Materials Handbook

Springer
London
Berlin
Heidelberg
New York
Barcelona
Hong Kong
Milan
Paris
Singapore
Tokyo

By the same author

Scientific Unit Conversion:
A Practical Guide to Metrication
ISBN 1-85233-043-0

François Cardarelli

Materials
Handbook

A Concise Desktop Reference

Springer

François Cardarelli, BSc, MSc, PhD, ChE
Materials Expert
Argo-Tech Productions, Inc.
1560, De Coulomb
Boucherville, J4B 7Z7, PQ, Canada
Tel: +450 655 3161 ext 243; Fax: +450 655 9297;
email: cardaref@argotech.qc.ca

Consultant in Electrodes Materials
44, rue de la Cosarde
F-94240 L'Hay-les-Roses, France
Tel: +33 1 47 40 89 95; Fax: +33 1 47 07 63 68;
email: cardarelli-f@calva.net

Cover: Based on a design by ANTONIO, Paris, France

ISBN 1-85233-168-2 Springer-Verlag London Berlin Heidelberg

British Library Cataloguing in Publication Data
Cardarelli, François, 1966–
    Materials handbook: a concise desktop reference
    1. Materials – Handbooks, manuals, etc.
    I. Title
    620.1'1
ISBN 1852331682

Library of Congress Cataloging-in-Publication Data
Cardarelli, François, 1966–
    Materials handbook: a concise desktop reference/François
Cardarelli.
    p. cm.
Includes bibliographical references and index.
ISBN 1-85233-168-2 (pbk.: alk. paper)
99–20194 CIP rev.
1. Materials. I. Title.
TA404.8 C37 1999
620.1'1–dc21

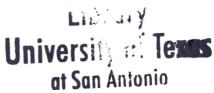

Typeset by EXPO Holdings, Malaysia
Printed and bound by Kyodo Printing Co. (S'pore) Pte. Ltd., Singapore
69/3830-543210 Printed on acid-free paper SPIN 10688664

Dedication

The *Materials Concise Reference Guide* is dedicated to my father Antonio and my mother Claudine, to my sister Elsa, and to Louise Saint-Amour, for their love and support. In addition, I want to express my thanks to my two parents and my uncle Consalvo Cardarelli, who in close collaboration provided valuable help when I was teenager to contribute to my first fully equipped geological and chemical laboratory, and to my personal scientific library. This was the starting point of my strong and extensive interest in both science and technology, and gluttony for scientific and technical literature.

François Cardarelli

Acknowledgments

Grateful thanks go to Mr Nicholas Pinfield (Engineering Editor, London), Mr Jean-Étienne Mittelmann (Editor, Paris), Mrs Alison Jackson (Editorial Assistant, London), and Mr Nicolas Wilson (Senior Production Controller, London), for their valued assistance, patience, and advice.

Units Policy

In this book all the units of measurement used for describing physical quantities are those recommended by the *Système International d'Unités* (SI). For accurate conversion factors between these units and the other non-SI units (e.g., cgs, fps, Imperial and US customary) please refer to Cardarelli, F., *Scientific Unit Conversion: A Practical Guide to Metrication*, 2nd edition, Springer-Verlag, London (1999), 504 pp.

Author Biography

Dr François Cardarelli
Born in Paris XV (France), 17 February 1966.

Academic Background

PhD in Chemical Engineering (University Paul Sabatier, Toulouse III),
MSc in Electrochemical Engineering (University Pierre and Marie Curie, Paris VI),
BSc in Physical Chemistry and Nuclear Engineering (University Pierre and Marie Curie, Paris VI/Conservatoire National des Arts et Métiers, CNAM),
Associate Degree in Geophysics and Geology (University Pierre and Marie Curie, Paris VI).

Areas of Work

The successive areas of work from 1990 to 2000 are in order: (1) Research scientist at the Laboratory of Electrochemistry (University Pierre & Marie Curie, Paris, France) for the development of a nuclear detector device for electrochemical experiments involving beta radiolabelled compounds; (2) Research scientist at the Institute of Marine Biogeochemistry (CNRS & École Normale Supérieure, Paris, France) for the environmental monitoring of heavy metals pollution by electroanalytical techniques; (3) Research scientist for the production of tantalum protective coatings for the chemical process industries by electrochemistry in molten salts; (4) Research scientist for the preparation and characterization of iridium-based industrial electrodes for oxygen evolution in acidic media at the Laboratory of Electrochemical Engineering (University Paul Sabatier, Toulouse, France); (5) Consultant in Chemical and Electrochemical Engineering (Toulouse, France); (6) Battery Product Leader at the Technology Department (Argo-Tech Productions, Inc., Boucherville, Québec, Canada) in charge of both the stationary and down-hole drilling applications of lithium polymer batteries; (7) Materials Expert at the Lithium Department (Argo-Tech Productions, Inc., Boucherville, Québec, Canada) working on both the processing of lithium metal anodes, and the recycling of lithium polymer batteries.

Contents

Introduction. xi

1 Ferrous Metals and Their Alloys 1
   1.1 Iron and Steels 1
   1.2 Nickel and Nickel Alloys. 32
   1.3 Cobalt Alloys 42

2 Common Nonferrous Metals 45
   2.1 Aluminum and Aluminum Alloys. 45
   2.2 Copper and Copper Alloys 51
   2.3 Zinc and Zinc Alloys 59
   2.4 Lead and Lead Alloys. 68
   2.5 Tin and Tin Alloys 69

3 Less Common Nonferrous Metals. 75
   3.1 Alkali Metals 75
   3.2 Alkaline-Earth Metals. 94
   3.3 Refractory and Reactive Metals (RMs). 113
   3.4 Precious and Noble Metals (NM) 192
   3.5 Platinum Group Metals (PGMs) 203
   3.6 Rare-Earth Metals (Sc, Y, and Lanthanides) 217
   3.7 Uranides (Th and U) 223

4 Semiconductors 245
   4.1 Band Theory of Bonding in Crystalline Solids. 245
   4.2 Electrical Classification of Solids 246
   4.3 Semiconductor Classes 247
   4.4 Semiconductor Physical Quantities 249
   4.5 Transport Properties 250
   4.6 Semiconductor Physical Properties 252
   4.7 Applications of Semiconductors 252
   4.8 Some Common Semiconductors. 252
   4.9 Semiconductor Wafer Processing 260
   4.10 The *P–N* Junction 263

5 Superconductors 265
   5.1 General Description 265
   5.2 Superconductor Types 266

	5.3	Basic Theory	269
	5.4	The Meissner–Ochsenfeld Effect	270
	5.5	History	271
	5.6	Industrial Uses and Applications	272

6   Magnetic Materials **275**

	6.1	Magnetic Physical Quantities	275
	6.2	Classification of Magnetic Materials	278
	6.3	Ferromagnetic Materials	281

7   Insulators and Dielectrics **291**

	7.1	Physical Quantities of Dielectrics	291
	7.2	Insulator Physical Properties	298
	7.3	Dielectric Behavior	300
	7.4	Dielectric Breakdown Mechanisms	302
	7.5	Electro-mechanical Coupling	303
	7.6	Piezoelectricity	303
	7.7	Ferroelectrics	303
	7.8	Aging of Ferroelectrics	305
	7.9	Industrial Dielectrics Classification	307
	7.10	Selected Properties of Insulators and Dielectric Materials	308

8   Miscellaneous Electrical Materials **313**

	8.1	Thermocouple Materials	313
	8.2	Electron-Emitting Materials	314
	8.3	Photocathode Materials	315
	8.4	Secondary Emission	318
	8.5	Electrode Materials	319

9   Ceramics and Glasses **337**

	9.1	Ceramics and Refractories	337
	9.2	Glasses	340

10 Polymers and Elastomers **371**

	10.1	Thermoplastics	371
	10.2	Thermosets	377
	10.3	Rubbers and Elastomers	378
	10.4	Polymer Physical Properties	381

11 Minerals, Ores, and Gemstones **395**

	11.1	Definitions	395
	11.2	Mineralogical, Physical, and Chemical Properties	396
	11.3	Strunz Classification of Minerals	407
	11.4	Mineral Properties Table	408
	11.5	Mineral Synonyms	409

12 Rocks and Meteorites **479**

	12.1	Introduction	479
	12.2	Types of Rocks	480
	12.3	Igneous Rocks	480
	12.4	Sedimentary Rocks	490
	12.5	Metamorphic Rocks	495
	12.6	Meteorites	499
	12.7	Physical Properties of Common Rocks	506

13 Timbers and Woods **509**

	13.1	General Description	509

13.2   Properties of Woods 510
13.3   Applications. 511
13.4   Wood Performance in Various Corrosives 511

14   **Building and Construction Materials** 517
14.1   Portland Cement. 517
14.2   Aggregates. 520
14.3   Mortars and Concrete 521
14.4   Ceramics for Construction 522
14.5   Building Stones. 522

15   **Appendices** 527
15.1   Periodic Chart of the Elements 527
15.2   Selected Physical Properties of the Elements. 528
15.3   Geochemical Classification of the Elements. 528
15.4   Historical Names of the Elements. 529
15.5   Cost of the Pure Elements. 529
15.6   Crystallography and Crystallochemistry 529
15.7   Properties of Liquid Metals. 559
15.8   Properties of Molten Salts. 559
15.9   Electrochemical Galvanic Series 559
15.10 Hardness Scales 563
15.11 UNS Standard Alphabetical Designation. 563
15.12 Fuel Energy Content 564
15.13 Natural Radioactivity. 564
15.14 Scientific and Technical Societies 569

16   **Bibliography** 579
16.1   General Desk References. 579
16.2   Dictionaries and Encyclopediae 581
16.3   Comprehensive Series 582

Index 584

Introduction

Despite the several comprehensive series available in Material Sciences and their related fields, it is a hard task to find grouped properties of metals and alloys, ceramics, polymers, minerals, woods, and building materials in a single volume source book. Actually, the scope of this practical handbook is to provide to scientists, engineers, professors, technicians, and students working in numerous scientific and technical fields ranging from nuclear to civil engineering, easy and rapid access to the accurate physico-chemical properties of all classes of materials. Classes used to describe the materials are: (i) metals and their alloys, (ii) semiconductors, (iii) superconductors, (iv) magnetic materials, (v) miscellaneous electrical materials (e.g., dielectrics, thermocouple and industrial electrode materials), (vi) ceramics, refractories, and glasses, (vii) polymers and elastomers, (viii) minerals, ores, meteorites, and rocks, (ix) timbers and woods, and finally (x) building materials. Particular emphasis is placed on the properties of the most common industrial materials in each class. Physical and chemical properties usually listed for each material are (i) mechanical (e.g., density, elastic moduli, Poisson's ratio, yield and tensile strength, hardness, fracture toughness), (ii) thermal (e.g., melting point, thermal conductivity, specific heat capacity, coefficient of linear thermal expansion, spectral emissivities), (iii) electrical (e.g., resistivity, dielectric permittivity, loss tangent factor), (iv) magnetic (e.g., magnetic permeability, remanence, Hall constant), (v) optical (e.g., refractive indices, reflective index), (vi) electrochemical (e.g., Nernst standard electrode potential, Tafel slope, specific capacity) and (vii) miscellaneous (e.g., corrosion rate, thermal neutron cross section, natural abundances, electron work function, Richardson constant). Detailed appendices provide additional information (e.g., properties and cost of the elements, molten salts and liquid metals, crystallographic calculations), and an extensive bibliography completes this comprehensive guide. The comprehensive index and large format of the book enable the reader to locate and extract the relevant information quickly and easily. Charts and tables are all referenced and tabs are used to denote the different sections of the book. It must be emphasized that the information presented here is taken from several various scientific and technical sources, and has been meticulously checked; every care has been taken to select the most reliable data.

1

Ferrous Metals and Their Alloys

The ferrous metals are defined as the three upper transition metals of the group VIIIB(8, 9, and 10) of Mendeleev's periodic chart, i.e., iron (Fe), cobalt (Co), and nickel (Ni).

1.1 Iron and Steels

1.1.1 Iron

1.1.1.1 General Properties and Description

Iron [7439-89-6] with the chemical symbol Fe, the atomic number 26 and the relative atomic mass of 55.845(2), is the first element of the upper transition metals of the group VIIIB(8) of Mendeleev's periodic chart. Its name comes from the Anglo-Saxon, *iren*, while the symbol Fe and words such as ferrous and ferric derive from the Latin name of iron, *ferrum*. Pure iron is a soft, dense (7874 kg.m^{-3}), silvery-lustrous, magnetic metal, with a high melting point (1538°C). In addition, when highly pure, iron has both a good thermal conductivity (78.2 W.m^{-1}.K^{-1}), a low coefficient of linear thermal expansion (12.1 μm.m^{-1}.K^{-1}), and is a satisfactory electric conductor (9.71 $\mu\Omega$.cm). At room temperature, highly pure iron crystallizes into a body-centered cubic (**bcc**) space lattice. From a mechanical point of view, pure iron exhibits a high Young's modulus of 208 GPa, with a Poisson's ratio of 0.291, but is malleable and can be easily shaped by hammering. Other mechanical properties such as yield and tensile strength strongly depend on interstitial impurity levels, and type of crystal space lattice structure. Natural iron is composed of four stable nuclides: ^{54}Fe (5.845at%), ^{56}Fe (91.754at%), ^{57}Fe (2.1191at%), and ^{58}Fe (0.2819at%), and the element has a thermal neutron cross-section of 2.6 barns. From a chemical point of view, pure iron is an active metal, and hence it rusts (i.e., oxidizes) when put in contact with moist air, forming a porous non-protective hydrated ferric oxide layer. In addition, pure iron readily dissolves in several diluted strong mineral acids such as hydrochloric and sulfuric acids with evolution of hydrogen. Various types of relatively pure or high-purity iron can be found on the market, although only a few of them are used as structural material. Most commercial irons, except **ingot iron** and **electrolytic iron**, contain perceptible quantities of carbon, which affects their properties. Other common high-purity iron types include **reduced irons** and **carbonyl iron** (powders). Cost (1998) – pure iron metal (i.e., 99.99wt% Fe) is priced at 2.205 $US.kg^{-1} (1.00 $US.lb^{-1}), while common iron (i.e. 99wt% Fe) is 9.92 ¢US.kg^{-1} (4.5 ¢US.lb^{-1}).

1.1.1.2 Natural Occurrence, Minerals, and Ores

Because nuclides of iron are particularly stable with the highest binding energy per nucleon (i.e., –8.79 MeV/nucleon for ^{56}Fe), its cosmic abundance is particularly high and it is thought to be the main constituent of the Earth's inner core as an iron–nickel alloy, called, after its chemical composition, *NiFe* by the Austrian geophysicist Suess. Moreover, iron is the major component of "siderite" class meteorites. The relative terrestrial abundance in the Earth's crust is about 5.63wt%, hence, it is the fourth most abundant element after oxygen, silicon, and aluminum and the second most abundant metal after aluminum. Iron, owing to its chemical reactivity, never occurs free in nature (except in meteorites) but it is present combined in a wide variety of mineral species found either in igneous, metamorphic, or sedimentary rocks and also in other geologic materials (e.g., soils). Amongst them, the most widely distributed iron-bearing minerals are oxides such as: **hematite** (Fe_2O_3, rhombohedral), **magnetite** (Fe_3O_4, cubic), and **limonite** ($Fe_2O_3 \cdot 3H_2O$, orthorhombic), the carbonate **siderite** ($FeCO_3$, rhombohedral), and the two sulfides **pyrite** (FeS_2, cubic), and **marcasite** (FeS_2, cubic). Amongst these minerals, only oxides are commonly used as iron ores. Commercially profitable extraction requires iron ore deposits providing a raw ore with more than 30wt% Fe. Although certain exceptional iron ores contain as much as 66wt% Fe, usually the major commercial iron ores contain 50–60wt% Fe. In addition the quality of the iron ore is influenced by the type of inert gangue materials. In addition to iron content, the amounts of silica, phosphorus, and sulfur-bearing compounds, are also important because they strongly affect the steelmaking process. China, Brazil, Australia, Russia, and the Ukraine are the five largest world producers of iron ore, but significant amounts are also mined in India, the United States, Canada, and Kazakstan. Together, these nine countries produce 80% of the world's iron ore.

1.1.1.3 History

Iron has been known since prehistoric times and no other element has played a more important role in man's material progress. Iron beads dating from around 4000 BC were no doubt of meteoritic origin, and later samples, produced by reducing iron ore with charcoal, were not cast because adequate temperatures were not attainable without use of some form of bellows. Instead the spongy material produced by low-temperature reduction would have had to be shaped by prolonged hammering. It seems that iron was first smelted by Hittites sometime in the third millennium BC, but the value of the process was so great that its secret was carefully guarded and it was only with the fall of the Hittite empire around 1200 BC that the knowledge was dispersed and the "Iron-Age" began. In more recent times the introduction of coke as the reductant had far-reaching effects, and was one of the major factors in the initiation of the industrial revolution.

1.1.1.4 Crystallography and Critical Points

The existence of pure iron element in more than one phase, i.e., a single crystallographic form, of space lattice structure is called **allotropism**. This should not be confused with the term extended to a pure compound (e.g., molecule, alloy, etc.) which exists in several space lattice structures and which is called **polymorphism**. The temperatures at which the changes of the crystallographic structure take place under constant pressure are called phase **transition temperatures** or **critical points**. These phase changes can be accurately determined by means of thermal analysis and dilatometry techniques. At atmospheric pressure, solid-state pure iron gives several allotropic crystallographic structures depending on temperature range. (i) Between room temperature and the critical point of 768°C, pure iron has a body-centered cubic space lattice structure (**bcc**, $a = 286.645$ pm at 25°C), and is known as **alpha-iron** (α-Fe). Alpha-iron is soft, ductile and ferromagnetic with a density of 7874 kg.m^{-3}. (ii) When heated above its Curie temperature of 768°C, alpha-iron loses its ferromagnetic properties, but retains its body-centered cubic structure. This form of iron is called **beta-iron** (β-Fe) which is considered as a different allotropic form owing to its nonmagnetic properties. However, because no changes to the crystal lattice structure occur, it is customary to consider it as nonmagnetic alpha-iron. (iii) At about 912°C, the crystallographic structure changes to the face-centered cubic form (**fcc**, $a = 364.68$ pm at 916°C) and iron assumes the allotropic form known as **gamma-iron** (γ-Fe). At this transition temperature, a considerable absorption of heat occurs due to the endothermic reaction, and the volume of the iron unit cell contracts by 1%. Gamma-iron is nonmagnetic and has slighter greater density than low-temperature phases having a body-centered cubic structure. At a temperature of 1394°C a third transformation occurs and the face-centered cubic lattice reverts to a body-centered cubic form, which again becomes magnetic and is named **delta-iron** (δ-Fe, $a = 293.22$ pm). Finally, at the melting point of 1539°C a final pause takes place in the rise of temperature as the iron absorbs the heat required for fusion and becomes liquid (i.e., molten iron). The crystallographic phase transformations are summarized in the following reaction schemes:

$$\underbrace{\alpha - Fe}_{bcc,\ magnetic} \quad \overset{T=768°C}{\longleftrightarrow} \quad \underbrace{\beta - Fe}_{bcc,\ nonmagnetic}$$

$$\underbrace{\beta - Fe}_{bcc,\ nonmagnetic} \quad \overset{T=912°C}{\longleftrightarrow} \quad \underbrace{\gamma - Fe}_{fcc,\ nonmagnetic}$$

$$\underbrace{\gamma - Fe}_{fcc,\ nonmagnetic} \quad \overset{T=1394°C}{\longleftrightarrow} \quad \underbrace{\delta - Fe}_{bcc,\ magnetic}$$

1.1.1.5 Mining and Mineral Dressing

Most iron ores are extracted by common open pit mining. Some underground mines exist, but, wherever possible, surface mining is preferred because it is less expensive. After mining, depending on the quality of the raw iron ore, two routes can be used to prepare the concentrated ore. For certain rich iron ore deposits the raw ore (above 66wt% Fe) is crushed to reduce the maximum particle size, and sorted into various fractions by passing it over sieves through which lump or rubble ore (i.e., 5 to 25 mm) is separated from the fines (i.e., less than 5 mm). Due to the high iron content, the lumps can be charged directly in the blast furnace without any further processing. Fines, however, must first be agglomerated, which means reforming them into lumps of suitable size by a process called **sintering**. In this agglomeration process, fines are heated in order to achieve partial melting during which ore particles fuse together. For this purpose, the elevated heat required is generated by burning of fine coke known as coke breeze. After cooling, the sinter is broken up and screened to yield blast-furnace feed and an undersize fraction that is recycled.

Common ore is crushed and ground in order to release the ore minerals from the inert gangue materials (e.g., silica). Gangue minerals are separated from iron ore particles by common ore refinement processes in order to decrease silica content to less than 9wt%. Most concentration processes use froth flotation, and gravity separation based on density differences to separate light minerals from heavier iron ores. Electromagnetic separation techniques are also used but hematite is not sufficiently ferromagnetic to be easily recovered. After refinement, the ore concentrate is in a powdered form unsuitable for direct use in the blast furnace. It has a much smaller particle size than ore fines and cannot be agglomerated by sintering. Instead, concentrates must be agglomerated by **pelletizing** (a process originated in Sweden and Germany in 1911, and optimized in the 1940s). In this process, first, humidified concentrates are fired in a rotary kiln, in which the tumbling action produces soft, spherical agglomerates. These agglomerates are then dried and hardened by firing in air at a temperature ranging between 1250°C to 1340°C, yielding spherical pellets with about 1 cm diameter.

1.1.1.6 Iron Making

Highly pure iron is prepared on a small scale by the reduction of the pure oxide or hydroxide with the hydrogen, or by the carbonyl process in which iron is heated with carbon monoxide under pressure and the $Fe(CO)_5$ so formed decomposed at 250°C to give off the powdered metal. By contrast, in industrial scale production of steel the first stage in the conversion of iron ore into steel is the blast furnace, which accounts for the largest tonnage of any metal produced by man.

Iron making produces iron metal from iron chemically combined with oxygen. The blast furnace process, which consists of the carbothermic reduction of iron oxides, is industrially the most efficient process, and is used to produce 95% of the iron made in the world. From a chemical engineering point of view, the **blast furnace** can be described as a countercurrent heat and oxygen exchanger in which rising combustion gas loses most of its heat on the way up, leaving the furnace at a temperature of about 200°C, while descending iron oxides are reduced to metallic iron. The blast furnace is a tall, vertical steel reactor lined internally with refractory ceramics such as high-alumina firebrick (45–63wt% Al_2O_3) and graphite. Five sections can be clearly identified: (i) At the bottom is a parallel-sided hearth where liquid metal and slag collect, and this is surmounted by (ii) an inverted truncated cone known as the **bosh**. Air is blown into the furnace through (iii) **tuyeres** (water-cooled copper nozzles), mounted at the top of the hearth close to its junction with the bosh. (iv) A short vertical section called the bosh parallel, or the barrel, connects the bosh to the truncated upright cone that is the stack. Finally, the fifth and top section, through which the charge is fed into the furnace, is the **throat**. The lining in the bosh and hearth, where the highest temperatures occur, is usually made of carbon bricks, which are manufactured by pressing and baking a mixture of coke, anthracite, and pitch. Actually, carbon exhibits excellent corrosion resistance to the molten iron and slag in comparison with aluminosilicate firebricks used for the remainder of the lining.

During the blast furnace process, the solid charge (mixture of iron ore, limestone, and coke) is loaded either by operated skips or by conveyor belts at the top of the furnace at a temperature ranging from 150°C to 200°C, while preheated air (900 to 1350°C) in hot-blast stoves, sometimes enriched up to 25vol% oxygen, is blown into the furnace through the tuyeres. During the process, the coke serves both as fuel and reducing agent, and a fraction combines with iron. The limestone acts as a fluxing agent, i.e., it reacts with both silica gangue materials and traces of sulfur to form a slag; sometimes fluorspar is also used as fluxing agent. During the carbothermic reduction operations, the ascending carbon monoxide, resulting from the exothermic combustion of coke at the tuyere entrance, begins to react with the descending charge, partially reducing the ore to ferrous oxide. At the same time the carbon monoxide is cooled by the descending charge and it reacts forming carbon dioxide and carbon black (soot). This soot dissolves in the iron forming a eutectic and hence decreases the melting temperature. At this stage the temperature is sufficiently high to decompose the limestone into lime and carbon dioxide. Carbon dioxide reacts with the coke to give off carbon monoxide, and the free lime combines with silica gangue to form a molten silicate slag floating upon molten iron. Slag is removed from the furnace by the same taphole as the iron, and it exhibits the following chemical composition: 30–40wt% SiO_2, 5–15wt% Al_2O_3, 35–45wt% CaO, and 5–15wt% MgO. As the partially reduced ore descends, it encounters both increasing high temperature and high concentration of carbon monoxide which accelerates the reactions. At this stage the reduction of ferrous oxide into iron is completed and the main product, called molten pig iron (i.e., hot metal or blast-furnace iron), is tapped from the bottom of the furnace at regular intervals. The gas exiting at the top of the furnace is composed mainly of 23vol% CO, 22vol% CO_2, 3vol% water, and 49vol% N_2, and after the dust particles have been removed using dust collectors, it is mixed with coke oven gas and burned in hot-blast stoves to heat the air blown in through the tuyeres. It is important to note that during the process, traces of aluminum, manganese, and silicon from the gangue are oxidized and are recovered into the slag, while phosphorus and sulfur dissolves in the molten iron.

1.1.1.7 Pure Iron Grades

Table 1.1 Pure iron grades

Pure iron grade	Purity (wt% Fe)	Description
Ingot iron	99.8–99.9	Ingot iron is a nearly chemically pure iron type (i.e., 99.8 to 99.9wt% Fe) that is used for construction work where a ductile, rust resistant metal is required. It is mainly applied to boilers, tanks, enameled ware, and galvanized culvert sheets, as well as to electromagnetic cores and as a raw material for producing specialty steels. A well-known commercial type is Armco® ingot iron (99.94wt% Fe). Typical ingot irons have as low as 0.02wt% carbon or less. The Armco® ingot iron, for example, typically has carbon concentrations of 0.013% and a manganese content around 0.017wt%. Ingot iron may also be obtained in grades containing 0.25 to 0.30wt% copper, which increases the corrosion resistance. Ingot iron is made by the basic open-hearth process and highly refined, remaining in the furnace 1 to 4 h longer than the normal time, and maintained at a temperature of 1600 to 1700°C.
Electrolytic iron	99.9	Electrolytic iron is a chemically pure iron (i.e., 99.9wt% Fe) produced by the cathodic deposition of iron in an electrochemical refining process. Bars of cast iron are used as soluble anodes and dissolved in an electrolyte bath containing iron (II) chloride ($FeCl_2$). The cathodic reduction of ferrous cations gives pure iron on the cathodes, which are often hollow steel cylinders. The deposited iron tube is removed by hydraulic pressure or by splitting, and then annealed and rolled into plates. The product is used for magnetic cores and, in general, applications where both high ductility and purity are required.
Reduced iron	99.9	Reduced iron is a fine gray amorphous powder made by reducing crushed iron ore by heating in hydrogen atmosphere. It is used for special chemical purposes.
Carbonyl iron	99.99	Carbonyl iron or carbonyl iron powder is metallic iron of extreme purity, produced as microscopic spherical particles by the reaction of carbon monoxide on iron ore. This reaction gives a liquid, called iron carbonyl $Fe(CO)_5$, which is vaporized and deposited as a powder. Carbonyl iron is mainly used for magnetic cores for high-frequency equipment and for pharmaceutical application of iron.
Wrought iron	99	Wrought iron, which is no longer commercially produced, is a relatively pure iron containing nonmetallic slag inclusions produced by the blast furnace. Modern wrought iron products are actually made of low-carbon steel.

1.1.1.8 The Major Carbon–Iron Phases

Plain carbon steel is an alloy of iron and carbon. Other alloying elements may be added to produce steels for special purposes. As a general rule, the structure of steel is determined by the amounts and nature of alloying materials, and by the rate of cooling from the molten state. Carbon is a strongly reactive element which induces stabilization of the gamma-iron phase. Indeed, carbon is always present owing to its use during the steelmaking process. First of all, it is necessary to describe the structural phases occurring in the structural Fe–C diagram.

Ferrite alpha (ferrite α, bcc) is a solid solution of an insertion of carbon in alpha-iron. For instance, maximum solubility of carbon in alpha-iron is 0.01wt% at room temperature (RT) and 0.02wt% at 727°C. Therefore, a solid solution of which alpha iron is the solvent is called ferrite. Because of the extremely small amount of carbon which it can contain in solid solution, ferrite in a steel that contains only iron and carbon may be considered substantially pure iron, and it is sometimes called pure iron in some textbooks. Because the ferrite of an alloy steel may contain in solid solution appreciable amounts of other elements, however, it is better to use the term only in its exact meaning: a solid solution of any elements in alpha-iron.

Ferrite delta (ferrite δ, bcc) is a solid solution of an insertion of carbon in delta-iron. For instance, maximum solubility of carbon in delta-iron is 0.1wt% at 1487°C.

Cementite (Fe_3C) which is a hard, brittle, and ferromagnetic compound at RT is a chemical combination of excess carbon as iron carbide with 6.68wt% of C. Actually, at RT under conditions of equilibrium, any carbon present in excess of that small amount must exist in a form other than that of a solute in a solid solution.

Austenite gamma (austenite γ, fcc), is a solid insertion solution of carbon in gamma-iron which represents the solvent. The solute consists of iron carbide only, or of any number of other elements. It is definitively established that the carbon atoms in austenite occupy interstitial positions in the fcc space lattice causing the lattice parameter to increase progressively with the carbon content.

Table 1.2 Stabilizing elements

Effect	Alloying elements or impurities
Ferrite stabilizers	C, Cr, Mo, Ti, W, V, Nb, and Si
Austenite stabilizers	Ni, Mn

The various temperatures at which pauses occur in the rise or fall of temperature when steel is heated from room temperature or cooled from the molten state are called **arrest points**. These are identical with the critical or transformation points in the pure iron phase diagram. The arrest points obtained on heating are designated *Ac* and those obtained on cooling are designated *Ar*. The suffixes *c* and *r* are respectively named after the French words chauffage (meaning heating) and refroidissement (meaning cooling).

1.1.2 Cast Irons

Cast irons contain much higher carbon and silicon levels than steels, theorically higher than 1.8wt%, but typically 3–5wt% Fe and 1–3wt% Si. These form another category of ferrous materials, which are intended to be cast from the liquid state to the final desired shape. Various types of cast iron are widely used in industry, especially for valves, pumps, pipes, filters and certain mechanical parts, including:

- Gray cast iron
- White cast iron

- Chilled iron (duplex)
- Malleable cast irons
- Ductile or nodular cast irons
- Alloy cast irons
- High-silicon cast irons
- Nickel cast irons

Cast iron can be considered as a ternary Fe–Si–C alloy. The carbon concentration ranges between 1.7 and 4.5%, most of which is present in insoluble form (e.g., graphite flakes or nodules). Such material is, however, normally called unalloyed cast iron and exists in four main types (Tables 1.3–1.5): (i) white iron, which is brittle and glass hard, (ii) unalloyed gray iron, which is soft but still brittle, and which is the most common form of unalloyed cast iron, (iii) the more ductile malleable iron, and (iv) nodular or ductile cast iron, the best modern form of cast iron, which has superior mechanical properties and equivalent corrosion resistance. In addition there are a number of alloyed cast irons, many of which have improved corrosion resistance and substantially modified mechanical and physical properties (Table 1.6). Generally, cast iron is not a particularly strong or tough structural material. Nevertheless it is one of the most economical and is widely used in industry. Its annual production is only exceeded by steel. Iron castings are used in many items of equipment in the chemical process industries, but its main use is in mechanical engineering industries, e.g., automobile, machine tools. Some of the most well-known classes include the high-silicon and nickel cast irons.

1.1.2.1 Gray or Graphitic Cast Iron

Gray cast iron contains 1.7–4.5wt% of carbon and other alloying elements such as Si, Mn, and Fe. Due to the slow cooling rate during casting, the carbon is precipitated as thin flakes of graphite dispersed throughout the metal. Therefore, gray cast irons are relatively brittle. Gray cast iron is the least expensive material, is quite soft, has excellent machinability and is easy to cast into intricate shapes. Various strengths are produced by varying size, amount, and distribution of the graphite. For instance, ultimate tensile strength typically ranges from 155 to 400 MPa and the Vickers hardness from 130 to 300 HV. Gray iron has excellent wear resistance and damping properties. However, it is both thermal and mechanical shock sensitive. Gray iron castings can be welded with proper techniques and adequate preheating of the components.

1.1.2.2 White Cast Iron

White cast iron is made by controlling the chemical composition (i.e., low Si and high Mn) and rate of solidification of the iron melt. Rapid cooling leads to an alloy that has practically all its carbon retained as dissolved cementite which is hard and devoid of ductility. The resulting cast is hard, brittle and virtually unmachinable and finishing must be achieved by grinding. Typical Vickers hardness ranges from 400 to 600 HV. Its main use is for wear- or abrasion-resistant applications. In this respect white irons are superior to manganese steel, unless deformation or shock resistance is required. The major applications of white cast iron include: pump impellers, slurry pumps, crushing and grinding equipment.

1.1.2.3 Malleable Cast Irons

Malleable iron exhibits a typical carbon content of 2.5wt%. It is made from white cast iron by prolonged heating of the casting. This causes the carbides to decompose and graphite aggregates are produced in the form of dispersed compact rosettes (i.e., no flakes). This gives a tough, relatively ductile material. There are two main types of malleable iron: (i) standard and

(ii) pearlitic. The latter contains both combined carbon and graphite nodules. Standard malleable irons are easily machined. This is less true for pearlitic iron. All malleable cast irons withstand cold working and bending without cracking.

1.1.2.4 Ductile (Nodular) Cast Irons

This is the best modern form of cast iron having superior mechanical properties and equivalent corrosion resistance. Ductility is much improved and may approach that of steel. Ductile iron is sometimes also called nodular cast iron or spheroidal graphite cast iron, as the graphite particles are approximately spherical in shape, in contrast to the graphite flakes in gray cast iron. Ductile cast iron exhibits a typical microstructure. This is achieved by the addition of a small amount of nickel–magnesium alloy, or by doping the molten metal with magnesium or cerium. Composition is nearly the same as gray iron, with some nickel, and with more carbon (3.7% C) than malleable iron. There are a number of grades of ductile iron. Some have maximum machinability and

Table 1.3 Cast iron classification[a]

Cast iron class	Average chemical composition range	Carbon rich phase	Matrix	Fracture
Gray cast iron (FG) (flake graphite cast iron)	2.5–4.0wt% C 1.0–3.0wt% Si 0.2–1.0wt% Mn 0.002–1.0wt% P 0.02–0.25wt% S	Lamellar graphite	Pearlite	Gray
Compacted graphite cast iron (CG)	2.5–4.0wt% C 1.0–3.0wt% Si 0.2–1.0wt% Mn 0.01–0.1wt% P 0.01–0.03wt% S	Compacted vermicular graphite	Ferrite, pearlite	Gray
Ductile cast iron (SG) (nodular or spheroidal graphite cast iron)	3.0–4.0wt% C 1.8–2.8wt% Si 0.1–1.0wt% Mn 0.01–0.1wt% P 0.01–0.03wt% S	Spheroidal graphite	Ferrite, pearlite, austenite	Silver gray
White cast iron	1.8–3.6wt% C 0.5–1.9wt% Si 0.25–0.8wt% Mn 0.06–0.2wt% P 0.06–0.2wt% S	Cementite Fe_3C	Pearlite, martensite	White
Malleable cast iron (TG)	2.2–2.9wt% C 0.9–1.9wt% Si 0.15–1.2wt% Mn 0.02–0.2wt% P 0.02–0.2wt% S	Temper graphite	Ferrite, pearlite	Silver gray
Mottled cast iron		Lamellar graphite and cementite Fe_3C	Pearlite	Mottled
Austempered ductile iron		Spheroidal graphite	Bainite	Silver gray

Note: types of graphite flakes in gray cast iron, A uniform distribution, random orientation, B rosette grouping, random orientation, C superimposed flake size, random orientation, D interdendritic segregation, random orientation, E interdendritic segregation, preferred orientation.
[a]From Stefanescu, D.M. *Classification and Basic Metallurgy of Cast Irons*. In: *ASM Metals Handbook, Volume 1: Ferrous Metals*. ASM, Metal Park, OH (1992) p. 3.

Table 1.4 Physical properties of gray cast irons

Type of iron	Grade	Density (ρ, kg.m^{-3})	Young's modulus (E, GPa)	Shear modulus (G, GPa)	Yield tensile strength (σ_{YS}, MPa)	Ultimate tensile shear strength (σ_{UTS}, MPa)	Compressive strength (MPa)	Brinell hardness (HB)	Coefficient of linear thermal expansion (α, 10^{-6} K^{-1})	Thermal conductivity (k, W.m^{-1}.K^{-1})
Gray cast iron (FG) (flake graphite cast iron)	20	7200–7600	66–97	27–39	152	179	572	156	13	46–49
	25	7200–7600	79–102	32–41	179	220	669	174	13	46–49
	30	7200–7600	90–113	36–45	214	276	752	210	13	46–49
	35 (coarse pearlite)	7200–7600	100–119	40–48	252	334	855	212	13	46–49
	40 (fine pearlite)	7200–7600	110–138	44–54	293	393	965	235	13	46–49
	50 (acicular iron)	7200–7600	130–157	50–55	362	503	1130	262	13	46–49
	60	7200–7600	141–162	54–59	431	610	1293	302	13	46–49

toughness, others have maximum oxidation resistance. Ductile iron castings can also be produced to have improved low-temperature impact properties. This is achieved by adequate thermal treatment, by control of the P and Si content and by various alloying processes.

1.1.2.5 High-Silicon Cast Irons

Cast iron with a high silicon level (13–16wt% Si) named Duriron® exhibits, for all concentrations of H_2SO_4 even up to boiling point, a constant corrosion rate of 130 μm (5 mil) per year. For these reasons, it is widely used in sulfuric acid service. Duriron® is a cheap material that does not contain any amount of strategic metal. Nevertheless, it is very hard and brittle and thermal shock sensitive, so it is not readily machined or welded.

1.1.3 Carbon steels (C–Mn Steels)

Iron containing more than 0.15wt% carbon chemically combined is normally termed **steel**. This 0.15wt% C is a somewhat arbitrary borderline and sometimes the nearly chemically pure "ingot iron" is referred to as **mild steel**. To make it even more confusing the term mild steel is often also used as synonym for **low-carbon steels**, which are materials with 0.15–0.30% carbon.

1.1.3.1 Carbon Steels

Carbon steels, also called **plain carbon steels** are primarily Fe and C, with small amounts of Mn. Specific heat treatments and slight variations in composition will lead to steels with varying mechanical properties. Carbon is the principal hardening and strengthening element in steel. Actually, carbon increases hardness and strength and decreases weldability and ductility. For plain carbon steels, about 0.2 to 0.25wt% carbon provides the best machinability. Above and below this level, machinability is generally lower for hot-rolled steels. They are sometimes divided into three groups: (i) low-carbon or mild steels, (ii) medium-carbon steels, and

Table 1.5 Physical properties of nodular and other cast irons

Type of iron	Grade	Density (ρ, kg.m^{-3})	Young's modulus (E, GPa)	Shear modulus (G, GPa)	Yield strength 0.2% (σ_{YS}, MPa)	Ultimate tensile strength (σ_{UTS}, MPa)	Elongation (Z, %)	Brinell hardness (HB)	Coefficient of linear thermal expansion (α, 10^{-6} K^{-1})	Thermal conductivity (k, W.m^{-1}.K^{-1})
Ductile cast iron (SG) (nodular or spheroidal graphite cast iron)	60–40–18	7200–7600	169–172	66–69	190–220	300–350	22	150	12	35
	80–60–03	7200–7600	169–172	66–69	190–220	300–350	18	160	12	35
	60–40–18	7200–7600	169–172	66–69	200–250	300–410	18	180	12	35
	65–45–12	7200–7600	169–172	66–69	270–332	420–464	12	212	12	35
	80–55–06	7200–7600	169–172	66–69	320–362	450–559	11	221	12	35
	100–70–03	7200–7600	169–172	66–69	320	500	7	241	12	35
	120–90–02	7200–7600	169–172	66–69	370–864	600–974	1–3	269	12	35
	D4018	7200–7600	169–172	66–69	420	700	1–3	302	12	35
	D4512	7200–7600	169–172	66–69	480	800	n.a.	352	12	35
	D5506	7200–7600	169–172	66–69	600	900	n.a.	359	12	35
Compacted graphite cast iron (CG)	Ferritic	7200–7600	162	68	290	365	4.5	140–155	n.a.	41
	Pearlitic	7200–7600	165	68	330	440	1.5	225–245	n.a.	41
Malleable cast iron (TG)	A602	7200–7600	168	67.5	180	300	6	150	n.a.	n.a.
	A603	7200–7600	168	67.5	190	320	10	150	n.a.	n.a.
	A604	7200–7600	168	67.5	230–280	400–480	12	150	n.a.	n.a.

(iii) high-carbon steels. **Low-carbon steels** (e.g. AISI grades 1005 to 1030) have up to 0.30wt% carbon. They are characterized by low tensile strength and high ductility. They are non-hardenable by heat treatment, except by surface hardening processes. **Medium-carbon steels** (e.g. AISI grades 1030 to 1055) have between 0.31wt% and 0.55wt% carbon. They provide a good balance between strength and ductility. They are hardenable by heat treatment, but hardenability is limited to thin sections or to the thin outer layer on thick parts. **High-carbon steels** (e.g. AISI grades 1060 to 1095) have between 0.56 and about 1wt% carbon. They are, of course, hardenable and are very suitable for wear-resistant and/or high-strength parts. Steels that have been worked or wrought while hot will be covered with a black so-called mill scale (i.e., magnetite, Fe_3O_4) on the surface, and are sometimes called **black iron**. Cold-rolled steels have a bright surface, accurate cross-section, and higher tensile and yield strengths. They are preferred for bar stock to be used for machining rods and shafts. Carbon steels may be specified either by chemical composition, mechanical properties, method of deoxidation, or thermal

Table 1.6 Properties of alloyed cast irons

Alloyed cast iron class	Alloyed cast iron type	Density (ρ, kg.m^{-3})	Ultimate tensile strength (σ_{UTS}, MPa)	Compressive strength (σ, MPa)	Brinell hardness (HB)	Coefficient of linear thermal expansion (α, 10^{-6} K^{-1})	Thermal conductivity (k, W.m^{-1}.K^{-1})	Electrical resistivity (ρ, $\mu\Omega$.cm)
Abrasion-resistant white iron	Low-carbon white iron	7600–7800	n.a.	n.a.	n.a.	12	22	53
	Martensitic Ni–Cr iron	7600–7800	n.a.	n.a.	n.a.	8–9	30	80
Corrosion-resistant irons	High-silicon iron (Duriron®)	7000–7050	90–180	690	480–520	12.4–13.1	n.a.	50
	High-chromium iron	7300–7500	205–380	690	250–740	9.4–9.9	n.a.	n.a.
	High-nickel gray iron	7400–7600	170–310	690–1100	120–250	8.1–19.3	38–40	100
	High-nickel ductile	7400	380–480	1240–1380	130–240	12.6–18.7	13.4	78
Heat-resistant gray iron	Medium-silicon iron	6800–7100	170–310	620–1040	170–250	10.8	37	210
	High-nickel iron	7300–7500	170–310	690–1100	130–250	8.1–19.3	37–40	140–170
	Ni–Cr–Si iron	7330–7450	140–310	480–690	110–210	12.6–16.2	30	150–170
	High-aluminum iron	5500–6400	180–350	n.a.	180–350	15.3	n.a.	240
Heat-resistant white iron	High-chromium iron	7300–7500	210–620	690	250–500	9.3–9.9	20	120
Heat-resistant ductile iron	Medium-silicon ductile iron	7100	n.a.	n.a.	140–300	10-8–13.5	n.a.	58–87
	High-Ni ductile (20wt% Ni)	7400	380–415	1240–1380	140–200	18.7	13	102
	High-Ni ductile (23wt% Ni)	7400	400–450	n.a.	130–170	18.4	n.a.	100
	High-Ni ductile (30wt% Ni)	7500	n.a.	n.a.	n.a.	12.6–14.4	n.a.	n.a.
	High-Ni ductile (36wt% Ni)	7700	n.a.	n.a.	n.a.	7.2	n.a.	n.a.

treatment and the resulting microstructure. However, wrought steels are most often specified by their chemical composition. No single element controls the characteristics of a steel; rather, the combined effects of several elements influence hardness, machinability, corrosion resistance, tensile strength, deoxidation of the solidifying metal, and microstructure of the solidified metal. Standard wrought-steel compositions for both carbon and alloy steels are designated by the SAE-AISI four-digit code, the last two digits of which indicate the nominal carbon content (see Table 1.7).

1.1.3.2 Low-Alloy Steels

Steels that contain specified amounts of alloying elements other than carbon and the commonly accepted amounts of manganese, copper, silicon, sulfur, and phosphorus, are known as alloy

Table 1.7 Carbon and low-alloy steel designation (SAE/AISI)

Series	Main group	Subgroups
1XXX	(Plain) carbon steel	10XX: Plain carbon (1.00wt% Mn, max) 11XX: Resulfurized 12XX: Resulfurized and rephosphorized 13XX: Manganese steels (1.75wt% Mn) 15XX: Nonresulfurized (1.00–1.65wt% Mn max)
2XXX	Nickel steels	23XX: 3.5wt% Ni 25XX: 5.0wt% Ni
3XXX	Nickel–chromium steels	31XX: 1.25wt% Ni and 0.65 to 0.80wt% Cr 32XX: 1.75wt% Ni and 1.07wt% Cr 33XX: 3.50wt% Ni and 1.50 to 1.57wt% Cr 34XX: 3.00wt% Ni and 0.77wt% Cr
4XXX	Molybdenum steels	40XX: 0.20 to 0.25wt% Mo 44XX: 0.40 to 0.52wt% Mo
	Chromium–molybdenum steels	41XX: 0.50 to 0.95wt% Cr, 0.12 to 0.30wt% Mo
4XXX	Nickel–molybdenum steels	46XX: 0.85–1.82wt% Ni, and 0.20–0.25wt% Mo 48XX: 3.50wt% Ni, 0.25wt% Mo
4XXX	Nickel–chromium–molybdenum steels	43XX: 1.82wt% Ni, 0.50 to 0.80wt% Cr, and 0.25wt% Mo 43BVXX: 1.82wt% Ni, 0.50wt%Cr, 0.12 to 0.25wt% Mo, and 0.03wt% 47XX: 1.05wt% Ni, 0.45wt% Cr, and 0.25 to 0.35wt% Mo
5XXX	Chromium steels	50XX: 0.25 to 0.65wt% Cr 51XX: 0.80 to 1.05wt% Cr 50XXX: 0.50wt% Cr, min. 1wt% C 51XXX: 1.02wt% Cr, min. 1wt% C 52XXX: 1.45wt% Cr, min. 1wt% C
6XXX	Chromium–vanadium steels	61XX: 0.6 to 0.95wt% Cr, 0.10 to 0.15wt% V
7XXX	Tungsten–chromium steels	72XX: 1.75 W, and 0.75 Cr
8XXX	Nickel–chromium–molybdenum steels	81XX: 0.30wt% Ni, 0.40wt% Cr, and 0.12wt% Mo 86XX: 0.55wt% Ni, 0.50wt% Cr, and 0.20wt% Mo 87XX: 0.55wt% Ni, 0.50wt% Cr, and 0.25wt% Mo 88XX: 0.55wt% Ni, 0.50wt% Cr, and 0.35wt% Mo
9XXX	Nickel–chromium–molybdenum steels	93XX: 3.25wt% Ni, 1.20wt% Cr, and 0.12wt% Mo 94XX: 0.45wt% Ni, 0.40wt% Cr, and 0.12wt% Mo 97XX: 0.55wt% Ni, 0.20wt% Cr, and 0.20wt% Mo 98XX: 1.00wt% Ni, 0.80wt% Cr, and 0.25wt% Mo
9XXX	Silicon–manganese steels	92XX: 1.40 to 2.00wt% Si, 0.65 to 0.85wt% Mn, and 0.65wt% Cr

Note: The letter L between the second and third digits indicates a leaded steel, while the letter B indicates a boron steel.

steels. Alloying elements are added to change mechanical or physical properties. A steel is considered to be an alloy when the maximum of the range given for the content of alloying elements exceeds one or more of these limits: 1.65wt% Mn, 0.60wt% Si, or 0.60wt% Cu, or when a definite range or minimum amount of any of the following elements is specified or required within the limits recognized for constructional alloy steels: aluminum, chromium (up to 3.99%), cobalt, columbium, molybdenum, nickel, titanium, tungsten, vanadium, zirconium or other element added to obtain an alloying effect. According to the previous definition, rigorously, tool and stainless steels are also considered alloy steels. However, the term alloy steel is reserved for those steels that contain a minute amount of alloying elements and that usually depend on thermal treatment to develop specific properties. Subdivisions for most steels in this family include: (i) through-hardening grades, which are heat treated by quenching and tempering and

Table 1.8 Physical properties of carbon and low-alloy steels

AISI type	UNS	Density (ρ, kg.m^{-3})	Yield strength 0.2% proof (σ_{YS}, MPa)	Ultimate tensile strength (σ_{UTS}, MPa)	Elongation (Z, %)	Brinell hardness (HB)	Coefficient of linear thermal expansion (α, 10^{-6} K^{-1})	Thermal conductivity (k, W.m^{-1}.K^{-1})	Specific heat capacity (c_p, J.kg^{-1}.K^{-1} (50–100°c)	Electrical resistivity (ρ, $\mu\Omega$.cm)
1008	G10080	7750	n.a.	n.a.	n.a.	n.a.	12.6	59.5	481	14.2
1010	G10100	7750	n.a.	310	25	121	12.2	n.a.	450	14.3
1015	G10150	7750	284	386	n.a.	n.a.	12.2	51.9	486	15.9
1020	G10200	7750	200–355	400–690	12–21	126–179	11.7	51.9	486	15.9
1025	G10250	7750	215–415	430–770	11–20	n.a.	12.0	51.9	486	15.9
1030	G10300	7750	230–415	460–775	10–20	126–207	11.7	51.9	486	16.6
1035	G10350	7750	280–480	490–775	9–18	179–229	11.1	50.8	486	16.3
1040	G10400	7750	245–530	510–770	7–17	152–255	11.3	50.7	486	16.0
1045	G10450	7750	280–525	540–770	7–16	n.a.	11.6	50.8	486	16.2
1050	G10500	7750	280–585	570–1000	8–14	n.a.	11.1	51.2	486	16.3
1137	G11370	7750	n.a.	n.a.	n.a.	n.a.	12.8	50.5	n.a.	17.0
1141	G11410	7750	340–495	540–850	7–20	152–255	12.6	50.5	461	17.0
1151	G11510	7750	n.a.	n.a.	n.a.	n.a.	12.6	50.5	502	17.0
1330	G13300	7750	n.a.	n.a.	n.a.	n.a.	12.0	n.a.	n.a.	n.a.
1335	G13350	7750	n.a.	n.a.	n.a.	n.a.	12.2	n.a.	n.a.	n.a.
1345	G13450	7750	n.a.	n.a.	n.a.	n.a.	12.0	n.a.	n.a.	n.a.
1522	G15220	7750	n.a.	n.a.	n.a.	n.a.	12.0	51.9	486	n.a.
2330	G23300	7750	689	841	19	248	10.9	n.a.	n.a.	n.a.
2515	G25150	7750	648	779	25	233	10.9	34.3	n.a.	n.a.
3120	G31200	7750	n.a.	n.a.	n.a.	n.a.	11.3	n.a.	n.a.	n.a.
3140	G31400	7750	n.a.	n.a.	n.a.	n.a.	11.3	n.a.	n.a.	n.a.
3150	G31500	7750	n.a.	n.a.	n.a.	n.a.	11.3	n.a.	n.a.	n.a.
4023	G40230	7750	586	827	20	255	11.7	n.a.	n.a.	n.a.
4042	G40420	7750	1448	1620	10	461	11.9	n.a.	n.a.	n.a.
4053	G40530	7750	1538	1724	12	495	n.a.	n.a.	n.a.	n.a.
4063	G40630	7750	1593	1855	8	534	n.a.	n.a.	n.a.	n.a.
4130	G41300	7750	1172	1379	16	375	12.2	42.7	477	22.3
4140	G41400	7750	1172	1379	15	385	12.3	42.7	475	22.0
4150	G41500	7750	1482	1586	10	444	11.7	41.8	n.a.	n.a.
4320	G43200	7750	1062	1241	15	360	11.3	n.a.	n.a.	n.a.
4337	G43370	7750	965	1448	14	435	11.3	n.a.	n.a.	n.a.
4340	G43400	7750	1379	1512	12	445	12.3	n.a.	n.a.	n.a.
4615	G46150	7750	517	689	18	n.a.	11.5	n.a.	n.a.	n.a.

Table 1.8 (continued)

AISI type	UNS	Density (ρ, kg.m^{-3})	Yield strength 0.2% proof (σ_{YS}, MPa)	Ultimate tensile strength (σ_{UTS}, MPa)	Elongation (Z, %)	Brinell hardness (HB)	Coefficient of linear thermal expansion (α, 10^{-6} K^{-1})	Thermal conductivity (k, W.m^{-1}.K^{-1})	Specific heat capacity (c_P, J.kg^{-1}.K^{-1})	Electrical resistivity (ρ, $\mu\Omega$.cm)
4620	G46200	7750	655	896	21	n.a.	12.5	44.1	335	n.a.
4640	G46400	7750	1103	1276	14	390	n.a.	n.a.	n.a.	n.a.
4815	G48150	7750	862	1034	18	325	11.5	n.a.	481	26.0
4817	G48170	7750	n.a.	n.a.	15	355	n.a.	n.a.	n.a.	n.a.
4820	G48200	7750	n.a.	n.a.	13	380	11.3	n.a.	n.a.	n.a.
5120	G51200	7750	786	986	13	302	12.0	n.a.	n.a.	n.a.
5130	G51300	7750	1207	1303	13	380	12.2	48.6	494	21.0
5140	G51400	7750	1172	1310	13	375	12.3	45.8	452	22.8
5150	G51500	7750	1434	1544	10	444	12.8	n.a.	n.a.	n.a.
6120	G61200	7750	648	862	21	n.a.	n.a.	n.a.	n.a.	n.a.
6145	G61450	7750	1165	1213	16	429	n.a.	n.a.	n.a.	n.a.
6150	G61500	7750	1234	1289	13	444	12.2	n.a.	n.a.	n.a.
8617	G86170	7750	676	841	n.a.	n.a.	11.9	45.0	481	30.0
8630	G86300	7750	979	1117	14	325	11.3	39.0	449	n.a.
8640	G86400	7750	1262	1434	13	420	13.0	37.6	460	n.a.
8650	G86500	7750	1338	1475	12	423	11.7	37.6	453	n.a.
8720	G87200	7750	676	841	21	245	14.8	37.6	450	n.a.
8740	G87400	7750	1262	1434	13	420	11.3	37.6	448	n.a.
8750	G87500	7750	1338	1476	12	423	14.8	37.6	448	n.a.
9255	G92550	7750	1482	1600	9	477	14.6	46.8	420	n.a.
9261	G92610	7750	1558	1779	10	514	14.6	46.8	502	n.a.

Other properties common to all carbon and low-alloy steels are: Young's modulus is 201–209 GPa, Coulomb's or shear modulus is 81–82 GPa, bulk or compression modulus is 160–170 GPa, and Poisson's ratio of 0.27–0.30.

are used when maximum hardness and strength must extend deep within a part, while (ii) carburizing grades are used where a tough core and relatively shallow, hard surface are needed. After a surface-hardening treatment such as carburizing or nitriding for nitriding alloys, these steels are suitable for parts that must withstand wear as well as high stresses. Cast steels are generally through hardened, not surface treated. Carbon content and alloying elements influence the overall characteristics of both types of alloy steels. Maximum attainable surface hardness depends primarily on the carbon content. Maximum hardness and strength in small sections increase as carbon content increases, up to about 0.7wt%. However, carbon contents greater than 0.3% can increase the possibility of cracking during quenching or welding. Alloying elements primarily influence hardenability. They also influence other mechanical and fabrication properties including toughness and machinability. Lead additions (0.15 to 0.35wt% Pb) greatly improve machinability of alloy steels by high-speed tool steels. For machining with carbide tools, calcium-treated steels are reported to double or triple tool life in addition to improving surface finish. Alloy steels are often specified when high strength is needed in moderate-to-large sections. Whether tensile or yield strength is the basis of design, thermally treated alloy steels generally offer high strength-to-weight ratios. In general, the wear resistance can be improved by increasing the hardness of an alloy or by specifying an alloy with greater carbon content, or using nitrided parts, which have better wear resistance than would be expected from the carbon content alone. Fully hardened-and-tempered, low-carbon (0.10 to 0.30wt% C) alloy steels have a good combination of strength and toughness, both at room and low temperature.

1.1.3.3 Cast Steels

Cast steels are steels that have been cast into sand molds to form finished or semi-finished machine parts or other components. Mostly, the general characteristics of steel castings are closely comparable to wrought steels. Cast and wrought steels of equivalent composition respond similarly to heat treatment and have fairly similar properties. A major difference is that cast steel has a more isotropic structure. Therefore, properties tend to be more uniform in all directions and do not vary according to the direction of hot or cold working as in many wrought steel products. Cast steels are often divided into the following four categories: cast plain carbon steels, (low-)alloy steel castings, heat-resistant cast steels, and corrosion-resistant cast steels, depending on the alloy content and intended service. Cast plain carbon steels can be divided into three groups similar to wrought steels: low-, medium-, and high-carbon steels. However, cast carbon steel is usually specified by mechanical properties (primarily tensile strength), rather than composition. Low-alloy steel castings are considered as steels with a total alloy content less than about 8%. The most common alloying elements are manganese, chromium, nickel, molybdenum, vanadium, and small quantities of titanium or aluminum (grain refinement) and silicon (improved corrosion and high-temperature resistance).

1.1.4 Stainless Steels

1.1.4.1 General Description

Stainless steel is a family of iron–chromium-based alloys, which are essentially low-carbon steel alloys that contain a percentage of chromium above 10.5wt%. This addition of chromium gives to the steel its unique corrosion resistance properties (i.e., stainless, or rust proof). The chromium content of the steel allows the formation onto the steel surface of a passivating layer of chromium oxide. This protective oxide film is impervious, adherent, transparent, and corrosion resistant in many chemical environments. If damaged mechanically or chemically, this film is self-healing when small traces of oxygen are present in the corrosive medium. It is important to note that in order to be corrosion resistant, the iron base alloy must contain at least 10.5wt% chromium and when this percentage is decreased, e.g., by chromium carbide precipitation when the alloy is heated, protection is lost and the rusting process occurs. Moreover, the corrosion resistance and other useful properties of stainless steels are largely enhanced by increased chromium content. In addition, further alloy additions (e.g., Mo, Ti) can be made to tailor the chemical composition in order to meet the needs of different corrosion conditions, operating temperature ranges, strength requirements, or to improve weldability, machinability, and work hardening. Generally, the corrosion resistance of stainless steels is, as a rule, improved by increasing the alloy content. The terminology heat resistant and corrosion resistant is highly subjective and somewhat arbitrary. The term heat-resistant alloys commonly refers to oxidation-resistant metals and alloys (see Ni–Cr–Fe alloys in Section 1.2), while corrosion-resistant commonly refers only to metals and alloys which are capable of sustained operation when exposed to attack by corrosive media at service temperatures below 315°C. They are normally Fe–Cr or Fe–Cr–Ni ferrous alloys and can normally be classified as stainless steels. There are roughly more than 60 commercial grades of stainless steel available. However, the entire group can be divided into five distinct classes. Each class is identified by the alloying elements which affect their microstructure and for which each is named. These classes are: (i) the austenitic stainless steels, (ii) the ferritic stainless steels, (iii) the martensitic stainless steels, (iv) the duplex or austeno-ferritic stainless steels, and finally (v) the precipitation-hardened (P-H) stainless steels. As a general rule, the Ni-bearing austenitic types have the highest general corrosion resistance, and higher-nickel alloys are more resistant than lower-nickel compositions. Galling and wear are failure modes that require special attention with stainless steels because these materials serve in many harsh environments. They often operate, for example, at high temperatures, in food-contact applications, and where access is limited. Such restrictions prevent the use of lubricants, leading to metal-to-metal contact, a condition that promotes galling and accelerated wear.

1.1.4.2 Martensitic Stainless Steels

Martensitic stainless steels (i.e., AISI 400 series) are typically iron–chromium–carbon alloys which contain at least 10.5–18wt% chromium; they may have small quantities of additional alloying elements, and the carbon content must be lower than 1.2wt%. They are called martensitic owing to the distorted body-centered cubic (bcc) crystal lattice structure in the hardened condition. Martensitic stainless steels exhibit the common following characteristics: (i) a martensitic crystal structure, (ii) they are ferromagnetic, (iii) they can be hardened by heat treatment, (iv) they have high strength and moderate toughness in the hardened-and-tempered condition, and (v) they have poor welding characteristics. Forming should be done in the annealed condition. Martensitic stainless steels are less resistant to corrosion than the austenitic or ferritic grades. Two types of martensitic steels, 416 and 420F, have been developed specifically for good machinability. Martensitic stainless steels are used where strength and/or hardness are of primary concern and where the environment is not too corrosive. These alloys are typically used for bearings, molds, cutlery, medical instruments, aircraft structural parts, and turbine components. The most common three grades are: the most used grade AISI 410, the grade AISI 420 used extensively in cutlery for making knife blades, and finally the grade AISI 440C used when very high hardness is required. Type 420 is used increasingly for molds for plastics and for industrial components requiring hardness and corrosion resistance. Physical properties of selected martensitic stainless steels are listed in Table 1.9.

1.1.4.3 Ferritic Stainless Steels

Ferritic wrought stainless steel alloys (i.e., AISI 400 series) (Table 1.10) exhibit a chromium content ranging from 10.5wt% to 27wt% Cr but have a lower carbon level (i.e., less than 0.2% C) than martensitic stainless steels. Ferritic wrought stainless steels exhibit the following common characteristics: (i) they exhibit a ferritic metallurgical crystal structure due to the high chromium content, (ii) they are ferromagnetic and retain their basic microstructure up to the melting point if sufficient Cr and Mo are present, (iii) they cannot be hardened by heat treatment, and they can be only moderately hardened by cold working, hence they are always used in the annealed condition, (iv) in the annealed condition, their strength is approximately 50% higher than that of carbon steels, and finally (v) like martensitic steels they have poor weldability. Ferritic stainless steels are typically used where moderate corrosion resistance is required and where toughness is not a major need. They are also used where chloride stress-corrosion cracking (SCC) may be a problem because they have high resistance to this type of corrosion failure. In heavy sections, achieving sufficient toughness is difficult with the higher-alloyed ferritic grades. Typical applications include automotive trim and exhaust systems and heat-transfer equipment for the chemical and petrochemical industries. The two common grades are: the grade AISI 409 used for high-temperature applications, and the most used grade AISI 430.

Table 1.9 Physical properties of martensitic stainless steels

AISI type	UNS	Average chemical composition (wt%)	Density (ρ, kg.m^{-3})	Yield strength 0.2% proof (σ_{YS}, MPa)	Ultimate tensile strength (σ_{UTS}, MPa)	Elongation (Z, %)	Rockwell hardness (HRB)	Coefficient of linear thermal expansion	Thermal conductivity (k, W.m^{-1}.K^{-1})	Electrical resistivity (ρ, $\mu\Omega$.cm)
403	S40300	Fe–12.25Cr–1Mn–0.5Si–0.15C	7800	275–620	485–825	12–20	88	n.a.	n.a.	n.a.
410	S41000	Fe–12.5Cr–1Mn–1.0Si–0.15C	7800	275–620	450–825	12–20	95	9.9	24.9	57
414	S41400	Fe–12.5Cr–1.88Ni–1Mn–1Si–0.15 C	7800	620	795–1030	15	n.a.	10.4	24.9	70
416	S41600	Fe–13Cr–1.25Mn–1Si–0.6Mo–0.15C	7800	275	845	20	n.a.	9.9	24.9	57
420	S42000	Fe–13Cr–1Mn–1Si–0.15C	7800	1480	690–1720	8–15	n.a.	10.3	24.9	55
422	S42200	Fe–12.5Cr–1Mo–1Mn–0.75Ni–0.75Si–0.22C–1W	7800	585–760	825–965	13–17	n.a.	11.2	23.9	59
431	S43100	Fe–16Cr–1.88Ni–1Mn–1.0Si–0.2C	7800	620–930	795–1210	13–15	n.a.	10.2	20.1	72
410 Cb	S41040	Fe–12.5Cr–1Mn–1Si–0.2Nb–0.15C	7800	275–690	485–860	12–13	n.a.	n.a.	24.9	57
410 S	S41008	Fe–12.5Cr–1Mn–1Si–0.6Ni–0.08C	7800	205–240	415–450	22	88	n.a.	n.a.	n.a.
416 Plus X	S41610	Fe–13Cr–2Mn–1Si–0.6Mo–0.15C	7800	300	585–1210	n.a.	n.a.	n.a.	n.a.	n.a.
416 Se	S41623	Fe–13Cr–1.25Mn–1Si–0.6Mo–0.15C 0.15Se	7800	275	845	20	n.a.	n.a.	n .a.	n.a.
420 FSe	S42023	Fe–13Cr–1Mn–1.25Si–0.15C–0.15Se	7800	1480	1720	8–15	96	n.a.	n.a.	n.a.
440 A	S44002	Fe–17Cr–1Mn–1Si–0.75Mo–0.675C	7800	415–1650	725–1790	5–20	95	10.2	24.2	60
440 B	S44003	Fe–17Cr–1Mn–1Si–0.75Mo–0.85C	7800	425–1860	740–1930	3–18	96	10.2	24.2	60
440 C	S44004	Fe–17Cr–1Mn–1Si–0.75Mo–1.1 C	7800	760–1970	450–1900	2–14	97	10.2	24.2	60
440 FSe	S44023	Fe–17Cr–1.25Mn–1Si–0.75Ni–0.95 C	7800	n.a.	n.a.	n.a.	n.a.	n.a.	n.a.	n.a.
CA6NM	S41500	Fe–12.75Cr–4Ni–0.75Mn–1Mo–0.6Si–0.05C	7800	620	795	15	n.a.	n.a.	n.a.	n.a.
E4	S41050	Fe–11.5Cr–0.85Ni–1Mn–1Si–0.1N–0.04C	7800	205	415	22	88	n.a.	n.a.	n.a.
Greek Ascoloy	S41800	Fe–13Cr–3W–2Ni–0.5Mn–0.5Si–0.15C	7800	760	965	15	n.a.	n.a.	n.a.	n.a.
Lapelloy	S42300	Fe–11.5Cr–3Mo–1.15Mn–0.5Si–0.5Ni–0.3C	7800	760	965	8	n.a.	n.a.	n.a.	n.a.

Table 1.9 (continued)

AISI type	UNS	Average chemical composition (wt%)	Density (ρ, kg.m^{-3})	Yield strength 0.2% proof (σ_{YS}, MPa)	Ultimate tensile strength (σ_{UTS}, MPa)	Elongation (Z, %)	Rockwell hardness (HRB)	Coefficient of linear thermal expansion	Thermal conductivity (k, W.m^{-1}.K^{-1})	Electrical resistivity (ρ, $\mu\Omega$.cm)
Trim -Rite	S42010	Fe–14Cr–1Mn– 1Mo–0.75Ni–1Si– 0.2C	7800	n.a.	690– 725	n.a.	n.a.	n.a.	n.a.	n.a.

Other properties common to all martensitic stainless steel types are: Young's modulus is 204–215 GPa, Coulomb's or shear modulus is 83.9 GPa, bulk or compression modulus is 166 GPa, and Poisson's ratio is 0.283, and their specific heat capacity is roughly 460 J.kg^{-1}.K^{-1}.

1.1.4.4 Austenitic Stainless Steels

Austenitic stainless steels are iron–chromium based alloys containing at least 18wt% Cr; in addition they contain sufficient nickel and manganese to stabilize and insure fully austenitic metallurgical crystal structure at all temperatures ranging from the cryogenic region to the melting point of the alloy. Carbon content is usually less than 0.15wt%. As a general rule, they further exhibit the following common characteristics: (i) by contrast with other classes they are not ferromagnetic even after severe cold working, (ii) they cannot be hardened by heat treatment, but (iii) they can be hardened by cold working, (iv) they have better corrosion resistance than other classes, (v) they can be easily welded, (vi) they possess an excellent cleanability and allow excellent surface finishing, and finally (vii) they exhibit excellent corrosion resistance to several corrosive environments at both room and high temperatures. However, the austenitic stainless steels have some limitations: (i) the maximum service temperature under oxidizing conditions is 450°C, above this temperature heat resistant steels are required, (ii) they are suitable only for low concentrations of reducing acids such as HCl; super austenitics are required for higher acid concentrations; (iii) in services and shielded areas, there might not be enough oxygen to maintain the passive oxide film and crevice corrosion might occur and they must be replaced by super austenitics, duplex and super ferritic, (iv) very high levels of halide ions, especially the chloride ion can lead to the breakdown of the passivating film. It is important to note that on heating, carbon combines with chromium to form chromium carbide. If the chromium content falls below the critical percentage of 10.5wt%, the corrosion resistance of the alloy is lost. Austenitic wrought stainless steels are classified in three groups: (i) the AISI 200 series, i.e., alloys of iron–chromium–nickel–manganese, (ii) the AISI 300 series, i.e., alloys of iron–chromium–nickel, and (iii) nitrogen-strengthened alloys (with the suffix N added to the AISI grade). Nitrogen-strengthened austenitic stainless steels are alloys of chromium–manganese–nitrogen; some grades also contain nickel. Yield strengths of these alloys in the annealed condition are typically 50% higher than those of the non-nitrogen-bearing grades. Like carbon, nitrogen increases the strength of a steel. But unlike carbon, nitrogen does not combine significantly with chromium in a stainless steel. The combination with carbon, which forms chromium carbide, reduces the strength and the corrosion resistance of an alloy. Because of their valuable structural and corrosion resistance properties this group is the most widely used alloy group in the process industry. Actually austenitic stainless steels are generally used where corrosion resistance and toughness are primary requirements. Typical applications include shafts, pumps, fasteners, and piping for servicing in seawater and equipment for processing chemicals, food, and dairy products. The most widely used grades are: AISI 304, the refractory grade AISI 310 for high-temperature applications, the grade AISI 316L for better corrosion resistance, and finally the AISI 317 for best corrosion resistance in chloride containing media. The largest single alloy in terms of total industrial usage is 304. Physical properties of austenitic stainless steels are listed in Table 1.11.

Table 1.10 Physical properties of ferritic stainless steels

AISI type	UNS	Average chemical composition (wt%)	Density (ρ, kg.m^{-3})	Yield strength 0.2% proof (σ_{YS}, MPa)	Ultimate tensile strength (σ_{UTS}, MPa)	Elongation (Z, %)	Rockwell hardness (HRB)	Coefficient of linear thermal expansion	Thermal conductivity (k, W.m^{-1}.K^{-1})	Electrical resistivity (ρ, $\mu\Omega$.cm)
405	S40500	Fe–13Cr–1Mn–1Si–0.2Al–0.08C	7800	170–280	415–480	20	88	10.8	27.0	60
409	S40900	Fe–11.13Cr–1Mn–1Si–0.5Ni–0.48Ti–0.08C	7800	205	380	20–22	80	11.7	n.a.	60
429	S42900	Fe–15Cr–1Mn–1Si–0.12C	7800	205–275	450–480	20–22	88	10.3	25.6	59
430	S43000	Fe–17Cr–1Mn–1Si–0.12C	7800	205–275	415–480	20–22	88	10.4	26.1	60
430F	S43020	Fe–17Cr–1.25Mn–1 Si–0.6Mo–0.12C	7800	275	485–860	20	n.a.	10.4	26.1	60
430FSe	S43023	Fe–17Cr–1.25Mn–1 Si–0.15Se–0.12C	7800	275	485–860	20	n.a.	10.4	26.1	60
430 Ti	S43036	Fe–18Cr–1Mn–1Si–0.75Ni–0.75Ti–0.1C	7800	310	515	30	n.a.	10.4	26.1	60
434	S43400	Fe–17Cr–1Mn–1Si–1Mo–0.12C	7800	365–315	530–545	23–33	83–90	10.4	26.3	60
436	S43600	Fe–17Cr–1Mn–1Si–1Mo–0.12C	7800	365	530	23	83	9.3	23.9	60
439	S43900	Fe–18Cr–1.1Ti–1Mn–1Si–0.5Ni–0.15Al–0.07C	7700	205–275	450–485	20–22	88	10.4	24.2	63
442	S44200	Fe–20.5Cr–1Mn–1Si–1Mo–0.2C	7800	275–310	515–550	20	90–95	n.a.	n.a.	n.a.
444	S44400	Fe–18.5Cr–1Ni–1Mn–1Si–2Mo–0.025C	7800	275	415	20	95	10.0	26.8	62
446	S44600	Fe–25Cr–1.5Mn–1Si–0.2C–0.25N	7500	275	480–515	16–20	95	10.8	20.9	67
E-Brite 26–1	S44627	Fe–26Cr–1Mo–0.5Ni–0.4Mn–0.4Si–0.2Cu–0.2Nb–0.01C	7800	275	450	16–22	90	n.a.	n.a.	n.a.
Monit 25-4-4	S44635	Fe–25Cr–4Ni–4Mo–1Mn–0.75Si–0.025C	7800	515	620	20	n.a.	n.a.	n.a.	n.a.
Sea-Cure (SC-1)	S44660	Fe–26Cr–2.5Ni–1Mn–1Si–3Mo–0.025C	7800	450	585	18	100	n.a.	n.a.	n.a.
AL 29-4C	S44735	Fe–29Cr–4Mo–1Ni–1Mn–1Si–0.8Ti–0.03C	7800	415	550	20	98	n.a.	n.a.	n.a.
Al 29-4-2	S44800	Fe–29Cr–4Mo–2Ni–0.3Mn–0.2Si–0.8Ti–0.15Cu–0.03C	7800	380–415	480–550	15–20	98	n.a.	n.a.	n.a.

Other properties common to all ferritic stainless steel types are: Young's modulus is 200–215 GPa, Coulomb's or shear modulus between 80 and 83 GPa, a Poisson's ratio of 0.27–0.29, and their specific heat capacity is roughly 460 J.kg^{-1}.K^{-1}.

Table 1.11 Physical properties of austenitic stainless steels (annealed)

AISI type	UNS	Average chemical composition (wt%)	Density (ρ, kg.m^{-3})	Yield strength 0.2% proof (σ_{YS}, MPa)	Ultimate tensile strength (σ_{UTS}, MPa)	Elongation (Z, %)	Rockwell hardness (HRB)	Coefficient of linear thermal expansion (α, 10^{-6} K^{-1})	Thermal conductivity (k, W.m^{-1}.K^{-1})	Electrical resistivity (ρ, $\mu\Omega$.cm)
201	S20100	Fe–17Cr–4.5Ni–0.25N–6.5Mn–1Si–0.15C	7800	275–965	515–1280	9–40	100	15.7	16.2	69
202	S20200	Fe–18Cr–5Ni–0.25N–8.75Mn–1Si–0.15C	7800	260–515	515–860	12–40	100	17.5	16.2	69
205	S20500	Fe–17.5Cr–1.5Ni–14.75Mn–1Si–0.15C	7800	450	790	40	100	17.9	16.2	69
301	S30100	Fe–17Cr–7Ni–2Mn–1Si–0.15C	8000	205–965	620–1280	9–40	95	17.9	16.2	72
302	S30200	Fe–18Cr–9Ni–2Mn–1Si–0.15C	8000	205–965	515–1275	4–40	92	17.2	16.2	72
302B	S30215	Fe–18Cr–9Ni–2Mn–2.5Si–0.15C	8000	205–310	515–620	30–40	95	16.2	15.9	72
303	S30300	Fe–18Cr–9Ni–2Mn–1Si–0.15C–0.6Mo	8000	205	515–1000	40	n.a.	17.2	16.2	72
303Se	S30323	Fe–18Cr–9Ni–2Mn–1Si–0.15C–0.15Se	8000	205	515–1000	40	n.a.	17.2	16.2	72
304	S30400	Fe–19Cr–9.25Ni–2Mn–1Si–0.08C	8000	205–760	515–1035	7–40	92	17.2	16.2	72
304 HN	S30409	Fe–19Cr–9.25Ni–2Mn–1Si–0.08C–0.12N	8000	275–345	585–620	30	100	n.a.	n.a.	n.a.
304 L	S30403	Fe–19Cr–9.25Ni–2Mn–1Si–0.03C	8000	170–310	450–620	30–40	88	17.2	16.0	n.a.
304 LN	S30453	Fe–19Cr–10Ni–2Mn–1Si–0.03C	8000	205	515	40	92	n.a.	n.a.	n.a.
304 N	S30451	Fe–19Cr–9.25Ni–2Mn–1Si–0.08C–0.13N	8000	240	550	30	92	n.a.	n.a.	74
305	S30500	Fe–18Cr–11.75Ni–2Mn–1Si–0.12C	8000	170–310	515–1690	30–40	88	17.8	16.2	72
308	S30800	Fe–20Cr–11Ni–2Mn–1Si–0.08C	8000	205–310	515–620	30–40	88	17.8	15.2	72
309	S30900	Fe–23Cr–13.5Ni–2Mn–1Si–0.20C	8000	205–310	515–620	30–40	95	17.2	15.6	78
310	S31000	Fe–25Cr–20.5Ni–2Mn–1.5Si–0.25C	8000	340–855	600–1280	46–58	n.a.	15.9	14.2	78
314	S31400	Fe–24.5Cr–20.5Ni–2Mn–2Si–0.25C	7800	205–310	515–620	30–40	n.a.	15.1	17.5	77
316	S31600	Fe–17Cr–12Ni–2.5Mo–2Mn–1Si–0.08C	8000	205–310	515–620	30–40	n.a.	15.9	16.2	74
316 L	S31603	Fe–17Cr–12Ni–2.5Mo–2Mn–1Si–0.03C	8000	170–310	450–620	30–40	95	15.9	15.6	74
316 N	S31651	Fe–17Cr–12Ni–2.5Mo–2Mn–1Si–0.08C–0.13N	8000	240	550	30–35	95	15.9	15.6	72

Table 1.11 (continued)

AISI type	UNS	Average chemical composition (wt%)	Density (ρ, kg.m^{-3})	Yield strength 0.2% proof (σ_{YS}, MPa)	Ultimate tensile strength (σ_{UTS}, MPa)	Elongation (Z, %)	Rockwell hardness (HRB)	Coefficient of linear thermal expansion (α, 10^{-6} K^{-1})	Thermal conductivity (k, W.m^{-1}.K^{-1})	Electrical resistivity (ρ, $\mu\Omega$.cm)
317	S31700	Fe–19Cr–13Ni–3.5Mo–2Mn–1Si–0.08C	8000	205–310	515–620	30–40	95	16.5	14.1	79
317 L	S31703	Fe–19Cr–13Ni–3.5Mo–2Mn–1Si–0.03C	8000	205–240	515–585	40–55	85–95	16.5	14.1	79
321	S32100	Fe–18Cr–10.5Ni–2Mn–1Si–0.4Ti–0.08C	8000	205–310	515–620	30–40	95	17.8	16.1	72
321 H	S32109	Fe–18Cr–10.5Ni–2Mn–1Si–0.35Ti–0.07C	8000	205–310	515–620	30–40	95	n.a.	n.a.	n.a.
347	S34700	Fe–18Cr–11Ni–2Mn–1Si–0.8Nb–0.08C	8000	205–310	515–620	30–45	92	16.6	16.2	73
384	S38400	Fe–16Cr–18Ni–2Mn–1Si–0.08C	8000	n.a.	550–585	n.a.	n.a.	17.8	16.2	79

Other properties common to all austenitic stainless steel types are: Young's modulus ranging between 192 and 200 GPa, a Coulomb's or shear modulus between 74 and 86 GPa, a Poisson's ratio of 0.25–0.29, and their specific heat capacity is roughly 500 J.kg^{-1}.K^{-1}.

1.1.4.5 Duplex Stainless Steels

When the chromium content is high (18 to 26wt%) and the nickel content is low (4 to 7wt%), the resulting structure is called duplex. In addition most grades contain 2 to 3wt% molybdenum. This results in a structure that is a combination of both ferritic and austenitic, hence the name duplex. The most common grade is the AISI 2205. They have the following characteristics: (i) high resistance to stress corrosion cracking, (ii) increased resistance to chloride ion attack, (iii) very weldable, (iv) higher tensile and yield strengths than austenitic or ferritic stainless steels.

1.1.4.6 Precipitation-Hardening Stainless Steels (PH)

Precipitation-hardening stainless steels, well known under the common acronyms, PH or P-H, develop very high strength through a low-temperature heat treatment that does not significantly distort precision parts. Compositions of most precipitation-hardening stainless steels are balanced to produce hardening by an aging treatment that precipitates hard, intermetallic compounds and simultaneously tempers the martensite. The beginning microstructure of PH alloys is austenite or martensite. The austenitic alloys must be thermally treated to transform austenite to martensite before precipitation hardening can be accomplished. These alloys are used where high strength, moderate corrosion resistance, and good fabricability are required. Typical applications include shafting, high-pressure pumps, aircraft components, high-temper springs, and fasteners.

Table 1.12 Physical properties of duplex stainless steels

Trade name	UNS	Average chemical composition (wt%)	Density (ρ, kg.m^{-3})	Yield strength 0.2% proof (σ_{YS}, MPa)	Ultimate tensile strength (σ_{UTS}, MPa)	Elongation (Z, %)	Rockwell hardness (HRC)
44LN	S31200	Fe–25Cr–6Ni–2Mn–1.6Mo–1Si–0.03C	7800	450	690	25	32
DP–3	S31260	Fe–25Cr–6.5Ni–3Mo–1Mn–0.75Si–0.03C	7800	450–485	690	20–25	31
3RE60	S31500	Fe–18.5Cr–4.75Ni–2.75Mo–1.6Mn–1.7Si–0.03C	7800	440	630	30	31
2205	S31803	Fe–22Cr–5.5Ni–3Mo–2Mn–1Si–0.03C	7800	450	620	25	32
2304	S32304	Fe–23Cr–4.25Ni–2.5Mn–1Si–0.5Mo–0.03C	7800	400	800	25	31
Ferralium 255	S32550	Fe–25.5Cr–5.5Ni–3Mo–1.5Mn–1Si–0.04C	7800	550	760	15	32
7–Mo Plus	S32950	Fe–27.5Cr–4.35Ni–2Mo–2Mn–0.6Si–0.03C	7800	480	690	15–20	31

1.1.4.7 Cast Heat-Resistant Stainless Steels

Cast stainless steels usually have corresponding wrought grades that have similar compositions and properties. However, there are small but important differences in composition between cast and wrought grades. Stainless-steel castings should be specified by the designations established by the Alloy Casting Institute (ACI), and not by the designation of similar wrought alloys. Service temperature provides the basis for a distinction between heat-resistant and corrosion-resistant cast grades. The C series of ACI grades designates the corrosion-resistant steels; the H series designates the heat-resistant steels, which can be used for structural applications at service temperatures between 650°C and 1200°C. Carbon and nickel contents of the H-series alloys are considerably higher than those of the C series. H-series steels are not immune to corrosion, but they corrode slowly, even when exposed to fuel-combustion products or atmospheres prepared for carburizing and nitriding. C-series grades are used in valves, pumps, and fittings. H-series grades are used for furnace parts and turbine components.

1.1.4.8 Processing and Melting Process

The feedstock used in the melting process is essentially made from stainless-steel scrap, i.e., scrap arising from sheet metal fabrication and disused plant and equipment. This approach enables the economical recycling of valuable alloys by the steel industry. After the chemical identification and analysis of the incoming steel scrap, scrap is sorted by grade, and a charge is prepared adding various alloys of chromium, nickel, and molybdenum depending on the stainless type to produce an alloy content which is as close as possible to the final grade required for the steel. The scrap charge is then fed into an electric arc furnace, where carbon

Table 1.13 Physical properties of PH stainless steels

AISI type	UNS	Average chemical composition (wt%.)	Density (ρ, kg.m^{-3})	Young's modulus (E, GPa)	Yield strength 0.2% proof (σ_{YS}, MPa)	Ultimate tensile strength (σ_{UTS}, MPa)	Elongation (Z, %)	Rockwell hardness (HRC)	Coefficient of linear thermal expansion (α, 10^{-6} K^{-1})	Thermal conductivity (k, W.m^{-1}.K^{-1})	Electrical resistivity (ρ, $\mu\Omega$.cm)
15–5PH	S15500	Fe–14.75Cr–4.5Ni–3.5Cu–1Mn–1Si–0.2Nb–0.07C	7800	196	515–1170	795–1310	10–18	40–48	10.8	17.8	77
17–4PH	S17400	Fe–16.5Cr–4Ni–4Cu–1Mn–1Si–0.2Nb–0.07C	7800	196	515–1170	795–1310	10–18	40–48	10.8	18.3	80
17–7PH	S17700	Fe–17Cr–7.125Ni–1Mn–1Al–1Si–0.09C	7800	204	965–1590	1170–1650	1–7	41–44	11.0	16.4	83
AM350 (Type 633)	S35000	Fe–16.5Cr–4.5Ni–1Mn–2.7Mo–0.5Si–0.07C	7800	n.a.	1000–1030	1140–1380	2–12	36–42	n.a.	n.a.	n.a.
AM355 (Type 634)	S35500	Fe–16.5Cr–4.5Ni–0.75Mn–2.7Mo–0.5Si–0.07C	7800	n.a.	1030–1070	1170–1310	10–12	37	n.a.	n.a.	n.a.
Custom 450	S45000	Fe–15Cr–6Ni–1Mn–1Si–0.7Mo–1.5Cu–0.05C	7800	n.a.	515–1100	860–1240	6–18	26–39	n.a.	n.a.	n.a.
Custom 455	S45500	Fe–11.75Cr–8.5Ni–0.5Mn–0.5Si–0.7Ti–1.5Cu–0.05C	7800	n.a.	1280–1520	1410–1620	4–10	44–47	n.a.	n.a.	n.a.
PH 13–8Mo	S13800	Fe–12.75Cr–8Ni–2Mo–1Al–1Mn–1Si–0.05C	7800	203	585–1410	860–1520	10–16	26–45	10.6	14.0	102

electrodes are placed in contact with the scrap. Under a high-voltage difference, a current is passed through the electrodes providing sufficient energy to melt the charge. The furnace is connected to a pot lined with refractory ceramic material which resists the high temperatures encountered in the melting process. The molten material from the electric furnace is then transferred into an argon-oxygen decarbonization vessel, where the carbon levels are reduced and the final alloy additions, i.e. nickel, ferro-chromium and ferro-molybdenum, are made to achieve the exact desired chemical composition of the final steel. After, the furnace is emptied into a tapping ladle by tilting the furnace forward. The ladle is an open-topped container lined with refractories. The melt is then transferred to a converter where the steel is refined, or purified, from impurities of mainly carbon, silicon and sulfur. This process involves blowing a mixture of oxygen and argon through the melt from the bottom of the converter. Samples are taken from the melt and analyzed and the chemical composition of the steel can, if necessary, be modified by the addition of alloying metals in the converter or in the ladle afterwards. Afterwards the desired molten metal is cast either into ingots or continuously cast into a slab or billet form. Then the material is hot rolled or forged into its final form. Some material receives cold rolling to further reduce the thickness as in sheets or is drawn into smaller diameters as in rods and wire. Most stainless steels receive a final annealing and acid pickling in order to remove furnace scale from annealing and this helps to promote the passive surface film that naturally occurs.

1.1.5 High-Strength Low-Alloy (HSLA) Steels

High-strength carbon and low-alloy steels, denoted usually under the acronym HSLA, must exhibit a minimum yield strength of 275 MPa or higher. They were developed primarily for the automotive industry to replace low-carbon steel in order to improve strength-to-weight ratio and meet the need for higher-strength construction-grade materials, particularly in the as-rolled condition. Strength is usually attained through the addition of small amounts of alloying elements, and hence several of these steels exhibit enhanced atmospheric corrosion resistance. Typically, HSLA steels are low-carbon steels containing up to 1.5wt% manganese, strengthened by small additions of elements, such as columbium, copper, vanadium or titanium and sometimes by special rolling and cooling techniques. Improved-formability HSLA steels contain additions such as zirconium, calcium, or rare-earth elements for sulfide-inclusion shape control. While additions of elements such as copper, silicon, nickel, chromium, and phosphorus improve atmospheric corrosion resistance, they also increase the cost. HSLA steels are particularly attractive for transportation-equipment components where weight reduction is important. HSLA steels are available in all standard wrought forms (i.e., sheet, strip, plate, structural shapes, bar-size shapes, and special shapes). In addition most HSLA steels exhibit directionally sensitive properties. For instance, formability and impact strength vary significantly for some grades depending on whether the material is tested longitudinally or transversely to the rolled direction. Major applications are typical passenger-car applications including door-intrusion beams, chassis members, reinforcing and mounting brackets, steering and suspension parts, bumpers, and wheels. Trucks, construction equipment, off-highway vehicles, mining equipment, and other heavy-duty vehicles use HSLA sheets or plates for chassis components, buckets, grader blades, and structural members outside the body. Structural forms are specified in applications such as offshore oil and gas rigs, single-pole power-transmission towers, railroad cars, and ship construction. Forming, drilling, sawing, and other machining operations on HSLA steels usually require 25 to 30% more power than do structural carbon steels. These alloys can be grouped into four classes: (i) as-rolled carbon–manganese steels, (ii) high-strength low-alloy steels, (iii) heat-treated carbon steels, and (iv) heat-treated low-alloy steels. Over 20 types of these commercial high-strength alloy steels are produced. Some have been developed to combine improved welding characteristics along with high strength. Most have good impact properties in addition to high strength. An example of the high-yield-strength grades are HY-80 and HY-100, which are used for naval vessels. This material combines high strength and toughness with weldability.

1.1.6 Ultrahigh-Strength Steels

Ultrahigh-strength structural steels must exhibit a minimum yield strength above 1380 MPa. Ultrahigh-strength steels start with grade 4340 and are modifications of this alloy. When these steels are used for aerospace components, they are usually produced by the vacuum-arc-remelt (VAR) process. They are classified into several broad categories based on chemical composition or metallurgical-hardening mechanisms. Medium-carbon alloy steels are generally modifications of grade 4330 or 4340 usually with increased molybdenum, silicon, and/or vanadium. These grades provide excellent hardenability in thick sections. Modified tool steels of the hot-work tool-steel varieties H11 modified and H13 provide the next step in increased hardenability and greater strength. Most steels in this group are air hardened in moderate to large sections and, therefore, are not likely to distort or quench crack. Structural uses of these steels are not as widespread as they once were, mainly because of the development of other steels costing about the same but offering greater fracture toughness.

Table 1.14 Mechanical properties of HSLA steels

Usual and trade name	Density (ρ, kg.m^{-3})	Min. yield strength 0.2% proof (σ_{YS}, MPa)	Min. ultimate tensile strength (σ_{UTS}, MPa)	Elongation (Z, %)
ASTM A242	7750	290–345	435–480	18
ASTM A517	7750	620–690	760–895	18
ASTM A572	7750	290–450	415–550	15–20
ASTM A588	7750	290–345	435–485	18
ASTM A606	7750	310–345	450–480	21–22
ASTM A607	7750	310–485	410–590	14–22
ASTM A618	7750	290–380	430–655	18–23
ASTM A633	7750	290–415	430–690	18–23
ASTM A656	7750	345–550	415–620	12–20
ASTM A715	7750	345–550	415–620	16–24
ASTM A808	7750	290–345	415–450	18–22
ASTM A871	7750	415–450	520–550	15–18
SAE 942X	7750	290	415	20
SAE 945C	7750	275–310	415–450	18–19
SAE 950 A	7750	290–345	430–483	18
SAE 955 X	7750	380	483	17
SAE 960 X	7750	415	520	16
SAE 970 X	7750	485	590	15
SAE 980 X	7750	550	655	10
HY-80	7750	550–690	n.a.	17–20
HY-130	7750	895–1030	n.a.	14–15

1.1.7 Tools and Machining Steels

Tool steels, owing to their relatively high hardness, developed from certain carbon, medium- and low-alloy steels through compositional adjustments or quenching and tempering at relatively low temperatures. These steels are used for applications that require: (i) resistance to wear/abrasion, (ii) thermal shock resistance, (iii) stability during heat treatment, (iv) strength at high temperatures, and (v) toughness. Tool steels are increasingly being used for machining tools and dies. Tool steels are melted at relatively low heat in electric furnaces and produced with careful attention to homogeneity. They can be further refined by techniques such as: argon/oxygen decarburization (AOD), vacuum-arc melting (VAM), or electroslag refining (ESR). Because of the high alloy content of certain groups, tool steels must be rolled or forged with care to produce satisfactory bar products. To develop their best properties, tool steels are always heat treated. Because the parts may distort during heat treatment, precision parts should be semifinished, heat treated, then finished. Severe distortion is most likely to occur during liquid quenching, so an alloy should be selected that provides the needed mechanical properties with the least severe quench. Tool steels are classified according to the American Institute of Steel and Iron (AISI) designation into several broad groups, some of which are further divided into subgroups according to alloy composition, hardenability, or mechanical similarities.

Table 1.15 Tool steels AISI designation

Class	AISI type	Description
Air-hardening medium-alloy tool steels (cold worked)	A	Air-hardening medium-alloy tool steels are best suited for applications such as machine ways, brick mold liners, and fuel-injector nozzles. The air-hardening types are specified for thin parts or parts with severe changes in cross-section, i.e., parts that are prone to crack or distort during hardening. Hardened parts from these steels have a high surface hardness; however, these steels should not be specified for service at elevated temperatures.
Air-hardening high-carbon and chromium steels (cold worked)	D	Air-hardening high-carbon and chromium tool steels possess high wear resistance and high hardenability and exhibit little distortion. They are best suited for applications such as machine ways, brick mold liners, and fuel-injector nozzles. The air-hardening types are specified for thin parts or parts with severe changes in cross-section, i.e., parts that are prone to crack or distort during hardening. Hardened parts from these steels have a high surface hardness; however, these steels should not be specified for service at elevated temperatures.
Hot-work steels	H	Hot-work tool steels, due to addition of tungsten and molybdenum, exhibit good heat and abrasion resistance from 315°C to 540°C. Hence, they serve well at elevated temperatures. However, although these alloys do not soften at these high temperatures, they should be preheated before and cooled slowly after service to avoid cracking. Note the chromium-containing grades are less expensive than the tungsten and molybdenum grades. For instance, the chromium grades H11 and H13 are used extensively for aircraft parts such as primary airframe structures, cargo-support lugs, catapult hooks, and elevon hinges. Subgroups are divided according to H10 to H19: chromium grades, H21 to H26: tungsten grades, H41 to H45: molybdenum grades.
Low-alloy tool steels	L	Low-alloy tool steels are often specified for machine parts when wear resistance combined with toughness is required.
High-speed tool steels (molybdenum alloy)	M	High-speed tool steels are the best known tool steels because they exhibit both abrasion and heat resistance with toughness. Hence they make good cutting tools because they resist softening and maintain a sharp cutting edge due to high hardness up to high service temperatures. This characteristic is sometimes called "red heat hardness". These deep-hardening alloys in which cobalt addition improves cutting are used for steady, high-load conditions rather than shock loads. Note that tempering at about 595°C increases toughness. Typical applications are pump vanes and parts for heavy-duty strapping machinery.
Oil-hardening cold-work tool steels	O	Oil-hardening cold-work tool steels are expensive but can be quenched less drastically than water-hardening types.
Mold tool steels	P	Special purpose tool steels containing chromium and nickel as major alloying elements. They exhibit low hardness and low resistance to work hardening when annealed.
Shock-resistant tool steels	S	Shock-resistant tool steels with Cr–W, Si–Mo, and Si–Mn as major alloys, are strong and tough, but they are not as wear resistant as many other tool steels. These steels resist sudden and repeated loadings. Applications include pneumatic tooling parts, chisels, punches, shear blades, bolts, and springs subjected to moderate heat in service.
High-speed tool steels (tungsten alloys)	T	High-speed tool steels are the best known tool steels because they exhibit both abrasion and heat resistance with toughness. Hence they make good cutting tools because they resist softening and maintain a sharp cutting edge due to high hardness up to high service temperatures. This characteristic is sometimes called "red heat hardness". These deep-hardening alloys in which cobalt addition improves cutting are used for steady, high-load conditions rather than shock loads. Note that tempering at about 595°C increases toughness. Typical applications are pump vanes and parts for heavy-duty strapping machinery.

Table 1.15 (continued)

Class	AISI type	Description
Water-hardening, or carbon, tool steels	W	Water-hardening tool steels containing 0.6 to 1.4wt% carbon are widely used because they combine low cost, good toughness, and excellent machinability. They are available as shallow, medium, or deep hardening, so the specific alloy selected depends on part cross-section and required surface and core hardnesses. Common applications include drills, shear knives, chisels, hammers and forging dies.

1.1.8 Iron-Based Superalloys

Iron-, nickel-, and cobalt-based alloys used primarily for high-temperature applications are known as superalloys. The iron-based grades, which are less expensive than cobalt or nickel-based grades, are of three types: (i) alloys that can be strengthened by a martensitic type of transformation, (ii) alloys that are austenitic and are strengthened by a sequence of hot and cold working, usually forging at 1100°C to 1150°C followed by finishing at 650°C to 880°C, and finally (iii) austenitic alloys strengthened by precipitation hardening. Some metallurgists consider the last group only as superalloys, the others being categorized as high-temperature, high-strength alloys. In general, the martensitic types are used at temperatures below 540°C, and the austenitic types above 540°C. The American Institute of Steel and Iron (AISI) designation defined the AISI 600 series, and divided superalloys in six subclasses of iron-based alloys:

- AISI 601 to 604: martensitic low-alloy steels,
- AISI 610 to 613: martensitic secondary-hardening steels,
- AISI 614 to 619: martensitic chromium steels,
- AISI 630 to 635: semiaustenitic and martensitic precipitation-hardening stainless steels,
- AISI 650 to 653: austenitic steels strengthened by hot/cold working,
- AISI 660 to 665: austenitic superalloys; all grades except alloy 661 are strengthened by second-phase precipitation.

Iron-based superalloys are characterized by both high-temperature and room-temperature strength and resistance to creep, oxidation, corrosion, and wear. Wear resistance increases with carbon content. Maximum wear resistance is obtained in alloys 611, 612, and 613, which are used in high-temperature aircraft bearings and machinery parts subjected to sliding contact. Oxidation resistance increases with chromium content. The martensitic chromium steels, particularly alloy 616 are used for steam-turbine blades. The superalloys are available in all conventional mill forms (i.e., billet, bar, sheet, and forgings) and special shapes are available for most alloys. In general, austenitic alloys are more difficult to machine than martensitic types, which machine best in the annealed condition. Crack sensitivity makes most of the martensitic steels difficult to weld by conventional methods. These alloys should be annealed or tempered prior to welding; even then, preheating and postheating are recommended. Welding drastically lowers the mechanical properties of alloys that depend on hot/cold working for strength. All of the martensitic low-alloy steels machine satisfactorily and are readily fabricated by hot working and cold working. The martensitic secondary-hardening and chromium alloys are all hot worked by preheating and hot forging. Austenitic alloys are more difficult to forge than the martensitic grades.

Table 1.16 Physical properties of tool steels

AISI type	UNS	Average chemical composition (wt%)	Density (ρ, kg.m^{-3})	Yield strength 0.2% proof (σ_{YS}, MPa)	Ultimate tensile strength (σ_{UTS}, MPa)	Elongation (Z, %)	Rockwell hardness scale C (HRC)	Coefficient of linear thermal expansion (α, 10^{-6} K^{-1})	Thermal conductivity (k, W.m^{-1}.K^{-1})	Quench medium	Annealing (°C)	Hardening (°C)	Tempering (°C)
Air-hardening tool steels													
A2	T30102	Fe–5Cr–1C–1Mn–1Mo–0.35V–0.5Si	7860	n.a.	n.a.	n.a.	57–62	10.7	n.a.	A	845–870	925–980	175–540
A6	T30106	Fe–2.2Mn–1.2Mo–1Cr–0.7C–0.5Si	7840	n.a.	n.a.	n.a.	54–60	11.5	n.a.	A	730–745	830–870	150–425
A7	T30107	Fe–5.4Cr–4.5V–2.4C–0.8Mn–1.1Mo–0.5Si–1W	7660	n.a.	n.a.	n.a.	58–66	12.0	n.a.	A	870–900	955–980	150–540
A8	T30108	Fe–5.1Cr–1.4Mn–1.25W–1Si–0.55C–0.5Mn	7870	n.a.	n.a.	n.a.	48–57	n.a.	n.a.	A	845–870	980–1010	175–595
A9	T30109	Fe–1.25Cr–1.9Mn–1.8Ni–1.5Mo–1.25Si	7780	n.a.	n.a.	n.a.	40–56	12.0	n.a.	A	845–870	980–1025	510–620
A10	T30110	Fe–1.25Cr–0.5Mn–1.8Ni–1.5Mo–1.0Si	7680	n.a.	n.a.	n.a.	52–62	12.8	n.a.	A	765–795	790–815	175–425
Air-hardening cold-work steels													
D2	T30402	Fe–12Cr–1.5C–1.1V–0.6Mn–0.9Mo	7700	n.a.	n.a.	n.a.	58–64	10.4	n.a.	A	870–900	980–1025	205–540
D3	T30403	Fe–12Cr–2.2C–1V–0.6Mn	7700	n.a.	n.a.	n.a.	58–64	12.0	n.a.	O	870–900	925–980	205–540

Designation	Composition											
D5	T30405	Fe–12Cr–1.5C–0.9Mo–1.1V–0.6Mn–0.6Si	7700	n.a.	n.a.	n.a.	58–64	11.0	n.a.	A	870–900	970–1010	205–540
D7	T30407	Fe–12Cr–2.25C–4.1V–0.6Mn–0.6Si	7700	n.a.	n.a.	n.a.	58–64	12.2	n.a.	A	870–900	1010–1065	150–540
Hot-work tool steels												
H11	T20811	Fe–5.1Cr–1.4Mo–0.45V–1Si	7750	n.a.	n.a.	n.a.	38–55	11.9	42	A	845–900	995–1025	540–650
H13	T20813	Fe–5.1Cr–1.4Mo–1V–0.3C	7760	1290–1570	1495–1960	13–15	40–53	10.4	29	A	845–900	995–1040	540–650
H21	T20821	Fe–9.25W–3.4Cr–0.45V–0.3Mn	8280	n.a.	n.a.	n.a.	40–55	12.4	27	A, O	870–900	1095–1205	595–675
H42	T20842	Fe–6.13W–4Cr–5Mo–2V–0.6C	8150	n.a.	n.a.	n.a.	50–60	11.0	n.a.	O, A, S	845–900	1120–1220	565–650
Low-alloy tool steels												
L2	T61202	Fe–1Cr–0.5Si–0.73C–0.5Mn–0.25Mo–0.2V	7860	510–1790	710–2000	5–25	30–54	11.3	n.a.	O, W	760–790	845–925	175–315
L6	T6106	Fe–1.6Ni–0.9Cr–0.5Si–0.7C–0.5Mn–0.5Mo	7860	380–1790	655–2000	4–25	32–54	11.3	n.a.	O	760–790	790–845	175–540
Molybdenum alloy high-speed tool steel												
M1	T11301	Fe–8.7Mo–3.75Cr–1.25W–1.15V–0.8C	7890	n.a.	n.a.	n.a.	60–65	10.1	n.a.	O, A, S	815–870	1175–1220	540–595
M2	T11302	Fe–5Mo–3.75Cr–6W–2V–0.8C	8160	n.a.	n.a.	n.a.	63–65	10.1	22	O, A, S	870–900	1190–1230	540–595
M3	T11303	Fe–5.5Mo–5.8W–2.5V–1C	8150	n.a.	n.a.	n.a.	62–66	10.1	n.a.	O, A, S	870–900	1205–1230	540–595
M48	T11348	Fe–10W–4Mo–3V–9Co–3.8Cr–1.5C–0.3Si	7960	n.a.	n.a.	n.a.	65–70	10.6	n.a.	O, A, S	870–900	1175–1200	540–595

Table 1.16 (continued)

AISI type	UNS	Average chemical composition (wt%)	Density (ρ, kg.m^{-3})	Yield strength 0.2% proof (σ_{YS}, MPa)	Ultimate tensile strength (σ_{UTS}, MPa)	Elongation (Z, %)	Rockwell hardness scale C (HRC)	Coefficient of linear thermal expansion (α, 10^{-6} K^{-1})	Thermal conductivity (k, W.m^{-1}.K^{-1})	Quench medium	Annealing (°C)	Hardening (°C)	Tempering (°C)
Oil-hardening tool steels													
O1	T31501	Fe–0.9C–1.2Mn–0.5W	7850	n.a.	n.a.	n.a.	57–62	11.2	n.a.	O	760–790	790–815	175–260
O2	T31502	Fe–0.9C–0.5Cr–1.2Mn–0.5Si	7660	n.a.	n.a.	n.a.	57–62	11.2	n.a.	O	745–775	760–800	175–260
O6	T31506	Fe–1.4C–0.7Mn–1Si–	7700	n.a.	n.a.	n.a.	58–63	11.2	n.a.	O	765–790	790–815	175–315
O7	T31507	Fe–1.2C–1Mn–1.5W–0.6Cr	7800	n.a.	n.a.	n.a.	58–64	11.2	n.a.	O, W	790–815	790–880	175–290
Mold steels													
P2	T51602	Fe–1Cr–0.3Ni–0.3Mo	7860	n.a.	n.a.	n.a.	58–64	13	n.a.	A, O	730–815	840–845	175–230
P6	T51606	Fe–3.5Ni–1.5Cr–0.5Mn	7850	n.a.	n.a.	n.a.	58–64	12	n.a.	A, O	845–870	800–830	175–260
Tungsten alloy high-speed tool steels													
T1	T12001	Fe–18W–1.1V–4.13Cr–0.75C	8670	n.a.	n.a.	n.a.	63–65	9.7	20.0	O, A, B	870–900	1260–1300	540–595
T5	T12005	Fe–18.5W–2V–4.38Cr–1Mo–0.8C	8750	n.a.	n.a.	n.a.	n.a.	11.2	n.a.	O, A, B	870–900	1275–1300	540–595
T15	T12015	Fe–12.4W–4.8V–4.88Cr–1.55C–5Co	8190	n.a.	n.a.	n.a.	64–68	9.9	21.0	O, A, B	870–900	1205–1260	540–650
Shock-resisting tool steels													
S1	T41901	Fe–1.4Cr–2.3W–0.75Si–0.48C–0.25Mn–0.25V	7880	415–1895	690–2070	4–24	48–58	12.4	n.a.	O	790–815	900–955	205–650

S2	T41902	Fe–1.1Si–0.48C–0.45Mo–0.40Mn–0.5V	7790	n.a.	n.a.	n.a.	n.a.	10.9	n.a.	W, B	760–790	845–900	175–425
S5	T41905	Fe–2Si–0.8Mn–0.58C–0.5Cr	7760	440–1930	725–2345	5–25	37–60	11.0	n.a.	O	775–800	870–925	175–425
S7	T41907	Fe–0.5C–1.5Mo–3.3Cr–0.55Mn–0.6Si	7760	380–1585	640–2170	7–25	45–57	10.5	n.a.	A, O	815–845	925–955	205–620
Water-hardening, or carbon tool steels													
W1	T72301	Fe–1.1C–0.25Si–0.25Mn	7840	n.a.	n.a.	n.a.	58–65	10.4	48	W, B	740–790	760–845	175–345
W2	T72302	Fe–1.2C–0.2V–0.25Si–0.25Mn	7850	n.a.	n.a.	n.a.	58–65	10.4	48	W, B	740–790	760–845	175–345
W5	T72305	Fe–1.1C–0.5Cr–0.25Si–0.25Mn	7850	n.a.	n.a.	n.a.	58–65	10.4	48	W, B	760–790	760–845	175–345

Quenching medium: A for air, W for water, B for brine, S for salt bath and O for oil.

1.2 Nickel and Nickel Alloys

1.2.1 Description and General Properties

Nickel [7440-02-0] with the chemical symbol Ni, the atomic number 28, and the relative atomic mass 58.6934(2), is the third element of the upper transition metal of the group VIIIB (10) of Mendeleev's periodic chart. It was named after the English Old Nick. Pure nickel is a dense (8902 kg.m^{-3}), tough, silvery white lustrous ferromagnetic metal, which exhibits both a high electrical (6.9 $\mu\Omega$cm) and thermal conductivity (90.7 W.m^{-1}.K^{-1}), and has a high melting point ($1455°$C). The face-centered cubic (fcc) crystal structure gives to the metal good ductility and nickel can be fabricated readily by the use of standard hot and cold working methods. From a chemical point of view, the pure metal is corrosion resistant to attack by moist air or water at room temperature, but dissolves in strong diluted mineral acids such as HCl. Nickel reacts only slowly with fluorine, due to the self-formation of a thin protective passivating layer of nickel fluoride. Therefore nickel and cupro-nickel alloys such as Monel$^{®}$ 400 and K-500 are extensively used for handling fluorine gas, hydrogen fluoride, and hydrofluoric acid. Nickel is extensively used in coinage but is more important either as the pure metal or in the form of alloys for its many domestic and industrial applications. Cost (1998) – pure nickel (99.99wt% Ni) is priced at 5.05 \$US.kg^{-1} (2.29 \$US.lb^{-1}).

1.2.2 History

Nickel was used industrially as an alloying metal almost 2000 years before it was isolated and recognized as a new element. As early as 200 BC, the Chinese made substantial amounts of a white alloy from zinc and a copper–nickel ore found in Yunnan province. The alloy, known as pai-t'ung, was exported to the Middle East and even to Europe. Later, miners in Saxony (Germany) encountered what appeared to be a copper ore but found that processing it yielded only a useless slaglike material. Earlier, an ore of this same type was called Kupfernickel because they considered it bewitched and ascribed it to the devil, Old Nick, and his mischievous gnomes because, though it resembled copper ore, it yielded a brittle, unfamiliar metal. It was from niccolite, studied by the Swedish chemist and mineralogist, Baron Axel Fredrik Cronstedt, that nickel was first isolated and recognized as a new element in 1751. In 1776 it was established that pai-t'ung, now called nickel–silver, was composed of copper, nickel, and zinc. Demand for nickel–silver was stimulated in England about 1844 by the development of silver electroplating, for which it was found to be the most desirable base. The use of pure nickel as a corrosion-resistant electroplated coating developed a little later. Small amounts of nickel were produced in Germany in the mid-19th century. More substantial amounts came from Norway, and a little from a mine at Gap, PA in the United States. In 1877, a new large ore deposit was discovered in New Caledonia and dominated the market until the development of the copper–nickel ores of the Copper Cliff, Sudbury, ON (Canada), which after 1905 became the world's largest source of nickel, until the discovery in the late 1970s of the Norylsk complex in the former Soviet Union.

1.2.3 Natural Occurrence, Minerals, and Ores

Nickel, with a relative abundance in the Earth's crust of 70 mg.kg^{-1} is twice as abundant as copper, and the inner core is supposed to be made of a Ni–Fe alloy. Nickel never occurs free in nature, but only as an alloy with iron in certain meteorites. However, due to its chalcophile geochemical character like copper, most nickel occurs primarily as minerals in combination with arsenic, antimony, and sulfur. Hence, the major minerals are the sulfides: **pentlandite** ((Ni, Fe)$_9$S$_8$, cubic), and **millerite** (NiS, hexagonal) and the sulfosalts: **niccolite** (NiAs, hexagonal), (NiSb, hexagonal) **rammelsbergite** (NiAs$_2$, orthorhombic), **gersdorfite** (NiAsS, cubic), and ullmanite (NiSbS, cubic). As traces, it often occurs in **pyrrhotite** (FeS_{1-x}, hexagonal), **chalcopyrite** ($CuFeS_2$, tetragonal), **cubanite** ($CuFe_2S_3$, orthorhombic), and nickel-bearing laterites, such as **garnierite** (($Ni,Mg)_3Si_2O_5(OH)_4$), or **nickeliferous limonite** (($Fe,Ni)O(OH).nH_2O$). Nickel is industrially recovered from nickel-bearing sulfides (e.g., Sudbury, Canada, and Norylsk, CIS) and laterites (e.g., New Caledonia). As a general rule, in sulfide ore deposits, precious metals (i.e., Au, Ag), the six PGMs (i.e., Ru, Rh, Pd, Os, Ir, and Pt), along with Co, Se, and Te are always present, and represent an important commercial by product. Laterites, which are residual sedimentary rocks such as bauxite (see Section 2.1.2) is the result of in-situ weathering of ultramafic igneous rocks (e.g., peridotites such as dunite). This alteration under subtropical climates exerts a leaching action of the host rock, and the nickel-soluble cations percolate downward and may reach a concentration sufficiently high to make mining economical. Owing to this method of formation, laterite deposits are found near the surface as a soft, frequently claylike material, with nickel concentrated in strata as a result of weathering. The world's largest nickel producers are Russia (CIS), Canada, New Caledonia, and Australia.

1.2.4 Processing and Industrial Preparation

The metallurgy of nickel depends on the type of ore processed. As a general rule, the sulfide ore is transformed into nickel(III) sulfide, Ni_2S_3, which is roasted in air to give nickel(II) oxide, NiO, while the laterite ore is fired to give off nickel oxide. In both processes, the metal is extracted by carbothermic reduction of the oxide. Some high-purity nickel is made by refining.

Nickel oxide from sulfide ores – sulfide ore deposits are usually mined by underground techniques in a manner similar to that of copper ores. Sulfide ores are crushed and ground in order to liberate nickel-bearing minerals from the inert gangue materials. Afterwards, raw ores are concentrated selectively by common beneficiation processes (e.g., both froth flotation, and magnetic separation). After separation from gangue minerals, the ore concentrate contains between 6wt% to 12wt% nickel. For high-copper-containing ores the concentrate is then subjected to a second selective flotation which produces: (i) a low-nickel copper concentrate, and (ii) a rich nickel concentrate, each processed in a separate smelting process. Nickel concentrates may be leached either with sulfuric acid or ammonia, or they may be dried and smelted in flash and bath processes, as is the case with copper. Nickel requires higher smelting temperatures of 1350°C in order to produce an artificial nickel–iron sulfide known as matte, which contains 25 to 45wt% Ni. In the next step, iron in the matte is converted in a rotating converter to iron oxide, which combines with a silica to form the slag. The slag is removed leaving a matte of 70–75wt% nickel. The conversion of nickel sulfide directly to metal is achieved at high temperature above 1600°C. The matte is then roasted in air to give the nickel oxide.

Nickel oxide from laterite ores – laterites are usually mined in an earth-moving operation, with large shovels, draglines, or front-end loaders extracting the nickel-rich strata and discarding large boulders and waste material. Recovery of nickel from laterite ores is an energy intensive process, requiring high energy input. In addition, laterites are difficult to concentrate by common ore beneficiation process, and hence a large amount of ore must be smelted to extract the metal. Because they contain large amounts of water (i.e. 35 to 40wt% H_2O), the major operation consists of drying in rotary-kiln furnaces, giving the nickel oxide.

Nickel metal – nickel metal is obtained by carbothermic reduction of the nickel oxide in an electric furnace operating between 1360°C and 1610°C.

Refining – The two common refining processes are: (i) the electrolytic refining, and (ii) the carbonyl process. The electrorefining process uses a sulfate or chloride electrolyte, and is performed in electrolyzers with two compartments separated by a diaphragm to prevent the passage of impurities from anode to cathode. During electrolysis, the impure nickel anode (+)

is dissolved and nickel electrodeposits onto pure nickel cathodes, while more noble metals (e.g., Au, Ag, and PGMs) are recovered in the slurries at the bottom of the reactor, and soluble metals (e.g., Fe, Cu) remain in the electrolyte. In the carbonyl refining process, carbon monoxide is passed through the matte, yielding nickel and iron carbonyls (i.e., $Ni(CO)_4$ and $Fe(CO)_5$). After separation, nickel carbonyl is decomposed onto pure nickel pellets to produce nickel shot.

1.2.5 Nickel Alloys

Nickel and the nickel alloys constitute a family of alloys with increasing importance in many industrial applications because they exhibit both a good corrosion resistance in a wide variety of corrosive environments, and an excellent heat resistance from low to elevated temperatures. Some types have an almost unsurpassed corrosion resistance in certain media, but nickel alloys are usually more expensive than, for example, iron-based or copper-based alloys or than plastic materials of construction. Nickel alloys refer to alloys in which nickel is present in greater proportion than any other alloying element. Actually, nickel content throughout the alloy families ranges from 32.5wt% to 99.5wt%. The most important alloying elements are Fe, Cr, Cu, and Mo, and a variety of alloy classes are commercially available. Two groups of alloy classes can be distinguished: (i) alloys which depend primarily on the inherent corrosion characteristics of nickel itself; (ii) alloys which greatly depend on chromium as the passivating alloying element such as for the stainless steels. Common nickel alloy families include: commercially pure nickel, binary systems (e.g., Ni–Cu, Ni–Si, and Ni–Mo), ternary systems (e.g., Ni–Cr–Fe, and Ni–Cr–Mo), more complex systems (e.g., Ni–Cr–Fe–Mo–Cu), and superalloys, and they are usually grouped in the following classes:

- Commercially pure and high nickel alloys,
- Nickel–molybdenum alloys,
- Nickel–copper alloys,
- Nickel–chromium alloys,
- Nickel–chromium–iron alloys,
- Nickel–chromium–molybdenum alloys,
- Nickel–chromium–iron–molybdenum–copper alloys,
- Nickel superalloys.

Structural applications that require specific corrosion resistance or elevated temperature strength receive the necessary properties from nickel and its alloys. Some nickel alloys are among the toughest structural materials known. When compared to steel, other nickel alloys exhibit both an ultrahigh tensile strength, high proportional limits, and large Young's moduli. At cryogenic temperatures, nickel alloys are strong and ductile. Several nickel-based superalloys are specific for high-strength applications at temperatures to 1090°C. High-carbon nickel-based casting alloys are commonly used at moderate stresses above 1200°C. Commercial nickel and nickel alloys are available in a wide range of wrought and cast grades; however, considerably fewer casting grades are available. Wrought alloys tend to be better known by trade names such as Monel, Hastelloy, Inconel, Incoloy. The casting alloys contain additional elements, such as silicon and manganese, to improve castability and pressure tightness.

1.2.6 Ni–Ti Shape Memory Nickel Alloys

The shape memory alloy nickel–titanium (55Ni–45Ti) has several attractive characteristics. Depending on its crystal structure, the alloy can be either superelastic or reversible in shape. To exhibit shape memory, an object is deformed at low temperatures (i.e., the martensitic

Table 1.17 Description of main nickel alloy classes

Nickel alloy class	Description
Commercially pure nickels and extrahigh nickel alloys	Major wrought alloys in this group are commercially pure nickel 200 and 201 grades. The cast grade is recommended for use at temperatures above 315°C owing to its lower carbon content which prevents graphitization and attendant ductility loss. These two grades are particularly suitable where corrosion resistance to caustic alkaline hydroxides (i.e. NaOH, KOH), high-temperature halogens and hydrogen halides (e.g., HF), and molten fluorides in non-oxidizing conditions is required. These alloys are particularly well suited to food-contact applications. Duranickel[®]301, a precipitation-hardened nickel alloy, has excellent spring properties up to 315°C, and corrosion resistance is similar to that of commercially pure wrought nickel. Commercially pure nickel has good electrical, magnetic, and magnetostrictive properties.
Cupro-nickels (Ni–Cu)	In this category the most common cupro-nickel alloys are the Monel[®]400 and Monel[®]K-500. The Ni–Cu alloys differ from Nickel 200 and 201 because their strength and hardness can be increased by age hardening. Ni–Cu alloys exhibit higher corrosion-resistance than commercially pure nickel, especially to sulfuric and hydrofluoric acids, and chloride brines. Handling of waters, including seawater and brackish water, is the major application of these two alloys in the CPI (e.g., desalination plants). In addition, Monel[®]400 and K-500 are immune to chloride-ion stress-corrosion cracking, which is often considered in their selection.
Ni–Mo	The Ni–Mo binary type, Hastelloy[®]B-2, offers superior resistance to hydrochloric acid, aluminum chloride catalysts, and other strongly reducing chemicals. It also has excellent high-temperature strength in inert atmospheres and in a vacuum. The Ni–Mo alloys are commonly used for handling hydrochloric acid in all concentrations at temperatures up to boiling point. These alloys are produced commercially under the trade names Hastelloy[®]B and Chlorimet 2.
Ni–Cr–Fe	Ni–Cr–Fe alloys are known commercially under the common trade names Haynes[®]214 and 556, Inconel[®]600, and Incoloy[®]800. Haynes[®]214 has excellent resistance to oxidation to 1200°C, and resists carburizing and chlorine-contaminated atmospheres. Haynes[®]556 combines effective resistance to sulfidizing, carburizing, and chlorine-bearing environments with good oxidation resistance, fabricability and high-temperature strength. Inconel[®]600 exhibits good resistance to both oxidizing and reducing environments. Incoloy[®]800 has good resistance to oxidation and carburization at elevated temperatures, and it resists sulfur attack, internal oxidation, scaling, and corrosion in many harsh atmospheres, and is suitable for severely corrosive conditions at elevated temperatures.
Ni–Cr–Mo	Ni–Cr–Mo alloys are known commercially under the common trade names Hastelloy[®]C-276 and C-22, and Inconel[®]625. Hastelloy[®]C-22 has better overall corrosion resistance and versatility than any other Ni–Cr–Mo alloy. In addition it exhibits outstanding resistance to pitting, crevice corrosion, and stress-corrosion cracking. Hastelloy[®]C-276 has excellent corrosion resistance to strong oxidizing and reducing corrosives, acids, and chlorine-contaminated hydrocarbons. It is also one of the few materials (along with titanium) that withstands the corrosive effects of wet chlorine gas, hypochlorite, and chlorine dioxide. Present applications include the pulp and paper industry, various pickling acid processes, and production of pesticides and various agrichemicals.
Ni–Cr–Fe–Mo–Cu	Ni–Cr–Fe–Mo–Cu alloys are known commercially under the common trade names Hastelloy[®]G-30 and H, Haynes[®]230, Inconel[®]617, 625, and 718, and Incoloy[®]825. Haynes[®]230 has excellent high-temperature strength, and heat and oxidation resistance, making it suitable for various applications in the aerospace, airframe, nuclear, and chemical-process industries. Hastelloy[®]G-30 has many advantages over other metallic and nonmetallic materials in handling phosphoric acid, sulfuric acid, and oxidizing acid mixtures. Hastelloy[®]H exhibits a localized corrosion resistance equivalent to, or better than, Inconel[®]625. In addition it has good resistance to hot acids and excellent resistance to stress-corrosion cracking. It is often used in flue gas desulfurization equipment. Inconel[®]617 resists cyclic oxidation at 1100°C, and has good stress-rupture properties above 990°C. Inconel[®]625 has high strength and toughness from cryogenic temperatures to 1100°C, good oxidation resistance, exceptional fatigue strength, and good resistance to many corrosives. It is extensively used in furnace mufflers, electronic parts, chemical and food-processing equipment, and heat-treating equipment. Inconel[®]718 has excellent strength from –250°C to 700°C. The alloy is age hardenable, can be welded in the fully aged condition, and has excellent oxidation resistance up to 1800°C.

Table 1.17 (continued)

Nickel alloy class	Description
Ni–Cr–Fe–Mo–Cu	Incoloy® 825 resists pitting and intergranular corrosion, reducing acids, and oxidizing chemicals. Applications include pickling-tank thermowell and bayonet heaters, spent nuclear-fuel-element recovery and radioactive-waste handling, chemical-tank trailers, evaporators, sour-well tubing, hydrofluoric-acid production, pollution-control equipment.
Nickel superalloys	Ni-based superalloys can be classified in three groups. (i) Those strengthened by intermetallic compound precipitation in an fcc matrix; these alloys are well known under the common trade names Astroloy, Udimet® 700, and Rene® 95. (ii) Another type of Ni-based superalloy is represented by Hastelloy® X. This alloy is essentially solid-solution strengthened. (iii) The final class consists of oxide-dispersion-strengthened (ODS) alloys such as MA-754 which is strengthened by dispersions of yttria coupled with gamma prime precipitation (e.g., MA-6000). Nickel-based superalloys are used in cast and wrought forms, although special processing (e.g., powder metallurgy, isothermal forging) often is used to produce wrought versions of the more highly alloyed compositions such as Udimet® 700, or Astroloy®.

Table 1.18 Physical properties of commercially pure and high nickel alloys (annealed)

Usual and trade name	UNS	Average chemical composition (wt%)	Density (ρ, kg.m^{-3})	Yield strength 0.2% proof (σ_{YS}, MPa)	Ultimate tensile strength (σ_{UTS}, MPa)	Elongation (Z, %)	Brinell hardness (HB)	Coefficient of linear thermal expansion (a, 10^{-6} K^{-1})	Thermal conductivity (k, W.m^{-1}.K^{-1})	Specific heat capacity (c_P, J.kg^{-1}.K^{-1})	Electrical resistivity (ρ, $\mu\Omega$.cm)
Nickel 200	N02200	99.5	8890	148	462	47	109	13.3	74.9	456	9.5
Nickel 201	N02201	99.6	8890	103	403	50	129	13.1	79.3	456	8.5
Nickel 205	N02205	99.6	8890	90	345	45	80	13.3	75.0	456	9.5
Nickel 211	N02211	93.7Ni–4.75Mn	8890	240	530	40	n.a.	13.3	44.7	532	16.9
Nickel 233	N02233	99.00	8890	150	400	40	100	13.3	n.a.	n.a.	n.a.
Nickel 270	N02270	99.95	8890	60–110	310–345	50	85	13.3	86	460	7.5
Nickel 290	N02290	99.95	8890	n.a.	n.a.	n.a.	n.a.	13.3	n.a.	n.a.	n.a.

condition) and maintained in this state. Upon heating, the object attempts to return to its original state (i.e., the austenitic condition). Superelastic 55Ni–45Ti tubes can be bent ten times more than steel tubes without kinking or collapsing. Owing to these properties combined with biocompatibility, 55Ni–45Ti shape memory alloy is used for manufacturing microsurgical instruments (e.g., catheters, stents). In addition, superelastic 55Ni–45Ti wire is currently used to make extremely flexible but essentially unbreakable spectacle frames, and flexible antennae for cellular phones.

Table 1.19 Physical properties of nickel alloys (annealed)

Usual and trade name	UNS	Average chemical composition (wt%)	Class	Density (ρ, kg.m^{-3})	Young's modulus (E, GPa)	Yield strength 0.2% proof (σ_{YS}, MPa)	Ultimate tensile strength (σ_{UTS}, MPa)	Elongation (Z, %)	Brinell hardness (HB)	Coefficient of linear thermal expansion	Thermal conductivity (k, W.m^{-1}.K^{-1})	Specific heat capacity (c_P, J.kg^{-1}.K^{-1})	Electrical resistivity (ρ, μW.cm)
904L	N08904	25Ni–21Cr–4.5Mo–2Mn–1Si	Fe–Ni–Cr alloys	n.a.	n.a.	220	490–830	35	n.a.	n.a.	n.a.	n.a.	n.a.
AL-6X	N08366	24Ni–21Cr–6.5Mo–2Mn–1Si	Fe–Ni–Cr alloys	n.a.	n.a.	210–240	515	10–30	n.a.	n.a.	n.a.	n.a.	n.a.
AL-6XN	N08367	24Ni–21Cr–3.5Mo–2Mn–1Si	Fe–Ni–Cr alloys	n.a.	n.a.	315	715	30	n.a.	n.a.	n.a.	n.a.	n.a.
Alloy® 20Mo-4	N08024	38Ni–32Fe–24Cr–4Mo–1Cu	Fe–Ni–Cr alloys	8106	186	262	615	41	155	14.9	12.1	458	105.6
Astroloy® M	N13017	54.8Ni–15Cr–17Co–5.3Mo–4Al–3.5Ti	Ni–Cr alloys	7910	n.a.	1050	1410	16	n.a.	n.a.	n.a.	n.a.	n.a.
Carpenter® 20Cb-3	N08020 W88021	35Ni–38Fe–20Cr–3Cu–2Mo–1Nb	Fe–Ni–Cr alloys	8080	n.a.	240	550	30	184	14.7	12.2	500	108.2
Carpenter® 20Mo-6	N08026	46Fe–35Ni–24Cr–6Mo–3Cu	Fe–Ni–Cr alloys	8133	186	275	607	50	n.a.	14.8	12.1	460	108.2
Cronifer® 1025	N08925	20Ni–25Cr–6.5Mo–2Mn–1Si–1Cu	Fe–Ni–Cr alloys	n.a.	n.a.	300	600	40	n.a.	n.a.	n.a.	n.a.	n.a.
Duranickel® 301	N03301	Ni–4.5Al–0.75Ti	Low alloy	8250	207	862	1170	25	346	13.0	23.8	435	42.5
Hastelloy® B	N10001 W80001	64Ni–28Mo–1Cr–5Fe–1Si	Ni–Cr–Fe alloys	9240	n.a.	n.a.	600–980	<60	100–230	10.3	11.1	n.a.	n.a.
Hastelloy® B2	N10665 W80665	68Ni–28Mo–1Cr–2Fe	Ni–Mo alloys	9220	n.a.	526	955	53	235	10.3	11.1	373	137

Table 1.19 (continued)

Usual and trade name	UNS	Average chemical composition (wt%)	Class	Density (ρ, kg.m^{-3})	Young's modulus (E, GPa)	Yield strength 0.2% proof (σ_{YS}, MPa)	Ultimate tensile strength (σ_{UTS}, MPa)	Elongation (Z, %)	Brinell hardness (HB)	Coefficient of linear thermal expansion	Thermal conductivity (k, W.m^{-1}.K^{-1})	Specific heat capacity (c_P, J.kg^{-1}.K^{-1})	Electrical resistivity (ρ, μW.cm)
Hastelloy® C22	N26022 W86022	51.6Ni–21.5Cr–13.5Mo–5.5Fe–2.5Co–4V	Ni–Cr–Fe alloys	n.a.	n.a.	372	785	62	209	n.a.	n.a.	n.a.	n.a.
Hastelloy® C276	N10276 W80276	57Ni–17Mo–15.5Cr–5.5Fe–4W	Ni–Cr–Fe alloys	8940	205	355–415	790	50–61	185	9.8–11.3	10.1–12.5	427	123
Hastelloy® C4	N06455 N26455	Ni–16Mo–16Cr	Ni–Mo alloys	8640	n.a.	416	768	52	184	10.8	10.1	406	125
Hastelloy® G3	N06985	49Ni–22.5Cr–19.5Fe–7Mo–2Cu	Ni–Cr–Fe alloys	8300	199	320	690	50	184	12.2	14.26	452	107.5
Hastelloy® HX	N06920	48.3Ni–22Cr18.5Fe–9Mo	Ni–Cr–Fe alloys	8230	205	358	793	46	184	13.3	11.6	461	116.0
Hastelloy® S	N06635	Ni–16Cr–15Mo–3Fe–2Co–1W	Ni–Cr–Fe alloys	8750	212	445	835	49	168	11.5	14.0	398	128.0
Hastelloy® X	N06002 W86002	52Ni–9Mo–21Cr–18Fe	Ni–Cr–Fe alloys	8250	196	360	785–1250	43	150–280	12.6–13.8	9.1–12	486	118.0
Haynes® 230	N06230	n.a.	Ni–Cr–Fe alloys	8830	211	390	860	48	200	12.6	8.9	397	125.0
Haynes® 556	n.a.	31Fe–20Ni–22Cr–18Co–3Mo–3W	Fe–Ni–Cr alloys	8230	205	410	815	47	91	14.6	11.1	464	95.2
Incoloy® 902	N09902	47Fe–42Ni–5Cr–2.5Ti–1Si–0.5Al	PH Ni superalloys	8050	n.a.	760	1210	25	300	7.6	12.1	502	101.0
Incoloy® 907	N19907	42Fe–38Ni–13Co–4.7Nb–1.5Ti	PH Ni superalloys	8330	n.a.	1100	1310	14	n.a.	7.7	14.8	431	69.7
Incoloy® 909	N19909	42Fe–38Ni–13Co–4.7Nb–1.5Ti	PH Ni superalloys	8300	159	1035	1275	15	n.a.	7.7	14.8	427	72.8

Incoloy® 800	N08800	46Fe–33Ni–21Cr	Fe–Ni–Cr alloys	7940	193	295–310	590–600	44–45	138–180	14.4	11.5	460	93
Incoloy® 800HT	N08811	46Fe–33Ni–21Cr	Fe–Ni–Cr alloys	7940	193	295–310	590–600	44–45	138–180	14.4	11.5	460	98.9
Incoloy® 825	N08825	42Ni–22Cr–30Fe–3Mo–2Cu–1Ti	Fe–Ni–Cr alloys	8140	206	310	690	45	150	14.0	11.1	440	113.0
Incoloy® 925	N09925	44Ni–28Fe–21Cr3Mo–2Ti	PH Fe–Ni superalloys	8140	n.a.	815	1210	24	315	13.2	13.9	435	116.6
Incoloy® 903	N19903	42Fe–38Ni–15Co–3Nb–1.5Ti	PH Ni superalloys	8250	n.a.	1100	1310	14	n.a.	7.65	16.7	435	61.0
Inconel® 600	N06600	72Ni–15.5Cr–8Fe	Ni–Cr–Fe alloys	8420	207	310	655	40–45	120–150	11.5–13.3	14.8	460	103
Inconel® 601	N06601	60.5Ni–23Cr–1.5Al–15.1Fe	Ni–Cr–Fe alloys	8110	207	210–340	550–790	40–70	110–150	13.75	11.2	448	119
Inconel® 617	N06617	55.7Ni–21.5Cr–12.5Co–9Mo–1.2Al–0.1C	Ni–Cr–Fe alloys	8360	211	350	760	45–58	173	11.6	13.6	419	122
Inconel® 625	N06625	58Ni–22Cr–4Nb–9Mo–0.3Ti–0.3Al	Ni–Cr–Fe alloys	8440	207	517	930	42–45	186	12.8	9.8	410	129
Inconel® 690	N06690	58Ni–22Cr–9Mo–5Fe–1Co–3.5Nb	Ni–Cr–Fe alloys	8190	211	348	725	41	175	14.1	13.5	450	114.8
Inconel® 718	N07718	52.5Ni–19Cr–18.8Fe–5.2Nb–3.1Mo–0.9Ti–0.5Al	PH Ni superalloys	8190	211	1036–1180	1240–1350	17	382	13.0	11.4	435	125
Inconel® MA 754	N07754	78Ni–20Cr–1Fe–0.6Y₂O₃	PH Ni superalloys	8300	n.a.	545–560	940–965	20–22	n.a.	12.2	14.3	n.a.	107.5
Inconel® X-750	N07750	70Ni–15Cr–7Fe–2.5Ti–0.6Al–0.8Nb	PH Ni superalloys	8250	207	690–900	1137–1240	20	382	12.6	12.0	425	122
Invar® 36	K93601	64Fe–36Ni	Fe–Ni low exp. alloys	8100	150	275–415	450–585	30–45	160	1	11	515	n.a.

Table 1.19 (continued)

Usual and trade name	UNS	Average chemical composition (wt%)	Class	Density (ρ, kg.m^{-3})	Young's modulus (E, GPa)	Yield strength 0.2% proof (σ_{YS}, MPa)	Ultimate tensile strength (σ_{UTS}, MPa)	Elongation (Z, %)	Brinell hardness (HB)	Coefficient of linear thermal expansion (α, 10^{-6} K^{-1})	Thermal conductivity (k, W.m^{-1}.K^{-1})	Specific heat capacity (c_P, J.kg^{-1}.K^{-1})	Electrical resistivity (ρ, μW.cm)
Invar® 42	K94100	48Fe–52Ni	Fe–Ni low exp. alloys	8100	144	235	538	32	160	1	11	515	n.a.
JS-700	N08700	25Ni–21Cr–5Mo–2Mn–1Si	Fe–Ni–Cr alloys	n.a.	n.a.	240	550	30	n.a.	n.a.	n.a.	n.a.	n.a.
Monel® 400	N04400	66(Ni+Co)–32Cu–1.5Fe–1.0Mn	Ni–Cu (cupronickel)	8830	180	240	550	40	110–150	13.9	21.7	423	51.0
Monel® 450	N04405	Cu–31Ni–1Zn–1Mn	Ni–Cu (cupronickel)	8910	180	165	385	46	90	15.5	29.4	380	41.2
Monel® K500	N05500	Ni–29Cu–2.8Al–0.5Ti	Ni–Cu (cupronickel)	8460	180	790	1100	30	290–300	13.7	17.4	419	61.4
Nimonic® 105	N07105	Ni–15Cr–20Co–5Mo–5Al–1.2Ti	PH Fe–Ni superalloys	8010	n.a.	780	1140	22	380	12.2	10.9	419	131
Nimonic® 115	N07115	Ni–14Cr–13Co–3Mo–5Al–4Ti	PH Fe–Ni superalloys	7850	n.a.	865	1240	27	400	12.0	10.6	444	139
Nimonic® 263	N07263	Ni–20Cr–20Co–6Mo–2Ti–0.5Al	PH Fe–Ni superalloys	8360	n.a.	580	970	39	320	11.1	11.7	461	115
Nimonic® 75	N06075	Ni–20Cr–0.4Ti	PH Fe–Ni superalloys	8370	n.a.	240	750	40	170	11.0	11.7	461	102
Nimonic® 80A	N07080	77Ni–20Cr–2Ti–1.5Al	PH Fe–Ni superalloys	8190	222	780	1220	30	370	12.7	11.2	460	117
Nimonic® 81	N07081	Ni–30Cr–1.8Ti–1Al	PH Fe–Ni superalloys	8060	n.a.	n.a.	n.a.	n.a.	n.a.	11.1	10.9	461	127

Name	UNS	Composition	Type										
Nimonic® 90	N07090	Ni–20Cr–17Co–2.4Ti–1.4Al	PH Fe-Ni superalloys	8180	n.a.	750	1175	30	380	12.7	11.5	445	114
Nimonic® 901	N09901	Ni–13Cr–35Fe–6Mo–3Ti	PH Fe-Ni superalloys	8160	n.a.	900	1220	15	n.a.	13.5	n.a.	419	119
Nimonic® PE16	n.a.	Ni–16Cr–32Fe–3Mo–1Ti–1Al	PH Fe-Ni superalloys	8020	n.a.	460	880	37	280	11.3	11.7	544	110
Nimonic® PK33	n.a.	Ni–18Cr–14Co–7Mo–2.25Ti–2.1Al	PH Fe-Ni superalloys	8210	n.a.	790	1170	30	n.a.	12.1	11.3	419	126
Rene® 41	N07041	55.4Ni–19Cr–11Co–11Mo–1Al–3Ti	PH Ni superalloys	8250	n.a.	1060	1420	14	n.a.	n.a.	9	n.a.	130.8
Rene® 95	n.a.	61.5Ni–14Cr–8Co–3.5Mo–3.5Nb–3.5Al–2.5Ti	PH Ni superalloys	n.a.	n.a.	1310	1620	15	n.a.	n.a.	8.7	n.a.	n.a.
Sanicro® 28	N08028	31Ni–27Cr–2Mn–1Si	Fe-Ni-Cr alloys	n.a.	n.a.	215	500	40	n.a.	n.a.	n.a.	n.a.	n.a.
Udimet® 500	N07500	53.7Ni–18Cr–18.5Co–4.0Mo–2.9Ti–2.9Al	PH Ni superalloys	8020	n.a.	840	1310	32	n.a.	n.a.	11.1	n.a.	120.3
Udimet® 700	N07700	55.5Ni–15Cr–17Co–5Mo–4Al–3.5Ti	PH Ni superalloys	7910	n.a.	965	1410	17	n.a.	n.a.	19.6	n.a.	n.a.
Waspaloy®	N07001	58.7Ni–19.5Cr–13.5Co–4.3Mo–3Ti–1.3Al–Fe	PH Ni superalloys	8190	n.a.	795	1275	25	n.a.	11.7	10.7	n.a.	124

Table 1.20 Properties of the 55Ni–45Ti shape memory alloy

Structure type	Density (ρ, kg.m^{-3})	Young modulus (E, GPa)	Yield strength 0.2% proof (σ_{YS}, MPa)	Ultimate tensile strength (σ_{UTS}, MPa)	Elongation (Z, %)	Thermal conductivity (k, W.m^{-1}.K^{-1})	Coefficient of linear thermal expansion (α, 10^{-6}K^{-1})
Austenitic	6450	83	195–690	895	25–50	18	11
Martensitic	6450	28–41	70–140	1900	5–10	8.6	6.6

1.3 Cobalt Alloys

Table 1.21 Properties of selected cobalt alloys

Usual and trade name	UNS	Density (ρ, kg.m^{-3})	Yield strength 0.2% proof (σ_{YS}, MPa)	Ultimate tensile strength (σ_{UTS}, MPa)	Elongation (Z, %)	Rockwell hardness C (HRC)	Melting point or liquidus range (°C)	Thermal conductivity (k, W.m^{-1}.K^{-1})	Specific heat capacity (c_p, J.kg^{-1}.K^{-1})	Coefficient of linear thermal expansion (a, 10^{-6} K^{-1})	Electrical resistivity (ρ, $\mu\Omega$.cm)
Co (99.9wt%)	n.a.	8900	758	944	10–25	65	1495	96	427	12.5	6.34
Haynes® 1233	n.a.	n.a.	558	1020	33	28	1333–1335	n.a.	n.a.	n.a.	n.a.
Haynes® 188	R30188	8980	464	945	53	n.a.	1302–1330	10.8	n.a.	11.9	101
Haynes® 25	R30605	9130	445	970	62	60–63	1329–1410	n.a.	n.a.	12.3	89
Haynes® 6B	n.a.	8390	619	998	11	37	1265–1354	n.a.	n.a.	13.9	91
MAR-M509	n.a.	8860	585	780	3.5	n.a.	1290–1400	8.8	n.a.	n.a.	100
MP35N	R30035	n.a.	380–414	895–931	65–70	90	1315–1440	11.2	n.a.	12.8	103
Stellite® 1	R30001	8690	n.a.	618	1	55	1255–1290	n.a.	n.a.	10.5	94
Stellite® 12	R30012	8560	649	834	1	48	1280–1315	n.a.	n.a.	11.4	88
Stellite® 21	R30021	8340	494	694	9	32	1186–1383	n.a.	n.a.	11	88
Stellite® 6	R30006	8460	541	896	1	40	1285–1395	n.a.	n.a.	11.4	84
Tantung G	n.a.	8300	n.a.	585–620	n.a.	60–63	1150–1200	26.8	n.a.	4.2	n.a.

Further Reading

Bringas JE (1995) The metals black book, vol. 1, Ferrous metals, 2nd edn. CASTI Publishing, Edmonton, Canada.

Harvey Ph (1982) Engineering properties of steels. ASM Books, Ohio Park.

Parr JG, Hanson A, Lula RA (1985) Stainless steel. ASM Books, Ohio Park.

Roberts GA, Cary RA (1980) Tool steels, 4th edn. ASM Books, Ohio Park.

Wegst CW (1995) Stahlschlussel (Key to steel), 17th edn. Verlag Stahlschlussel.

2 Common Nonferrous Metals

2.1 Aluminum and Aluminum Alloys

2.1.1 General Properties and Description

Aluminum (aluminium in the UK) [7429-90-5] with the chemical symbol Al, the atomic number 13, and the relative atomic mass 26.981538(5) is the second element of the group IIIA(13) of Mendeleev's periodic chart. Pure aluminum is a light (2698 kg.m^{-3}), silvery white metal with a low melting point (660.3°C) and a boiling point approximately 2519°C. From a mechanical point of view, pure aluminum metal is ductile, malleable, and exhibits good formability, but the mechanical strength of the metal can be largely improved by either (i) cold working, or (ii) adding alloying elements such as manganese, silicon, copper, magnesium, or zinc. At low temperatures, aluminum is stronger than at room temperature and is no less ductile. Actually, the mechanical strength of aluminum increases under very cold temperatures, making it especially useful for cryogenic applications and in the extreme cold of outer space, as well as for aircraft and for construction in high latitudes. Hence, aluminum is an attractive structural material for applications requiring high strength-to-weight ratios such as aerospace, high-rise construction, and automotive design. Due to the high thermal conductivity (237 W.m.$^{-1}$.K^{-1}) of aluminum, this element and its alloys provide better heat transfer capabilities than other common metals. This makes aluminum ideal for applications requiring heat exchangers, especially because extrusion, as a metal-forming process, is well-suited to produce shapes that make optimal use of thermal conduction properties. In addition, aluminum is nonsparking, and hence it is appropriate for applications involving explosive materials or taking place in highly flammable environments. On the other hand, aluminum exhibits a good electrical conductivity (2.6548 $\mu\Omega$.cm), i.e. 60% IACS, and is paramagnetic. Hence aluminum does not acquire a magnetic charge, and it is useful for high-voltage applications, as well as for electronics, especially where magnetic field interference occurs. The high reflective index of polished aluminum can be used to shield products or areas from light, radio waves, or infrared radiation. Aluminum is a mononuclidic element with the nuclide ^{27}Al. From a chemical point of view, aluminum, due to its excellent valve action property, when put in contact with oxidizing environments, develops spontaneously onto the exposed surface area a thin passivating film of impervious aluminum oxide (alumina). Moreover, the thickness of this passivating film can be further enhanced artificially by anodizing or other finishing techniques. Therefore, aluminum by contrast with iron and steels is corrosion resistant to normal atmospheric conditions (i.e., air, water), and it does not rust. However, aluminum is readily attacked by both diluted strong mineral acids and alkaline solutions evolving hydrogen, and reacts vigorously with chlorinated organic solvents. **Cost** (1998) – pure aluminum (99.99wt%) is priced at 1.48 $US.kg^{-1} (0.67 $US.lb^{-1}).

2.1.1.1 History

There is evidence of use of aluminum compounds such as **alum** (potassium aluminum sulfate dodecahydrate, $KAl(SO_4)_2 \cdot 12H_2O$) from as early as 300 BC, but it was not until 1888 that an economically feasible process was developed for modern, commercial production of aluminum by the Hall–Heroult electrolytic process. Sir Humphrey Davy first named the metal aluminum in 1805. Even though he could not isolate it, he was convinced that it existed and named it anyway. Later the name was changed to aluminium to be consistent with other metal Latin names. North Americans still use the old spelling. Aluminium was once considered so valuable that the earliest aluminum metal was made into jewelry and luxury cookware, and a sample was displayed along side the British Crown jewels at the Paris Exposition of 1855. Prior to the discovery of a method of reducing aluminum electrolytically, aluminum metal was produced from 1850 until 1888, by the metallothermic reduction of the aluminum (III) chloride with pure sodium metal:

$$AlCl_3(s) + 3Na(s) \rightarrow Al(s) + 3NaCl(s)$$

However, in 1886, the American engineer, Charles Martin Hall, and the French engineer, Paul Heroult, simultaneously but independently announced that they had discovered how to electrowin aluminum metal from alumina. Because aluminum oxide melts at such a high temperature (i.e., 2030°C) and is a nonionic liquid (i.e., nonconductive), electrolysis of the molten oxide is not feasible. What both Hall and Heroult discovered was that a natural fluoride mineral called **cryolite** (Na_3AlF_6, monoclinic) naturally occurring in Greenland, which melts at only 1009°C would act as fluxing agent and easily dissolve purified aluminum oxide. This molten salt mixture could then be electrolyzed using carbon electrodes.

2.1.2 Natural Occurrence, Minerals, and Ores

Aluminum is the third most abundant chemical element in the Earth's crust with a relative natural abundance of 8.3wt%, just after oxygen, and silicon. Owing to its high chemical reactivity with oxygen aluminum never occurs as a native element in nature. Actually, the element occurs in numerous minerals (e.g., oxides, hydroxides, sulfates, fluorides, and silicates) amongst which aluminosilicates represent the major aluminum-bearing minerals. However, only the sedimentary rock bauxite is used industrially as a primary aluminum ore because the recovery of aluminum from silicate minerals is more energy intensive. Nethertheless, some alternative extractive processes were developed at the pilot scale to recover aluminum from silicate ores (e.g., feldspars such as **orthoclase**, $KAlSi_3O_8$, and feldpathoids such as **nepheline**, $Na_3KAl_4Si_4O_{16}$) in the worst case of depletion of bauxite ore deposits. **Bauxite** is defined as a residual sedimentary rock originated from in-situ superficial alteration under moist tropical conditions of clays, clayey limestones, or high-alumina-content igneous rocks. Afterwards alteration products are depleted from soluble cations (e.g., Na, K, Ca, Mg) by the leaching action of water. Hence, insoluble cations such as iron (III) and aluminum (III) associated with clays and silica remain in the materials. As a general rule, bauxite is mainly composed of hydrated alumina minerals such as **gibbsite** ($Al_2O_3 \cdot 3H_2O$, monoclinic) in tropical and equatorial bauxite ore deposits, while **boehmite** ($Al_2O_3 \cdot H_2O$, orthorhombic) is the major mineral in subtropical and temperate bauxite ore deposits. The average chemical composition of bauxite is 45–60wt% Al_2O_3, and 22–30wt% Fe_2O_3; the remainder consists of silica, iron oxides, titanium dioxide and water. According to the US Geological Survey, the world bauxite resources were estimated to be 55 to 75 billion tonnes located as follows: South America (33%), Africa (27%), Asia (17%), Oceania (13%), and elsewhere (10%).

2.1.3 Processing and Industrial Preparation

The aluminum metal production from bauxite ore is a three-step process. (1) **Pure anhydrous alumina preparation** – first the alumina is extracted from bauxite ore concentrate usually using the **Bayer process**. (2) **Aluminum electrowinning/scrap recycling** – pure and anhydrous alumina previously obtained from the Bayer process is then reduced to aluminum metal usually using the **Hall–Heroult process**, while aluminum scrap is remelted. (3) **Alloying and refining** – the molten aluminum can be further electrorefined or purified, or is mixed with desired alloying elements in order to obtain required mechanical characteristics, and cast into ingots.

(1) **The Bayer process** – first the raw bauxite ore is crushed and washed with water in order to remove clay minerals and silica-forming waste by-products. Afterwards, the washed concentrated ore is ground with a concentrated solution of caustic soda (sodium hydroxide, NaOH) providing a suspension of bauxite with 90% of particle sizes less than 300 μm. The suspension is then heated in an autoclave at 235–250°C under a pressure of 35 to 40 bar. During the digestion the hydrated alumina is dissolved by the hot concentrated alkaline solution. During the dissolution reaction, the sodium hydroxide reacts with both alumina and silica but not with the other impurities such as calcium, iron, and titanium oxides, which remain as insoluble residues and form a sludge at the bottom of the digester (called **red mud**). The insoluble residues sink gradually to the bottom of the tank and the resulting slurry is removed by filtration. Slowly heating the solution causes the $Na_2Si(OH)_6$ to precipitate out. The solution is then filtered to leave only the aluminum-containing $NaAl(OH)_4$. The clear sodium tetrahydroxyaluminate solution is pumped into a huge tank called a precipitator. The solution is next acidified by bubbling carbon dioxide gas through the solution. Carbon dioxide forms a weak acid solution of carbonic acid which neutralizes the sodium hydroxide from the first treatment. This neutralization selectively precipitates the aluminum hydroxide ($Al(OH)_3$), but leaves the remaining traces of silica in solution. Fine particles of aluminum hydroxide are usually added to seed the precipitation process of pure alumina particles as the liquor cools. The particles of aluminum hydroxide crystals sink to the bottom of the tank, are removed, are then vacuum dewatered, and a small fraction is reused as seed crystals while the major fraction passes through a rotary-kiln or fluidized calciner at 1100–1300°C to drive off the chemically combined water. The result is a white powder made of pure alumina, Al_2O_3. The caustic soda is returned to the start of the process and used again.

(2) **Aluminum metal** – aluminum metal can either be produced from bauxite ore or from recycling aluminum scrap. Electrowinning of aluminum ore is sufficiently expensive that the secondary production industry commands much of the market. Because electrowinning aluminum from alumina is an highly energy intensive process, the world's major primary aluminum smelters are located in areas which have access to low-cost energy and abundant power resources (e.g., hydroelectric, natural gas, coal or nuclear), while secondary producers tend to locate near industrial centers where aluminum scrap is largely available.

(2a) **The Hall–Heroult process**[1] – because aluminum oxide (i) is a refractory compound that only melts at high temperature (2030°C), (ii) is highly stable chemically (seen from Ellingham's diagram, which may be found in textbooks), and (iii) on melting it gives a nonionic molten liquid (i.e., insulator), fused salt electrolysis of this compound is not feasible. Hence, purified

---

[1] Hall–Heroult in North America, and Heroult–Hall in France.

anhydrous alumina must be dissolved as solute in an electrolyte made of molten cryolite (Na_3AlF_6). Actually, alumina forms with the cryolite a eutectic (935°C) containing 18.5wt% alumina, but common industrial baths contain no more than 7.5wt% in order to maintain fluidity of the melt. Cryolite, which was at one time extracted from natural ore deposits in Greenland, is today synthesized industrially.

The electrolytic cell consists of a large carbon- or graphite-lined steel container (pot). The carbon anode is made of prebaked petroleum coke and pitch (Soderberg anodes are today discontinued), and the cathode is formed by the thick carbon or graphite lining of the pot. The electrolytic bath is heated by passing current to about 940–980°C to produce the molten salt electrolyte. Then the molten bath mixture is electrolyzed at a low voltage of 4–5 V, but a high current of 50–150 kA. During electrolysis, which is a continuous smelting process, aluminum cations are reduced to aluminum metal at the cathode (−) forming the molten aluminum pool at the bottom of the cell which is siphoned off periodically, while at the carbon anode (+), oxygen is produced from the oxide anions and combines with carbon to form carbon dioxide with some traces of fluorinated hydrocarbons. Hence, the anode material is permanently consumed and must be replaced quite frequently. The molten aluminum, siphoned from the bottom of the smelter, is placed in a crucible, then formed into ingots or transferred to an alloying furnace.

Although the electrochemistry of the Hall-Heroult process is more complex and involves several different kinds of ions containing aluminum, oxygen, and fluorine, the overall reaction can be simplified to the following:

$$2Al_2O_3 \text{ (dissolved in } Na_3AlF_6) + 3C(s) \rightarrow 4Al(l) + 3CO_2(g)$$

Because the average specific energy consumption of 13 to 15.7 kW.h.kg^{-1} is relatively high, electricity used to produce aluminum represents 25% of the cost of the metal. A typical aluminum smelter consists of around 250–300 pots. These will produce some 125,000 tonnes of aluminum annually. However, some of the latest generation of smelters are in the 350,000–400,000 tonnes range. As a general rule, 1 tonne of aluminum metal requires about 1.9 tonnes alumina, and hence 5–6 tonnes bauxite. Most smelters produce aluminum of 99.7wt% purity, which is acceptable for most applications. However, high-purity aluminum (i.e., 99.99wt%) required for some special applications, typically those where high ductility and/or electrical conductivity is required, can be obtained by electrorefining.

(2b) **Secondary aluminum production** – in the secondary aluminum production industry, scrap aluminum is melted in gas- or oil-fired reverberatory or hearth furnaces. Impurities are removed using chlorine bubbling or other fluxes until the aluminum reaches the desired purity. Other aluminum production plants use **dross** in addition to scrap. Dross is a by-product of primary aluminum smelting. This process further reduces the pollution resulting from primary aluminum production. It contains fluxes and varying concentrations of aluminum. The dross is crushed, screened and melted in a rotary furnace where the molten aluminum is collected in the bottom.

(3) **Alloying and refining** – for production of primary aluminum alloys, the molten aluminum may be transferred from the smelter to the alloying furnace or previously produced aluminum ingot may be melted in the furnace. Alloying metals can then be mixed with the molten aluminum. Molten aluminum may be further heated to remove oxides, impurities, and other active metals such as sodium and magnesium, before casting. Chlorine may also be bubbled through the molten aluminum to further remove impurities.

2.1.4 Aluminum Alloys

Aluminum and its alloys are available in all common commercial forms. Aluminum-alloy sheet can be formed, drawn, stamped, or spun. Many wrought or cast aluminum alloys can be welded, brazed, or soldered, and aluminum surfaces readily accept a wide variety of finishes, both mechanical and chemical. Because of their high electrical conductivity, aluminum alloys are used as electrical conductors. Aluminum reflects radiant energy throughout the entire spectrum, is nonsparking, and nonmagnetic. The mechanical properties of aluminum may be improved by alloying, by strain hardening, by thermal treatment, or by combinations of all three techniques. Copper, magnesium, manganese, silicon, and zinc are used as the major constituents in aluminum alloys. Chromium, lead, nickel, and other elements are used for special purposes as minor alloy constituents. Impurities such as iron affect the performance of aluminum alloys and must be considered. Pure aluminum can be strengthened by alloying with small amounts of Mn (up to 1.25wt%) and Mg (up to 3.5wt%). The addition of larger percentages of magnesium produces still higher strengths, but precautions are needed for satisfactory performance. These alloys and pure aluminum can be further hardened by cold working up to tensile strengths of 200 or even 300 MPa. Higher strengths are achieved in alloys which are heat-treatable.

2.1.4.1 Aluminum Alloy Standard Designations

The most widely used standard designation for the aluminum alloys is that introduced by the Aluminum Association, Inc. (AA) and now extensively used in several countries. According to these standard designations, aluminum based alloys are divided into two main categories: (1) **wrought alloys** (i.e. mechanically worked), and (2) **cast alloys**.

Wrought aluminum alloys have a systematic identification according to the type of alloying elements. These designations are reported in Table 2.1. For instance, according to the designation the first series 1XXX corresponds to 99.00% pure aluminum or greater. In this series, the second digit indicates special purity controls and the last two digits indicate the minimum aluminum content beyond 99.00% (e.g., a 1030 aluminum has 99.30% Al and no special control of individual impurities). For other series, the second digit refers to alloying modifications, and the last two digits identify the alloy (usually from its former commercial designation). For instance, the former aluminum alloy 75S is now referred to as 7075. This number is followed by a temper designation that identifies thermal and mechanical treatments.

Cast aluminum alloys have a similar designation to the wrought materials. They are normally identified as 2XX.0, 3XX.0, 4XX.0, 5XX.0, etc.

Table 2.1 Wrought aluminum alloy AA standard designation[a]

Series	Designation
1XXX	Commercially pure aluminum (i.e., 99.00wt% or greater)
2XXX	Al–Cu and Al–Cu–Mg alloys
3XXX	Al–Mn alloys
4XXX	Al–Si alloys
5XXX	Al–Mg alloys
6XXX	Al–Mg–Si alloys
7XXX	Al–Zn–Mg and Al–Zn–Mg–Cu alloys
8XXX	Other alloying elements
9XXX	Unused series

First digit: principal alloying element. **Second digit**: variations of initial alloy. **Third** and **fourth digits**: individual alloy variations.

[a]Source: The Aluminum Association Inc.

Table 2.2 Cast aluminum alloy standard designation[a]

Series	Designation
1XX.X	Commercially pure aluminum (i.e., 99.00wt% or greater)
2XX.X	Al–Cu alloys
3XX.X	Al–Si–Cu–Mg alloys
4XX.X	Al–Si alloys
5XX.X	Al–Mg alloys
6XX.X	Al–Mg–Si alloys
7XX.X	Al–Zn alloys
8XX.X	Al–Sn alloys
9XX.X	Other alloying element

First digit: principal alloying element. **Second and third digits**: specific alloy designation. **Fourth digit**: casting (0), or ingot (1,2).

[a]Source: The Aluminum Association Inc.

There is also a system of temper designations used for all forms of wrought and cast aluminum.

Table 2.3 Aluminum alloy temper designation[a]

Letter	Description
F	As manufactured or fabricated
O	Annealed
H	Strain hardened (wrought products only); H1x: Strain hardened only H2x: Strain hardened only and partially annealed to achieve required temper H3x: Strain hardened only and stabilized by low-temperature heat treatment to achieve required temper H12, H22, H32: Quarter hard, equivalent to about 20–25% cold reduction H14, H24, H34: Half hard, equivalent to about 35% cold reduction H16, H26, H36: Three quarter hard, equivalent to about 50–55% cold reduction H18, H28, H38: Fully hard, equivalent to about 75% cold reduction
W	Solution heat treated
T	Thermally treated to produce stable tempers other than F. H and O. Usually solution heat treated, quenched, and precipitation hardened T1: Cooled from elevated temperature shaping process and aged naturally to a substantially stable condition T2: Cooled from elevated temperature shaping process, cold worked and aged naturally to a substantially stable condition T3: Solution heat treated, cold worked and aged naturally to a substantially stable condition T4: Solution heat treated, and aged naturally to a substantially stable condition T5: Cooled from elevated temperature shaping process, and then aged artificially T6: Solution heat treated, and then aged artificially T7: Solution heat treated, and then stabilized (over-aged) T8: Solution heat treated, cold worked, and then aged artificially T9: Solution heat treated, aged artificially, and then cold worked T10: Cooled from an elevated temperature shaping process, artificially aged and then cold worked

Note: a large number of numeric additions have been introduced to indicate specific variations.

[a]Source: The Aluminum Association, Inc.

2.1.4.2 Wrought Aluminum Alloys

See **Table 2.4** (pages 52–55).

2.1.4.3 Cast Aluminum Alloys

See **Table 2.5** (pages 56–57).

2.1.5 Industrial Uses and Applications

Aluminum and aluminum alloys are used extensively in applications requiring high strength-to-weight ratios, corrosion resistance to atmospheric conditions, and good electrical and thermal conductivity. Since 1994, the automotive industry has represented the major market for aluminum and aluminum alloys, followed by the packaging and container industry. Alclad[R] (registered trademark of Texas Instruments) aluminum alloys are aluminum structural alloys covered with a thin skin of pure aluminum. Other uses of aluminum are sprayed or hot-dipped coatings on steel substrates (i.e., aluminized steel) and paint pigments in marine applications. For process industries applications, the most widely used wrought aluminum alloys are from the 1XXX, 3XXX, 4XXX, 5XXX, and 6XXX series.

2.2 Copper and Copper Alloys

2.2.1 General Properties and Description

Copper [7440-50-8], with the chemical symbol Cu, the atomic number 29, and the relative atomic mass 63.546(3), is the first element of the group IB(11) of Mendeleev's periodic chart. Its chemical symbol comes from the Latin, Cuprum, meaning the island of Cyprus. Highly pure copper is a dense (8960 kg.m^{-3}), reddish, bright, malleable and ductile metal, which melts at 1084°C. It exhibits a very low electrical resistivity (1.6730 $\mu\Omega$ cm) which is only 3% more than that of pure silver; this makes it the second most conductive element at room temperature. Hence, the resistivity of the **International Annealed Copper Standard** (IACS)[2] was used as a reference for the comparison of the electrical conductivity of metals and alloys. Moreover, copper exhibits a high thermal conductivity (397 W.m^{-1}.K^{-1}) combined with good corrosion resistance, and ease of forming and joining. Therefore copper and copper alloys are extensively used in heat exchangers. In addition, however, copper and its alloys have relatively low strength-to-weight ratios and low strengths at elevated temperatures. Some copper alloys are also susceptible to stress-corrosion cracking unless they are stress relieved. In air copper forms a passivating adherent film that is relatively impervious to corrosion at room temperature and that protects the base metal from further attack. Nevertheless, on heating copper oxidizes readily forming a black nonprotective spalling oxide, and under most outdoor conditions copper surfaces develop a blue-green patina due to formation of a complex scale made of both carbonates and hydroxide of copper. It resists against corrosion by fresh water and steam, and hydrofluoric acid. But copper exhibits poor resistance to inorganic acids, ammonia, and ammonium salt. Natural copper has two stable isotopes: ^{63}Cu (69.17at%), and ^{65}Cu (30.83at%). **Cost** (1998) – highly pure copper (99.999wt% Cu) is priced at 2.80 $US.kg^{-1} (1.27 $US.lb^{-1}).

---

[2] Defined as a conductive material which the electrical resistance of a wire of 1 m long and weighing 0.001 kg is exactly 0.15328 Ω at 20°C (i.e., 100%IACS = 58.00 MS.m^{-1} = 1.72413793 $\mu\Omega$.cm^{-1}).

Table 2.4 Physical properties of selected wrought aluminum alloys

AA designation	Average chemical composition (x, wt%)	Density (ρ, kg.m^{-3})	Temper	Young modulus (E, GPa)	Yield strength 0.2% proof (σ_{YS}, MPa)
1199	99.992	2710	H18	69	12
1050	99.50	2705	O	69	30
1100	99.00	2710	O	69	35
			H18		150
2014	Al–4.5Cu–0.85Si–0.7Fe–0.5Mg–0.25Zn	2800	O	73	95
			T6		415
2017	Al–4.0Cu–0.7Fe–0.6Mg–0.5Si–0.25Zn	2790	O	72	70
			T4	72	315
2024	Al–4.4Cu–1.5Mg–0.6Mn–0.5Si–0.5Fe–0.25Zn	2780	O	73	95
			T6		415
2219	Al–6.3Cu–0.3Mn–0.2Si–0.3Fe–0.25Zr	2840	O	73	75
			T62		290
3003	Al–1.5Mn–0.6Si–0.7Fe–0.2Cu	2730	O	69	40
			H12		125
3004	Al–1.25Mn–1.1Mg–0.7Fe–0.3Si–0.25Zn	2720	O	69	70
3105	Al–0.35Mn–0.7Fe–0.6Si–0.6Mg–0.3Cu–0.25Zn	2730	O	69	55
4032	Al–12Si–1Mg–1Ni–1Fe–0.9Cu–0.25Zn	2680	O	79	n.a.
			T6		315
4043	Al–5Si–0.8Fe–0.3Cu	2690	O	71	75
5052	Al–2.5Mg–0.4Fe	2680	O	70	90
5083	Al–4.5Mg–0.4Mn–0.5Si 0.25Zn	2660	O	71	145
6061	Al–1Mg–0.6Si–0.3Cu	2700	O	69	55
			T6		275

Ultimate tensile strength (σ_{UTS}, MPa)	Elongation (Z, %)	Brinell hardness (HB)	Liquidus range (°C)	Thermal conductivity (k, W.m^{-1}.K^{-1})	Specific heat capacity (c_p, J.kg^{-1}.K^{-1})	Coefficient of linear thermal expansion (α, 10^{-6} K^{-1})	Electrical resistivity (ρ, $\mu\Omega$.cm)
85	43	28	660	234	917	23.6	2.70
70	43	19	645–655	222–230	920	23.5	2.80
90	35–45	23	643–655	222	917	23.6	3.0
165	5–15	44		170			2.80
185	18	45	507–638	193	962	23.0	3.50
485	13	135		154			4.30
180	22	45	513–640	193	920	23.6	3.5
420	12	120		134			5.15
185	22	47	503–638	193	920	23.2	3.50
495	13	135		151			4.50
175	18	n.a.	543–643	172	920	22.3	4.00
415	10	n.a.		121			5.80
110	30–40	28	643–655	193	920	23.2	3.50
130	10–20	35		163			4.15
180	20–25	45	630–655	163	920	23.9	4.15
115	24	85	635–655	172	920	23.6	3.8
n.a.	n.a.	n.a.	532–570	154	950	19.4	4.30
380	9	120		138			5.00
130	20	n.a.	575–632	163	920	22.1	4.15
195	25–30	47	607–650	138	962	23.75	5.00
290	22	77	590–638	117	962	23.75	6.00
125	25–30	30	580–650	180	962	23.6	3.65
310	12–17	95		167			4.00

Table 2.4 (continued)

AA designation	Average chemical composition (x, wt%)	Density (ρ, kg.m^{-3})	Temper	Young modulus (E, GPa)	Yield strength 0.2% proof (σ_{YS}, MPa)
6063	Al–0.5Mg–0.5Si–0.35Fe	2700	O	69	50
			T6		215
7075	Al–5.7Zn–2.5Mg–1.6Cu	2810	T6	72	105
7178	Al–6Zn–2.6Mg–2Cu–0.3Cr	2830	T6	n.a.	n.a.
8017	Al–0.6Fe–0.2Cu	2710	H12	n.a.	n.a.
			H212	n.a.	n.a.
8176	Al–0.1Si–0.7Fe–0.1Zn	2710	H24	n.a.	n.a.
8081	Al–1Cu–0.7Si–0.7Fe	2700	H24	n.a.	n.a.

2.2.2 Natural Occurrence, Minerals, and Ores

Copper's relative abundance in the Earth's crust is about 50 mg.kg^{-1} (i.e., ppm wt), which is less than nickel and zinc. It occurs as a native element (4%), but the major part of its occurrence is as oxide minerals (10%) such as **cuprite** (Cu_2O, cubic); carbonates (5%); **malachite** ($CuCO_3 \cdot Cu(OH)_2$, monoclinic) and **azurite** ($2CuCO_3 \cdot Cu(OH)_2$, monoclinic), and sulfide minerals (80%): **chalcocite** (Cu_2S, monoclinic), **chalcopyrite** ($CuFeS_2$, tetragonal), and **bornite** (Cu_5FeS_4, cubic), and in other rare minerals (1%) such as **atacamite** ($Cu_2Cl(OH)_3$, orthorhombic). But only the oxide and sulfide minerals are used industrially as copper ores. Chile is the world's largest producer of copper, followed by the United States.

2.2.3 Processing and Industrial Preparation

After mining the sulfide and oxide ores through digging or blasting, they are crushed into walnut-sized pieces. Crushed ore is then milled in a large ball or rod mill until it becomes a powder usually containing less than 1wt% copper. Sulfide ores are concentrated from inert gangue material by froth flotation, while oxide ores are routed directly to leaching tanks. After froth flotation the copper concentrate slurry contains about 15–30wt% Cu. Waste inert gangue is removed and water is recycled. Tailings containing copper oxide are routed to leaching tanks or are returned to the surrounding terrain for disposal. Depending on the ore type (i.e., sulfide or oxide) copper is recovered as pure copper metallic cathode in two different ways: (1) **hydrometallurgical**: leaching and electrowinning, or (2) **pyrometallurgical**: smelting and electrorefining.

(1) **Hydrometallurgical process** (i.e., leaching and electrowinning) (oxide ores) – oxide ores and tailings are leached by a dilute sulfuric acid solution, producing a diluted copper sulfate

Ultimate tensile strength (σ_{UTS}, MPa)	Elongation (Z, %)	Brinell hardness (HB)	Liquidus range (°C)	Thermal conductivity (k, W.m^{-1}.K^{-1})	Specific heat capacity (c_p, J.kg^{-1}.K^{-1})	Coefficient of linear thermal expansion (α, 10^{-6} K^{-1})	Electrical resistivity (ρ, $\mu\Omega$.cm)
90	n.a.	25	615–655	218	962	23.4	3.20
240	12	73		200			3.30
230	16–17	60	475–635	130	962	23.6	5.15
n.a.	n.a.	n.a.	475–630	125	962	23.4	5.50
n.a.	n.a.	n.a.	645–655	n.a.	n.a.	23.6	2.8
n.a.	n.a.	n.a.		n.a.	n.a.		3.0
n.a.	n.a.	n.a.	645–655	230	962	23.6	2.8
n.a.	n.a.	n.a.	n.a.	230	962	23.6	2.8

aqueous solution. The mother liquor is then treated and transferred into an electrolytic cell. During electrolysis, copper metal electrodeposits at the cathode (–), made from pure copper foil, while oxygen is evolved at the anode (+). During the electrolytic process traces of precious metals (i.e., Ag, Au) and platinum group metals (e.g., Pt) can be recovered from the spent electrolytic bath.

(2) **Pyrometallurgical** process (i.e., smelting and electrorefining) (sulfide ores) – sulfide ore concentrate is roasted in air at 1100°C in a reverberatory furnace. The sulfides of copper and iron melt together to give the **matte**, while silica and calcium oxide form a molten slag above the bath. The matte is oxidized by flowing dry air above the melt. Iron sulfide is oxidized and combines with the slag, while copper sulfide gives off molten copper, called **blister**. The blister, which contains about 1wt% impurities, is melted under pure oxygen, and cast into rectangular anodes and must be further electrorefined.

Pure electrorefined copper cathodes (i.e., 99.9wt% Cu, with maximum 30 ppm wt oxygen) are remelted under air or inert atmosphere and cast for making commercial products (e.g., bars, rods, ingots, billets).

2.2.4 Copper Alloys

2.2.4.1 UNS Copper Alloy Designation

The three-digit system developed by the US copper and brass industry was expanded to five digits following the prefix letter C and made part of **Unifed Numbering System for Metals and Alloys.** UNS designations are simply expansions of the former designations of the **Copper Development Association, Inc.** (CDA). For example, the copper alloy CDA 377 (forging brass) in the original three-digit system became C37700 in the UNS designation. Because these old

Table 2.5 Physical properties of selected cast aluminum alloys

AA designation	Average chemical composition (x, wt%)	Density (ρ, kg.m^{-3})	Young modulus (E, GPa)	Yield strength 0.2% proof (σ_{YS}, MPa)	Ultimate tensile strength (σ_{UTS}, MPa)
100.1	99.9	2700	69	30	80
201.0	Al–4.6Cu–0.7Ag–0.35Mg–0.35Mn–0.25Ti	2750	71	170–215	225–295
204.0	Al–4.6Cu–0.25Mg–0.35Fe–0.17Ti	2750	70	200	225
356.0	Al–7Si–0.35Mg–0.2Fe–0.2Cu–0.35Zn	2685	73	195–210	240–290
359.0	Al–9Si–0.6Mg	2685	72	180	230
360.0	Al–9.5Si–2Fe–0.6Cu–0.5Zn–0.5Ni	2740	71	170	305–320
383.0	Al–10.5Si–2.5Cu–0.5Mn	2740	71	172	310
390.0	Al–17Si–4.5Cu–0.6Mg	2731	88	248	317
413.0	Al–11.5Si–2Fe–1Cu–0.5Ni–0.5Zn	2657	71	145–280	200–297
443.0	Al–5.5Si–0.8Fe	2670	71	60–70	125–155
512.0	Al–4Mg–2Si–0.3Fe	2600	70	n.a.	n.a.
514.0	Al–4Mg–0.5Fe	2650	71	95	145
518.0	Al–8Mg–1.8Fe	2570	n.a.	180–190	295–340
535.0	Al–6.5Mg–0.2Mn	2620	71	100	160–215
712.0	Al–5.8Zn–0.6Mg–0.5Cr–0.2Ti	2810	71	170	220

numbers are embedded in the new UNS numbers, no confusion need result. The designation system is an orderly method of defining and identifying coppers and copper alloys; it is not a specification. It eliminates the limitations and conflicts of alloy designations previously used and at the same time provides a workable method for the identification marking of mill and foundry products. The designation system is administered by the Copper Development Association Inc. New designations are assigned as new coppers and copper alloys come into commercial use, and designations are discontinued when an alloy composition ceases to be used commercially. In the designation system, numbers from C10000 through C79999 denote wrought copper alloys. Cast copper alloys are numbered from C80000 through C99999. Within these two categories, the compositions are grouped into the following families of coppers and copper alloys:

Elongation (Z, %)	Brinell hardness (HB)	Melting point or liquidus range (°C)	Thermal conductivity (k, W.m^{-1}.K^{-1})	Specific heat capacity (c_p, J.kg^{-1}.K^{-1})	Coefficient of linear thermal expansion (α, 10^{-6} K^{-1})	Electrical resistivity (ρ, $\mu\Omega$.cm)
30	25	n.a.	218	n.a.	24.0	3.0
8	90	535–650	121	920	22.5	5.4–6.4
26	118–137	570–650	121	920	19.3	5.40
6	90	555–615	167	963	21.5	4.01
1	105	555–615	138	963	20.9	4.00
2.5 – 3.5	55–60	555–595	93	963	20.88	6.16
3.5	75	515–580	96–100	n.a.	21.1	6.6
1	120	505–650	134	n.a.	18.0	8.6
2.5	80–125	650–760	121–142	963	20.34	5.3
5–6	n.a.	575–630	159	963	21.0	4.1
n.a.	n.a.	n.a.	134	963	22.0	5.1
3.0	50	585–630	146	963	24	4.93
12–18	85–95	535–620	96.2	n.a.	24.1	6.89
6–10	60–65	550–630	130	n.a.	23.6	5.6
5	70	570–615	138	963	24.7	4.93

2.2.4.2 Wrought Copper Alloys (Table 2.7, pages 60–63)

- Unalloyed copper
- High-copper alloys
- Brasses (Cu–Zn)
- Leaded brasses (Cu–Zn–Pb)
- Tin brasses (Cu–Zn–Sn)
- Phosphor bronzes (Cu–Sn–P)
- Copper–phosphorus–silver alloys (Cu–P–Ag)
- Aluminum–silicon bronzes (Cu–Al–Ni–Fe–Si–Sn)
- Silicon bronzes (Cu–Si–Sn)

Table 2.6 Copper alloy categories

Copper metal	Alloys which have a designated minimum copper content of 99.3% or higher
High-copper alloys	For the wrought products, these are alloys with designated copper contents less than 99.3wt% but more than 96wt% which do not fall into any other copper alloy group. The cast high-copper alloys have designated copper contents in excess of 94wt% to which silver may be added for special properties.
Brasses (Cu–Zn)	These alloys contain zinc as the principal alloying element with or without other designated alloying elements such as iron, aluminum, nickel, and silicon. The wrought alloys comprise three main families of brasses: (i) copper–zinc alloys (i.e., **brasses** senso-stricto, Cu–Zn), (ii) copper–zinc–lead alloys (i.e., **leaded brasses**, Cu–Zn–Pb), and (iii) copper–zinc–tin alloys (i.e., **tin brasses**, Cu–Zn–Sn). The cast copper alloys comprise four main families of brasses: (i) copper–tin–zinc alloys (i.e., red, semi-red, and yellow brasses), (ii) manganese bronze alloys (i.e., high-strength yellow brasses), (iii) leaded manganese bronze alloys (i.e., leaded high-strength yellow brasses), (iv) copper–zinc–silicon alloys (i.e., **silicon brasses** and bronzes), and cast copper–bismuth and copper–bismuth–selenium alloys.
Bronzes (Cu–Zn–Sn)	Broadly speaking, bronzes are copper alloys in which the major alloying element is not zinc or nickel. Originally bronze described alloys with tin as the only or principal alloying element. Today, the term is generally used not by itself but with a modifying adjective. For wrought alloys, there are four main families of bronzes: (i) copper–tin–phosphorus alloys (i.e., **phosphor bronzes**), (ii) copper–tin–lead–phosphorus alloys (i.e., **leaded phosphor bronzes**), (iii) copper–aluminum alloys (i.e., **aluminum bronzes**), and (iv) copper–silicon alloys (i.e., **silicon bronzes**). The cast alloys have four main families of bronzes: (i) copper–tin alloys (i.e., **tin bronzes**), (ii) copper–tin–lead alloys (i.e., **leaded** and **high-leaded tin bronzes**), (iii) copper–tin–nickel alloys (i.e., **nickel–tin bronzes**), and (iv) copper–aluminum alloys (i.e., **aluminum bronzes**).
Copper–nickels (Cu–Ni)	These are alloys with nickel as the principal alloying element, with or without other designated elements.
Leaded coppers	These comprise a series of cast alloys of copper with 20wt% or more lead, sometimes with a small amount of silver, but without tin or zinc.
Special alloys	Alloys whose chemical compositions do not fall into any of the above categories are combined in special alloys.

- Copper–nickel (Cu–Ni–Fe)
- Nickel silvers (Cu–Ni–Zn)

2.2.4.2 Cast Copper Alloys (Table 2.8, page 64)

- Unalloyed coppers
- High-copper alloys
- Red brasses and leaded red brasses (Cu–Zn–Sn–Pb)
- Yellow and leaded yellow brasses (Cu–Zn–Sn–Pb)
- Manganese bronzes and leaded manganese bronzes (Cu–Zn–Mn–Fe–Pb)
- Silicon bronzes, silicon brasses (Cu–Zn–Si)
- Copper–phosphorus–silver alloys (Cu–P–Ag)
- Tin bronzes and leaded tin bronzes (Cu–Sn–Zn–Pb)
- Nickel–tin bronzes (Cu–Ni–Sn–Zn–Pb)
- Copper–nickel alloys (Cu–Ni–Fe)
- Nickel silvers (Cu–Ni–Fe)
- Leaded coppers (Cu–Pb)
- Miscellaneous alloys
- G-Bronzes (gunmetal)

2.3 Zinc and Zinc Alloys

2.3.1 General Properties and Description

Zinc [7440-66-6] with the chemical symbol, Zn, the atomic number 30, and the relative atomic mass 65.39(2), is the first element of the group IIB(12) of Mendeleev's periodic chart. The element is named after the German, Zink. Pure zinc is a hard, dense (7133 kg.m^{-3}), and low melting point (419.5°C) metal with a bluish-white luster when freshly cut. It exhibits a good electrical resistivity (5.916$\mu\Omega$.cm) and a good thermal conductivity (113 W.m^{-1}.K^{-1}). Zinc crystal structure is hexagonal close packed (hcp) which explains its hardness at room temperature. Zinc has five stable isotopes: ^{64}Zn (48.6at%), ^{66}Zn (27.9at%), ^{67}Zn (4.1at%), ^{68}Zn (18.8at%), and ^{70}Zn (0.6at%). Zinc exhibits a relatively high corrosion resistance in natural environments, but is readily attacked by strong mineral and organic acids, with evolution of hydrogen. Furthermore, zinc is now known to be an essential nutrient which is indispensable for human health as well as for many species of animals and plants. **Cost** (1998) – pure zinc is priced at 1.6 $US.kg^{-1} (0.48 $US.lb^{-1}).

2.3.2 History

Zinc has been known since ancient times in the form of alloys with tin and copper, and served mankind long before it was identified as a separate metal. The modern history of zinc began in 1743 when the first commercial zinc smelter was built in Bristol, UK.

2.3.3 Natural Occurrence, Minerals, and Ores

The relative abundance of zinc in the Earth's crust is about 75 mg.kg^{-1} (i.e., ppm wt), less than vanadium, nickel, and chromium. Zinc occurs chiefly as minerals: the sulfide **sphalerite** or **blende** (ZnS, cubic), **smithsonite** (ZnCO$_3$, trigonal), and **hemimorphite** or **calamine** (Zn$_4$Si$_2$O$_7$(OH)$_2$·H$_2$O, monoclinic), and **franklinite** (ZnFe$_2$O$_4$, cubic). Amongst them, mainly sphalerite is used as zinc ore. Moreover, sphalerite is often associated in ore deposits with galena and pyrite, and because it contains non-negligible amounts of impurities such as Cd, In, and Ge, sphalerite is the principal source of these metals as by-products. In 1994, world zinc production was 6.81 × 10^6 tonnes. Australia, Canada, China, Peru, and United States are the world's largest zinc producers accounting for 60% of the world total.

2.3.4 Processing and Industrial Preparation

(1) **Roasting process** – the raw sulfide ores (4 to 8wt% Zn) are concentrated from inert gangue materials by common ore beneficiation processes (e.g., froth flotation). The concentrated sulfide ore (60wt% Zn) is then roasted in air at a temperature ranging from 900°C to 1100°C. During the oxidation reaction the sulfide reacts with oxygen forming zinc oxide also called **calcine** (ZnO), and giving off sulfur dioxide (SO$_2$) which is used for the synthesis of sulfuric acid (e.g., about 2 tonnes of acid per tonne of zinc produced).

(2) **Preparation of zinc metal** – at this stage, the zinc metal can be extracted from the calcine, ZnO, according to two routes: (a) the hydrometallurgical process, and (b) the pyrometallurgical process. The second route, which is the older process, provides roughly 58% of the world zinc metal production.

(2a) **Hydrometallurgical process** – in this process, zinc oxide is dissolved at 55–65°C in diluted sulfuric acid (200 kg.m^{-3} H$_2$SO$_4$). After dissolution is completed, the mother liquor is then purified, especially from interfering cations such as Fe(III), which is removed by precipitating the hydroxide, Fe(OH)$_3$, while other more noble cations such as Cu(II), and Ni(II) are reduced by a redox reaction, adding zinc powder to the solution. The electrowinning of zinc from the mother liquor is performed in an electrolyzer made of PVC, using silver-bearing lead

Table 2.7 Physical properties of wrought copper alloys

Usual and trade names	UNS	Average chemical composition (x, wt%)	Density (ρ, kg.m⁻³)	Yield strength 0.2% proof (σ_{YS}, MPa)	Ultimate tensile strength (σ_{UTS}, MPa)	Elongation (Z, %)	Brinell hardness (HB)	Melting point or liquidus range (°C)	Thermal conductivity (k, W.m⁻¹.K⁻¹)	Coefficient of linear thermal expansion (a)	Electrical resistivity (ρ, μΩ.cm)
Unalloyed copper											
Pure copper (oxygen free electronic copper)	C10100	99.99+	8941	69–365	221–455	4–55	42–96	1084	392	17.7	1.741
Pure copper (oxygen free)	C10200	99.95	8941	69–365	221–455	4–55	49–87	1084	397	17.7	1.741
Electrolytic tough pitch copper	C11000	99.90Cu–0.04O	8920	69–365	224–314	4–55	49–87	1084	397	17.7	1.707
Phosphorus deoxidized non-arsenical copper	C10800	99.95Cu–0.009P	8940	69–345	221–379	4–50	54–82	1082	177	17.7	2.028
Phosphorus deoxidized arsenical copper	C14200	99.68Cu–0.35As–0.02P	8940	69–345	221–379	8–45	50	1082	177	17.4	3.831
High-copper alloys											
Beryllium copper	C17200	99.5Cu–1.85Be–0.25Co	8250	172–1344	469–1462	1–48	100–363	865–980	84	17.8	4.009
Beryllium copper	C17000	99.5Cu–1.7Be–0.2Co	8250	221–1172	483–1310	3–45	100–363	1000	84	17.0	2.053
Cadmium copper	C16200	99.2Cu–0.8Cd	8940	600	649	n.a.	n.a.	1080	376	17.0	2.028
Chromium copper	C18200	99.4Cu–0.6Cr	8890	479–531	232–593	14–60	58–140	1081	188	17.0	3.831
Cobalt beryllium copper	C17500	99.5Cu–0.5Be–2.5Co	8750	172–758	310–793	5–28	67–215	1060	84	17.0	7.496

Lead copper	C18700	99.0Cu–1Pb	8950	69–345	221–379	8–45	n.a.	n.a.	n.a.	n.a.	n.a.
Silver-bearing copper	C11300	99.9Cu–0.05Ag–0.02O	8920	69–365	221–455	4–55	55–90	1079	397	17.7	1.741
Sulfur copper	C14700	99.65Cu–0.4S	8920	69–379	221–393	8–52	55–85	1075	373	17.0	1.815
Tellurium copper	C14500	99.5Cu–0.5Te–0.008P	8940	69–352	221–386	8–45	49–80	1082	382	17.7	1.759
Zirconium copper	C15000	99.8Cu–0.15Zr	8900	41–496	200–524	2–54	n.a.	n.a.	n.a.	n.a.	n.a.
Brasses (i.e., copper–zinc)											
Admiralty brass	C44300	71Cu–28Zn–1Sn	n.a.	124–152	331–379	60–65	105	n.a.	n.a.	n.a.	n.a.
Aluminium brass	C68700	77.5Cu–20.5Zn–2Al–0.1As	8350	186	414	55	75	1010	101	18.5	7.496
Cartridge brass	C26000	70Cu–30Zn	8550	76–448	303–896	3–66	65–132	965	121	19.1	6.152
Free-cutting brass	C36000	61.5Cu–35.5Zn–3Pb	8500	124–310	338–469	18–53	105	900	109	20.9	6.631
Gilding metal (cap copper)	C21000	95.0Cu–5.0Zn	8850	69–400	234–441	8–45	65–105	1065	234	18.1	3.079
High tensile brass (architectural bronze)	C38500	57Cu–40Zn–3Pb	8350	138	414	30	95	990	88–109	21.0	8.620
Hot stamping brass (forging)	C37700	59Cu–39Zn–2Pb	8450	138	359	45	95	910	109	20.9	6.631
Low brass	C24000	80Cu–20Zn	8650	83–448	290–862	3–55	65–130	1000	138	18.7	4.660
Muntz metal	C28000	60Cu–40Zn	8400	145–379	372–510	10–52	85–120	900	126	20.8	6.157

Table 2.7 (continued)

Usual and trade names	UNS	Average chemical composition (x, wt%)	Density (ρ, kg.m^{-3})	Yield strength 0.2% proof (σ_{YS}, MPa)	Ultimate tensile strength (σ_{UTS}, MPa)	Elongation (Z, %)	Brinell hardness (HB)	Melting point or liquidus range (°C)	Thermal conductivity (k, W.m^{-1}.K^{-1})	Coefficient of linear thermal expansion (a)	Electrical resistivity (ρ, $\mu\Omega$.cm)
Naval brass	C46400	60Cu-39.25Zn-0.75Sn	8400	172–455	379–607	17–50	100	915	117	21.2	6.631
Red brass	C23000	85Cu-15Zn	8750	69–434	269–724	3–55	65–125	1020	159	18.2	3.918
Yellow brass	C26800	65Cu-35Zn	8500	97–427	317–883	3–65	65–137	940	121	19.9	6.631
Bronzes (i.e., copper-tin-zinc)											
Phosphor bronze	C51100	98.75Cu-3.5Sn-0.12P	8850	345–552	317–710	2–48	70–195	1070	85	18.8	9.171
Phosphor bronze A	C51000	95.0Cu-5Sn-0.09P	8850	131–552	324–965	2–64	71–205	1060	75	18.0	10.26
Phosphor bronze C	C52100	92.0Cu-7Sn-0.12P	8800	165–552	379–965	2–70	80–210	1050	67	18.5	12.32
Phosphor bronze D	C51100	90.0Cu-10Sn-0.05P	8800	193	455–1014	3–70	n.a.	1040	63	18.0	12.32
Silicon bronzes (i.e., copper-silicon-tin)											
Silicon bronze A	C65500	97.0Cu-3.0Si	8520	145–483	386–1000	3–63	n.a.	1028	50	18.0	21.29
Silicon bronze B	C65100	98.5Cu-1.5Si	8700	103–476	276–655	11–55	n.a.	n.a.	n.a.	n.a.	n.a.
Aluminium–silicon bronzes (i.e., Cu–Al–Ni–Fe–Si–Sn)											
Aluminum bronze D	C61400	91Cu-7Al-2Fe	7800	228–414	524–614	32–45	90–160	1045	n.a.	17.0	12.32

Aluminum bronze	C60800	95Cu–5Al	8150	186	414	55	85	1065	85	18.0	9.741
Aluminum bronze	C63000	Cu–9.5Al–4Fe–5Ni–1Mn	7570	345–517	621–814	15–20	200	1060	62	17.0	13.26
Copper–nickel alloys											
Copper nickel	C70400	92.4Cu–5.5Ni–1.5Fe–0.6Mn	8940	276–524	262–531	2–46	60–65	1121	67	17.5	13.79
Copper nickel	C70600	87.7Cu–10Ni–1.5Fe–0.6Mn	8940	110–393	303–414	10–42	65–70	1150	42	17.1	21.55
Copper nickel	C71500	67Cu–31Ni–1Fe–1Mn	8900	138–483	372–517	15–45	90	1238	21	16.6	38.31
Nickel silvers											
Nickel silver 10%	C74500	65Cu–25Zn–10Ni	8600	124–524	338–896	1–50	66–155	1010	37	16.4	20.75
Nickel silver 12%	C74500	65Cu–23Zn–12Ni	8640	124–545	359–641	2–48	65–210	1025	30	16.2	22.36
Nickel silver 15%	C74500	65Cu–20Zn–15Ni	8690	124–545	365–634	2–43	70–210	1060	27	16.2	24.59
Nickel silver 18%	C745200	65Cu–17Zn–18Ni	8720	172–621	386–710	3–45	77–166	1100	28	16.0	27.37
Nickel silver 25%	C74500	65Cu–10Zn–25Ni	8820	103–579	345–758	1–37	75–201	1160	21	17.0	33.81

2

Common Nonferrous Metals

Table 2.8 Physical properties of selected cast copper alloys

Usual and trade names	UNS	Average chemical composition (x, wt%)	Density (ρ, kg.m^{-3})	Young's modulus (E, GPa)	Yield strength 0.2% proof (σ_{YS}, MPa)	Ultimate tensile strength (σ_{UTS}, MPa)	Elongation (Z, %)	Brinell hardness (HB)	Melting point or liquidus range (°C)	Thermal conductivity (k, W.m^{-1}.K^{-1})	Specific heat capacity (c_P, J.kg^{-1}.K^{-1})	Coefficient of linear thermal expansion (a)	Electrical resistivity (ρ, $\mu\Omega$.cm)
Admiralty gun metal	C90500	88Cu–10Sn–2Zn	8720	105	140–152	305–310	20–310	75–80	854–1000	74	376	20.0	15.7
Aluminum bronze	C95200	88Cu–9Al–3Fe	7640	105	186	552	35	125	1040–1045	50	380	16.2	14.4
Aluminum bronze	C95500	81Cu–11Al–4Fe–4Ni	7530	110	275–303	620–689	6–12	192–230	1040–1055	42	418	16.2	20.3
Beryllium copper cast	C81300	98.5Cu–1Co–0.15Be–	8810	110	250	365	11	39	1066–1093	260	390	18	2.87
Beryllium copper cast	C81400	99Cu–0.8Cr–0.06Be	8810	110	83–250	205–365	11	62–69 HRB	1065–1095	259	389	18	3.00
Beryllium copper 20C	C82500	97.2Cu–2Be–0.5Co–0.25Si	8260	128	275–1035	515–1105	2–35	30–43 HRC	855–980	105	420	17	8.62
Cast copper	C81100	99.70Cu	8940	115	62	170	40	44	1065–1083	346	380	16.9	1.87
Chromium copper	C81500	99Cu–1Cr	8820	115	275	350–352	17	105	1075–1085	315	376	17.1	3.83
High tensile brass	C86500	58Cu–39Zn–1.3Fe–1Al–0.5Mn	8300	105	193	490	28	100–130	862–880	87	373	20.3	8.41
Hydraulic bronze	C83800	83Cu–7Zn–6Pb–4Sn	8640	92	110	240	25	60	845–1005	72.5	380	18.0	11.5
Leaded gun metal	C83600	84.8Cu–4.8Pb–5.1Sn–4.8Zn	8830	83	117	255	30	60–65	855–1010	72	380	18.0	11.5
Silicon brass	C87500	82Cu–14Zn–4Si	8410	106–138	207–310	462–585	21–25	115–134	903–917	84	375	19.6	25.7
Tin bronze	C90700	89Cu–11Sn	8770	105	150–207	241–310	25	80–110	832–1000	71	376	18.0	15.7

anodes (+), and pure aluminum cathodes (−). During electrolysis, the zinc electrodeposits onto aluminum cathodes while oxygen evolves at the anodes. The cell voltage ranges between 3.2 V to 3.7 V, while the cathodic current density is between 0.4 to 0.7 kA.m^{-2}. The specific energy consumption is 3 to 3.5 kWh.kg^{-1} of zinc. After electrolysis zinc is detached from aluminum anodes, melted, and cast into ignots. The zinc purity obtained is above 99.995wt%.

(2b) **Pyrometallurgical process** – this process consists of the carbothermic reduction of the zinc oxide at temperature above 900°C, according to the chemical reaction

$$ZnO(s) + C(s) \rightarrow Zn(g) + CO(g).$$

In the **New Jersey process**, the solid mixture of zinc oxide and carbon is heated at 1300°C in a large vertical crucible, the zinc vapor and carbon monoxide evolve and the zinc is recovered in the condenser. The purity of the zinc condensate is about 99.5wt%. In the **imperial smelting process**, the reactor vessel is similar to a blast furnace, and a preheated mixture of zinc oxide and coke is poured into the top of the furnace, while air at 750°C under pressure is introduced at the bottom. In the condenser, a spray of molten lead droplets collect the zinc forming a two-phase liquid mixture, from which the zinc is easily separated by decantation in a settler. The impure zinc (99.8wt% Zn) contains several impurities. Afterwards, high-purity zinc (99.993wt% Zn) is obtained by fractional distillation under inert atmosphere, and cadmium is an important by-product.

2.3.5 Industrial Applications and Uses

Zinc is used extensively in protective coatings on steel and sometimes on aluminum. Its excellent protective properties, along with its relatively high corrosion resistance in a natural environment, ensure that the metal will have wide industrial applications. When used as a coating either electroplated or hot dip galvanized, it produces an anodic coating that not only protects mechanically by shielding, but also electrochemically by galvanic coupling. Zinc and its alloys are used extensively as galvanic anodes (i.e., sacrificial anodes) in cathodic protection systems (e.g., buried pipelines). Large amounts of zinc are also used to produce Cu–Zn alloys (i.e., brasses). However, zinc-based alloys find few applications.

Hot dip galvanizing is a method distinct from galvanizing which involves the electrodeposition of a zinc layer usually achieved in an aqueous electrolyte. Hot dip galvanizing (or hot galvanizing) consists of producing a thick coating of zinc by immersing the steel workpiece in a molten bath of zinc maintained at temperatures ranging from 440 to 465°C. The thickness of the coating varies greatly with bath temperature, immersion time, and withdrawal rate from the bath. The mass surface density of the zinc coating ranges between 1.22 to 2.14 kg.m^{-2} for steel sheet counting both sides, which corresponds to actual thicknesses of 86 to 150 μm for a single side. The coating is actually formed in several distinct layers. The layers closest to the steel surface are composed of iron–zinc intermetallic compounds, while the outer layers consist mainly of pure zinc. The best steels for galvanizing are those containing less than 0.15wt% carbon. Cast iron can be galvanized but should be low in silicon and phosphorus to avoid brittleness in the zinc–iron layer closest to the surface of the cast iron. There are many uses for galvanized steel, including many mill products, fasteners of all descriptions, pipes and fittings, structural members, heat exchanger coils, highway guard rails, etc. The zinc on the surface of steel, like zinc and cadmium platings, cathodically protects the underlying steel forming a sacrificial anode, especially if breaks occur in the coating. The corrosion protection is best in atmospheres that do not contain sulfur gases and other industrial pollution; however, galvanized steel is widely used in industrial atmospheres because galvanizing is so economical.

2.3.6 Zinc Alloys

See **Table 2.9** (pages 66-67).

Table 2.9 Physical properties of zinc and zinc alloys

Usual and trade names (cast)	UNS	Average chemical composition (x, wt%)	Density (ρ, kg.m^{-3})	Young's modulus (E, GPa)	Ultimate tensile strength (σ_{UTS}, MPa)	Elongation (Z, %)	Brinell hardness (HB)	Melting point or liquidus range (°C)	Thermal conductivity (k, W.m^{-1}.K^{-1})	Specific heat capacity (c_P, J.kg^{-1}.K^{-1})	Coefficient of linear thermal expansion (α, 10^{-6} K^{-1})	Electrical resistivity (ρ, $\mu\Omega$.cm)
Pure zinc	Z13000	99.9993	7133	104.5	n.a.	n.a.	n.a.	419.5	119.5	382	39.7	5.96
AC41A (Zamak 5)	Z35531	Zn–4Al–1Cu–0.05Mg	6700	n.a.	328	7	91	380–386	109.0	394	27.4	6.50
AC43A (Zamak 2)	Z35541	Zn–4Al–2.5Cu–0.04Mg	6600	n.a.	358	7	100	379–390	105.0	419	27.8	6.89
AG40A (Zamak 3)	Z33521	Zn–4Al–0.04Mg	6600	n.a.	283	10	82	381–387	113.0	419	27.4	6.40
AG40B (Zamak 7)	Z33523	Zn–4Al–0.015Mg	6600	n.a.	283	13	80	381–387	113.0	419	27.4	6.39
Copper hardened rolled zinc	Z44330	Zn–1.0Cu	7170	n.a.	170–210	35–50	52–60	419–422	104.7	402	21.1–34.7	6.20
ILZRO 16	n.a.	Zn–1.25Cu–0.2Ti–0.15Cr	7100	97.0	230	6	113	416–418	104.7	402	27.0	8.4
Korloy 2684	n.a.	Zn–22Al	5200	68–93	310–380	25–27	70–85	n.a.	n.a.	n.a.	22.0	6.00

Rolled zinc	Z21210	Zn–0.08Pb	7140	n.a.	134–159	50–65	42	419	108.0	395	32.5	6.20
Rolled zinc	Z21220	Zn–0.06Pb–0.06Cd	7140	n.a.	145–173	32–50	43	419	108.0	395	32.5	6.06
Rolled zinc	Z21540	Zn–0.3Pb–0.03Cd	7140	n.a.	160–200	32–50	47	419	108.0	395	39.9	6.06
Rolled zinc alloy	Z45330	Zn–1.0Cu–0.01Mg	7170	n.a.	200–276	10–20	61–80	419–422	104.9	401	21.1–34.8	6.30
Slush casting alloy A	Z34510	Zn–4.75Al–0.25Cu	n.a.	n.a.	193	1.0	n.a.	380–390	n.a.	n.a.	n.a.	n.a.
Slush casting alloy B	Z30500	Zn–5.5Al	n.a.	n.a.	172	1.0	n.a.	380–395	n.a.	n.a.	n.a.	n.a.
ZA-12 (ILZRO 12)	Z35631	Zn–11Al–1Cu–0.025Mg	6030	82.7	328–404	2–5	100	377–432	116.0	450	24.1	6.10
ZA-27	Z35841	Zn–27Al–2Cu–0.015Mg	5000	77.9	426	2.5	119	375–484	125.5	525	26.0	5.8
ZA-8	Z35636	Zn–8Al–1Cu–0.02Mg	6300	85.5	240–374	1–8	103	375–404	115.0	435	23.2	6.20
Zn–Cu–Ti alloy	Z41320	Zn–0.8Cu–0.15Ti	7170	63.5–88.0	221–290	21–38	61–80	419–422	105.0	402	19.4–24.0	6.24

2.4 Lead and Lead Alloys

2.4.1 Description and General Properties

Lead [7439-92-1] with the chemical symbol Pb, the atomic number 82, and the relative atomic mass 207.2(1), is the heaviest element of the group IVA(14) of Mendeleev's periodic chart. The symbol Pb is the abbreviation of the Latin name of the metal, Plumbum. Pure lead is a soft, dense (11,680 kg.m^{-3}), malleable, fusible (mp 327.46°C), bluish-gray metal, but tarnishes upon exposure to air. It exhibits a low electrical resistivity (20.648 $\mu\Omega$.cm), a low thermal conductivity (34.9 W.m^{-1}.K^{-1}), and a high coefficient of thermal expansion (29.1 μm.m^{-1}.K^{-1}). From a mechanical point of view, lead exhibits both a low Young's modulus (16.1 GPa), and low tensile strength, and it is sensitive to creep even at room temperature. Therefore, lead structures are unable to support their own weight and are subject to creep, which forbids their use as structural materials. The combination of its high density with low stiffness and good damping capacity makes lead an excellent material for absorbing sound and vibrations. Lead has four stable isotopes, amongst them ^{204}Pb (1.4at%) is nonradiogenic, while ^{206}Pb (24.1at%), ^{207}Pb (22.1at%), and ^{208}Pb (52.4at%) are the end products of the three natural radioactive series. Owing to rapid build-up of a passivating protective film when put in contact with numerous corrosive oxidizing environments (e.g. chromates, sulfates, carbonates, phosphates, fluorides), lead and lead alloys were extensively used for handling these chemicals in the CPI. For this main reason, it was widely used in the chemical industry at the beginning of the century and especially in the historical lead chamber process for production of concentrated sulfuric acid. Actually, corrosion rate in cold 50% H_2SO_4 is about 130 μm per year.

Sometimes lead is clad to another base metal. There are a number of lead alloys developed specifically to increase durability and hardness or strength of the metal. Small amounts of silver and copper are present in many natural ores and such elements are believed to increase the corrosion resistance and improve creep and fatigue behavior. There are, however, various proprietary alloys to which copper and other elements have deliberately been added for improved corrosion and creep resistance (e.g., Nalco$^{®}$, which is a specialty product for anodes with improved corrosion resistance in chromic acid plating baths). Chemical lead, acid lead, and copper lead are the grades usually specified in the chemical industries. Other common grades include antimonial lead (also called hard lead), and tellurium lead. Chemical lead contains traces of silver and copper from the original ores left in, basically since it is not considered economical to recover the silver, while the copper is thought to improve the general corrosion resistance. Antimonial lead has been alloyed with 2 to 6wt% Sb. Antimony hardens lead and improves its physical characteristics up to about 100°C. However, it lowers the melting point, and while it may not detract from lead's corrosion resistance, it seldom improves it. Above 100°C both strength and corrosion resistance rapidly decrease. Lead can be strengthened by the addition of a fraction of a percentage of tellurium. Tellurium in extruded lead products retards grain growth and tellurium lead will work-harden under strain. Therefore it has better fatigue resistance. Corrosion resistance is comparable with chemical lead. In lead-cladded products, the corrosion resistance of lead is combined with specific properties of the substrate or base metal, e.g. the strength of steel or the excellent heat transfer of copper. Vessels designed for operation at high temperatures, fluctuating temperatures or in a vacuum are usually constructed of steel with a lining of lead bonded directly to the steel. Steel pipes and valves with internal lead cladding are quite common, especially in the manufacture of nitroglycerine. Also heating coils of copper with an external lead cladding are in use. Sheet lead is further used as a membrane in acid-brick-lined vessels.

History – lead has been known since ancient times; lead pipes bearing the insignias of Roman emperors were used as drains from the baths during the Roman Empire. Moreover, lead is mentioned in Exodus. Alchemists believed lead to be the oldest metal and associated it with the planet Saturn.

Natural occurrence, minerals, and ores – the relative abundance of lead in the Earth's crust is roughly 15 mg.kg^{-1} (i.e., ppm wt) which is below that of nitrogen (19) and lithium (20). Lead never occurs free in nature, and amongst the 60 mineral varieties known containing lead the chief minerals are the sulfide **galena** (PbS, cubic), the carbonate **cerussite** (PbCO$_3$, orthorhombic), the oxide **minium** (Pb$_3$O$_4$, tetragonal), and the sulfate **anglesite** (PbSO)$_4$, orthorhombic). Amongst them, galena is the main lead ore used industrially to recover the metal. Because in ore deposits, galena is always associated with copper and zinc ores, it is not normally mined independently. Australia (19%) is the world's largest lead producer, followed by the United States (13%), China (12%), Peru (8%), and Canada (6%).

Industrial uses and applications – today, major applications of lead and lead alloys are in order of importance: (i) battery grids for electrode manufacture in lead–acid rechargeable batteries, (ii) type metals in the printing industry, (iii) cable sheathing in electrical engineering, (iv) piping for handling corrosive chemicals in the chemical process industries, (v) solders and bearing alloys, (vi) ammunitions, (vii) anodes in industrial electrolyzers (see electrode materials in Chapter 8). **Cost** (1998) – pure refined lead (99.99wt% Pb) is priced at 0.53 \$US.kg^{-1} (0.24 \$US.lb^{-1}).

2.4.2 Lead Alloys

See **Table 2.10** (pages 70–71).

2.5 Tin and Tin Alloys

See **Table 2.11** (pages 72–73).

References

[1] Jarrett N, Frank WB, Keller R (1981) Advances in the smelting of aluminum. In Tien JK, Elliott JF (eds) Metallurgical treatises. The Metallurgical Society of AIME.

Further Reading

Aluminum and Aluminum Alloys

AA/ANSI (1997) Standard alloy and temper designation systems for aluminum. The Aluminum Association, Inc. and American National Standardization Institute.
ASM International (1991) ASM metal handbook, vol. 2, Properties and selection: nonferrous alloys and special-purpose materials. ASM, Ohio Park, OH.
Gerard G (ed.)(1963) Extractive metallurgy of aluminum, vol. 1: alumina. Interscience Publishers.
Hatch JE (ed.) (1984) Aluminum: properties and physical metallurgy. American Society for Metals, Ohio Park, OH.
Keeffe J (ed.) (1982) Aluminum: profile of the industry. McGraw-Hill, New York.
Pearson TG (1955) The chemical background of the aluminium industry. The Royal Institute of Chemistry, London.
Serjeanston R (ed.) (1996) World aluminium: a metal bulletin databook, 3rd edn. Metal Bulletin Books Ltd, London.

Lead and Lead Alloys

Hoffmann W (1960) Lead and lead alloys, 2nd edn. Springer-Verlag, Berlin.
Properties of lead and lead alloys (1984) Lead Industries Association.
Worcester AW, O'Reilly JT (1991) Lead and lead alloys. In: American Society of Metals, ASM handbook of metals, 10th edn, vol. 2: Properties and selection: nonferrous alloys and special-purpose materials. ASM, Ohio Park, OH, pp 543–556.

Table 2.10 Physical properties of lead and lead alloys

Usual and trade names (ASTM grade)	UNS	Average chemical composition (x, wt%)	Density (ρ, kg.m⁻³)	Ultimate tensile strength (σUTS, MPa)	Elongation (Z, %)	Brinell hardness (HB)	Melting point or liquidus range (°C)	Thermal conductivity (k, W.m⁻¹.K⁻¹)	Specific heat capacity (cP, J.kg⁻¹.K⁻¹)	Coefficient of linear thermal expansion (a)	Electrical resistivity (ρ, μΩ.cm)
Pure lead											
Pure lead (zone refined)	L50001	99.9999	11,680	16.8	35	3–4	327.5	34.9	129.8	29.0	20.6
Pure lead (soft, refined)	L50005	99.999	11,680	12–13	30	3–4	327.5	34.9	129.8	29.0	20.6
Pure lead (Soft refined)	L50011	99.99	11,680	12–13	30	3–4	327.5	34.9	129.8	29.0	20.6
Pure lead (corroding)	L50042	99.94	11,350	12–13	30	3.2–4.5	327.5	34.9	129.8	29.0	20.6
Lead–copper alloys											
Copper lead	L51110	99.9Pb–0.05Cu	n.a.	16–19	30–60	n.a.	n.a.	n.a.	n.a.	n.a.	n.a.
Lead–tellurium copper	L51123	99.85Pb–0.055Te–0.06Cu	n.a.	21.1	n.a.	5.8	n.a.	n.a.	n.a.	n.a.	n.a.
Silver-copper lead (chemical lead)	L51120	99.9Pb–0.02Ag–0.08Cu	n.a.	16–19	30–60	4–6	n.a.	n.a.	n.a.	n.a.	n.a.
Copper-bearing alloy (Alloy 485)	L51180	51Pb–44Cu–3Sn	n.a.	n.a.		n.a.	n.a.	n.a.	n.a.	n.a.	n.a.
Lead–antimony alloys											
Antimonial lead	L52700	Pb–2.0Sb	n.a.	n.a.	n.a.	n.a.	n.a.	n.a.	n.a.	n.a.	n.a.
Antimonial lead	L52900	Pb–4.0Sb	n.a.	30	n.a.	8–12	n.a.	n.a.	n.a.	n.a.	n.a.
Antimonial lead	L53200	Pb–8.0Sb	n.a.	37.7	n.a.	9–16	n.a.	n.a.	n.a.	n.a.	n.a.

Electrowinning alloys	L50122	98Pb–1Ag–1As	n.a.	n.a.	n.a.	n.a.	n.a.	n.a.	n.a.	n.a.
Lead tin	L55000	52Pb–48Sn	n.a.	n.a.	n.a.	n.a.	n.a.	n.a.	n.a.	n.a.
Cathodic protection anode	L50150	98Pb–2Ag	n.a.	n.a.	n.a.	n.a.	n.a.	n.a.	n.a.	n.a.
Lead cadmium eutectic	L50940	83Pb–17Cd	n.a.	n.a.	n.a.	n.a.	n.a.	n.a.	n.a.	n.a.
Battery grid	L50735	99.9Pb–0.06Ca	n.a.	n.a.	n.a.	n.a.	n.a.	n.a.	n.a.	n.a.
Type metals										
Electrotype	L52730	95Pb–2.5Sb–2.5Sn	n.a.	n.a.	n.a.	12.5	246–303	n.a.	n.a.	n.a.
Stereotype	L53530	80Pb–14Sb–6Sn	n.a.	n.a.	n.a.	22–25	239–256	n.a.	n.a.	n.a.
Linotype	L53420	86Pb–11Sb–3Sn	n.a.	n.a.	n.a.	19–22	239–247	n.a.	n.a.	n.a.
Monotype	L53558	78Pb–15Sb–7Sn	n.a.	n.a.	n.a.	24–33	239–262	n.a.	n.a.	n.a.
Lead-based Babitt alloys										
Lead alloy 7	L53465	77.5Pb–12.5Sb–10Sn	9730	n.a.	n.a.	22.5	240–268	n.a.	n.a.	n.a.
Lead alloy 8	L53560	Pb–15Sb–5Sn–0.5Cu	10040	n.a.	n.a.	20.0	237–272	n.a.	n.a.	n.a.
Lead alloy 15	n.a.	Pb–16Sb–1Sn–1.1As–0.5Cu	10040	n.a.	n.a.	21.0	248–281	n.a.	n.a.	n.a.
Lead alloy B	n.a.	83.3Pb–12.45Sb–0.84Sn–0.1Cu	n.a.	n.a.	n.a.	n.a.	n.a.	n.a.	n.a.	n.a.

Table 2.11 Physical properties of selected tin alloys (chillcast)

Usual and trade names	Average chemical composition (x, wt%)	Density (ρ, kg.m⁻³)	Young's modulus (E, GPa)	Ultimate tensile strength (σ_{UTS}, MPa)	Elongation (Z, %)	Brinell hardness (HB)	Melting point or liquidus range (°C)	Thermal conductivity (k, W.m⁻¹.K⁻¹)	Specific heat capacity (c_P, J.kg⁻¹.K⁻¹)	Coefficient of linear thermal expansion (a)	Electrical resistivity (ρ, μΩ.cm)
Electrolytic (AAA)	99.98wt% Sn (extra high purity)	7280	49.9	14.5	n.a.	3.9	231.9	66.8	222	23.5	12.6
Electrolytic (AA)	99.95wt% Sn (high purity)	7280	49.9	14.5	n.a.	n.a.	231.9	66.8	222	23.5	n.a.
Tin Grade A	99.80wt% Sn	7280	49.9	14.5	n.a.	n.a.	231.9	66.8	222	23.5	n.a.
Tin Grade B	99.80wt% Sn	7280	49.9	14.5	n.a.	n.a.	231.9	66.8	222	23.5	n.a.
Tin Grade C	99.65wt% Sn	7280	49.9	n.a.	n.a.	n.a.	n.a.	n.a.	n.a.	n.a.	n.a.
Tin Grade D	99.50wt% Sn	7280	49.9	n.a.	n.a.	n.a.	n.a.	n.a.	n.a.	n.a.	n.a.
Tin Grade E	99.00wt% Sn	7280	49.9	n.a.	n.a.	n.a.	n.a.	n.a.	n.a.	n.a.	n.a.
Antimonial tin	95Sn–5Sb	7250	49.99	31	25	15	234–240	n.a.	n.a.	n.a.	14.5
Bearing alloy	75Sn–12Sb–10Pb–3Cu	7530	n.a.	27	n.a.	27	184–306	n.a.	n.a.	n.a.	n.a.
Casting alloy	65Sn–18Pb–15Sb–2Cu	7750	n.a.	22.5	n.a.	22.5	181–296	n.a.	n.a.	n.a.	n.a.

Material	Composition										
Hard tin	99.6Sn–0.4Cu	n.a.	n.a.	23	n.a.	n.a.	227–230	n.a.	n.a.	n.a.	n.a.
Soft solder	70Sn–30Pb	n.a.	n.a.	46.9	n.a.	12	183–192	50.0	n.a.	21.6	14.6
Soft solder	62Sn–36Pb–2Ag	8420	22.96	43	7	17	177–189	50.0	n.a.	27.0	14.5
Soft solder	60Sn–40Pb	8520	29.99	19	135	16	183–188	50.0	150	23.9	14.9
Tin Babbitt alloy 1 (White metal)	91Sn–4.5Sb–4.5Cu	7340	n.a.	64	9	17	223–371	n.a.	n.a.	n.a.	n.a.
Tin Babbitt alloy 2	89Sn–7.5Sb–3.5Cu	7390	n.a.	77	18	24	241–354	n.a.	n.a.	n.a.	n.a.
Tin Babbitt alloy 3	84Sn–8Sb–8Cu	7450	n.a.	69	1	27	240–422	n.a.	n.a.	n.a.	n.a.
Tin die casting alloy	82Sn–13Sb–5Cu	7750	n.a.	69	1	29	181–296	n.a.	n.a.	n.a.	n.a.
Tin foil	92Sn–8Zn	n.a.	n.a.	60	40	n.a.	200	n.a.	n.a.	n.a.	n.a.
Tin–silver solder eutectic	96.5Sn–3.5Ag	7290	n.a.	37	31	15	221	n.a.	n.a.	n.a.	12.31
White metal	92Sn–8Sb	7280	53	64.7	24	23.8	244–295	n.a.	n.a.	n.a.	15.5

Tin and Tin Alloys

Barry BTK, Thwaites JC (1983) Tin and its alloys and compounds. Ellis Horwood, Chichester.

Hedges ES (1964) Tin and its alloys. Edward Arnold, London.

Manko HH (1964) Solders and soldering. McGraw-Hill, New York.

Mantell CL (1959) Tin, 2nd edn. Hafner Publishing, New York.

Wright PA (1982) Extractive metallurgy of tin, 2nd edn. Elsevier, Amsterdam.

Less Common Nonferrous Metals

3.1 Alkali Metals

The alkali metals are represented by the six chemical elements of the group IA(1) of Mendeleev's periodic chart. These six elements are, in increasing atomic number: lithium (Li), sodium (Na), potassium (K), rubidium (Rb), cesium (Cs), and francium (Fr). The name alkali metals is given owing to the fact they form strong alkaline hydroxides (MOH, with M = Li, Na, K, etc.) when they combine with water (i.e., strong bases capable of neutralizing acids). The only members of the alkali metal family that are relatively abundant in the Earth's crust are sodium and potassium. Amongst the alkali metals only lithium, sodium, and to a lesser extent potassium are widely used in industrial applications. Hence, only these three metals will be reviewed in detail in this section. Nevertheless, a short description of the main properties and industrial uses of the last three alkali metals (i.e., Rb, Cs, and Fr) will be presented at the end of the section. Some physical, mechanical, thermal, electrical, and optical properties of the five chief alkali metals (except francium which is radioactive with a short half-life) are listed in Table 3.1.

3.1.1 Lithium

3.1.1.1 Description and General Properties

Lithium [7439-93-2] with the atomic number 3, and a relative atomic mass (atomic weight) of 6.941(2), is the lightest of the alkali metals, i.e., group IA(1) of Mendeleev's periodic chart. It has the chemical symbol Li and is named after the Greek word, lithos, meaning stone. Highly pure lithium is soft, ductile (like lead) and malleable metal having a Mohs hardness of 0.6; hence, it is the hardest of the alkali metals. Nevertheless, small amounts of interstitial impurities (e.g., H, C, O, N) or solid inclusions (e.g., Li_2O, Li_3N) strongly modify its mechanical properties. Moreover, it is the lightest of all the metals and solid elements with the lowest density, 534 kg.m^{-3}, which is roughly half that of pure water. The pure metal has two allotropes. The alpha phase corresponds to a crystalline structure

Table 3.1 Selected properties of five alkali metals

Properties (at 298.15 K, unless otherwise specified)		Lithium	Sodium	Potassium	Rubidium	Cesium
General	Chemical symbol [IUPAC]	Li	Na	K	Rb	Cs
	Chemical abstract registered number [CAS RN]	[7439-93-2]	[7440-23-5]	[7440-09-7]	[7440-17-7]	[7440-46-2]
	Unified numbering system [UNS]	[L06990]	[L11001]	[L08001]	[L09001]	[L02001]
Mineral occurrence and economics	Earth's crust abundance (mg.kg^{-1})	20	23,600	20,900	90	3
	Seawater abundance (mg.kg^{-1})	0.18	10,800	399	0.12	0.003
	World estimated reserves (R, 10^3 kg)	7.3×10^6	Unlimited	$>10^{10}$	n.a.	n.a.
	World production (1997) (P, 10^3 kg.yr^{-1})	12,000 (metal)	90,000 (metal)	200 (metal)	n.a.	20 (metal)
	Cost of pure metal (1998) (C, $US.kg^{-1}) (purity, wt%)	95.40 (99.8)	250 (99.95)	650 (99.95)	20,000 (99.8)	20,283 (99.98)
Atomic properties	Atomic number (Z)	3	11	19	37	55
	Relative atomic mass A_r ($^{12}C = 12.000$)[1]	6.941(2)	22.989770(2)	39.0983(1)	85.4678(3)	132.90545(2)
	Electronic configuration	[He] 2s^1	[Ne] 3s^1	[Ar] 4s^1	[Kr] 5s^1	[Xe] 6s^1
	Fundamental ground state	$^2S_{1/2}$	$^2S_{1/2}$	$^2S_{1/2}$	$^2S_{1/2}$	$^2S_{1/2}$
	Electronegativity χ_a (Pauling)	0.98	0.93	0.82	0.82	0.79
	Electron work function (W_S, eV)	2.93	2.36	2.29	2.261	1.95
	X-ray absorption coefficient CuK$_{a1.2}$ ((μ/ρ), cm^2.g^{-1})	0.716	30.1	143	117	318
Nuclear properties	Thermal neutron cross-section (σ_n, 10^{-28} m^2)	0.045	0.53	2.1	0.38	29
	Isotopic mass range	4–11	17–35	32–54	72–102	112–151
	Natural isotopes (including isomers)	8	21	24	38	56
Crystallographics properties (at 293.15 K)	Crystal structure at room temperature (phase α or β)	bcc (β-Li)	bcc (β-Na)	bcc	bcc	bcc
	Strukturbericht designation	A2	A2	A2	A2	A2
	Space group (Hermann–Mauguin)	Im3m	Im3m	Im3m	Im3m	Im3m
	Pearson symbol	cI2	cI2	cI2	cI2	cI2
	Space lattice parameter (a, pm)	350.89	429.06	532.10	570.50	614.10
	Miller's indices of slip plane (hkl)	(111)	(111)	(111)	(111)	(111)
	Phase transition temperature α-β (T,K)	77 (−196°C)	5 (−268°C)	nil	nil	nil
Mechanical properties (annealed)	Density (293 K) (ρ, kg.m^{-3})	534	971	862	1532	1873
	Young or elastic modulus (300 K) (E, GPa)	4.91	6.80–10.00	3.53	2.35	1.70
	Coulomb or shear modulus (G, GPa)	4.22–4.24	2.53–3.34	1.27–1.30	0.91	0.67
	Bulk or compression modulus (K, GPa)	11.402	6.967	4.201	2.985	2.693
	Mohs hardness (HM)	0.6 (HV < 5)	0.5	0.5	0.3	0.2
	Brinell hardness (HB)	n.a.	0.690	0.363	0.216	0.140
	Ultimate tensile strength (σ_{UTS}, MPa)	1.156	n.a.	n.a.	n.a.	n.a
	Longitudinal velocity of sound (V_S, m.s^{-1})	5830	3200	2000	1300	n.a.
	Poisson's ratio ν (dimensionless)	0.362	0.340	0.350	0.300	0.295–0.356

Table 3.1 (continued)

Properties (at 298.15 K, unless otherwise specified)		Lithium	Sodium	Potassium	Rubidium	Cesium
Thermal properties (293.15 K)	Melting point (mp, K)	453.69 (180.54°C)	370.96 (97.82°C)	336.80 (63.65°C)	312.2 (39.05°C)	301.55 (28.40°C)
	Boiling point (bp, K)	1620 (1346.85°C)	1156.1 (883°C)	1047 (773.85°C)	961 (687.85°C)	951.6 (678.45°C)
	Volume expansion on melting (%)	+1.65	+2.70	+2.55	+2.50	+2.60
	Thermal conductivity (k, $W.m^{-1}.K^{-1}$)	84.7	141	102.4	58.2	35.9
	Coefficient of thermal linear expansion (α, 10^{-6} K^{-1})	56	70.6	83	90	97
	Molar heat capacity (C_p, $J.mol^{-1}.K^{-1}$)	24.770	28.21903	29.97749	30.89858	31.40100
	Specific heat capacity (c_p, $J.kg^{-1}.K^{-1}$)	3568.65	1227.460	766.721	361.523	236.266
	Vapor pressure at melting point (p_v, Pa)	1.82×10^{-10}	1.43×10^{-5}	1.06×10^{-4}	1.56×10^{-4}	2.50×10^{-5}
	Enthalpy molar fusion (ΔH_{fus}, $kJ.mol^{-1}$) (Δh_{fus}, $kJ.kg^{-1}$)	4.60 (662.73)	2.64 (114.83)	2.39 (61.12)	2.198 (25.74)	2.09 (15.73)
	Enthalpy molar vaporization (ΔH_{vap}, $kJ.mol^{-1}$) (Δh_{vap}, $kJ.kg^{-1}$)	147.80 (21,293)	98.00 (4261)	79.53 (2034)	75.8 (887)	65.90 (496)
	Enthalpy molar sublimation (ΔH_{sub}, $kJ.mol^{-1}$)(Δh,$kJ.kg^{-1}$)	161.6 (23.3)	108.90 (4740)	90.00 (2302)	87.5 (1020)	78.70 (600)
	Enthalpy molar combustion (ΔH_{comb}, $kJ.mol^{-1}$)(Δh,$kJ.kg^{-1}$)	−597.9 (86,140)	−414.2 (18,020)	−361.5 (9250)	−339.0 (3970)	−345.8 (2600)
Electrical, electrochemical, and magnetic (293.15 K)	Electrical resistivity (ρ, $\mu\Omega.cm$)	8.55–9.29	4.2	6.15	12.5	20.0
	Temperature coefficient of resistivity (0–100°C) (10^{-3} K^{-1})	4.271–4.350	4.34–5.50	5.70–5.81	4.80	6.00
	Hall coefficient (R_H, $a.\Omega.m.T^{-1}$)	−2.2	−2.3	−4.2	−5.9	−7.8
	Seebeck absolute coefficient (e_S, $\mu V.K^{-1}$)	+14.37	−4.4	−12	−8.26	+0.2
	Thermoelectric potential versus platinum (mV)	+1.82	+0.29	−0.83	n.a.	+1.50
	Electrochemical equivalence (E_q, $A.h.kg^{-1}$)	3860	1166	685	314	202
	Nernst standard electrode potential (E_0, V_{SHE})	−3.040	−2.713	−2.924	−2.924	−2.923
	Mass magnetic susceptibility (χ_m, 10^{-9} $kg^{-1}.m^3$)	+25.6	+8.8	+6.7	+2.49	+2.8
Optical	Wavelength maximum intensity atomic spectra line (bunsen flame color) (λ, nm)	670 (deep red)	589 (bright yellow)	766 (purple-red)	424 (violet)	460 (bluish purple)
	Reflective index (at 650 nm)	0.913	0.975	0.950	n.a.	n.a.

[1]Standard atomic masses from atomic weights of the elements 1995. Pure Appl Chem 68(1996):2339.

down to a very low temperature (to −196°C), which is hexagonal close-packed (hcp). Above 25°C, the crystallographic structure changes slowly to the room temperature beta phase, which is body-centered cubic (bcc). Lithium thermal properties outline the highest specific heat capacity of all the elements (3552 $J.kg^{-1}K^{-1}$), a low vapor pressure of the liquid metal and a high coefficient of linear thermal expansion (56 $\mu m.m^{-1}.K^{-1}$).

Freshly cut lithium has a silvery lustrous appearance, like Na and K. However, owing to its strong chemical reactivity with oxygen, nitrogen, and water, it tarnishes readily in moist air and becomes yellowish, forming a mixture of several compounds (LiOH, Li_2CO_3, and Li_3N). Hence, it should be stored in air-tight containers, under inert gas atmosphere or better totally immersed in benzene, heptane or a mineral oil such as petrolatum, or Nujol[R], totally free from traces of oxygen or water. Actually, lithium reacts vigorously with water, forming a corrosive cloud of lithium hydroxide (LiOH) particles, and evolving hydrogen gas. Nevertheless, this hydrolysis is less vigorous than with sodium or potassium, probably due to the fair solubility and strong adherence of the LiOH to metal surfaces in water. Lithium also reacts violently with concentrated inorganic acids and reactive gases such as chlorine. Nevertheless, it does not react with oxygen at room temperature, and lithium oxide, Li_2O only forms when the metal is heated above 100°C. Lithium ignites spontaneously in air near its melting point (i.e. 180.5°C). Lithium reacts with nitrogen, even at room temperature, to form the reddish-brown nitride, Li_3N. Like Na and K, lithium is entirely soluble in liquid ammonia giving a deep blue solution; the saturated solution has, at 20°C, a density of 477 kg.m^{-3}.

From an electrochemical point of view, it has the highest negative standard electrode potential (–3.045 V/SHE), a high electrochemical equivalence (3860 A.h.kg^{-1}), and a good electronic conductivity, which makes it the most attractive anode material available for high specific energy and energy density electrochemical power sources,[1] both primary and rechargeable batteries. Moreover, the small ionic radius of lithium (60 pm) explains the ability of the lithium ion to pass through its own passivation layer and this advantage is extensively used in primary cells with liquid cathodes (e.g., SO_2, $SOCl_2$). Lithium colors the flame of a bunsen gas burner with a characteristic crimson color (670.8 nm). Moreover, extremely thin foils of lithium are transparent to far-UV radiation.

Natural lithium contains the two stable nuclide isotopes ^6Li (7.42at%) and ^7Li (92.58at%). However, samples with modified isotopic compositions may be found in commercially available material because it has been subject to an undisclosed or inadvertant isotopic fractionation. Therefore, substantial deviations in atomic weight of the element from that given in the literature can occur. For these reasons, in commercially available materials, lithium has atomic relative weights that range between 6.94 and 6.99; if a more accurate value is required, it must be determined for the specific material by high-resolution mass spectrometry. For instance, the less abundant lithium-6 isotope, owing to its high thermal neutron cross-section (940 barns) is an interesting material to serve as breeder blanket for producing tritium gas by the following neutron capture nuclear reaction: ^6Li (n, α)^3H. Tritium gas, so produced, is suitable for thermonuclear fusion power systems, while lithium-7 with less than 0.01at% ^6Li was proposed as a high-temperature coolant for thermonuclear reactor heat exchanger loops. Hence, many methods have been used to achieve partial separation of the two natural lithium isotopes on a small scale. Amongst them, the countercurrent liquid–liquid exchange of lithium isotopes between aqueous lithium hydroxide and lithium–mercury amalgam[2] was developed with a separation factor approaching 1.072.

Lithium in the molten state is a very corrosive medium, like liquid sodium, and readily attacks aluminum, copper, lead, platinum, silicon, silver, and zinc. Nevertheless, below 550°C, common ferrous alloys such as pure iron (e.g., Armco[R]), or stainless steels (AISI 304L or 316L series)[3] with a carbon content below 0.12wt%, are satisfactory for handling,[4] and containing the molten metal.[5] Above 600°C, corrosion resistant materials for handling and containing liquid lithium with less than 100 ppm wt free oxygen are, in order of decreasing resistance:[1] molybdenum, tungsten, and rhenium (up to 1650°C), pure tantalum and tantalum alloy, grades such as Ta–10Hf, Ta–8.5W–2.5Hf, T–111 (up to 1000–1200°C),[6] niobium, and Nb–1Zr alloy (up to 1300°C), titanium, zirconium, and hafnium (up to 820°C), but their corrosion resistance strongly depends on the amount of trace impurities, especially dissolved oxygen,[7] carbon, and nitrogen in the molten lithium.[8] Regarding ceramics, containment materials,[9] silica, SiO_2, and alumina, Al_2O_3, are strongly attacked and hence readily dissolve in liquid lithium. By contrast, alkaline-earth oxides such as: beryllia, BeO, magnesia, MgO, calcium oxide, CaO, rare-earth oxides such as ceria, CeO_2, or yttria, Y_2O_3, chromite spinel, $MgAl_2O_4$, and yttrium–aluminum garnet, $Y_3Al_2O_{12}$, seem to be satisfactory below 500°C, while aluminum, titanium and

---

[1]The maximum working temperatures mentioned correspond to molten lithium with an extra low level of impurities.

zirconium nitrides, or titanium and zirconium carbides seem to be quite inert in this medium below 1000°C. From a safety point of view, fires caused by reaction of solid lithium with water require special treatment owing to the low density of the metal, which floats. Lithium fires can only be efficiently stopped with special and efficient extinguishing agents such as copper or graphite powder, or the commercial product with the common trade name Lith-X®.

3.1.1.2 History

Lithium was first discovered in a mineral by the Swedish mineralogist Arfvedson, in 1817. One year later, the pure metal was first prepared independently by the British chemist Sir Humphrey Davy and the French chemist Brandé by molten salt electrolysis in 1818.[10] Lithium metal was first industrially produced by Metallgesellschaft AG (Germany) in 1925 and soon after by The Maywood Chemical Company in New Jersey (USA). The Foote Mineral Company began commercial production in the US at the end of the 1930s. Since World War II, especially after the development of processes for preparing enriched lithium-6 for nuclear fusion, lithium became available in large stock quantities, especially in the US. Therefore, large quantities of depleted lithium-7 were available for other applications. At the same time, electrochemists started to consider lithium as a potential anode material since the original patent[11] of the French engineer Hajek who was the first in 1949 to suggest lithium metal as an anode material in primary batteries. This led to the point of entry of lithium in industrial applications. Actually, several years later, the lithium battery concept was claimed in the French patent of Herbert and Ulam.[12] During the 1960s several American laboratories began R&D in this field. At present, as the consequence of these great developments, lithium takes today an important place in electrochemical generators, either in primary batteries such as Li/SO_2, $Li/SOCl_2$, and Li/SO_2Cl_2 or in secondary (i.e., rechargeable) batteries such as Li-Ion, and lithium solid polymer electrolyte. The future of lithium development is closely linked with (i) the development of lightweight lithium–aluminum alloys, (ii) the secondary batteries for both electric vehicles and stationary applications, and finally (iii) the nuclear fusion technology.

3.1.1.3 Natural Occurrence, Minerals, and Ores

Owing to its high reactivity with water and air, lithium metal obviously never occurs free in nature and hence it only appears as a definite combined form. Moreover, despite its relatively high and widespread natural abundance in the Earth's crust, estimated to be 20 mg.kg^{-1} (i.e., ppm wt) in comparison with that of lead (14 mg.kg^{-1}), large ore deposits are extremely rare. Nevertheless, lithium is widespread in small amounts, especially in nearly all the igneous and metamorphic rocks, or sedimentary such as clays, in spring waters and finally in seawater, natural brines, oil-field brines, and geothermal brines (see Table 3.2). The chief lithium minerals are the two phyllosilicates **lepidolite** $(K(Li,Al)_3(Si,Al)_4O_{10}(F,OH)_2$, monoclinic) and **petalite** $(LiAlSi_4O_{10}$, monoclinic), the inosilicate **spodumene** $(LiAlSi_2O_6$, monoclinc), and finally the two phosphates **triphylite** $(LiFePO_4$, orthorhombic), and **amblygonite** $((Li,Na)Al(PO_4)(F,OH)$, triclinic), but only spodumene, lepidolite, and to a lesser extent petalite are used as industrial minerals for recovering lithium from ore deposits. These minerals are usually found in pegmatite veins, which are coarse-grained granitic igneous rocks composed largely of quartz, feldspars, and micas (e.g., King Mountains, North Carolina, USA, and Rhodesia, South Africa). However, presently for economic reasons, spodumene recovery is discontinued and lithium is mainly recovered from natural chloride brines found in some particular places such as: USA (e.g., Clayton Valley (Nevada), Searles Lakes (California), Silver Peak (Western Nevada), and Great Salt Lake (Utah)), China[13] (e.g., provinces of Qinghai, Sichuan, and Hubei), Tibet, Northern Chile (e.g., Salar de Atacama) and Argentina (e.g., Salar de Hombre Muerto). Since the 1990s, a new commercial lithium resource has provided the recycling of lithium primary and secondary batteries by Toxco Inc. in British Columbia, Canada. Moreover, from 1997, Chile surpassed the USA and became the world's largest lithium carbonate producer.

Table 3.2 Lithium abundances in different geological materials

Source		Mass fraction (mg.kg^{-1})
Igneous rocks	Granites and granodiorites	35
	Peridotites and other ultramafic rocks	< 1
Sedimentary rocks	Shales	70
	Clays	70
	Limestones and dolomites	8
Water	Ocean	0.18
	Brines^a	100–1000

^{a}Natural, oil-field, or geothermal.

3.1.1.4 Processing and Industrial Preparation

Lithium carbonate from silicate ores (sulfuric acid roast process) – winning lithium from silicate minerals such as spodumene or to a lesser extent lepidolite and petalite is a highly energy-demanding chemical process compared with the recovery of lithium from natural brines and it is therefore expensive and recently discontinued in the US. It was performed where large spodumene ore deposits were found (e.g., Bessemer City, and King Mountains, North Carolina in the USA or in Rhodesia, South Africa). After mining the pregmatite veins, the raw ore, sometimes hand sorted, undergoes a comminution process, i.e., the ore is finely crushed, then ground to a final particle size smaller than 200 μm. Nevertheless, the fraction having particle sizes below 15 μm is rejected as slimes owing to interference during the flotation process. The desliming process consists of treating the ore with sodium hydroxide, and applying sedimentation–decantation principles in order to eliminate the slimes. Afterwards, the clean ore powder is impregnated with surfactant additives such as sodium xanthate, or fatty acids and undergoes a froth flotation beneficiation process to concentrate and separate spodumene crystals from inert gangue particles and other common by-product minerals (e.g., quartz, feldspars plagioclases/orthoclases, and micas). The flotation process is commonly achieved in three or four stages until a spodumene rich concentrate is obtained. Then, the enriched concentrate is calcinated during a decrepitation process in a brick-lined rotary kiln between 1075°C and 1100°C. Actually, the α-spodumene occurring in nature is chemically inert, and has to be transformed into the tetragonal β-spodumene crystals which are more chemically reactive. During decrepitation, the phase transformation results in volume expansion which introduces cracks into the spodumene crystals. Therefore, this irreversible phase transition gives a material of high specific area, straightforward to dissolve into sulfuric acid. However, owing to the flux properties of remaining gangue materials, the temperature must be carefully controlled below 1400°C in order to avoid the formation of eutectics between α-spodumene and other silicate minerals. After decrepitation and cooling, the β-spodumene is gently ground in a rubber lined ball-mill to 150 μm particle size in order to enhance the specific area of the powder. The acid dissolution or leaching process consists of mixing the finely powdered decrepitated β-spodumene with concentrated sulfuric acid (98wt% H_2SO_4) and then roasting the blend in a small kiln up to 250–300°C, to form the water-soluble lithium sulfate, Li_2SO_4. The mixture is leached with water at room temperature in order to dissolve the soluble salts giving an impure lithium sulfate solution containing traces of iron, aluminum, and other alkali metal cations (e.g., Ca^{2+} and Mg^{2+}). The excess sulfuric acid is neutralized with powdered natural calcium carbonate (ground limestone) and the resulting insoluble slurries formed during operation (e.g., $CaSO_4$, $FeCO_3$, $Al(OH)_3$) are removed by filtration, giving a purified solution of lithium sulfate. Afterwards, the liquor is treated with soda ash (sodium carbonate, Na_2CO_3) and spent lime (calcium hydroxide, $Ca(OH)_2$), in order to precipitate the traces of calcium and magnesium, which are removed by filtration. Later the solution is pH-adjusted between 7 and 8 and concentrated by an evaporation process to a liquor containing between 200 and 250 $g.dm^{-3}$ lithium sulfate. Then, the slightly soluble lithium carbonate, Li_2CO_3 is precipitated at 90–100°C with a dilute solution of sodium carbonate (20wt% Na_2CO_3), and separated by centrifugation, washed, and dried for sale or used as feedstock. However, the remaining mother liquor contains 15wt% lithium with a large amount of sodium sulfate. Sodium sulfate decahydrate (i.e., Glauber salts) is recovered as a by-product within the circuit to maximize lithium recovery.

Apart the previously discontinued sulfuric roast acid process which has been operated on a commercial scale for several years by FMC Corp. at Bessemer City, and the former Cyprus Foote at King Mountains, both located in North Carolina, several other routes for recovering lithium from silicates ores have been developed on a pilot scale. These processes are (i) the ion-exchange processes, and (ii) the alkaline processes. In the ion-exchange processes, the β-spodumene obtained by calcination and ground is leached at moderate temperature by an aqueous solution or a molten salt of a strong mineral acid, a sodium or a potassium salt (e.g., chloride, sulfate). During the leaching/wetting between reactants, the ion-exchange process occurs between the cations in the solution (e.g., H^+, Na^+, or K^+) and lithium cations in spodumene. After reaction, the lithium-enriched liquor is filtered and serves to recover lithium carbonate. In the alkaline process, spodumene is mixed with ground limestone and some additives such as calcium chloride or sulfate. The mixture undergoes pyrolysis and after reaction the clinker formed is crushed and leached with water giving an impure liquor containing lithium hydroxide. Finally, in the case of lepidolite-bearing ores, the ore concentrate follows the same beneficiation sequence as for spodumene, then the lepidolite lamellar crystals react with dry hydrogen chloride (HCl) at 935°C forming at a high-yield volatile lithium chloride. Lithium chloride may be also volatilized from carbo-chlorination of lepidolite mixed with carbon powder and heated in a stream of chlorine gas (Cl_2).

Lithium carbonate from brines (solar evaporation process)[14] – the recovery of lithium from brines is a less energy-intensive process and it is therefore extensively used where natural brines are found. Nevertheless, the methods of recovery used vary with the nature of the brines, especially the lithium concentration and the concentration of interfering cations such as magnesium and calcium. Brines are pumped either from natural ponds (e.g., Chile, Argentina, US), geothermal fluids (e.g., New Zealand), or oil-field reservoirs (e.g., Texas). The pumped brine undergoes a series of solar evaporation steps in artificial ponds of decreasing sizes. The process is very similar to the original recovery of sodium chloride from seawater in Mediterranean regions. Actually, gypsum, $CaSO_4 \cdot 2H_2O$, rock salt, NaCl, and carnallite, $KMgCl_3 \cdot 6H_2O$, precipitate as the evaporation progresses. Over the course of 12 to 18 months, the lithium concentration of the brine increases up to 6000 $mg. kg^{-1}$ (i.e., ppm wt). When the lithium chloride reaches the optimum concentration of 3.1wt%, the liquor is pumped to a recovery plant and purified from the residual magnesium and calcium by adding spent lime and soda ash (calcium hydroxide and sodium carbonate). During this step, calcium precipiates as calcium carbonate, while magnesium forms insoluble magnesium hydroxide. Then, the purified liquor is treated with additional soda ash to precipitate the insoluble lithium carbonate, Li_2CO_3. It is then purified and dried for sale or feedstock. However, concentration of lithium chloride in brines may vary widely in composition, and the economical recovery of lithium from such sources depends not only on the lithium content but also the concentrations of interfering ions, especially calcium and magnesium. Actually, when brines contain high levels of magnesium it is difficult to remove it. Hence, some countercurrent liquid–liquid extraction or ion-exchange processes have been proposed.

Preparation of pure lithium metal (molten lithium chloride electrolysis) – even if in some older processes, the metal was prepared by direct metallothermic reduction of the lithium oxide with magnesium or aluminum,[15,16] today the lithium metal is essentially obtained directly by molten salt electrolysis of LiCl–KCl according to a process developed from an old original process devised in 1893.[17,18] Hence, after preparation the lithium carbonate is converted by hydrogen chloride into highly pure anhydrous lithium chloride which serves as feed during the

Table 3.3 Lithium metal molten salt electrowinning

Parameter	Description
Electrolytic bath composition	LiCl–KCl eutectic (44.2–65.8wt%), mp 352°C
Anode material	Graphite
Cathode material	Low-carbon mild steel
Operating temperature	420–460°C
Operating cell voltage	4.86 to 8.19 V (theoretical 3.6 V)
Cathodic current density (max.)	20 kA.m^{-2}
Cathodic current	3000 A per cell
Specific energy consumption	$35–40 \text{ kW.h.kg}^{-1}$
Energy efficiency	80%
Power input	15–25 kW per cell
6 kg of raw LiCl gives 1 kg of Li metal (99.8wt%)	

electrolytic process. However, there exist four electrolytic cell designs worldwide, grouped into two main classes: open swept cells and closed cells. Open air swept cells are used by FMC Corporation in the United States, while modified Downs cells commonly used for sodium electrolysis are used both by Péchiney Électrométallurgie/Métaux Spéciaux SA in France and E.I. DuPont de Nemours/ Cyprus Foote Mineral Company in the United States. Finally, closed cells are used by Chemetall Gmbh (Metallgesellschaft AG/Degussa AG[19]) in Germany. But as a general rule, the electrolysis characteristics in all electrolytic cells are similar to those listed in Table 3.3.

The electrolytic cells are made of a low-carbon steel shell, in which graphite anodes are used, while the cathodes are made of low-carbon mild steel. During the electrochemical reduction, owing to their low density, both chlorine gas evolved at the anode, and the molten lithium droplets at the cathode rise by buoyancy to the surface of the electrolyte bath. Liquid lithium floats around the cathode, forming, after coalescence, a molten pool, while lithium chloride is continuously fed to the cell top. The lithium metal is protected from both oxidation and nitriding by an inert atmosphere in combination with a tight closed hood and it is pumped off in the closed cell design, while in the open, air-swept cells, lithium is protected by a thin film of the liquid electrolyte, and is collected by ladling and directly cast by pouring molten metal into rectangular molds, giving when solidified 2 lb trapezoidal bars, so-called lithium traps. Pure lithium ingots (purity >99.95wt% Li) are later obtained by purifying the remelted lithium traps (99.8wt% Li). Purification processes are, in order of importance: (i) addition of alumina for removing both calcium and nitrogen, (ii) hot gettering of O, C, and N with Ti, Zr or Y, (iii) filtration of lithium metal, (iv) cold trapping the impurities in trapping devices such as for sodium, (v) vacuum distillation, and finally (vi) containerless zone refining. The remelting and purifying step reduces the sodium content to less than 100 ppm wt. Four commercial purities or grades of lithium metal ingots are now produced and available. These four common grades are: (i) the standard or catalyst grade traps with a high sodium content (i.e., Na content between 1200 (low grade) and 8000 ppm wt (high grade), used for producing n-butyllithium, (ii) traps with low sodium content, (iii) the technical grade ingot with low sodium but high potassium, and finally (iv) the battery grade ingot with low sodium, calcium, and lithium nitride (Na content less than 100 ppm wt) for manufacturing lithium anodes for the battery industry.

3.1.1.5 Industrial Uses and Applications

According to J. Ober in the 1997 annual report presented by the US Geological Survey (USGS), the main uses of lithium are in order of importance: (i) lithium carbonate and lithium ore concentrates as additives in the ceramic and the glass manufacturing process, (ii) lithium carbonate as fluxing agent added to potlime in the Héroult–Hall aluminum melting process, (iii) in the chemical industry for the preparation of the catalyst n-butyllithium for production of synthetic rubbers, and in pharmaceutical production of various drugs, (iv) for water insoluble grease-lubricants, and finally, (v) as anode material used in primary and secondary batteries. More details about lithium end uses are listed in Table 3.4 with the percentage of lithium market in brackets when available. However, the only lithium market which has shown

Table 3.4 Lithium applications and industrial uses

Lithium product	Applications
Lithium ores and concentrates (46%)	Lithium oxide, Li_2O, sometimes called lithia, is extensively used in the glass industry (31%), as a flux during glass making. It is also used (15%) as a minor component in container glass, television tube glass, glasses with high-strength, fiber glass, glass ceramics (e.g., Vitroceram®), enamels and glazes.
Lithium chemicals (36%)	Lithium carbonate, Li_2CO_3, is extensively used in (i) the aluminum Héroult-Hall electrowinning process and also (ii) as an active component in pharmaceuticals (16%). Lithium hydroxide, LiOH or lithine, forms by saponification with stearate esters, the lithium stearate. This Li-stearate is used as a thickener and gelling agent to transform oils into high-temperature lubricating greases. Lithium fluoride, LiF, owing to its etching properties in the molten state is extensively used in welding and brazing fluxes. Lithium chloride, LiCl, is one of the most hygroscopic compounds known, hence it is used in temperature and humidity control in heat, ventilation, and air conditioning (HVAC) engineering. Lithium hypochlorite, LiClO, is used in water sanitizers although most of this market is currently held by calcium or sodium hypochlorite, both of which are cheaper though not as effective. Finally, Li-compounds are also used in the production of monosilane gas for the semiconductor industry, as Li-salts in primary and secondary battery electrolytes and in dye pigments. Lithiated intercalation compounds were specifically developed for serving as positive electrode materials (i.e., cathode) in secondary or rechargeable Li-batteries (3%). Lithium is also utilized to a considerable extent in chemical organic synthesis. Organolithium compounds act somewhat like organomagnesium compounds and undergo reactions similar to the Grignard reaction. Lithium compounds also serve as catalysts in synthetic rubber manufacture.
Lithium metal (18%)	Lithium has the lowest density of the solid elements, and so is used as an alloying element in aluminum-based alloys, in magnesium–lithium alloys for armor plate and aerospace components and to a lesser extent in zinc and lead alloys for improving densities, strength, and toughness (5%). Because lithium has highest standard electrode potential, high electrochemical equivalence, and good electronic conductivity, it is the most attractive anode material available for high-energy-density electrochemical power sources (i.e., primary and secondary batteries) (5%). Lithium metal is also used as a catalyst or chemical intermediate in organic synthesis, and as an efficient scavenger (i.e., remover of impurities) in the refining of such metals as iron, nickel, copper, and zinc, and their alloys or for maintaining dry inert gases. Actually, a large variety of nonmetallic elements are scavenged by metallic lithium, including oxygen, hydrogen, nitrogen, carbon, sulfur, and the four halogens. Owing to it possessing the highest specific heat capacity of all the elements, combined with a low vapor pressure and because the lithium-7 isotope, the more common stable isotope, has a low neutron cross-section, lithium has been proposed as a heat-transfer fluid for high-power-density nuclear reactors such as breeder reactors in which coolant temperatures above roughly 800°C are required. Finally, during the 1950s and 1960s, the less abundant lithium-6 isotope served as breeder blanket for producing tritium gas (i.e., ^3H or T). Tritium was produced according to the nuclear reaction by neutron capture $^6Li(n, \alpha)^3H$. Tritium gas, so produced, is employed in the manufacture of hydrogen bombs, among other uses. Finally, many lithium alloys are produced directly by the electrolysis of molten salts containing lithium chloride in the presence of a second chloride, or by the use of cathode materials that interact with the electrodeposited lithium, introducing other elements into the melt. Metallic lithium is used in the preparation of compounds such as lithium hydride.

Table 3.5 Lithium metal world producers

Company	Address
Chemetall Gmbh. (subsidiary of Metallgesellschaft AG)	Trakehner Strasse 3, D-60441 Frankfurt am Main, Germany Telephone: (+49) 69 71650 Facsimile: (+49) 69 7165 3018 E-mail: lithium @ chemetall.com Internet: http://www.chemetall.com/
FMC Corporation, Lithium Division (former Lithium Corporation of America), world's largest producer of lithium compounds	449 North Cox Road, Box 3925, Gastonia, NC 28054, USA Telephone: (704) 868 5300 Facsimile: (704) 868 5370 E-mail: lithium-info@fmc.com Internet: http://www.fmc.com/
Foote Chemetall Company (former Cyprus Foote Company, subsidiary of Cyprus Amax Minerals Co.), first company to produce lithium from brines	348 Holiday Inn Drive, Kings Mountain, NC 28086, USA Telephone: (704) 739 2501 (800) 523 7116 (Toll free USA, and Canada) Facsimile: (704) 734 0208 Internet: http://www.cyprusamax.com/
Métaux Spéciaux SA (Péchiney-Électrometallurgie)	Usine de Plombière-Saint-Marcel, Moutiers F-73600, France Telephone: 33 04 79 09 40 40 Fax: 33 04 79 09 40 50

sustained growth through the recession is demand for batteries; however, this is unlikely to make any major impact on total lithium demand in the next few years and the chief demand for lithium remains the glass and ceramic industry.

3.1.1.6 Lithium Metal World Producers

Some chief world producers and suppliers of lithium metal and its compounds are listed in Table 3.5. The world annual production of lithium carbonate for 1997 is estimated to be 42,000 tonnes.

3.1.2 Sodium

3.1.2.1 Description and General Properties

Sodium [7440-23-5] with the chemical symbol Na, the relative atomic mass (atomic weight) 22.989770(2) and the atomic number 11, is the second element of the alkali metals, i.e. of the group IA (1) of Mendeleev's periodic chart. The modern name of the element is derived from the Medieval English name soda, itself coming from the old Latin name, sodanum, a headache remedy. Nevertheless, its chemical symbol, Na, is the abbreviation of the Latin name, natrium, meaning soda ash, a natural sodium carbonate occurring in some warm desert regions. Pure sodium is a soft, malleable, and light metal with a low density ($971 \ kg.m^{-3}$). It has a silvery lustrous appearance when freshly cut but readily tarnishes in air containing moisture, becoming dull and gray. Sodium crystallizes in a body-centered cubic (bcc) crystal lattice structure. It is a strongly more reactive alkali metal than lithium. Actually, it reacts vigorously with water evolving hydrogen gas and forming a corrosive cloud of sodium hydroxide (caustic soda) particles. The explosive hazards of the reaction are only associated primarily with the hydrogen gas that is generated. Moreover, sodium ignites spontaneously in air above 120–125°C giving highly hazardous aerosols of sodium peroxide, Na_2O_2. Sodium readily dissolves in anhydrous liquid ammonia to form unstable deep blue solutions with a slight reaction evolving hydrogen and producing sodamide, $NaNH_2$. Other liquids in which sodium is soluble are, for instance, ethylenediamine, naphtalene in dimethyl ether, forming a dark green complex. Owing to its reactivity it must be stored in an air-tight container containing an oxygen and moisture free atmosphere or immersed under a mineral oil, benzene, naphtha or kerosene. Nevertheless, sodium does not react with nitrogen but combines directly with halogens and phosphorus. Sodium readily dissolves in mercury giving sodium amalgam with a highly exothermic reaction (-20.570 kJ.mol^{-1}). From a nuclear point of view, sodium is a monoisotopic element with the only stable nuclide ^{23}Na. Hence, it does not exhibit natural radioactivity, unlike potassium. Finally, sodium vapor is essentially a monoatomic gas.

Sodium is miscible with the alkali metals below it in the periodic table (i.e., K, Rb, and Cs). Sodium forms a eutectic alloy with potassium (Na, 22wt%, K 78wt%) commercially known as "NaK" which melts at $-10°$C. The eutectics formed in the Na–Rb and Na–Cs binary systems melt respectively at $-4.5°$C and $-30°$C. Sodium is the minor component with potassium and cesium of the ternary alloy Na–K–Cs. The composition of this ternary alloy is 3wt% Na, 24wt% K, and 73wt% Cs. This fluid has the lowest melting point of any liquid alloy yet isolated, melting at $-78°$C.

Sodium colors the flame of a bunsen gas burner a characteristic yellow owing to the highly intense D line of its atomic spectra (589 nm). Sodium is ordinarily quite reactive with air, and its chemical reactivity is a function of the moisture content of air. The reactivity of solid sodium with pure oxygen depends on traces of impurities in the metal. In ordinary air, sodium metal forms a sodium hydroxide film (NaOH), which rapidly absorbs exothermically traces of carbon dioxide and moisture always present in air, forming sodium hydrogenocarbonate, or simply bicarbonate ($NaHCO_3$). Sodium is more reactive in air as a liquid than as a solid, and the liquid can ignite spontaneously at about 125°C. When burning in dry air, sodium burns quietly, giving off a dense, white aerosol, strongly caustic smoke of sodium peroxide, Na_2O_2. The temperature of burning sodium increases rapidly to more than 800°C, and under these conditions the fire is extremely difficult to extinguish. Pure sodium begins to absorb hydrogen gas at about 100°C, with an absorption rate increasing with temperature. Sodium reacts vigorously with halogen vapors producing chemiluminescence. With molten sulfur it reacts violently to produce polysulfides; under more controlled conditions it reacts with organic solutions of sulfur. Liquid selenium and tellurium both react vigorously with solid sodium to form selenides and tellurides. Sodium shows relatively little reactivity with carbon, although the existence of lamellar materials prepared from graphite with the formula NaC_{64} has been reported. At 625°C carbon monoxide reacts with sodium to form sodium carbide and sodium carbonate. Ammonia also serves as a solvent for reactions of sodium with arsenic, tellurium, antimony, bismuth, and a number of other low-melting metals. Sodium also forms alloys with the alkaline-earth metals. Beryllium is soluble in sodium only to the extent of a few atomic percent at approximately 800°C. Liquid sodium and magnesium are only partially miscible. The degree of solubility in sodium of the alkaline-earth metals increases with increasing atomic weight, with the result that the solubility of calcium is 10wt% at 700°C. In the sodium–strontium system, there is a considerable degree of miscibility. Sodium forms a number of compounds with barium, and several eutectics exist in the system. The precious metals, silver, gold, platinum, palladium, and iridium, and the white metals, such as lead, tin, bismuth, and antimony, alloy to an appreciable extent with liquid sodium. Cadmium and mercury also react with sodium, and a number of compounds exist in both binary systems. Seven sodium–mercury compounds, or amalgams, exist (e.g., $HgNa_3$, Hg_2Na, Hg_4Na, Hg_2Na_3, Hg_2Na_5), with Hg_2Na having the highest melting point (354°C). Sodium amalgams are used chiefly for carrying out reactions in situations in which pure elemental sodium would be violently reactive and difficult to control. The solubility of transition metals in alkali metals is generally very low, often in the 1 to 10 ppm wt range even at temperatures in excess of 500°C. Sodium is essentially nonreactive with carbon monoxide. Sodium fires can be extinguished with special extinguishing agents such as those developed in France at the Superphenix breeder reactor facilities and called Marcalina$^®$ composed of Na_2CO_3, Li_2CO_3, and graphite.

Cost (1998) – sodium is priced at 3.95 $US.kg^{-1} (1.79 $US.lb^{-1}) for regular grade (99.9wt%).

3.1.2.2 History

Despite the fact that sodium compounds were known in ancient times (e.g., natron, rock salt), the metal was first isolated by the famous British scientist Sir Humphrey Davy, in 1807.[20] It was recovered by molten salt electrolysis of the fused caustic soda (sodium hydroxide, NaOH).

3.1.2.3 Natural Occurrence, Minerals, and Ores

Sodium is the most abundant of the alkali metals and it is the sixth most abundant element in the Earth's crust (roughly 2.36wt%). Owing to its high chemical reactivity with water and to a lesser extent with air, sodium metal never occurs free in nature; however, the element is ubiquitous and occurs naturally in a wide variety of compounds. Sodium chloride is the most common compound of sodium known either dissolved in seawater or in the crystalline form of **halite** or **rock salt** (NaCl, cubic). However, it is widely present in numerous complex silicates, such as feldspars and micas and other nonsilicate minerals such as **cryolite** (Na_3AlF_6, monoclinic), **natrolite** or **soda ash** (Na_2CO_3, monoclinic), **borax** ($Na_2B_4O_7 \cdot 10H_2O$, monoclinic), **sodium hydroxide** or **caustic soda** (NaOH), **Chilian salpeter, nitratite**, or **soda niter** ($NaNO_3$, rhombohedral). Obviously, the chief ore is sodium chloride either recovered from brines or rock salt ore deposits. There are large ore deposits of rock salt in various parts of the world, and sodium nitrate ore deposits exist in South America (e.g., Chile, Peru). The average sodium content of the sea is approximately 1.05wt%, corresponding to an average concentration of approximately 3.5wt% of sodium chloride. On the other hand, sodium, owing to its high-intensity D line, has been identified in both the atomic and ionic forms in the solar spectrum, the spectra of other stars and in the interstellar medium.

3.1.2.4 Processing and Industrial Preparation

The first and now obsolete industrial processes for producing raw sodium metal were based on the carbon reduction of sodium carbonate or sodium hydroxide. The first industrial production of pure sodium metal was performed by molten salt electrolysis of the pure sodium hydroxide, NaOH, in the so-called Castner cells. But most modern processes for the production of sodium involve now the molten salt electrolysis of highly pure sodium chloride. Actually, since 1921 when the process was invented by J.C. Downs, the electrolysis has been performed in Downs electrolytic cells at the DuPont de Nemours US facilities at Niagara Falls, New York. The electrolytic cell consists of four cylindrical anodes made of graphite surrounded at the bottom

Table 3.6 Sodium molten salt electrowinning (Downs cells)

Parameter	Description
Electrolytic bath composition	Mixture of NaCl (28wt%), $CaCl_2$ (26wt%) and $BaCl_2$ (46wt%)
Anode material	Graphite
Cathode material	Low-carbon mild steel
Temperature of operation	600°C
Operating cell voltage	7 V
Cathodic current (max.)	50 kA
Specific energy consumption	10 kW.h.kg^{-1}
Power input	350 kW per cell

of the cell by steel cathodes and a fine steel mesh acts as a separator between anodic and cathodic compartments. Each cell contains a batch of 8 tonnes of a molten salts mixture with the following chemical composition: NaCl (28wt%), CaCl$_2$ (26wt%), and BaCl$_2$ (46wt%).

The feedstock sodium chloride is purified after dissolution by aqueous precipitation from traces of other interfering ions such as sulfate anions, SO$_4^{2-}$ (max. 30 ppm wt), and magnesium cations, Mg^{2+} (max. 1 ppm wt). The operating conditions are: a temperature of 600°C, a cell voltage of 7 V (owing to high anodic and cathodic overpotentials), an electric current of 50 kA, and a specific energy consumption of 10 kW.h. kg^{-1} with a total power requirement of 350 kW. Liquid Na is cathodically formed at the bottom of the cell but owing to its lower density it is collected at the top of the cell by a skimmer. Chlorine gas which evolves at the anodes is removed by a nickel collector. The preparation is a continuous process and hence, the cell is continuously fed by pure sodium chloride. The melting is initiated and sustained by direct Joule effect and each electrolytic cell produces roughly 800 kg per day of sodium metal. A complete electrolysis plant consists of roughly 50 electrolytic cells. However, sodium is relatively contaminated with calcium metal (0.5 to 1wt% Ca), but it is readily purified by filtration on stainless steel meshes at 100°C since the calcium remains as solid particles at this temperature and it is easy to filter. This purification process decreases the calcium below 300 ppm wt to give sodium of a technical grade. However, to obtain sodium of nuclear grade (Ca < 10 ppm wt) extensively used in large liquid-metal reactor cooling systems, it is necessary to used the cold trapping technique. In this second purification technique, calcium is oxidized to calcia, CaO, and to remove this calcium oxide, the cold trapping involves running the molten sodium through a cooled, packed bed of material, upon which the oxide can precipitate. Filtration and cold trapping are also effective in removing large amounts of carbonate, hydroxide, and hydride impurities.

In conclusion, two grades of metallic sodium are commercially available. These two chief grades are: (i) the regular or technical grade, and (ii) the reactor or nuclear grade sodium as a high-temperature coolant extensively used in fast neutron breeder nuclear reactors.

3.1.2.5 Industrial Uses and Applications

Sodium is the alkali metal of most commercial importance by far, and some principal sodium compounds of industrial and commercial importance are: the chloride (NaCl), the anhydrous carbonate (soda ash, Na$_2$CO$_3$), and hydrogenocarbonate (baking soda, NaHCO$_3$), the hydroxide (caustic soda, NaOH) and the sulfate (Glauber salt, Na$_2$SO$_4 \cdot$10H$_2$O). Large amounts of sodium chloride are, for instance, used in the production of bulk quantities of other industrial chemicals. To a lesser extent, sodium cyanide, sodium peroxide, sodium sulfide, sodium borates, sodium phosphates, and sodium alkylsulfates are also industrially produced. Nevertheless, the aim of this book is not to detail the uses of sodium compounds, which are extensively described in comprehensive industrial inorganic chemistry textbooks, and hence, only the uses and applications of metallic sodium are listed in Table 3.7 (page 88).

3.1.2.6 Sodium Metal World Producers

Table 3.8 Sodium metal world producers

Producer	Address
E.I. DuPont de Nemours (Speciality Chemicals)	E.I. Du Pont de Nemours Canada Inc., Box 2200, Streetsville, Mississauga L5M 2H3, Ontario ON, Canada
Métaux Spéciaux S.A. (Péchiney Électrométallurgie)	Usine de Plombière-Saint-Marcel, Moutiers F-73600, France Telephone: 33 04 79 09 40 40 Fax: 33 04 79 09 40 40

Table 3.7 Sodium industrial applications and uses

Application	Description
Production of lead tetraethyl	Sodium metal is used chiefly to produce tetraethyl lead ($Pb(C_2H_5)_4$ or tetramethyl lead ($Pb(CH_3)_4$) by means of the reaction of lead–sodium alloy (e.g., 90Pb–10Na) with monochloroethane (ethyl chloride, C_2H_5Cl) or monochloromethane (methyl chloride, CH_3Cl) forming NaCl. Lead tetraethyl is used to increase the anti-knock rating of gasoline.
Heat-transfer fluid	Owing to its good thermal conductivity (141 $W.m^{-1}.K^{-1}$), its low density (971 $kg.m^{-3}$), low absolute viscosity (0.71 mPa.s), and both low melting point (98°C), and boiling point (883°C), molten sodium is a liquid metal having a low Prandtl number (e.g., 0.005 at 371°C) suited for use as an excellent heat-transfer fluid at atmospheric pressure. Therefore, it has potentially large-scale industrial uses. Moreover, owing to the low thermal neutron cross-section, it is extensively used as heat-transfer liquid in large fast-breeder nuclear reactors. Nevertheless, due to the formation of the two radionuclides: ^{24}Na with a short half-life (14.97 h) and ^{22}Na (2.6019 years) during the neutron irradiation in the nuclear reactor core, a sodium-cooled reactor must have a second heat-transfer loop in order to avoid the direct contact of radioactive sodium with the external environment. For instance, the Superphenix French breeder reactor used 5600 tonnes of liquid sodium as coolant. Because sodium forms a low melting point (–10°C) eutectic alloy with potassium (Na 22wt% and K 78wt%) commercially known as "NaK", it was considered for use as the heat-transfer fluid in submarine nuclear reactor systems and also in the secondary coolant system in a high power-density reactor system for use in spacecraft. Finally, the alloy also is used as a cooling liquid for crucibles in consumable arc-melting processes for preparing titanium ingot from the titanium sponge.
Metals processing as a reducing agent	Owing to its high chemical affinity for oxygen and halogens, sodium metal is extensively used in pyrometallurgy as a deoxidant and as a strong reducing agent. With the exception of the oxides of the refractory metals of subgroup IVB (i.e., Ti, Zr, Hf), the oxides of the other transition metals are all reduced to the respective metals with sodium metal. Sodium also reacts with a large number of metallic halides, displacing the metal from the salt and forming a sodium halide in the process. This reaction is used in the preparation of several pure reactive metals and refractory metals such as K, Ca, Zr, Ti, and other transition metals. For instance, commercial production of silicon, titanium by the Hunter process, and tantalum according to the Marignac process involves the reduction respectively of titanium tetrachloride or tantalum pentachloride with molten sodium metal. Sodium is also currently used in refining of metallic lead, silver, and zinc.
Chemical manufacture	Preparation of alkolates, amide, sodium hydride and borohydride, azotide (NaN_3) used in air-bags, sodium oxides (i.e., Na_2O, Na_2O_2, and NaO_2), dyes, and pigments (e.g., indigo).
Pharmaceutical manufacture	Pharmaceutical uses include vitamins A and C, barbiturates, ibuprophen, and sulfamethoxizane.
Electrical conductor	Owing to its excellent electrical conductivity (4.2 $\mu\Omega.cm$), sodium metal has also been considered as a substitute for copper in electrical conductors for applications requiring the supply of high currents.
Other	Sodium, owing to the high intensity of its monochromatic yellow D line (589 nm) is extensively used as vapor in high-pressure lamps for outdoor lighting (e.g., highways). Moreover, due to its low work function sodium is also used as a photocathode in photoelectric cells.

3.1.3 Potassium

3.1.3.1 Description and General Properties

Potassium [7440-09-7], with the atomic number 19, and the relative atomic mass (i.e., atomic weight) 39.0983(1) is the third element of the alkali metals, i.e., group IA(1) of Mendeleev's periodic chart. Potassium was named after the old English word potash meaning potassium carbonate, K_2CO_3, found in pot ashes after combustion of wood. Its chemical symbol, K, comes from its old chemical name kallium which is a consequence of the the Arabic word qali meaning alkali. Potassium is a soft, ductile, and malleable metal with a low density (856 $kg.m^{-3}$) and having a silvery white appearance with a body-centered cubic (bcc) crystal lattice structure. Nevertheless, it tarnishes on exposure to moist air, and becomes brittle at low temperature.

Potassium metal reacts with water, even with ice at −100°C, evolving hydrogen gas and forming the hydroxide, KOH. Owing to its reactivity it must be stored in an air-tight container containing an oxygen and moisture free atmosphere or immersed under a mineral oil, petrolatum, benzene, naphtha or kerosene. It reacts vigorously with pure oxygen and ignites spontaneously with bromine and iodine. It reacts with hydrogen at approximately 350°C to form the hydride. Potassium, like lithium and sodium, is readily dissolved by liquid ammonia giving a deep blue solution. Other solvents which dissolve potassium are, for instance, ethylenediamine and aniline. At elevated temperatures, potassium reduces carbon dioxide to carbon monoxide and carbon. Solid carbon dioxide and potassium react explosively when subjected to mechanical shock. Oxidation of potassium amalgam with carbon dioxide results in the formation of potassium oxalate.

The chemical properties of potassium are similar to those of sodium, although the former is considerably more reactive. Potassium differs from sodium in a number of respects. Whereas sodium is essentially unreactive with graphite, potassium, rubidium, and cesium react to form a series of interlamellar compounds, the richest in metal having the formula MC_8. Compounds are formed with carbon–alkali metal atomic ratios of 8, 16, 24, 36, 48, and 60 to 1. The graphite lattice is expanded during penetration of the alkali metal between the layers. Potassium reacts with carbon monoxide at temperatures as low as 60°C to form an explosive carbonyl ($K_6C_6O_6$), a derivative of hexahydroxybenzene. Potassium colors the flame of a bunsen gas burner purple-red (766 nm).

Natural potassium contains three isotope nuclides: ^{39}K (93.2581at%), ^{40}K(0.0117at%), and ^{41}K(0.7302at%). Potassium-40 is a slightly radioactive nuclide which is a pure beta emitter (i.e., emitting electrons without gamma rays and having a maximum kinetic energy of 1.32 MeV) with a half-life of 1.277 billion years. Actually, its specific activity (i.e., radioactivity per unit mass) is roughly 32 kBq.kg^{-1} (i.e., 0.87 μCi.kg^{-1}). Therefore, the natural element is slightly radioactive and it is the main source of the background radioactivity of the human body. Moreover, the disintegration of ^{40}K is commonly used in georadiochronology for age calculations of minerals and rocks.

From a biological point of view, potassium is an essential element in all forms of life. Each organism has a closely maintained potassium level and a relatively fixed potassium–sodium ratio. For instance, in the human body, the ratio of potassium between the cell and plasma is approximately 27/1. Potassium is the primary inorganic cation, within the living cell, and sodium is the most abundant cation in extracellular fluids. Moreover, potassium is a chief element of fertilizers.

Potassium forms alloys with all the alkali metals. Complete miscibility exists in the K–Rb and K–Cs binary systems. The latter system forms an alloy eutectic melting at approximately −38°C. Modification of the system by the addition of sodium results in a ternary eutectic melting at approximately −78°C. Potassium is essentially immiscible with all of the alkaline-earth liquid metals, as well as with molten zinc, aluminum, and cadmium.

3.1.3.2 History

Potassium was first prepared by the British chemist Sir Humphrey Davy, in 1807. Davy used molten salt electrolysis of the molten hydroxide.

3.1.3.3 Natural Occurrence, Minerals, and Ores

Potassium is the seventh most abundant element in the Earth's crust and is roughly as abundant as sodium. The chief potassium-containing minerals are the two tectosilicates: feldspars (**orthoclase** $KAlSi_3O_8$, monoclinic) and **microcline** ($KAlSi_3O_8$, triclinic), the two chlorides: **sylvinite** (KCl, cubic) and the **carnallite** ($MgCl_2 \cdot KCl \cdot 6H_2O$, rhombohedral) and finally the nitrate: **niter** or **saltpeter** (KNO_3, orthorhombic). The main mineral deposits are found in igneous rocks (i.e., felsites such as granite, pegmatite, syenite, etc.), metamorphic rocks (i.e., gneiss, micaschistes), and sedimentary rocks (i.e., shales, clays). Owing to the solubility of its chloride, KCl, potassium is always present as a dissolved cation, K^+, in seawater and also in numerous natural brines and many lake deposits (e.g., Searles Lakes, California). For instance, the potassium content of the Dead Sea (Israel) is approximately 1.7wt% KCl. The waste liquors obtained after evaporation of certain brines may contain up to 4wt% KCl, and hence, are used as a primary source of potassium. On the other hand, the salt deposits in Stassfurt (Germany), containing potassium sulfate and large quantities of carnallite, are one of the richest and most important sources of potassium in the world. These deposits became so important with respect to caustic potash (potassium hydroxide) production that Germany had a virtual monopoly on production of that substance up to the time of World War I. The other chief sources of potash are found in France, Austria, Spain, India, and Chile.

On the other hand, and from a biological aspect, the potassium content of muscle tissue is approximately 0.3wt%, whereas that of blood serum is about 0.01 to 0.02wt%. Finally, the potassium content of plants varies considerably, but it is ordinarily in the range of 0.5 to 2wt%. Therefore, the ashes of calcined vegetation are sometimes used as secondary sources of potash.

3.1.3.4 Processing and Industrial Preparation

The techniques used for extracting potassium salts depend to a considerable extent on the nature of the ore constituent and the types of other salts mixed with the potassium salt. Potassium chloride can be easily separated from carnallite ore by fractional crystallization of potassium chloride brines during evaporation of the mother liquor. Brines are evaporated in vacuum pans, and after the sodium chloride and sulfate are crystallized and settled, the resulting liquor is boiled to reach a fixed concentration where potassium chloride begins to crystallize, producing a crude potash deposit containing about 66wt% potassium chloride. Potassium metal is produced using the metallothermic reduction of molten potassium chloride with sodium metal at 870°C. Molten KCl is continuously fed into a packed distillation column, while sodium vapor is passed up through the column. By condensation of the volatile potassium at the top of the distillation tower, the reaction is forced to the reduction. Despite its historical impact, the electrolytic production of potassium from potassium hydroxide has been unsuccessful because there are few fluxes that can decrease the melting point of potassium chloride to temperatures where electrolysis is efficient.

3.1.3.5 Industrial Uses and Applications

Table 3.9 Potassium metal industrial applications and uses

Application	Description
Preparation of potassium superoxide	Most of the pure potassium metal produced is converted to the superoxide (KO_2), a yellow solid, directly by combustion in dry air or burning the potassium amalgam with oxygen. Potassium superoxide is extensively used as a source of oxygen in respiratory equipment, since it generates oxygen and also absorbs carbon dioxide forming the carbonate.
Heat-transfer fluid	Sodium-potassium eutectic alloy commercially known as "NaK" (Na 22wt% and K 78wt%) (see also sodium section) owing to its low melting point (–10°C) is used only to a limited extent as a heat-transfer coolant in nuclear reactors, although it has significant potential for this purpose. It is presently being evaluated for use in crucible cooling in titanium arc melting in replacement of pure sodium.

3.1.4 Rubidium

3.1.4.1 Description and General Properties

Rubidium [7440-17-7], with the chemical symbol Rb, the atomic number, 37, and the relative atomic mass (atomic weight) 85.4678(3) is the fourth alkali metal of the group IA(1) of Mendeleev's periodic chart. Rubidium is named after the Latin word, rubidus, owing to the deep red color of its chief atomic spectra line. Rubidium is a soft, ductile, and malleable silvery white metal when freshly cut with a low density (1532 kg.m^{-3}) between that of sodium and magnesium (it is the fourth lightest element). It has a body-centered cubic (bcc) crystal lattice structure. Owing to its low melting point (39.05°C), it can be liquid at room temperature. It decomposes vigorously in water, and also reacts with ice even at –108°C evolving hydrogen gas and forming the caustic rubidium hydroxide, RbOH. It ignites spontaneously in oxygen, hence, rubidium is difficult to handle. Like the other alkali metals, it combines vigorously with mercury forming a stable rubidium amalgam. Rubidium and cesium are miscible in all proportions and have complete solid solubility; a melting-point minimum of +9°C is reached. Because of the higher specific volume of rubidium, compared with the lighter alkali metals, there is less tendency for it to form alloy systems with other metals. Owing to its more negative standard electrode potential (–2.924 V/SHE), it is the second most electropositive element after lithium. It colors the flame of a bunsen gas burner violet (424 nm).

Rubidium has two natural isotopes: the stable nuclide ^{85}Rb (72.165at%) and the radionuclide ^{87}Rb (27.835at%) with a half-life of 47.5 billion years, which is a pure beta emitter (i.e., it emits electrons, without gamma rays, with a maximum kinetic energy of 273 keV). Hence, rubidium is sufficiently radioactive to expose a photographic emulsion. Actually, its specific radioactivity is roughly 907 Bq.kg^{-1} (24.5 μCi.kg^{-1}).

3.1.4.2 History

Rubidium was first discovered in 1861, by the two German chemists R.W. Bunsen and G. Kirchhoff at the University of Heidelberg, who identified the new element in the lithium mineral lepidolite using the spectroscope. The pure metal was prepared in 1928 by Hackspill.

3.1.4.3 Natural Occurrence, Minerals, and Ores

The rubidium abundance in the Earth's crust is established as 90 mg.kg^{-1} (i.e., ppm wt), hence, it is the fifteenth most abundant metal on Earth. Rubidium, like the other alkali metals, due to its chemical reactivity, never occurs free in nature. Despite its relative abundance, there are no specific chief rubidium-containing minerals, and it is only dispersed in common minerals in which it occurs as traces. Moreover, it often occurs together with cesium, ranging in content up to 5wt%. These particular minerals are the three phyllosilicates **lepidolite** (K(Li,Al)$_3$(Si,Al)$_4$O$_{10}$(F,OH)$_2$, monoclinic), **zinnwaldite** (KLiFeAl(AlSi$_3$)O$_{10}$(F,OH)$_2$, monoclinic) and **pollucite** ((Cs,Na)$_2$Al$_2$(Si$_4$O$_{12}$).H$_2$O, cubic), the tectosilicate **leucite** (KAlSi$_2$O$_6$, tetragonal), the chloride **carnallite** (MgCl$_2$·KCl·6H$_2$O, rhombohedral), and finally the **rhodizite** ((K,Cs)Al$_4$Be$_4$(B,Be)$_{12}$O$_{28}$, cubic). Apart from minerals, rubidium also naturally occurs in seawater, brines, spring water and salt lake deposits and in some biological products such as tea, coffee, and tobacco. For instance, some brines has contain up to 6 mg.kg^{-1} of rubidium.

3.1.4.4 Processing and Industrial Preparation

The principal commercial source of rubidium is accumulated stocks of a mixed carbonate produced as a by-product in the extraction of lithium salts from lepidolite. Primarily potassium

3
Less
Common
Nonferrous
Metals

carbonate, the by-product also contains approximately 23wt% rubidium and 3wt% cesium carbonates. The primary difficulty associated with the production of either pure rubidium or pure cesium is that these two elements are always found together in nature and also are mixed with other alkali metals; because these elements have very close ionic radii, their chemical separation encounters numerous issues. Before development of procedures based on thermochemical reduction and fractional distillation, the elements were purified in the salt form through laborious fractional crystallization techniques. Once pure salts have been prepared by precipitation methods, it is a relatively simple task to convert them to the free metal. This end is ordinarily accomplished by metallothermic reduction with calcium metal in a high-temperature vacuum system in which the highly volatile alkali metal is distilled from the solid reaction mixture. Today, direct reduction of the mixed carbonates from lepidolite purification, followed by fractional distillation, is perhaps the most important of the commercial methods for producing rubidium. The mixed carbonate is treated with excess sodium at approximately 650°C, and much of the rubidium and cesium passes into the metal phase. The resulting crude alloy is vacuum distilled to form a second alloy considerably richer in rubidium and cesium. This product is then refined by fractional distillation in a tower to produce elemental rubidium more than 99.5wt% pure.

3.1.4.5 Industrial Uses and Applications

Rubidium metal has few commercial uses, and hence it is of very minor economic significance. The high price (20,000 $US.kg^{-1}) and uncertain and limited supply of the metal discourage the development of any uses. Only a few commercial uses have been developed for the metal itself such as photocathode materials in photoelectric cells for photomultoplier tubes, and as getter in radio vacuum tubes to fix residual gases such as oxides, nitrides, and hydrides.

3.1.4.6 World Chief Producers

Cabot Performance Materials, 144 Holly Road, P.O. 1607, Boyertown, PA 19512-1607, USA.

3.1.5 Cesium

3.1.5.1 Description and General Properties

Cesium or caesium [7440-46-2] with the atomic number 55, and the relative atomic mass (atomic weight) 132.90545(2) is the fifth alkali metal of the group IA(1) of Mendeleev's periodic chart. It has the chemical symbol Cs. Cesium is named after the Latin word, caesius, meaning blue sky owing to the deepest blue color of its chief spectra line. Cesium is a soft, malleable silvery white element when freshly cut with the highest density (1873 kg.m^{-3}) of the alkali metals, just slightly greater than that of magnesium. Like the other alkali metals, it has a body-centered cubic (bcc) crystal lattice structure. Owing to its low melting point (28.4°C), it can be liquid at room temperature. It decomposes vigorously in water, evolving hydrogen and forming the caustic cesium hydroxide, CsOH, and it reacts even with ice at temperatures above −116°C. It ignites spontaneously in oxygen and hence, cesium is difficult to handle because it reacts spontaneously in air. Therefore, it must be kept immersed in mineral oil. It is readily soluble in liquid and anhydrous ammonia giving deep blue solutions. Moreover, cesium hydroxide is the strongest base known. Like the other alkali metals, it combines vigorously with mercury forming a stable cesium amalgam. Cesium is the most reactive of the alkali metals with nitrogen, carbon, and hydrogen. Cesium salt colors the nonluminous flame of a bunsen gas burner reddish violet or bluish purple (460 nm). Cesium is a monoisotopic element, and hence the only natural occurring nuclide,[133]Cs.

3.1.5.2 History

Cesium was the first element discovered spectroscopically in the spring water from Durkheim, by the two German chemists R.W. Bunsen and G. Kirchhoff at the University of Heidelberg. Cesium salts were not reduced to metal until 1880 and had no significant utility until the 1920s, when cesium was used as a coating for tungsten filaments in lighting.

3.1.5.3 Natural Occurrence, Minerals, and Ores

The relative abundance of cesium in the Earth's crust is estimated to be 3 mg.kg^{-1} (ppm wt), and cesium is widespread at very low concentration in igneous rocks (e.g., pegmatites), and sedimentary rocks. The only chief mineral containing cesium is the phyllosilicate **pollucite** (($Cs,Na)_2Al_2(Si_4O_{12})\cdot H_2O$, cubic) and in a lesser extent cesium also occurs in the rare mineral called **rhodizite** (($K,Cs)A_{14}Be_4(B,Be)_{12}O_{28}$, cubic). Pollucite, which is by far the major source of the metal, theoretically contains 40.1wt% Cs, and impure samples are ordinarily separated by hand-sorting methods to greater than 25wt% Cs. Large pollucite ore deposits have been found in Zimbabwe and in the lithium-bearing pegmatites at Bernic Lake (Manitoba, Canada). Cesium also occurs as traces in lepidolite, in salt brines and salt deposits. On the other hand, cesium-137 radionuclide is an important by-product of the spent nuclear fuel of pressurized water reactors (PWR).

3.1.5.4 Processing and Industrial Preparation

The direct metallothermic reduction of pollucite ore with sodium metal is the primary commercial source of cesium metal. In the process raw pollucite ore is reduced with sodium molten metal at approximately 650°C to form a sodium cesium alloy containing some rubidium as impurity. Fractional distillation of this alloy in a distillation column at approximately 700°C produces 99.9wt% pure cesium metal. Cesium can also be obtained pyrometallurgically, reducing the chloride CsCl with calcium metal or the hydroxide CsOH with magnesium metal. Nevertheless, the electrolytic recovery of a cesium amalgam from an aqueous solution of cesium chloride can be achieved in a similar process to the chlor-alkali production with a mercury cathode. The cesium is after removed from the amalgam by vacuum distillation. However, cesium metal is produced in rather limited amounts because of its relatively high cost (40,800 $US.kg^{-1}).

3.1.5.5 Industrial Uses and Applications

Cesium, owing to its low electronic work function, is extensively used for manufacturing photocathode materials for photoelectric cells used in photomultiplier tubes. Cesium salts are used in moderate quantities in the manufacture of spectrometer prisms made of high purity CsCl, CsBr, CsI, in infrared signaling lamps, as X-ray screen phosphors, in gamma scintillation counters such as a single monocrystal made of thallium-doped CsI(TI), in spectrophotometer lamps, and finally, as a getter for the final evacuation of radio and television vacuum low voltage electron tubes to fix residual gases such as oxides, nitrides, and hydrides. At one time the second, unit of time, was defined on the vibrational frequency 9,192,769 MHz of hyperfine transition between electronic levels of the cesium-133 atom and it is therefore used in atomic clocks. The radionuclide, cesium-137, with an half-life of 33 years, is a beta emitter, extensively used in radiotherapy as a gamma source. It is produced by fission in power nuclear reactors, and therefore, essentially recovered from spent nuclear fuel and wastes. On the other hand, cesium has been recently evaluated for power systems for spacecraft applications as a potential rocket fuel for interplanetary travel. Actually, owing to its low work function the cesium atom can be easily ionized thermally. The cations can then be accelerated to great speeds, hence cesium fuel can provide extraordinarily high specific impulses for rocket propulsion. Cesium also has application in thermionic converters that generate electricity directly within nuclear reactors. Another potentially large application of cesium metal is in the production of the lowest melting-point Na–K–Cs eutectic alloy (see sodium, Section 3.1.2).

3.1.5.6 World Cesium Metal Producers

(1) Chemetall Gmbh, Trakehner Strasse 3, D-60441 Frankfurt am Main, Germany, and (2) Cabot Performance Materials, 144 Holly Road, PO 1607, Boyertown, PA 19512-1607, USA.

3.1.6 Francium

Francium occurs in nature as the ^{223}Fr radionuclide in uranium minerals. It was first discovered in 1939, by Mademoiselle Marguerite Perey of The Institut Curie (Paris, France). It is named after the Latin, Francium, France, the country of its discovery. Francium is the heaviest known member of the alkali metals series, and occurs as a result of alpha disintegration of actinium. It can also be made by artificially by bombarding thorium with protons. Thirty-three isotopes of francium are recognized. The longest lived, ^{223}Fr (Ac, K), a daughter of ^{227}Ac, has a half-life of 22 min. This is the only isotope of francium occurring in nature. While it occurs naturally in uranium minerals, there is probably less than an ounce of francium at any time in the total crust of the earth. It has the highest equivalent weight of any element, and is the most unstable of the first 101 elements of the periodic chart. Because all known isotopes of francium are highly unstable, knowledge of the chemical properties of this element only comes from radiochemical techniques. No weighable quantity of the element has been prepared or isolated. The chemical properties of francium most resemble cesium. Owing to the minute quantities of the element that have been produced, it is doubtful that it has metallurgical or other industrial applications at present.

3.2 Alkaline-Earth Metals

The alkaline-earth metals are represented by the six chemical elements of the group IIA(2) of Mendeleev's periodic chart. These six elements are, in the order of increasing atomic number: beryllium (Be), magnesium (Mg), calcium (Ca), strontium (Sr), barium (Ba), and radium (Ra). The designation "earth" for these metals derives from the Middle Ages when alchemists referred to substances that were insoluble in water and unchanged by calcination as earths. Those earths, such as lime (CaO), that bore a resemblance to the alkalines (e.g., soda ash and potash) were called alkaline-earths. Amongst the alkaline-earth metals, magnesium and calcium are the only abundant alkaline-earth elements in the Earth's crust. They also are the most commercially important members of the family, and to a lesser extent beryllium is also commonly used as pure metal and alloys in large industrial applications. Hence, only these three metals will be reviewed in detail in this section. Nevertheless, a short description of the main properties and industrial uses of the last three alkali metals (i.e., Sr, Ba, and Ra) will be presented at the end of the section. Some physical, mechanical, thermal, electrical, and optical properties of the chief alkaline-earth metals are listed in Table 3.10.

Table 3.10 Selected properties of five alkaline-earth metals

Properties (at 298.15 K, unless otherwise specified)		Beryllium	Magnesium	Calcium	Strontium	Barium
General	Chemical symbol (IUPAC)	Be	Mg	Ca	Sr	Ba
	Chemical abstract registry number [CAS RN]	[7440-41-7]	[7439-95-4]	[7440-70-2]	[7440-24-6]	[7440-39-3]
	Unified numbering system [UNS]	[R19920]	[M19995]	[M03001]	[M06001]	[M02002]
Mineral occurrence and economics	Earth's crust abundance (mg.kg^{-1})	2.6	23,000	41,000	370	500
	Seawater abundance (mg.kg^{-1})	3.5–22	1200	390–440	7.6	4.7–20
	World estimated reserves (R, 10^3 kg)	0.4×10^6	$> 10^{10}$	Unlimited	n.a.	450×10^6
	World production of ore (1998) (P, 10^3 kg.yr^{-1})	364 (metal)	536,000 (metal)	2000 (metal)	137,000	6×10^6
	Cost of pure metal (1998) (C, \$US.kg^{-1}) (purity, wt%)	720.91 (99.5)	4.07 (99.8)	8.86 (99.8)	10,000 (99.95)	400 (99.7)
Atomic properties	Atomic number (Z)	4	12	20	38	56
	Relative atomic mass A_r (^{12}C = 12)	9.012182(3)	24.3050(6)	40.078(4)	87.62(1)	137.327(7)
	Electronic configuration	[He] 2s^2	[Ne] 3s^2	[Ar] 4s^2	[Kr] 5s^2	[Xe] 6s^2
	Fundamental ground state	$^1S_0$	$^1S_0$	$^1S_0$	$^1S_0$	$^1S_0$
	Electronegativity (Pauling)	1.57	1.31	1.00	0.95	0.89
	Electron work function (W_S, eV)	4.98	3.66	2.87	2.59	2.52
	X-ray absorption coefficient CuK$_{\alpha1,2}$ ((μ/ρ), cm^2.g^{-1})	1.50	38.6	162	125	330
Nuclear	Thermal neutron cross-section (σ_n, 10^{-28} m^2)	0.0092	0.063	0.43	1.2	1.3
	Isotopic mass range	6–14	19–36	34–53	74–102	114–151
	Natural isotopes (including isomers)	9	18	20	32	45
Crystallographic properties	Crystal structure at RT (phase α)	hcp	hcp	fcc	fcc	bcc
	Strukturbericht designation	A3	A3	A1	A1	A2
	Space group (Hermann–Mauguin)	P6$_3$/mmc	P6$_3$/mmc	Fm3m	Fm3m	Im3m
	Pearson's symbol	hP2	hP2	cF4	cF4	cI2
	Lattice parameters (pm) (@293 K)	a = 228.58, c = 358.42	a = 320.94, c = 521.3	a = 558.84	a = 608.49	a = 502.5
	Miller's indices of slip plane at RT($hkil$)	(0002), (10$\bar{1}$0)	n.a.	n.a.	n.a.	n.a.
	Phase transformation temperature (T,K)	1527 (1254°C)	nil	573 (300°C)	508,813 (235°C, 540°C)	643 (370°C)
Mechanical properties (annealed)	Density (ρ, kg.m^{-3}) (293 K)	1847.7	1738	1550	2540	3594
	Young or elastic modulus (E, GPa) (300 K)	318	44.7	19.6	15.7	12.8
	Coulomb or shear modulus (G, GPa)	156	34.2	7.36	6.08	4.86
	Bulk or compression modulus (K, GPa)	110	35.6	17.2	11.61	10.30
	Mohs hardness (HM)	n.a.	2.5	1.75	1.5	n.a.
	Brinell hardness (HB)	589–637	26.0	16.7	n.a.	n.a.
	Vicker's microhardness (μHV) (100 g load)	150	30–35	17	n.a.	n.a.
	Yield strength proof 0.2% (σ_{YS}, MPa)	117–158	69	37.8	n.a.	n.a.

Table 3.10 (continued)

Properties (at 298.15 K, unless otherwise specified)		Beryllium	Magnesium	Calcium	Strontium	Barium
Thermal properties (293.15 K)	Ultimate tensile strength (ρ_{UTS}, MPa)	420–503	176	53.8	49.0	12.8
	Elongation (Z, %)	2–5	n.a.	7	n.a.	n.a.
	Longitudinal velocity of sound (V_S, m.s^{-1})	12,600	4602	3560–4060	n.a.	1620
	Poisson's ratio ν (dimensionless)	0.039	0.291	0.310	0.304	0.280
	Melting point (mp, K)	1556 (1283°C)	922 (649°C)	1112 (839°C)	1042 (769°C)	1002 (729°C)
	Boiling point (bp, K)	3243 (2969°C)	1380 (1107°C)	1768 (1495°C)	1657 (1384°C)	1910 (1637°C)
	Vapor pressure at mp (ρ,Pa)	4.18	361	254	246	98
	Thermal conductivity (k, W.m^{-1}.K^{-1})	210	156	200	35.3	18.4
	Coefficient of thermal linear expansion (α, 10^{-6}K^{-1})	11.6	26.1	22.3	23	18.1
	Molar heat capacity (C_p, J.mol^{-1}.K^{-1})	16.65412	24.91146	26.37684	27.30277	27.17233
	Specific heat capacity (c_p, J.kg^{-1}.K^{-1})	1848	1024.952	674.629	311.604	197.865
	Volume expansion on melting (%)	n.a.	+4.12	n.a.	n.a.	n.a.
	Enthalpy mol. fusion (ΔH_f, kJ.mol^{-1}) (Δh_f, kJ.kg^{-1})	12.22 (1356)	9.04 (372)	9.33 (232.8)	9.16 (104.5)	7.66 (55.8)
	Enthalpy molar vaporization (ΔH_v, kJ.mol^{-1}) (Δh_v, kJ.kg^{-1})	308.80 (34,265)	128.70 (5295)	149.95 (3741)	154.5 (1736)	177.2 (1290)
	Enthalpy molar sublimation (ΔH_s, kJ.mol^{-1}) (Δh_s, kJ.kg^{-1})	324.4 (35,996)	146.5 (6028)	176.2 (4396)	177.1 (2021)	192.0 (1398)
	Gibbs free enthalpy of oxide formation (ΔG_{oxide}, kJ.mol^{-1})	−580.1	−569.3	−603.3	−561.9	−520.3
Electrical, electrochemical, and magnetic (293.15 K)	Superconductive critical temperature (T_c K)	0.026	n.a.	n.a.	n.a.	1–1.8(5.5 GPa)
	Electrical resistivity (ρ 10^{-8}Ω.m)	4.266	4.38	3.43	23.0	50
	Temperature coefficient of resistivity (0–100°C) (K^{-1})	0.00667	0.00390	0.00402	0.00320	0.00360
	Hall coefficient (R_H, aΩm.T^{-1})	+7.7	−0.9	−0.228	n.a.	n.a.
	Thermoelectric potential versus platinum (emf, mV)	n.a.	+0.44	−0.51	n.a.	n.a.
	Electrochemical equivalence (A.h.kg^{-1})	5.948	2205	1337	611	390
	Standard electrode potential (E,V vs. SHE)	−1.970	−2.356	−2.840	−2.890	−2.920
	Mass magnetic susceptibility (χ_m, 10^{-8} kg^{-1}.m^3)	−1.3	+0.68	+1.40	+1.32	+0.19
Optical	Wavelength maximum intensity atomic spectra line (λ, nm) (bunsen flame color)	n.a.	518	467 (yellow-red)	408 (red)	455 (yellow-green)
	Spectral hemispherical emissivity (650 nm)	0.51–0.61	0.07	n.a.	n.a.	n.a.

3.2.1 Beryllium

3.2.1.1 Description and General Properties

Beryllium [7440-41-7], with the atomic number 4 and a relative atomic mass (i.e., atomic weight) of 9.012182(3) is the first element of the alkaline-earth metals, i.e., group IIA(2) of Mendeleev's periodic chart. It has the chemical symbol Be and is named after the Greek word, beryllos, meaning the mineral beryl in which it was first discovered. Nevertheless, in the past, especially in France, it was called glucinium from the Greek word, glykys, owing to the sweetness of its salts.[2] Highly pure beryllium has a steel gray aspect, and is a brittle and tough metal which scratches glass, probably due to the hard protective layer of beryllia always present. It has a hexagonal close-packed (hcp) crystalline structure with a strong anisotropy as indicated by its extremely low Poisson's ratio of 0.039. Beryllium with a density of 1847.7 kg.m^{-3} is the seventh lightest metal. Moreover, it has the highest melting point of the light metals (1283°C). Its Young's modulus is about three-fold greater than that of steel. It has a high thermal conductivity (210 W.m^{-1}.K^{-1}) comparable of that of aluminum, hence it is used as a heatsink material. At room temperature, beryllium does not react with air. Owing to the passivation by a thin protective layer of its oxide BeO (also called beryllia), it is not dissolved by concentrated nitric acid. Nevertheless, finely divided or amalgamated metal reacts readily with diluted hydrochloric, sulfuric and nitric acids. Beryllium is attacked by strong alkalis with evolution of hydrogen. Due to its low mass attenuation coefficient for a wide range of wavelengths of X-rays, it is transparent, and hence, extensively used as window material in X-ray tubes or detectors. Another interesting property of beryllium it is its high normal reflectivity in the long wave infrared (LWIR). Natural beryllium is mononuclidic, i.e., it is composed only of the ^{9}Be nuclide. The water soluble salts are strongly hazardous and dermal contact can cause contact dermatitis and in the worst cases granulomatous skin ulceration. Furthermore, inhalation exposure to dusts or fines containing Be-compounds (e.g., during machining) can cause acute pulmonary diseases and beryllosis, which is a serious chronic lung disease.

Cost (1998) the pure beryllium metal (98.5wt% Be) sold as vacuum arc cast ingots is priced 720.92 \$US.kg^{-1} (327 \$US.lb^{-1}).

3.2.1.2 History

Beryllium was first discovered in 1798 by the French chemist N.L. Vauquelin who extracted the oxide from the cyclosilicate mineral beryl. The pure metal was first isolated in 1828 independently by the German chemist F. Wöhler and the French chemist A.A.A. Bussy by reducing beryllium chloride with molten potassium metal.

3.2.1.3 Natural Occurrence, Minerals, and Ores

Beryllium is present in the Earth's crust at a level roughly of 2.6 mg.kg^{-1} (ppm wt) Amongst the 45 mineral species identified, the chief minerals are: the nesosilicate **phenakite** (Be$_2$[SiO$_4$], rhombohedral), the sorosilicate **bertrandite** (Be$_4$[Si$_2$O$_7$(OH)$_2$], orthorhombic), the cyclosilicate **beryl** (Be$_3$Al$_2$[Si$_6$O$_{18}$], hexagonal), and finally, the binary oxide **chrysoberyl** (BeAl$_2$O$_4$, orthorhombic). It is important to note that certain precious forms of beryl such as the emerald (Cr^{3+} impurities), aquamarine (Fe^{3+} impurities), and heliodore (U^{6+} impurities) are gemstones known for centuries. Beryl occurs as small and isolated pockets in pegmatites, which are coarse-grained granitic igneous rocks composed largely of quartz, feldspars, and micas. Therefore, beryllium ore deposits are closely linked with pegmatite occurrence for instance in North America, especially USA, Africa (e.g., Madagascar, Zimbawe), South America (e.g., Brazil) and Eastern Europe (e.g., Russia and Kazakstan). Amongst the chief minerals, only beryl and bertrandite are used as ores for extracting beryllium on an industrial scale.

[2] Be extremely careful and never taste beryllium compounds, because beryllium salts are highly poisonous.

3.2.1.4 Mining and Mineral Dressing

Beryl is a by-product of other minerals occurring in pegmatites such as spodumene and lepidolite, and is normally recovered by hand sorting. Commercial beryl contains roughly 10wt% BeO (3.6wt% Be metal). Bertrandite ore deposits are mined by an open-pit method, such as US ore deposits of Delta in Utah, processed by Brush Wellman.

3.2.1.5 Processing and Industrial Preparation

Preparation of beryllium hydroxide – beryllium is extracted from bertrandite ore by a leaching process. In this process the crushed and ground bertrandite concentrate is directly dissolved in sulfuric acid by leaching to produce the crude beryllium sulfate liquor, while extraction of beryllium hydroxide from beryl ore is achieved through a more energy-intensive process. Actually, the beryl concentrate is treated by arc melting at 1650°C, and then quenched into cold water to form dispersed particles in order to make beryllium easily accessible by sulfuric acid. The next stage, called sulfatation, consists of dissolving the powder in sulfuric acid at a temperature of 400°C. However, these two routes produce a crude beryllium sulfate liquor and silica as a by-product. The crude beryllium sulfate liquor either obtained by the first or the second process follows a solvent extraction process. Then, the beryllium hydroxide, $Be(OH)_2$, is precipitated by adding calcium hydroxide to the reactor. The beryllium hydroxide is an essential intermediate product which is the starting material for the manufacture of beryllium metal.

Preparation of beryllium metal – at the pilot plant scale most of the methods for winning beryllium metal from its compound are based on the molten salt electrolysis of the chloride or the fluoride. The molten salt electrolysis is achieved from a eutectic bath, e.g., $KCl–LiCl–BeCl_2$. However, electrowinning of beryllium at the industrial scale has never been competitive with the metallothermic reduction of beryllium fluoride with magnesium. In this last process, the beryllium hydroxide is dissolved in ammonium bifluoride, purified and crystallized from aqueous solution as ammonium fluoroberylate, $(NH_4)_2BeF_4$. The salt undergoes thermal decomposition by a pyrolysis process to form anhydrous beryllium fluoride, BeF_2, with simultaneous evolution of ammonium fluoride gas, NH_4F. The beryllium fluoride is reacted with magnesium to produce beryllium and magnesium fluoride. Beryllium pebbles previously obtained are then vacuum remelted into a magnesia crucible in order to remove slags and volatilize residual magnesium, and cast into graphite molds. This vacuum cast ingot obtained is the starting point for the manufacture of beryllium powder. For applications requiring high-purity beryllium (i.e., above 99.99wt% Be), the metal can also be further refined by the following three methods: (i) vacuum distillation, (ii) containerless zone refining, and (iii) molten salt electrorefining. Two beryllium metal commercial grades are available: vacuum cast beryllium (99.5wt% Be) and sintered beryllium (99.4wt% Be).

3.2.1.6 Industrial Uses and Applications

As previously discussed, beryllium is considerably tough and its largest application (65%) is as an alloying element with copper. The two general classes of copper–beryllium alloys are: (i) the high-strength alloys containing 1.6–2.0wt% Be and 0.25wt% Co which are used for small electrical contacts, springs, clips, switches, Bourdon pressure gages and dies for plastics and (ii) the high-conductivity alloys used for electrical applications. Nickel–beryllium alloys find limited applications as electrical connectors and in the glass industry. A new application for the metal is the beryllium–aluminum alloys used in some military helicopter electro-optical systems. Small additions of beryllium as an alloying element are made to magnesium and aluminum alloys for improving fluidity during casting and to decrease oxidation losses. Due to its low mass attenuation coefficient for a wide range of X-ray energies, it is transparent, and hence, extensively used as window material in X-ray tubes or detectors such as those used in energy dispersive analysis of X-ray equipment. Moreover, owing to its high normal reflectivity in the long-wave infrared (LWIR), its stiffness and low density, and the fact that it can be polished to a mirror finish, it is used for manufacturing IR-mirrors in satellites for surveillance and deep-space observatories. Beryllium can be machined to extremely close tolerances; this leads, in combination with its excellent dimensional stability, to its extensive use for manufacturing highly precise and stable components and devices for optical apparatus, instrumentation (e.g., gyroscopic systems), or in guidance or navigation for ships and aircrafts. It is used in nuclear industry (i) as a source of thermal neutrons, when irradiated by an alpha-emitter source such as ^{226}Ra, according to the nuclear reaction ^{9}Be$(\alpha,n)^{12}$C, (ii) as a neutron reflector, and (iii) as a neutron moderator in nuclear reactors in the form of beryllium carbide Be$_2$C core material,[21] as a missile part and in other weapons. Owing to both its high thermal conductivity and high electrical resistivity, it is also used as heatsink material in electronics devices requiring good electrical insulation properties. In conclusion, some 60% of beryllium consumption is as alloy and oxide in electronic components and some 20% in the same form for electrical components. Approximately 13% is consumed as an alloy, oxide and metal in aerospace and defense applications, while the balance is used as an alloy, metal and oxide for other purposes.

3.2.1.7 Beryllium Metal World Producers

Table 3.11 Beryllium metal world producers

Producer	Address
Brush Wellman Inc., Beryllium Metals Products Division	14710 W. Portage River Road South, Elmore, OH 43416, USA Telephone: (419) 862 4173 Facsimile: (419) 862 4174 Internet: http://www.brushwellman.com
NGK Metals Corp.	Temple, PA, USA Telephone: (909) 340 0190

3.2.2 Magnesium and Magnesium Alloys

3.2.2.1 Description and General Properties

Magnesium [7439-95-4] with the chemical symbol Mg, the atomic number 12 and the relative atomic mass of 24.3050, is the second element of the alkaline-earth metals, i.e., group IIA(2) of Mendeleev's periodic chart. It is a silvery white metal quite similar in appearance to bright aluminum but with a lower density of only 1738 kg.m^{-3}. Hence, it is the lightest structural metal known. It has a hexagonal close-packed (hcp) crystalline structure, therefore, like most metals having this crystal lattice structure, it lacks ductility when worked at low temperatures. In addition, in its pure form, it lacks sufficient strength for most structural applications. However, the addition of alloying elements greatly improves these properties to such an extent that both cast and wrought magnesium alloys are widely used, particularly where strength-to-weight ratio is an important requirement. Magnesium derives its name from magnesite, a magnesium carbonate mineral (MgCO$_3$, rhombohedral), and this mineral in turn is said to owe its name to magnesite deposits found in Magnesia, a district in the ancient region of Thessaly (Greece). Magnesium is strongly reactive with oxygen at high temperatures; actually, above its ignition temperature of 645°C in dry air, it burns with a dazzling bright white light and intense heat generation. For this reason, magnesium powders were used in the first generation of photographic flash bulbs, but today this property is only used in pyrotechnics.

Owing to its more negative standard electrode potential (–2.687 V/SHE) magnesium is an active element and as a general rule most magnesium alloys range at or near the top of the galvanic or electrochemical series. At room temperature, magnesium and magnesium alloys may be sufficiently resistant to atmospheric corrosion because a protective anodic film of water-insoluble magnesium hydroxide, $Mg(OH)_2$, forms in a process similar to the formation of passivating film in the active metal aluminum. When corrosion does occur, it is only the result of the breakdown of this protective layer. Being a strong reducing agent that forms stable compounds with chlorine, oxygen, and sulfur, magnesium has several pyrometallurgical applications, such as in the production of titanium or zirconium from metallothermic reduction of their tetrachlorides and in the desulfurization of blast-furnace iron. Its chemical reactivity is also evident in the magnesium compounds that have wide application in industry, medicine, and agriculture.

Cost (1998) – pure magnesium metal (i.e., 99.8wt% Mg) sold as ingots is priced at 3.1–4.5 $US.kg^{-1} (1.41–2.04 $US.lb^{-1}).

3.2.2.2 History

The British chemist Humphry Davy is said to have produced an amalgam of magnesium in 1808 by electrolyzing moist magnesium sulfate, using a liquid mercury cathode. The first metallic magnesium, however, was produced 20 years later in 1828 by the French scientist A.A.B. Bussy. He achieved the reduction of molten magnesium chloride by metallic potassium. In 1833, the British scientist Sir Michael Faraday was the first to produce magnesium by the electrolysis of molten magnesium chloride ($MgCl_2$). His experiments were successfully repeated by the German chemist Robert Bunsen. The first successful industrial production was begun in Germany in 1886 by Aluminum und Magnesiumfabrik Hemelingen Gmbh, based on the electrolysis of molten carnallite. Hemelingen later became part of the industrial complex IG Farben Industrie, which, during the 1920s and 1930s, developed a process for producing large quantities of molten and essentially water-free magnesium chloride, now known as the IG Farben process, as well as the technology for electrolyzing this product to magnesium metal and chlorine. Other contributions by IG Farben were the development of numerous cast and malleable alloys, refining and protective fluxes, wrought magnesium products, and a large number of aircraft and automobile applications. During World War II, the Dow Chemical Company of the United States and Magnesium Elektron Limited of the United Kingdom began the electrolytic reduction of magnesium from seawater pumped from Galveston Bay, at Freeport, Texas, and the North Sea at Hartlepool, UK. At the same time in Ontario, Canada, L.M. Pidgeon's process of thermally reducing magnesium oxide with silicon in externally fired retorts was introduced. Following the war, military applications lost prominence. Dow Chemical broadened civilian markets by developing wrought products, photoengraving technology, and surface treatment systems. Extraction remained based on electrolysis and thermal reduction. To these processes were made such refinements as the internal heating of retorts (the Magnetherm process, introduced in France in 1961), extraction from dehydrated magnesium chloride pills introduced by the Norwegian company Norsk-Hydro in 1974, and improvements in electrolytic cell technology from about 1970. Magnesium is the lightest of all machinable metals. Casting characteristics are excellent, since the molten metal has a low heat content and low viscosity. Magnesium alloys have limited cold-forming capabilities, owing to the hexagonal crystalline structure of magnesium, but they are readily hot-worked at temperatures ranging from 150°C to 400°C.

3.2.2.3 Natural Occurrence, Minerals, and Ores

Magnesium is the eighth most abundant element in the Earth's crust; magnesium constitutes 2.5wt% of the lithosphere, and an average of 1.3 kg.m^{-3} of ocean water with a maximum of

35 kg.m^{-3} for certain seas. Owing to its strong chemical reactivity, it does not occur in the native state, but rather it is found in a wide variety of compounds in seawater, brines, and rocks. Among the ore minerals, the most common are the carbonates such as **dolomite** (Mg Ca(CO$_3$)$_2$, rhombohedral) and **magnesite** (MgCO$_3$, rhombohedral). Less common are the hydroxide mineral **brucite** (Mg(OH)$_2$), the three sulfates: **kieserite** (MgSO$_4$·H$_2$O), **kainite** (MgSO$_4$·KCl·3H$_2$O), and **langbeinite** (2MgSO$_4$·H$_2$O), and the halide mineral **carnallite** (MgCl$_2$KCl·6H$_2$O). Magnesium chloride is recoverable from naturally occurring brines such as the Great Salt Lake (typically containing 1.1wt% Mg) and the Dead Sea (3.4wt% Mg), but by far the largest sources are the oceans of the world. Although seawater is only approximately 0.13wt% Mg, it represents an almost inexhaustible source. Both dolomite and magnesite are mined and concentrated by conventional methods. Carnallite is dug as ore or separated from other salt compounds that are brought to the surface by solution mining. Naturally occurring magnesium-containing brines are concentrated in large ponds by solar evaporation. Magnesium is produced for profit in such places as the United States and Canada, Western Europe, South America and Asia.

3.2.2.4 Processing and Industrial Preparation

Owing to its strong chemical reactivity, magnesium combines with oxygen and chlorine to form stable compounds. This means that the extraction of the metal from raw materials is always an energy-intensive process requiring both low-cost electricity, and well-tuned technologies. All commercial producers of magnesium metal use either (i) the **electrolytic reduction** of molten anhydrous (e.g., Norsk Hydro process) or hydrated (e.g., Dow Chemical process) magnesium chloride, or (ii) the **nonelectrolytic processes** which consist mainly of the thermo-chemical reduction of either calcinated magnesite (i.e., magnesia, MgO) or dolomite (i.e., a mixture of magnesium and calcium oxides, called dolime). These nonelectrolytic processes can be sorted according to the reducing agent: ferrosilicon in the silicothermic process (e.g., Pidgeon's and magnetherm processes), carbon in the carbothermic, or aluminum powder in the aluminothermic process. Where power costs are low, electrolysis is the cheaper method and, indeed, it accounts for approximately 75–80% of world magnesium production, while the nonelectrolytic routes form the balance. Hence, the chief raw materials for magnesium metal preparation are dolomite (CaMg(CO$_3$)$_2$), magnesite (MgO), carnallite (MgKCl$_3$·6H$_2$O), seawater, and brines.

   Electrolytic reduction processes – the electrolytic processes comprise two steps: (i) the preparation of a feedstock containing anhydrous (essentially water-free) magnesium chloride, MgCl$_2$ (e.g., Norsk Hydro process) or partially dehydrated magnesium chloride, MgCl$_2$·H$_2$O (e.g., Dow Chemical process), or sometimes anhydrous carnallite (e.g., Russian processes), and (ii) the dissociation of this compound into magnesium metal and chlorine gas into an electrolytic cell. In order to avoid impurities present in carnallite ores, dehydrated artificial carnallite is produced by controlled crystallization from heated magnesium- and potassium-containing solutions. Partly dehydrated magnesium chloride can be obtained by the Dow process, in which seawater is mixed in a flocculator with lightly burned reactive dolomite (dolime). An insoluble magnesium hydroxide precipitates to the bottom of a settling tank, whence it is pumped as a slurry, filtered, converted to magnesium chloride by reaction with hydrochloric acid, and dried in a series of evaporation steps to 25wt% water content. Final dehydration takes place during smelting. Finally, anhydrous magnesium chloride is produced either by (i) dehydration of magnesium chloride brines or (ii) direct chlorination of magnesium oxide with HCl. In the later method, exemplified by the IG Farben process, lightly burned dolomite is mixed with seawater in a flocculator, where magnesium hydroxide is precipitated out, filtered, and calcined in a kiln to dry magnesium oxide. This is mixed with charcoal, formed into globules with the addition of magnesium chloride solution, and dried. The globules are charged into a chlorinator, a brick-lined shaft furnace where they are heated by carbon electrodes to approximately 1000–1200°C. Chlorine gas introduced through portholes in the furnace reacts with the magnesium oxide to produce molten magnesium chloride, which is tapped at intervals and sent to the electrolytic cells. Dehydration of magnesium brines is conducted in stages. In the Norsk Hydro process, impurities are first removed by precipitation and filtering. The purified brine, which contains approximately 8.5wt% magnesium, is concentrated by evaporation to 14wt% and converted to particulates in a prilling tower. This product is further dried to water-free particles and conveyed to the electrolytic cells. In the Norsk-Hydro (resp. Dow Chemical) process, electrolytic cells comprise a refractory semi-wall separator (resp. no wall), and are essentially refractory brick-lined (resp. carbon steel) vessels equipped with multiple cast mild steel cathodes (–) and graphite anodes (+). These are mounted vertically through the cell hood and partially submerged in a molten salt electrolyte composed of a mixture of alkali chlorides (e.g., 20wt% $MgCl_2$, 20wt% $CaCl_2$, and 60wt% NaCl) to which the anhydrous (resp. hydrous) magnesium chloride produced in the processes described above is added in concentrations of 18wt% (resp. 6%). The cell is internally (resp. externally) heated by the Joule effect, with an operating temperature which is 750°C (resp. 700°C), and specific energy consumption is 18 kW.h.kg^{-1} (resp. 12 kW.h.kg^{-1}) of magnesium metal produced. In the two electrolytic processes, chlorine is generated at the graphite anodes, leading to a certain consumption of the anode material, due to a formation of chlorinated hydrocarbons, while molten magnesium metal forms into droplets, which coalesce, and due to buoyancy rise the surface of the fused salt bath forming a liquid metal pool, where it is collected. The chlorine can be reused in the dehydration process.

Thermochemical reduction processes – these metallothermic reductions are commonly achieved on either calcined magnesite (magnesia, MgO) or calcined dolomite (mixture of magnesium and calcium oxides, called dolime). In the **silicothermic reduction** (**Pidgeon's process**[3]), a mixture of ground, and briquetted, dolime and ferrosilicon (reducing agent, containing 75wt% Si), and fluorspar (fluxing agent) is charged into in a horizontal tubular stainless steel retort. Because the reaction is endothermic, heat must be supplied to initiate and sustain the reduction. Hence retorts are externally heated by an oil- or gas-fired furnace. Moreover, because magnesium reaches a vapor pressure of 1 atm only at 1800°C, heat requirements can be quite high, and in order to lower reaction temperatures, the process operates under primary vacuum (13 Pa) at 1200°C. After the reduction is completed, magnesium crystals (called crowns) are removed from the condensers (cold walls), and the silicate slag ($CaSiO_3$) with iron particles is removed as a spent solid, and the retort is recharged. In the **Bolzano process**, dolime–ferrosilicon briquettes are stacked on a special charge support system through which internal electric heating is conducted to the charge. A complete reaction takes 24 h at 1200°C below 400 Pa. In the **aluminothermic reduction**, the thermal reduction of dolime is performed using aluminum as reducing agent, and alumina as fluxing agent. By adding alumina (aluminum oxide, Al_2O_3) to the charge, the melting point can be reduced to 1550–1600°C. This technique has the advantage that the liquid slag can be heated directly by electric current through a water-cooled copper electrode. The reduction reaction occurs at 1600°C and 400–670 Pa pressure. Vaporized magnesium is condensed in a separate system attached to the reactor, and molten slag and depleted ferrosilicon (20wt% Si) are tapped at intervals.

Refining – after extraction by the different processes described above, crude magnesium metal is transported to cast shops for removal of impurities, addition of alloying elements, and transformation into ingots, billets, and slabs. During melting and handling, molten magnesium metal and alloys are protected from burning by a layer of flux or of a gas such as sulfur hexafluoride or sulfur dioxide. For shipping and handling under severe climatic conditions, suitable ventilated plastic or paper wrappings are required to prevent corrosion. Primary magnesium is available in grades of 99.90wt%, 99.95wt%, and 99.98wt%, but, in practice, grades 99.95 and 99.98wt% have only limited use in the uranium and nuclear industries. For bulk use, commercial grades 99.90 and 99.80wt% are supplied.

---

[3] Usually ferrosilicon is prepared by direct melting of quartzite and iron scrap in an arc melting furnace.

3.2.2.5 Magnesium Alloys

Table 3.12 Standard ASTM designation of magnesium alloys

First letters	Second code	Third code	Fourth code
Two-letter code which corresponds to the two major alloying elements ordered according to their mass fraction	Rounded percentage by mass of the two major alloying elements arranged in the order defined for the first code	One-letter code for distinguishing two different alloys having the same first and second part code. All the letters of the Roman alphabet can be used except I and O	One-letter, one-number code which corresponds to the tempering of the alloy
A = Al M = Mn B = Bi N = Ni C = Cu P = Pb D = Cd Q = Ag E = Rare R = Cr earths S = Si F = Fe T = Sn G = Mg W = Y H = Th Y = Sb K = Zr Z = Zn L = Li			F = As fabricated O = Annealed H10, H11 = Slightly strain hardened H23, H24 = Strain hardened and partially annealed T4 = Solution heat treated T5 = Artificially aged T6 = Solution heat treated T8 = Solution heat treated, cold worked, and artificially aged

Example: AZ31B O magnesium alloys having the following average chemical composition: Mg–2.5Al–1.6Zn annealed.

3.2.2.6 Industrial Uses and Applications

Magnesium applications are motivated by the light weight, high strength, high damping capacity, close dimensional tolerance, and ease of fabrication of its alloys (See physical properties shown in Table 3.13, page 104). The chief uses and applications are listed in Table 3.14 (page 107).

3.2.2.7 Magnesium World Producers

- Dow Chemical, Freeport, Texas, USA (stopped activity in 1999)
- Norsk Hydro, Porsgrunn, Norway
- Northwest Alloys, Addy, Washington, USA
- Magcorp Bowley, Salt Lake City, USA
- Péchiney Électrométallurgie, France
- Magnola Asbestos, Québec, Canada (should start in 2001)

3.2.3 Calcium

3.2.3.1 Description and General Properties

Calcium [7440-70-2] with the chemical symbol Ca, the atomic number 20 and the relative atomic mass (atomic weight) 40.078(4), is the third element of the alkaline-earth metals of the main group IIA(2) of Mendeleev's periodic chart. It is named after the Latin words, calc, calcis, meaning quicklime. Calcium is a bright silvery white metal when freshly cut. But on exposure to moist air, it reacts slowly with oxygen, water vapor, and nitrogen of the air to form a bluish gray to yellow tarnish coating of a mixture of oxide, hydroxide, and nitride. Calcium has a face-centered cubic crystalline structure below 300°C. The metal is much harder than sodium but softer than aluminum and magnesium. Calcium ignites and burns in air or pure oxygen above

Table 3.13 Physical properties of selected magnesium alloys

Usual and trade names	UNS	Average chemical composition (x, wt%)	Category[a]	Density (ρ, kg.m^{-3})	Yield strength 0.2% proof (σ_{YS}, MPa)	Ultimate tensile strength (σ_{UTS}, MPa)	Elongation (Z, %)	Brinell hardness (HB)	Liquidus range (°C)	Thermal conductivity (k, W.m^{-1}.K^{-1})	Specific heat capacity (c$_P$, J.kg^{-1}.K^{-1})	Coefficient of linear thermal expansion (α, 10^{-6} K^{-1})	Electrical resistivity (ρ, $\mu\Omega$.cm)
Pure magnesium	n.a.	99.97wt% Mg	C	1738	n.a.	n.a.	n.a.	n.a.	649	156	1050	26.0	4.2
AM100A F	M10100	Mg–9.5Al–0.3Zn	C	1830	83	150	2	53	493–595	73	1050	25.0	15.0
AM60A F	M10600	Mg–6Al–0.5Si–0.35Cu	C	1800	130	220	6	n.a.	540–615	62	1000	25.6	n.a.
AM60B F	M10600	Mg–6Al–0.25Mn	C	1800	130	220	6	n.a.	540–615	62	1000	25.6	n.a.
AS41A F	M10410	Mg–4Al–1Si–0.35Mn	C	1770	140	210	6	63	565–620	68	1000	26.1	n.a.
AS41XB F	n.a.	Mg–4Al–1Si–0.35Mn	C	1770	140	210	6	63	565–620	68	1000	26.1	n.a.
AZ10A	M11100	Mg–1.5Al–0.5Zn	W	1760	145–150	240	10	n.a.	630–645	110	1050	26.6	6.4
AZ31B T1	M11312	Mg–3Al–1Zn	W	1780	150–220	241–290	12–21	46–73	605–630	84	1050	26.0	9.2
AZ61 AF	M11610	Mg–6Al–1Zn	W	1800	165–220	285–305	8–16	50–60	525–620	79	1050	27.3	12.5
AZ63 T1	M11630	Mg–6Al–3Zn–0.3Si	C	1830	97	200	6	50	455–610	77	1050	26.1	11.5
AZ80A T4	M11800	Mg–8Al–0.5Zn	W	1810	230–275	330–380	6–11	67–80	475–610	84	1050	27.2	14.5
AZ81A T1	M11810	Mg–8Al–0.5Zn	C	1800	83	275	15	55	490–610	79	1000	25.5	14.3
AZ91A F	M11910	Mg–9Al–1Zn–0.1Mn	C	1810	97–150	165–230	3	60–63	470–595	72	1050	27.0	17
AZ91C T6	M11914	Mg–9Al–1Zn–0.1Mn	C	1810	90–145	275	6–15	62–77	470–595	72	1050	27.0	15.2

Alloy	UNS	Composition												
AZ92A T6	M11920	Mg–9Al–2Zn–0.3Si	C	1830	150	275	3	81	445–595	72	1050	26.0	14.0
EQ21 T6	M16330	Mg–1.5Ag–2RE	C	1810	170	235	2	65–85	540–640	113	1000	26.7	6.85
EZ33A T5	M12330	Mg–3RE–2.5Zn–0.6Zr	C	1830	110	160	3	50	545–645	100	1050	26.1	7.0
HK31 A T6	M13310	Mg–3Th–0.6Zr–0.3Zn	C	1800	105	220	8	55	590–650	92	1050	n.a.	7.7
HK31A H24	M13310	Mg–3Th–1Zr	W	1830	205	260	9	n.a.	590–647	105	960	26.7	7.2
HM21A T8	M13210	Mg–2Th–0.5Mn	W	1780	170	235	10	n.a.	605–650	134–138	1050	27.0	5.2
HM31A O	M13312	Mg–3Th–1.2Mn	W	1800	230	283	10	n.a.	605–650	104	1050	26.0	6.6
HZ32A T5	M13320	Mg–2Zn–3Th–1Zr	C	1830	90	185	4	55	550–650	110	960	26.7	6.5
K1A F	M18010	Mg–0.5Zr	C	1740	55	180	19	n.a.	650	122	1000	27.0	5.7
M1A	M15100	Mg–1Mn–0.3Cr	W	1770	125–180	230–255	7–12	42–54	648–649	138	1050	26.0	5.0
QE22A T6	M18220	Mg–2.5Ag–2Re–0.6Zr	C	1820	195	260	3	65–85	550–643	113	1000	26.7	6.85
QH21A T6	n.a.	Mg–2.5Ag–1RE–1Th–0.7Zr	C	1830	200–207	275–285	4–8	n.a.	535–640	113	1005	26.7	6.85
WE43 T6	M18430	Mg–4Y–3.4RE–0.6Zr	C	1840	162	250	2	75–95	545–640	51	966	26.7	14.8
WE54 T6	M18410	Mg–5.1Y–3RE–0.6Zr	C	1840	172	250	2	75–95	545–640	52	960	24.6	17.3
ZC63 T6	M16631	Mg–5.5Zn–2.5Cu–0.5Mn	C	1870	125	210	3–5	55–65	465–635	122	n.a.	n.a.	5.4
ZC71 T6	M16710	Mg–6.5Zn–1Cu–0.8Mn	W	1830	340–345	360–375	4–6	n.a.	455–635	122	962	26.0	5.4
ZE41A T5	M16410	Mg–4.5Zn–1.5RE	C	1820	140	205	3.5	62	525–645	113	n.a.	n.a.	6.0

Table 3.13 (continued)

Usual and trade names	UNS	Average chemical composition (x, wt%)	Category[a]	Density (ρ, kg.m^{-3})	Yield strength 0.2% proof (σ_{YS}, MPa)	Ultimate tensile strength (σ_{UTS}, MPa)	Elongation (Z, %)	Brinell hardness (HB)	Liquidus range (°C)	Thermal conductivity (k, W.m^{-1}.K^{-1})	Specific heat capacity (c_P, J.kg^{-1}.K^{-1})	Coefficient of linear thermal expansion (α, 10^{-6} K^{-1})	Electrical resistivity (ρ, $\mu\Omega$.cm)
ZE63A T6	M16630	Mg–6Zn–2.5RE	C	1870	190	300	10	60–85	510–635	109	960	26.5	5.6
ZH62A T5	M16620	Mg–5.5Zn–2Th–0.65Zr	C	1860	150	240	4	70	520–630	110	960	27.1	6.5
ZK21A	M16210	Mg–2.3Zn–0.65Zr	W	1800	180–195	235–260	4	n.a.	600–635	125	960	27.0	5.5
ZK40A T5	M16400	Mg–4Zn–0.45Zr	W	1830	250–255	275	4	n.a.	530–630	117	1050	26.0	6.0
ZK51A T5	M16510	Mg–4.5Zn–0.45Zr	C	1830	140	205	3.5	62	560–640	110	1020	26.0	6.2
ZK60A T5	M16600	Mg–5.5Zn–0.45Zr	W	1830	215–285	305–350	11–16	65–82	520–635	117	1050	26.0	6.0
ZK61A T6	M16610	Mg–6Zn–0.8Zr	C	1830	195	310	10	n.a.	530–635	117	1050	27.0	6.0
ZM21 T1	n.a.	Mg–2Zn–1Mn	C	1780	n.a.	n.a.	n.a.	n.a.	n.a.	n.a.	n.a.	27.0	n.a.
ZW10 T1	n.a.	Mg–1.3Zn–0.6Zr	C	1800	n.a.	n.a.	n.a.	n.a.	625–645	134	1000	27.0	5.3

[a] W = wrought alloys, C = cast alloys.

Table 3.14 Magnesium industrial applications and uses

Application	Description
Metallurgical applications (69%)	By far the greatest use of pure magnesium is as an alloying element in aluminum, zinc, lead, and other nonferrous metals and alloys. In the particular case of aluminum alloys (48% of the market), magnesium additions ranging from less than 1wt% to approximately 10wt%, enhance the mechanical properties as well as the corrosion resistance. Similarly, pure aluminum is also used as an alloying element in many magnesium-based alloys. In the iron and steel industry, small quantities of magnesium and magnesium-containing alloys as ladle addition agents are introduced just the casting is poured. In the particular case of white cast iron, magnesium transforms the graphite flakes into spherical nodules, thereby significantly improving the strength, the toughness and ductility of the iron (4% of the market). In addition, owing to its oxygen scavenger properties, particulate magnesium blended with lime or other fillers is injected into liquid blast-furnace iron, where it improves mechanical properties of steel by combining with sulfur and oxygen (15% of the market). Other metallurgical applications are based on the metallothermic reducing properties of the metal, these include the production of pure titanium, zirconium, hafnium, and uranium from their halides (2% of the market). By far the most important of these is in the pyrometallurgical Kroll process for reducing titanium tetrachloride to titanium metal sponge.
Electrochemical applications (3%)	The strong electronegative nature of magnesium (i.e., its readiness to give up electrons) makes it useful in dry-cell batteries and as an efficient sacrificial anode in the cathodic protection of steel structure. Magnesium dry cells, mostly used in military and rescue equipment, combine light weight, long storage life, and high energy content. The batteries consist of a magnesium anode and a cathode of silver chloride or cuprous chloride. When activated by water, they rapidly build up voltages of 1.3 to 1.8 V per cell and operate at a constant potential between -55 and $95°C$. On the other hand, when magnesium comes into electrical contact with steel in the presence of water, owing to its position in the galvanic series, the magnesium dissolves anodically (i.e., corrodes) sacrificially, while the steel is polarized cathodically, leaving the steel material uncorroded. Ship hulls, water heaters, storage tanks, bridge structures, pipelines, and a variety of other steel products are protected in this manner. Magnesium corrosion can be accelerated by galvanic coupling, high levels of certain impurities (e.g., especially Ni, Cu, and Fe), or contamination (especially of castings) by salts. Magnesium alloys are anodic to all other structural metals and will undergo galvanic attack if coupled to them. The attack is especially severe if the other metal in the couple is passive or fairly inert as, for example, stainless steels or copper-based alloys. For this reason, magnesium and alloys are widely used in cathodic protection of subterranean pipelines. Of the three main metals used in cathodic protection, i.e., aluminium, zinc, and magnesium, magnesium is the most efficient even in low-humidity environments.
Pyrotechnics	In finely divided form magnesium, both as pure magnesium and alloyed with 30wt% Al, has been used in military pyrotechnics for many years and has found numerous uses in incendiary devices and flares. In the form of finely divided particles, it has been used as a fuel component, particularly in solid rocket propellants.
Structural applications (25%)	The mechanical properties of magnesium improve when it is alloyed with small amounts of other metals. In most cases, the alloying elements form intermetallic compounds that permit heat treatment for enhanced mechanical properties. Magnesium alloys can be divided into two types. General-purpose alloys, suitable for applications at temperatures up to $150°C$, contain 3–9wt% Al, 0.5–3wt% Zn, and about 0.2wt% Mn. Special alloys are used at temperatures up to approximately $250°C$; these contain various amounts of Zn, Zr, Th, Ag, and Y and other rare-earth metals. In addition, high-purity alloys, with low contents of Fe, Ni, and Cu, have greater corrosion resistance than conventional alloys. Structural applications include automotive, industrial, materials-handling, commercial and aerospace. The automotive applications include clutch and brake pedal support brackets, gear boxes, steering column lock housings, and manual transmission housing. In industrial machinery, such as textile and printing machines, magnesium alloys are used for moving parts that operate at high speeds and hence must be lightweight to minimize inertial forces. Material-handling equipment includes dock-boards, grain shovels, and gravity conveyors. Commercial applications include hand-held tools, sporting goods, luggage frames, cameras, household appliances, business machines, computer housing, and ladders. The aerospace industry employs magnesium alloys in the manufacture of aircraft, rockets, and space satellites. Magnesium is also used in tooling plates and, because of its rapid and controlled etching characteristics, in photoengraving.
Agricultural applications (1%)	The use of dolomite as a fertilizer in areas with acid soil and the use of magnesium oxide as a mineral addition to cattle feed at the start of the grazing season in early spring.

Table 3.14 (continued)	
Application	Description
High temperature thermal insulation (2%)	The predominant industrial application of magnesium compounds is in the use of magnesite and dolomite in refractory bricks. Bricks of high-purity magnesia are exceptionally wear- and temperature-resistant, with high heat capacity and conductivity. The more expensive fused magnesia serves as an insulating material in electrically heated stoves and ovens, while the less expensive caustic magnesia is a constituent in leaching lyes for the paper industry, where it reduces losses and allows for the processing of both coniferous and deciduous wood. Another nonstructural use of magnesium is in the Grignard reaction in organic chemistry.

300°C to form calcium oxide, CaO (calcia, or quicklime), and reacts rapidly with warm water and more slowly with cold water evolving hydrogen gas and calcium hydroxide (spent lime). Calcium metal reacts strongly with cold fuming sulfuric acid (Nordhausen's acid) to give off elemental sulfur and sulfur dioxide. It colors the flame of a bunsen gas burner crimson (brick red). It readily dissolves in liquid anhydrous ammonia giving a deep blue solution. Owing to its reactivity, it should be stored in air-tight containers, under inert gas atmosphere or totally immersed in benzene, heptane or a mineral oil such as petrolatum, free from traces of oxygen or water. Naturally occurring calcium consists of a mixture of six stable nuclide isotopes: ^{40}Ca (96.941at%), ^{44}Ca (2.086at%), ^{42}Ca (0.647at%), and smaller proportions of ^{48}Ca (0.187at%), ^{43}Ca (0.135at%), and ^{46}Ca (0.004at%).

3.2.3.2 History

Calcium was first isolated by Sir Humphry Davy, in 1808. Davy produced the calcium amalgam by electrolyzing an aqueous solution of the chloride, CaCl$_2$, using a liquid mercury cathode such as in the chlor-alkali process employing a mercury cathode. After distilling the mercury from the amalgam formed, he obtained the pure calcium metal. His discovery showed lime to be an oxide of calcium. Later Moissan reduced calcium diiodide with sodium. But the first industrial production of calcium metal was reported in 1904 and attributed to Brochers and Stockem who prepared it by electrolysis of the molten chloride. This process was discontinued in 1940 and replaced by aluminothermic reduction of the oxide.

3.2.3.3 Natural Occurrence, Minerals, and Ores

Calcium, with a concentration of 4.1wt% in the Earth's crust, is therefore, the fifth most abundant element in the lithosphere. However, owing to its chemical reactivity with oxygen, calcium never occurs naturally in the free state, but the compounds of the element are widely distributed amongst the geological materials. The chief calcium bearing minerals are the three carbonates **calcite** (CaCO$_3$, rhombohedral), the **aragonite** (CaCO$_3$, orthorhombic), the **dolomite** ((Ca, Mg)(CO$_3$)$_2$, rhombohedral), the sulfate **gypsum** (CaSO$_4$·2H$_2$O, monoclinic), the **anhydrite** (CaSO$_4$, orthorhombic), the fluoride, **fluorite** or **fluorspar** (CaF$_2$, cubic) and to a lesser extent the phosphate **apatite** (Ca$_5$(PO$_4$)$_3$(OH, F, Cl), hexagonal), and other tectosilicates such as **feldspars** (e.g., Ca-plagioclases) and **zeolites**. Calcite is the major constituent of sedimentary rocks such as limestone, chalk, marble, dolomite, eggshells, pearls, coral, and the shells of many marine animals, while aragonite is the main component of stalactites, and stalagmites. As calcium phosphate, it is the principal inorganic constituent of teeth and bones. The human body contains 2wt% of calcium and hence, it is the most abundant metallic element in the human body.

3.2.3.4 Processing and Industrial Preparation

Formerly produced by molten salt electrolysis of anhydrous calcium chloride with a specific energy consumption of 30 kW.h.kg^{-1} until the 1940s, the pure calcium metal is now made by metallothermic reduction on an industrial scale by reducing calcium oxide (calcia or lime) with molten silicon or aluminum. Sometimes high-purity (>99.9wt% Ca) calcium metal, required for certain demanding applications, is obtained by redistillation of the metal. In the aluminothermic reduction, crushed pure limestone or calcite (min. 98wt% CaCO$_3$), is calcinated in a kiln at 1000°C to produce calcium oxide giving off carbon dioxide. Then calcium oxide is ground and mixed with aluminum powder. The mixture is heated and the reduction begins producing calcium metal and calcium aluminate. The calcium is removed from the slag by distillation under vacuum and condensed in a cold mold. Calcium alloys are produced industrially by various techniques such as direct alloying, or chemical reduction of the raw components.

3.2.3.5 Industrial Uses and Applications

Calcium was extensively used by the ancients as the compound quicklime. The metal is used as an alloying agent for aluminum, lead, magnesium, and other base metals. It is used in metallurgy as a desulfurizer for ferrous metals, and as deoxidizer for certain high-temperature alloys such as copper, beryllium, and for nickel, steel, and tin bronzes. It is used in electronics as a getter in vacuum electron tubes to fix residual gases such as oxides, nitrides, and hydrides. It is used in pyro-metallurgy as a strong reducing agent in the metallothermic preparation of the refractory metals such as Cr, Th, U, Ti, Zr, and Hf, from their tetrachlorides and other metals from their oxides (Be, Sc, and Y). Finally, calcium is used as a dehydrating agent for organic solvents. Alloyed with lead (0.04wt% Ca), it is employed as sheaths for telephone cables and as grids for storage batteries of the lead–acid stationary type. It is alloyed with cerium to make flints for cigarette and gas lighters. Limelights, formerly used in staged lighting, emit a soft, very brilliant white light upon heating a block of calcium oxide to incandescence in an oxyhydrogen flame.

Calcium-based chemicals – the most important of the calcium compounds are calcium carbonate, CaCO$_3$, the major constituent of limestones, marbles, chalks, oyster shells, and corals. Calcium carbonate obtained from its natural sources is used as a filler in a variety of products, such as ceramics and glass, and as a starting material for the production of calcium oxide. Synthetic calcium carbonate, called "precipitated" calcium carbonate, is employed when high purity is required, as in pharmacy (e.g., anti-acid and dietary calcium supplement), in the food industry (e.g., baking powder), and for manufacturing pure laboratory reagents. Calcium oxide, also known as lime, or quicklime, CaO, is a white or grayish white solid produced in large quantities by calcinating (roasting) calcium carbonate so as to drive off carbon dioxide. Lime, one of the oldest products of chemical reaction known, is used extensively as a building material (e.g., mortar) and as a fertilizer. Large quantities of lime are utilized in various industrial neutralization reactions. A large amount also is used as starting material in the production of calcium carbide, CaC$_2$. Also known as carbide, or calcium acetylenide, this grayish black solid decomposes in water, forming flammable acetylene gas and spent lime or calcium hydroxide, Ca(OH)$_2$. The decomposition reaction is used for the production of acetylene, which serves as an important fuel for welding torches. Calcium carbide also is used to make calcium cyanamide, CaCN$_2$, a fertilizer component and starting material for certain plastic resins. Calcium hydroxide, also called slaked or spent lime Ca(OH)$_2$, is obtained by the reaction of water with calcium oxide. When mixed with water, a small proportion of it dissolves, forming a solution known as limewater, the rest remaining as a suspension called milk of lime. Calcium hydroxide is used primarily as an industrial alkali and as a constituent of mortars, plasters, and cement. Another important compound is calcium chloride, CaCl$_2$, a colorless or white solid produced in large quantities either as a by-product of the manufacture of sodium carbonate by the Solvay process or by the action of hydrochloric acid on calcium carbonate. The anhydrous solid is used as a drying agent. Calcium hypochlorite, $Ca(ClO_2)$, widely used as bleaching powder, is produced by the action of chlorine on calcium hydroxide. The hydride CaH_2, formed by the direct action of the elements, liberates hydrogen when treated with water. Calcium sulfate, $CaSO_4$, is a naturally occurring calcium salt. It is commonly known in its dihydrate form, $CaSO_4 \cdot 2H_2O$, a white or colorless powder called gypsum. When gypsum is heated and loses three-quarters of its water, it becomes the hemi-hydrate $CaSO_4 \cdot 0.5H_2O$, plaster of Paris. If mixed with water, plaster of Paris can be molded into shapes before it hardens by recrystallizing to dihydrate form. Calcium sulfate may occur in groundwater, causing hardness that cannot be removed by boiling. Calcium phosphates occur abundantly in nature in several forms. For example, the tribasic variety (precipitated calcium phosphate), $Ca_3(PO_4)_2$, is the principal inorganic constituent of bones and bone ash. The acid salt $Ca(H_2PO_4)_2$, produced by treating mineral phosphates with sulfuric acid, is employed as a plant food and stabilizer for plastics.

3.2.4 Strontium

3.2.4.1 Description and General Properties

Strontium [7440-24-6], with the chemical symbol Sr, the atomic number 38, and the relative atomic mass (i.e., atomic weight) 87.62(1) is the fourth alkaline-earth metal, i.e. elements of the group IIA(2) of Mendeleev's periodic chart. Strontium is named after the word Strontian, a Scottish town. Stronium is a hard, silvery-white metal when freshly cut, but it readily tarnishes in moist air, beginning yellowish due to the formation of the oxide. It ressembles barium in its properties but it is harder and less reactive. It has a low density (2540 kg.m^{-3}), and a relatively high melting point (1042 K). It has a face-centered cubic (fcc) crystalline structure. It decomposes in water evolving hydrogen and forming the hydroxide, $Sr(OH)_2$. Strontium fines such as powder, dust and turnings, are pyrophoric, i.e., they ignite spontaneously in air.

3.2.4.2 History

The element was first identified by the British chemist Adair Crawford in 1790 (Edinburgh, Scotland). Actually, he recognized a new heavy mineral (later named strontianite) which differed from the heavy barium sulfate barite. Later in 1809, the metal was first isolated by the British chemist Sir Humphrey Davy (London, UK) using the molten salt electrolysis of the strontium chloride.

3.2.4.3 Natural Occurrence, Minerals, and Ores

The strontium content in the lithosphere is approximately 370 mg.kg^{-1} (ppm wt), but owing to its chemical reactivity, the metal does not occur free in nature. The chief strontium containing minerals are the sulfate **celestite** or **celestine** ($SrSO_4$, orthorhombic) and the carbonate **strontianite** ($SrCO_3$, orthorhombic) but strontium traces can also be found in calcium- and barium-containing minerals. Nevertheless, strontium minerals rarely concentrate in large ore deposits, and the chief ore is only represented by celestite because of strontianite no economically workable deposits are known. However, strontium widely occurs dispersed in seawater, and in igneous rocks as a minor constituent of rock-forming minerals.

3.2.4.4 Preparation and Processing

After mining the raw ore is finely crushed, ground and undergoes a froth flotation beneficiation process to concentrate and separate celestine from by-product minerals (e.g., barite). Afterwards, the concentrated ore is reduced by pyrolysis in a kiln to strontium sulfide (i.e., SrS or black ash). The black ash is then dissolved in pure water, and the aqueous solution is treated with sodium carbonate to precipitate the strontium carbonate crystals. After the strontianite crystals are removed and dried, the strontianite undergoes calcination, evolving carbon dioxide and giving the anhydrous strontium oxide (SrO, strontia). The strontium metal can be obtained either by thermal reduction of the previously obtained strontium oxide with molten aluminum in vacuum or by fused strontium chloride electrolysis.

3.2.4.5 Industrial Uses and Applications

Strontium, owing to the flame color is extensively used in pyrotechnics for fireworks blends and also as a component in red emergency signal flares, and on tracer bullets. It is also used as strontium carbonate for colored glass for cathode ray, television (CRT) tubes. Strontium ferrite is used for manufacturing small magnets for electric motors. Strontium titanate, $SrTiO_3$, owing to its high refractive index and an optical dispersion greater than that of diamond, is used as optical device. Strontium compounds have also been used as deoxidizers of nonferrous metals and alloys. The radionuclide ^{90}Sr with a half-life of 29 years is a by-product of nuclear fission, and easily recovered from spent nuclear fuel. This radionuclide is the best long-lived high-energy pure beta emitter known (max. kinetic energy of the electrons is 546 keV). Hence it is used as a heat generation source converted by the Peltier effect (i.e., thermoelectric) into electrical energy in SNAP (Systems Nuclear Auxiliary Power). Strontium metal is also sometimes used as a getter in vacuum electron tubes.

3.2.5 Barium

3.2.5.1 Description and General Properties

Barium [7440-39-3], with the chemical symbol Ba, the atomic number 56, and the relative atomic mass (atomic weight) 137.327(7) is the fifth alkaline-earth metal, i.e., elements of the group IIA(2) of Mendeleev's periodic chart. Barium is named after the Greek word, baryos, meaning heavy owing to the high density of its main mineral, barite. Barium, resembling calcium chemically, is a relatively dense compared to other alkaline-earth metals, soft, malleable, silvery-white metal like lead when freshly cut, but it readily tarnishes in moist air, beginning yellowish due to the formation of the oxide. It has a body-centered cubic (bcc) crystalline structure. It decomposes in water evolving hydrogen and forming the hydroxide, $Ba(OH)_2$. Barium fines such as powder, dust and turnings, are pyrophoric, i.e., they ignite spontaneously in air. The metal should be kept under mineral oil such as petroleum or other suitable oxygen-free liquids to exclude air. It is also decomposed by ethanol. Naturally occurring barium is a mixture of seven stable isotopes.

3.2.5.2 History

Barium was first identified in the mineral barite by the Swedish chemist Scheele in 1774. The pure element was first prepared by the British chemist Sir Humphrey Davy in 1808 who produced the barium amalgam by electrolyzing an aqueous solution of barium chloride using a liquid mercury cathode. After distilling mercury from the barium amalgam formed, he obtained the pure barium metal.

3.2.5.3 Natural Occurrence, Minerals and Ores

The barium content in the lithosphere is approximately 500 mg.kg^{-1} (ppm wt), but owing to its chemical reactivity, the metal does not occur free in nature. The chief barium-containing minerals are the sulfate **barite** or **heavy spar** (BaSO$_4$, orthorhombic) and the carbonate **witherite** BaCO$_3$, orthorhombic.

3.2.5.4 Processing and Preparation

After mining the barite is finely crushed, ground and undergoes a flotation beneficiation process to concentrate and separate barite from by-product minerals (e.g., quartz). After, the concentrate ore is reduced by pyrolysis in a kiln to barium sulfide (BaS or black ash). The black ash is then dissolved in pure water, and the aqueous solution is treated with sodium carbonate to precipitate the barium carbonate crystals. After the witherite crystals are removed and dried, the carbonate undergoes calcination, evolving carbon dioxide and giving the anhydrous barium oxide, BaO. The barium metal can be obtained either by thermal reduction of the previously obtained barium oxide with molten aluminum in a vacuum or by fused barium chloride electrolysis.

3.2.5.5 Industrial Uses and Applications

The metal is used as a getter in vacuum tubes. Lithopone, a pigment containing barium sulfate and zinc sulfide, has good covering power, and does not darken in the presence of sulfides. The sulfate, as permanent white is also used in paint, in X-ray diagnostic work, and in glassmaking. Barite is extensively used as a weighting agent in oilwell drilling fluids, and is used in making rubber. The carbonate has been used as a rat poison, while the nitrate and chlorate give colors in pyrotechnics. The impure sulfide phosphoresces after exposure to the light. All barium compounds that are water or acid soluble are poisonous.

3.2.6 Radium

3.2.6.1 Description and General Properties

Radium [7440-14-4], with the chemical symbol Ra, the atomic number 88, and the relative atomic mass 226 for the longest-life isotope, is the heaviest element of the alkaline-earth metals. It is named after the Latin, radius (ray) owing its radioactivity. The pure metal is brilliant white when freshly prepared, but tarnishes on exposure to air, probably due to formation of the nitride. It exhibits luminescence, as do its slats; it decomposes in water and is somewhat more volatile than barium. Radium imparts a carmine red color to the flame of a bunsen gas burner. Radium emits alpha, beta, and gamma rays and when mixed with beryllium produces neutrons according to the nuclear reaction $^9\text{Be}(\alpha,n)^{12}\text{C}$. Twenty-five isotopes are known; amongst them, radium-226, the most common isotope, has a half-life of 1620 years. Radium loses about 1% of its activity in 25 years, being transformed into elements of lower atomic weight. Lead-206 is the final product of disintegration. Stored radium should be ventilated to prevent build-up of radon.

3.2.6.2 History

Radium was discovered in 1898 in the uranium ore, pitchblende (uraninite) found in Joachimsthal (North Bohemia, Europe), by the two French physicists and chemists Pierre and

Marie Curie in Paris. Actually, there was about 1 g of radium in 7 tonnes of pitchblende. The pure element was isolated in 1911 by Marie Curie and Debierne by the electrolysis of an aqueous solution of pure radium chloride, $RaCl_2$, employing a mercury cathode; on distillation in an atmosphere of hydrogen this amalgam yielded the pure metal.

3.2.6.3 Natural Occurrence

Radium is an extremely rare element in the Earth's crust with an abundance of 90 ng.kg^{-1} (ppt wt). However, it is always present in uranium minerals and ores, and could be extracted, if desired, from the extensive wastes of uranium processing. Today, the carnotite sands of Colorado in the United States furnish some radium, but richer ore deposits are found in the Republic of Zaire. Other large uranium deposits are located in Canada (e.g., Ontario), in the United States (e.g., New Mexico, Utah), Australia, and elsewhere.

3.2.6.4 Industrial Preparation

Radium is obtained commercially as the bromide and chloride; it is doubtful if any appreciable stock of the isolated element now exists.

3.2.6.5 Industrial Uses and Applications

At the beginning of the century, radium was used to establish the standard prototype of the curie. Actually, one curie is equal exactly to the radioactivity of a source which has the same radioactivity as one gram of the radionuclide radium-226 in secular equilibrium with its derivative radon-222 (or emanation).[22] In spite of the new mandatory SI unit of radioactivity, the becquerel, with the symbol Bq, the curie, Ci, is sometimes still in use (1 Ci = 37 GBq). Twenty-five isotopes are now known; amongst them, radium-226, the most common isotope, has a half-life of 1620 years. This is purged from the radium and sealed in minute tubes, which are used in the treatment of cancer and other diseases. Radium is used in the production of self-luminous paints, neutron sources, and in medicine for the treatment of disease. Some of the more recently discovered radioisotopes, such as ^{60}Co or ^{137}Cs, are now being used in place of radium. Some of these sources are much more powerful, and others are safer to use. Inhalation, injection, or body exposure to radium can cause cancer and other body disorders. The maximum permissible burden in the total body for ^{226}Ra is 7400 Bq.

3.3 Refractory and Reactive Metals (RMs)

3.3.1 General Overview

The following subgroups of inner transition metals of Mendeleev's periodic chart such as IVB(4) (i.e., Ti, Zr, and Hf), VB(5) (i.e., V, Nb, and Ta) and VIB(6) (i.e., Cr, Mo, and W) are usually classified as high-melting-point or refractory metals. Sometimes, other elements such as rhenium, iridium, and osmium are also taken into account in this group but their use is still negligible industrially. On the other hand, properties of iridium and osmium will be discussed in Section 3.5 which deals with the six platinum group metals (PGMs).

3.3.1.1 Common Properties

As a general rule, these metals share the following properties in common: (i) they are **refractory metals** where refractory means that their melting point (i.e., temperature of fusion) is higher than the setting point of pure iron, 1539°C. Actually, their melting point ranges from 1668°C for titanium to 3410°C for tungsten. (ii) They exhibit a strong chemical reactivity, i.e., they combine strongly with oxygen, nitrogen, carbon, and other metals and non-metals outside their group to form highly stable compounds. For this reason, sometimes particularly metals of group IVB(4) are named **reactive metals**. (iii) Many of their mechanical properties are extremely sensitive to relatively minute amounts of interstitial atomic species (e.g., H, C, O, and N). (iv) In oxygen-containing environments such as air and moisture, owing to their high chemical reactivity with oxygen, they spontaneously form a tenacious, strongly adherent, and protective oxide film. This impervious passivating barrier exhibits excellent dielectric and insulating properties and protects very efficiently the underlying base metal from the corrosion. Moreover, for this reason, all the refractory metals exhibit a strong valve action (VA) property which can be defined as follows: when acting as cathodes these metals allow electric current to pass but when acting as anodes they prevent passage of current owing to a rapid build-up of this insulating oxide layer. Since the beginning of the century, refractory metals have been used for fabricating special devices such as high-temperature crucibles, X-ray targets, wires for lamp filaments, electron tube grids, heating elements, and electrical contacts, However, owing to the industrialization of processing, refractory metals are now also used for engineering special equipment and devices in a wide range of industrial areas such as the nuclear power industry and high-energy physics, pharmaceutical industry, chemical process industries, foodstuffs industries, marine engineering, and also civil engineering. Moreover, each refractory metal is assigned to a particular application according to its particular physical properties and chemical inertness.

3.3.1.2 Cleaning, Descaling, Pickling, and Etching

Owing to their tenacious and strongly adherent protective oxide films, refractory metals should be compulsorily etched, when interesting properties of the base metal should be enhanced. Actually, joining such as welding and brazing, plating, coating or electrical contact is difficult or impossible to achieve when a thick passivating film is present on the surface of the metal. In order to remove efficiently this passivating film, there exist several methods such as: (i) mechanical (e.g., blasting, or grinding), (ii) chemical (e.g., etching, descaling, and pickling) by alkaline or acid solutions, or (iii) electrochemical (e.g., electropolishing). For instance, some efficient etching procedures for the main refractory metals are reported in Table 3.15.

3.3.1.3 Machining of Pure Reactive and Refractory Metals

See **Table 3.16** (page 116).

3.3.1.4 Pyrophoricity of Refractory Metals

Owing to their strong chemical reactivity with oxygen to form highly stable oxides (e.g., see their Gibbs free energy as a function of temperature in Ellingham's diagram), reactive and refractory metals in finely divided form exhibit highly hazardous properties. Amongst these, pyrophoricity is the main issue encountered when handling reactive and refractory metal powders. Hence, great care should be taken when handling powder, hot sponge, turnings or other divided form of the raw metals. Appropriate extinguishers should be used in order to fight

Table 3.15 Selected etching and descaling procedures for reactive and refractory metals

Refractory metals	Type	Operating conditions
Group IV(4): titanium	Acid cleaning	Boiling in oxalic acid solution (10wt% $H_2C_2O_4$), hydrochloric acid, or diluted sulfuric acid (30wt% H_2SO_4) for 10 min (discard the solution after each operation due to the interference of inhibiting titanium(IV) cations)
Group IVB(4): zirconium and hafnium	Acid cleaning	Etching bath at 25°C (ratio HNO_3/HF 10:1) (during 2 s): 25–59vol% HNO_3 ($d = 1.42$) 2–7vol% HF ($d = 1.16$) remainder deionized water Stopping bath (during 5 s): 70vol% HNO_3 ($d = 1.42$) 30vol% deionized water Rinsing bath (during 1 min): deionized water
Group VB(5): niobium and tantalum	Acid cleaning	Etching bath at 50 to 65°C: 40–60vol% HNO_3 ($d = 1.42$) 10–30vol% HF ($d = 1.16$) remainder deionized water Rinsing bath: deionized water
Group VIB(6): molybdenum and tungsten	Acid cleaning	Etching bath at 50 to 65°C: 50–70vol% HNO_3 ($d = 1.42$) 10–20vol% HF ($d = 1.16$) remainder deionized water Stopping bath: 750–900 $g.dm^{-3}$ HNO_3 ($d = 1.42$) 100–200 $g.dm^{-3}$ H_2SO_4 ($d = 1.84$) Rinsing bath: deionized water

Important note: acid cleaning is preferred to alkaline cleaning to avoid hydrogen embrittlement of the refractory metals.

fires. Some parameters of combustion of dusts of refractory metals are listed in Table 3.17. More detailed information related to pyrophoricity of both reactive and refractory metals with a particular emphasis of titanium, zirconium, and tantalum powders used in chemical plant equipment are given by McIntyre and Dillon.[23]

3.3.1.5 Physical Properties and Corrosion Resistance

Selected properties are given in Table 3.18 (pages 117–119), and a comparison of corrosion rates for common chemicals is given in Table 3.19 (page 120).

3.3.2 Titanium and Titanium Alloys

3.3.2.1 Description and General Properties

Titanium [7440-32-6] is the first metallic element in subgroup IVB(4) of Mendeleev's periodic chart with the atomic number 22 and a relative atomic mass (atomic weight) of 47.867. It has the chemical symbol, Ti, and is named after the latin word, Titans, the names of the powerful mythological first sons of the Earth goddess. Titanium is a hard, lustrous, gray silvery and light metal with a low density 4540 $kg.m^{-3}$, which falls between those of aluminum and iron and give very attractive strength-to-weight ratios. For instance, the yield strength 0.2% proof of titanium and titanium alloys ranges between 470 to 1800 MPa. However, highly pure titanium (e.g., obtained by molten fluoride electrolysis or from the iodide processes) is ductile owing to the

Table 3.16 Machining characteristics of pure reactive and refractory metals

Pure reactive and refractory metals		Ti	Zr	Hf	Nb	Ta	Mo	W
Tool type	High-speed steel	T15, M42	None	None	S9, S11, M2, T15, M42	S9, S11, M2 T15, M42	None	None
	Cemented carbide	C2, C3	C1, C2, C10, C11	C1, C2, C10, C11	M20, C2, K20	M20, C2,	K20, M20, C2	K01,C4
Tool shape (°)	Approach angle	None	None	None	15–20	30	30	None
	Side rake	0–15	12–15	10	30–35	5	5	0
	Side and end clearances	5–7	10	10	5	5	5	8–10
	Plane relief angle	5–7	None	None	15–20	5	7	7
	Nose radius (mil)	20–30	31.25	31.25	5–30	5–30	5–30	90
Cutting speed (sfm)	High-speed steel tools	95–250	80–100	150–200	80–120	40–60	None	None
	Carbide cutting tools	120–560	100–150	200–250	300–350	60–110	290–325	80–130
Feed (mil per rev.)	Rough	5–15	3–20	3–20	8–120	8–150	8–150	
	Finish	5	1–3	1–3	5	5	5–100	7–100
	Depth of cut (mil)	40–300	40–250	50–100	30–125	40–150	40–150	40–150

Conversion factors: 1 mil $= 25.4$ μm; 1 surface foot per minute (sfm) $= 5.08$ mm s^{-1}.

Table 3.17 Pyrophoric properties of reactive and refractory metals

Refractory metal powder		Ignition temperature of the dust cloud (T_{ign}, °C)	Minimum ignition energy (E_{ign}, mJ)	Specific enthalpy of combustion for the oxide (ΔH, kJ.kg^{-1})	Minimum explosive concentration (c, g.m^{-3})	Maximum explosion gage pressure (P_g, kPa)	Maximum rate of pressure rise (kPa.s^{-1})	Final oxygen volume fraction (vol%)
Group IVB(4)	Ti	460	10	n.a.	45.05	552	44,500	0
	Zr	20	5	5940	45.05	448	40,000	0
	Hf	n.a.	n.a.	5772	n.a.	n.a.	n.a.	0
Group VB(5)	V (86wt%)	500	60	n.a.	220.25	331	2670	13
	Nb	n.a.	n.a.	n.a.	n.a.	n.a.	n.a.	n.a.
	Ta	630	120	n.a.	200.23	352	16,500	n.a.
Group VIB(6)	Cr	580	140	n.a.	230.26	386	22,241	14
	Mo	720	n.a.	7580	n.a.	n.a.	n.a.	n.a.
	W	>950	n.a.	n.a.	n.a.	n.a.	n.a.	n.a.

Table 3.18 Selected properties of reactive and refractory metals

Properties (at 298.15 K, unless otherwise specified)		Titanium	Zirconium	Hafnium	Niobium	Tantalum	Molybdenum	Tungsten
General properties	Chemical symbol (IUPAC)	Ti	Zr	Hf	Nb	Ta	Mo	W
	Chemical abstract registered number [CARN]	[7440-32-6]	[7440-67-7]	[7440-58-6]	[7440-03-1]	[7440-25-7]	[7439-98-7]	[7440-33-7]
	Unified numbering system [UNS]	[R50250]	[R60702]	[R02001]	[R04210]	[R05200]	{R03600}	[R07004]
Natural occurrence and economics	Earth's crust abundance (mg.kg^{-1})	5600	190	5.3	20–24	2	1.5	1
	Seawater abundance (mg.kg^{-1})	4.8×10^{-4}	9×10^{-6}	7×10^{-6}	9×10^{-7}	2×10^{-6}	0.01	9.2×10^{-6}
	World estimated reserves (R, 10^3 kg)	650×10^6	21×10^6	n.a.	n.a.	n.a.	5×10^6	2.80×10^6
	World production of metal in 1998 (P, 10^3 kg.yr^{-1})	98,000	40,000	50–90	15,000	840–900	170,000	45,100
	Cost of pure metal 1998 (C, $US.kg^{-1}) (purity, wt%)	48.50 (99.8)	190 (99.8)	1614 (9.8)	221 (99.85)	461 (99.5)	215 (99.95)	620 (99.95)
Atomic properties	Atomic number (Z)	22	40	72	41	73	42	74
	Relative atomic mass A_r (^{12}C = 12.000)^a	47.867(1)	91.224(2)	178.49(2)	92.90638(2)	180.9479(1)	95.94(1)	183.84(1)
	Electronic configuration	[Ar] $3d^24s^2$	[Kr} $4d^25s^2$	[Xe] $4f^{14}5d^26s^2$	[Kr] $4d^45s^1$	{Xe] $4f^{14}5d^36s^2$	[Kr] $4d^55s^1$	[Xe] $4f^{14}5d^46s^2$
	Fundamental ground state	$^3F_2$	$^3F_2$	$^3F_2$	$^6D_{1/2}$	$^4F_{3/2}$	$^7S_3$	$^5D_0$
	Electronegativity χ_a (Pauling)	1.54	1.33	1.3	1.6	1.5	2.16	2.36
	Electron work function (W_S, eV)	4.17	4.10	3.53	4.01	4.12	3.9	4.40
	X-ray absorption coefficient CuK$_{a,1,2}$ ((μ/ρ), cm^2.g^{-1})	208	143	159	153	166	162	172
Nuclear properties	Thermal neutron cross-section (σ_n, 10^{-28}m^2)	6.1	0.184	104	1.15	20.5	2.60	18.4
	Isotopic mass range	38–58	80–107	154–185	82–110	156–187	84–113	158–190
	Natural isotopes (including isomers)	21	32	41	44	39	33	37
Crystallographic properties	Crystal structure (phase α)	hcp	hcp	hcp	bcc	bcc	bcc	bcc
	Strukturbericht designation	A3	A3	A3	A2	A2	A2	A2
	Space group (Hermann–Mauguin)	P6$_3$/mmc	P6$_3$/mmc	P6$_3$/mmc	Im3m	Im3m	Im3m	Im3m
	Pearson's notation	hP2	hP2	hP2	cI2	cI2	cI2	cI2
	Lattice parameters (pm) (293.15 K)	$a =$ 295.030 $c =$ 468.312 $c/a =$ 1.5873	$a =$ 323.115 $c =$ 514.770 $c/a =$ 1.5931	$a =$ 319.460 $c =$ 505.100 $c/a =$ 1.5811	$a =$ 329.860	$a =$ 330.290	$a =$ 314.680	$a =$ 316.522
	Miller's indices of slip plane at room temperature (hkl)	n.a.	n.a.	n.a.	110	110	110	n.a.
	Latent heat transition (L_t kJ.kg^{-1})	69.8	42.2	38.7	nil	nil	nil	n.a.
	Phase transition temperature α–β (T, K)	1155 (882°C)	1136 (863°C)	2033 (1760°C)	nil	nil	nil	903 (630°C)

Table 3.18 (continued)

Properties (at 298.15 K, unless otherwise specified)	Titanium	Zirconium	Hafnium	Niobium	Tantalum	Molybdenum	Tungsten
Density (ρ, kg.m^{-3}) (293.15 K)	4540	6506	13,310	8570	16,654	10,220	19,300
Young's or elastic modulus (E, GPa) (polycrystalline)	120.2	97.1	137–141	104.9	185.7	324.8	411
Coulomb or shear modulus (G, GPa) (polycrystalline)	45.6	36.5	56	37.5	69.2	125.6	160.6
Bulk or compression modulus (K, GPa) (polycrystalline)	108.4	89.8	109	170.3	196.3	261.2	311
Mohs hardness (HM)	6.0	5.0	5.0	6.0	n.a.	n.a.	n.a.
Brinell hardness (HB)	1810	638–687	1680–1800	736	411–1230	1370–1815	3540
Vickers hardness (HV) (hardened)	65–74 (400)	80 (110)	150–180	60–80 (160)	90 (200)	200 (250)	350 (500)
Yield strength proof 0.2% (σ_{YS}, MPa)	140	207	230	105	172	345	550
Ultimate tensile strength (σ_{UTS}, MPa)	235	379	445	195 (585)	285 (650)	435 (540)	620 (1920)
Elongation (Z, %)	54	16	25	26	40	25	2
Creep strength (MPa) (hardened)	550	250–310	240	240 (550)	310–380 (705)	415–450 (550)	n.a.
Longitudinal velocity of sound (V_s, m.s^{-1})	4970	4262	3000	3480	3400	6370	5174
Static friction coefficient (vs air)	0.47	0.63	n.a.	n.a.	n.a.	0.46	0.51
Poisson ratio ν (dimensionless)	0.340	0.345–0.361	0.260	0.397	0.342	0.293	0.280
Melting point (mp, K)	1941 (1668°C)	2125 (1852°C)	2506 (2233°C)	2503 (2230°C)	3269 (2996°C)	2883 (2610°C)	3680 (3407°C)
Boiling point (bp, K)	3560 (3287°C)	4650 (4377°C)	4876 (4603°C)	5015 (4742°C)	5698 (5427°C)	5883 (5560°C)	6000 (5727°C)
Thermal conductivity (k, W.m^{-1}.K^{-1})	21.9	22.6	22.3	53.7	57.5	139	174
Coefficient of thermal linear expansion (α, 10^{-6} K^{-1})	8.35	5.78	5.90	7.1	6.6	5.43	4.59
Specific heat capacity (c_p, J.kg^{-1}.K^{-1})	536.30	250.57	145.00	265.86	150.70	250.78	132.00
Enthalpy molar fusion (ΔH_{fus}, kJ.mol^{-1}) (Δh_{fus}, kJ.kg^{-1})	20.9 (437)	23.0 (252)	255 (143)	27.2 (290)	36.6 (174)	27.6 (288)	35.2 (191)
Enthalpy molar vaporization (ΔH_{vap}, kJ.mol^{-1}) (Δh_{vap}, kJ.kg^{-1})	428.9 (8958)	581 (6376)	661.1 (3704)	689.9 (7490)	732.8 (4270)	594.1 (6192)	799.15 (4345)
Enthalpy molar sublimation (ΔH_{sub}, kJ.mol^{-1})	469.4	610.08	618.46	726.20	778.6	571.25 (3200)	860.37 (4680)
Enthalpy molar combustion (ΔH_{comb}, kJ.mol^{-1})	–944.1	–1101.3	–1113.7	–1900.8	–2047.3	–746.1	–838.6

Mechanical properties (annealed) spans the rows from Density through Poisson ratio.

Thermal properties (298.15 K) spans the rows from Melting point through Enthalpy molar combustion.

Table 3.18 (continued)

Properties (at 298.15 K, unless otherwise specified)		Titanium	Zirconium	Hafnium	Niobium	Tantalum	Molybdenum	Tungsten
Electrical and electrochemical properties	Superconductive critical Temp. (T_c, K)	0.37–0.56	0.61	0.128	9.25	4.47	0.915	0.0154
	Electrical resistivity (ρ, $\mu\Omega.cm^{-1}$)	42.00	42.10	35.5	15.22	13.15	5.70	5.65
	Temperature coefficient of resistivity (0–100°C) (10^{-3} K^{-1})	0.0038	0.0044	0.0044	0.002633	0.003820	0.00435	0.00480
	Hall coefficient (R_H, nV.m.A^{-1}.T^{-1})	n.a.	n.a.	n.a.	0.09	0.095	n.a.	n.a.
	Thermoelectronic emission constant (A, kA.m^{-2}. K^{-2})	n.a.	3300	220	1200	1200	550	600
	Thermoelectric potential versus platinum (mV)	n.a.	+1.17	n.a.	n.a.	+0.33	+1.45	+1.12
	Nernst standard electrode potential (E, VSHE)	−1.21	−1.55	−1.70	−1.10	−0.81	n.a.	n.a.
	Hydrogen overvoltage (η, mV) ([H^+] = 1M, 200 A.m^{-2})	754	n.a.	n.a.	803	n.a.	343	363
Magnetic and optical properties	Mass magnetic susceptibility (χ_m, 10^{-8} m^3.kg^{-1})	+4.21	+1.22	+0.53	+2.81	+1.56	+1.17	+0.39
	Spectral emissivity (650 nm)	n.a.	n.a.	n.a.	n.a.	0.49	0.37	0.45
	Reflective index under normal incidence (650 nm)	0.540	0.500	0.489	0.505	0.460	0.576	0.518

[a] Standard atomic masses from: Atomic weights of the elements 1995. Pure Appl Chem 68 (1996): 2339.

very low concentration of interstitial alloying elements (i.e., H, C, O, N). The metal has two allotropes; the alpha phase corresponds to a low temperature crystal structure which has a hexagonal close-packed (hcp) crystal lattice. When the metal reaches the transition point of 882°C, the crystallographic structure changes slowly to the high-temperature beta phase which is body-centered cubic (bcc). Like many other elements, this behavior is strongly influenced (i.e., raised or decreased) by the type and amount of impurities or alloying elements. Moreover, the addition of alloying elements splits this temperature into two temperatures: (i) the alpha transus point below which the alloy contains only alpha-phase crystals, and (ii) the beta transus point above which all the crystals are bcc. Between these two temperatures, both alpha and beta phases coexist. On the other hand, from a thermal point of view, titanium is the first member of the refractory metals owing to its high melting point of 1668°C. Industrially, titanium is the world's sixth most abundant structural metal after Fe, Al, Cu, Zn, and Mg.

Corrosion resistance – pure titanium metal, although highly chemically reactive (i.e., electropositive), readily forms a stable, impervious, and highly adherent protective oxide layer of rutile (TiO_2) when put in contact with oxygen-containing media, such as air, or moisture. At room temperature, a clean titanium surface exposed to air immediately forms a passivation oxide film which is 1.2–1.6 nm thick. After 70 days it is about 5 nm and it continues to grow slowly reaching a thickness of 8–9 nm in 545 days and 25 nm in 4 years. The film growth is accelerated under strongly oxidizing conditions, such as heating in air, anodic polarization in an electrolyte or exposure to oxidizing agents such as HNO_3, and CrO_3. The composition of this film varies from stoichiometric rutile TiO_2 at the surface to Ti_2O_3 with TiO at the metal interface. However, strong oxidizing conditions promote chiefly the formation of rutile TiO_2 film. This film is transparent in its normal thin configuration and not detectable by visual

Table 3.19 Comparison of corrosion rates (μm/year) of reactive and refractory metals for common corrosive chemicals[a,b]

Corrosive chemical	Ti	Zr[c]	Hf	Nb[d]	Ta[e]	Ir
HCl 37wt% (fuming) at 60°C	Poor (>1250)	Excellent (<25)	Excellent (<25)	Poor (250)	Excellent (<2.54)	Excellent (<2.54)
H$_2$SO$_4$ 80wt% (boiling)	Poor (>1250)	Poor (>500)	Poor (>500)	Poor (5000)	Excellent (<2.54)	Excellent (<2.54)
HNO$_3$ 70wt% (boiling)	Good (<125)	Excellent (<25)	Excellent (<25)	Excellent (<25)	Excellent (<2.54)	Excellent (<2.54)
KOH 50wt% (boiling)	Poor (2700)	Excellent (<25)	Excellent (<25)	Poor (>300)	Poor (>300)	Excellent (<2.54)
H$_2$O$_2$ 30wt% (boiling)	Poor (n.a.)	Excellent (nil)	Excellent (nil)	Poor (500)	Excellent (<2.54)	Excellent (<2.54)
H$_2$C$_2$O$_4$ 10wt% (boiling)	Poor (>1250)	Excellent (<25)	Excellent (<25)	Poor (1250)	Excellent (<2.54)	Excellent (<2.54)
Aqua regia (3 HCl: 1 HNO$_3$)	Excellent (0)	Poor (>1250)	Poor (>1250)	Excellent (<25)	Excellent (<2.54)	Excellent (<2.54)
HF 10% at room temperature	Poor (>1250)	Poor (>1250)	Poor (>1250)	Poor (>1250)	Poor (>1250)	Excellent (<2.54)
Strong mineral acids (with 200 ppm F)	Poor (>1250)	Poor (>1250)	Poor (>1250)	Poor (>500)	Poor (>500)	Excellent (<2.54)

To convert μm. year^{-1} to mpy divide by 25.4, and note that iridium is used as a comparative example.

Source: Cardarelli F, Taxil P, Savall A (1996) Tantalum protective thin coating techniques for the chemical process industry: molten salts electrocoating as a new alternative. Int J Refract Metals Hard Mater 14: 365–881.
[a] Yau TL, Bird KW (1995) Manage corrosion with zirconium. Chem Eng Progr 91: 42–46.
[b] Wortly JPA (1963) Corrosion resistance of titanium, zirconium, and tantalum. Corrosion Prevent Control 10: 21–26.
[c] Zircadyne® corrosion data. Teledyne Wah Chang Albany, OR, USA (1995).
[d] Bishop CR (1963) Corrosion test at elevated temperatures and pressures. Corrosion 19: 308t–314t.
[e] Technical Note on KBI® tantalum. Cabot Performance Material Inc., Boyertown, FL (1995).

means but when grown, anodically formed optical interferences give a common dark blue to black color. Hence, as a general rule the corrosion resistance of titanium is identical to the corrosion resistance of the oxide film which gives to underlying metal an excellent corrosion resistance in many harsh environments, including oxidizing acids and chlorides, and good elevated temperature properties up to about 440°C in some cases. The oxide film on titanium is very stable and titanium immunity is only lost when the metal is in contact with reagents which prevent or slow down the oxide layer formation or damage this natural barrier. For instance, hydrogen fluoride, dilute hydrofluoric acid and fluoride ions, hot strong mineral acids such as HCl H$_3$PO$_4$, H$_2$SO$_4$ or strongly reducing or complexing reagents such as hot oxalic acid readily dissolve titanium metal. Nevertheless, in many environments titanium is capable of self-healing this film almost instantly in any media where a trace of moisture or oxygen is present. Therefore, anhydrous conditions in the absence of a source of oxygen should be avoided since the protective film may not be regenerated if damaged. Furthermore, titanium is also susceptible to hydrogen pick-up and embrittlement and is hence attacked by nascent hydrogen produced by galvanic coupling. Therefore it is necessary to protect the metal by anodic protection in order to prevent this phenomenon. Titanium and its alloys provide excellent resistance to general localized attack under most oxidizing, neutral and inhibited reducing conditions. On the other hand, titanium and alloys have excellent corrosion resistance and thus good dimensional stability in wet chlorine, media containing chloride anions and in oxidizing acidic brines. However, titanium reacts readily with oxygen at red heat and with chlorine above $550°C$. Titanium in a divided form such as powder, or hot sponge is hazardous owing to its high pyrophoric character and hence must be stored under an inert gas. Moreover, it is the only element that burns in pure nitrogen.

Cost (1998) – titanium sponge is roughly priced at 8.81 $US.kg^{-1} (4.00 $US.lb^{-1}) and commercially pure titanium (e.g., ASTM grades 2 and 4) are priced at 48.50 $US.kg^{-1} (22 $US.lb^{-1}) and the cost reaches 550 $US.kg^{-1} (249.48 $US.lb^{-1}) for highly pure titanium with extra-low interstitial content (ELI grade). Titanium alloys such as the well-known Ti-6Al-4V (ASTM grade 5) is priced at 50 $US.kg^{-1} (22.70 $US.lb^{-1}). According to Roskill Information Services Ltd, titanium metal production in 1997 reached 98,000 tonnes per year.

3.3.2.2 History

It was almost two hundred years ago that titanium was first isolated and named in 1791 by Rev William Gregor (1761–1817), Bishop of Creed (Cornwall, UK) and amateur chemist. After examination of the minerals found in alluvion sands from the Helford river, he separated with a permanent magnet a dense black mineral (ilmenite). After dissolution of the ilmenite crystals in concentrated boiling hydrochloric acid in order to remove iron, he obtained an insoluble powder which was the first impure titanium dioxide. Independently, in 1795 in Berlin, the German chemist M.H. Klaporth discovered the same oxide prepared from rutile ores. Titanium metal was first isolated in an impure form by J.J. Berzelius in 1825 in Stockholm (Sweden) and later by Nilson and Pettersson in 1887. However, pure titanium was first prepared by metallothermic reduction in 1910 by Hunter, by heating $TiCl_4$ with sodium in a steel bomb. This slow progress in the production of the pure metal is the result of the strong chemical reactivity of the element for oxygen, which requires complex and energy-intensive processes to win the pure element from highly stable compounds. From 1930 and 1950, the only method allowing the metal to be processed into useful shapes was the powder-metallurgy technique applied to titanium sponge. Hence, it was not until 1947 with the development by Dr Wilhem Justin Kroll of the metallothermic reduction of $TiCl_4$ with magnesium at the US Bureau of Mines that it began to be available as a commercial material. Moreover, improvement of melting methods was a compulsory step to overcome in order to provide homogeneous structural material. In the 1950s, the first melting techniques were resistance, induction and tungsten arc heating, but the development of skull melting by the US Bureau of Mines allowed the production of large quantities of low-interstitial titanium as ingots or complex shapes. Hence, the industry as we know it today is over 40 years old. This was initially stimulated by aircraft applications. Although the aerospace industry still provides the major market where titanium is most commonly associated with jet engines and airframes, titanium and titanium alloys are finding increasingly widespread use in other industries, such as desalination plants and chlorine production. The most recent media attention has been given to equipment in the chemical process industries and to fittings for prosthetic devices and an artificial heart.

3.3.2.3 Natural Occurrence, Minerals, and Ores

The lithospheric relative abundance of titanium which comprises 5.65 g.kg^{-1} (5650 ppm wt) of Earth's crust makes it the ninth most abundant element and the seventh most abundant metallic element on Earth. Titanium is also present in seawater (0.00048 ppm wt), ash coal, in plants, and in the human body. Owing to its strong chemical reactivity with oxygen, titanium never occurs as a native element and it is chiefly present as an oxide in minerals, such as **rutile** (TiO_2, tetragonal), **brookite** (TiO_2, orthorhombic), **anatase** (TiO_2, tetragonal), **ilmenite** ($FeTiO_3$, trigonal), **sphene** or **titanite** (CaTi [SiO$_4$|(O, OH, F)], monoclinic), **perowskite** (CaTiO$_3$, tetragonal), and **titanomagnetite** ((Ti, Fe)FeO$_4$, cubic). Moreover, titanium is also present as complex titanates. The metal is extracted commercially from the three chief ore minerals: **rutile**, **leucoxene** and **ilmenite**, the first, which is by far the most common being richest in titanium content. Actually, these ores are ubiquitous minerals of igneous (e.g., syenites) and metamorphic rocks, but they are difficult to extract because segregation is rarely important. Therefore, the main ore deposits are found in quaternary sedimentary rocks such as alluvion deposits (e.g., river, sea beaches) known as black sand beach. Titanium ore deposits are found in North America, South America, Europe, Africa, CIS, India, China and Australia in the forms of ilmenite, rutile and other ores. In particular, US domestic deposits are found in California, Florida, Tennessee, and New York. Australia, Canada, Norway, and South Africa are the major producing countries of titanium ores and concentrates.

3.3.2.4 Mining and Mineral Dressing

The black sand deposits usually contain about 4wt% heavy minerals such as rutile, ilmenite, leucoxene, monazite, xenotime, and zircon. Other heavy minerals such as staurolite, tourmaline, sillimanite, corundum, and magnetite may be recovered as the local situation warrants. Initially sand is mined and excavated from beach deposits using front-end loaders, or sand dredges. Typically, the overburden is bulldozed away, the excavation is flooded, and the raw sand is handled by a floating sand dredge capable of dredging to a depth of 18 m. The material is broken up by a cutter head to the bottom of the deposit and the sand slurry is pumped to a wet-mill concentrator mounted on a floating barge behind the dredge. Secondly, sand is wet concentrated by the classical mineral dressing operations using screens, Reichert cones, spirals, and cyclones in order to remove the coarse sand, slimes, and light density sands to produce a 40wt% heavy mineral concentrate. The tailings are returned to the back end of the excavation and used for rehabilitation of worked-out areas. The concentrate is then dried and iron oxide and other surface coatings are removed by means of various operation units such as dense media or gravity, magnetic, and electrostatic separation.

3.3.2.5 Processing and Industrial Preparation

The classical processes involved in the recovery of metals from their oxides are not suited for preparing the metals of the group IVB(4) (i.e., Ti, Zr, and Hf), owing to two main reasons: (i) the carbothermic reduction of the oxides by carbon leads to the formation of highly stable refractory carbides, and (ii) the metallothermic reduction by the molten alkali metals (e.g., Na) or alkaline-earth metals (e.g., Ca, Mg) does not completely remove the oxygen from the ores. Furthermore, at high temperature these three metals are too reactive with air and hence, the process should be achieved under an inert atmosphere in order to prevent some unsuitable reactions (e.g., nitriding and oxidation). Historically, the first industrial process was invented by Wilhem Justin Kroll in Luxembourg in 1932. The titanium sponge was prepared by metallothermic reduction of the titanium tetrachloride (TiCl$_4$) in the gas phase using molten calcium metal as reducing agent. Later, in 1940 the reduction was achieved with magnesium and sodium.[24] Today, the major commercial extraction process involves treatment of the ore with chlorine gas to produce titanium tetrachloride (TiCl$_4$) which is purified and reduced to a metallic titanium sponge by reaction with magnesium or sodium (e.g., Kroll process, see below).

Kroll process – the most widely used means of winning the metal from the ore is the batch metallothermic reduction with magnesium as a reducing agent. To produce titanium, the raw ore, usually ilmenite (FeTiO$_3$) or rutile (TiO$_2$) is converted to titanium sponge in two distinct steps. Firstly, the ground ore such as FeTiO$_3$ or TiO$_2$ is mixed with carbon powder, coke or tar and charged in a chlorinator vessel. The reactor is heated up to 900°C and chlorine gas is passed through the fluidized bed charge. The titanium ore reacts with the chlorine to form volatile titanium tetrachloride ($TiCl_4$) and the oxygen is removed as CO and CO_2. The reaction scheme is the following:

$$2FeTiO_3 + 7Cl_2 + 6C \xrightarrow{900°C} 2TiCl_4 + 2FeCl_3 + 6CO.$$

The resultant crude $TiCl_4$ called "Ticle" by titanium producers, is a volatile (bp 136°C), colorless liquid which is purified by continuous fractional distillation (i.e., rectification) in order to separate the iron chloride. The pure gaseous titanium tetrachloride is then reduced with molten magnesium under an inert atmosphere of argon according to the chemical reaction:

$$TiCl_4(g) + 2Mg(l) \xrightarrow{950-1150°C} Ti(s) + 2MgCl_2(l).$$

3
Less
Common
Nonferrous
Metals

The chemical reactor in which the reduction reaction occurs is a clean dry retort made of stainless steel AlSI 304 grade or low-carbon steel. The reactor bottom is filled with pure magnesium ingots and with liquid titanium tetrachloride. In order to decrease the amount of intermediate titanium chloride species (i.e., $TiCl_2$ and $TiCl_3$) the total amount of magnesium is in excess of 25% of the stoichiometric quantity. Owing to the exothermic reaction route the process is controlled by the feed of titanium tetrachloride at 820°C. The molten magnesium chloride produced is removed continuously from the reactor bottom. It is then recycled by molten salt electrolysis for producing magnesium metal and the chlorine is reused for the ore chlorination. The preparation follows a batch process and classically produces 1 to 7 tonnes of titanium sponge per turn. The titanium raw sponge contains large amounts of impurities (about 30wt%). The magnesium chloride and magnesium metal are removed by leaching with a mixture of diluted hydrochloric and nitric acids. Hence, the cleaning process is energy-intensive and produces a large waste rejection per unit mass of titanium sponge produced. The leached and dried titanium sponge is then crushed and rinsed with aqua regia mixture ($HCl + 3HNO_3$) before being formed into ingots by melting under argon or a vacuum. Melting is the second step; titanium is converted from sponge to solid ingot by first blending the crushed sponge with the desired alloying elements (and reclaimed scrap) to insure uniformity of composition, and then pressing into briquets which are welded together to form an electrode. The electrode is melted in a consumable electrode vacuum arc furnace where an arc is struck between the electrode and a layer of titanium in a water-cooled copper crucible. The molten titanium on the outer surface solidifies on contact with the cold wall, forming a shell or skull to contain the molten pool (i.e., containerless melting). The ingot is not poured, but solidifies under vacuum in the melting furnace. Several meltings may be necessary to achieve a homogeneous ingot; classically, a second or sometimes a third melting operation is applied. Then the ingot is ready for processing into useful shapes typically by forging followed by rolling. Although this process is expensive, it has allowed the production of high-purity titanium needed in some particular applications. Titanium sponge is currently produced by the Kroll process in China, France, Japan, Kazaksthan, CIS, and in the United States, while titanium ingot is produced in France, Germany, Japan, Russia, the UK and the US.

Hunter process – this metallothermic reduction process, developed in 1910, is similar to the Kroll process in which the reducing agent, i.e., magnesium, is replaced by sodium. These two processes are chemically similar, but differ in operating details. Actually, the overall basic reaction is as follows:

$$4Na(s) + TiCl_4(l) \rightarrow Ti(s) + 4NaCl(s).$$

However, in some Hunter processes, the total reduction is split into two mechanisms. The initial step involves only a partial reduction of Ti(IV) to TI(II), followed by a second step for complete reduction of Ti(IV) to titanium metal and formation of sodium chloride. In order to insure complete reduction and to provide a coarse sponge particle by sintering, the final temperature is high as 1400°C. By contrast with the Kroll process, the Hunter process offers the following main advantages: (i) the frozen sodium chloride is easily removed by hot water leaching, (ii) the titanium sponge is easy to grind, while sponge produced by the Kroll process is harder, (iii) the titanium metal is less contaminated by impurities, and finally (iv) this winning process is 30% less energy demanding. However, it exhibits also some important drawbacks, due to the monovalent sodium; in theory, a tonne of titanium tetrachloride is required to reduce 485 kg of sodium metal, while in the Kroll process it requires only 256 kg of magnesium metal. Moreover, at the end of reduction the retort is mainly filled with sodium chloride and titanium is highly dispersed.

Refining – for particular applications requiring a high-purity titanium (i.e., above 99.99wt% Ti), the pure titanium can be further refined by the Van Aerkel–De Boer or iodine process, electrotransport, or by zone refining of ingots.

3.3.2.6 Commercially Pure Titanium

The classical commercially pure grades of titanium, designated by the common acronym CP, correspond to unalloyed titanium where the mechanical properties are strongly influenced by small additions of oxygen and iron occurring during metal processing. By careful control of these additions, the various grades of commercially pure titanium are produced to give properties suited to different applications. Designation of these commercial types of pure unalloyed titanium are more commonly known by ASTM grades than by their UNS numbers. Hence, according to ASTM B265 standard,[26] four grades of CP titanium are commercially available: grade 1 contains the lowest oxygen and iron levels, producing the most formable grade of material, while grades 2, 3, and 4, have progressively higher oxygen contents and correspondingly higher strength levels.

Table 3.20 Chemical composition (%wt.) of CP titanium grades (Ti balance)

ASTM grades (B265)	UNS	N	C	H	Fe	O	Individual	Total
ASTM Gr. 1	R50250	0.03	0.08	0.015	0.20	0.18	0.05	0.3
ASTM Gr. 2	R50400	0.03	0.08	0.015	0.30	0.25	0.05	0.3
ASTM Gr. 3	R50550	0.05	0.08	0.015	0.30	0.35	0.05	0.3
ASTM Gr. 4	R50700	0.05	0.08	0.015	0.50	0.40	0.05	0.3

Table 3.21 Corresponding designations of CP titanium in several countries

ASTM (B265)	UNS	ASM	AECMA	MIL-T-9046J	DIN		BS	AFNOR
	USA			NATO	Germany		UK	France
Engineering		Aerospace		Military	Engi-neering	Aero-space	Aero-space	Engineer-ing & aerospace
Grade 1	R50250	4941	Ti-PO1	CP-4	3.7025	3.7024	BS TA 1, DTD 5013	NFT-35
Grade 2	R50400	4902, 4941, 4942, 4951	Ti-PO2	CP-3	3.7035	3.7034	BS TA 2, 3, 4, 5	NFT-40
Grade 3	R50550	4900	n.a.	CP-2	3.7055	3.7064	DTD 5023, 5273 (5283)	NFT-50
Grade 4	R50700	4901, 4921	Ti-PO4	CP-1	3.7065	3.7064	BS TA.6	NF T-60
Timet® 100		4921	n.a.	n.a.	3.7065	3.7064	BS TA 7, 8, 9	NF T-60

Table 3.22 Selected trade names of CP titanium

ASTM (B265)	Titanium Industries, Inc.	Cabot Performance Materials, Inc.	IMI Limited	RMI Titanium Company	TIMET (Titanium Metal Corporation)
	Fairfield, NJ, USA)	(Boyertown, PA, USA)	(Birmingham, UK)	(Niles, OH, USA)	(Denver, CO, USA)
Grade 1	Grade 1	Cabot$^\circledR$ T170	IMI$^\circledR$ 115	RMI$^\circledR$ 70	Timetal$^\circledR$ 35A
Grade 2	Grade 2	Cabot$^\circledR$ T155	IMI$^\circledR$ 125	RMI$^\circledR$ 55	Timetal$^\circledR$ 50A
Grade 3	Grade 3	Cabot$^\circledR$ T140	IMI$^\circledR$ 130	RMI$^\circledR$ 40	Timetal$^\circledR$ 65A
Grade 4	Grade 4	Cabot$^\circledR$ T125	IMI$^\circledR$ 155	RMI$^\circledR$ 25	Timetal$^\circledR$ 75A
n.a.	n.a.	n.a.	IMI$^\circledR$ 160	n.a.	Timetal$^\circledR$ 100A

Apart from ASTM grades, the world's largest titanium producers proposed their own commercial designation for trades. For example, the commercial designations according to several of the chief titanium producers are listed in Table 3.22.

3.3.2.7 Titanium Alloys

General description and properties – there are several reasons for using titanium alloys according to their properties and characteristics. Chief advantages of titanium alloys which are important to design engineers are, in order: (i) their high strength-to-weight ratio, (ii) superior erosion-corrosion resistance, and (iii) their excellent corrosion resistance in harsh media. Actually, first of all, the densities of titanium-based alloys range between 4430 kg.m^{-3} (light alloying elements) and 4850 g.cm^{-3} when alloyed with dense elements. Moreover, yield strength varies from 172 MPa for commercially pure ASTM grade 1 to above 1380 MPa for heat-treated beta alloys. Hence, this combination of high strengths and low densities results in exceptionally favorable strength-to-weight ratios. These ratios for titanium-based alloys are superior to almost all other structural metals and allow the use of thinner-walled equipment, and hence become important in such diverse applications as deep-well tube strings in the oil industry and surgical implants in the medical field. Secondly, titanium alloys offer superior resistance to erosion, cavitation or impingement attack. Actually, titanium alloys are at least 20 times more erosion resistant than the common copper–nickel alloys in chloride solutions, and hence permit significantly higher operation velocities. Thirdly, titanium alloys have excellent corrosion resistance in brines, seawater or marine atmospheres. Moreover, they also exhibit exceptional resistance to a broad range of acids, alkalis, natural waters and industrial chemicals. This relative absence of corrosion in media where titanium is generally used leaves the surface bright and smooth for improved lamellar flow.

Manufacturing – depending on the alloy, titanium alloys may be industrially produced by vacuum arc remelting (VAR), electron-beam melting (EB), or plasma melting. Ingots are commonly 60 to 125 cm outside diameter, weighing 2.3 to 18 tonnes. Wrought products are produced by conventional metallurgical processing in air. All standard mill products are available. Casting may also be produced using investment casting technologies and rammed graphite molding technologies. Although it is available in all conventional forms, titanium is not very easy to shape and form; it has a high spring-back and tends to gall, while welding must be carried out in an inert atmosphere. Nevertheless, recent manufacturing processes have employed an emerging technology, which makes it possible to mold some sheet titanium in a manner similar to one way in which plastics are molded. This manufacturing process is called superplastic forming.

Table 3.23 Physical properties of CP titanium grades

ASTM Grades (UNS)	Density (ρ, kg.m^{-3})	Young's modulus (E, GPa)	Poisson's ratio (ν)	Yield strength 0.2% proof (σ_{YS}, MPa)	Ultimate tensile strength (σ_{UTS}, MPa)	Elongation (Z, %)	Reduction in area (%)	Bend radius 2 mm sheet (ton)	Charpy V-notch impact (J^2)	Brinell hardness (HB)	α-β transition temperature (°C)	Melting range (°C)	Short temperature creep (1000 h, 250°C) (MPa)	Thermal conductivity (k, W.m^{-1}.K^{-1})	Specific heat capacity (c_P, J.kg^{-1}.K^{-1})	Coefficient of linear thermal expansion (20–100°C) (α, 10^{-6} K)	Coefficient of linear thermal expansion (21–538°C) (α, 10^{-6} K)	Electrical resistivity (ρ, $\mu\Omega$.cm)
Grade 1	4512	103	0.34	172–310	241–390	25–37	35	2.0	109–302	120	888	16–70	103	17.30	528	8.7	9.8	48.2
Grade 2	4512	102	0.34	275–450	345–483	20–28	35	2.5	41–114	160	913	1677	103	16.3	540	8.7	9.8	56
Grade 3	4512	103	0.34	379–550	448–593	18–25	35	2.5	20–38	200	921	1677	131	16.3	540	8.7	9.8	56
Grade 4	4512	104	0.34	379–485	550–640	16–25	35	3.0	20–30	265	950	1677	131	16.3	540	8.7	9.8	56

Table 3.24 Types of phase stabilizers

Stabilizer type	Alloying element (or impurities)
Alpha (hcp)	B, Al, C, O, and N
Alpha and beta	V, Nb, Ta, and Mo
Near alpha	Sn, and Zr
Beta (bcc)	Si, Mn, Cr, Fe, Co, Ni, Cu, and H

Metallurgical classification – the crystallographic structure of titanium exhibits a phase transformation from a low-temperature close-packed hexagonal arrangement (α-hcp, alpha-titanium) to a high-temperature form body-centered cubic crystal lattice (β-bcc, beta-titanium) at 882°C. This transformation can be considerably modified by the addition of alloying elements (see Table 3.24) to produce at room temperature alloys which have all alpha, all beta or alpha–beta structures.

Therefore, the basic properties of titanium and its alloys strongly depend on their basic metallurgical structure and the way in which this is manipulated during their mechanical and thermal treatment during manufacture. Four main types of titanium alloy have been developed and hence titanium alloys fall into the four categories: alpha, near alpha, alpha plus beta, and beta.

Alpha-titanium alloys – these alloys range in yield tensile strength from 173 MPa to 483 MPa. Variations are generally achieved by alloy selection and not heat treatment. They usually contain alpha stabilizers and have the lowest strengths. However, they are formable and weldable. Some contain beta stabilizers to improve strength. Alpha-titanium alloys are generally in the annealed or stress-relieved condition. They are considered fully annealed after heating to 675–790°C for 1–2 h. Alpha alloys are generally fabricated in the annealed condition. All fabrication techniques used for stainless steels are generally applicable. Weldability is considered good under a proper shielding from oxygen.

Alpha–beta-titanium alloys – these alloys range in yield tensile strength from 862 MPa to more than 1200 MPa. Strength can be varied both by alloy selection and heat treatment. Water quenching is required to attain higher strength levels. Section thickness requirements should be considered when selecting these alloys. Alpha–beta alloys are widely used for high-strength applications and have moderate creep resistance. Alpha–beta titanium alloys are generally used in the annealed or solution-treated and aged condition. Annealing is generally performed in a temperature range 705 to 845°C for 0.5 to 4 h. Solution treating is generally performed in a temperature range 900 to 955°C followed by a water quench. Aging is performed between 480 to 593°C for 2 to 24 h. The precise temperature and time is chosen to achieve the desired mechanical properties. Generally, alpha–beta alloys are fabricated at elevated temperatures, followed by heat treatment. Cold forming is limited in these alloys.

Near-alpha-titanium alloys – near-alpha alloys have medium strength but better creep resistance. They can be heat treated from the beta phase to optimize creep resistance and low cycle fatigue resistance, and some are weldable. Palladium stabilized grades of these materials are also available for enhanced corrosion resistance. For example, the ASTM grade 12 is a highly weldable, near-alpha alloy, exhibiting improved strength and temperature capability over CP grades combined with superior crevice corrosion resistance and excellent resistance under oxidizing to mildly reducing conditions especially chlorides.

Beta-titanium alloys – beta alloys range in yield strength from 795 MPa to far more than 1380 MPa which are attainable through cold working and direct aging treatments. Beta phase alloys are usually metastable, formable as quenched and can be aged to the highest strengths, but then lack ductility. Fully stable beta alloys need large amounts of stabilizers and are therefore too dense. In addition, the Young's modulus is low (below 100 GPa) unless the beta phase structure is decomposed to precipitate alpha phase. They have poor stability at 200–300°C, have low creep resistance and are difficult to weld without embrittlement. Metastable beta alloys have some application as high-strength fasteners. Beta-titanium alloys are generally used in the solution-treated and aged condition. The annealed condition may also be employed for service temperature less than 205°C. Annealing and solution treating are performed in a temperature range of 730 to 980°C with temperatures around 815°C most common. Aging between 482 to 593°C for 2 to 48 h is chosen to obtain the desired mechanical properties. Duplex aging is often employed to improve age response; the first age cycle is performed between 315 and 455°C for 2 to 8 h followed by the second age cycle between 480°C and 595°C for 8 to 16 h. Beta alloys may be fabricated using any of the techniques employed for alpha alloys including cold forming in the solution-treated condition. Forming pressure will increase because the yield strength is high compared to alpha alloys. The beta alloys are weldable and may be aged to increase strength after welding. The welding process will produce an annealed condition exhibiting strength at the low end of the beta alloy range.

ASTM designation (Table 3.25, pages 130–132) – as for CP titanium grades the commercial types of titanium alloys are more commonly known by ASTM grades than by their UNS numbers. Titanium grades 7 and 11 contain 0.15wt% Pd to improve resistance to crevice corrosion and to reducing acids. Actually, the noble alloy additions enhance passivation. Titanium grade 12 contains 0.3wt% Mo and 0.8wt% Ni and is known for its improved resistance to crevice corrosion and its higher design allowances than unalloyed grades. It is available in many product forms. Other alloying elements (e.g., V, Al) are used to increase strength (e.g., grades 5 and 9).

3.3.2.8 Corrosion Resistance

Corrosion resistance has been an important consideration in the selection of titanium alloys as engineering structural materials, since titanium first became an industrial reality in 1950. Actually, titanium has gained acceptance in many harsh media where its corrosion resistance and engineering properties have provided the corrosion and design engineer with a reliable and economic material. Although many titanium alloys have been developed for aerospace applications where mechanical properties are the primary consideration, in the chemical process industries, however, corrosion resistance is the most important property. Finally, titanium has the ability to resist erosion by high-velocity seawater. Velocities as high as 37 $m.s^{-1}$ cause only a minimal rise in erosion rate. The presence of abrasive particles, such as sand, has only a small effect on the corrosion resistance of titanium under conditions that are extremely detrimental to copper-based alloys. The corrosion resistance of titanium and titanium alloys to specific environments are listed in Table 3.31 (pages 140–143) with an explanation of the types of corrosion that can affect it. However, discussion of corrosion resistance in this table is limited to commercially pure (CP) and alloy grades typically used in the chemical process industries. The data given should be used with caution, as a guide to the applications of titanium, because in many cases, they were obtained in the laboratory. Actually, in-plant environments often contain impurities which can exert their own effects. Moreover, in particular, heat transfer conditions or unanticipated deposited residues can also alter results. Such factors may require in-plant corrosion tests. In general, commercially pure ASTM grade 7 extends the usefulness of unalloyed titanium to more severe conditions. On the other hand, Ti-6Al-4V provides less corrosion resistance than unalloyed titanium, but is still outstanding in many environments compared to other structural metals. Recently, ASTM incorporated a series of new titanium grades containing 0.05wt% Pd. These new grades exhibit nearly identical corrosion resistance to the old ASTM grade 7 with 0.15% Pd grades, yet offer considerable cost savings.

3.3.2.9 Titanium Metalworking

Titanium can be formed into intricate shapes by common forming techniques such as bending, shearing, pressing, deep drawing, and expanding. Titanium work hardens significantly during forming such as stainless or low-carbon steels. However, the titanium alloys exhibit a greater tendency to galling than stainless steels. This implies a close attention to lubrication in any forming operations in which the metal is in direct contact with dies or other forming equipment.

Bending – minimum bend-radius rules are nearly the same as for stainless steels, although springback is greater for titanium due to its lower Young's modulus. Commercially pure titanium grades of heavy plate are cold formed or, for more severe bending or stretch forming, hot forming at temperatures to about 425°C is required. However, titanium alloys can also be formed at temperatures as high as 760°C under an inert gas atmosphere (e.g., Ar). Tubes can be cold bent to radii three times the tube outside diameter, provided that both inside and outside surfaces of the bend are in tension at the point of bending. In some cases, tighter bends can be made.

Superplastic forming – despite their high strength, some alloys of titanium exhibit a superplastic behavior in the range 815 to 925°C. The titanium alloy most often used in superplastic forming is the standard Ti–6Al–4V grade. Several aircraft manufacturers are producing components formed by this method. Some applications involve assembly by diffusion bonding.

Shearing, punching – titanium plates or sheets can be sheared, punched, or perforated on standard equipment suitable for steel. Titanium and Ti–Pd alloy plates can be sheared subject to equipment limitations similar to those for stainless steel. The harder alloys are more difficult to shear, so thickness limitations are generally about two-thirds those for stainless steel.

Casting – titanium castings can be produced by investment or graphite-mold methods. Casting must be done in a vacuum furnace, however, because of the highly chemical reactivity of molten titanium with oxygen.

Annealing – residual stresses can be removed by annealing under inert gases or vacuum at temperatures ranging between 500°C to 600°C, while complete annealing must be performed at 700°C.

3.3.2.10 Titanium Machining

Titanium and titanium alloys can be machined and abrasive ground; however, sharp tools and continuous feed are required to prevent work hardening. Tapping is difficult because the metal galls. Coarse threads should be used where possible. As a general rule it is necessary to use low cutting speed, high feed rates, and use large amounts of cutting fluid, and due to its chemical reactivity it causes galling, smearing, or in the worst case galling with tools.

3.3.2.11 Titanium Joining

Apart from mechanical fastening, titanium and its alloys can be usually joined by fusion, resistance, flash butt, electron beam, diffusion bonding, and pressure welding techniques. Nevertheless, production of joints by fusion welding is restricted to commercially pure titanium or weldable titanium alloys. However, brazing and friction welding can only be performed for making joints between two nonweldable titanium alloys or titanium and dissimilar metals. Several titanium characteristics make the welding of titanium different from that of other structural alloys. Actually, its high melting point implies use of a high-temperature technique, and owing to its strong chemical reactivity with oxygen, titanium must be welded under inert atmosphere to avoid oxidation, and nitriding, and the work piece must be carefully pickled before welding. Hence welding of titanium alloys can be performed by gas–tungsten arc welding (GTAW) or plasma-arc techniques. Sometimes, when appropriate, particular techniques such as gas–metal arc welding (GMAW), laser welding, and electron beam welding are used. Metal–inert-gas processes can be used under special conditions. Thorough cleaning and shielding and

Table 3.25 Chemical composition (wt%) of ASTM titanium alloy grades (Ti balance)

ASTM grades (B 265)	Alloy composition	UNS	N	C	H	Fe	O	Other	Individual	Total
Grade 5	Ti-6Al-4V	R56400	0.05	0.08	0.015	0.40	0.20	5.5–6.75 Al 3.5–4.5 V	0.05	0.3
Grade 6	Ti-5Al-2.5Sn	R54520	0.03	0.08	0.020	0.50	0.20	4.0–6.0 Al 2.0–3.0 Sn	0.05	0.3
Grade 7	Ti-0.15Pd	R52400	0.03	0.08	0.015	0.30	0.25	0.12–0.25 Pd	0.05	0.3
Grade 9	Ti-3Al-2.5V	R56320	0.03	0.05	0.015	0.30	0.12	2.5–3.5 Al 2.0–3.0 V	0.05	0.3
Grade 10	Ti-11.5Mo-6Zr-4.5Sn	R58030	0.05	0.08	0.020	0.35	0.18	3.75–5.25 Sn 4.50–7.50 Zr	0.05	0.3
Grade 11	Ti-0.15Pd	R52250	0.03	0.08	0.015	0.20	0.18	0.12–0.25 Pd	0.05	0.3
Grade 12	Ti-0.3Mo-0.8Ni	R53400	0.03	0.08	0.015	0.30	0.25	0.2–0.4 Mo 0.6–0.9 Ni	0.05	0.3
Grade 13	Ti-0.5Ni-0.05Ru	R53413	0.03	0.08	0.015	0.20	0.10	0.04–0.06 Ru 0.40–0.60 Ni	0.05	0.3
Grade 14	Ti-0.5Ni-0.05Ru	R53414	0.03	0.08	0.015	0.30	0.15	0.04–0.06 Ru 0.40–0.60 Ni	0.05	0.3
Grade 15	Ti-0.5Ni-0.05Ru	R53415	0.05	0.08	0.015	0.30	0.25	0.04–0.06 Ru 0.40–0.60 Ni	0.05	0.3
Grade 16	Ti-0.6Pd	R52402	0.03	0.08	0.015	0.30	0.25	0.04–0.08 Pd	0.05	0.3
Grade 17	Ti-0.6Pd	R52252	0.03	0.08	0.015	0.20	0.18	0.04–0.08 Pd	0.05	0.3
Grade 18	Ti-3.5V-2Al-0.05Pd	R56322	0.03	0.08	0.015	0.25	0.15	0.04–0.08 Pd 2.5–3.5 Al 2.3–3.0 V	0.05	0.3
Grade 19	Ti-8V-6Cr-4Mo-4Zr-3Al	R58640	0.03	0.05	0.02	0.30	0.12	3.0–4.0 Al 7.5–8.5 V 5.5–6.5 Cr 3.5–4.5 Zr	0.15	0.4

Grade	Alloy	UNS						Other elements		
Grade 20	Ti-8V-6Cr-4Zr-4Mo-3Al-0.05Pd	R58645	0.05	0.05	0.02	0.30	0.12	3.0-4.0 Al / 7.5-8.5 V / 3.5-4.5 Mo / 5.5-6.5 Cr / 3.5-4.5 Zr / 0.04-0.08 Pd	0.15	0.4
Grade 21	Ti-15Mo-03Al-2.7Nb-0.25Si	R58210	0.03	0.05	0.015	0.40	0.17	2.5-3.5Al / 14.0-16.0Mo / 2.2-3.2 Nb / 0.15-0.25Si		0.4
Grade 23	Ti-6Al-4V (ELI)	R56401	0.03	0.08	0.0125	0.25	0.13	5.5-6.75 Al / 3.5-4.5 V	0.1	0.4
Grade 24	Ti-6Al-4V-0.05Pd	R56405	0.05	0.08	0.015	0.40	0.20	5.5-6.75 Al / 3.5-4.5 V / 0.04-0.08Pd	0.1	0.4
Grade 25	Ti-6Al-4V-0.5Ni-0.05Pd	R56403	0.05	0.08	0.0125	0.40	0.20	5.5-6.75 Al / 3.5-4.5 V / 0.04-0.08Pd / 0.3-0.8Ni	0.1	0.4
Grade 26	Ti-0.1Ru	n.a.	0.03	0.08	0.015	0.30	0.25	0.08-0.14 Ru	0.1	0.04
Grade 27	Ti-0.1Ru	n.a.	0.03	0.08	0.015	0.20	0.18	0.08-0.14 Ru	0.1	0.4
Grade 28	Ti-3Al-2.5V-0.5Ru	n.a.	0.03	0.08	0.015	0.25	0.15	2.5-3.5 Al / 2.0-3.0 V / 0.08-0.14Pd	0.1	0.4
Grade 29	Ti-6Al-4V-0.1Ru	n.a.	0.03	0.08	0.015	0.25	0.13	5.5-6.5 Al / 3.5-4.5 V / 0.08-0.14Pd	0.1	0.4
n.a.	Ti-10V-3Al-2Fe	R56410	0.05	0.05	0.015	1.6-2.2	0.13	2.6-3.4 Al / 9.0-11.0 V	0.10	0.3
n.a.	Ti-6Al-6V-2Sn-Ti	R56620	0.04	0.05	0.015	0.35-1.0	0.20	5.0-6.0 Al / 5.0-6.0 V / 1.5-2.5 Sn / 0.35-1.0 Cu	0.10	0.3
n.a.	Ti-8Al-1Mo-1V	R54810	0.05	0.08	0.015	0.30	0.12	7.35-8.35 Al / 0.75-1.25 Mo / 0.75-1.25 V	0.10	0.3

Table 3.25 (continued)

ASTM grades (B 265)	Alloy composition	UNS	N	C	H	Fe	O	Other	Individual	Total
n.a.	Ti–6Al–4Zr–2Sn–2Mo	R56210	0.05	0.05	0.150	0.250	0.15	5.5–6.5 Al 1.8–2.2 Sn 3.6–4.4 Zr 1.8–2.2 Mo 0.06–0.010 Si	0.10	0.3
n.a.	Ti–2.5Cu	n.a.	0.05	0.08	0.01	0.2	0.2	2.0–3.0 Cu	–	0.4

Table 3.26 Uses of common titanium alloys

Corrosion resistant	High strength	High temperature
Commercially Pure Grade 1	Ti–6Al–4V Grade 5	Ti–6Al–2Sn–4Zr–2Mo
Commercially Pure Grade 2	Ti–5Al–2.5Sn Grade 6	Ti–6Al–2Sn–4Zr–6Mo
Commercially Pure Grade 3	Ti–2.5Cu	Ti–11Sn–5Zr–2.5Al–1Mo
Commercially Pure Grade 4	Ti–6Al–7Nb	Ti–5.5Al–3.5Sn–3Zr–1Nb
Ti–Pd Grade 7 and Grade 16	Ti–4Al–4Mo–2Sn	Ti–5.8Al–4Sn–3.5Zr–0.7Nb
Ti–3Al–2.5V Grade 9 and Grade 18	Ti–6Al–6V–2Sn	
Ti–Pd Grade 11 and Grade 17	Ti–10V–2Fe–3Al	
Ti–0.3Mo–0.8Ni Grade 12	Ti–15V–3Cr–3Sn–3Al	
Ti–3Al–8V–6Cr–4Zr–4Mo (Beta C)	Ti–5.5Al–3Sn–3Zr–0.5Nb	
Ti–15Mo–3Nb–3Al–0.2Si	Ti–5Al–2Sn–4Mo–2Zr–4Cr (Ti 17)	
	Ti–8Al–1Mo–1V	
	Ti–6Al–5Zr–0.5Mo–0.25Si	

supplying argon gas to the surface of the work piece are essential because molten titanium reacts readily with nitrogen, oxygen, and hydrogen, and will dissolve large quantities of these gases, which embrittles the metal. In all other respects, GTA welding of titanium is similar to that of stainless steel. Normally, a sound weld appears bright silver with no discoloration on the surface or along the heat-affected zone. The weldability is strongly influenced by the chemical composition and microstructure of the alloy. Actually, the presence of a beta phase leads to a deleterious effect.

3.3.2.12 Titanium Etching, Descaling, and Pickling

Grinding – owing to the pyrophoricity of titanium dust, grinding must be achieved with a lubrication liquid to avoid explosion hazards.

Blasting – owing to the intrinsic hardness of rutile which is higher than that of quartz, blasting can be performed using harder minerals such as corundum or zircon rather than silica as abrasive medium. After operation, the workpiece must be carefully pickled for a few seconds in an HF–HNO_3 mixture in order to remove the abrasive particles embedded into the metal, washed with deionized water, and dried.

Cleaning and degreasing – these operations can be performed in alkaline solutions, organic solvents, or emulsions and the procedures are similar to those used for other metals, but careful attention must be exerted to avoid hydrogen embrittlement in alkaline solutions.

Pickling and descaling – there exist several pickling chemicals for titanium and titanium alloys. When thick oxide films, e.g. due to thermal treatments are present on the surface, pickling for 5 s in the ternary corrosive mixture HNO_3–HF–H_2O (20vol%–2vol%–78vol%) at room temperature gives good results. For thick scale deposits, hot pickling for 5–10 minutes in boiling oxalic acid (10wt%), or diluted sulfuric acid (30wt%) is efficient.

3.3.2.13 Industrial Uses and Applications

Although 95% of titanium metal is consumed in the form of rutile (titanium dioxide, TiO_2), the use of titanium metal products increases owing to its corrosion resistance in harsh media, and high strength-to-weight ratio. Examples of industrial applications in which titanium-based alloys are currently utilized are given in Table 3.32 (page 144).

Table 3.27 Description of common titanium alloys (source TIMET Corp.)

Class	Tradename	Chemical composition (wt%)	Description
Medium- and high-strength titanium alloys	Timetal® 230	Ti–2.5Cu	Binary age-hardening alloy, ease of formability and weldability of commercially pure titanium with improved mechanical properties up to 350°C
	Timetal® 62S	Ti–6Al–2Fe–0.1Si	Properties and processing characteristics equivalent to or better than Timetal® 6–4, but with significantly higher Young's modulus. Due to the use of iron as the beta stabilizer, the alloy has lower formulation costs than Timetal® 6–4. The combination of reasonable cost and excellent mechanical properties make it a practical substitute for many engineering materials
	Timetal® 6–4	Ti–6Al–4V	This titanium alloy is a versatile medium-strength titanium alloy, which exhibits good tensile properties at room temperature, a creep resistance up to 325°C and excellent fatigue strength. It is often used in less critical applications up to 400°C. It is the alloy most commonly used in wrought and cast forms
	Timetal® 3–2.5	Ti–3Al–2.5V	Cold formable and weldable, this alloy is used primarily for honeycomb foil and hydraulic tubing applications. Industrial applications such as pressure vessels and piping also utilize this alloy. Available with Pd stabilization to enhance corrosion resistance
	Timetal® 367	Ti–6Al–7Nb	Medium-strength titanium alloy dedicated for surgical implants
	Timetal® 10–2–3	Ti–10V–2Fe–3Al	Readily forgeable alloy that offers excellent combinations of strength, ductility, fracture toughness and high cycle fatigue strength. Typically used for critical aircraft structures, such as landing gear
	Timetal® 550	Ti–4Al–4Mo–2Sn–0.5Si	Readily forgeable and is generally used in a heat treated condition. It has superior room and elevated temperature tensile strength and fatigue strength compared to Timetal® 6–4, and is creep resistant up to 400°C
	Timetal® 551	Ti–4Al–4Mo–4Sn–0.5Si	High strength and creep resistant up to 400°C. It has a similar composition to Timetal® 550, apart from an increase in tin content, which gives increased strength at room and elevated temperatures
	Timetal® 6–6–2	Ti–6Al–6V–2Sn–0.5Fe–0.5Cu	Improved strength properties and greater depth hardenability compared with Timetal® 6–4
	Timetal® 15–3	Ti–15V–3Cr–3Sn–3Al	Cold formable and weldable, this strip alloy is primarily used for aircraft ducting, pressure vessels and other fabricated sheet metal structures up to 300°C
	Timetal® 21S	Ti–15Mo–3Nb–3Al–0.2Si	Good cold formability and weldability of a beta strip alloy, but with greatly improved oxidation resistance and creep strength. Aerospace applications include engine exhaust plug and nozzle assemblies
High-temperature titanium alloys	Timetal® 6–2–4–2	Ti–6Al–2Sn–4Zr–2Mo–0.08Si	Good tensile creep and fatigue properties up to 540°C. It is the most commonly used high-temperature alloy in jet engine compressors and airframe structures
	Timetal® 17	Ti–5Al–2Sn–4Mo–2Zr–4Cr	High-strength, deep hardenable forging alloy primarily for jet engines. Allows heat treatment to a variety of strength levels in sections up to 15 cm. Offers good ductility and toughness, as well as good low-cycle and high-cycle fatigue properties
	Timetal® 6–2–4–6	Ti–6Al–2Sn–4Zr–6Mo	Stronger derivative of Timetal® 6–2–4–2 offering higher strength, depth hardenability and elevated temperature properties up to 450°C
	Timetal® 679	Ti–11Sn–5Zr–2.2Al–1Mo–0.2Si	Excellent tensile strength and creep resistant up to 450°C
	Timetal® 685	Ti–6Al–5Zr–0.5Mo–0.25Si	Excellent tensile strength and is creep resistant up to 520°C. It is weldable and has good forging characteristics

Table 3.27 (continued)

Class	Tradename	Chemical composition (%wt.)	Description
High-temperature titanium alloys	Timetal® 8–1–1	Ti–8Al–1Mo–1V	Designed for creep resistance up to 450°C, used primarily in engine applications such as forged compressor blades and disks. This alloy has a relatively high tensile modulus to density ratio compared to most commercial titanium alloys
	Timetal® 829	Ti–5.6Al–3.5Sn–3Zr–1Nb–0.25Mo–0.3Si	Combines creep resistance up to 540°C with good oxidation resistance. It is weldable and like Timetal® 685, and has good forgeability
	Timetal® 834	Ti–5.8Al–4Sn–3.5Zr–0.7Nb–0.5Mo–0.35Si–0.06C	Near-alpha titanium alloy offering increased tensile strength and creep resistance up to 600°C together with improved fatigue strength when compared with established creep resistant alloys such as Timetal® 6-2-4-2, Timetal® 829 and Timetal® 685. Like these alloys, it is weldable and has good forgeability
	Timetal® 1100	Ti–6Al–2.7Sn–4Zr–0.4Mo–0.45Si	Near-alpha, high-temperature creep resistant alloy developed for elevated temperature use in the range of 600°C that offers the highest combination of strength, creep resistance, fracture toughness and fatigue crack growth resistance
Developmental titanium alloys	Timetal® 21SRA	n.a.	Development from the alloy Timetal® 21S with the aluminum additions removed and targeted at biomedical applications
	Timetal® LCB	n.a.	Metastable beta alloy produced in bar or rod form and targeted at titanium spring and other high-strength requirement applications
	Timetal® 5111	Ti–5Al–1Sn–1Zr–1V–0.8Mo	Near alpha alloy with excellent weldability, seawater stress corrosion cracking resistance and high dynamic toughness

3.3.2.14 Titanium Metal World Producers

See **Table 3.33** (page 145).

3.3.3 Zirconium and Zirconium Alloys

3.3.3.1 Description and General Properties

Zirconium [7440-67-7] is the second metallic element in subgroup IVB(4) of the periodic chart with the atomic number 40 and a relative atomic mass (atomic weight) of 91.224. It has the chemical symbol Zr and is probably named after the Arabic word zargun, which, in the ancient times, described the gold or dark amber color of the gemstone now known today as **zircon** $Zr[SiO_4]$. Zirconium is a hard, grayish-white lustrous metal similar to stainless steel in appearance. It has a moderate density of 6506 kg.m^{-3} which is lower than that of pure iron or nickel. Moreover, it has a low coefficient of thermal expansion (5.78 μm.m^{-1}.K^{-1}) which is about one-third that of AISI type 316L stainless steel. Finally, its thermal conductivity is slightly greater than that of stainless steel AISI type 316L. Owing to its high melting point of 1852°C, zirconium is obviously classified as a refractory metal, like titanium or tantalum. The main physical properties are listed in Table 3.36 (pages 149–150). From a chemical point of view, as the result of the lanthanide contraction, the ionic radii of cations Zr^{4+} (87 pm) and Hf^{4+} (84 pm) are virtually identical and the separation is the most difficult of all the elements and could

Table 3.28 Mechanical properties of selected titanium alloys (annealed)

Common and trade name	Average chemical composition (wt%)	Density (ρ, kg.m^{-3})	Young's modulus (E, GPa)	Yield strength proof 0.2% (σ_{YS}, MPa)	Ultimate tensile strength (σ_{UTS}, MPa)	Elongation (Z, %)	Reduction in area (%)	Bend radius 2 mm sheet (ton)	Charpy V-notch impact (J)	Brinell hardness (HB)
Grade 5 Timetal® 6.4)	Ti–6Al–4V	4420	106–114	808–1075	897–1205	10–18	20–30	5.0	24	330 (HR C36)
Grade 6 (Timetal® 5.3)	Ti–5Al–2.5Sn	4480	110	779–897	828–972	10–16	25	4.5	26	(HR C36)
Grade 7	Ti–0.2Pd	4510	103	276–352	345–483	20–28	50	3.0	43	160–200
Grade 9 (Timetal® 3.2.5)	Ti–3Al–2.5V	4500	91–107	483–607	621–740	15–17	–	2.5	48	(HR C15)
Grade 10 (Ti–alloy $\beta3$)	Ti–11.5Mo–6Zr–4.5Sn	5760	n.a.	n.a.	690	10	n.a.	n.a.	n.a.	n.a.
Grade 11	Ti–0.2Pd	4510	103	175–221	241–345	24–37	35	2.0	109	120
Grade 12	Ti–0.3Mo–0.8Ni	4510	103	345–462	483–607	18–22	25	2.5	14	180
Grade 13	Ti–0.5Ni–0.05Ru	n.a.	n.a.	170	275	24	n.a.	n.a.	n.a.	n.a.
Grade 14	Ti–0.5Ni–0.05Ru	n.a.	n.a.	275	410	20	n.a.	n.a.	n.a.	n.a.
Grade 15	Ti–0.5Ni–0.05Ru	n.a.	n.a.	380	483	18	n.a.	n.a.	n.a.	n.a.
Grade 16	Ti–0.06Pd	4510	103	276–352	345–483	20–28	50	3.0	43	160–200
Grade 17	Ti–0.06Pd	4510	103	175–221	241–345	24–37	35	2.0	109	120
Grade 18	Ti–3Al–2.5V–0.06Pd	4500	91–107	483–607	621–740	15–17	n.a.	2.5	48 (35)	(HR C15)
Grade 19	Ti–8V–6Cr–4Mo–4Zr–3.5Al	4810	103	759	793	15	n.a.	n.a.	n.a.	n.a.
Grade 20 (Beta C)	Ti–8V–6Cr–4Zr–4Mo–3.5Al–0.05Pd	4810	103	1104–1152	1172–1248	6–10	19	n.a.	n.a.	n.a.
Grade 21	Ti–15Mo–3Al–2.7Nb–0.25Si	n.a.	n.a.	759	793	15	n.a.	n.a.	n.a.	n.a.
Grade 23	Ti–6Al–4V (ELI)	4420	106–114	759	828	10	n.a.	n.a.	n.a.	n.a.
Grade 24	Ti–6Al–4V–0.05Pd	n.a.	n.a.	828	895	10	n.a.	n.a.	n.a.	n.a.
Grade 25	Ti–6Al–4V–0.5Ni–0.05Pd	n.a.	n.a.	828	895	10	n.a.	n.a.	n.a.	n.a.
Grade 26	Ti–0.1Ru	4510	n.a.	275	345	20	n.a.	n.a.	n.a.	n.a.
Grade 27	Ti–0.1Ru	4510	n.a.	170	240–310	24	n.a.	n.a.	n.a.	n.a.
Grade 28	Ti–3Al–2.5V–0.5Ru	n.a.	n.a.	483	620	15	n.a.	n.a.	n.a.	n.a.
Grade 29	Ti–6Al–4V–0.1Ru	n.a.	n.a.	759	828	10	n.a.	n.a.	n.a.	n.a.
Timetal® 1.7	Ti–5Al–4Mo–2Sn–2Zr–4Cr	4650	109	1055–1193	1125	7–10	n.a.	n.a.	n.a.	n.a.
Timetal® 10–2–3	Ti–10V–3Al–2Fe	4650	112	970–1228	1040–1310	8–15	42	n.a.	n.a.	n.a.

Table 3.28 (continued)

Common and trade name	Average chemical composition (wt%)	Density (ρ, kg.m^{-3})	Young's modulus (E, GPa)	Yield strength proof 0.2% (σ_{YS}, MPa)	Ultimate tensile strength (σ_{UTS}, MPa)	Elongation (Z, %)	Reduction in area (%)	Bend radius 2 mm sheet (ton)	Charpy V-notch impact (J.m^{-2})	Brinell hardness (HB)
Timetal® 11.0.0	Ti–6Al–4Zr–2.7Sn–0.4Mo–0.45Si	4500	120	910	1030	6	n.a.	6	n.a.	n.a.
Timetal® 15.3	Ti–15V–3Al–3Sn–3Cr	4780	63–107	890–1172	703–1241	5–7	n.a.	4	n.a.	n.a.
Timetal® 21S	Ti–15Mo–3Nb–3Sn–0.2Si	4940	83	890	945	12	n.a.	4	n.a.	n.a.
Timetal® 3.6.7	Ti–7Nb–6Al	4520	105–120	800	900	10	n.a.	n.a.	n.a.	n.a.
Timetal® 5.5.0	Ti–4Al–4Mo–2Sn–0.5Si	4600	110–130	920–960	1050–1100	9–14	n.a.	n.a.	n.a.	n.a.
Timetal® 5.5.1	Ti–4Al–4Mo–4Sn–0.5Si	4600	113–130	1065–1140	1205–1300	8–12	40	n.a.	n.a.	n.a.
Timetal® 6.2.1	Ti–6Al–2Nb–1Ta–0.8Mo	n.a.	n.a.	n.a.	n.a.	n.a.	n.a.	n.a.	n.a.	n.a.
Timetal® 6.2.4.2	Ti–6Al–4Zr–2Sn–2Mo–0.08Si	4540	115	830–917	931–1100	8–18	n.a.	4.5	–	(HRC32)
Timetal® 6.2.4.6	Ti–6Al–6Mo–4Zr–2Sn	4640	115	725–1000	850–1100	6–10	n.a.	n.a.	n.a.	n.a.
Timetal® 6.2.5	Ti–6Al–2.5V	n.a.	n.a.	n.a.	n.a.	n.a.	n.a.	n.a.	n.a.	n.a.
Timetal® 6.2.S	Ti–6Al–2Fe–0.1Si	4440	128	960	1000	16	n.a.	4.5	n.a.	n.a.
Timetal® 6.6.2	Ti–6Al–6V–2Sn–0.5Fe–0.5Cu	4530	115	950–1021	1035–1150	8–17	15–30	4.5	18 (14)	(HRC38)
Timetal® 6.7.9	Ti–11Sn–5Zr–2.25Al–1Mo–0.2Si	4840	105–110	970	1100	8	n.a.	n.a.	n.a.	n.a.
Timetal® 6.8.5	Ti–6Al–5Zr–0.5Mo–0.25Si	4450	125	850–900	990–1030	6–12	n.a.	n.a.	n.a.	n.a.
Timetal® 8.1.1	Ti–8Al–1Mo–1V	4360	124	931	1020	10	28	4.5	33	HRC 35
Timetal® 8.2.9	Ti–5.6Al–3.5Sn–3.5Zr–0.7Nb–0.25Mo–0.35Si–0.06C	4510	120	820–860	950–980	10	n.a.	n.a.	n.a.	n.a.
Timetal® 8.3.4	Ti–5.8Al–4Sn–3.5Zr–0.7Nb–0.5Mo–0.35Si–0.06C	4590	120	910–930	1030–1050	6	n.a.	6	n.a.	n.a.
Timetal® 230	Ti–2.5Cu	4560	105–125	510–520	620–760	20–25	n.a.	2.5	n.a.	n.a.
Timetal® 5.3 (ELI)	Ti–5Al–2.5Sn (ELI)	4480	110	n.a.	n.a.	n.a.	n.a.	n.a.	n.a.	n.a.
Titanium niobium	Ti–45Nb	5600	62.05	480	546	23	n.a.	n.a.	n.a.	n.a.

Table 3.29 Thermal and electrical properties of selected titanium alloys

Common and trade name	Average chemical composition (wt%)	α–β transition temperature (T, °C)	Melting range (T, °C)	Short-term creep (1000 h, 250°C) (MPa)	Specific heat capacity (c_p, Jkg⁻¹)	Thermal conductivity (k, Wm⁻¹.K⁻¹)	Coefficient of linear thermal expansion (20–100°C) (α, 10⁻⁶ K)	Coefficient of linear thermal expansion (21–538°C) (α, 10⁻⁶ K)	Electrical resistivity (ρ, μΩ.cm)
Grade 5 (Timetal® 6.4)	Ti–6Al–4V	993	1649	n.a.	560	7.2	8.8	9.2	170
Grade 6 (Timetal® 5.3)	Ti–5Al–2.5Sn	1038	1571	n.a.	540	7.7	9.4	9.5	157
Grade 7	Ti–0.2Pd	913	1677	n.a.	540	20.6	8.6	9.2	56
Grade 9 (Timetal® 3.2.5)	Ti–3Al–2.5V	935	1704	100	540	7.6	n.a.	7.9	124
Grade 11	Ti–0.2Pd	888	1670	n.a.	540	20.6	8.6	9.2	56
Grade 12	Ti–0.8Ni–0.3Mo	888	n.a.	221	540	22.7	9.5	–	51
Grade 16	Ti–0.06Pd	913	1677	n.a.	540	20.6	8.6	9.2	56
Grade 17	Ti–0.06Pd	888	1670	n.a.	540	20.6	8.6	9.2	56
Grade 18	Ti–3Al–2.5V–0.06Pd	935	1704	100	540	7.6	n.a.	7.9	124
Ti-alloy (Beta C)	Ti–8V–6Cr–4Mo–4Zr–3Al	793	1649	n.a.	n.a.	8.4	9.4	9.7	n.a.
Ti-alloy 6.2.1	Ti–6Al–2Nb–1Ta–0.8Mo	1015	n.a.	n.a.	n.a.	n.a.	n.a.	n.a.	n.a.
Timetal® 1.7	Ti–5Al–2Sn–4Mo–2Zr–4Cr	800	n.a.	n.a.	n.a.	n.a.	n.a.	n.a.	n.a.
Timetal® 10.2.3	Ti–10V–3Al–2Fe	796	1649	n.a.	n.a.	n.a.	9.7	n.a.	n.a.
Timetal® 11.0.0	Ti–6Al–2.7Sn–4Zr–0.4Mo–0.45Si	1015	n.a.	n.a.	n.a.	n.a.	n.a.	n.a.	n.a.
Timetal® 15.3	Ti–15V–3Cr–3Sn–3Al	760	n.a.	n.a.	n.a.	n.a.	n.a.	n.a.	n.a.
Timetal® 2.3.0	Ti–2.5Cu	895	n.a.	n.a.	n.a.	n.a.	n.a.	n.a.	n.a.
Timetal® 21.s	Ti–15Mo–3Cr–3Sn–0.2Si	800	n.a.	n.a.	n.a.	n.a.	n.a.	n.a.	n.a.
Timetal® 3.6.7	Ti–7Nb–6Al	1015	n.a.	n.a.	n.a.	n.a.	n.a.	n.a.	n.a.
Timetal® 5.5.0	Ti–4Al–4Mo–2Sn–0.5Si	975	n.a.	n.a.	n.a.	n.a.	n.a.	n.a.	n.a.
Timetal® 5.5.1	Ti–4Al–4Mo–4Sn–0.5Si	1050	n.a.	n.a.	n.a.	n.a.	n.a.	n.a.	n.a.
Timetal® 6.2.4.2	Ti–6Al–4Zr–2Sn–2Mo–0.08Si	996	1649	n.a.	420	6.0	9.9	n.a.	191
Timetal® 6.2.4.6	Ti–6Al–6Mo–4Zr–2Sn	940	n.a.	n.a.	n.a.	n.a.	n.a.	n.a.	n.a.
Timetal® 6.2.5	Ti–6Al–2.5V	1024	n.a.	n.a.	n.a.	n.a.	n.a.	n.a.	n.a.
Timetal® 6.6.2	Ti–6Al–6V–2Sn–0.5Fe–0.5Cu	946	1704	n.a.	650	7.2	9.0	9.4	n.a.
Timetal® 6.8.5	Ti–6Al–5Zr–0.5Mo–0.25Si	1020	n.a.	n.a.	n.a.	n.a.	n.a.	n.a.	n.a.
Timetal® 8.1.1	Ti–8Al–1Mo–1V	1038	n.a.	n.a.	n.a.	n.a.	n.a.	n.a.	n.a.

Table 3.29 (continued)

Common and trade name	Average chemical composition (%wt.)	α-β transition temperature (T°C)	Melting range (T°C)	Short-term creep (1000 h, 250°C) (MPa)	Specific heat capacity (c_p, Jkg^{-1})	Thermal conductivity (k, Wm^{-1}.K^{-1})	Coefficient of linear thermal expansion (20–100°C) (α, 10^{-6} K)	Coefficient of linear thermal expansion (21–538°C) (α, 10^{-6} K)	Electrical resistivity (ρ, $\mu\Omega$.cm)
Timetal® 8.2.9	Ti–5.6Al–3.5Sn–3.5Zr–0.7Nb–0.25Mo–0.35Si–0.06C	1015	n.a.	n.a.	n.a.	n.a.	n.a.	n.a.	n.a.
Timetal® 8.3.4	Ti–5.8Al–4Sn–3.5Zr–0.7Nb–0.5Mo–0.35Si–0.06C	1045	n.a.	n.a.	n.a.	n.a.	n.a.	n.a.	n.a.

Table 3.30 Some typical uses of titanium alloys in the CPI

Titanium ASTM grades	Most common CPI applications
Grade 1	Chemically pure titanium, relatively low strength and high ductility. Plate heat exchangers
Grade 2	The CP titanium most used. The best combination of strength, ductibility and weldability. Piping systems and heat exchanger tubing
Grade 3	High-strength titanium used for matrix-plates in shell and tube heat exchangers
Grade 4	Exchangers
Grade 7 (high Pd)	Superior corrosion resistance in reducing and oxidizing environments
Grade 16 (low Pd)	Use in chemical industries
Grade 9	Very high strength and corrosion resistance. Used for hydraulic piping, marine technology
Grade 11	Applications as for Gr 7. Suitable for deep drawing
Grade 12	Better heat resistance than pure titanium. Applications as Gr 7 and 11. Good for shell and tube heat exchangers

be compared to the separation of two isotopes. This very close chemical similarity leads to their parallel association in natural ores and minerals where hafnium is invariably found in zirconium ores in quantities around 2wt%. As a consequence, the commercial grades of zirconium always contain from 1 to 4.5wt% maximum of hafnium. Zirconium forms anhydrous compounds in which the valence may be 1, 2, 3, or 4. But the chemistry of zirconium is characterized by the difficulty of reduction to oxidation states less than four. Zirconium is a highly reactive metal, as evidenced by its standard redox potential of –1.55 V/SHE. Chemical inertness of zirconium is due to its tenacious, strongly adherent and protective oxide layer made of zirconia (ZrO_2). This passivating layer spontaneously forms when metal is exposed to oxygen-containing media such as air or moisture. Even when minute amounts of oxygen are present, this film is impervious and also self-healing and hence protects efficiently the base metal from further chemical attack at temperatures up to 300°C in some cases. Therefore, zirconium is very corrosion-resistant in most strong mineral and organic acids, strong alkalis, saline solutions, some molten salts, and liquid metals above their melting points. Nevertheless, there exist a few chemical reagents which readily dissolve zirconium metal. Amongst these chemicals, hydrofluoric acid, ferric chloride, cupric chloride, aqua regia, concentrated sulfuric acid, and wet chlorine gas strongly attack zirconium. In a finely divided form such as hot

Table 3.31 Corrosion resistance of titanium alloys

Corrosive chemical	Description
Chlorine, chlorine chemicals and chlorides	Titanium is unique among metals in handling these environments. The corrosion resistance of titanium to moist chlorine gas and chloride-containing solutions is the basis for the largest number of titanium applications in the chlor-alkali cells, such as dimensionally stable anodes; bleaching equipment for pulp and paper; heat exchangers, pumps, piping and vessels used in the production of organic intermediates; pollution control devices; human body prosthetic devices; seawater desalination plants.
Chlorine gas	Titanium is widely used to handle moist chlorine gas and has earned a reputation for out-standing performance in this service. The strongly oxidizing nature of moist chlorine passivates titanium resulting in low corrosion rates in moist chlorine. Dry chlorine can cause rapid attack on titanium and may even cause ignition if moisture content is sufficiently low. However, 1–1.5wt% of water is generally sufficient for passivation at 200°C or repassivation after mechanical damage to titanium in chlorine gas under static conditions at room temperature. Factors such as gas pressure, gas flow, and temperature as well as mechanical damage to the oxide film on the titanium, influence the actual amount of moisture required. Caution should be exercised when employing titanium in chlorine gas where moisture content is low.
Chlorinated chemicals	Titanium is fully resistant to solutions of chlorites, hypochlorites, chlorates, perchlorates, and chlorine dioxide. Titanium equipment has been used to handle these chemicals in the pulp and paper industry for many years with no evidence of corrosion. Titanium is used today in nearly every piece of equipment handling wet chlorine or chlorine chemicals in a modern bleach plant, such as chlorine dioxide mixers, piping, and washers. In the future it is expected that these applications will expand including use of titanium in equipment for CO_2 generators and waste water recovery.
Chlorides	Titanium has excellent resistance to corrosion by neutral chloride solutions even at relatively high temperatures. Titanium generally exhibits very low corrosion rates in chloride environments. The limiting factor for application of titanium and its alloys to aqueous chloride environments appears to be crevice corrosion. When crevices are present, unalloyed titanium will sometimes corrode under conditions not predicted by general corrosion rates. Corrosion in sharp crevices in near neutral brine is possible with unalloyed titanium at about 90°C and above. Lowering the pH of the brine lowers the temperature at which crevice corrosion is likely, whereas raising the pH reduces crevice corrosion susceptibility. However, crevice corrosion on titanium is not likely to occur below 70°C. The presence of high concentrations of cations other than sodium such as Ca^{2+} or Mg^{2+}, can also alter this relationship and cause localized corrosion at lower temperatures. ASTM grade 7 and 12 offer considerably improved resistance to crevice corrosion compared to unalloyed titanium. These alloys have not shown any indication of any kind of corrosion in laboratory tests in neutral saturated brines to temperatures in excess of 316°C. ASTM grade 12 maintains excellent resistance to crevice corrosion down to pH values of about 3.
Bromine and iodine	The resistance of titanium to bromine and iodine is similar to its resistance to chlorine. It is attacked by the dry gas but is passivated by the presence of moisture. Titanium is reported to be resistant to bromine water.
Fluorine	Titanium is not recommended for use in contact with fluorine gas. The possibility of formation of hydrofluoric acid even in minute quantities can lead to very high corrosion rates. Similarly, the presence of free fluorides in acid aqueous environments can lead to formation of hydrofluoric acid and, consequently, rapid attack on titanium. On the other hand, fluorides chemically bound or fully complexed by metal ions, or highly stable fluorine-containing compounds (e.g. fluorocarbons), are generally noncorrosive to titanium.
Oxidizing acids	Titanium is highly resistant to oxidizing acids over a wide range of concentrations and temperatures. Common acids in this category include nitric, chromic, perchloric, and hypochlorous (i.e., wet chlorine) acids. These oxidizing compounds insure oxide film stability. Low, but finite, corrosion rates from continued surface oxidation may be observed under high-temperature and highly oxidizing conditions. Titanium has been extensively utilized for handling and producing nitric acid in applications where stainless steels have exhibited significant uniform or intergranular attack. Titanium offers excellent resistance over the full concentration range at sub-boiling temperatures. At higher temperatures, however, titanium's corrosion resistance is highly dependent on nitric acid purity. In hot, very pure solutions or vapor condensates of nitric acid, significant general corrosion (and trickling acid condensate attack) may occur in the 20 to 70wt% range. Under marginal high-temperature conditions, higher purity unalloyed grades of titanium (i.e., ASTM grade 2) are preferred for curtailing accelerated corrosion of

Table 3.31 (continued)

Corrosive chemical	Description
Oxidizing acids	weldments. On the other hand, various metallic species such as Si, Cr, Fe, Ti or various precious metal ions (i.e., Pt, Ru) in very minute amounts tend to inhibit high-temperature corrosion of titanium in nitric acid solutions. Titanium often exhibits superior performance to stainless steel alloys in high-temperature metal-contaminated nitric acid media, such as those associated with the Purex Process for U_3O_8 recovery. Titanium's own corrosion product Ti(IV), is a very potent inhibitor. This is particularly useful in recirculating nitric acid process streams, such as stripper reboiler loops, where effective inhibition results from achievement of steady-state levels of dissolved Ti(IV).
Other inorganic acids	Titanium offers excellent resistance to corrosion by several other inorganic acids. It is not significantly attacked by boiling 10wt% solutions of boric or hydriodic acids. At room temperature, low corrosion rates are obtained on exposure to 50wt% hydriodic and 40wt% hydrobromic acid solutions. But hydrochloric acid attacks readily titanium.
Mixed acids	The addition of nitric acid to hydrochloric or sulfuric acids significantly reduces corrosion rates. Titanium is essentially immune to corrosion by aqua regia ($3HCl:1HNO_3$) at room temperature. ASTM grades 7 and 12 show respectable corrosion rates in boiling aqua regia. Corrosion rates in mixed acids will generally rise with increases in the reducing acid component concentration or temperature.
Organic acids	Titanium is generally quite resistant to organic acids. Its behavior is dependent on whether the environment is reducing or oxidizing. Only a few organic acids are known to attack titanium. Among these are hot non-aerated formic acid, hot oxalic acid, concentrated trichloroacetic acid and solutions of sulfamic acid. Aeration improves the resistance of titanium in most of these nonoxidizing acid solution. In the case of formic acid, it reduces the corrosion rates to very low values. Unalloyed titanium corrodes at a very low rate in boiling 0.3wt% sulfamic acid and at a rate of over 2.54 mm.yr^{-1} in 0.7wt% boiling sulfamic acid. Addition of ferric chloride (0.375 g.l^{-1}) to the 0.7wt% solution reduces the corrosion rate to 0.031 mm.yr^{-1}. Boiling solutions containing more than 3.5 g.l^{-1} of sulfamic acid can rapidly attack unalloyed titanium. For this reason, extreme care should be exercised when titanium heat exchangers are descaled with sulfamic acid. The pH of the acid should not be allowed to go below 1 to avoid corrosion of titanium. Consideration should also be given to inhibiting the acid with ferric chloride. Titanium is resistant to acetic acid over a wide range of concentrations and temperatures well beyond the boiling point. It is being used in terephthalic acid and adipic acid up to 200°C and at 67wt% concentration. Good resistance is observed in citric, tartaric, stearic, lactic and tannic acids. ASTM grade 7 and 12 may offer considerably improved corrosion resistance to organic acids which attack unalloyed titanium. Similarly, the presence of multivalent metal ions in solution may result in substantially reduced corrosion rates.
Organic chemicals	Titanium generally shows good corrosion resistance to organic media and is steadily finding increasing application in equipment for handling organic compounds. Titanium is a standard construction material in the Wacker process for the production of acetaldehyde by oxidation of ethylene in an aqueous solution of metal chlorides. Successful application has also been established in critical areas of terephthalic and adipic acid production. Generally, the presence of moisture (even trace amounts) and oxygen is very beneficial to the passivity of titanium in organic media. In certain anhydrous organic media, titanium passivity can be difficult to maintain. For example, methyl alcohol can cause stress corrosion cracking in unalloyed titanium when the water content is below 1.5wt%. At high temperatures in anhydrous environments where dissociation of the organic compound can occur, hydrogen embrittlement of the titanium may be possible. Since many organic processes contain either trace amounts of water and or oxygen, titanium has found successful application in organic process streams.
Alkaline media	Titanium is generally very resistant to alkaline media including solutions of sodium hydroxide, potassium hydroxide, calcium hydroxide and ammonium hydroxide. In concentrations of up to approximately 70wt% NH_4OH, for example, titanium exhibits corrosion rates of less than or equal to 0.127 mm.yr^{-1} up to the boiling point. Near nil corrosion rates are exhibited in boiling calcium hydroxide, and magnesium hydroxide solutions up to saturation. Despite low corrosion rates in alkaline solution, hydrogen pickup and possible embrittlement of titanium can occur at temperatures above 75°C when solution pH is greater than or equal to 12.
Inorganic salt solutions	Titanium is highly resistant to corrosion by inorganic salt solutions. Corrosion rates are generally very low at all temperatures to the boiling point. The resistance of titanium to chloride solutions is excellent. However, crevice corrosion is a concern. Other acidic salt solutions, particularly those formed from reducing acids, may also cause corrosion of unalloyed titanium

Table 3.31 (continued)

Corrosive chemical	Description
Inorganic salt solutions	at elevated temperatures. For instance, a boiling solution of 10wt% sodium sulfate, pH 2.0, causes crevice corrosion on ASTM grade 2. The ASTM grades 12 and 7, on the other hand, are resistant to this environment.
Liquid metals	Titanium exhibits good corrosion resistance to many liquid metals at moderate temperatures. In some cases at higher temperatures it dissolves rapidly. It is used successfully in some applications up to 900°C, for instance with molten aluminum for pouring nozzles, skimmer rakes and casting ladles. Rapidly flowing molten aluminum, however, can erode titanium; and some metals such as cadmium can cause stress corrosion cracking.
Fresh water, steam	Titanium resists all forms of corrosive attack by fresh water and steam to temperatures in excess of 315°C. The corrosion rate is very low or a slight weight gain is experienced. Titanium surfaces are likely to acquire a tarnished appearance in hot water steam but will be free of corrosion. Some natural river waters contain manganese which deposits as manganese dioxide on heat exchanger surfaces. Chlorination treatments used to control sliming results in severe pitting and crevice corrosion on stainless steel surfaces. Titanium is immune to this form of corrosion and is an ideal material for handling all natural waters.
Seawater	Titanium resists corrosion by seawater as high as 260°C. Titanium tubing, exposed for 16 years to polluted seawater in a surface condenser, was slightly discolored but showed no evidence of corrosion. Titanium has provided over 30 years of trouble-free seawater service for the chemical, oil refining and desalination industries. Exposure of titanium for many years to depths of over a mile below the ocean surface has not produced any measurable corrosion. Pitting and crevice corrosion are totally absent, even if marine deposits form. The presence of sulfides in seawater does not affect the resistance of titanium to corrosion. Exposure of titanium to marine atmospheres or splash or tide zone does not cause corrosion.
Oxygen	Titanium has excellent resistance to gaseous oxygen and air at temperatures up to about 371°C. At 350°C it acquires a light straw color. Further heating to 425°C in air may result in a heavy oxide layer because of increased diffusion of oxygen through the titanium lattice. Above 650°C, titanium lacks oxidation resistance and will become brittle. Scale forms rapidly at 925°C. Titanium resists atmospheric corrosion in a marine atmosphere in a similar rate as in industrial and rural atmospheres. Caution should be exercised in using titanium in high oxygen atmospheres. Under some conditions, it may ignite and burn. Ignition cannot be induced even at very high pressure when the oxygen content of the environment is less than 35%. However, once the reaction has started, it will propagate in atmospheres with much lower oxygen levels than are needed to start it. Steam as a diluent allows the reaction to proceed at even lower O_2 levels. When a fresh titanium surface is exposed to an oxygen atmosphere, it oxidizes rapidly and exothermically. Rate of oxidation depends on O_2 pressure and concentration. When the rate is high enough so that heat is given off faster than it can be conducted away, the surface may begin to melt. The reaction becomes self-sustaining because, above the melting point, the oxides diffuse rapidly into the titanium interior, allowing highly reactive fresh molten titanium to react at the surface.
Hydrogen	The surface oxide film on titanium acts as an effective barrier to penetration by hydrogen. Disruption of the oxide film allows easy penetration by hydrogen. When the solubility limit of hydrogen in titanium (about 100–150 ppm for Timetal 50A) is exceeded, hydrides begin to precipitate. Absorption of several hundred ppm of hydrogen results in embrittlement and the possibility of cracking under conditions of stress. Titanium can absorb hydrogen from environments containing hydrogen gas. At temperatures below 80°C hydrogen pickup occurs so slowly that it has no practical significance, except in cases where severe tensile stresses are present. In the presence of pure hydrogen gas under anhydrous conditions, severe hydriding can be expected at elevated temperatures and pressures. The surface condition is important to hydrogen penetration. Titanium is not recommended for use in pure hydrogen because of the possibility of hydriding if the oxide film is broken. Laboratory tests have shown that the presence of as little as 2% moisture in hydrogen gas effectively passivates titanium so that hydrogen absorption does not occur. This probably accounts for the fact that titanium is being used successfully in many process steams containing hydrogen with very few instances of hydriding being reported. A more serious situation exists when cathodically impressed or galvanically induced currents generate nascent hydrogen directly on the surface of titanium. The presence of moisture does not inhibit hydrogen absorption of this type. Laboratory experiments have shown that three conditions usually exist simultaneously for hydriding to occur: (i) The pH of the solution is less than 3 or greater than 12; the metal surface must be damaged by abrasion;

Table 3.31 (continued)

Corrosive chemical	Description
Hydrogen	or impressed potentials are more negative than –0.70 V. (ii) The temperature is above 80°C or only surface hydride films will form which, experience indicates, do not seriously affect the properties of the metal. Failures due to hydriding are rarely encountered below this temperature. (There is some evidence that severe tensile stresses may promote hydriding at low temperatures.) (iii) There must be some mechanism for generating hydrogen. This may be a galvanic couple, cathodic protection by impressed current, corrosion of titanium, or dynamic abrasion of the surface with sufficient intensity to depress the metal potential below that required for spontaneous evolution of hydrogen. Most of the hydriding failures of titanium that have occurred in service can be explained on this basis. In seawater, hydrogen can be produced on titanium as the cathode by galvanic coupling to a dissimilar metal such as zinc or aluminium which are very active (low) in the galvanic series. Coupling to carbon steel or other metals higher in the galvanic series generally does not generate hydrogen in neutral solutions, even though corrosion is progressing on the dissimilar metal. The presence of hydrogen sulfide, which dissociates readily and lower pH, apparently allows generation of hydrogen on titanium if it is coupled to actively corroding carbon steel or stainless steel. Within the range pH 3 to 12, the oxide film on titanium is stable and presents a barrier to penetration by hydrogen. Efforts at cathodically charging hydrogen into titanium in this pH range have been unsuccessful in short-term tests. If pH is below 3 or above 12, the oxide film is believed to be unstable and less protective. Breakdown of the oxide film facilitates access of available hydroge to the underlying titanium metal. Mechanical disruption of the film (i.e. iron is smeared into the surface) permits of hydrogen at any pH level. Impressed currents involving cathodic potentials more negative than – 0.7 V in near neutral brines can result in hydrogen pickup in long-term exposures. Furthermore, very high cathodic current densities (more negative than –1.0 V/SCE) may accelerate hydrogen absorption and eventual embrittlement of titanium in seawater even at ambient temperatures. Hydriding can be avoided if proper consideration is given to equipment design and service conditions in order to eliminate detrimental galvanic couples or other conditions that will promote hydriding.
SO_3 and H_2S	Titanium is resistant to corrosion by gaseous sulfur dioxide and water saturated with sulfur dioxide. Sulfurous acid solutions also have little effect on titanium. Titanium has demonstrated superior performance in wet SO_2 scrubber environments of power plant systems. Titanium is not corroded by moist or dry hydrogen sulfide gas. It is also highly resistant to aqueous solutions containing hydrogen sulfide. The only known detrimental effect is the hydriding problem discussed in the previous section. In galvanic couples with certain metals such as iron, the presence of H_2S will promote hydriding. Hydriding, however, does not occur in aqueous solutions containing H_2S if unfavorable galvanic couples are avoided. Titanium is highly resistant to general corrosion and pitting in the sulfide environment to temperatures as high as 260°C. Sulfide scales do not form on titanium, thereby maintaining good heat transfer.
N_2 and NH_3	Titanium reacts with pure nitrogen to form surface films having a gold color above 538°C. Above 816°C, diffusion of the nitride into titanium may cause embrittlement. Nevertheless, titanium is not corroded by liquid anhydrous ammonia at room temperature. Low corrosion rates are obtained at 40°C. Titanium also resists gaseous ammonia. However, at temperatures above 150°C, ammonia will decompose and form hydrogen and nitrogen. Under these circumstances, titanium could absorb hydrogen and become embrittled. The high corrosion rate experienced by titanium in the ammonia-steam environment at 220°C is believed to be associated with hydriding. The formation of ammonium chloride scale could result in crevice corrosion of Timetal 50A at boiling temperatures. Timetal Code-12 and 50A Pd are totally resistant under these conditions. This crevice corrosion behavior is similar to that shown for sodium chloride.

Sources: TIMET Corp./ Titanium Industries.

sponge or turnings, powdered zirconium is highly hazardous due to its pyrophoricity. Zirconium contains five naturally occurring stable isotopes: ^{90}Zr, ^{91}Zr, ^{92}Zr, ^{94}Zr, and ^{96}Zr. Owing to the low thermal neutron cross-section of hafnium-free zirconium, zirconium alloys such as Zircaloys [R] are extensively used as nuclear fuel container materials in nuclear power reactors.

Corrosion resistance – zirconium exhibits a better chemical resistance than titanium when subject to corrosive attack in harsh conditions. The corrosion resistance of zirconium is due to

Table 3.32 Titanium industrial uses and applications

Application	Description
Aerospace industry	The aerospace industry is the single largest market for titanium products primarily due to the exceptional strength-to-weight ratio, elevated temperature performance and corrosion resistance. Titanium applications are most significant in jet engine and airframe components that are subject to temperatures up to $600°C$ and for other critical structural parts. Usage is widespread in most commercial and military aircraft. Titanium is also used in spacecraft where the many benefits of titanium are effectively utilized. The largest single use of titanium is in the aircraft gas turbine engine.
Gas turbine engines	Highly efficient gas turbine engines are possible only through the use of titanium-based alloys in components like fan blades, compressor blades, disks, hubs and numerous non-rotor parts. The key advantages of titanium-based alloys in this application include high strength-to-weight ratio, strength at moderate temperatures and good resistance to creep and fatigue. The development of titanium aluminides will allow the use of titanium in hotter sections of a new generation of engines.
Heat transfer	A major industrial application for titanium remains in heat transfer applications in which the cooling medium is seawater, brackish water or polluted water. Titanium condensers, shell and tube heat exchangers, and plate and frame heat exchangers are used extensively in power plants, refineries, air conditioning systems, chemical plants, offshore platforms, surface ships and submarines. The life span and dependability of titanium are demonstrated by the fact that of the millions of feet of welded titanium tubing in power plant condenser service, there have been no reported failures due to corrosion on the cooling water side.
Electro-chemistry	Titanium electrodes coated with precious metal or precious metal oxides named DSA®, dimensional stable anodes, are used for: (i) impressed current cathodic protection, (ii) chlorine production, (iii) electroplating, (iv) hydrometallurgy and metal recovery, and (v) electrophoresis and electro-osmosis and other applications where long-term electrode stability is required. The unique electrochemical properties of the titanium-based DSA® make it the most energy efficient unit for the production of chlorine, chlorate, and hypochlorite. Hydrometallurgical extraction, and electrowinning of metals from ores in titanium reactors is an environmentally safe alternative to smelting processes. Extended lifespan, increased energy efficiency, and greater product purity are factors promoting the usage of titanium electrodes in electrowinning and eletrorefining of metals like copper, gold, manganese, and manganese dioxide.
Desalination	Excellent resistance to corrosion in seawater and chloride solutions, erosion, and high condensation efficiency, make titanium the most cost-effective and dependable material for critical segments of desalination plants. Increased usage of very thin-walled welded tubing makes titanium competitive with copper–nickel.
Chemical processing	Titanium vessels, heat exchangers, tanks, agitators, coolers, and piping systems are utilized in the processing of aggressive compounds, like nitric acid, organic acids, chlorine dioxide, inhibited reducing acids, and hydrogen sulfide.
Hydrocarbon processing	The need for longer equipment life, coupled with requirements for less downtime and maintenance favor the use of titanium in heat exchangers, vessels, columns, and piping systems in refineries, and offshore platforms. Titanium is immune to general attack and stress corrosion cracking by hydrocarbons, hydrogen sufhide, brines, and carbon dioxide.
Marine applications	Owing to its high toughness, high strength, and exceptional erosion/corrosion resistance, titanium is currently being used for submarine ball valves, fire pumps, heat exchangers, castings, hull material for deep sea submersibles, water jet propulsion systems, shipboard cooling and piping systems.
Medical	Titanium is widely used for implants, surgical and prosthetic devices, pacemaker cases and centrifuges. Titanium is the most bio-compatible of all metals due to its total resistance to attack by body fluids, high strength and low modulus.

Sources: TIMET Corp. / Titanium Industries.

the formation of a dense, tenaciously adherent, chemically inert oxide film of zirconia (ZrO_2) which forms spontaneously at ambient temperature on the metal surface exposed to the corrosive media. Any breakdown in the film reforms instantly and spontaneously in most oxidizing environments. Actually, zirconium is a highly reactive metal. It is normally covered with a layer of oxide film resulting from the spontaneous oxidation reaction in air or water at ambient temperatures or below. This film will form on the fresh surfaces of zirconium created

Table 3.33 Titanium world producers

Producer	Address
Oremet-Wah Chang	600 N.E. Old Salem Road, P.O. Box 460, Albany, OR 97321-0460, USA Telephone: 1 (541) 926 4211 Facsimile: 1 (541) 967 6994 Telex: 360-741 Internet: http://www.wahchang.com
TIMET (former Titanium Metals Corporation)	1999 Broadway, Suite 4300, Denver, CO 80202, USA Telephone: 1 (303) 296 5600 Facsimile: 1 (303) 296 5640 E-mail: danielle.nelson@timet.com Internet: http://www.timet.com
Titanium Industries	48 South Street, Morristown, NJ 07960, USA Telephone: 1 (973) 984 8200 Facsimile: 1 (973) 984 8206 E-mail: titanium@titanium.com Internet: http://www.titanium.com
RMI Titanium Company	1000 Warren Avenue, P.O. Box 269, Niles, OH 44446-0269, USA Telephone: 1 (330) 544 7700 Facsimile: 1 (330) 544 7701

by operations like cutting, machining, and pickling. It is very thin, ranging from 10 nm to 100 nm, but is still more protective than most oxide films in a broad range of corrosive media. This thin film will suppress the reactivity of zirconium and grows to a steady state when zirconium equipment is exposed to a compatible environment.

Zirconium is one of very few metals that can take oxygen from water to form a protective oxide film even in highly reducing acids, such as hydrochloric acid. In an incompatible medium, such as hydrofluoric acid, concentrated sulfuric acid and aqua regia, corrosive breakdown occurs because fluoride anions avoid the self-healing process. This property, known as the valve action (VA), is not peculiar to zirconium but concerns all the refractory metals. This oxide film protects the base metal from both chemical and mechanical attack at temperatures up to about $400°C$.

Zirconium has a high resistance to localized forms of corrosion pitting, crevice corrosion, and stress corrosion cracking. Zirconium is exceptionally resistant to corrosion by most strong mineral acids (e.g., HCl, HNO_3, H_2SO_4, H_3PO_4). It is resistant to most organic acids (e.g., formic, acetic, lactic, oxalic). However, very few materials also exhibit resistance to strong alkali hydroxides (e.g., NaOH, KOH), and alkalis (e.g., NH_3) as well as zirconium. Therefore, zirconium is almost unique in this regard and can be used interchangeably between acid and alkaline conditions. Zirconium also resists seawater and chloride solutions, except ferric and cupric chlorides, and some molten salts. Furthermore, owing to its hydride-forming capabilities, zirconium has a much higher capability than titanium, niobium, and tantalum to form a protective oxide film under reducing conditions and it is not subjected to hydrogen embrittlement with material blistering when it is cathodically polarized undergoing the hydrogen evolution reaction.

Nevertheless zirconium is readily attacked by acid solutions containing fluorides. As little as 3 ppm wt fluoride ion in 50wt% boiling sulfuric acid corrodes zirconium at 1.25 mm per year. Therefore, solutions of $NH_4H_2F_2$ or KHF_2 have been used for pickling and electropolishing zirconium. For instance, commercial pickling is conducted with HNO_3–HF mixtures (sometimes called fluorhydric aqua regia). On the other hand, on boiling in 80wt% sulfuric acid, zirconium exhibits depleted chemical resistance with a corrosion rate above 56 mm per year.

In addition, anodizing of zirconium equipment or components for service in corrosive media can be improved by artificially increasing the impervious film thickness. However, a coherent, homogeneous, and adherent film can only form on zirconium with a clean surface. Common oxide film formation methods include autoclaving in hot water (360°C for 14 days) or steam (400°C for 1 to 3 days), heating in air or oxygen (560°C for 4–6 h), and immersing in molten salts (600°C to 800°C for 6 or more hours). Thick oxide films of high quality should be shiny, dark blue to black, and approximately 20 μm in thickness. Zirconium with a rough or contaminated surface may prematurely get into the breakaway oxidation stage and form white oxide films spottily or extensively. The white oxide film is thicker than the black film and may be porous. It is only adequate for non-demanding services.

Cost (1998) – chemically pure zirconium is priced at 190 $US.kg^{-1}(86.18 $US.lb^{-1}).

3.3.3.2 History

The minerals **jargon**, **hyacinth**, and **jacinth** also are varieties of zircon and these have been known since the ancient times and are mentioned in the Bible in several places. The existence of a new element within these minerals was not suspected until an analysis of a zircon crystal from Ceylon (Sri Lanka) conducted by Martin Heinrich Klaproth in Berlin (Germany) in 1789. Klaproth announced the discovery of an unknown earth which he called zirkonerde.[27] In 1797, the French chemist Nicolas Louis Vauquelin studied this new earth, to which the name zirconia was given, and published the preparation and properties of some of its compounds.[28] The impure metal, a black powder, was first isolated by the Swedish chemist Jöns Jakob Berzelius in 1824 who heated a mixture of potassium metal and potassium hexafluorozirconate IV (K_2ZrF_6) in a small closed iron tube.[29] Nevertheless, the first relatively pure zirconium was not prepared until 1914 by reducing zirconium tetrachloride ($ZrCl_4$) with sodium in a bomb. High-purity zirconium was first produced by A.E. Van Arkel and J.H. DeBoer (Netherlands) in 1925 by their tetraiodide decomposition process.[30] They vaporized zirconium tetraiodide (ZrI_4) into a bulb containing a hot tungsten filament which caused the tetraiodide to dissociate, depositing pure zirconium crystals on the filament. The successful commercial production of the pure ductile zirconium via the magnesium reduction of zirconium tetrachloride vapor in an inert gas atmosphere was the result of the intense research efforts of Kroll and co-workers at the US Bureau of Mines in 1945–1950.

3.3.3.3 Natural Occurrence, Minerals, and Ores

Zirconium comprises 0.019wt% (i.e., 190 ppm wt) of the Earth's crustal rocks but it is also found in abundance in S-type stars, and has been identified in the sun and meteorites. Hence it is more abundant than many common metals, such as chromium, nickel and cobalt. More recently, analyses of the lunar rock samples show a surprisingly high zirconium oxide content, compared with terrestrial rocks. Zirconium is found at least 37 different mineral forms but the predominant ore of commercial significance is the nesosilicate **zircon** ($Zr[SiO_4]$, tetragonal). Other current mineral sources are the naturally occurring zirconium oxide **baddeleyite** (ZrO_2, trigonal) and **eudialyte** ($(Na,Ca)_6Zr[(Si_3O_9)_2|OH]$). Chief properties of these minerals are briefly listed in Table 11.4. Zircon occurs worldwide as an accessory mineral in igneous (e.g., granites, syenites), metamorphic and sedimentary rocks. Weathering has resulted in segregation and concentration of the heavy mineral sands in layers or lenses of placer deposits in river bedrocks and ocean beaches. Hence all commercial zircon ore deposits are derived from the mining of the ancient unconsolidated beach deposits, in the largest of which are in Kerala State in India, Sri Lanka, the East and West Coast of Australia, on the Trail Ridge in Florida, and at

Richards Bay in the Republic of South Africa and in the State of Minas Gerais, Brazil. Nevertheless, these heavy mineral sands are primarily processed for the recovery of the titanium-bearing minerals such as ilmenite, rutile, and leucoxene, and the zircon is only obtained as a by-product. Therefore, the output of zircon depends largely then on the market for these titanium minerals used in producing titanium oxide white pigment and titanium metals. On the other hand, some forms of zircon have excellent gemstone qualities.

3.3.3.4 Mining and Mineral Dressing

The deposits usually contain ca. 4wt% heavy minerals such as **rutile** (TiO_2, tetragonal), **ilmenite** ($FeTiO_3$), **leucoxene, monazite** (Ce, La, Th)$[PO_4]$), **xenotime** (Y$[PO_4]$), and **zircon** (Zr$[SiO_4]$). Other heavy minerals such as staurolite, tourmaline, sillimanite, corundum, and magnetite may be recovered as local situations warrant. Initially zirconium sand is mined and excavated from beach deposits using front-end loaders, or sand dredges. Typically, the over-burden is bulldozed away, the excavation is flooded, and the raw sand is handled by a floating sand dredge capable of dredging to a depth of 18 m. The material is broken up by a cutter head to the bottom of the deposit and the sand slurry is pumped to a wet-mill concentrator mounted on a floating barge behind the dredge. Secondly, sand is wet concentrated by the classical mineral dressing operations using screens, Reichert cones, spirals, and cyclones are used in order to remove the coarse sand, slimes, and light density sands to produce a 40wt% heavy mineral concentrate. The tailings are returned to the back end of the excavation and used for rehabilitation of worked-out areas. The concentrate is then dried and iron oxide and other surface coatings are removed by means of various operation units such as gravity, magnetic, and electrostatic separation.

3.3.3.5 Processing and Industrial Preparation

The concentrate of zircon is then mixed with coke and introduced in a ball mill. The ground mixed powder is poured into a closed reactor vessel filled with dry pure chlorine gas and heated. Then the carbo-chlorination reaction of zircon occurs giving a mixture of volatile chlorides ((Zr + Hf) Cl_4, $SiCl_4$, $TiCl_4$) and carbon monoxide (CO). A raw distillation separates silicon tetrachloride from the zirconium and hafnium tetrachlorides. After rectification, the crude $ZrCl_4$ is sparged in a tank containing an aqueous solution of ammonium thiocyanate (NH_4SCN). At this stage, zirconium is separated from is very close chemical neighbor hafnium by means of a liquid–liquid extraction operation unit. The selected solvent is methyl-iso-butylketone (MIBK) dissolved in kerosene. After extraction all the hafnium is virtually recovered in the organic phase and the Zr-enriched aqueous solution is mixed with ammonia in a tank where zirconia precipitated. The filtration gives a wet zirconia cake which after washing is calcined in a rotary kiln. The pure zirconia is then submitted to a second carbo-chlorination operation in order to give a high-purity zirconium tetrachloride. Winning of zirconium from $ZrCl_4$ is commonly performed by metallothermic reduction by magnesium metal in an argon inert atmosphere according to the Kroll process (see titanium, Section 3.3.2). The Kroll process is extensively used in the US by Oremet-Wah Chang and Western Zirconium, and in France by the former CEZUS, while in the former Soviet Union (CIS) the metal is obtained by molten salt electrolysis of a molten chloride bath containing K_2ZrF_6 as solute. The zirconium sponge produced is then chopped, crushed, and graded by hand sort. In order to remove $MgCl_2$ the crushed sponge is vacuum heated. After cooling, it is agglomerated by pressing into briquettes. These briquettes are stacked end to end and electron beam welded (EBM) to form a consumable electrode for the melting process. The consumable electrode is loaded into a vacuum arc furnace (VAR) and remelted. The zirconium ingot is machined and heated prior to forging, extruding, or rolling according to the desired mill products.

For applications requiring high-purity zirconium, refining of the metal can be achieved by (i) the Van Arkel–DeBoer process which consists of chemical vapor deposition of pure zirconium from the volatile iodide, (ii) electron-beam melting of the ingot in order to volatilize low vapor pressure elements, and finally (iii) electrotransport which consists of applying a high current density (several $MA.m^{-2}$) to the zirconium ingot in order to electromigrate impurities.

3.3.3.6 Zirconium Alloys

Table 3.34 Zircadyne[®] chemical composition[a]

Trade designation (ASTM grade)	Zircadyne[®] 702	Zircadyne[®] 704	Zircadyne[®] 705	Zircadyne[®] 706
UNS designation	R60702	R60704	R60705	R60706
Zr + Hf (wt%min)	99.2	97.5	95.5	95.5
Hf (wt%max)	4.5	4.5	4.5	4.5
Fe + Cr (wt%max)	0.20	0.2–0.4	0.2	0.2
Sn (wt%max)	–	1.0–2.0	–	–
H (wt%max)	0.005	0.005	0.005	0.005
N (wt%max)	0.025	0.025	0.025	0.025
C (wt%max)	0.05	0.05	0.05	0.05
O (wt%max)	0.16	0.18	0.18	0.16
Nb (wt%)	–	–	2.0–3.0	2.0–3.0

[a]Source: Oremet-Wah Chang.

For each nuclear power use, each reactor vendor issues particular, detailed specifications which usually include the pertinent ASTM nuclear specifications. Zirconium metal, hafnium free, is being produced in volume by Oremet-Wah Chang, and Western Zirconium in the US, CEZUS (Péchiney) in France, and in the CIS for use in nuclear energy programs.

Table 3.35 Nuclear grades of zirconium (i.e., hafnium free)

Alloying element (wt%)	Zircaloy[®] 2	Zircaloy[®] 4	Zr–2.5Nb	Excel	ATR	Ozhennite 0.5
	US Naval Nuclear Propulsion Program (PWR)		Atomic Energy of Canada, Ltd. (CANDU)	Canada	UK Nuclear Program	Former USSR Nuclear Reactor (RMBK)
Sn	1.5	1.5	<0.02	3.5	<0.01	0.02
Fe	0.14	0.22	<0.08	<0.08	<0.05	0.1
Cr	0.1	0.1	<0.02	<0.02	<0.01	<0.02
Ni	0.05	<0.004	<0.007	<0.007	<0.004	0.1
Nb	–	–	2.5	0.8	–	0.1
Cu	–	–	–	–	0.55	–
Mo	–	–	–	0.8	0.55	–

Table 3.36 Physical properties of zirconium and zirconium alloys

Usual trade name	UNS	Chemical composition (wt%)	Density (ρ, kg.m⁻³)	Young's modulus (E, GPa)	Poisson's ratio (ν)	Yield strength 0.2% proof (σ_{YS}, MPa)	Ultimate tensile strength (σ_{UTS}, MPa)	Elongation (Z, %)	Vickers hardness (HV)	Thermal conductivity (k, W.m⁻¹.K⁻¹)	Specific heat capacity (c_P, J.kg⁻¹.K⁻¹)	Coefficient of linear thermal expansion (α, 10⁻⁶ K⁻¹)	Electrical resistivity (ρ, $\mu\Omega$.cm)
Zirconium unalloyed reactor grade (iodide process)	R60001	>99.9 Zr	6510	n.a.	n.a.	100–130	170–210	40–45	85–100	n.a.	n.a.	n.a.	n.a.
Zirconium unalloyed (Kroll process)	R60701	99.6 (Zr + Hf)	6510	n.a.	n.a.	250–310	350–390	23–31	195–215	21.1	n.a.	5.04	n.a.
Zr sponge	R60703 (Zr + Hf)	98.0 (Zr + Hf)	n.a.	n.a.	n.a.	250–310	350–390	23–31	195–215	n.a.	n.a.	n.a.	n.a.
Zircadyne® 702	R60702 (Zr + Hf)	99.2 (Zr + Hf)	6510	99.28	0.35	30	55	16	n.a.	22.48	285	5.89	39.7
Zircadyne® 704	R60704	(Zr + Hf)-1.2Sn	6570	n.a.	n.a.	241	413	14	n.a.	13.84	280	n.a.	n.a.
Zircadyne® 705	R60705	(Zr + Hf)-2.5Nb	6640	96.53	0.33	379	552	16	n.a.	17.1	280	6.30	55.0
Zircadyne® 706	R60706	Zr-2.5Nb	6640	n.a.	n.a.	345	510	20	n.a.	17.1	280	6.3	55.0
ERZr4	R60707 (Zr + Hf)	95.5 (Zr + Hf)	n.a.	n.a.	n.a.	n.a.	n.a.	n.a.	n.a.	16.60	280	n.a.	n.a.
Zircaloy® 2	R60802	99.2Zr	6550	n.a.	n.a.	340–490	450–520	29	205–220	14.01	n.a.	5.67	n.a.
Zircaloy® 4	R60804	97.5Zr	6560	n.a.	n.a.	n.a.	n.a.	n.a.	n.a.	n.a.	n.a.	n.a.	n.a.
Zr-alloy reactor grade	R60901	Zr-2.5Nb	6640	n.a.	n.a.	470	590	21	150–600	17.1	n.a.	6.3	55

Table 3.36 (continued)

Usual trade name	UNS	Chemical composition (wt%)	Density (ρ, kg.m^{-3})	Young's modulus (E, GPa)	Poisson's ratio (ν)	Yield strength 0.2% proof (σ_{YS}, MPa)	Ultimate tensile strength (σ_{UTS}, MPa)	Elongation (Z, %)	Vickers hardness (HV)	Thermal conductivity (k, W.m^{-1}.K^{-1})	Specific heat capacity (c_P, J.kg^{-1}.K^{-1})	Coefficient of linear thermal expansion (α, 10^{-6} K^{-1})	Electrical resistivity (ρ, $\mu\,\Omega$.cm)
Zr-alloy reactor grade	R60902	Zr–2.5Nb	6640	n.a.	n.a.	470	590	24	230–250	17.1	n.a.	6.3	55
Zirconium 30 (IMI)	n.a.	Zr–0.55Cu–0.55Mo	6550	n.a.	n.a.	220–320	470–550	20–31	130–180	25.3	n.a.	5.93	n.a.
Zirconium-0.65Cb	R60904	Zr–0.65Nb	n.a.	n.a.	n.a.	n.a.	n.a.	n.a.	n.a.	n.a.	n.a.	n.a.	n.a.

3.3.3.7 Corrosion Resistance

Table 3.37 Corrosion resistance of zirconium and zirconium alloys (source Oremet-Wah Chang)

Corrosive media	Description
Hydrochloric acid	Zirconium has corrosion rates of less than 0.0127 mm.yr^{-1}. Zirconium is totally immune to attack by hydrochloric acid at all concentrations to temperatures well above boiling. Aeration has no effect, but oxidizing agents such as cupric or ferric ions may cause pitting. Furthermore, zirconium also has excellent corrosion resistance to hydrobromic and hydroiodic acid.
Sulfuric acid	As a general rule zirconium is completely resistant to sulfuric acid up to boiling temperatures, at concentrations up to 70wt%, except that the heated affected zones at welds have lower resistance in more than 55 wt% concentration acid. Fluoride ions must be excluded from the sulfuric acid. Cupric, ferric, or nitrate ions significantly increase the corrosion rate of zirconium in 65–75wt%.
Nitric acid	Zircadyne® 702 exhibits strong corrosion resistance in a variety of conditions. The metal is particularly well suited to high-temperature nitric acid applications. Zirconium resists attack by nitric acid at concentrations up to 70wt% and up to 250°C. Above a concentration of 70wt%, zirconium is susceptible to stress-corrosion cracking in welds and points of high sustained tensile stress. Otherwise, zirconium is resistant to nitric acid concentrations of 70–98wt% up to the boiling point.
Oxygen	Owing to its high reactivity with oxygen, when finely divided the zirconium powder may ignite spontaneously in air, especially at elevated temperatures. For instance, powder (size below 4.4 μm) prepared in an inert atmosphere by the hydride–dehydride process ignites spontaneously upon contact with air unless its surface has been conditioned, i.e., preoxidized by slow addition of air to the inert atmosphere. However, the solid metal is much more difficult to ignite. Nevertheless, pure cleaned zirconium plate ignites spontaneously in oxygen of about 2 MPa; the autoignition pressure drops as the metal thickness decreases.
Nitrogen	Zirconium reacts more slowly with nitrogen than with oxygen. Heating at 700°C in nitrogen during 3 min gives a 0.3 μm layer of zirconium nitride or a 1–2 μm at 900°C. The nitriding rate is enhanced by the presence of oxygen in the nitrogen on the metal surface. Clean zirconium in ultrapure nitrogen reacts more slowly. Through the nitride reaction occurs at 900°C or higher, diffusion of nitrogen into zirconium is slow, and the temperatures of 1300°C are needed to fully nitride the material.
Chlorine and chlorides	Heated zirconium is readily chlorinated by ammonium chloride, molten $SnCl_2$, $ZnCl_2$, and chlorinated hydrocarbons and the common chlorinating agents. It is slowly attacked by molten magnesium chloride in the absence of free magnesium metal.

3.3.3.8 Zirconium Machining

Zirconium alloys can be machined by conventional methods, but they have a tendency to gall and work harden during machining. Consequently, tools with higher than normal clearance angles are needed to penetrate previously work-hardened surfaces. Results can be satisfactory, however, with cemented carbide or high-speed steel tools. Carbide tools usually provide better finishes and higher productivity. Mill products such as grades 702, 704, 705, and 706 can be formed, bent, and punched on standard equipment with a few modifications and special techniques. Grades 702 and 704 sheet and strip can be bent on conventional press-brake or roll-forming equipment to a 5t bend radius at room temperature and to 3t at 200°C. Grades 705 and 706 can be bent to a 3t and 2.5t radius at room temperature and to about 1.5t at 200°C. Zirconium has better weldability than some of the more common construction metals including some alloy steels and aluminum alloys. Low distortion during welding stems from a low coefficient of thermal expansion. Zirconium is most commonly welded by the gas–tungsten-arc (GTAW) method, but other methods can also be used, including gas–metal-arc (GMAW), plasma-arc, electron-beam, and resistance welding. Welding zirconium requires proper shielding because of the metal reactivity to gases at welding temperatures. Actually, welding without proper shielding (i.e., argon or helium) causes absorption of oxygen, hydrogen, and nitrogen from the atmosphere, resulting in brittle welds. Although a clean, bright weld results from the use of a proper shielding system, discoloration of a weld is not necessarily an indication of its unacceptability. However, white deposits or a black color in the weld area are not acceptable. A bend test is usually the best way to determine acceptability of a zirconium weld.

Cleaning, pickling, etching and descaling – aqueous solutions of $NH_4H_2F_2$ or KHF_2 have been used for pickling and electropolishing zirconium. Nevertheless, commercial pickling or chemical etching is usually conducted with an aqueous solution of fluorhydric aqua regia containing 25–59% HNO, and 3–7% HF with the balance being water. The ratio of nitric acid to hydrofluoric acid should be a minimum 10 to 1 to minimize pick-up during pickling. The pickling should be immediately followed by a water rinse in a stopping bath to prevent etching and staining of the metal surfaces. Descaling, involving the removal of some scale and lubricant residue, can be accomplished using caustic-based compounds, molten alkaline-based salt baths, abrasive methods or pickling solutions.

3.3.3.9 Industrial Uses and Applications

Owing to these good properties zirconium and zirconium alloys are used extensively in a wide range of industrial applications.[31]

Table 3.38 Zirconium industrial application and uses

Application	Description
Nuclear reactor applications	A major use of zirconium and Zr-based alloys is for structural material in nuclear powder reactors where it serves as a tubular container for nuclear-fuel pellets (enriched UO_2) in the core of pressurized water reactors (PWR). Zirconium is particularly useful for this application because of its ready availability, good ductility, resistance to radiation damage, and low-thermal neutron cross-section (0.18 barn). The thermal neutron absorption cross-section is the ability of a nuclide to absorb thermal neutrons (i.e., neutron in thermal equilibrium with energy around 0.025 eV). Hence, zirconium was originally developed as a nuclear material by Teledyne Wah Chang. Nevertheless, the reactor-grade zirconium should contain very little amounts of the hafnium element. Actually, the hafnium nuclides increase dramatically its thermal neutron cross-section and therefore increase strongly its neutron-absorbing properties.[a] Therefore in this application, reactor grade (i.e., low hafnium) material must be used. These zirconium alloys with an extra-low hafnium content are referred to as Zircaloys®. Owing to its low thermal neutron absorption cross-section and good strength, Zircaloy® is ideal in this application. The major nuclear grade zirconium based alloys are: Zircaloy®-2, Zircaloy®-4, Zr-2.5Nb which have excellent corrosion resistance to high-temperature steam and water, and good mechanical strength.
Chemical processing applications	Zirconium–hafnium alloys offer a significant economic advantage over many other construction materials. Their thermal conductivity, corrosion resistance, formability, strength, and minimum creep characteristics under high operational temperatures make these alloys the logical choice to replace many other materials and they are commonly used as construction materials for the Chemical Processing Industries (CPI). Since the early 1960s, zirconium has been used in the manufacture of urea. Zirconium has been used in acetic acid manufacturing since the early 1970s in replacement of Hastelloy® equipment and this is the number one commercial market for zirconium worldwide. Zirconium is also widely used in the manufacture of formic acid (Leonard–Kemira process). Since the early 1980s, zirconium has been specified for heat exchanger use in nitric acid production facilities where Zr-distillation columns are used to nitric acid concentration rectification. Gas scrubbers and pickling tanks, resin plants, chlorination systems, batch reactors, coal degasification reactors and prilling tanks are but a few of the applications in which zirconium alloys will function with superior efficiency compared to many other common metals.
Miscellaneous applications	Zirconium and zirconium alloys are non-toxic and compatible with body fluids and thus are used in making surgical implants and prosthetic devices. The high oxygen affinity of the divided metal powder allows Zr to be used as a getter in vacuum tubes or used in photoflash bulbs or explosive primers. It is also used as an alloying agent in steel. Other applications include making of rayon spinnerets and lamp filaments.

[a] Hafnium nuclide has a thermal neutron cross-section 565 times higher than zirconium nuclide.

3.3.3.10 Zirconium Metal World Producers

Table 3.39 Zirconium metal world producers

Company	Address
Oremet-Wah Chang	1600 N.E. Old Salem Road, P.O. Box 460, Albany, OR 97321-0460, USA Telephone: 1 (541) 926 4211 Facsimile: 1 (541) 967 6994 Internet: http://www.wahchang.com

3.3.4 Hafnium and Hafnium Alloys

3.3.4.1 Description and General Properties

Hafnium [7440-58-6] with the atomic number 72 and the relative atomic mass (atomic weight) 178.49 is the last element of the subgroup IVB of the periodic chart. It has the chemical symbol Hf and it is named after the Latin word Hafnia meaning Copenhagen. Pure hafnium is a shiny, ductile and soft metal with a gray steel color and has a brilliant silver luster. It has a high density (13,310 $kg.m^{-3}$) and owing to its high melting point of 2230°C it is classified as a refractory metal. The natural element contains six isotopes among which the isotope ^{174}Hf is slightly radioactive with an alpha decay. As a general rule, its physical properties are strongly influenced by the amount of zirconium present and particularly nuclear properties. Actually, of all the natural elements, zirconium and hafnium are two of the most difficult to separate. As the result of the lanthanide contraction the ionic radii of cations Zr^{4+} (87 pm) and Hf^{4+} (84 pm) are virtually identical and the separation is the most difficult of all the elements and could be compared to the separation of two isotopes. Therefore, this very close chemical similarity leads to their parallel association in natural ores and minerals where hafnium is invariably found in zirconium ores in quantities between 1 to 2wt%. It has excellent mechanical properties, and is extremely corrosion resistant owing to the valve action properties of its tenacious and impervious protective oxide (hafnia, HfO_2). The pure metal readily dissolves in HF but hafnium has greater resistance in water, steam, molten alkali metals, and air than zirconium and its alloys. Nevertheless, finely divided hafnium such as hot sponge, powder, or turnings is highly pyrophoric and can ignite spontaneously in air. Therefore, during the machining of the metal, great care should be taken to avoid hazardous conditions. Hafnium contains six naturally occurring stable isotopes: ^{174}Hf, ^{176}Hf, ^{177}Hf, ^{178}Hf, ^{179}Hf, and ^{180}Hf. Moreover, hafnium has a high thermal neutron cross-section roughly 565 times that of zirconium, hence, this capability has made it a primary material for nuclear power reactor control rods. Finally, the combination of its high thermal neutron cross-section and excellent corrosion resistance to strong and concentrated mineral acids also makes hafnium attractive for the reprocessing of spent nuclear fuels.

3.3.4.2 History

In 1911, the French chemist G. Urbain is said to have discovered the chemical element with atomic number 72 in a rare earth residue and he gave it the Latin name Celtium. Later, hafnium was thought to be present in various zirconium minerals and concentrations many years prior to its discovery, in 1923, which was credited to Dirk Coster and George Charles von Hevesey (Denmark). It was finally identified in zircon (a zirconium ore) from Norway, by means of X-ray analysis (it was the first application of Moseley's rule). It was named in honor of the city in which the discovery was made. As previously discussed, most zirconium minerals contain 1 to

5wt% hafnium. It was originally separated from zirconium by repeated recrystallization of the double ammonium or potassium fluorides by Von Hevesy and Jantzen. Pure hafnium metal was first prepared in 1925 by Anton Eduard Van Arkel, and Jan Hendrick DeBoer by decomposing by chemical vapor deposition hafnium tetrachloride, $HfCl_4$, on a heated tungsten filament (process known as Van Arkel–DeBoer). Later in the 1950s, development programs at Oak Ridge National Laboratory allowed the introduction of a liquid–liquid separation process for separating the hafnium fraction from zirconium and recovering the pure metal by the Kroll process applied to the tetrachloride.

3.3.4.3 Natural Occurrence, Mineral, and Ores

Hafnium is more abundant than uranium and tin in the Earth's crust with 5.3 ppm wt. As previously discussed, the close chemical similarity between hafnium and zirconium leads to their parallel association in natural ores and minerals where hafnium is invariably found in zirconium ores in quantities between 1 and 2wt%. Apart from specific zirconium ores such as **zircon** or **baddeleyite** where hafnium is always present, chief and specific hafnium-bearing minerals are rare – the nesosilicates **hafnon** ($HfSiO_4$) and **alvite** (($Hf, Zr, Th)SiO_4 xH_2O$).

3.3.4.4 Processing and Industrial Preparation

This is described under zirconium (Section 3.3.3.5).

3.3.4.5 Industrial Uses and Applications

Because hafnium has a high absorption cross-section for thermal neutrons (almost 565 times that of zirconium) it is extensively used for producing nuclear reactor control rods. On the other hand, hafnium carbide is the most refractory binary composition known, and the nitride is the most refractory of all known metal nitrides (mp $3310°C$). To a lesser extent, hafnium is used in gas-filled and incandescent lamps, as an efficient "getter" for scavenging oxygen and nitrogen, and alloying with iron, titanium, niobium and other refractory metal alloys.

3.3.4.6 Hafnium World Producers

Table 3.40 Hafnium metal world producers

Oremet-Wah Chang	1600 N.E. Old Salem Road, P.O. Box 460, Albany, OR 97321-0460, USA Telephone: 1 (541) 967 4211 Facsimile: 1 (541) 967 6994 Internet: http://www.wahchang.com

3.3.5 Vanadium and Vanadium Alloys

3.3.5.1 Description and General Properties

Vanadium [7440-62-2] with the atomic number 23 and the relative atomic mass of 50.9415(1) is the first metal of the subgroup VB(5) of Mendeleev's periodic chart. It was named after the Scandinavian goddess, Vanadis, owing to its beautiful multicolored compounds. Pure vanadium is a mildly dense (6110 $kg.m^{-3}$) bright white refractory metal (mp 1887°C), and is soft and ductile. It has good corrosion resistance to bromine water, alkalis, cold sulfuric and hot hydrochloric acid, and salt water, but the metal dissolves in aqua regia and hydrochloric acid, and it oxidizes readily above 660°C. The metal has good structural strength and a low fission neutron cross-section, making it useful in nuclear applications. Natural vanadium is a mixture of two isotope nuclides: ^{50}V(0.2497at%) and ^{51}V (99.7503at%) but geologically exceptional specimens (i.e., minerals and ores) are known in which the element has an isotopic composition outside the reported values. Vanadium-50 is slightly radioactive, having a half-life of $> 3.9 \times 10^{17}$ years, this lead to an extra-low specific radioactivity of 1.66 $mBq.kg^{-1}$ (i.e., 0.045 $pCi.kg^{-1}$) for natural vanadium. Vanadium and its compounds are toxic and should be handled with care. The maximum allowable concentration of V_2O_5 dust in air is about 0.05 $mg.m^{-3}$ (8 h time-weighted average – 40-h week).

Cost (1998) – commercial vanadium metal (i.e., 95wt% V) is priced at 44.09 $US.kg^{-1}$ (20 $US.lb^{-1}$) while highly pure vanadium (99.95wt% V) costs about 3215 $US.kg^{-1}$ (1458.33 $US.lb^{-1}$).

3.3.5.2 History

Vanadium was first discovered by Andres Manuel del Rio at Mexico City in 1801. He prepared a number of salts from this material contained in "brown lead" (vanadite), from a mine near Hidalgo in Northern Mexico whose colors del Rio found reminiscent of those shown by chromium, so he called the element panchromium (i.e. akin to chromium). He later renamed the element erythronium, from the Greek, erythros, red, after noting that most of these salts turned red upon heating. Unfortunately, del Rio did not get proper credit for his discovery because a French chemist incorrectly declared del Rio's new element was only impure chromium; del Rio thought himself to be mistaken and accepted the French chemist's statement. The element was rediscovered in 1830 by Selström, who named the element in honor of the Scandinavian goddess. Later Friedrich Wohler came into possession of del Rio's "brown lead" and confirmed del Rio's discovery of vanadium, although the name vanadium still stands rather than del Rio's suggestion of erythronium. It was isolated in nearly pure form by Henry Enfield Roscoe, in 1869, who reduced the vanadium trichloride (VCl_3) with hydrogen gas to give off vanadium metal and HCl. However, vanadium of 99.3 to 99.8wt% purity was not produced until 1922.

3.3.5.3 Natural Occurrence, Minerals, and Ores

Vanadium is found in about 65 different minerals among which are **carnotite** ($KU_4V_2O_{12} \cdot 3H_2O$), **roscoelite** ($K_4(Mg, Fe)SiAl_4V_6O_{26} \cdot 4H_2O$), **vanadinite** ($Pb_{10}V_6O_{16}Cl_2$) and **patronite** ($V_xS_y$), important sources of the metal. Vanadium is also found in phosphate rocks and certain iron or uranium ores, and is present in some crude oils in the form of organic complexes. It is also found in small percentages in meteorites. Commercial production from petroleum holds promise as an important source of the element.

3.3.5.4 Industrial Preparation and Processing

High-purity ductile vanadium can be obtained by metallothermic reduction of vanadium trichloride with magnesium or with magnesium–sodium mixtures. But most of the vanadium metal being produced industrially is now made by calciothermic reduction of vanadium pentoxide, V_2O_5, while aluminothermic or carbothermic reductions have been discontinued. **Calciothermic reduction** is the earliest technique reported for preparing vanadium metal and today is the chief commercial process. It was first described by Marden and Rich,[32] and later by McKechnie and Seybolt.[33] The reduction of vanadium pentoxide by calcium metal takes place in a pressure vessel adapted from the original design. The vanadium produced by this process is used in applications not requiring high purity such as alloying elements in steel. In the **aluminothermic reduction**, a 99.9wt% vanadium is produced reducing V_2O_5 with aluminum powder, followed by electron beam melting to remove aluminum from the metal. Finally, **carbothermic reduction** achieved at high temperature in a dynamic vacuum leads to the production of impure vanadium (95wt%) containing up to 2wt% C and a large amount of unreacted vanadium pentoxide. Apart from vanadium metal, the largest production of vanadium is reported as ferrovanadium for addition to tool steels and titanium alloys.

3.3.5.5 Industrial Applications and Uses

Vanadium, due to its structural strength is used in producing rust-resistant, spring, and high-speed tool steels. It chemical reactivity with carbon allows to use it as an important carbide stabilizer in making steels. About 80% of the vanadium now produced is used as ferrovanadium or as a steel additive. Vanadium foil is used as a bonding agent in cladding titanium to steel. Its low thermal neutron cross-section makes it useful in nuclear applications. It is also a mordant in dyeing and printing fabrics and in the manufacture of aniline black. Vanadium pentoxide is used in ceramics, as cathode intercalation compounds in rechargeable batteries, and as a catalyst. Vanadium–gallium tape is used in producing a superconductive magnet with a field of 17.5 T. Ductile vanadium is commercially available. The major uses for vanadium during are: carbon steel (15%), full alloy steel (17%), high-strength, low-alloy steel (10%), and tool steel (36%).

3.3.6 Niobium and Niobium Alloys

3.3.6.1 Description and General Properties

Niobium [7440-03-1], with the chemical symbol, Nb, the atomic number 41, and the relative atomic mass (atomic weight) 92.90638(2) is the second element of the group VB(5) of Mendeleev's periodic chart. It is named after the Greek, Niobe, goddess of tears and the daughter of Tantalus because tantalum is closely related to niobium in the naturally occurring ores and minerals. Niobium is mildly dense (8570 kg.m^{-3}), shiny white, with a gray steel color, and is a refractory metal (mp 2468°C). When pure it is soft, malleable, and ductile, and takes on a bluish cast when exposed to air at room temperatures for a long time. The metal starts to oxidize in air at temperatures above 230°C, and when processed at even moderate temperatures must be placed in a protective atmosphere or protected by a thermal barrier coating. Niobium is a chemically reactive metal which forms spontaneously a thin protective oxide layer of Nb_2O_5 in oxygen-containing environments. Niobium is mononuclidic with only one stable naturally occurring isotope, ^{93}Nb.

Corrosion resistance – from a chemical point of view, niobium, despite its resemblance to tantalum is less corrosion resistant in concentrated strong mineral acids. However, it remains chemically inert to several inorganic and organic corrosive chemicals below 100°C. Like the other reactive and refractory metals, niobium corrosion resistance is a direct consequence of enhancement of its valve action (VA) properties. Niobium pentoxide formed in oxidizing environments has an amorphous (i.e., vitreous or glassy) structure, is impervious, strongly adherent to metal, extremely thin (few nanometers) and is self-limiting of its own thickness growth. It exhibits more singular properties than the other reactive metal's protective oxide films (e.g., TiO_2, ZrO_2, and HfO_2) but it is less impervious than Ta_2O_5. This oxide has very good dielectric properties and it was used in highly effective electronic capacitors.

Niobium is resistant to most mineral acids at all concentration below $100°C$, such as concentrated HCl, HNO_3, and aqua regia; in addition it is not corroded by solutions containing 10wt% $FeCl_3$. However, niobium immunity is lost when the underlying substrate metal comes in contact with a corrosive chemical which prevents or slows down oxide layer formation or damage this natural barrier. For instance, hydrogen fluoride, HF, and hydrofluoric acid, aqueous solutions containing fluoride anions, sulfur trioxide, SO_3, concentrated sulfuric and orthophosphoric acids, sulfonitric mixtures, concentrated oxalic acid, concentrated strong alkali hydroxides (e.g., NaOH, KOH), sodium hypochlorite, sodium metasilicate, and fused alkali carbonates (e.g., Na_2CO_3 and K_2CO_3) readily attack the niobium metal.

Pure niobium is highly corrosion resistant to molten metals such as (i) the alkali metals Li $(1000°C)$, Na $(1000°C)$, and K $(1000°C)$, and their eutectic NaK $(1000°C)$, (ii) Th–Mg eutectic $(850°C)$, (iii) heavy metals such as Pb $(850°C)$, Hg $(600°C)$, Zn $(450°C)$, Ga $(400°C)$, and Bi $(510°C)$, while it is readily dissolved by liquid metals such as Al and U. The corrosion resistance strongly depends of the oxygen and nitrogen traces into the liquid metal. The niobium alloys such as Nb–1Zr are extremely resistant to molten alkali metals such as lithium and sodium up to $1000°C$ and were extensively used in sodium vapor lamps or for handling liquid alkali metals in heat transfer loops in nuclear reactors.[34]

Finally, niobium metal combines with air above $230°C$ to form the porous pentoxide; with hydrogen the reaction begins above $205°C$, while with nitrogen it reacts above $300°C$. It also combines with most common gases above $250°C$ (e.g., O_2, CO_2, and CO) and with several corrosive gases such as halogens: fluorine at room temperature, chlorine $(200°C)$, and bromine $(250°C)$. Niobium, like tantalum, is susceptible to hydrogen pick-up and extremely sensitive to embrittlement by nascent hydrogen gas evolved during galvanic coupling or cathodic polarization of the metal.[35] Therefore, for certain applications, it is necessary to protect the metal by anodic protection (e.g., impressed current).

Cost (1998) – pure niobium metal (99.85wt% Nb) is priced at 221 \$US.kg^{-1} (100.4 \$US.lb^{-1}).

3.3.6.2 History

Niobium was discovered in 1801 by Charles Hatchett in an ore called columbite (niobite) sent to England more than a century before by John Winthrop the Younger, the first governor of Connecticut, USA. Hatchett called the new element (**columbium**) with the symbol Cb in honor to the country or its origin Columbia, synonym for America. He was not able to isolate the free element. There was then some confusion concerning tantalum and niobium which was resolved by Heinrich Rose, who named niobium, and Marignac in 1846 who separated tantalum from niobium by means of the differences in solubilities of the potassium heptafluorotantalate and niobiate. The name niobium is now preferred over the original name columbium. The niobium metal was first prepared by chemical vapor deposition in 1864 by Blomstrand, who reduced the pentachloride, $NbCl_5$, by heating it in a hydrogen atmosphere. It was first produced on an industrial basis by Dr Von Bolton[36] in 1905. The name niobium was adopted in 1949 by the International Union of Pure and Applied Chemistry (IUPAC) held in Amsterdam (Netherlands), after 100 years of controversy. Today, many leading US chemical societies and government organizations refer to it by this name. However most metallurgists, leading US commercial producers, however, still refer to the metal as its American name columbium and its symbol Cb.

3.3.6.3 Natural Occurrence, Minerals, and Ores

Niobium is relatively rare in the Earth's crust with an abundance of roughly 20 mg.kg^{-1} (ppm wt), similar to cobalt and lithium. The element, owing to its strong chemical reactivity with oxygen, never occurs free in nature, and is found in a combined form in **niobite** (columbite) $((Fe,Mn)(Ta,Nb)_2O_6$, orthorhombic), **tapiolite** $(Fe(Ta,Nb)_2O_6$, tetragonal), niobite–tantalite,

pyrochlore $((Na,Ca)_2(Nb,Ta,Ti)_2O_4(OH,F)\cdot H_2O)$, and **euxenite** $((Y,Er,Ce,U)(Ta,Nb)TiO_6,$ orthorhombic). Large deposits of niobium have been found associated with carbonatites (carbonate–silicate rocks), as a constituent of pyrochlore. Extensive ore reserves are found in Canada, Brazil, Nigeria, Zaire, and in Russia. The metal can be isolated from tantalum, and prepared in several ways. According to Roskill Information Services Ltd, Brazil is the world's principal source of niobium. The country hosts extensive reserves of pyrochlore ores, and the operations of two producers, Cia. Brasileira de Metalurgia e Mineração (CBMM) and Mineração Catalão de Goias account for approximately 85% of all niobium produced worldwide. Although no new sources of niobium production have appeared recently, a number of attractive deposits are under investigation, and feasibility studies are being undertaken at sites around the world. In the former Soviet Union there were three main producers of niobium metal: Silmet in Estonia, Donetsk in Ukraine and Ulva in Kazakhstan. According to Roskill Information Services Ltd approximately 500 tonnes per year of high-purity niobium pentoxide is used worldwide, mainly in the production of capacitors, lithium niobate and optical glass. Of this total, Japan accounts for approximately 60% and the rest of the world, 40%.

3.3.6.4 Processing and Industrial Preparation

Niobium–tantalum concentrates – niobium can be isolated from tantalum, and prepared in several ways from different niobium-containing materials. (i) The primary source is the columbo-tantalite ore, $(Fe,Mn)(Ta,Nb)_2O_6$, obtained after concentration of the raw ore by means of common ore-beneficiation processes (e.g., froth flotation, gravity separation). (ii) A secondary important Nb–Ta source consists of treatment of tin slag resulting from the smelting of niobium–tantalum bearing cassiterite. Actually, the tin slag usually contains between 1.5wt% to 10wt% Ta_2O_5 and 1 to 3wt% Nb_2O_5. Afterwards, the tin slag is leached either by an acid or alkaline solution providing an insoluble Nb–Ta-bearing residue, principally comprising niobium and tantalum pentoxides. (iii) Finally, a third recent source of niobium and tantalum comes from recycling of hard metal containing tantalum and niobium carbides. After recycling, the process provides a Nb–Ta bearing oxide residue. These three Nb–Ta sources can be processed to produce niobium–tantalum concentrates.

Processing – the Nb–Ta concentrate, after size reduction by grinding into a ball mill, follows a digestion operation in which the material is dissolved in the mixture of hydrofluoric and sulfuric acids ($HF + H_2SO_4$). The dissolution gives respectively an aqueous solution of tantalum and niobium hydrogenofluorides (i.e., H_2NbF_7 and H_2TaF_7). The separation of niobium and tantalum is efficiently performed by means of a liquid–liquid solvent extraction operation unit to which niobium hydrogenofluoride is an important by-product. The aqueous phase is a mixture of HCl and HF containing metal ions, while the solvent selected is the methyl isobutyl ketone (MIBK) dissolved in kerosene or the 2-octanol and the process is achieved in a counterflow extraction column. The liquid–liquid solvent extraction process essentially depends on promoting ion exchange between the aqueous phase (i.e., mother liquor) and the immiscible organic phase (i.e., MIBK or 2-octanol) in intimate contact. This is typically achieved by varying the concentration of components. Owing to the solubility in organic phase of both niobium and tantalum fluorides, they are therefore easily separated from impurity cations such as iron, manganese, titanium, and zirconium, which remain in the aqueous mother liquor. Adjusting the acidity of the aqueous phase, it is then easy to separate niobium from tantalum in the stripping solution. Actually, tantalum fluoride is soluble over a wide range of acidity, while niobium is only soluble within high acidity regions. Hence MIBK is placed in contact with aqueous solutions of various acidities in order to achieve the separation of the two metals. At the exits of the extraction column, the two streams, respectively of tantalum and niobium fluorides are recovered for the further recovery of the two metals. Two niobium chemicals are used industrially as feedstock to prepare the niobium metal: (i) niobium heptafluorotantalate (K_2NbF_7), or niobium pentoxide (Nb_2O_5). Niobium heptafluorotantalate is obtained by adding potassium fluoride, KF to the purified stripping solution in order to precipitate the insoluble crystals. The settled crystals of K_2NbF_7 are easily removed from the solution by centrifugation and filtration. Once separated, the crystals are dried. Niobium pentoxide is prepared by precipitation of the niobium hydroxide, $Nb_2O_5 \cdot xH_2O$, adding ammonia gas, NH_3 into the stripping solution containing niobium. The settled precipitate is then filtrated, washed with deionized water, dried, and calcinated, giving off water to obtain the anhydrous niobium pentoxide.

Preparation of niobium metal – the niobium metal is further obtained either by thermal reduction of the niobium pentoxide (Nb_2O_5) or the thermal reduction of the niobium heptafluorotantalate (K_2NbF_7). There are two processes for producing niobium metal by thermal reduction of the niobium pentoxide. The **carbothermic reduction** consists of reducing Nb_2O_5 by carbon. The reaction is carried out in two steps: the first step involves preparation of niobium carbide by vacuothermal reduction of Nb_2O_5 with carbon at 1650°C under 133 mPa. The carbide is then mixed with additional niobium pentoxide and reacted at 2020°C under a diminished pressure of 1.33 mPa. After reaction completion, the reacted mass is cooled under a vacuum. The **metallothermic** reduction uses aluminum powder as a reducing agent. The reduction takes place in a magnesia-lined reactor, and after reaction completion the reacted mass is allowed to cool and the slag separated from the metal. In the metallothermic reduction of the niobium heptafluorotantalate, sodium is used as reducing agent.

3.3.6.5 Niobium Alloy Properties

See **Table 3.41** (pages 160–161).

3.3.6.6 Niobium Metalworking

The cold-working properties of niobium are excellent. Because of its body-centered cubic (bcc) crystal lattice structure, pure niobium is a very ductile metal that can undergo cold reductions of more than 95% without failure. The metal can be easily forged, rolled or swaged directly from ingot at room temperature. Annealing is necessary after the cross-sectional area has been reduced by approximately 90%. Heat treating under a vacuum at 1200°C for 1 h causes complete recrystallization of material cold worked over 50%. Note that due to its high reactivity with oxygen and air (above 260°C) the annealing process must be performed either in an inert gas or in a high vacuum at pressures below 10^{-4} Torr. Of the two methods, the use of a vacuum is preferred. Niobium is well suited to deep drawing. The metal may be cupped and drawn to tube but special care must be taken with lubrication. Sheet metal can also easily be formed by general sheet metalworking techniques. The low rate of work-hardening reduces springback and facilitates these operations.

3.3.6.7 Niobium Machining (Source: Rembar Company, Inc.)

Niobium may be machined using standard techniques. However, due to the tendency of the material to gall, special attention needs to be given to tool angles and lubrication. Niobium also has a tendency to stick to tools during metal-forming operations. To avoid this, specific lubricant and die material combinations are required in high-pressure forming operations.

Turning – machining on a lathe is best performed with high-speed steel tools. For cooling and lubricating, air, soluble oil, or other suitable products may be used. The metal turns very much like lead or soft copper. It must be sheared with the chip allowed to slide off the tool's surface. If any buildup of material occurs, the pressure will break the cutting edge, ruining the tool. Carbide tooling should be used only for fast, light cuts to work efficiently (254 to 381 μm deep).

Drilling – standard high-speed drills, ground to normal angles, may be used. However, the peripheral lands wear badly so that care must be exercised to ensure the drill has not worn undersize.

Table 3.41 Physical properties of selected niobium and niobium alloys

Usual and trade names	UNS	Chemical composition (x, wt%)	Density (ρ, kg.m^{-3})	Young's modulus (E, GPa)	Yield strength 0.2% proof (σ_{YS}, MPa)	Ultimate tensile strength (σ_{UTS}, MPa)	Elongation (Z, %)	Vickers hardness (HV)	Liquidus temperature (T, °C)	Thermal conductivity (k, W.m^{-1}.K^{-1})	Specific heat capacity (c_p, J.kg^{-1}.K^{-1})	Coefficient of linear thermal expansion (α, 10^{-6} K^{-1})	Electrical resistivity (ρ, $\mu\Omega$.cm)
Nb high purity	n.a.	99.97wt% (EB-melted, swaged, and recrystallized at 1100°C)	8570	103	170	240	50	65	2467	54.1	268	7.20	16.0
Nb (soft)	n.a.	99.9wt% (arc-melted, cold forged, recrystallized for 4 h at 1200°C)	8570	99	240	330	50	115	2467	54.1	268	7.20	n.a.
Nb (hard)	n.a.	99.9wt% (cold rolled)	8570	n.a.	550	585	5	160	2467	54.1	268	7.20	n.a.
Nb (unalloyed RG)	R04200	Reactor grade unalloyed niobium	8570	103	207	275–585	5–30	80–120	2468	52.3	270	7.31	12.5
Nb (unalloyed CG)	R04210	Commercial grade unalloyed niobium	8570	103	207	275–585	5–30	80–120	2468	52.3	270	7.31	12.5
Nb(unalloyed	R04211	Commercial grade	8570	103	207	275–585	5–30	80–120	2468	52.3	270	7.31	12.5

Grade	UNS number	Nominal composition											
WC 103		0.5Ta		90	670	725	26						
WC-3009	n.a.	Nb–30Hf–9W	10,100	123	752	862	24	n.a.	n.a.	n.a.	n.a.	7.5	n.a.
C–129Y	R04271	Nb–10W–10Hf–0.15Y	9500	112	515	620	25	220	2400	69.6	268	6.88	n.a.
Cb–752, WC–752	R04271	Nb–10W–2.5Zr	9030	110	400	540	20	180	2425	48.7	281	7.4	n.a.
SCb–291	n.a.	Nb–10Ta–10W	n.a.	n.a.	n.a.	n.a.	n.a.	n.a.	n.a.	n.a.	n.a.	n.a.	n.a.
FS–85	n.a.	Nb–28Ta–10W–1Zr	10,610	140	462–730	570–830	11–23	n.a.	2590	52.8	255	7.1–9.0	n.a.
Nb–Mo–V–Zr	n.a.	Nb–5Mo–5V–1Zr	n.a.	n.a.	n.a.	n.a.	n.a.	n.a.	n.a.	n.a.	n.a.	n.a.	n.a.
Nb–Zr (CG), WC–1Zr	R04261	Nb–1(Zr + Hf) commercial grade	8590	68.9–80	138–255	241–345	15	n.a.	2407	59	270	7.54	14.7
Nb–Zr (RG), KBI® 1	R04251	Nb–1Zr reactor grade	8570	101.3	73–125	125–195	30–40	n.a.	2410	43.9	270	7.54	12.6

Screw cutting – niobium may be screw-cut using a standard die-cutting head provided that an ample amount of lubricant is used. The use of sufficient lubricant prevents galling on the die resulting in the tearing of the thread. Roll threading is an alternate, and preferred, method.

Spinning – with some minor modifications, normal techniques of metal spinning may be applied successfully to niobium. It is generally better to work the metal in stages. For example, when spinning a right-angled cup from flat sheet, several formers should be used to perform the operation in steps of approximately $10°$. Wooden formers may be used for the rough spinning, but a brass or bronze former is essential for finishing. This is because niobium is soft and readily accepts the contour of the former. For small work, aluminum, bronze or narite tools should be used with a radius of approximately 9.525 mm. Note that if sharp angles are required, the tool must be shaped accordingly. Suitable lubrication for this process may be either yellow soap or tallow, both of which must be continually cold-worked. The peripheral speed of the workpiece should be approximately 2.54 m.s^{-1} (500 surface ft.min^{-1}). Niobium is prone to thinning during this process. This is avoided by working the tool in successive, long, sweeping strokes with light pressure instead of a few heavy strokes.

3.3.6.8 Niobium Joining and Welding

Welding – niobium is a highly reactive metal. It reacts at temperatures well below its melting point with all the common gases, e.g. nitrogen, oxygen, hydrogen, and carbon dioxide. At the melting point and above, niobium will react with all the known fluxes. This severely restricts the choice of welding methods. Niobium can be welded to several metals, one of which is tantalum. This can be readily accomplished by resistance welding, tungsten–inert gas (TIG), plasma welding and electron-beam welding. Formation of brittle intermetallic phases is likely with many metals and must be avoided. Surfaces to be heated above $300°C$ should be protected by an inert gas shielding such as argon or helium to prevent embrittlement.

Fusion welding – niobium can be welded satisfactorily by applying standard gas–tungsten-arc (GTAW) heli-arc procedures. The resulting welds are superior to those made under similar conditions with an alternating current. The argon from the torch seems to provide better protection for a small pool. The TIG method is the recommended procedure for welding niobium. However, some modifications to this method are required. It is essential to completely cover the area of the molten pool and the heated zone with inert gas to avoid contamination of the weld metal. This protection must be given to both the back of the weld and the face. For a small workpiece, the torch provides sufficient coverage to the face of the weld. The back of the weld may be protected with a stream of argon from a manifold positioned just below the weld bead. A trailing shield will provide further protection to the hot metal after the main shield has passed. In welding large workpieces, the current required for full penetration now becomes high enough to cause a spread of the molten pool outside the protection of the argon shield. The pool also becomes too large for the argon shield when welding with a filler rod. The solution in this case is to ensure complete protection with the use of an argon-filled box. When this metal is exposed to air at reaction temperatures, it acquires a relatively thick and adherent oxide film that is extremely difficult to remove. Vacuum annealing will cause the oxide film to diffuse rapidly into the metal. This results in a hardening of the weld bead and the heat-affected zone. Note that contamination-free welds can be produced under totally inert atmospheres compared to welds produced employing only inert shielding.

3.3.6.9 Niobium Cleaning, Pickling, and Etching

Cleaning and degreasing – to properly clean and degrease niobium, the following steps are recommended: degrease in a chlorinated solvent such as trichloroethylene, or immerse in commercial alkaline cleanser for 5–10 min. Rinse with water and afterwards immerse in 35–40% HNO_3 for 2–5 min at room temperature. Rinse thoroughly with deionized water.

Pickling – it is critical to ensure that the metal is clean and free of a thick passivating oxide film prior to welding or other joining techniques such as cladding. In these cases, an acid pickling operation is recommended. For ambient temperatures pickling for a few seconds is achieved in a typical solution of 25%–35% HF, and 25%–33% HNO_3 with the balance deionized water. The workpiece must be thoroughly rinsed with deionized water after pickling. In order to optimize the procedure before immersing the part, sample coupons must be used to check the etchant corrosion rate. Removal of approximately 2.54 μm is generally acceptable.

Anodic electroetching – this can be performed using the niobium workpiece polarized as anode (+), while a large platinum cathode is used (–). The electrolytic bath consists of an aqueous solution of 40wt% HF ($d = 1.16$). The electrocleaning is performed between 20–25°C, with an anodic current density of 20–105 A.dm^{-2} during 1 to 2 min.

3.3.6.10 Industrial Applications and Uses

See **Table 3.42** (page 164).

3.3.6.11 Niobium Metal World Producers

Table 3.43 Niobium metal producers

Company	Address
Cabot Performance Materials	P.O. Box 1607, 144 Holly Road, Boyertown, PA 19512, USA Telephone: (610) 367 1500; Facsimile: (610) 369 8259 Internet: http://www.cabot-corp,com/cpm
Oremet-Wah Chang	1600 N.E. Old Salem Road, P.O. Box 460, Albany, OR 97321–0460, USA Telephone: 1 (541) 926 4211 Facsimile: 1 (541) 967 6994 Internet: http://www.wahchang.com
Plansee AG	A–6600 Reutte Tyrol, Austria Telephone: +43 0 56 72 600 0 Facsimile: +43 0 56 72 600 500
COMETEC Gmbh	Lagerhausstrasse 7–9 Linsengericht D–63589, Germany Telephone: (49) 6051 71037 Facsimile: (49) 6051 72030 Internet: http://www.cometec.com
H.C. Starck Gmbh & Co. KG	Postfach 25 40, Im Schleeke 78–91, D-38615 Goslar, Germany Telephone: (49) 5321 751 0 Facsimile: (49) 5321 751 192 Internet: http://www.hcstarck.com
The Rembar Company, Inc.	P.O. Box 67, 67 Main Street, Dobbs Ferry, NY 10522, USA Telephone: 1 (914) 693 2620 Facsimile: 1 (914) 693 2247 Internet: http://www.rembar.com

3.3.7 Tantalum and Tantalum Alloys

3.3.7.1 Description and General Properties

Tantalum [7440-25-7], with the chemical symbol Ta, the atomic number 73, and the relative atomic mass (atomic weight) 180.9479 is the third and last element of the group VB(5) of Mendeleev's periodic chart. Tantalum is named after the Greek word, Tantalos, son of Zeus and

Table 3.42 Niobium industrial applications and uses

Industrial applications	Description
Metallurgy	Niobium is extensively used as an alloying element in carbon and alloy steels, Ni- and Co-base superalloys, and in several nonferrous metals. Niobium addition improves strength either by solid solution strengthening, or by forming intermetallic compounds, and hard carbides. Moreover, minute additions of niobium also improve thermal shock resistance of the alloy. These alloys have improved strength and other desirable properties. Hence, the niobium consumption is largely driven by the increasing demand for superalloys, for use in commercial aircraft, stationary gas turbines and corrosion-resistant applications such as equipment. Niobium is used in arc-welding rods for stabilized grades of stainless steel. It was used in advanced airframe systems such as were used in the Gemini space program.
Chemical process industries (CPI)	Even if niobium and niobium alloys are less corrosion resistant than tantalum and tantalum alloys, owing to its lower cost and its density which is half that of tantalum, it can be used efficiently in applications handling corrosive chemicals at lower temperature and concentration. On the other hand, the niobium alloy Nb–1Zr is extremely corrosion resistant to molten alkali metals such as lithium and sodium up to 1000°C and owing to the low capture cross-section for thermal neutrons, it was extensively used for tubing, and handling liquid alkali metals in heat transfer loops used in nuclear fast neutron breeder reactors.[a]
High-temperature applications	Combination of suitable physical properties such as light-weight, high melting point, good mechanical strength at elevated temperature, good thermal conductivity, and ease of fabrication allow the use of niobium and its alloys for applications requiring high operating temperature under an inert atmosphere or vacuum. Therefore, niobium is extensively used as thermocouple sheaths, thermowells, crucibles for melting speciality glasses, evaporating vessels for refractory compounds, heat shields, rocket nozzles, and finally containers for handling liquid alkali metals and their vapors.
Super-conductivity	Pure niobium, niobium alloys (e.g., Nb–7.5Ta, and Nb–46.5Ti), and the compound Nb_3Sn, exhibit a superconductive state at low temperature, and retain their superconductivity in strong magnetic fields, and hence are extensively used as superconducting materials. These niobium alloys are used in energy storage devices, or for manufacturing powerful electromagnets used either in magnetic resonance imaging (MRI), particle accelerators, or thermonuclear fusion power reactors. Actually, apart their superconductive state, niobium alloys exhibit ductility and can be easily extruded into thin wires, which can be further successfully embedded in a copper metal matrix forming a multifilament bundle.
Corrosion engineering	Owing to its protective oxide film, Nb_2O_5, which has excellent dielectric and insulating properties, niobium is used as base metal coated with platinum group metals or oxides for making noble metal coated industrial anodes. Actually, niobium is preferred to other common titanium base metal owing to its higher dielectric breakdown voltage (120 V) than the latter (10 V) in aqueous chloride solutions. These dimensionally stable anodes (i.e., DSA[R]) are used for cathodic protection of structures in harsh operating conditions: high chloride concentration, and high anodic current densities. Because of its low consumption rate, the anode may be employed in a cathodic protection system to achieve a design life of 20 years or more. These anodes are usually composed of a copper core acting as current collector, onto which a niobium protective layer is cladded. Metallurgically bonded to the niobium substrate is a platinum metal coating, or iridium dioxide deposit obtained by thermal treatment of a suitable precursor. By using these composite anodes, superior protection characteristics are produced (long service life, and corrosion resistance). Due to the copper core, the anode is highly conductive and can be operated at a maximum current density of 1 kA.m^{-2}. In addition, these anodes are also lightweight, flexible, and strong. Because the niobium is highly resistant to anodic corrosion, the anode remains dimensionally stable over its operating life, and consumption of the platinum electrocatalyst is extremely low (40 to 80 mg.A.yr^{-1}). These anodes have proven to operate effectively in fresh water, brackwish water, and seawater, and are not adversely affected by high chloride concentrations. They are extensively used in waterworks facilities (e.g., water storage tanks, waste water clarifiers), in marine technologies (e.g., oil rigs) and other industrial plants where cathodic protection is performed to overcome corrosion.

[a]Webster RT (1984) Niobium in industrial applications. In: Smallwood RE (ed.) Refractionary metals and their industrial applications. ASTM STP 849, ASTM, Philadelphia, pp 18–27.

the father of Niobe in the Greek mythology, because in nature, tantalum always occurs with its chemical neighbor niobium. Highly pure tantalum is a gray-blue, dense, ductile, malleable and very soft metal. Hence, it is very easy to fabricate and can be worked into intricate forms and drawn into fine wire. The physical properties of tantalum are quite similar to mild steel, except that tantalum has a much higher melting point (2996°C) exceeded only by tungsten and rhenium and a high density (16,654 kg.m^{-3}) exceeded only by tungsten, rhenium, the dense platinum group metals, gold, and uranium. Nevertheless, tantalum becomes very hard when traces of interstitial impurities such as H, C, N, and O are present in the body-centered cubic (bcc) crystal lattice structure. The ultimate tensile strength is about 345 MPa, which can be approximately doubled by cold working. It can be welded by a number of techniques but requires completely inert conditions during welding in order to prevent metal oxidation. Moreover, tantalum exhibits a very low ductile–brittle transition temperature (DBTT) which is below 25 K.

Corrosion resistance – from a chemical point of view, tantalum is chemically inert to most inorganic and organic chemicals below 150°C, and the broadest range of its corrosion resistance is a direct consequence of enhancement of its valve action (VA) properties. For instance solid tantalum is totally inert to hydrochloric and nitric acids until boiling point, resistant to aqua regia, perchloric and chromic acids, oxides of nitrogen, chlorine and bromine, organic acids, hydrogen peroxide and aqueous solutions of chlorides (e.g., $FeCl_3$, $AlCl_3$). This outstanding tantalum chemical inertness is due to the formation of a protective passivating film of tantalum pentoxide, Ta_2O_5. This protective barrier forms spontaneously in oxidizing conditions, and exhibits more singular properties than the other reactive metal protective oxide films (e.g. TiO_2, ZrO_2, HfO_2, and Nb_2O_5). Actually, anodic tantalum pentoxide film has an amorphous (i.e., vitreous or glassy) structure, is strongly adherent to the base metal, is extremely thin (about 1 to 4 nm) and self-limiting of its own thickness growth. Moreover, this oxide has very good insulating and dielectric properties which is the primary reason for tantalum's commercial development for highly efficient electronic capacitors.

Tantalum pentoxide is formed and persists even in extremely oxygen-deficient environments and exhibits excellent self-healing properties. Nevertheless, tantalum immunity is lost when the underlying substrate metal becomes in contact with a corrosive chemical which prevents or slows down oxide layer formation or damages this natural barrier. For instance, corrosive breakdown of tantalum occurs when the metal is in contact with hydrogen fluoride and hydrofluoric acid, HF, fluoride salts or aqueous solutions containing fluoride anions in excess of about 5 ppm wt, orthophosphoric acid above 190°C, sulfur trioxide, SO_3, and a fortiori hot concentrated and fuming sulfuric acid (Nordhausen's acid), concentrated strong alkalis, hydroxides (e.g. NaOH, and KOH), ammoniac, ammonium hydroxide, molten sodium and potassium pyrosulfates, and fused carbonates (e.g., Na_2CO_3, and K_2CO_3).[37] In this way tantalum's corrosion resistance is similar to that of borosilicated glasses (e.g., Pyrex$^{®}$) despite it being nonbrittle and a better thermal conductor. Tantalum is susceptible to hydrogen pick-up and extremely sensitive to embrittlement by nascent hydrogen gas evolved during galvanic coupling or cathodic polarization of the metal.[38] This phenomenon leads rapidly to the metal blistering. Therefore, for certain applications, it is necessary to protect the metal by anodic protection (e.g. impressed current). Pure tantalum is highly corrosion resistant to molten metals such as (i) the alkali metals: Li (1000°C), Na (1000°C), and K (1000°C), and their eutectic NaK (1000°C), (ii) alkaline-earths such as Mg (1150°C), (iii) heavy metals such as Pb (1000°C), Hg (600°C), Ga (450°C), and Bi (900°C), while it is readily dissolved by liquid metals such as Al, Zn, Sn, and U.

Finally, tantalum metal combines with most common gases above 250°C (e.g., O_2, N_2, H_2, CO_2, and CO) and with several corrosive gases such as halogens: fluorine at room temperature, chlorine (250°C), bromine (300°C), and with air above 260°C to form the nonprotective and porous thermal oxide. In conclusion, despite its cost and high density, tantalum is an inescapable corrosion resistant material for building chemical process equipment submitted to harsh media such as hot and concentrated strong acids (e.g., nitric, hydrochloric, and hydrobromic) and when no corrosion products are tolerated.

Cost (1998) – pure tantalum metal (99.5wt% Ta) sold as rods is priced at 461 $US.kg^{-1} (i.e., 209 $US.lb^{-1}).

3.3.7.2 History

Historically, tantalum was discovered in 1802 by the Swedish chemist Anders Gustaf Ekenberg in Upsala, but many chemists thought niobium and tantalum were identical elements (some thought that perhaps tantalum was an allotrope of niobium) until Rose, in 1844, and Marignac, in 1866, indicated and showed that niobic and tantalic acids were two different acids. The first relatively pure tantalum was produced by Dr Von Bolton[39] in 1905. In 1922, the first industrial production was started by Fanstell Inc. in Chicago (Illinois), USA. Development of tantalum electrochemistry in molten salts is closely linked with refractory metals electrochemistry. The first studies on tantalum recovery in fused salt by electrolysis appeared in 1931[40] and concerned only metal electrowinning. During the 1930s–1940s, electrowinning was the only way of recovering tantalum metal from ores. After World War II, due to US nuclear, aerospace, and electronics programs, tantalum electrochemistry in molten salt underwent rapid expansion in the period 1950–1965. The greater number of references on the subject concern academic studies on fundamental aspects, and numerous industrial patents (e.g., Norton Company,[41−44] Union Carbide Corp.,[45,46] Horizon Titanium Corp.,[47−52] Timax Corp.,[53] Ciba Ltd,[54−57] Péchiney,[58] SOGEV,[59] Mitsubishi Heavy Industries[60]). This large development of tantalum metallurgy is the result of major military nuclear, space and aeronautical programs which have involved high-refractory material studies for ballistic equipment used to handle liquid metals or molten salts in nuclear power reactor systems, thermal shields for aerospace engines and selection of high-efficiency electronic devices. A wide range of melts and mixtures has been explored for tantalum electrorecovery ranging from tantalum pentoxide diluted in cryolithe melt[61] to tantalum fluoride dissolved in chloride or fluoride melts. However, for electroplating applications, the selection of molten alkaline fluorides was highlighted by industrial studies, especially those performed by the two precursors S. Senderoff and G.W. Mellors[62] from Union Carbide Corp. They showed that in molten alkaline fluorides, refractory metal electrodeposits were dense, coherent and adherent.

3.3.7.3 Natural Occurrence, Minerals, and Ores

The abundance of tantalum in the Earth's crust is 2 mg.kg^{-1} (ppm wt). The chief tantalum-containing minerals are niobio-tantalates of complex general formula $(Fe,Mn)(Nb,Ta)_2O_6$ which is called tantalite when the tantalum pentoxide, Ta_2O_5, content exceeds the niobium pentoxide content, and columbite (or niobite) in the reverse situation. Another important mineral containing both tantalum and niobium is **pyrochlore** $((Na,Ca)_2(Nb,Ta,Ti)_2O_4(OH, F)\cdot H_2O)$ which forms important ore deposits in Brazil and Canada. Tantalum is also often found as an impurity in tin ores such as cassiterite, therefore it is often recovered as by-product slags from tin smelting. Hence, tantalum is industrially recovered where large tin ore deposits are naturally found such as in Thailand, Malaysia, and Australia. Other minerals containing tantalum are: the rare native tantalum, struverite, skobolite, tapiolite, mangano-tantalite, mossite, microlite, nioccalite, calogerasite, thoreaulite, euxenite, wodginite, and samarskite. The main tantalum ore deposits are found in Australia, Brazil, Mozambique, Thailand, Portugal, CIS, Nigeria, Zaire, and Canada.

3.3.7.4 Processing and Industrial Preparation

Tantalum can be prepared in several ways from different tantalum-containing materials. The primary source is the columbo-tantalite ore, obtained after concentration of the raw ore by means of common ore-beneficiation processes (e.g., froth flotation, gravity separation). A secondary important source consists of tin slag resulting from the smelting of niobium–tantalum-bearing cassiterite. The tin slag usually contains between 1.5wt% to 10wt% Ta_2O_5 and 1 to 3wt% Nb_2O_5. Afterwards tin slags are leached either by an acid or alkaline solution providing an insoluble Nb–Ta bearing residue, principally comprising niobium and tantalum pentoxides. Finally, a third recent source of niobium and tantalum comes from recycling of hard metal containing tantalum and niobium carbides. After recycling, the process provides a Nb–Ta-bearing oxide residue. Tantalum is separated from other metals either by the precipitation of the hydroxide $Ta(OH)_5$ dried and calcined to tantalum pentoxide, Ta_2O_5, or crystallized with potassium fluoride, KF, to potassium heptafluorotantalate (K_2TaF_7). Nevertheless, today industrially the fractional crystallization process of fluoride salt has been largely replaced by the solvent extraction procedure. For the preparation of tantalum metal, although several industrial processes have been historically developed for producing tantalum metal from tantalum compounds, electrolysis in molten salts, and reduction of tantalum pentoxide with tantalum carbide, today the majority of tantalum metal is prepared from thermal reduction of potassium heptafluorotantalate by molten sodium as described below.

Preparation of potassium heptafluorotantalate – all the three Nb–Ta residues can be processed by the same digestion operation in which the material is dissolved in a mixture of hydrofluoric and sulfuric acids giving respectively an aqueous solution of tantalum and niobium hydrogenofluorides (i.e., H_2NbF_7 and H_2TaF_7). The separation of niobium and tantalum fluorides is efficiently performed by mean of a liquid–liquid solvent extraction operation unit to which niobium fluoride is an important by-product. The aqueous phase is a mixture of HCl and HF containing metal cations, while the solvent selected is methyl isobutyl ketone (MIBK) dissolved in kerosene. The liquid–liquid extraction process is performed in a counterflow extraction column. The extraction process essentially depends on promoting ion exchange between aqueous phase (i.e., mother liquor) and the immiscible organic phase (i.e., MIBK) in intimate contact. This is typically achieved by varying the concentration of components. Owing to the solubility in the organic phase of both niobium and tantalum fluorides, they are therefore easily separated from impurity cations such as iron, manganese, titanium and zirconium, which remain in the aqueous mother liquor. Adjusting the acidity of the aqueous phase, it is then easy to separate niobium from tantalum. Actually, tantalum fluoride is soluble over a wide range of acidity, while niobium is only soluble within high-acidity regions. Hence, MIBK is placed in contact with aqueous solutions of various acidities in order to achieve the separation of the two metals. At the exits of the extraction column, the two streams, respectively of tantalum and niobium fluoride are recovered for the further recovery of the two metals. The potassium fluoride, KF, is added to the purified tantalum fluoride solution in order to precipitate the potassium heptafluorotantalate (K_2TaF_7). The crystals of K_2TaF_7 are easily removed from the solution by a centrifugation and a filtration operation unit. Once separated, the crystals are dried.

Preparation of tantalum metal – tantalum metal is directly obtained from the metallothermic reduction of potassium heptafluorotantalate with pure sodium metal. After reduction and cooling, the frozen mixture of both sodium fluoride and tantalum particles is crushed. The salt-encased tantalum powder is recovered by a leaching operation unit with fresh water followed by acidified water. Tantalum particles are usually spherical in shape, with a tendency to form grape-like clusters during reduction. After drying and size selection by sieving, large tantalum particles are pressed, sintered and electron-beam (EB) melted under argon to give a bar and other mill products, while the fines with high specific surface area are directly used as capacitor anode powder.

Tantalum powder by the hydriding–dehydriding process – on the other hand, tantalum capacitor powder can also be prepared by the hydriding–dehydriding process. The EB ingot obtained previously is placed in an evacuated and purged furnace into which highly pure hydrogen is introduced. The tantalum is fully hydrided on slow cooling from $800°C$ under hydrogen. The brittle hydride is crushed, ground, and classified to yield powder with an average particle size of 5 μm. The final tantalum powder is obtained after dehydriding (i.e., reversible degassing) the tantalum hydride powder.

3.3.7.5 Tantalum Alloy Properties

See **Table 3.44** (page 169).

3.3.7.6 Tantalum Metalworking (Sources: Rembar Company, Inc./ Cometech Gmbh)

General considerations – because of its body-centered cubic (bcc) crystal structure, highly pure tantalum is a very ductile metal that can undergo cold reductions of more than 95% without failure, and hence is an extremely workable metal with a behavior between that of pure copper and austenitic stainless steels (e.g., AISI 316L). Actually, it can be easily cold-worked, rolled, forged, blanked, formed and drawn with common metalworking equipment. Tantalum is also machinable with high-speed and carbide tools using a suitable coolant. Tantalum does have a tendency to stick to tools during metal-forming operations and hence, it exhibits a strong tendency to seize, tear, and gall. To avoid this, specific lubrication and die material combinations are required in high-pressure forming operations. Most procedures used in working and fabricating tantalum are conventional and can be mastered without very much difficulty. However, all forming, bending, stamping, or deep drawing operations are normally performed cold to avoid oxidation of the metal. Heavy sections can be heated for forging to approximately 300°C. Tantalum can also be joined by: (i) mechanical techniques (i.e. riveting and explosion cladding), (ii) brazing, and finally (iii) welding using techniques such as resistance welding, electron-beam (EB) or tungsten–inert-gas (TIG) welding.

    Forming and stamping – most sheet metal work in tantalum is performed on metal with a thickness ranging from 0.1 to 1.5 mm. The instructions given here apply to metal in this thickness range.

    Punching – punching presents no special difficulties. Steel dies are recommended for use. The clearance between the punch and die should approximate 6% of the thickness of the metal being worked. Close adherence to this clearance is important. The use of light oil is recommended to prevent scoring of the dies. A suitable lubricant is necessary.

    Stamping – stamping techniques are similar to those used with mild steel except that precautions should be taken to prevent seizing or tearing of the metal. Dies may be made of steel except where there is considerable slipping of the metal. In this case, aluminum–bronze or beryllium–copper alloys are recommended. Low-melting-point alloys such as Kirksite® may be used for experimental work or short runs. Rubber or pneumatic die cushions should be used where required. Annealed tantalum takes a permanent set in forming and does not spring back from the dies.

    Deep drawing – deep drawing is an operation where the depth of the draw in the finished part is equal to, or greater than, the diameter of the blank. For deep-drawing operations, only annealed tantalum sheet should be used owing to its appropriate ductility. Note that tantalum does not work-harden as rapidly as most metals, and that work-hardening begins to appear at the top, rather than at the deepest part of the draw. If the piece is to be drawn in one operation, a draw in which the depth is equal to the diameter of the blank can be made. If more than one drawing operation is to be performed, the first draw should have a depth of not more than 40% to 50% of the diameter. Dies should be made of aluminum–bronze alloy, although the punch may be made of steel if not too much slippage is encountered. Sulfonated tallow, chlorinated oil, castor oil or Johnsons No. 150 drawing wax seem to be suitable lubricating agents.

    Spinning – spinning can be accomplished using conventional techniques. Steel roller wheels may be used as tools, although yellow brass may be used for short runs. Yellow soap or Johnsons No. 150 drawing wax may be used as a lubricant.

    Annealing – annealing tantalum is performed by heating the metal at a temperatures above 1100°C, in a high vacuum in order to avoid nitriding, oxidation or hydrogen embrittlement.

Table 3.44 Physical properties of selected tantalum alloys

Usual and trade names	UNS	Average chemical composition (x, wt%)	Density (ρ, kg.m^{-3})	Young's modulus (E, GPa)	Yield strength 0.2% proof (σ_{YS}, MPa)	Ultimate tensile strength (σ_{UTS}, MPa)	Elongation (Z, %)	Micro-Vickers hardness (HV)	Melting or liquidus temperature (T, °C)	Thermal conductivity (k, W.m^{-1}.K^{-1})	Specific heat capacity (c_p, J.kg^{-1}.K^{-1})	Coefficient of linear thermal expansion (α)	Electrical resistivity (ρ, $\mu\Omega$.cm)
Ta (unalloyed, EB-cast)	R05200	99.98wt% (electron-beam melted, cold rolled, and annealed at 1200°C)	16,656	179–185	165–180	205–276	35–50	70–80	2996	57.55	142	6.5	12.5
Tantalum (soft)	n.a.	99.95wt% (recrystallized)	16,656	179	310–380	310–485	25–40	90	2996	57.55	142	6.5	n.a.
Tantalum (hard)	n.a.	99.95wt% (cold worked)	16,656	185.	705	760	3	200	2996	57.55	142	6.5	n.a.
Ta (sintered, cast)	R05210	Sintered unalloyed	16,656	185	220	310	30	120	2996	57.55	139	6.5	12.5
Ta (sintered, cast)	R05400	Sintered unalloyed tantalum	16,656	185	220	310	30	120	2996	57.5	139	6.5	12.5
KBI®2, Tantaloy®63	R05252	Ta-2.5W (electron beam cast)	16,700	179–195	230–241	345–379	40	130	3005	n.a.	n.a.	n.a.	n.a.
KBI®6	n.a.	Ta-6W	n.a.	n.a.	n.a.	n.a.	n.a.	n.a.	n.a.	n.a.	n.a.	n.a.	n.a.
KBI®10, Tantaloy®60, Cabot 10	R05255	Ta-10W (electron beam cast)	16,900	205–207	460–482	550–620	30	245	3030	n.a.	n.a.	n.a.	17.6
KBI®40	R05240	Ta-40Nb (electron beam cast)	12,100	152	193–207	275–310	25–40	n.a.	2705	n.a.	n.a.	n.a.	n.a.
Tantaloy 61 (P/M)	n.a.	Ta-7.5W (sintered)	16,800	200	875	1165	7	400	3025	n.a.	n.a.	n.a.	n.a.
T-111®	n.a.	Ta-8W-2Hf-0.02C	n.a.	n.a.	n.a.	n.a.	n.a.	n.a.	n.a.	n.a.	n.a.	n.a.	n.a.
T-222®	n.a.	Ta-10W-10Hf-0.01C	n.a.	n.a.	n.a.	n.a.	n.a.	n.a.	n.a.	n.a.	n.a.	n.a.	n.a.

3.3.7.7 Tantalum Machining

Turning and milling – in lathe operations, cemented carbide tools such as Vascoloy® Ramet grade 2A-5, with high cutting speeds have been found to be satisfactory. The tools should be kept sharp and should be ground with as much positive rake as the strength of the tool can withstand. The same rakes and angles used with soft copper will usually give satisfactory results with tantalum. A minimum speed of 508 mm.s^{-1} (100 surface ft.min^{-1}) will be found to be correct for most turning operations. Slower speeds will cause the metal to tear, especially if annealed metal is being cut. A suitable lubricant is recommended as a cutting medium and the work must be kept well-flooded at all times. When filing or using emery cloth, the file or cloth must be kept well-wetted. The same general procedures should be followed when milling, drilling, threading, or tapping tantalum. Milling cutters should be of the staggered-tooth type, having substantial back and side relief. In drilling, the point of the drill should be relieved so that it does not rub the work. In threading larger diameters, it is preferable to cut the threads on a lathe rather than with a threading die. When dies or taps are used, they must be kept free of chips and cleaned frequently. Extremely light finishing cuts should be avoided. It is better to use sharp tools and light feeds to finish the work in one cut rather than to take the usual roughing cut followed by a finish cut.

    Grinding – grinding tantalum is difficult and should be avoided if at all possible. Grinding of annealed tantalum is nearly impossible, but cold-worked tantalum can be ground with fair results by using aluminum oxide Norton 38A-60 wheels or equivalent. Most other wheels load rapidly when grinding tantalum.

3.3.7.8 Tantalum Joining

Welding – tantalum may be welded to several other metals. This can be readily accomplished by resistance welding, tungsten–inert gas (TIG), plasma welding, and electron-beam welding. Formation of brittle intermetallic phases is likely with many metals and must be avoided. Surfaces to be heated above 300°C should be protected by an inert gas such as argon or helium to prevent rapid oxidation and embrittlement. Tantalum may also be welded to itself by inert gas arc welding. Note that acetylene torch welding is destructive to tantalum. Resistance welding can be performed with conventional equipment. The methods applied are not substantially different from those used in welding other materials. Because its melting point is higher than that of SAE 1020 steel and its resistivity is only two-thirds that of SAE 1020 steel, tantalum requires a higher power input to accomplish a sound weld. The weld duration should be kept as short as possible, i.e., in the range of one to ten cycles at 60 Hz. This is to prevent excessive external heating. Where possible, the work should be flooded with water for cooling and reduction of oxidation. RWMA Class 2 electrodes are recommended with internal water cooling. As in all resistance welding, the work must be cleaned and be free of scales and oxides. The electrode contours should maintain a constant area and contour to prevent lowering of current and pressure densities. A common mistake in welding tantalum is to apply too much electrode force. This causes so little interface resistance that no weld is made. Strong, ductile welds can be made by the TIG method. Extreme care must be taken to cover all surfaces that are raised above 300°C by the welding heat with an inert gas. Helium, argon, or a mixture of the two gases creates an atmosphere that prevents embrittlement by absorption of oxygen, nitrogen, or hydrogen into the heated metal. Where a pure, inert atmosphere is provided, the fusion and adjacent area will be ductile. Extreme high ductility can be obtained in a welding chamber that can be evacuated and purged with inert gas. When the use of a welding chamber is not practical, the heated surfaces can be protected by proper gas-backed fixturing. This usually serves two purposes: to hold the work in alignment and to chill the work in order to limit the heat area. Weld ductilities in the order of a 180°C bend over one metal thickness can be consistently accomplished when backup gas fixtures and gas-filled trailing cups are used.

3.3.7.9 Tantalum Cleaning and Degreasing

Cleaning and degreasing of tantalum parts must be achieved after metalworking and machining in order to remove adherent greases and scales. It presents no special problems and conventional methods and materials may be used. However, owing to hydrogen embrittlement and corrosion by akaline solutions, hot caustics (e.g., NaOH and KOH) must be avoided.

Grit blasting – the first step consists of blasting tantalum parts with steel grit. The recommended procedure is a blast of a few seconds with sharp particles of No. 90 steel grit, at a pressure ranging between 1.2 to 3 bar. To achieve the best results, the blasting nozzle should be held at an angle nearly tangential to the work, rather than perpendicular to the work to avoid indentation of the metal. Sand (SiO_2, silica), corundum (Al_2O_3, alumina), or carborundum (SiC, silicon carbide) abrasives should not be used because they become embedded in the tantalum and cannot be removed with any chemical treatment that would not also damage the tantalum (i.e., HF).

Cleaning – then the workpieces are immersed in hot hydrochloric acid in order to remove the particles of steel grit embedded in the metal. Owing to the complete inertness of tantalum metal versus hydrochloric acid, the acid can be used concentrated and hot. The parts should then be thoroughly rinsed with distilled or deionized water. Actually, tap water which often contains calcium cations that may be converted to insoluble sulfates in the subsequent cleaning process must be avoided.

Degreasing – the second step consists of a chemical degreasing process. A hot chromic acid cleaning solution, commonly used for cleaning glass, may be applied. A sulfochromic acid solution, which consists of a saturated solution of potassium dichromate in hot concentrated sulfuric acid (98wt% H_2SO_4) may be used for this purpose. However, chromium trioxide (CrO_3) is preferred to potassium dichromate ($K_2Cr_2O_7$) because its use avoids the presence of potassium salt residues in crevices or elsewhere on the tantalum parts. The cleaning solution should be applied at approximately 110°C and should maintain its red color at all times. Actually, when the solution is reduced it becomes muddy or turns green, and should be discarded. After the chromic acid wash, it is necessary to rinse the parts thoroughly with hot distilled water. The parts should be dried in clean, warm air, free from dust. The parts should not be wiped with paper or cloth and they should be handled with great care.

Etching, pickling, and descaling – when the tantalum metal surface must be free from either thick tantalum oxide scale, or embedded abrasive particles of corundum or silica, it is necessary to etch the surface with chemicals which prevent or slow down the passivation process, or dissolve the abrasive. The best acid pickling solution consists of a mixture of 40-60vol% of concentrated HNO_3 combined with 10-30wt% HF; the balance is deionized water. The operating temperature ranges between 50 and 60°C. After pickling the workpiece must be rinsed thoroughly with deionized water.

Anodic electroetching – this can be performed using the tantalum workpiece as anode (+), while a large platinum cathode is used (–). The electrolytic bath consists of a solution of 90wt% H_2SO_4 ($d = 1.8$), with 10wt% HF ($d = 1.16$). The electrocleaning is performed between 25°C and 40°C, with an anodic current density of 10–50 A.dm^{-2} for 1 to 2 min.

3.3.7.10 Tantalum Cladding and Coating Techniques

Although it was pointed out that solid tantalum metal exhibits singular properties, it has two main drawbacks: a high density (16,656 kg.m^{-3}) combined with high cost (461 \$US.kg^{-1}). Therefore, despite its excellent corrosion resistance in harsh environments, tantalum remains too expensive for manufacturing large industrial equipment (e.g., piping, pumps, and reactor vessels). However, it has been demonstrated that a tantalum coating more than 100 μm thickness[63] provides excellent corrosion protection for the underlying metal. Therefore, the only alternative to lowering the amount of metal in order to reduce the cost of using solid tantalum is to manufacture a composite material made of: (i) a base metal or substrate with proper mechanical strength acting as structural material, coated or cladded with (ii) a thin tantalum protective layer. The **duplex material** obtained exhibits both the corrosion resistance of tantalum and the mechanical strength of the underlying metal. Actually, today, most industrial tantalum vessels used in the chemical process industries (CPI) and manufactured by specialized suppliers (e.g., Cabot Performance Materials, Cometec GmbH, Rembar Company Inc., and H.C. Starck GmbH) are made of tantalum-clad ordinary base metal (e.g., steel, copper).[64,65] A detailed comparison of the several techniques suitable for coating/cladding tantalum was presented by Cardarelli et al.[66] with a particular emphasis on electroplating performed in molten salts. In the following paragraphs a brief description of the main industrial techniques is presented. The tantalum cladding/coating techniques can be classified into five categories: (i) mechanical, (ii) thermal, (iii) physical, (iv) chemical and (v) electrochemical.

(i) Mechanical cladding process – mechanical cladding techniques are straightforward to achieve industrially which explains their widespread use in the manufacture of large clad vessels for the CPI.

Loose lining – the loose-lined construction provides thick tantalum liners and was historically the first approach to a duplex system. The tantalum liner is manufactured separately and inserted into the reactor vessels without bonding with the structural base metal. This loose-lined construction is the most economical and most widely used fabrication method throughout the industry. Liner thicknesses of 0.5 to 1 mm are satisfactory against corrosion. It is also possible to improve this technique by welding the liners to the base metal (i.e., weld overlay). Although economical, loose-lined construction has some intrinsic drawbacks: (i) unsuitability for vacuum use due to indentation or collapse of the liner owing to its mechanical instability, (ii) limitations with regard to temperature and pressure, (iii) poor heat transfer coefficient due to the air gap located between the inner and base metal, (iv) difficulty in failure inspection and control, (v) large thickness of the liner. However it continues to be used.

Roll bonding and **hot rolling** – in this method the base metal and tantalum plates are joined together by applying pressure sometimes combined with heat, with or without the use of an intermediate filler material (to ensure good bonding). When rolling is completed, the sandwich cladding plates give the duplex material. The minimum cladding thickness is about 1.5 mm. For certain special procedures, tantalum-lined vessels are being produced with 0.3 mm thick elastomer-bonded tantalum sheet on steel plate. Good ductility of tantalum is the main property required for good bonding. Hence, amounts of interstitial elements (e.g., H, C, N, O), which decrease ductility, must be kept low and usually highly pure tantalum is required (99.99wt% Ta). In order to increase the plastic deformation, high-temperature rolling (e.g., above $1000°C$) can be performed, but it requires an inert atmosphere[67] or a vacuum[68] in order to prevent thermal oxidation of tantalum.

Explosive bonding[69] and **cladding** – in this method, the controlled energy of a detonating explosive[70] is used to create a metallurgical bond between tantalum and the base metal.[71] An intermediate copper layer is used when tantalum is clad onto carbon steel in order to avoid brittle intermetallic formation at the interface. Limitations of the explosive bonding are: (i) difficulties in obtaining a high-quality bonded interface without a filler material because of the large difference in density between common base metals and tantalum, (ii) complex geometries cannot be cladded, (iii) the process is not amenable to automated production techniques, and requires considerable manual labor, and finally (iv) coating thicknesses approach millimeters and need non-negligible amounts of tantalum.[72]

(ii) Thermal spraying methods – the tantalum coating is obtained by projection of molten metal droplets which are carried by compressed gas toward the workpiece. Droplets of liquid metal are obtained by melting metal powder. Particles are carried by an inert gas (e.g., argon) to a heating source. For tantalum only a high-temperature source and inert atmosphere are suitable owing its high melting point and high chemical reactivity with oxygen and nitrogen. Thus arc-spraying, plasma-spraying, and the use of a detonation gun in a protective gas overcome the limitations of the flame-spraying process and give adherent coatings. Nevertheless, due to the mosaic structure having some defects (e.g., amounts of oxide, vacancies), a deposit thickness of 1 mm is required to protect the base metal against corrosion attack. This large coating thickness requires large amounts of tantalum, and makes this process costly, which is unsuitable for complex workpieces and heavy equipment.

(iii) Physical coating processes – only two **physical vapor deposition** (PVD) methods are suitable for thin tantalum coatings but both are restricted to small devices.

Vacuum deposition – the vacuum deposition or evaporation is performed in a chamber under high vacuum, and the vapor of tantalum is produced by heating the metal. The vapor is expanded into the vacuum toward the surface of precleaned base metal. Tantalum must be heated above $3350°C$ in order to get a sufficient vapor pressure. Tantalum vapor condenses in a solid phase onto the cold base metal. Two kinds of heating source are used: (i) direct by resistance heating, i.e., tantalum is contained in a crucible heated by a spiral coil or is deposited onto a refractory filament (e.g. Mo, W) subjected to a high current; (ii) indirect heating, where the temperature is raised by: electron beam, laser beam, electric arc or induction. This technique gives coherent tantalum deposits, and has a high deposition rate ($75~\mu m.h^{-1}$) combined with a simple design. However, deposits are poorly adherent and extremely thin ($5~\mu m$), and protection of the underlying metal against corrosion cannot be guaranteed. Moreover, vacuum deposition is directional; so only the front of the workpiece is coated and it leads to poor throwing power, and is restricted to small items.

Cathodic sputtering deposition – this is performed in an inert gas chamber, under reduced pressure. A spark discharge is produced by application of a high voltage of several kilovolts between the cathode (i.e., target) and the anode (i.e., base metal). Under these conditions, argon atoms are ionized. A beam which is produced by argon ions accelerated by the high field strength arrives at the cathode with a kinetic energy up to 10 keV. Under incident ion impacts, the tantalum atoms are extracted and deposit onto the anode. The higher energy than vacuum evaporation allows greater metallic bonding between coating and base metal. Cathodic sputtering exhibits both a low deposition rate ($2~\mu m.h^{-1}$)[73] and a low throwing power which prohibits its use for complex geometry workpieces and industrial uses.

(iv) Chemical coating processes – **chemical vapor deposition** (CVD) allows the production of coherent and adherent tantalum coatings 20–30 μm thick, with good corrosion resistance. The coating is obtained by reduction of tantalum pentachloride, $TaCl_5$, with pure hydrogen[74] at a temperature of the base metal surface roughly of $1100°C$.[75] The deposition rate is between 1 and 10 $\mu m.min^{-1}$.[76] Sometimes, with carbon steel, a filler metal such as titanium or copper is inserted between the base metal and tantalum to prevent formation of brittle tantalum carbide in the boundary region. CVD has been used to deposit tantalum onto the inner surface of long carbon steel pipes.[77] Nevertheless, CVD exhibits several drawbacks: the structure, the thickness, and the adherence of the coating are very sensitive to the substrate temperature and the gaseous stream. In addition, complex geometry workpieces lead to difficulty in the control of temperature and gaseous stream flow rate, and require large reactor vessels.

(v) Electrochemical coating processes – because tantalum is highly electropositive, it cannot be electrodeposited in aqueous solutions, though some attempts were made.[78] Actually, for more positive cathodic potential, the hydrogen evolution prevents cathodic deposition of tantalum. However, owing to their wide potential span between decomposition limit,[79] high ionic conductivity, molten alkaline-fluoride electrolytes are the most appropriate baths for electrodepositing tantalum.[80] In addition,[81] these molten salts exhibit: a high melting temperature favorable to reaction kinetics, a good throwing power, and a high solute content.

Moreover, the coating adherence is enhanced by etching of the base metal and/or interdiffusion phenomena. As a general rule, two routes are used to perform molten salt electrodeposition of tantalum: (i) coherent deposition, or (ii) alloyed deposition with base metal.

The **coherent deposition process (or electroplating process)** gives a dense, smooth and adherent coating. It is performed by classic electrolysis, under galvanostatic control with two electrodes: a soluble tantalum anode, and base metal as cathode. The main characteristics of the coherent deposition process are: (i) the concentration of tantalum cations is maintained constant by the anodic dissolution process, (ii) the rate of electrochemical reduction is controlled by the current density, and (iii) as a general rule, there is no limitation on the coating thickness. However the formation of dendrites increases roughness, and leads to short circuits in the case of a narrow gap between electrodes.

The **metalliding process (electrolytic cementation, surface alloying, or diffusion coating)** in molten salt is a process in which metallic cations are transferred to the cathode surface where they give a surface alloy with base metal which is uniform, dense, nonporous, and smooth.[82] Metalliding can be performed in two ways: (i) the first method consists of maintaining a galvanic coupling between the tantalum (anode), and the workpiece (cathode). When the short circuit is well established, the redox reaction occurs in which tantalum dissolved anodically and tantalum cations are then reduced at the cathode by outer circuit electrons and give metal alloying. The sufficient condition for ensuring good metalliding reaction is that tantalum must be more electropositive (i.e., less noble) than the base metal. Thus, the metalliding process is self-supporting by galvanic electromotive force, with cathodic current densities ranging from 1 to 20 mA.cm^{-2}. (ii) In the second method, metalliding is obtained by electrodepositing tantalum, then disconnecting the power supply, and maintaining the cathode immersed in the high-temperature melt (800–1100°C) for several hours to allow interdiffusion of the two metals. Stabilized alloyed deposits could be solid solutions or intermetallic compounds. The precursors for metalliding in molten salts are Cook[83–86] and Ilyushchenko et al.[87]

The optimum conditions for the electrodeposition of tantalum in molten fluoride were established by Senderoff and Mellors,[88–90] using the ternary eutectic mixture LiF–NaF–KF (FLiNaK) with 15 to 40wt% K$_2$TaF$_7$ as solute, under inert atmosphere, at 800°C, with a cathodic current density of 400 A.m^{-2}. The electroplating process was later largely improved by Balikhin et al.[91,92] using LiF–NaF–30% K$_2$TaF$_7$ at 800°C, with a cathodic current density of 600 A.m^{-2}.

3.3.7.11 Industrial Applications and Uses

Close examination of tantalum properties reveals that besides its excellent corrosion resistance, tantalum metal exhibits numerous physical properties of interest to chemical engineers (e.g., high melting point, good electrical and thermal conductivities, high elastic modulus and yield tensile strength). These properties are suitable for its industrial use. These additional assets bring tantalum closer to other high-performance construction metals and alloys used in industrial applications. Therefore, tantalum is the construction material to consider in any application where corrosion is a factor and the long-term benefits of reduced downtime, increased equipment life, and profitability are important.

3.3.7.12 Tantalum Metal World Producers

See **Table 3.46** (page 176).

Table 3.45 Tantalum industrial applications

Application	Description
Chemical process industries (CPI) equipment[a]	Tantalum's good thermal conductivity (57.5 W.m^{-1}.K^{-1} at $20°$C) gives a suitable construction material when corrosion resistance has to be combined with good heat transfer conduction. Thus, tantalum is widely used for heat transfer devices working in concentrated acidic media (e.g., plate and tube heat exchangers, spiral coil, U-tubes, spargers, bayonet heaters and thermowells, condensers, boilers, etc.). For instance, tube heat exchangers inserted into boilers for vaporization of concentrated strong mineral acids (e.g., HCl) and for rectification, are entirely built in solid tantalum. Its good Young's modulus (185 GPa) and yield tensile strength (172 MPa) allow the use of tantalum when good mechanical strength is required for the device (e.g., rupture disks, impellers, column, and reactor vessels). For example, some distillation columns for concentration of hydrochloric acid or high boiling organic acids required tower internals made of solid tantalum (e.g., Ta-Intalox$^®$). In comparison with other corrosion resistant materials extensively used in the CPI, the main advantages of tantalum as a consequence of its inertness combined with above suitable physical properties are: no contamination of chemicals due to absence of corrosion products, low maintenance costs due to negligible corrosion rates (<25.4 μm.yr^{-1}), nonfouling, excellent heat transfer, and less downtime. It is also often used in pharmaceutical, biotechnological and foodstuff processes. Moreover, owing to its similarity with glass, it is sometimes used for assembling pipes or repair in glass-lined technology.
Pharmaceutical industry	Tantalum has many applications in pharmaceutical plants. Tantalum alloys are preferred to other corrosion resistant materials because they exhibit higher corrosion resistance, and hence zero contamination, preserving product purity and offering the versatility to reuse the same processing equipment for many applications.
High-temperature applications	Combination of suitable physical properties such as high melting point, good mechanical strength at elevated temperature, good thermal conductivity, and ease of fabrication allow the use of tantalum and its alloys for applications requiring high operating temperature under an inert atmosphere or vacuum. Therefore, tantalum is extensively used as resistance heaters for vacuum furnaces, thermocouple sheaths, crucibles for melting speciality glasses, crucibles for evaporation of refractory compounds, heat shields, and finally containers for handling liquid alkali metals and their vapors. As a general rule, tantalum alloys such as Ta–2.5W and Ta–10W, owing to their higher yield strength (241 and 482 MPa respectively) at high temperatures are preferred when structural properties are important. It is sometimes used as a getter in high-temperature vacuum furnaces, and as targets in X-ray tubes.
Metallurgy	Tantalum is used as an alloying element up to 12wt% in speciality superalloys for particular aircraft applications (e.g., turbine blades). Actually, tantalum provides solid solution strengthening, reacts with interstitial carbon to form stable carbides, and improves the thermal stability of intermetallic compounds.
Surgical implants	Because tantalum is completely inert to body fluids and is a nonirritating metal, it has therefore found wide use in making surgical appliances.
Aerospace and aircraft industries	Tantalum is used as rocket nozzles, hot gas tubing, nose caps for supersonic airplanes, heat shields, and caesium vapor inlets in ion engines.
Electrochemical and corrosion engineering	A better electrical conductivity and higher corrosion resistance in harsh environments compared to other common reactive and refractory metals (e.g., titanium, zirconium, niobium) used as electrode base metal is responsible for its use in association with niobium as base metal for platinized anodes in replacement of titanium substrates. These anodes are widely used for cathodic current protection in seawater. Actually, these anodes (e.g., Protectodes$^®$ from Heraeus) are suited to large surface area plants and vessels (e.g., tankers, oil-rigs) when localized anodic current densities are very high (i.e., several kA.m^{-2}). Moreover, tantalum is also employed as base metal in dimensionally stable anodes (DSA$^®$ type electrodes) in some electrochemical processes working in harsh conditions: i.e., requiring both a high anodic current density, a high temperature, and a concentrated acidic media or brines.

[a]Gramberg U, Renner M, Diekmann H (1995) Tantalum as a material of construction for the chemical processing industry: a critical survey. Int Symp Tantalum and Niobium (TIC) Brussels, 27–29 September.

Table 3.46 Tantalum metal world producers

Company name	Address
Cabot Performance Materials	P.O. Box 1607, 144 Holly Road, Boyertown, PA 19512, USA Telephone: 1 (610) 367 1500 Facsimile: 1 (610) 369 8259 Internet: http://www.cabot-corp.com/cpm
Cometec Gmbh	Lagerhausstrasse 7–9 Linsengericht D-63589, Germany Telephone: (49) 6051 71037 Facsimile: (49) 6051 72030 Internet: http://www.cometec.com
H.C. Starck Gmbh & Co. KG	Postfact 25 40, Im Schleeke 78–91, D-38615 Goslar, Germany Telephone: (49) 5321 751 0 Facsimile: (49) 5321 751 192 Internet: http://www.hcstarck.com
Plansee AG	A–6600 Reutte Tyrol, Austria Telephone: (43) 0 56 72 600 0 Facsimile: (43) 0 56 72 600 500
The Rembar Company, Inc.	P.O. Box 67, 67 Main Street, Dobbs Ferry, NY 10522, USA Telephone: 1 (914) 693 2620 Facsimile: 1 (914) 693 2247 Internet: http://www.rembar.com

3.3.8. Molybdenum and Molybdenum Alloys

3.3.8.1 Description and General Properties

Molybdenum [7439-98-7] with the chemical symbol, Mo, the atomic number 42 and the relative atomic mass (atomic weight) 95.94(1), is the second element, between chromium and tungsten, of the group VIB(6) of Mendeleev's periodic chart. The element is named after the Greek word, molybdos, meaning lead, owing to the resemblance of the mineral molybdenite to lead. Molybdenum is a mildly dense (10,220 $kg.m^{-3}$), hard refractory metal, silvery-white, with a gray steel color, having a high melting point (2610°C). It has a high Young's modulus (325 GPa) and both exceptional strength and stiffness at high temperatures but it is softer than tungsten. Moreover, molybdenum has a low coefficient of linear thermal expansion (5.43 $\mu m.m^{-1}.K^{-1}$) allowing it to be used for sealing hard glass. Its low electrical resistivity (5.7 $\mu\Omega.cm$) combined with high melting point are suited to making electrical contacts. Natural molybdenum contains seven stable isotopes: $^{92}Mo(14.84at\%)$, $^{94}Mo(9.25at\%)$, $^{95}Mo(15.92at\%)$, $^{96}Mo(16.68at\%)$, $^{97}Mo(9.55at\%)$, $^{98}Mo(24.13at\%)$, and $^{100}Mo(9.63 at\%)$.

Chemically, molybdenum is an extremely versatile element, forming chemical compounds in a range of readily interconvertible oxidation states, complexes with many inorganic and organic ligands, and finally compounds in which the molybdenum coordination number ranges from four (tetrahedral) to eight (octahedral). Molybdenum has a good corrosion resistance in strong mineral acids but vigorously dissolves in hydrofluoric and nitric acid mixture (hydrofluoric aqua regia, HF–HNO$_3$). Molybdenum is resistant to alkaline aqueous solutions but readily dissolves in molten caustic alkali hydroxides (e.g., NaOH and KOH), molten alkali carbonates (e.g., Na$_2$CO$_3$ and K$_2$CO$_3$) and molten alkali nitrates (e.g., NaNO$_3$ and KNO$_3$). The metal is also inert in gases such as carbon monoxide, hydrogen, ammonia, and nitrogen at temperatures up to 1095°C. Molybdenum oxidizes only at high temperatures which allow it to be used as a thermal shield in air at high temperatures. Moreover, molybdenum is highly corrosion-resistant to halogens such as iodine, bromine, and chlorine vapors. Molybdenum and its alloys are also suitable container materials for handling strongly corrosive liquid metals such as molten alkali metals (e.g., Li, Na, and K), and also bismuth and magnesium, while pure molybdenum is strongly attacked by molten tin, aluminum, and cobalt. Actually, along with rhenium, molybdenum is the most resistant pure metal to oxygen-free molten lithium up to $1200°C$.

In biological systems, molybdenum is an essential constituent of enzymes which catalyse redox reactions, e.g., oxidation of aldehydes, xanthine, and other purines, and reduction of nitrates and molecular nitrogen. The biochemical importance of molybdenum is due to two main factors: (i) its various oxidation states provide easy and various electron-transfer pathways, and (ii) its ability to form strong bonds with oxygen, sulfur, and nitrogen donors which allow both the existence of stable complexes with straightforward ligand exchange reactions and changes in molybdenum coordination number.

Cost (1998) – pure molybdenum (99.95wt% Mo) obtained by powder metallurgy (i.e., sintered) is priced at 130.65 $US.kg^{-1} (59.26 $US.lb^{-1}), while pure molybdenum (99.95wt%) arc cast and vacuum melted is priced at 214.95 $US.kg^{-1} (97.5 $US.lb^{-1}).

3.3.8.2 History

In 1778, the Swedish chemist Carl Welhelm Scheele conducted research on a sulfide mineral now known as molybdenite (MoS_2). It was often confused with graphite and lead ore. He concluded that it did not contain lead as was suspected at the time and reported that the mineral contained a new element that he called molybdenum after the mineral. Molybdenum metal was prepared in an impure form in 1782 by Peter Jacob Hjelm. During World War II a German artillery piece called "Big Bertha" contained molybdenum as an essential alloying element of the steel. Commercial production of molybdenum and ferromolybdenum began in the 1920s.

3.3.8.3 Natural Occurrence, Minerals, and Ores

Molybdenum abundance in Earth's crust is about 1.5 mg.kg^{-1} (ppm wt). Moreover, molybdemun, owing to its high chemical reactivity with oxygen, never occurs in the native state (i.e., free) in nature. The chief minerals containing molybdenum are the sulfide, **molybdenite** (MoS_2, hexagonal), the molybdate, **wulfenite** ($PbMoO_4$, tetragonal), and the molybdo-wolframate, **powellite** ($Ca(MoW)O_4$, tetragonal). Because molybdenum is a chalcophile element (i.e., is often combined with sulfur in geological materials), the main molybdenum ore is the molybdenite which is largely found in igneous rocks such as granites, syenites and pegmatites, and their associated metamorphic belt. From a biological point of view, molybdenum is an essential trace element in plant nutrition and some soils are barren for lack of this element in the soil.

3.3.8.4 Processing and industrial Preparation

The pure metal powder is prepared by hydrogen reduction of the powdered molybdenic trioxide. The molybdenum trioxide is itself obtained by the calcination of the ammonium molybdate.

3.3.8.5 Molybdenum Alloy Properties

There are three broad classes of molybdenum alloys: (i) those strengthened by reactive metal carbides, (ii) those strengthened by substitutional elements, and finally (iii) those stabilized by a mechanically dispersed second phase. Alloys designed to take advantage of a combination of these different approaches also can be found. The carbide-strengthened alloys are normally

Table 3.47 Properties of selected molybdenum alloys

Usual and trade name	UNS	Average chemical composition (x, wt%)	Density (ρ, kg.m⁻³)	Young's modulus (E, GPa)	Yield strength 0.2% proof (σYS, MPa)	Ultimate tensile strength (σUTS, MPa)	Elongation (Z, %)	Vickers hardness (HV)	Coefficient of linear thermal expansion (α, 10⁻⁶ K⁻¹)	Thermal conductivity (k, W.m⁻¹.K⁻¹)	Electrical resistivity (ρ, μΩ.cm)
Mo (high purity)	R03600	99.99wt% Mo (EB-melted, hot rolled, recrystallized at 1100°C)	10,200	325	345	435	5–25	n.a.	5.1	137	5.7
Mo (soft)	n.a.	99.95wt% Mo (arc cast, hot worked, and recrystallized at 1100°C for 1 h)		325	415–450	485–550	30–40	200	5.1	137	5.7
Mo (hard)	n.a.	99.95wt% Mo (rolled at 1000°C)	10,200	325	550	620–690	10–20	250	5.1	137	5.7
Mo-alloy 362	R03620	Mo–0.5Ti	10,200	315	825	895	10	n.a.	6.6	n.a.	n.a.
Mo-alloy 363 (arc cast), TZM	R03630	Mo–0.5Ti–0.1Zr	10,160	315	380–860	550–965	10–20	n.a.	4.9	n.a.	n.a.
Mo-alloy 364 (P/M), TZM	R03640	Mo–0.5Ti–0.1Zr	10,160	315	380–860	550–965	10–20	n.a.	4.9	n.a.	n.a.
TZC	n.a.	Mo–1Ti–0.3Zr	n.a.	n.a.	640–725	725–995	22	n.a.	n.a.	n.a.	n.a.
HCM	n.a.	Mo–1.1Hf–0.07C	n.a.	n.a.	n.a.	n.a.	n.a.	n.a.	n.a.	n.a.	n.a.
HWM-25	n.a.	Mo–25W–1Hf–0.07C	n.a.	n.a.	n.a.	n.a.	n.a.	n.a.	n.a.	n.a.	n.a.
HWM-45	n.a.	Mo–45W–0.9Hf	n.a.	n.a.	n.a.	n.a.	n.a.	n.a.	n.a.	n.a.	n.a.
Mo–Re alloys	n.a.	Mo–47.5Re	13,700	357	848	980–1034	22–25	n.a.	n.a.	n.a.	n.a.

employed where high-temperature strength is required. The high strength at elevated temperatures combined with high thermal diffusivity inherent to molybdenum make the carbide-strengthened alloys attractive for hot-work tooling applications.

3.3.8.6 Molybdenum Metalworking

Forming and metalworking of molybdenum and its alloys can be achieved by all common metalworking practices such as punching, shearing, drawing, stamping, spinning, and bending. Note that operations that employ shearing, such as stamping, punching, and blank shearing, are particularly sensitive to the formation of planar cracks in the sheet being formed. These defects are commonly called delaminations; they are in fact intergranular cracks which propagate along the planar grain boundaries which develop during the rolling of sheet and plate. Tool clearances and edge condition are the major contributors to this phenomenon. Dull and damaged tool blades are invitations to delamination. Clearances between blades, or between punch and die in stamping operations, should be in the range 5–8% per side to minimize delamination.

Punching and shearing – conventional equipment is normally satisfactory for this operation and moderate heating is recommended. Sharp tools with close tool clearances of approximately 5% of the sheet thickness are essential to clean cutting action without the sheet cracking or delamination occurring. Sheet up to 0.5 mm thick can be successfully sheared at ambient temperature. Preheat temperatures of 65–95°C are recommended for sheet between 0.5 and 1.2 mm thick. In the range of 1.5–3.2 mm, the preheat temperature should be increased to about 350°C, and 600°C preheat is necessary to shear plate of 6.3 mm thickness. Linear gas burners, infrared lights, air furnaces, hand-held torches, and hot plates have all been successfully employed as heat sources for shearing operations.

Deep drawing and stamping – wall reductions of as much as 20% between heat treatments have been achieved with the deep drawing process. Heating of both sheet and dies is suggested for best results on sheets over 0.5 mm thick. Conventional equipment, tooling, and lubricants normally produce acceptable results.

Spinning – the use of stress-relieved material and the continuous application of heat are the only precautions to observe. Otherwise, molybdenum can be routinely fabricated into a variety of shapes.

Bending – in bending operations, the bend radius which can be successfully bent without cracking will be a function of the sheet thickness. Thicker sections may require heating above room temperature. In addition, molybdenum and its alloys are typically anisotropic in their ductility properties, unless special processing has been employed to equalize the directionality of deformation in the material. When bending sheet, for instance, orienting the bend axis of a blank perpendicular to the dominant rolling direction will result in better performance. Heated to the proper temperature, molybdenum sheet can be accurately formed into complex shapes. Sheets under 0.5 mm thick will normally take a 180° bend at room temperature. Red heat may be required for forming heavy thick plate to remain in the ductile regime, due to the greater triaxiality of stress present during the forming operation. If necessary, dies and tools can be warmed with infrared lamps or strip heaters to assist in the bending process.

3.3.8.7 Molybdenum Joining

Mechanical joining – mechanical joining methods such as bolting, riveting, and lock seams are the simplest ways of joining molybdenum where fluid-tight joints are not required. It is recommended that rivets be heated in place to 200°C to 760°C, depending on the section size. Lacing with molybdenum wires is often employed for parts such as furnace shields.

Welding – joining of molybdenum parts can be accomplished using conventionally accepted welding techniques except for gas but welding is normally employed only for applications not subjected to great stress. Indeed, the weld and surrounding recrystallized zone in the base metal have significantly lower strength, and a much higher ductile–brittle transition temperature than the surrounding material which is unaffected by the welding process. This tends to concentrate the deformation in the weld zone, and the triaxial stresses produced by the constraint of the base metal can result in brittle fracture. Amongst welding techniques, helium-arc welding is most common and usually provides satisfactory results. Complex welding operations may require more sophisticated or special techniques. For instance, electron-beam welds, with their narrow weld and heat-affected zones, are less susceptible to failure than GTA welds which require large amounts of heat input. Careful cleaning of the joint surfaces is essential. Owing to its chemical reactivity with oxygen, most welding of molybdenum components is performed inside high-purity inert gas chambers in order to minimize oxygen pickup. Therefore, controlled weld atmospheres, such as a dry box, are recommended. In designing fixtures, all clamping forces should be compressive and should be released immediately after welding to permit unstressed cooling. Oxygen is also a bad actor in welded components. It tends to segregate to grain boundaries, further reducing ductility. For this reason, the arc-cast alloys which generally contain higher carbon levels, are somewhat more readily welded than the powder metallurgy analogs. The carbide-strengthened alloys are also more forgiving than pure molybdenum for the same reason. The doped alloys generally do not weld as successfully as the other alloys, because the volatile alloy elements in the materials produce gassy welds. Rhenium alloys are quite weldable. The well-known rhenium ductilizing effect renders these alloys ductile at cryogenic temperatures even in the as-solidified or recrystallized condition. As noted earlier, this property has been utilized to design and fabricate large chemical pressure vessels by weld cladding Mo–Re to inexpensive plate steel alloys.

Brazing – this is commonly used for joining molybdenum and its alloys. Usually, copper-based alloys are normally acceptable in creating a relatively low-strength joint. However, higher-strength joints can be achieved by using commercial brazing alloys that have flow points ranging from 630°C to 1400°C. Most of these alloys contain noble and precious metals such as gold, platinum, or other nickel-based alloys. With proper temperature precautions, brazing will normally produce a more ductile joint than welding. In most cases, the brazing temperature should maintained below the recrystallization temperature of the alloy to be brazed. In this manner, the improvement in strength and ductile–brittle transition behavior which accrues with mechanical working can be retained.

3.3.8.8 Molybdenum Machining (Source: Rembar Company, Inc.)

Machining of pressed and sintered or recrystallized molybdenum is similar to that of medium-hard cast iron, while the machining of wrought molybdenum is similar to stainless steel, and it can be machined with conventional tools and equipment. However, the machining characteristics of molybdenum differ basically from those of medium-hard cast irons or cold-rolled steels in two ways: (i) it has a tendency to break out on the edges when cutting tools become dull, and (ii) it is very abrasive and causes tools to wear out much faster than steel.

Turning – most turning operations on molybdenum and molybdenum alloys are consistent with machining practices on steel. However, the cost of machining can be a concern. The only obvious differences are that: cutting speeds are 50% faster than high-strength steels and almost three times as fast as nickel-based alloys. Greater attention must be paid to tool geometry and tool replacement. In general, any of the straight tungsten carbide tools are suggested for use. General purpose high-speed steel tools may be used for rough turning. Recommendations for the turning of molybdenum are as follows: the lathe should be rigid and well powered. The workpiece should be well clamped and rigidly supported. Because molybdenum is shock-sensitive, care must be taken when mounting the workpiece. Avoid excessive chucking pressure, which could distort the workpiece or cause a fracture even before machining begins. A useful practice is to apply copper shims at all chucking and workpiece contact points. Tool overhang should be kept to a minimum. Tools must be kept sharp to avoid buildup that can cause failure of the tool or the workpiece. Tool geometry similar to that used for cast iron is generally suitable. Liberal rake and clearance angles can be used. Cutting fluids are required to control the tool tip temperature. Water-soluble cutting oils work well.

Threading – this is usually performed by thread grinding, single point turning or chasing operations. Carbide tools should be used and best results are obtained when: the back rake is 10°, the side rake is 5°, there are clearances of 10° on the leading edge and 3° on the trailing edge. Rough threading is done best at speeds from 356 to 635 mm.s^{-1} (70 to 125 ft.min^{-1}). For making fine threads or shallow threads, grinding is usually more practical than turning. The use of dies is not recommended for threading because they have a tendency to tear or pull threads from the workpiece. If tapping must be performed, care should be taken that the tap is very sharp, perfect alignment is maintained, and a tapping compound is used. Thread rolling is also possible to produce the strongest threads.

Face milling – for face milling operations, the conventional carbide or carbide-tipped face mills designed for use on cast iron are employed. High-speed steel will also work, but cutter wear is rapid and tool life is very short. Tool angles are the same as used on cast iron. Cutter conditions should be checked frequently and workpiece backup plates should be used, particularly for heavy stock removal. Rough cutting of molybdenum in depths of 1.27 mm to 2.54 mm calls for speeds of 508–813 mm.s^{-1} (100–160 ft.min^{-1}) at an average feed of 127 μm per tooth. Finish cuts are made mainly at speeds of 1778 to 2032 mm.s^{-1} (350–400 ft.min^{-1}) with a range of the depth of cut from 25.4 to 76.2 μm and a feed in a range 102 to 127 μm per tooth. Care must be taken to eliminate corner and edge breakout, especially when using multiple cutters. This can be minimized by in-feeding. Breakdown of one cutter can cause vibration from eccentric loading. This will result in a poor surface finish and accelerated breakdown of the other cutters. A cutting fluid of soluble oil should be used since it has a decided influence on the effective tool life.

Drilling – drilling of molybdenum and molybdenum alloys presents no special problems. Standard high-speed steel drills are used. Because high temperatures will be generated, extra care must be exercised in cooling the drill. Variations of heat expansion between the drill and the workpiece can cause excessive binding that can result in tool failure or damage to the workpiece. Heavy duty machines with substantial power, absolute rigidity, and a true running spindle with no end play are necessary for successful drilling. The workpiece should be adequately supported at the point of thrust to forestall vibrations. When small parts are to be drilled, this may require fixturing for support. Drill rigidity is important. In addition to using the shortest drills possible, the use of a drill bushing should also be considered. For the drilling process itself, the following considerations need to be observed: standard high-speed steel drills are used with a 118° angle to provide maximum drilling efficiency. A generous flow of soluble oil coolant should be maintained. Maintain a positive, consistent feed rate to avoid work-hardening and loss of tool life. Light feed rates normally will provide significantly longer tool life. Although solid carbide drills have been used for drilling holes of up to 9.525 mm in diameter, high-speed steel drills give better performance. Standard points ground to 118° angles with clearance angles of approximately 10° are the most widely used. Crankshaft points may also be considered since they reduce the area of contact and minimize heat build-up. All drills should be carefully checked for sharpness and proper geometry before being put to use. Positive drill feed should always be maintained. Any riding of the drill inside the hole without cutting causes excessive heating and reduces tool life. Also, the drills should be examined periodically during production and re-sharpened or replaced at the first sign of wear. Once operating performance has been established, a drill replacement schedule should be established. A copious flow of soluble drill oil coolant is recommended. When drilling through holes, the workpiece should be backed up to prevent edge breakout.

Grinding – grinding of molybdenum should be considered primarily for finishing, not for major stock removal. Grinding can be handled on conventional machines with standard feeds and speeds. As long as the machines are in good condition and vibration free, standard practices produce good results. Molybdenum, like some steels, is susceptible to surface heat checking. Therefore, soft grade wheels are used and they should be sharply dressed. Carborundum wheels No. GA-463-J6-V-10 can be used for rough grinding and No. PA-60-H8-

V40 wheels can be used for finish and contour grinding. Copious amounts of standard grinding coolant should always be used. Soluble oil mixtures are recommended over highly chlorinated or highly sulfurized fluids. Actually, grinding also has the potential to cause overheating and surface cracking in these materials if sufficient amounts of coolant are not employed.

Sawing – molybdenum saws readily with high-speed steel bands or hacksaws. No coolant is required, although it may be used. Approximately 3.175 mm should be allowed for the kerf and 4.763 mm for the camber on heavier sections. Abrasive cut-off operations can also be used. Flame cutting, however, is not recommended.

Electrical discharge machining (EDM) is also commonly performed on molybdenum and its alloys. Care must be exercised when EDMing molybdenum and its alloys because the surface zone frequently contains a resolidified layer. This structure is susceptible to microcracking and should be removed by mechanical or chemical polishing prior to placing the part in service.

3.3.8.9 Molybdenum Cleaning, Etching, and Pickling

Cleaning of molybdenum is necessary to remove the following: (i) surface scale, (ii) general contamination, and (iii) basis metal. Amongst the potential contaminants in wrought products, iron is of primary concern. Others, such as Al, C, Ca, Cu, and Ni, may also be present as elements, but they are more frequently present in the form of oxides. Removal of a controlled amount of basis metal may be desired to insure complete removal of contaminants. There are three main methods for cleaning molybdenum and molybdenum alloys.

3.3.8.10 Industrial Applications and Uses

Molybdenum and its alloys are widely used in industrial applications. The properties that have made molybdenum and molybdenum alloys most attractive are: (i) their high strength

Table 3.48 Pickling, descaling and etching procedures for molybdenum and alloys

Cleaning type	Pickling or etching bath composition	T (°C)	Procedure
Acid cleaning	95wt% H_2SO_4 4.5wt% HNO_3 0.5wt% HF	20–25	Immersion of the workpiece in the bath for 5 s, followed by thorough rinsing with deionized water.
Alkaline cleaning	Bath (1): 10wt% NaOH 5wt% $KMnO_4$ 85wt% H_2O Bath (2): 15wt% H_2SO_4 15wt% HCl 70wt% H_2O 6–10wt% chromic acid	65–80	Soak for 5 to 10 min in the first alkaline bath. When immersion in the alkaline bath is complete, immersion in the second bath is required to remove smut that may be formed by the first treatment. The second bath should also provide a soak duration of 5–10 min.
Electrochemical cleaning	80wt% H_2SO_4	55	This method is generally performed on the molybdenum alloy TZM. The recommended procedure is: (i) solvent degreasing for 10 min, (ii) immersion in a commercial alkaline cleaner for 2–3 min, (iii) rinsing with cold water, (iv) buff and vapor blasting, (v) second immersion in a commercial alkaline cleaner, (vi) rinsing with cold water, (vii) electrochemical polishing performed at a current density of 50 mA.cm^{-2}.

Table 3.49 Molybdenum industrial applications and uses

Applications	Description
Metallurgy	Molybdenum is widely used as an alloying element in cast irons, steel, heat-resistant alloys, and corrosion-resistant alloys such as stainless steel and Hastelloy®. Actually, it increases toughness, strength, stiffness, creep resistance, abrasion, and corrosion resistance. Indeed molybdenum is a valuable alloying agent which contributes to the hardenability and toughness of quenched and tempered steels. Almost all ultrahigh-strength steels contain molybdenum in amounts from 0.25 to 8wt% It also improves the strength of steel at high temperatures.
Electrical engineering	This market is probably the largest for molybdenum and its alloys. It includes applications such as mandrel wire for manufacturing lamp filaments, wire leads and support structures for lighting and electronic tube manufacture, powders for specially formulated circuit inks and the tooling used to apply them to multi-layer circuit boards, internal components for microwave devices, high-performance electronic packaging, and heat sinks for solid state power devices. Molybdenum serves also in the production of electrical and electronic equipment used in the medical industry. For instance, many of the internal components of X-ray tubes, from the target itself to support structures and heat shields, are manufactured from molybdenum and molybdenum alloys. Molybdenum also finds its way into X-ray detectors, where sheet with precisely controlled gage is used. Molybdenum finds application as a buffer between the relatively low-expansion materials used in integrated circuit (IC) packages and the copper normally used to supply electrical power to the devices and to remove heat from them as well. It is even finding application as a replacement for the silicon substrates used in some devices. Power rectifiers use large quantities of molybdenum sheet that is stamped and plated with nickel, copper, or rhodium to provide both thermal expansion control and heat management. These devices find application in diesel-electric and electric railroad motors and industrial motor power supplies and controls. Pressed and sintered heat sinks for small electrical devices are ubiquitous. Cladding molybdenum with copper results in a material (Cu/Mo/Cu, or CMC) whose properties can be tailored to the application at hand. The copper increases the thermal expansion coefficient of the composite allowing a good match with ceramic substrate materials such as alumina (Al_2O_3), beryllia (BeO), and aluminum nitride AlN). CMC brings the added benefit of high elastic modulus to the assembly, resulting in reduced susceptibility to vibration-induced failures. These materials are available commercially in various compositions and shapes. Powder composites offer the potential advantage of isotropic properties and less hysteresis in thermal expansion, probably due to the triaxiality of the internal stress distribution. They also offer greater flexibility in tailoring thermal properties because varying powder blends is significantly less cumbersome than manufacturing different cladding ratios on rolled sheet.
Lubrication	Molybdenum disulfide (MoS_2) is a good lubricant, especially at high temperatures where normal lubricant oils readily decompose.
Materials processing	Aerospace forgers employ tooling made of molybdenum alloys to forge engine materials at high temperature. Extrusion houses found molybdenum alloys to be ideal for certain applications in the brass industry. The processing of many electronic components, whether it be by sintering the ceramic material used in high-performance circuit boards or the metallization of silicon wafers, requires molybdenum metal components. The gatorizing®, or isothermal forging, process is used to forge titanium alloys or superalloy engine disks. In this process, the tooling and workpiece are both heated to the forging temperature, and the disk is formed superplastically. The entire tooling stack and workpiece are contained in a vacuum chamber, in order to avoid formation of volatile oxides of molybdenum. This technique is capable of producing highly defined disk forgings that require much less machining than those produced by conventional techniques. This practice has produced integrally bladed disks on an experimental basis. These alloys also find application in conventional hot-work tooling. A primary reason for this is their resistance to thermal shock and cracking. While the steel has a distinct advantage in strength, and both steel and nickel alloys have an advantage in modulus, the high thermal conductivity and low coefficient of expansion of molybdenum make it the preferred material for thermal shock applications. TZM and MHC alloys both find application in the extrusion of copper and copper alloys. Extrusion die design requirements are somewhat different from those normally used by tool designers, primarily because of molybdenum's low coefficient of thermal expansion. Significantly more shrink fit is required for molybdenum than for steel or nickel alloy dies in nickel alloy cases, so that the die does not loosen as the assembly heats up to its normal operating temperature. Once this is accounted for, the molybdenum dies perform very well.
Molten metal processing	Molybdenum–tungsten alloys are used in the handling of molten zinc, due to their chemical compatibility with that material. Aluminum die casters use molybdenum to solve thermal checking and cracking problems that otherwise cannot be eliminated. In this case, TZM inerts, cores and pins are used in areas prone to hot checking. Rapid solidification equipment using rotating disc

Table 3.49 (continued)

Applications	Description
	and rotating drum technology benefits from the use of TZM and MHC alloys. Here again, the high-temperature strength of these materials and their resistance to thermal shock permits the processing of higher-melting-point materials than would be otherwise possible. Another unique application for molybdenum alloys is in the handling of molten zinc. At one point in time, tungsten was thought to be the only material resistant to corrosion by molten zinc. Alloying molybdenum with tungsten resulted in an equally resistant material at a greatly reduced cost. The Mo–25W and Mo–30W alloys evolved from this work, and are widely for impellers, pump components, and piping that handle molten zinc.
Thermal spraying	A significant amount of molybdenum powder is consumed by thermal spray applications. In this technology, molybdenum metal powder is blended with binders rich in chromium and nickel, then plasma-sprayed on piston rings and other moving parts where wear is a critical performance issue. The older wire spray process still accounts for a significant amount of molybdenum consumed in the market. In both cases, the material to be sprayed is fed through a high-temperature gas jet. This jet may be generated by a plasma torch or a high-velocity gas torch. The feed material is melted in the flame and droplets are carried by the jet to the surface of a substrate, where they impact the surface and freeze rapidly. With time, a coating is built up on the substrate's surface. Composite or graded coating can be produced by controlling the composition of the feed material to the jet. Piston rings are coated with pure molybdenum, or alloy powder blends. The paper and pulp industry also employs the coatings for powder blends. The paper and pulp industry also employs the coatings for wear and corrosion resistance. A variety of compositions is possible by blending with other powder components. The most common alloys blends contain varying amounts of Ni, Cr, B, and Si. The powders used in spray applications are markedly different from those used to produce mill products. Because most mill products start as pressed and sintered billets, great attention is paid to producing a powder that will press to high density and produce green billets that have strength enough to be handled in industrial operations. This means that the powders that work best for mill products tend to be agglomerates of fine particles that provide easy mechanical interlocking. Spray applications require just the opposite characteristics – good flowability. Thermal spray powders are generally processed by spray drying to produce spherical or nearly spherical powders that flow easily through spray equipment. In addition to these powders, prealloyed powder grades are also available, in which the molybdenum and alloy blend powders are themselves densified together in a plasma jet. These powders have been reported to give improved wear resistance in laboratory evaluation, and are useful where corrosion is a concern.
Chemical processing	Although tantalum is by far the most widely used of the refractory metals to impart corrosion resistance to chemical process vessels and components, there are some applications where molybdenum has been used with great success. Molybdenum support structures have replaced graphite in the processing of high-purity alcohols. Molybdenum–rhenium alloys, first developed because of their vastly improved ductility at low temperatures and in the recrystallized condition, have been employed as vessel lining and piping components for the manufacture of Freon® replacements.
Glass manufacturing	Because of its compatibility with many molten glass compositions, molybdenum has found application in handling equipment, tooling, and furnace construction. The most common use for molybdenum is as electrodes for the melting of glass. Because glasses are electrically conductive when molten, molybdenum electrodes can be used to increase the energy input by direct resistance heating in conventionally fired furnaces and thereby increase the throughput of the furnaces. There are as many electrode designs as there are design firms, but all immerse the molybdenum electrode into the furnace where it is protected from oxidation by the glass itself. Molybdenum is used for making electrodes for electrically heated glass furnaces in the glass manufacturing industry, in nuclear energy applications, missile and aircraft parts.
Heat	Molybdenum's strength and stability at elevated temperatures make it an attractive material for construction of high-temperature furnaces and the fixtures and tooling associated with them. Molybdenum's high melting point means that at typical operating temperatures for vacuum furnaces, volatilization of internal components made from molybdenum or molybdenum alloys will be negligible. Metal hot zones offer the utmost in vacuum cleanliness for those heat treating applications that cannot tolerate carbon or oxygen contamination. Actually, titanium, niobium, and tantalum are all metals that require environments free of oxygen and carbon when heated above 500°C. The increasing use of hot isostatic pressing (HIP) to consolidate powder materials and improve the integrity of cast metals has also boosted the need for molybdenum products. Molybdenum and its alloys are widely used as materials of construction for HIP vessels, being

Table 3.49 (continued)

Applications	Description
	found in their heating elements, mantles, and support structures. The ceramic processing industry also makes extensive use of molybdenum components for fixtures and sintering boats. Oxide ceramics processed by the electronics industry are nearly universally sintered in hydrogen on molybdenum carriers. Molybdenum metal and its alloys are used in electrical and electronic devices, such as filament material, e.g. $MoSi_2$ in heating coils in high-temperature furnaces.
Aerospace and defense applications	Compatibility with hot gases and strength at high temperatures are the typical properties that result in molybdenum being used in this market area. Molybdenum's poor oxidation resistance prevents it from being used in a wider variety of applications that could use its high strength, but in rocket and reactive gas valves, where high performance is required for a relatively short time, it finds application. For certain of these components, the metal injection molding process is being developed because of its potential for significant material and machining savings. Molybdenum is also being employed in ammunition applications, a relatively new application. It is also used in glass-to-metal seals because molybdenum has a coefficient of linear thermal expansion similar to that of hard glass combined with solubility.

Source: CSM.

combined with a high stiffness up to high temperatures, (ii) good thermal conductivity, (iii) a low coefficient of linear thermal expansion, (iv) low electrical resistivity, (v) low vapor pressure at high temperature, (vi) good resistance to abrasion/wear, and oxidation combined with corrosion-resistance in many harsh environments, and finally, (vii) good ductility, machinability, and workability. Moreover, today a considerable base of fabrication knowledge and manufacturing allows these alloys to be fabricated into useful components. Therefore, combinations of these valuable properties and characteristics predict increasing uses in the electronics and aerospace industries owing to the requirement for materials that maintain reliability under oxidizing and high temperature conditions.

3.3.8.11 Molybdenum Metal World Producers

Table 3.50 Molybdenum metal world producers

Company	Address
CSM Industries Inc.	21801 Tungsten Road, Cleveland, Cleveland, OH 44117–1117, USA Phone: (216) 692 3990 Facsimile: (216) 692 0031 Internet: http://www.csm-moly.com
Molycorp Inc.	P.O. Box 469, Questa, NM 87556, USA Telephone: + 1 505 586 7606 Facsimile: + 1 505 586 0811 E-mail: irstark@questa.unocal.com
Osram Sylvania Products Inc.	Hawes Street, Towanda, PA 18848, USA Telephone: + 1 570 268 5000 Facsimile: + 1 570 268 5113 E-mail: dunns@osi.sylvannia.com
Plansee AG	A-6600 Reutte Tyrol, Austria Telephone: +43 0 56 72 600 0 Facsimile: +43 0 45 72 600 500

3.3.9 Tungsten and Tungsten Alloys

3.3.9.1 Description and General Properties

Tungsten (synonym wolfram) [7440-33-7], with the atomic number 74, the chemical symbol W, and the relative atomic mass (atomic weight) of 183.84(1) is the heaviest metal of the group VIB (6) of Mendeleev's periodic chart. Its chemical symbol, W, is named after the German, wolfram, from the mineral wolframite, said to be named from wolf rahm or spumi lupi, because the ore interfered with the smelting of tin and was supposed to devour the tin, while tungsten is named after the Swedish, tung sten, meaning heavy stone. Pure tungsten is a highly dense (19,300 kg.m^{-3}) steel-gray to shiny tin-white metal with a body-centered cubic (bcc) crystal space lattice structure. In its highly pure form it is ductile, and can be cut with a hacksaw, and can be forged, spun, drawn, and extruded. Nevertheless, traces amount of interstitial impurities such as carbon of oxygen give to the metal its considerable hardness and brittleness. It has the highest melting point (3422°C) of the four common refractory metals and the lowest vapor pressure of any metal and at temperatures over 1650°C. Mechanically, it has a high Young's and bulk modulus and the highest ultimate tensile strength and creep resistance at high temperature. For this reason, it is used in very-high-temperature vacuum furnaces, i.e., those that operate above 2000°C, and in arc lamps for both the cathode and anode. Moreover, it has the lowest thermal expansion coefficient of all the metals which is quite similar to that of hard glass. Hence this allows tungsten to be used for making hermetic glass-to-metal seals used in electronic and military applications. The powdered metal is pyrophoric and may ignite spontaneously on contact with air or oxidants (e.g., F_2, CIF_3, NO_x, IF_5, N_2O, and N_2O). It has an excellent corrosion resistance to most chemicals and is only slightly attacked by concentrated strong mineral acids such as nitric acid or aqua regia. However, it is slightly attacked by aerated molten salts such as KOH, or Na_2CO_3, and readily soluble in a molten mixture of $NaNO_3$ and NaOH. Moreover, tungsten oxidizes in air at elevated temperatures and must be protected by a thermal impervious coating. Natural tungsten element contains five stable isotopes: ^{180}W (0.120at%), ^{182}W (26.498at%), ^{183}W (14.314at%), ^{184}W (30.642at%), and ^{186}W (28.426at%).

Cost (1998) – pure tungsten metal (99.95wt% W) is priced at 620 \$US.kg^{-1} (281 \$US.lb^{-1}).

3.3.9.2 History (Source: ITiA, 1997)

During the 17th century, the miners in the Erz Mountains of Saxony (Germany) noticed that certain ores interfered during tin-smelting disturbing the reduction of cassiterite (tin ore) associated with the formation of slags. The particular ore, known today as wolframite, was named from the German wolf rahm or spumi lupi, meaning wolf froth, because it was supposed to devour the tin. In 1758, the Swedish chemist and mineralogist, Axel Fredrik Cronstedt, discovered and described an unusually dense mineral which is known today as scheelite. He named it after the Swedish, tung sten, meaning heavy stone. Later, in 1779, Peter Woulfe examined the mineral now known as wolframite and concluded it must contain a new substance. In 1781, 23 years after the discovery of the so-called tung sten (scheelite), the Swedish pharmacist Carl Wilhem Scheele in Uppsala, found that a new "acid", i.e., tungsten oxide, could be isolated from the ore. Later Scheele and Berman suggested the possibility of obtaining a new metal by reducing this oxide. Independently, the two Spanish brothers Fausto and Juan Jose Elhuyar de Suvisa obtained a tungsten oxide in wolframite in 1783 that they succeeded in reducing to the elemental metal with charcoal. In 1816, the Swedish chemist Jöns Jacob Berzelius and later in 1824 the German chemist Friedrich Wohler described the tungsten oxides and bronzes and adopted the name wolfram for the element, while Anglo-Saxon scientists preferred the name tungsten. In 1821, K.C. von Leonhard suggested the name scheelite. The first attempt to use tungsten as an alloying element in steelmaking was made in

1855 but the high cost of the metal has rapidly led to the discontinuation of the process. Industrial application of tungsten as an alloying element to harden steel appeared in the late 19th century and subsequently applications of tungsten encountered a rapid growth. In 1903, the British scientist W.D. Coolidge prepared the first ductile tungsten wire by doping tungsten oxide before reduction with carbon. The reduced metal powder was first sintered and then forged into rods. A thin wire was then obtained after the drawing of the rods. The development of the incandescent lamp invented by Edison several years before, requiring tungsten filaments, was instrumental in the rapid consumption of tungsten wire. Moreover, in 1923, the German K. Schröter invented the first metal matrix composite know today as cemented carbides (cermets or hardmetal) combining tungsten monocarbide (WC) particles embedded inside a cobalt matrix acting as binder by liquid phase sintering giving the so-called hardmetal used for machining tools.

3.3.9.3 Natural Occurrence, Minerals, and Ores

Tungsten is a rare occurring element in the Earth's crust with an abundance of 1.5 mg.kg^{-1} (1.5 ppm wt). Tungsten, owing to its strong affinity for oxygen, never occurs free in nature and is found chiefly as wolframates in certain minerals, such as: **wolframite** ((Fe,Mn)WO$_4$, monoclinic), **scheelite** (CaWO$_4$, tetragonal), **huebnerite** (MnWO$_4$, monoclinic), and **ferberite** (FeWO$_4$, monoclinic). By far, wolframite and scheelite are its chief ores. Tungsten ore deposits are of magmatic or hydrothermal origin. Actually, scheelite and wolframite essentially form during fractional crystallization of magmas (magma differentiation) and concentrate into veins around batholithes. Moreover, tungsten ore deposits are strongly associated with recent orogenic activity such as in the Alps, Himalayas, and circum-Pacific belt. Hence, tungsten occurs in important ore deposits in the western United States (e.g., Alaska, California, and Colorado), in Asia (e.g., China and South Korea), in South America (e.g., Bolivia and Mexico), and Europe (e.g., Russia and Portugal). However, China is reported to have about 75% of the world's tungsten resources (3.7×10^6 tonnes of W) followed by Canada (0.57×10^6 tonnes of W), United States (0.45×10^6 tonnes of W), and the Commonwealth of Independent States (CIS). The concentration of workable ore (Clarke index) usually ranges between 0.3wt% and 1.0wt% WO$_3$. However, today apart the above natural resources, recycling of tungsten-containing scrap and other residues is another important source of tungsten and it is estimated to supply approximately 25–30% of the world demand. Actually, hardmetal is recycled by the zinc process. Tungsten carbide tools are immersed in a molten zinc bath at 900°C and the high volume expansion of the Co–Zn alloy destroys the binder and releases the tungsten carbide particles. After removal of zinc by distillation, the carbide particles are reprocessed.

3.3.9.4 Processing and Industrial Preparation

Tungsten ores such as scheelite and wolframite are recovered by underground mining techniques of rich bearing veins. Actually, even if the open pit process exists is economically negligible. After mining the raw ore, containing a mean average of 1.3wt% WO$_3$, undergoes a classic ore beneficiation process. The raw ore is crushed and ground in a ball mill. This size reduction operation is followed by a gravity separation and froth flotation in order to remove inert minerals of the gangue from valuable ore minerals, while electromagnetic separation is sometimes used, especially for wolframite. The concentrated tungsten ore (e.g., wolframite concentrate) is then processed to produce the important chemical intermediate ammonium paratungstate ((NH$_4$)$_{10}$W$_{12}$O$_{41}$·5H$_2$O). Calcination of this intermediate leads to the formation of tungsten oxides WO$_3$ (yellow) and W$_{20}$O$_{28}$ (blue). Ferrotungsten can be obtained commercially by carbothermic reduction of tungsten oxides with coal or coke in an electric arc furnace, while highly pure tungsten metal can be obtained commercially by two routes, either by the reduction of tungsten oxide by hydrogen at 700–1000°C in a rotary furnace, or chemical vapor deposition

(CVD) by reduction of volatile tungsten hexachloride (WCl_6) with hydrogen which deposits the metal onto a tungsten heated filament. Preparation of the tungsten monocarbide is performed by the carburization process which requires heating a mixture of tungsten powder and carbon black at 900–2000°C.

3.3.9.5 Tungsten Alloy Properties

See **Table 3.51** (page 189).

3.3.9.6 Industrial Uses and Applications

Cemented carbides, or hardmetal, since 1923 represent the major use of tungsten (60%). They are prepared by melting together graphite and tungsten metal in an electric arc furnace. The reaction leads to the formation of a eutectic mixture of WC and W_2C. The cooled material, owing to its brittleness is ground to fine particles and added to molten cobalt bath. Tungsten is useful for glass-to-metal seals since the linear thermal expansion of the metal is about the same as borosilicate glass. Tungsten and its alloys are used extensively for filaments for electric lamps, electron and television tubes (CRTV), and for metal evaporation work. Other applications include electrical contact points for car distributors, X-ray targets, windings and heating elements for electrical furnaces, missile and high-temperature applications, alloying element in high-speed tool steels (10%) and many other alloys. The two carbides (WC and W_2C) are important to the metalworking, mining, and petroleum industries. Calcium and magnesium tungstates are widely used in fluorescent lighting, tungsten salts are used in the chemical and tanning industries (10%). Tungsten chalcogenide, WS_2, like MoS_2, is a dry, high-temperature lubricant, stable up to 500°C. Tungsten bronzes ($NaWO_3$) and other tungsten compounds are used in paints.

3.3.9.7 Tungsten Metal World Producers

See **Table 3.52** (pages 190–191).

3.3.10 Rhenium and Rhenium Alloys

3.3.10.1 Description and General Properties

Rhenium [7440-15-5], with the symbol Re, the atomic number 75, and the relative atomic mass 186.207 is the last metal of the group VIIB(7) of Mendeleev's periodic chart. The metal is named after the Latin word, Rhenus, meaning the German river Rhine. Rhenium is silvery grayish white with a metallic luster; its density (21,010 kg.m^{-3}) is only exceeded by that of platinum, iridium, and osmium, and its melting point (3270°C) is only exceeded by that of tungsten and carbon. Its crystal lattice structure is hexagonal close packed (hcp). From a mechanical point of view, rhenium is highly ductile and exhibits a high Young's modulus (460 GPa) only exceeded by that of iridium and osmium, and a high tensile strength (1170 MPa). Like the other refractory metals it has a low coefficient of linear thermal expansion (6.7 μm.m^{-1}.K^{-1}). Rhenium does not exhibit a ductile–brittle transition temperature (DBTT). It is expensive but useful as a trace alloying agent in some refractory alloys. It has two natural isotopes: the stable nuclide ^{185}Re (37.07at%), and the radioactive nuclide ^{187}Re (62.93at%) with a half-life of roughly 4.5×10^{10} years. Hence the specific radioactivity of the natural element is 4.47 kBq.kg^{-1} (0.12 μCi.kg^{-1}). At room temperature, rhenium reacts with alkalis, and it is readily dissolved in concentrated nitric acid (1.5 mg.min^{-1}), but resists up to 50wt%

Table 3.51 Physical properties of selected tungsten alloys

Usual and trade name (or ASTM B459-67)	UNS	Average chemical composition (x, wt%)	Density (293.15 K), (ρ, kg.m^{-3})	Young's modulus (E, GPa)	Yield strength 0.2% proof (σ_{YS}, MPa)	Ultimate tensile strength (σ_{UTS}, MPa)	Elongation (Z, %)	Hardness, Rockwell C (HRC)	Coefficient of linear thermal expansion (α, 10^{-6} K^{-1})	Thermal conductivity (k, W.m^{-1}.K^{-1})	Electrical resistivity (ρ, $\mu\Omega$.cm)
W (soft, unalloyed)	R07030	99.5wt% W (sintered, hot swaged, and annealed 1590°C)	19,293	415	550	620	2	32 (360HV)	4.50	174	5.50
W (hard, unalloyed)	R07030 (sintered, cold worked)	99.95wt% W	19,293	n.a.	760	1862	0	500HV	4.50	174	5.50
W-Ni-Cu Cl. 1	n.a.	W-6Ni-4Cu	16,995	276	605	755	6	24-36	5.40	23	12.36
W-Ni-Fe Cl. 1	n.a.	W-7Ni-3Fe	16,995	310	615	895	16	25-32	4.61	18	17.31
W-Ni-Fe Cl. 2	n.a.	W-5.25Ni-2.25Fe	17,494	324	579	786	7	26-33	4.62	20	13.32
W-Ni-Cu Cl. 3	n.a.	W-3.5Ni-1.5Cu	17,992	310	586	758	7	27-33	4.43	33	10.81
W-Ni-Fe Cl. 3	n.a.	W-3.5Ni-1.5Fe	17,992	345	621	827	7	27-34	4.60	26	13.32
W-Ni-Cu Cl. 4	n.a.	W-2.1Ni-0.9Cu	18,490	365	586	848	5	28-35	4.50	30	10.18
W2Mo	n.a.	W-2Mo	n.a.	400	750	965	3	31	n.a.	n.a.	n.a.
W15Mo	n.a.	W-15Mo	n.a.	390	740	980	7	30	n.a.	n.a.	n.a.
W1Re	R07031	W-1.5Re	19,400	n.a.	n.a.	n.a.	10	n.a.	n.a.	n.a.	n.a.
W3Re	n.a.	W-3Re	19,550	403	n.a.	1242-2208	10	n.a.	n.a.	n.a.	9.14
W5Re	n.a.	W-5Re	19,570	400-411	n.a.	1380-2208	10	n.a.	n.a.	n.a.	11.63
W25Re	n.a.	W-25Re	19,800	370-430	n.a.	1370-1551	10-20	n.a.	n.a.	n.a.	27.43
AKS with thoria	R07911	W-1ThO$_2$	n.a.	n.a.	n.a.	n.a.	n.a.	n.a.	n.a.	n.a.	n.a.
AKS with thoria	R07005	W-1.5ThO$_2$	n.a.	n.a.	n.a.	n.a.	n.a.	n.a.	n.a.	n.a.	n.a.
AKS with thoria	R07912	W-2ThO$_2$	n.a.	n.a.	n.a.	n.a.	n.a.	n.a.	n.a.	n.a.	n.a.
AKS with zirconia	R07620	W-0.3ZrO$_2$	n.a.	n.a.	n.a.	n.a.	n.a.	n.a.	n.a.	n.a.	n.a.
AKS with lanthania	R07941	W-1La$_2$O3	n.a.	n.a.	n.a.	n.a.	n.a.	n.a.	n.a.	n.a.	n.a.

Table 3.52 Tungsten metal and hardmetal world producers

Company	Address
Fansteel Inc.	1 Tantalum Place, North Chicago, Illinois 60064, USA Telephone: + 1 847 689 4900 Facsimile: + 1 847 689 0307
Kennametal Inc.	P.O. Box 231, Latrobe, PA 15650, USA Telephone: + 1 724 539 5000 Facsimile: + 1 724 539 5079 E-mail: adl@kennametal.ibmmail.com http://www.kennametal.com
Kulite Tungsten Corp.	160 E Union Avenue, East Rutherford, New Jersey 07073, USA Telephone: + 1 201 438 9000 Facsimile: + 1 201 438 0891 E-mail: info@kulitetungsten.com
Avocet Mining PLC.	9th Floor New Zealand House, 80 Haymarket, London SW1Y 4TE, UK Telephone: + 44 171 389 8200 Facsimile: + 44 171 925 0888 E-mail: avocetmining@compuserve.com http://www.avocet.co.uk
Boart Longyear GmbH & Co. KG	Städeweg 18, D-36151 Burghaun, Germany Telephone: + 49 6652 82 300 Facsimile: + 49 6652 82 390 E-mail: 100.28930@germanynet.de http://www.boartlongyear-eu.com
H.C. Starck GmbH & Co. KG	Postfach 25 40, Im Schleeke 78–91, D-38615 Goslar, Germany Telephone: + 49 5321 751 0 Facsimile: + 49 5321 751 192 http://www.hcstarck.com
Nippon Tungsten Co. Ltd	2–8, Minoshima 1-chome, Hakata-ku, Fukuoka, 812 Japan Telephone: + 81 92 415 5507 Facsimile: + 81 92 415 5513 E-mail: sumikura@nittan.co.jp
North American Tungsten Corp.	11–1155 Melville Street, Vancouver, BC, V7K 2H4, Canada Telephone: + 1 604 682 1333 Facsimile: + 1 604 682 1324 E-mail: westpac@intergate.bc.ca
OM Group Inc.	50 Public Square, Suite 3800, Cleveland, OH 44113-2204, USA Telephone: + 1 216 781 0083 Facsimile: + 1 216 781 1502 E-mail: tacowowego@aol.com
Osram Sylvania Products Inc.	Hawes Street, Towanda, PA 18848, USA Telephone: + 1 570 268 5000 Facsimile: + 1 570 268 5113 E-mail: dunns@osi.sylvania.com
Plansee AG	A-6600 Reutte Tyrol, Austria Telephone: + 43 0 56 72 600 0 Facsimile: + 43 0 56 72 600 500
Sandvik AB	S-126 80 Stockholm, Sweden Telephone: + 46 8 726 6700 Facsimile: + 46 8 726 9096 http://www.sandvik.com
Sogem USA Inc.	Magnolia Building, Suite 110, 3120 Highwoods Boulevard, Raleigh, NC 27604, USA Telephone: + 1 919 874 7171 Facsimile: + 1 919 874 7195 E-mail: rick.holden@sogemnet.com

Table 3.52 (continued)

Company	Address
Sumitomo Electric Industries Ltd	Hardmetal Division, 1-1 Koyakita 1-chome, Itami, Hyogo 664, Japan Telephone: + 81 727 72 4535 Facsimile: + 81 727 71 0088 E-mail: dw800327@jnet.sei.co.jp
Teledyne Advanced Materials	1 Teledyne Place, Lavergne, TN 37086, USA Telephone: + 1 615 641 4245 Facsimile: + 1 615 641 4268 E-mail: jim_oakes@teledyne.com
Toho Kinzoku Co., Ltd	Osaka-Shinko Building, 6–17 Kitahama-2, Chuo-Ku, Osaka 541, Japan Telephone: + 81 6 202 3376 Facsimile: + 81 6 202 1390
Tokyo Tungsten Co., Ltd	24–8, 5-chome, Higashi-Ueno, Taito-Ku, Tokyo 110, Japan Telephone: + 81 3 5828 5632 Facsimile: + 81 2 5828 5517 E-mail: wzd16986@biglobe.ne.jp
Toshiba Corp.	1–1 Shibaura 1-Chome, Minato-Ku, Tokyo 105–01, Japan Telephone: + 81 3 3457 3311 Facsimile: + 81 3 5444 9341 E-mail: yoshiro.suzuka@toshiba.co.jp
Valenite Inc.	31700 Research Park Drive, P.O. Box 9636, Madison Heights, MI 48071-9636, USA Telephone: + 1 248 589 6310 Facsimile: + 1 248 597 4990 E-mail: jim esdale@valenite.com
Widia GmbH	Münchener Str. 90, D-45145 Essen, Germany Telephone: + 49 201 725 3353 Facsimile: + 49 201 725 3500

sulfuric (0.0015 mg.min^{-1}), and is quite inert to aqua regia (0.04 mg.min^{-1}) and concentrated hydrochloric acid (0.008 mg.min^{-1}). It reacts with boron, but unlike other refractory metals it does not form stable carbide. Rhenium has poor oxidation resistance. Hence, it should be protected by an iridium coating in high-temperature applications. Finally, rhenium is strongly corroded by molten metals such as Mo, W, Fe, Ni, and Co.

Cost (1998) – pure rhenium metal powder (99.95wt% Re) is priced at 900 $US.kg^{-1} (1984 $US.lb^{-1}).

3.3.10.2 History

In 1869, the famous Russian chemist Dimitri Mendeleev, establishing the periodic chart, predicted two new elements below manganese in the group VIIB. He gave them the names 43 eka-manganese (technetium) and 75 dwi-manganese (rhenium). However, due to lack of data regarding neighbor elements, he did not extrapolate their properties. The discovery of rhenium is generally attributed to the three German chemists, Walter Noddack, Ida Tacke-Noddack, and Otto Berg, who announced in 1925 they had detected the element in platinum ores and columbite.[93] In 1926, by treating 660 kg of molybdenite concentrate, they were able to extract and prepare the first gram of rhenium metal. Industrial production of rhenium began in 1928, but because it was an uneconomical process which produced an expensive metal, the production was discontinued in 1930, and rhenium remained a laboratory curiosity until the 1950s. In the 1950s, rhenium was industrially produced in Europe, the USA, and the USSR for alloying element and catalyst purposes.

3.3.10.3 Natural Occurrence, Minerals and ores

Rhenium is an extremely rare element in the Earth's crust with an average abundance of $1 \mu g.kg^{-1}$ (ppb wt). Rhenium does not occur free in nature or as a compound in a distinct mineral species. Actually, the element has no specific mineral and only occurs as trace impurities in gadolinite, molybdenite, columbite, rare earth minerals, and some sulfide ores. The total estimated free world reserve of rhenium metal is 3500 tonnes.

3.3.10.4 Processing and Industrial Preparation

Commercial rhenium is obtained from molybdenum roaster-flue dusts obtained from the processing of copper sulfide ores. Actually, some molybdenum ores contain from 0.002wt% to 0.2wt% rhenium. More than 4.67 tonnes per year of rhenium are now being produced in the United States. After roasting, the volatile rhenium heptoxide, carried off in flue gases, is absorbed in scrubbing liquors, refined and converted to ammonium perrhenate (APR). Afterwards, pure rhenium metal powder (99.99wt% Re) is obtained by thermal reduction of ammonium perrhenate with hydrogen at elevated temperatures. The preparation is usually performed in a two-stage process. In the first stage, the ammonium perrhenate is thermally decomposed into ammonia and rhenium heptoxide. In the second stage, the rhenium heptoxide is reduced to powdered metal by hydrogen gas. Afterwards, the rhenium powder is usually cold-isostatic pressed at 170–200 MPa, and sintered in an hydrogen atmosphere at 75% of its melting point (2480°C). When high-purity rhenium is required (99.995wt% Re), sintered rhenium ingots are remelted by an electron beam and zone refined. The world's largest resources of rhenium-bearing ores are located in countries with large copper ore deposits, such as in Chile, the Commonwealth of Independent states (CIS), and the USA.

3.3.10.5 Uses and Applications

Rhenium is used as an additive to tungsten- and molybdenum-based alloys to impart useful properties. It is also used to manufacture filaments for mass spectrophotometers and ion gages. Rhenium–molybdenum alloys are superconductors at very low temperatures, e.g., 10 K. Rhenium is also used in electrical contact material, as it has good wear resistance and withstands arc corrosion. In instrumentation, thermocouples made of Re–W are extensively used for measuring high temperatures up to 2200°C. Moreover, Rhenium wire is used in flash lamps for photography. Finally, rhenium catalysts are exceptionally resistant to pollution by nitrogen, sulfur, and phosphorus, and are used for hydrogenation of fine chemicals, hydrocracking, reforming, and the disproportionation of alkenes.

3.4 Precious and Noble Metals (NM)

3.4.1 Silver and Silver Alloys

3.4.1.1 Description and General Properties

Silver [7440-22-4] with the chemical symbol Ag, the atomic number 47 and the relative atomic mass (atomic weight) 107.8682(2) is the second metal of the group IB(11) of Mendeleev's periodic chart. Silver is named after the Anglo-Saxon, Seolfor, sulfur, while its chemical symbol, Ag, is named first of all from the Greek, argyros, and later after the Latin, argentum, silver. Pure silver is a midly dense metal (10,501 kg.m^{-3}) which has a brilliant white metallic luster. Silver crystals have a face-centered cubic (fcc) space lattice structure with a melting point of 961.78°C. From a mechanical point of view, silver is slightly harder than gold (its Mohs hardness is roughly between 2.5 to 3.0) and is very ductile and malleable, being exceeded only by gold and perhaps palladium. Pure silver has the lowest electrical resistivity (1.47 $\mu\Omega$.cm) and the highest thermal conductivity (450 W.m^{-1}.K^{-1}) of all the pure metals, and possesses the lowest contact resistance. When freshly electrodeposited, it has the highest reflective index (albedo) of all metals for visible light known (i.e., roughly 95% between 400 and 800 nm), but is rapidly tarnished and loses much of its reflectance. However, it is a poor reflector of ultraviolet radiation (8% at 320 nm). From a chemical point of view, silver is stable in pure air and water in comparison with copper, but it readily tarnishes when exposed to ozone, O_3, hydrogen sulfide, H_2S, or air containing sulfur dioxide, SO_2. Strong halogenated mineral acids (e.g., HCl, HBr, and HI), nitric acid, NHO_3, and sulfuric acid, H_2SO_4, readily dissolve silver but it is not corroded by hydrofluoric, HF, and orthophosphoric, H_3PO_4, acid, and other common sulfur-free organic acids (e.g., formic, and acetic acids). Moreover, it is highly resistant to molten alkali-metal hydroxides (e.g., NaOH and KOH) and is extensively used for making labware for alkali fusion. However, silver forms a eutectic with molten metals such as low-melting point metals: Hg, Pb, Sn, Bi, and alkali metals such as Na, and K. Silver, like copper, can form explosive acetylenides (AgCH≡CHAg) when put directly in contact with compressed acetylene gas. Moreover, it is attacked by sulfur even at room temperature because the sulfide layer is not protective. While silver itself is not considered to be toxic, most of its soluble compounds are poisonous. Exposure to silver metal and its compounds in air should not exceed 10 μg.m^{-3} Ag, on a basis of 8 h time weighted average and 40 h week. Silver compounds can be absorbed in the circulatory system and reduced silver deposited in the various tissues of the body. A condition, known as argyria, results, with a grayish pigmentation of the skin and mucous membranes. Silver soluble compounds have germicidal effects and kill many lower organisms effectively without harm to higher animals.

Cost (1998) – pure silver (99.9wt% Ag) is priced at 167 $US.kg^{-1} (5.19 US/troy oz).

3.4.1.2 History

Like copper and gold, silver has been known since ancient times. Actually, it is mentioned in Genesis and moreover slag dumps were discovered by archeologists in Asia Minor and on islands in the Aegean Sea (Europe). These traces indicate clearly that man learned to separate silver from lead as early as 3000 BC.

3.4.1.3 Natural Occurrence, Minerals, and Ores

Silver is a relatively rare metal in the Earth's crust with an average abundance of 75 μg.kg^{-1} (ppt wt). The element occurs free in nature as a native element sometimes alloyed with mercury giving the so-called **amalgam** (γ-Ag$_3$Hg$_4$, rhombohedral), or alloyed with gold as **electrum** (AgAu, cubic) and owing to its strong geochemical affinity for sulfur (i.e., chalcophile element), it is often found in a combined form in sulfide ores such as **acanthite** (Ag$_2$S, monoclinic), **argentite** (Ag$_2$S, cubic), and halides such as **chloroargyrite** or **horn silver** (AgCl, cubic), **bromoargyrite** (AgBr, cubic) and **iodoargyrite** (AgI, hexagonal). Moreover, it always occurs as trace impurities in lead, lead–zinc, copper, gold, and copper–nickel ores which are its principal commercial primary sources. Secondary source silver is recovered either as a by-product during electrorefining of copper as silver powder in slurries at the bottom of the electrolytic cell or from scrap recycling. The overall world production according the US Geological Survey (USGS) was estimated in 1996 as 25,000 tonnes from which the direct mine production is approximately 15,200 tonnes. The four largest silver-producing countries in the western hemisphere are, in decreasing order: in South America (1) Mexico (2600 tonnes), (2) Peru (1968 tonnes); in North America (3) the United States (1570 tonnes), and finally (4) Canada (1296 tonnes).

3.4.1.4 Processing and Industrial Preparation

Raw materials for the production of silver are, for primary sources: rich silver galena and all the by-products arising from gold bullion smelting in the mines containing small amounts of gold, such as slag, flue dust, furnace linings, and crucibles. Secondary sources include activated carbon, sweeps, ashes, concentrates, and computer scrap. By-products of the smelting process are leady copper matte and flue dust which are sold overseas, and barren residue slag which is sold for use in shot-blasting and glass coloring. When products from the mine are delivered to the smelter, they are first sampled and analyzed. They are then mixed thoroughly with fluxes such as limestone (impure $CaCO_3$) and iron mill-scale to prepare a feed of consistent composition for smelting in an electric-arc furnace. Included in the feed is lead which acts as a collector for gold and silver during smelting. This lead is tapped out of the arc furnace into a ladle and transferred molten into a top-blown rotary converter in which the lead is oxidized to slag. The end product of the smelting process is doré bullion, a mixture of approximately 30wt% gold and 70wt% silver, which is poured into bars and sent to the refining section. The chlorides produced in the gold refining process contain approximately 50wt% silver and are treated to recover this. The molten chlorides previously poured off the refining furnaces are stripped of gold in separate furnaces to speed up the recycling of traces of gold dispersed in the melt. They are then quenched in water and pumped to a reaction tank. In the reaction tank, base metals are removed by washing and then metallic silver is precipitated by the addition of zinc dust. This silver (99wt% Ag) is dried in a centrifuge, then melted in a resistance furnace, from which silver anodes are cast. These slab anodes are refined by electrorefining to produce silver crystals containing at least 99.9wt% Ag, which are either vacuum dried and sold, or cast into roughly 31.103477 kg (exactly 1000 troy oz) bars for sale.

3.4.1.5 Silver Alloys

See **Table 3.53** (page 195).

3.4.1.6 Industrial Applications and Uses

The uses and applications of silver alloys are important and various. Actually, the high thermal and electrical conductivities, associated with the high reflectivity and ductility of silver and silver alloys govern their uses despite their high costs.

Jewelry and silverware – because of its high reflectivity, sterling silver (a silver–copper alloy) is extensively used for jewelry, silverware, and tableware where appearance is paramount. Sometimes, a protective rhodium coating is electrodeposited onto thin nickel-plated silver objects in order to avoid tarnishing. Sterling silver alloy contains roughly 92.5wt% Ag, the remainder being copper or some other metals.

Photography – silver is of the utmost importance in photography, about 30% of the industrial consumption going into this application. Actually, the use of silver in photography is due to the ability of silver halide crystals (especially silver bromide), to become photosensitive and undergo a secondary image amplification called development.

Dentistry – silver is also used in dentistry either as Ag–Sn–Hg or Ag–Sn–Cu–Hg silver alloys owing to their small expansion during setting which is suitable for making dental amalgams.

Brazing filler metals – owing to its low melting point and its good wettability of several base metals, silver is a major component for making solder and brazing alloys.

Usual and trade name	UNS	Average chemical composition (wt%)	Density (ρ, kg.m^{-3})	Young's modulus (E, GPa)	Yield strength 0.2% proof (σ_{YS}, MPa)	Ultimate tensile strength (σ_{UTS}, MPa)	Elongation (Z, %)	Vickers hardness (HV)	Melting range (°C)	Coefficient of linear thermal expansion (α, 10^{-6} K^{-1})	Specific heat capacity (c_p, J.kg^{-1}.K^{-1})	Thermal conductivity (k, W.m^{-1}.K^{-1})	Electrical resistivity (ρ, $\mu\Omega$.cm)
Commercially pure silver	P07020	Ag (99.90 wt%min)	10,490	71	54	125	3–5	25	961.78	19.1	234	419	1.47
Refined silver ASTM 413	P07015	Ag (99.95 wt%min)	10,490	71	54	125	3–5	25	961.78	19.1	234	425	1.47
Refined silver ASTM 413	P07010	Ag (99.99 wt%min)	10,490	71	54	125	3–5	25–27	961.78	19.68	234	425	1.59
Silver magnesium alloy	n.a.	99.5Ag–0.25Ni–0.25Mg	10,500	83	130	250	28	n.a.	n.a.	n.a.	n.a.	n.a.	n.a.
Sterling silver (standard grade)	P07931	92.5Ag–7.5Cu	n.a.	n.a.	255	n.a.	30	n.a.	n.a.	n.a.	n.a.	n.a.	2.10
Sterling silver (silversmith grade)	P07932	92.5Ag–7.5Cu	n.a.	n.a.	n.a.	n.a.	n.a.	n.a.	n.a.	n.a.	n.a.	n.a.	n.a.
Eutectic alloy	P07720	72Ag–28Cu	n.a.	n.a.	275	380	20	n.a.	779–1435	n.a.	n.a.	n.a.	2.25
Silver-palladium	n.a.	61Ag–27Pd–2Au	10,800	n.a.	241	586	10	165	960–1055	n.a.	n.a.	n.a.	n.a.
Silver-palladium	n.a.	59Ag–25Pd	10,500	n.a.	660	751	9	235	900–980	n.a.	n.a.	n.a.	n.a.
Silver-palladium	n.a.	50Ag–30 Pd–3Au	10,500	n.a.	427	607	20	170	965–1030	n.a.	n.a.	n.a.	n.a.
Silver brazing filler metals	P07453	45Ag–30Cu–25Zn	n.a.	n.a.	n.a.	n.a.	n.a.	n.a.	n.a.	n.a.	n.a.	n.a.	n.a.
Dental amalgam	n.a.	45Hg–25Ag–15Sn–14Cu–1Zn	11,000	60	n.a.	n.a.	n.a.	90	n.a.	22–28	n.a.	n.a.	n.a.

Electronics – due to its high conductivity, both electrical and thermal, and low surface contact resistance, electrodeposited silver is also used for making electrical contacts.

Primary batteries – silver compounds such as AgO are used as cathode materials for making high-capacity silver–zinc and silver–cadmium primary batteries. Silver paints are used for making printed circuits.

Mirrors – it is used in mirror production and may be deposited on glass or metals by thermal decomposition, chemical vapor deposition, electroless plating and electrodeposition, or by vacuum evaporation.

Explosives – silver fulminate, AgCNO, is a powerful primary explosive.

Meteorology – hexagonal crystals of silver iodide, AgI, are used in seeding clouds to produce artificial rain.

Other – silver chloride, AgCl, has interesting optical properties as it can be made transparent; it is also a cement for glass. Silver nitrate, $AgNO_3$, also called lunar caustic, the most important silver compound, is used extensively in photography.

Coinage – for centuries silver has been used traditionally for coinage by many countries of the world. But today, silver coins of other countries have largely been replaced with coins made of other metals.

3.4.2 Gold and Gold Alloys

3.4.2.1 Description and General Properties

Gold [7440-57-5], with the chemical symbol Au, the atomic number 79, and the relative atomic mass (i.e., atomic mass) 196.96655(7) is a precious metal, along with copper and silver, of the group IB(11) of Mendeleev's periodic chart. The symbol comes from the Latin word, aurum, while the name gold comes from the Anglo-Saxon. Gold is a highly dense (19,320 kg.m^{-3}), soft, lustrous yellow, ductile, and malleable metal. In addition to its softness, it is both the most malleable and most ductile of all the metals. Actually, it can be hammered into extremely thin sheets, approaching a small number of atoms and can be drawn into extremely fine wire. For instance, one gram of pure gold can give a foil of 0.9 m^2 surface area, which corresponds to roughly 60 nm thickness. Gold in the form of very thin sheets is transparent and is called gold leaf. Moreover, it has a melting point of 1064.18°C (as defined by the ITS-90) and a boiling point of 2856°C. It has a very low electrical resistivity (2.35 $\mu\Omega$.cm) and a high thermal conductivity (317 W.m^{-1}.K^{-1}), surpassed only by the other members of group IB, i.e., copper and silver. The gold crystal space lattice structure, like copper, is face-centered cubic (fcc) which explains its intrinsic ductility and malleability. Gold is a mononuclidic element with one isotope, ^{197}Au. Gold has a high reflective index in the far infrared, and it is used as mirror. From a chemical point of view, gold has an excellent corrosion resistance, and hence it is inert to most corrosive chemicals. Therefore, it is unaffected by air and moisture and remains tarnish-free indefinitely. However, it is readily dissolved in various solutions of cyanides (e.g., NaCN, KCN) that are used in ore extraction process or by aqua regia (a mixture of three parts conc. HCl and one part conc. HNO_3 by volume), named from the Latin, regis, because it dissolves the king of the metals. Gold also forms amalgams with mercury and as such is used in dentistry. Gold usually forms chemical compounds and stable complexes. When used in jewelry, it is commonly alloyed with other metals in order to increase strength and yield colors. For instance, the alloy of gold, silver, and copper, in which the amount of silver predominates, is called green old, while the alloy of the same three elements in which copper predominates is called red gold. Finally, the alloy of gold and nickel is called white gold.

Cost (1998) – pure refined gold (99.995wt%) is priced at 10,224 $US. kg^{-1} (318 $US/troy oz)

Caratage – the caratage (with the symbols, k, K, or Kt) is the mass fraction of gold in a gold alloy, expressed in 24ths. For instance, 24 carat (24 Kt) gold corresponds to pure gold (i.e., 100wt% Au), and 18 carat (18 Kt) gold alloy contains 18/24ths, i.e., 75wt% of gold.

100wt% Au = 24/24ths 24 Kt
91.67wt% Au = 22/24ths 22 Kt
75.00wt% Au = 18/24ths 18 Kt
58.33wt% Au = 14/24ths 14 Kt
41.67wt% Au = 10/24ths 10 Kt

Fineness – the fineness is the mass fraction of gold in a gold alloy, expressed in 1000ths ($wt^o/_{oo}$) or 10,000ths. For instance, 9999 gold contains 9999 parts per 10,000 fine gold by weight. **Troy weight measures:**[94]

1 troy ounce (troy oz) = 31.1034768 g
1 troy pound (troy lb) = 12 troy ounces = 0.3732417216 kg

3.4.2.2 History

Gold has been for many thousands of years a prized metal throughout the world, which has fascinated man who has bestowed great value on it because of its beauty, its purity, and its scarcity. Gold was probably used in decorative arts before 9000 BC. Even civilizations that developed little or no use of other metals prized gold for its beauty. The ancients found quantities of gold in Ophir, Sheba, Uphaz, Parvaim, Arabia, India, and Spain. By the time of Christ, written reports were made of deposits in Thrace, Italy, and Anatolia. Gold is also found in Wales, in Hungary, in the Ural Mountains of Russia, and, in large quantities, in Australia. The greatest early surge in gold recovery followed the discovery of the Americas. From 1492 to 1600, Central and South America, Mexico, and the islands of the Caribbean Sea contributed significant quantities of gold to world commerce. Colombia, Peru, Ecuador, Panama, and Hispaniola contributed 61% of the world's newfound gold during the 17th century. In the 18th century they supplied 80%. Actually, discovery of large gold deposits in 1721 in Mato Grosso and later in Goias (Brazil) led to large gold rushes. Later in 1838, a gold placer deposit was discovered in alluvial deposits of the Tchara river (Siberia). In 1848, following the discovery of gold in California, North America became the world's major supplier of the metal. From 1850 to 1875 more gold was discovered than in the previous 350 years. In 1850 gold deposits were found in Western Australia. By 1890 the gold fields of Alaska and the Yukon edged out those in the western United States, and soon in 1884 the African Transvaal exceeded even these. South Africa burst on to the international scene over 100 years ago with the discovery of the largest and most productive gold fields (e.g. Witwatersrand near Johannesburg). Demand for gold has grown and changed somewhat from that of 100 years ago. The US Geographical Survey (USGS) has estimated that 123,000 tonnes of gold was mined since historical times through 1997, and about 15% is thought to have been lost, used in dissipative industrial applications, or otherwise unaccounted for or unrecoverable. Of the remaining 105,000 tonnes, about 34,000 tonnes is official stocks held by central banks, and about 71,000 tonnes is privately held as bullion, coin, and jewelry.

3.4.2.3 Natural Occurrence, Minerals, and Ores

Although the gold abundance in the Earth's crust averages 4 $\mu g.kg^{-1}$ (i.e., ppb wt) seawater which contains depending on location between 0.1 to 10 mg per tonne of gold is a large reserve of the metal, but owing to its dispersion the process of recovery (at one time suggested by the German chemist Haber) of the metal is not economical. Commercial concentrations of gold ranging between 2 and 20 $mg.kg^{-1}$ are found in areas distributed widely over the globe. Gold occurs free as a **native gold** metal (Au, cubic) or alloyed with silver as **electrum** (AuAg, cubic), in association with ores of copper and lead, in quartz veins, in the gravel of stream beds, and as a trace impurity in sulfide ores such as **pyrite** (FeS_2, cubic), **pyrrhotite** ($Fe_{1-x}S_2$, cubic), **arsenopyrite** or **mispickell** (FeAsS, monoclinic), and **stibine** (Sb_2S_3, orthorhombic). The distribution of gold seems to validate the theory that gold was carried toward the Earth's surface from great depths by magmatic activity, perhaps with other metals as a solid solution within magmas. After this solid solution cooled, its gold content was spread through such a great volume of rock that large fragments were unusual; this theory explains why much of the world's gold is in small, often microscopic particles. The theory also explains why small amounts of gold are widespread in all igneous rocks; they are rarely chemically combined and seldom in quantities rich enough to be called an ore. Because of its poor chemical reactivity, gold was one of the first two or three metals, along with copper and silver, to be used by humans in these metals' elemental states. Because it is relatively unreactive, it was found uncombined and required no previously developed knowledge of refining. Native gold is found in a few magmatic segregations in pegmatites and pyrosomatic lodes and recent or burred placer deposits derived from them such as in the Witwatersrand (Republic of South Africa). The world's unmined reserves were estimated in 1996 by the US Geological Survey as 86,000 tonnes, of which 15–20% was by-product resources. About one half of these resources are found in the Witwatersrand area of the Republic of South Africa. In 1997, the world mine production of gold was estimated by the USGS to 2300 tonnes. The four largest producer countries are, in decreasing order: (i) the Republic of South Africa (490 tonnes), (ii) the United States (325 tonnes), (iii) Australia (290 tonnes), and (iv) Canada (150 tonnes).

3.4.2.4 Mineral Dressing and Mining

Gold is obtained by two principal mining methods: (i) placer and (ii) vein mining, and also (iii) as a by-product of the mining of other metals.

Placer mining – this extraction process is used when the metal is found in unconsolidated deposits of sand and gravel from which gold can be easily separated due to its high density. The sand and gravel are suspended in moving water; the much heavier metal sinks to the bottom and is separated by hand. The simplest method, called gold panning, is to swirl the mixture in a pan rapidly enough to carry the water and most of the gravel and sand over the edge while the gold remains on the bottom. Much more efficient is a sluice box, a U-shaped trough with a gentle slope and transverse bars firmly attached to the trough bottom. The bars, which extend from side to side, catch the heaviest particles and prevent their being washed downslope. Sand and gravel are placed in the high end, the gate to a water supply is opened, and the lighter material is washed through the sluice box and out the lower end. The materials caught behind the bars are gleaned to recover the gold. A similar arrangement catches the metal on wool, and may have been the origin of the legend of Jason's search for the golden fleece. Another variation of the placer method is called hydraulic mining. A very strong stream of water is directed at natural sand and gravel banks, causing them to be washed away. The suspended materials are treated much as if they were in a giant sluice box. Today's the most important placer technique is dredging. In this method a shovel of several cubic meters capacity lifts the unconsolidated sand and gravel from its resting place and starts the placer process.

Vein, or lode mining – this process is the most important of all gold recovery methods. Although each kilogram of gold recovered requires the processing of about 100,000 kg of ore, there is so much gold deposited in rock veins that this method accounts for more than half of the world's total gold production today. The gold in the veins may be of microscopic particle size, in nuggets or sheets, or in gold compounds. Regardless of how it is found, the ore requires extensive extraction and refining.

3.4.2.5 Processing and Industrial Preparation

Gold extraction strongly depends on the nature of the ore and type of deposits. The main routes for gold recovery are: (i) the **placer method** or **gravity separation method**, and (ii) the **cyaniding process**.

The **placer process** or **gravity separation method** – this process which represents approximately 15% of the world production is only used to recover native gold from placer deposits with gold particle size larger than 75 μm. This extraction process allows mining of low-level deposits (less than 1 mg.kg^{-1} Au). Gold particles are separated by gravity from inert gangue material (e.g., quartz, micas) and from other heavy minerals by amalgamation with mercury. The gold amalgam is then distilled at 400–500°C to release a gold–silver alloy (e.g., 25–50wt% Au).

The **cyaniding process** – this process, introduced in 1888, is today the most used worldwide (approximately 85wt% of the world production). The gold-containing ore (e.g., quartzite) is first crushed in rod or ball mills to a size less than 100 μm. This size reduction operation unit reduces the ore to a powdery substance from which the gold can be easily separated from inert minerals and comes into contact with chemicals. Then the powdered ore undergoes a beneficiation froth flotation process in order to remove waste products such as gangue and sulfide minerals. About 70% of the gold is recovered at this point. The ore concentrate is then leached for 12 to 48 h with a dilute and alkaline aerated solution of sodium cyanide (0.5 g.dm^{-3} NaCN) or calcium cyanide, to dissolve the gold. In order to prevent evolution of the poisonous hydrogen cyanide, HCN, the leaching solution is strictly maintained at pH above 10. The dissolution reaction is given by:

$$4Au + 8CN^- + O_2 + 2H_2O \rightarrow 4Au(CN)_2^- + 4OH^-$$

After complete dissolution, the leaching solution contains several ppm wt Au. At this stage there exist at least two main methods for recovering the gold: (i) the **Merryl–Crowe process** or **cementation method** – in this process, which represents 40wt% of world production, the addition of metallic zinc powder to the leaching solution causes the dissociation of the auricyanide complex with precipitation of metallic gold onto zinc particles forming a cemented compound containing 50wt% Zn, 15wt% Cu, 15wt% Ag and 2 to 6wt% Au. The zinc is later dissolved by addition of sulfuric acid, while a nitric acid solution dissolves copper and silver. The gold residue is filtered and dried and refined by smelting. (ii) The **carbon in pulp process** (CiP) – in this process, the cyanide leaching solution is filtered through activated carbon which strips the gold from the solution. One tonne of activated carbon is required to recover about 70 kg of gold. The gold-impregnated activated carbon is then leached with a hot alkaline solution containing 1wt% NaOH and 0.1wt% NaCN. The carbon is reactivated by pyrolysis at 600–700°C, while gold is recovered from solution by electrolysis on a steel wool cathode. Sometimes carbon can be burned in a furnace releasing gold bullion. For instance, the South African bullion contains approximately 84wt% Au, 10wt% Ag, and 6wt% miscellaneous metals.

Refining process – on arrival at the refinery, the bullion bars are first dried and weighed and the results are verified against those of the mines. If the weights agree, the bars are loaded in lots of up to 350 kg into an induction furnace where they are melted and samples are taken for analysis of the precious metal contents. This analysis is done by X-ray fluorescence spectrometry. If the analyses agree, the molten metal is transferred to a refining induction furnace in quantities of up to 75 kg. Chlorine gas is blown into the melt (**Miller process**), coverting base metals and silver to their chlorides which either rise to the surface and are poured off or, being volatile, are removed with the off-gas. It takes 60 to 90 min depending on impurity content, before the remaining gold reaches the required minimum of 99.5wt%. The refined gold is transferred in ladles to a 1 tonne capacity holding/casting furnace. The gold is sampled and mechanically cast into 12.44 kg (exactly 400 troy oz) bars, which are subsequently quenched, dried, stamped, weighed, and packed all on a mechanized conveyor assembly. Gold refined to a purity of 99.5wt% is not always pure enough for certain industrial and technical applications or for the production of small investment bars. In order to attain a purity of 99.99wt%, the gold is further purified by electrorefining. The 99.5wt% gold is cast into rectangular slabs (i.e., anodes). These gold anodes (+) are then immersed in a bath containing hydrochloricauric acid (HAuCl$_4$) and slowly dissolved anodically by electrolysis. Pure gold electrodeposits in a crystalline sponge form onto titanium cathode (–) plates. This process removes most of the silver and traces of metallic impurities, especially platinum group metals (PGMs), not removed by chlorination. Actually, metal impurities more noble in the electrochemical series than gold such as PGMs are not dissolved anodically and the solid particles sink in the sludge at the bottom of the electrolyzer. On the contrary, less noble impurities than gold, such as silver and copper, are readily dissolved but are not cathodically reduced and remain as soluble cations in the electrolytic bath. At the end of the electrolysis, the cathodes are removed from the bath and rinsed, and deposits are then melted and cast into billets from which the refinery's final products, such as 1 kg bars, are subsequently made.

Copper electrorefining by-product – one-third of all gold is produced as a by-product of copper, lead, and zinc production (500–600 tonnes per year). Copper, for example, must be electrolytically refined to raise its purity from 99wt% to more than 99.99wt% as required for many industrial purposes. In the refining process an anode of impure copper is electrolyzed in a bath in which the cathode is a very thin sheet of highly refined copper. As the process continues, copper ions leave the impure anode and are electrodeposited onto the cathode. Because noble impurities are not anodically dissolved, and as the anode is consumed, the noble impurities sink to the bottom as a sludge. This anode sludge contains gold in quantities sufficient to make recovery profitable. One-third of all gold is obtained from such by-products. Silver and platinum are also recovered from the copper anode sludge in quantities large enough to more than pay for the total refining process.

3.4.2.6 Gold Alloys

Gold alloys and some of their physical properties are shown in Table 3.54.

3.4.2.7 Industrial Applications and Uses

Jewelry – when used in jewelry, it is commonly alloyed with other metals in order to increase strength and yield colors. For instance, the alloy of gold, silver, and copper, in which silver predominates, is called **green gold**. The alloy of the same three elements in which copper predominates is called **red gold**. Finally, the alloy of gold and nickel is called **white gold**.

Electronics – its relatively high electrical conductivity, low friction coefficient and extremely high resistance to corrosion make the metal critically important in microelectrical circuitry particularly in printed circuit boards, connectors, keyboard contacts, and miniaturized circuitry.

Dentistry and medicine – owing to its chemical inertness, its lack of toxicity and its compatibility with all the body fluids, gold is extensively used in dentistry in inlays, crowns, bridges, and orthodontic appliances.

Thermal barrier and IR reflectors – owing to its high reflective index in the far infrared region of the electromagnetic spectrum, gold is used for manufacturing radiant heating devices, thermal barrier, heat shields, and reflective coating for solar radiation.

Mechanics – because of its chemical stability, gold plated on surfaces exposed to corrosive fluids or vapors is in demand for bearings used in corrosive atmospheres.

3.4.2.8 Gold World Suppliers

See **Table 3.55** (page 203).

Table 3.54 Physical properties of selected gold alloys

Usual and trade name	UNS	Chemical composition (average)	Density (ρ, kg.m^{-3})	Young's modulus (E, GPa)	Yield strength 0.2% proof (σ_{YS}, MPa)	Ultimate tensile strength (σ_{UTS}, MPa)	Elongation (Z, %)	Vickers hardness (HV)	Melting range (°C)	Coefficient of linear thermal expansion (α, 10^{-6} K^{-1})	Specific heat capacity (c_p, J.kg^{-1}.K^{-1})	Thermal conductivity (k, W.m^{-1}.K^{-1})	Electrical resistivity (ρ, $\mu\Omega$.cm)
Refined gold (cast)	P00010	Au (99.995wt%)	19,320	78.5	n.a.	127	30	33HB	1064	14.2	130	300	2.2
Refined gold (wrought)	P00015	Au (99.99wt%)	19,320	79.9	n.a.	131	45	25HB	1064	14.2	130	300	2.2
Refined gold (wrought)	P00020	Au (99.95wt%)	n.a.	n.a.	n.a.	n.a.	n.a.	n.a.	n.a.	n.a.	n.a.	n.a.	n.a.
Refined gold (wrought)	P00025	Au (99.50wt%)	n.a.	n.a.	n.a.	n.a.	n.a.	n.a.	n.a.	n.a.	n.a.	n.a.	n.a.
Gold 18K (White gold)	P00275	75.0Au–17.8Ni–1.7Cu	n.a.	n.a.	n.a.	n.a.	n.a.	n.a.	n.a.	n.a.	n.a.	n.a.	n.a.
Gold 14K (White gold)	P00160	58.3Au–12.2Ni–23.5Cu	n.a.	n.a.	n.a.	n.a.	n.a.	n.a.	n.a.	n.a.	n.a.	n.a.	n.a.
Gold 10K (White gold)	P00125	41.7Au–1.51Ni–46.7Cu	n.a.	n.a.	n.a.	n.a.	n.a.	n.a.	n.a.	n.a.	n.a.	n.a.	n.a.
Gold 18K (Yellow gold)	P00255	75.0Au–10Cu–15Ag	n.a.	n.a.	n.a.	n.a.	n.a.	n.a.	n.a.	n.a.	n.a.	n.a.	n.a.
Gold 14K (Yellow gold)	P00180	58.3Au–26.0Cu–16.5Ag	n.a.	n.a.	n.a.	n.a.	n.a.	n.a.	n.a.	n.a.	n.a.	n.a.	n.a.
Gold 10K (Yellow gold)	P00115	41.7Au–43.8Cu–5.5Ag	n.a.	n.a.	n.a.	n.a.	n.a.	n.a.	n.a.	n.a.	n.a.	n.a.	n.a.
Gold 18K (Green gold)	P00280	75.0Au–2.5Cu–22.5Ag	n.a.	n.a.	n.a.	n.a.	n.a.	n.a.	n.a.	n.a.	n.a.	n.a.	n.a.
Gold 14K (Green gold)	P00180	58.3Au–6.5Cu–35Ag	n.a.	n.a.	n.a.	n.a.	n.a.	n.a.	n.a.	n.a.	n.a.	n.a.	n.a.
Gold 10K (Green gold)	P00140	41.7Au–48.9Ag–9.1Cu	n.a.	n.a.	n.a.	n.a.	n.a.	n.a.	n.a.	n.a.	n.a.	n.a.	n.a.

Table 3.54 (continued)

Usual and trade name	UNS	Chemical composition (average)	Density (ρ, kg.m^{-3})	Young's modulus (E, GPa)	Yield strength 0.2% proof (σ_{YS}, MPa)	Ultimate tensile strength (σ_{UTS}, MPa)	Elongation (Z, %)	Vickers hardness (HV)	Melting range (°C)	Coefficient of linear thermal expansion (α, 10^{-6} K^{-1})	Specific heat capacity (c_p, J.kg^{-1}.K^{-1})	Thermal conductivity (k, W.m^{-1}.K^{-1})	Electrical resistivity (ρ, $\mu\Omega$.cm)
Gold 18K (Red gold)	P00285	75.0Au–20Cu–5Ag	n.a.	n.a.	n.a.	n.a.	n.a.	n.a.	n.a.	n.a.	n.a.	n.a.	n.a.
Gold 14K (Red gold)	P00170	58.3Au–39.6Cu–2.1Ag	n.a.	n.a.	n.a.	n.a.	n.a.	n.a.	n.a.	n.a.	n.a.	n.a.	n.a.
Gold 10K (Red gold)	P00145	41.7Au–55.5Cu–2.8Ag	n.a.	n.a.	n.a.	n.a.	n.a.	n.a.	n.a.	n.a.	n.a.	n.a.	n.a.
Gold–platinum	n.a.	70Au–30Pt	19,920	113.8	200	245	50	130 HB	1,228–1,450	n.a.	n.a.	n.a.	34

Table 3.55 Gold metal and alloy world producers

Producer	Address
Engelhard-CLAL	28, rue Michel-Le-Comte, F-75003 Paris, France Telephone: (33) 01 44 61 30 00 Facsimile: (33) 01 44 61 30 01 E-mail: webadmin@engelhard-clal.fr Internet: http://www.engelhard-clal.ff
Degussa	Degussa AG E-mail: preciousmetals@degussa.de Internet: http://www.degussa.de
W.C. Heraeus GmbH & Co. KG	P.O. Box 1553 D-63405 Hanau, Germany Telephone: + 49 (6181) 35 1 Facsimile: + 49 (6181) 35 658 E-mail: public-relations@heraeus.de Internet: http://www.heraeus.com
Johnson Matthey Noble Metals	Orchard Road, Royston, Herts SG8 5HE, UK Telephone: + 44 (0) 1763 253000 Facsimile: + 44 (0) 1763 253313 E-mail: nobleuk@matthey.com Internet: http://www.noble.matthey.com

3.5 Platinum Group Metals (PGMs)

3.5.1 General Overview

The platinum group metals (PGMs) comprise six closely related metals of the group VIIIB (8, 9, and 10) of Mendeleev's periodic chart: ruthenium (Ru), rhodium (Rh), palladium (Pd), osmium (Os), iridium (Ir), and platinum (Pt). Owing to their close chemical properties, they commonly occur together free in nature as native metals and alloys and are amongst the scarcest of the metallic elements in the Earth's crust. Along with gold and silver, owing to their strong chemical inertness they are commonly known as precious or noble metals. They occur as native alloys in recent and burred placer ore deposits or, more commonly, in lode mine deposits associated with nickel and copper ores. Nearly all the chief world ore deposits of PGMs are extracted from placers and lode deposits found on the following three continents: Africa (e.g., Republic of South Africa), Eurasia (e.g., Russia), and North America (e.g., Canada, and the United States). However, the Republic of South Africa is the only country that produces all the six PGMs in substantial quantities.

PGMs have become critical to industry because of their extraordinary physical and chemical properties, the most important of which is their chemical inertness and high catalytic activity. Since the mid-1970s and continuing today, the automobile industry uses catalytic converters containing platinum, palladium, and rhodium to reduce pollutant exhaust gas emissions and the modern chlor-alkali industry uses extensively for evolving chlorine dimensionally stable anodes, DSA$^®$, made of a titanium base metal coated with a thin electrocatalytic layer of ruthenium dioxide (Ti/TiO$_2$–RuO$_2$). Similarly, the chemical and petroleum-refining industries have relied on PGM catalysts to produce a wide variety of chemicals and petroleum products. On the other hand, the remarkable chemical inertness in highly corrosive environments of three of the six platinum group metals (Pt, Ir, and Rh) allows them to be suitable materials for handling strong concentrated minerals acids at high temperature. Hence, they are indispensable materials for manufacturing devices and apparatus employed in the laboratory. However, their high cost combined with a high density forbid their use in the chemical process industries (CPI) although platinum was used as a linear in the lead chamber process one century ago.[95] Actually, since the mid-1970s when processes used harsh environments involving handling of highly corrosive chemicals, chemical engineers used to utilize refractory metals such as titanium, zirconium, niobium, tantalum, and molybdenum, either as solid metals or as protective coatings and liners. According to the US Geological Survey (USGS) identified resources at year-end 1996 were estimated at 100,000 tonnes. The reserve base was estimated at 66,000 tonnes and reserves at 54,000 tonnes. Of the reserve base and reserves, Republic of South Africa had nearly 90% of each, Russia had 9% and 11% respectively, and the USA had 1% and 0.4% respectively.

3.5.2 Natural Occurrence, Minerals, and Ores

The magmatic origin of PGM deposits has genetic affinities to both Ni–Cu sulfides and chromites. Actually, there exist several possible magmatic processes during which the concentration of PGMs occurs: (i) during high-temperature deposition of chromites, (ii) incorporation into immiscible magmatic liquids, (iii) remobilization and reconcentration during metasomatic and hydrothermal activity. The significant PGM production has come from: (i) the well-known deposit of Merenski Reef of the Bushveld Complex in the Republic of South Africa, (ii) the Ni–Cu deposits of the Noril'sk-Talnakh District in the CIS (ex-USSR), (iii) as by-products of several Ni–Cu deposits at Sudbury, Ontario (Canada), (iv) placer deposits derived from zoned (i.e., Alaskan-type) ultramatic intrusions (Columbia, Goodnews Bay, Tulameen), and (v) metasomatic dunite pipes of the Bushveld complex. The bulk of present world production comes from the Republic of South Africa and to a lesser extent, Russian deposits and most presently known reserves are within Merenski-type environments (Bushveld and Stillwater Complexes).

   The ores of the "Merenski" reef form thin but laterally persistent, disseminated, sulfide-poor horizons within polycyclic mafic–ultramafic cumulate sequences one-third of the way up from the base of the Bushveld intrusion. Principal ore minerals are pyrrhotite, chalcopyrite, pentlandite, PGM sulfides, arsenides and tellurides. The Noril'sk-Talnakh orebodies are essentially typical Ni–Cu deposits containing anomalously high concentrations of PGMs (6 mg.kg^{-1} of ore). They occur at or near the base of complexity differentiated gabbro-dolerite intrusions emplaced during the late Permian to Triassic period during rifting of the Siberian platform. The sills are considered to be feeders to overlying plateau basalts. The mineralogy of the ores includes pyrrhotite, chalcopyrite, pentlandite, and a great variety of PGM minerals. Placers derived from Alaskan-type intrusions are the results of the breakdown, transport, and concentration of Pt–Fe alloys mainly associated with Fe-rich chromite layers from the dunitic portions of these complexes. Finally, the metasomatic dunite pipes of the Bushveld Complex consist of central zones of Fe-rich dunite enveloped by shells of dunite and pyroxenite. The ores are pegmatitic and may contain slabs of chromite. Spot assays as high as 1990 mg.kg^{-1} Pt were recorded from dunite pipes.

   As a general rule, most disseminated sulfide deposits carrying appreciable PGM values are characterized by both pegmatitic textures, and the presence of hydrous minerals within otherwise anhydrous layered successions. These features, which point to high fluid activity during magmatic segregation, are important prospecting guides for these rare metals.

3.5.3 Common Properties

Physical and chemical properties of the PGMs are shown in **Table 3.56** (pages 206–208).

3.5.4. The Six PGMs

3.5.4.1 Ruthenium

Description and general properties – ruthenium [7440-18-8] has the chemical symbol Ru, the atomic number 44, and the relative atomic mass 101.07(2). Ruthenium is a white, lustrous, very hard metal, with a hexagonal close packed (hcp) crystal lattice structure. It has a high Young's modulus (414 GPa) and a high melting point (2310°C). Ruthenium is corrosion resistant to most strong minerals acids including hot aqua regia up to 100°C, but when potassium chlorate, $KClO_3$, is added to the solution, it oxidizes explosively. Nevertheless, it does not tarnish in air at room temperatures, but it oxidizes rapidly above 800°C forming the unprotective volatile tetroxide, RuO_4. However, in an oxygen-free atmosphere, ruthenium crucibles resist corrosion by molten alkali metals (i.e., Li, Na, and K), and group IB precious metals (i.e., Cu, Ag, and Au). Ruthenium has six naturally occurring stable isotopes: ^{96}Ru (5.48at%), ^{98}Ru (1.87at%), ^{99}Ru (12.63at%), ^{100}Ru (12.53at%), ^{101}Ru (17.02at%), ^{102}Ru (31.60at%), and ^{104}Ru (18.87at%).

Natural occurrence – ruthenium is an extremely rare element with an average abundance in the Earth's crust of 0.0004 mg.kg^{-1} (0.4 ppb wt). It is found in the minerals such as osmiridium, laurite, and some platinum ores.

History – ruthenium was first isolated as metal in 1844 by the German chemist Karl Karlovich Klaus, who obtained ruthenium from the part of crude osmiridium that is insoluble in aqua regia. Nevertheless, it is possible that the Polish chemist Jedrzej A. Sniadecki had in fact isolated ruthenium from some platinum ores rather earlier than this in 1807 but his work was not ratified, apparently as he withdrew his claims. He called it vestium.

Industrial uses and applications – ruthenium combines with platinum and palladium as an effective hardener, creating alloys that are extremely wear resistant to abrasion and are used to make electrical contacts. It also improves, by addition of 0.15–0.25wt%, the corrosion resistance of titanium alloys in hydrochloric acid by several times (e.g., ASTM grades 13, 14, and 15). When combined with molybdenum it gives superconductive alloys. Ruthenium dioxide, RuO_2, with the rutile type structure, is extensively used as an anodic electrocatalyst for diminishing the overpotential of chlorine evolution in chloride brines. The oxide, obtained by the thermal decomposition of a precursor, is coated onto a titanium base metal plate. These composite anodes, Ti/TiO_2-RuO_2, owing to their low overvoltage, long service life and mechanical stability, are named dimensionally stable anodes in comparison with graphite anodes. They are commercially registered under the common acronym DSA®. Therefore, the RuO_2-DSA® are extensively used in the chlor–alkali process. Moreover, ruthenium dioxide coated onto CdS particles in an aqueous suspension allows, under visible light irradiation, splitting of hydrogen sulfide. This may have application in removal of hydrogen sulfide from oil and in other industrial chemicals.

Cost (1999) – pure ruthenium (99.95wt% Ru) is priced at 1318 $US.kg^{-1} (41 $US/troy oz).

3.5.4.2 Rhodium

Description and general properties – rhodium [7440-16-6] has the chemical symbol Rh, the atomic number 45, and the relative atomic mass 102.90550(2). Rhodium is a slivery-white, hard metal, fairly ductile when heated and having a high melting point (1963°C). It has a face-centered cubic (fcc) crystal lattice structure. Rhodium is corrosion resistant to most strong mineral acids including hot aqua regia up to 100°C. Moreover, it is also oxidation resistant in air at high temperatures allowing solid rhodium or rhodium coatings to be used for high-temperature device manufacture. However, upon heating it turns to the oxide when red and at higher temperatures turns back to the element. Moreover, it absorbs oxygen on melting, releasing it on solidification. Rhodium has the highest specular reflectivity, the whitest luster and the highest electrical and thermal conductivities of the PGMs. Rhodium is a mononuclidic element with only one stable isotope, the nuclide ^{103}Rh.

Table 3.56 Selected physical and chemical properties of the six PGMs

Physical and chemical properties (@ 298.15 K unless otherwise indicated)		Ruthenium	Rhodium	Palladium	Osmium	Iridium	Platinum
General	Chemical symbol [IUPAC]	Ru	Rh	Pd	Os	Ir	Pt
	Chemical abstract registry number [CARN]	[7440-18-8]	[7440-16-6]	[7440-05-3]	7440-02-4]	[7439-88-5]	[7440-06-4]
	Unified numbering system [UNS]	[P06999]	[P05995]	[P03980]	[P02001]	[P01999]	[P04980]
Mineral occurrence and economics	Earth's crust abundance (mg.kg^{-1})	0.001	0.0002	0.0006	0.0001	0.000006	0.001
	Seawater abundance (mg.kg^{-1})	n.a.	n.a.	n.a.	n.a.	n.a.	n.a.
	World reserves (R,10^3 kg)	5000	3000	24,000	200	950	27,000
	World production of ore (P, 10^3 kg.yr^{-1})	0.12	3	111	0.06	3	146
	Cost of pure metal 1999 (C, \$US.kg^{-1}) (\$US.tr.oz^{-1})	1318 (41)	28,614 (890)	11,574 (360)	n.a. (n.a.)	13,343 (415)	11,446 (356)
Atomic properties	Atomic number (Z)	44	45	46	76	77	78
	Relative atomic mass A_r (^{12}C = 12.000)	101.07	102.90550	106.42	195.078	192.217	195.078
	Electronic configuration (ground state)	[Kr]4d^{7}5s^1	[Kr]4d^{8}s^1	[Kr]4d^{10}5s^0	[Xe]4f^{14}5d^{6}6s^2	[Xe]4f^{14}5d^{7}6s^2	[Xe]4f^{14}5d^{9}6s^1
	Fundamental ground state	$^5F_5$	$^4F_{9/2}$	$^1S_0$	$^5D_4$	$^4F_{9/2}$	$^3D_3$
	Electronegativity (Pauling)	2.2	2.3	2.2	2.2	2.2	2.2
	Electron work function (W_S, eV)	4.71	4.98	5.12	4.83	5.27	5.65
	Electronic affinity (E_A, kJ.mol^{-1})	101	109.7	53.7	106	151	205.3
	Atomic radius Goldschmidt (r, nm)	0.134	0.1345	0.137	0.135	0.1355	0.1385
	Covalent radius (r, nm)	0.124	0.125	0.128	0.126	0.126	0.129
	X-ray mass attenuation coefficient	183	194	206	186	193	200

(σ_{nth}, barn)						
Natural isotope range (including isomers)	87–118	89–121	91–123	162–196	166–198	168–202
Natural isotopes (including isomers)	34	51	40	41	50	41
Crystal structure (low temp. α-phase)	hcp	fcc	fcc	hcp	fcc	fcc
Stukturbericht designation	A3	A1	A1	A3	A1	A1
Space group (Hermann–Mauguin)	P6$_3$/mmc	Fm3m	Fm3m	P6$_3$/mmc	Fm3m	Fm3m
Pearson's notation	hP2	cF4	cF4	hP2	cF4	cF4
Miller's indices of slip plane (hkil)	(1$\bar{1}$00)	(111)	(111)	(1$\bar{1}$00)	(111)	(111)
Space lattice parameters (pm)	$a = 268.87$ $c = 428.11$	$a = 380.36$	$a = 389.08$	$a = 273.42$ $c = 431.97$	$a = 383.92$	$a = 391.58$
Density (293 K) (ρ, kg.m^{-3})	12,370	12,410	12,020	22,590	22,650	21,450
Young's modulus (E, GPa)	413.8–432	344.8–379	117.2–121	558.6	524–528	172.4
Shear modulus (G, GPa)	173	147	43.6	223	210	60.9
Bulk modulus (K, GPa)	286	276	187	373	371	276
Mohs hardness (HM)	6.5	n.a.	4.8	7.0	6–6.5	4.3
Vickers hardness (HV) (hardened)	38–42 (98)	350 (410)	37–42 (220)	110–120 (650)	200–240 (1000)	37–42 (210)
Yield strength (σ_{YS}, MPa) (0.2% proof) (hardened)	38	68	65 (400)	n.a.	234	70 (290)
Ultimate tensile strength (σ_{UTS}, MPa) (hardened)	496	758.6 (2068)	227.5 (480)	n.a.	550–1103	137.9 (330)
Elongation (Z, %)	40	9	35	n.a.	6.8	40
Creep strength (MPa)	372	69–275	205	n.a.	14–35 (185)	185
Poisson's ratio (ν, dimensionless)	0.250	0.260	0.394	0.250	0.262	0.397

clear properties | Crystallographic properties (293.15K) | Mechanical properties (annealed)

Table 3.56 (continued)

Physical and chemical properties (@ 298.15 K unless otherwise indicated)	Ruthenium	Rhodium	Palladium	Osmium	Iridium	Platinum
Thermal properties						
Melting point (T_{fus}, K) (mp, °C)	2583 (2310°C)	2239 (1966°C)	1825 (1552°C)	3327 (3054°C)	2683 (2410°C)	2045 (1772°C)
Boiling point (T_{vap}, K) (bp, °C)	4173 (3900°C)	4000 (3727°C)	3413 (3980°C)	5300 (5027°C)	4403 (4130°C)	4100 (3800°C)
Thermal conductivity (k, W.m⁻¹.K⁻¹)	117	150	71.8	87.6	146.5	71.6
Coefficient of linear thermal expansion (α, 10^{-6} K⁻¹)	9.6	8.5	11.76	4.57	6.4	9.1
Specific heat capacity (c_p, J.kg⁻¹.K⁻¹)	238	243	245	129.73	129.8	134
Enthalpy molar fusion (ΔH_{fus}, kJ.mol⁻¹) (Δh_{fus}, kJ.kg⁻¹)	23.7 (235)	21.55 (209)	17.2 (162)	29.3 (154)	26.4 (137)	19.7 (101)
Enthalpy molar vapor (ΔH_{vap}, kJ.mol⁻¹) (Δh_{vap}, kJ.kg⁻¹)	567.8 (5620)	495.4 (4814)	393.3 (3696)	627.6 (3300)	563.6 (2932)	510.5 (2617)
Electrical, magnetic, and optical properties						
Superconductive critical temperature (T_c, K)	2.04	n.a.	nil	0.66	–	nil
Electrical resistivity (ρ, $\mu\Omega$cm)	7.6	4.51	10.8	8.12	5.3	10.58
Temperature coefficient of resistivity (0–100°C) (K⁻¹)	0.0041	0.0043	0.00377	0.0041	0.00427	0.003927
Electromotive force versus platinum (0–100°C) (emf, mV)	n.a.	+0.70	−0.57	n.a.	+3.626 (400°C)	0.00 (defined)
Mass magnetic susceptibility (χ_m, 10^{-8} m³.kg⁻¹)	+0.542	+1.31	+6.574	+0.6	+0.256	+1.22
Spectral emmissivity (650 nm)	n.a.	0.24	0.33	n.a.	0.30	0.30
Reflective index under normal incidence	0.63 (Vis)	0.8	0.628 (Vis)	n.a.	n.a.	0.594

History – William Hyde Wollaston discovered rhodium in 1804 in crude platinum ore from South America rather soon after his discovery of another element, palladium. He dissolved the ore in aqua regia, neutralized the acid with sodium hydroxide (NaOH), and precipitated the platinum by treatment with salmiac (ammonium chloride, NH_4Cl), as ammonium hexachloroplatinate (NH_4PtCl_6). Palladium was then removed as palladium cyanide by treatment with mercuric cyanide. The remaining material was a red material containing rhodium chloride salts from which rhodium metal was obtained by reduction with hydrogen gas.

Natural occurrence – rhodium is one of the rarest elements in the Earth's crust with an abundance of 1 $\mu g.kg^{-1}$ (ppb wt). Rhodium occurs in nature as a native metal along with other platinum group metals in the native mineral iridosmine, or in sulfide ores such as **rhodite**, **sperrylite**, and in some copper–nickel ores.

Industrial uses and applications – it is a major component of industrial catalytic systems such as the BP–Monsanto process or as part of the catalytic system in car catalytic converters, used to clean up exhaust gases. It is also used as an alloying element in order to harden platinum and palladium alloys. Rhodium coating achieved by electroplating is used to protect silverware from tarnishing. Such alloys are used for furnace windings, thermocouple elements, bushings for glass fiber production, electrodes for aircraft spark plugs, and laboratory crucibles. In electrical engineering it is used as an electrical contact material as it has a low electrical resistance, a low and stable contact resistance, and is highly resistant to corrosion. Rhodium coatings produced by electroplating or vacuum evaporation are exceptionally hard and are used for optical instruments such as mirrors.

Cost (1999) – pure rhodium (99.95 wt%) is priced at 28,614 $US.kg^{-1}$ (890 $US/troy oz).

3.5.4.3 Palladium

Description and general properties – palladium [7440-05-3] has the chemical symbol Pd, the atomic number 46, and the relative atomic mass 106.42(1). It was named after the discovery by astronomers of the asteroid Pallas. Palladium is a steel-white, very ductile and malleable metal but cold working increases its strength and hardness. It has physical properties similar to those of platinum except a lower density. In fact, palladium has the lowest density (12,020 $kg.m^{-3}$) and melting point (1552°C) of the PGMs and it is the least corrosion resistant. Palladium crystals have a face-centered cubic (fcc) space lattice structure. Actually, palladium resists tarnishing by moist air, but tarnishes slightly upon exposure to sulfur-contamined environments. Moreover, it is slightly attacked by most strong concentrated mineral acids (e.g., HCl, H_2SO_4, and HNO_3) including aqua regia and readily corroded by halogen gases. In air, it oxidizes between 400°C and 800°C forming a thin oxide layer which decomposes readily above 800°C. However, palladium absorbs hydrogen efficiently (up to 900 times its volume) which diffuses at a high rate when the metal is heated. This particular property allows extensive use of palladium and palladium–silver alloys as membranes for purifying hydrogen. Therefore, finely divided palladium is an efficient catalyst for hydrogenation and dehydrogenation reactions. Palladium has six naturally occurring stable isotopes: ^{102}Pd (1.0at%), ^{104}Pd (11.0at%), ^{105}Pd (22.33at%), ^{106}Pd (27.33at%), ^{108}Pd (26.46at%), and ^{110}Pd (11.72at%).

History – William Hyde Wollaston discovered palladium in 1803 in crude platinum ore from South America. He dissolved the ore in aqua regia, neutralized the acid with sodium hydroxide (NaOH), and precipitated the platinum by treatment with ammonium chloride, NH_4Cl, as ammonium hexachloroplatinate (NH_4PtCl_6). Palladium was then removed as palladium cyanide by treatment with mercuric cyanide. The free metal was produced from this cyanide by heating.

Natural occurrence – palladium is a rare element in the Earth's crust with an average abundance of 10 $\mu g.kg^1$ (10 ppb wt) and is hence the most abundant of the PGMs. It is found as a native element alloyed with platinum or gold, or in the combined form in minerals such as **stibiopalladinite, braggite, porpezite**, and some nickel sulfide ores.

Industrial uses and application – palladium is used in some watch springs or alloyed for use in jewelry. White gold is an alloy of gold decolorized by the addition of palladium, and can be beaten into leaf as thin as 0.1 μm. It is also used in dentistry for making crowns in replacement of gold. Finally, it is also used to make surgical instruments.

Cost (1999) – pure palladium (99.95 wt%) is priced at 11,574 $US.kg^{-1} (360 $US/troy oz).

3.5.4.4 Osmium

Description and general properties – osmium [7440-02-4] has the chemical symbol Os, the atomic number 76, and the relative atomic mass 190.23(3). Osmium has the highest melting point (3054°C) and the lowest vapor pressure of the platinum group PGMs and the second highest density (22,590 kg.m^{-3}) after iridium. Its properties make it ideal for combining with other platinum metals to produce very hard alloys. Finely divided osmium (e.g., powder or sponge) oxidizes readily in air at room temperature forming a highly toxic and volatile tetroxide, OsO_4, which has a strong smell, boils at 130°C at atmospheric pressure and is a powerful oxidizing agent, while solid osmium metal is relatively inert in ambient conditions. The metal is very difficult to fabricate, but the powder can be sintered in a hydrogen atmosphere at a temperature of 2000°C. As previously indicated, the tetroxide is highly toxic, and concentrations in air as low as 10^{-7} g.m^{-3} can cause lung congestion, skin damage, or eye damage.

History – osmium was discovered in 1803 by Smithson Tennant in the dark colored residue left when crude platinum is dissolved by aqua regia. This dark residue contains both osmium (named after the Greek, osme, meaning odor) and iridium. It is a bluish–white silvery, extremely hard, brittle metal not malleable even at high temperatures.

Natural occurrence – osmium is an extremely rare element in the Earth's crust with an average abundance of 1 μg.kg^{-1} (1 ppb wt). It naturally occurs as native element alloyed with iridium and other PGMs in minerals such as **osmiridium**, and in all platinum ores.

Industrial uses and applications – osmium metal is almost entirely used to produce very hard alloys with other metals of the platinum group, for fountain pen tips, instrument pivots, and electrical contacts. The Pt90–Os10 alloy is used in implants such as pacemakers and replacement valves. On the other hand, OsO_4 is used in forensic sciences to detect fingerprints and to stain fatty tissue for microscope slides.

3.5.4.5 Iridium

Description and general properties – iridium [7439-88-5] has the chemical symbol Ir, the atomic number 77, and the relative atomic mass 192.217(3). It is a silvery-white, extremely hard and brittle metal not malleable at room temperature but which can be hot worked at high temperature. These properties make iridium most difficult to machine, form, or work. Iridium has a high melting point (2410°C), a high Young's modulus (517 GPa) and the highest density (22,650 kg.m^{-3}) of the PGMs. Moreover, it oxidizes in air between 600°C and 1000°C forming a thin layer of iridium dioxide, IrO_2, which decomposes readily above 1100°C releasing the metal. Iridium is the most corrosion resistant metal known. Actually, in the highly pure state, it resists hot concentrated strong mineral acids (e.g., HF, HCl, H_2SO_4, H_3PO_4, and HNO_3) including hot aqua regia but is readily attacked by molten alkali cyanides, such as KCN and NaCN.

History – iridium was discovered in 1803 by Smithson Tennant in the dark colored residue left when crude platinum is dissolved by aqua regia. This dark residue contains both osmium and iridium metals. It was named after the Latin, iris, rainbow, in reference to its compounds which are highly colored.

Industrial uses and applications – its principle use is as a hardening agent for platinum alloys, though it is frequently found in crucibles and other high-temperature labware. Because it is the most resistant metal known, it was used for making the second prototype of the standard meter bar of Paris, which is a 90Pt–10Ir alloy. To a lesser extent, it is used as electrical contacts. It forms an alloy with osmium used for tipping pens and compass bearings. Finally Pt–Ir alloy is used in spark plugs.

World production (1997) – 116 tonnes.

Cost (1999) – pure iridium is priced at 13,343 $US.kg^{-1} (415 $US/troy oz).

3.5.4.6 Platinum

Description and general properties – platinum [7440-06-4] has the chemical symbol Pt, the atomic number 78, and the relative atomic mass 195.078(2). Pure platinum is a silvery-white, highly dense (21,450 kg.m^{-3}) and ductile lustrous metal with a high electrical resistivity (9.85 $\mu\Omega$.cm) and a high thermal conductivity (71.6 W.m^{-1}.K^{-1}). It has a high melting point (1772°C) and does not oxidize in air until its melting point. Its coefficient of linear thermal expansion (9.1 μm.m^{-1}.K^{-1}) is similar to that of soda–lime–silica glass, and hence it is used to make glass-to-metal seals for high-temperature applications. Platinum resists corrosion to practically all chemicals at room temperature but readily dissolves in aqua regia forming hydrogen hexachloroplatinate(IV) (H$_2$PtCl$_6$). It does not corrode or tarnish in air and is not affected by water. In its finely divided form, platinum is an excellent catalyst. Nevertheless, platinum is readily attacked by the following hot chemicals: (i) aqua regia (a mixture of three parts conc. HCl and one part conc. HNO$_3$ by volume) (except iridium–rhodium alloys), (ii) mixtures of strong halogenated mineral acids and oxidizing agents, (iii) liquid metals: Pb, Zn, Sn, Bi, Au, Cu, (iv) molten oxides, hydroxides and peroxides of alkali metals, (v) concentrated sulfuric and phosphoric acids, and finally (vi) metal chlorides in air above 700°C. Moreover, pure platinum metal is sensitive to the following elements forming low-melting-point eutectics: P, As, Si, C, Se, B, Te, and is strongly corroded in hot gaseous atmospheres containing ammonia, NH$_3$, sulfur trioxide, SO$_3$, chlorine, Cl$_2$, carbonaceous vapors, and volatile chlorides. Like its neighbor palladium, platinum metal absorbs large volumes of hydrogen gas. In order to avoid the recrystallization of platinum crystal high service temperatures, it is possible to add to the pure metal or refractory oxide particles (e.g. zirconia) either solid solutions with small amount of iridium, rhodium, or gold. Platinum–rhodium and platinum–iridium alloys have higher strength and chemical resistance to corrosive chemicals. For instance, Pt–30Rh and Pt–30Ir are resistant to aqua regia. Platinum–gold alloys (Pt–5Au) have interesting nonwetting properties for molten glass. Platinum becomes ferromagnetic when alloyed with cobalt.

Natural occurrence – platinum metal occurs free in nature as a native metal contaminated with small amounts of all the platinum group metals such as iridium, osmium, palladium, ruthenium, and rhodium. These native minerals are found in placer ore deposits.

History – the metal was used by pre-Columbian Indians but platinum was rediscovered in South America by Ulloa in 1735 and later by Wood in 1741. In 1822 large amounts of platinum were discovered in the Ural Mountains in Russia. It was named after the Spanish, platina, meaning silver because the first conquistadores thought that the platinum grains found in South America placers were silver.

Industrial uses and applications – platinum is used in jewelry, for making wire and vessels for laboratory use, thermocouple elements, electrical contacts, corrosion–resistant apparatus, and in dentistry. Platinum–cobalt alloys have magnetic properties and are used for coating missile nose cones and jet engine fuel nozzles. The metal, like palladium, absorbs large volumes of hydrogen, giving it up at red heat. In the finely divided state platinum is an excellent catalyst, such as in the contact process for producing sulfuric acid, also for cracking oil and in fuel cells and in catalytic converters for cars. Platinum anodes are extensively used in cathodic protection systems for large ships and ocean-going vessels, pipelines, and steel piers. Platinum wire glows red hot when placed in the vapor of methanol – acting as a catalyst to convert the alcohol into formaldehyde. This phenomenon has been used commercially to produce cigarette lighters and hand warmers, and sealed electrodes in glass system laboratory vessels. Cis-platin $PtCl_2(NH_3)_2$ is an effective drug for certain types of cancer such as leukemia or testicular cancer. Platinum–osmium 90/10 alloy is used in implants such as pacemakers and replacement valves.

World production (1997) – 148 tonnes.

Cost (1999) – pure platinum is priced at 11,446 \$US.kg^{-1} (356 \$US/troy oz).

3.5.5 PGM Alloys

Properties of PGM alloys are shown in Tables 3.57 and 3.58.

3.5.6 PGM Corrosion Resistance

Corrosion properties of PGMs are shown in Tables 3.59 and 3.61.

Cleaning of platinum labware – several chemicals can be used for cleaning platinum crucibles. The first step consists of chemical etching with chemicals such as: (i) concentrated nitric acid (HNO_3) containing a small amount of oxidizing agent such as hydrogen peroxide (H_2O_2), (ii) concentrated hydrochloric acid (37wt% HCl), (iii) molten potassium hydrogenosulfate ($KHSO_4$, pyrosulfate), (iv) molten sodium hydrogenocarbonate ($NaHCO_3$), or finally (v) molten sodium tetraborate ($Na_2B_4O_7 \cdot 10H_2O$, borax). Afterwards, a slight mechanical polishing or lapping is achieved with a moist mixture of fine silica sand with talc, in order to remove thicker scale deposits not dissolved by etching. Then the crucible is rinsed with concentrated hydrofluoride acid (50wt% HF) in order to remove completely all the abrasive silica particles embedded in the walls. The crucible is then rinsed with deionized water followed by absolute ethanol and dried in an oven. Precautionary note: it is important to never heat platinum labware in the reducing flame of a bunsen burner (i.e., blue flame).

Table 3.57 Physical properties of selected platinum alloys at 273.15 K (annealed)

Pt alloy	Density (ρ, kg.m^{-3})	Melting point (mp, °C)	Electrical resistivity (ρ, $\mu\Omega$.cm)	Temperature Coefficient of resistivity (0–100°C) (10^{-4} K)	Ultimate tensile strength (σ_{UTS}, MPa)	Elongation (Z, %)	Vickers hardness (HV)
95Pt–5Rh	20,650	1825	17.5	20	240	31	69
90Pt–10Rh	19,970	1850	19.4	17	320	29	90–95
80Pt–20Rh	18,740	1900	20.8	14	450	27	120
70Pt–30Rh	17,620	1925	19.4	14	500	23	132
60Pt–40Rh	16,630	1940	17.5	13	550	22	150
95Pt–5Ir	21,490	1780	18.95	18.8	275	28	90
90Pt–10Ir	21,566	1800	24.9	13	380	24	130
85Pt–15Ir	21,570	1815	28.5	10	513	22	160
80Pt–20Ir	21,610	1825	31.9	8	674	20	190
75Pt–25Ir	21,680	1845	32.9	7	863	18	252
70Pt–30Ir	21,740	1885	35	5.8	1120	16	280
95Pt–5Au	21,000	1670	17	21	340	18	95

Table 3.58 Ultimate tensile strength and elongation of platinum alloys vs temperature

Pt alloy	Ultimate tensile strength (σ_{UTS}, MPa)			Elongation (Z, %)		
	900°C	1000°C	1200°C	900°C	1000°C	1200°C
Pt	110	75	50	56	63	63
95Pt–5Rh	150	102	68	50	63	70
90Pt–10Rh	205	133	96	47	60	66
80Pt–20Rh	330	205	109	10	43	60
90Pt–10Ir	n.a.	192	75	n.a.	n.a.	n.a.

Table 3.59 Corrosion rate of platinum in molten salts (after 1 h immersion)

Molten salt	Temperature ($T°$,C)	Corrosion rate (v, g.m^{-2}.day^{-1})
KNO$_3$	350	nil
NaNO$_3$	350	nil
Na$_2$O$_2$	350	nil
KHSO$_4$	440	7.2
Na$_2$CO$_3$	920	7.2[a]
KCN + 2 NaCN	550	84
NaCN	700	745
KCN	700	2800

[a]Under inert atmosphere.

Table 3.60 Eutectics with low-melting point metals (°C)

Element	Pt	Rh	Ru	Ir
B	825	1131	1370	1046
P	588	1254	1425	1262
Si	890	1389	1488	1470
Sn	1070	n.a.	n.a.	n.a.
Pb	327	n.a.	n.a.	n.a.
Bi	730	n.a.	n.a.	n.a.
Sb	633	n.a.	n.a.	n.a.
As	597	n.a.	n.a.	n.a.

3.5.6.1 Industrial Applications and Uses

See **Table 3.62** (page 217).

3.5.6.2 PGM World Suppliers

See **Table 3.63** (page 218).

Table 3.61 Corrosion properties[a] of the PGMs in several aerated corrosive media between 20°C and 100°C[b]

Corrosive chemical	Formula	T (°C)	Ag	Au	Pd	Pt	Rh	Ir	Ru	Os
Acetic acid conc. (glacial)	CH$_3$COOH	100	nil	nil	nil	nil	nil	nil	nil	n.a.
Aluminum sulfate	Al$_2$(SO$_4$)$_3$	100	nil	nil	nil	nil	nil	nil	nil	n.a.
Ammonia	NH$_4$OH	RT	attacked	nil	nil	nil	nil	nil	nil	n.a.
Ammonium chloride	NH$_4$Cl	300	attacked	n.a.	n.a.	n.a.	n.a.	n.a.	n.a.	n.a.
Aqua regia	HNO$_3$ + 3HCl	RT	attacked	attacked	attacked	attacked	nil	nil	nil	slight
Aqua regia	HNO$_3$ + 3HCl	100	attacked	attacked	attacked	attacked	nil	nil	nil	attacked
Bromine (moist)	Br$_2$	RT	nil	attacked	attacked	poor	nil	nil	nil	slight
Bromine anhydrous	Br$_2$	RT	nil	attacked	attacked	poor	slight	slight	slight	attacked
Chloric acid conc.	HClO$_3$	RT	attacked	nil	nil	nil	nil	nil	nil	nil
Chlorine (dry)	Cl$_2$	RT	slight	attacked	poor	slight	nil	nil	nil	nil
Chlorine (moist)	Cl$_2$	RT	nil	attacked	attacked	slight	nil	nil	nil	poor
Chlorosulfonic acid	CH$_2$ClSO$_3$	Boil.	nil	nil	nil	nil	nil	nil	nil	n.a.
Copper (II) sulfate	CuSO$_4$	100	attacked	nil	nil	nil	nil	nil	nil	n.a.
Copper (II) chloride	CuCl$_2$	100	attacked	nil	slight	nil	nil	n.a.	n.a.	n.a.
Iron (III) chloride	FeCl$_3$	100	nil	nil	attacked	attacked	n.a.	nil	n.a.	n.a.
Fluorine (dry)	F$_2$	RT	slight	nil	n.a.	slight	n.a.	n.a.	n.a.	n.a.
Formic acid conc.	HCHO	100	nil	nil	nil	nil	nil	nil	nil	n.a.
Hydrobromic acid 60wt%	HBr	RT	attacked	nil	attacked	slight	slight	nil	nil	nil
Hydrobromic acid 60wt%	HBr	100	attacked	nil	attacked	attacked	poor	nil	nil	poor
Hydrochloric acid 37wt%	HCl	RT	attacked	nil	nil	nil	nil	nil	nil	nil
Hydrochloric acid 37wt%	HCl	100	attacked	nil	slight	slight	nil	nil	nil	poor
Hydrochloric acid	HCl	RT	attacked	nil	attacked	slight	nil	nil	nil	poor

65wt%

Reagent	Formula	Conc./Temp.							
Hydrofluoric acid 50wt%	HF	100	nil	nil	nil	nil	nil	nil	poor
Hydrogen peroxide (30 vol.)	H$_2$O$_2$	100	dec.	nil	n.a.	attacked	n.a.	n.a.	n.a.
Hydrogen sulfide	H$_2$S	RT	attacked	attacked	nil	nil	nil	nil	nil
Hydrogen selenide	H$_2$Se	RT	attacked	attacked	nil	nil	nil	nil	nil
Hydroiodic acid conc.	HI	RT	attacked	attacked	attacked	nil	nil	nil	slight
Iodine (dry)	I$_2$	RT	attacked	nil	attacked	nil	nil	nil	n.a.
Iodine (moist)	I$_2$	RT	attacked	attacked	slight	slight	nil	nil	nil
Mercury (II) chloride	HgCl$_2$	100	poor	attacked	nil	nil	nil	poor	n.a.
Nitric acid 62wt%	HNO$_3$	RT	attacked	nil	attacked	nil	nil	nil	attacked
Nitric acid 62wt%	HNO$_3$	100	attacked	nil	attacked	nil	nil	nil	attacked
Nitric acid 95wt% (fuming)	HNO$_3$	100	attacked	nil	attacked	nil	nil	nil	attacked
Nitric oxide	NO$_2$	RT	attacked	attacked	slight	nil	nil	n.a.	n.a.
Orthophosphoric acid conc.	H$_3$PO$_4$	100	nil	nil	slight	nil	nil	nil	nil
Ozone	O$_3$	RT	nil	nil	n.a.	nil	n.a.	n.a.	n.a.
Potassium cyanide aerated	KCN	RT	attacked	attacked	poor	poor	n.a.	n.a.	n.a.
Potassium cyanide	KCN	100	attacked	attacked	attacked	poor	poor	n.a.	n.a.
Potassium bisulfate	KHSO$_4$	500	n.a.	nil	slight	nil	poor	nil	n.a.
Potassium hydroxide	KOH	400	nil	nil	slight	poor	poor	nil	attacked
Potassium iodide + iodine	KI$_3$	RT	n.a.	attacked	n.a.	nil	n.a.	attacked	n.a.
Potassium nitrate	KNO$_3$	335	attacked	n.a.	n.a.	attacked	n.a.	n.a.	n.a.
Potassium permanganate	KMnO$_4$	Boil.	attacked	n.a.	n.a.	n.a.	n.a.	n.a.	n.a.

Table 3.61 (continued)

Corrosive chemical	Formula	T (°C)	Ag	Au	Pd	Pt	Rh	Ir	Ru	Os
Potassium peroxodisulfate	$K_2S_2O_8$	RT	attacked	n.a.	n.a.	attacked	n.a.	n.a.	n.a.	n.a.
Potassium peroxide	K_2O_2	380	attacked	attacked	n.a.	attacked	n.a.	n.a.	n.a.	n.a.
Selenic acid	H_2SeO_4	RT	n.a.	n.a.	poor	nil	n.a.	n.a.	n.a.	n.a.
Selenic acid	H_2SeO_4	100	n.a.	n.a.	attacked	poor	n.a.	n.a.	n.a.	n.a.
Sodium hydroxide	NaOH	500	nil	nil	slight	slight	slight	n.a.	attacked	attacked
Sodium hypobromite	NaBrO	RT	n.a.	n.a.	slight	nil	nil	nil	nil	n.a.
Sodium hypochlorite	NaClO	RT	nil	n.a.	poor	nil	slight	slight	attacked	attacked
Sodium hypochlorite	NaClO	100	nil	n.a.	attacked	nil	slight	slight	attacked	attacked
Sodium perchlorate	$NaClO_4$	480	attacked	n.a.	n.a.	attacked	n.a.	n.a.	n.a.	n.a.
Sulfur dioxide (moist)	SO_2	RT	nil	nil	nil	nil	nil	nil	nil	nil
Sulfuric acid conc. 96wt%	H_2SO_4	20	attacked	nil	nil	nil	nil	nil	nil	nil
Sulfuric acid conc. 96wt%	H_2SO_4	100	attacked	nil	poor	nil	slight	nil	nil	nil
Sulfuric acid conc. 96wt%	H_2SO_4	300	attacked	nil	attacked	slight	poor	n.a.	n.a.	n.a.

[a] **nil**: uncorroded, normally excellent, undiscernable, i.e., corrosion rate less than 0.05 mpy; **slight**: good, suitable for particular uses; **poor**: fair; **attacked**: unsatisfactory, unsuitable, and high dissolution rate.

[b] Data compiled and arranged from technical specification sheets supplied from the following PGMS producers: Engelhard, CLAL, Heraeus Gmbh, and Johnson Mattey Ltd.

Table 3.62 PGM applications and uses

Application field	Description
Chemical process industry (CPI)	Platinum–rhodium alloy is used as long service life catalyst in the industrial production of nitric acid by direct oxidation of air–ammonia mixtures. Owing to its permeability to molecular hydrogen, palladium and palladium alloys, such as Pd–40Ag, are extensively used as a selective membrane for the purification of hydrogen gas. The alloy is also used as a catalyst in the manufacturing of nitrogen-based fertilizers. Because of its chemical inertness, platinum is used as a protective liner for vessels handling hot concentrated hydrochloric acid.
Dental and medical	Platinum is used to manufacture medical and surgical devices like endoscopes and catheters. Owing to its unmatched and stable electrical conductivity, it is also used extensively in the manufacturing of pacemakers.
Electrochemical engineering	Owing to its catalytic activity, ruthenium dioxide (RuO_2) is extensively used in industrial anodes for the chlor-alkali industry, and the production of perchlorates. These ruthenium dioxide-based anodes consist of a thin catalytic layer coated onto a titanium base metal while iridium dioxide-based anodes are used for the production of persulfates, in electroplating, and in hydro-metallurgy for evolving oxygen. These composite anodes, because of their corrosion resistance in chloride-containing media or concentrated acids, and ability to decrease the overpotential of chlorine and oxygen evolution, are called by the trade name dimensionally stable anodes (DSA®). Uses include fuel cells, electrodes, and electrocatalysts.
Electronics	Platinum is part of the coating process for hard drives and other high-density computer data storage devices. It is also an integral part of the communications network like the telephone and internet fiber optics system. It is the key component in the manufacturing of liquid crystal displays (LCD) used in laptop computers and other small display electronic devices.
Glass and ceramics industry	Pure platinum, owing its high melting point and chemical inertia, is used as crucibles for melting optical glass and optical salt crystals for scientific instruments.
Jewelry industry	The jewelry industry is the second largest user of Pt and Rh.
Laboratory	Platinum and iridium are commonly used for manufacturing crucibles and labware for handling highly corrosive reagents. Platinum is used in extremely sensitive scientific devices like light and oxygen sensors. Owing to its stable temperature relation and high temperature coefficient, ultrapure Pt is extensively used as resistance thermometer device (RTD). For the same reasons, Rh and Pt are extensively used for manufacturing thermocouples.
Petrochemical industry	Platinum is a key catalyst in the processing of low-lead and unleaded gasoline and jet fuel, in petroleum refining, and catalytic reforming. Petrochemical usages of Pt include the manufacturing of thermoplastics and polyester.
Pharmaceutical industry	The platinum metals are used extensively in the development of anti-cancer and many other drugs. Rhodium, the rarest of the PGMs, is the key catalyst in the processing of Tylenol®.
Pollution devices	Pollution devices are the primary usage of Pt, Pd, Rh to manufacture catalytic converters, industrial smoke stack scrubbers, and other combustion catalysts for gasoline engines and diesel engines.

3.6 Rare-Earth Metals (Sc, Y, and Lanthanides)

3.6.1 Description and General Properties

The rare-earth metals are usually defined as a group of chemical elements composed of the three inner transition metals: scandium (Sc), yttrium (Y), and lanthanum (La) of the group IIIA(3) of Mendeleev's periodic chart, and the lanthanide elements. The lanthanides are a group of 15 chemically similar elements with atomic numbers ranging from 57 for lanthanum to 71 for lutetium (i.e., lanthanum (La), cerium (Ce), praseodymium (Pr), neodymium (Nd), promethium (Pm), samarium (Sm), europium (Eu), gadolinium (Gd), terbium (Tb), dysprosium (Dy), holmium (Ho), erbium (Er), thullium (Tm), ytterbium (Yb), and lutetium (Lu)). Sometimes promethium is excluded because it is a synthetic radioactive element

Table 3.63 PGM world producers

Producer	Address
Engelhard-CLAL	28, rue Michel-Le-Comte, F-75003 Paris, France Telephone: (33) 01 44 61 30 00 Facsimile: (33) 01 44 61 30 01 E-mail: webadmin@engelhard-clal.fr Internet: http://www.engelhard-clal.fr
Degussa AG	E-mail: preciousmetals@degussa.de Internet: http://www.degussa.de
W.C. Heraeus GmbH & Co. KG	P.O. Box 1553 D-63405 Hanau, Germany Telephone: +49 (6181) 35 1 Facsimile: +49 (6181) 35 658 E-mail: public-relations@heraeus.de Internet: http://www.heraeus.com
Johnson Matthey Noble Metals	Orchard Road, Royston, Herts SG8 5HE, UK Telephone: +44 (0) 1763 253000 Facsimile: +44 (0) 1763 253313 E-mail: nobleuk@matthey.com Internet: http://www.noble.matthey.com

produced during neutron-induced fission of the ^{235}U radionuclide and hence it is only found as a fission by-product in spent nuclear reactor wastes. Although scandium and yttrium are not lanthanide elements, senso stricto, they are also included in the rare-earth elements because: (i) their chief containing ores always occur in nature in association with lanthanide minerals, (ii) Y and Sc occur exclusively in the trivalent oxidation state (i.e., Sc^{3+} and Y^{3+}), and finally (iii) they form stable complexes of high coordination number with chelating O-donor ligands. Also, historically, thorium was, at one time, considered as a rare-earth element because it always occurs in their ores, particularly monazite. Therefore, the rare-earth group with 17 elements comprises about 17% of the naturally occurring elements and hence, it is the major group of the periodic table. The rare-earth metals could be arbitrarily split into two main subgroups: the first four elements (i.e., La, Ce, Pr, Nd) are referred to as **light** or **ceric rare earths**, while the remaining 11 including yttrium, are called **heavy** or **yttric rare earths**. Sometimes, inorganic chemists employ the three general chemical symbols RE, Ln, or R for representing chemical compounds of rare earths (e.g., LnF_3, RCl_3).

As a general rule, the overall chemical properties of rare-earth elements are due to the particularity of their electronic configuration. Indeed, lanthanide elements differ from one another by the number of inner core electrons in the 4f subshell, from f^0 (lanthanum) to f^{14} (lutetium) while the number of outer electrons remains the same in the outer shell $5d^6s^2$ (i.e., $3d^4s^0$ for Sc, and $4d^5s^0$ for Y). Since the energy of 4f-electrons lies below the energy of outer electrons, core electrons are completely shielded from the crystal electric field induced by surrounding atoms. Hence the 4f-electrons do not contribute to the valence shell. This important characteristic of lanthanide elements, called **lanthanic contraction**, results in two particular features. Firstly, there are only minor differences in the chemical properties as the atomic number increases due to the similar ionic radii and valence state. For instance, the normal valence state is mainly the trivalent oxidation state, Ln(III), even if cerium occurs in the tetravalent oxidation state Ce(IV) and europium and ytterbium in the divalent oxidation states, Eu(II) and Yb(II) respectively. Secondly, there are major differences in atomic spectra, and magnetic properties. The rare-earth metals are amongst the electropositive elements forming ionic bonds in solids, and therefore, they are extremely reactive with hydrogen and electronegative elements such as halogens, oxygen, nitrogen, and sulfur forming respectively stable hydrides, halides, oxides, nitrides, and sulfides. This property is extensively used for gettering impurities from high-purity atmospheres or for the preparation of hydrogen storage compounds used in rechargeable secondary batteries (e.g., $LaNi_5$ in Ni–MH). Like the other reactive metals, physical properties of pure rare-earth metals are strongly influenced by the amount of interstitial impurities such as O, N, C, and H present in the metal lattice structure.

3.6.2 Physical Properties

See Table 15.2a, properties of the elements, in the appendices.

3.6.3 History

In 1751, the Swedish mineralogist A.F. Crondstedt discovered a new heavy mineral **ytterbite** from Ytterby. Later, in 1788, oxides of the heavy rare-earth elements where first discovered in ytterbite (gadolinite) by the Swedish chemist J. Gadolin who gave to them the name rare earth owing to the scarcity of these elements in nature, and the chemical similarity of their oxides to the earthy oxides of alkaline-earth metals. He called it **yttria** because he thought it was pure yttrium oxide. Moreover, at that time, owing to their similar chemical properties, chemists first considered them as one element that could be isolated from ore as an oxide. During the period 1794–1907, two new ores: **monazite** in the Urals (Russia) and **bastnaesite** in Sweden were discovered, and the individual rare-earth elements were isolated, identified, and finally named,

Table 3.64 Discovery milestones of the lanthanides

Rare earth	Date	Discoverer (country)
Yttrium (after Ytterby, Sweden)	1794	J. Gadolin, Abo (Finland)
Cerium (after asteroid, Ceres, discovered in 1801)	1803	J.J. Berzelius, and W. Hisinger, Vestmanland (Sweden) isolated by G.C. Mosander, Stockholm (Sweden)
Lanthanum (from Greek, lanthanein, hidden)	1839	C.G. Mosander, Stockholm (Sweden)
Erbium (after the Swedish town of Ytterby)	1842	C.G. Mosander, Stockholm (Sweden)
Terbium (after the Swedish town of Ytterby)	1843	G.C. Mosander, Stockholm (Sweden)
Holmium (from Latin, Holmia, Stockholm)	1878	P.T. Cleve, Uppsala (Sweden), M. Delafontaine and J.L. Soret, Geneva (Switzerland)
Ytterbium (after the Swedish town of Ytterby)	1878	J.-C. Galissard de Marignac, Geneva (Switzerland)
Scandium (from Latin, Scandia, Scandinavia)	1879	L.F. Nilson, Uppsala (Sweden)
Samarium (after the mineral, samarskite)	1879	P.-E. Lecoq de Boisbaudran, Paris (France)
Thulium (after Thule, ancient Scandinavia)	1879	P.T. Cleve, Uppsala (Sweden)
Gadolinium (after the Swedish chemist J. Gadolin)	1880	J.-C. Galissard de Marignac, Geneva (Switzerland)
Praseodymium (from Greek, prasios didymos, green twin)	1885	Baron C. Auer von Welsbach, Vienna (Austria)
Neodymium (from Greek, neo dydimos new twin)	1885	Baron C. Auer von Welsbach, Vienna (Austria)
Dysprosium (from Greek, dysprositos, hard to obtain)	1886	P.-E. Lecoq de Boisbaudran, Paris (France)
Europium (from Latin, Europa, Europe)	1901	E.-A. Demarçay, Paris (France) and C. James, New Hampshire (USA)
Lutetium (from Latin, Lutecia, Paris)	1907	G. Urbain, Paris (France)
Prometheum (from Greek, Prometheus, stole fire from Gods)	1945	J.A. Marinsky, L.E. Glendenin, and C.D. Coryell, Oak Ridge, TN (USA)

but at the same time, about a hundred claims of new elements appeared owing to the difficult chemical separation of the rare earths and lack of efficient analytical techniques to clearly identify new elements. In 1913, the British physicist W. Moseley using for the first time X-ray spectroscopy demonstrated the existence of 14 elements between lanthanum and lutecium. Between 1918 and 1921, the physicist Niels Bohr, with the new quantum theory of electronic atom configuration, interpreted and recognized lanthanides as 4f-elements. In 1939, after the discovery of neutron-induced nuclear fission of uranium by the two German scientists Hahn and Strassmann, rare-earth elements were identified in fission products. Hence, after World War II, owing to the recovery of lanthanide elements in fission products during reprocessing of spent nuclear wastes the separation of rare earths was greatly improved and this led to the large commercial scale solvent extraction process now widely used to recover lanthanides for industrial applications. Their first major industrial application must be credited to the German engineer Carl Auer von Welsbach, who invented in 1866 the gas mantle called the Auer gas mantle or incandescent Auer nozzle, which involves the heating of a cotton sock soaked in a doped lanthanum oxide solution until it reaches thermal luminescence. Later, ceria-doped thoria was selected for improving the brightness of the lighting system. Moreover, in 1903 the efficiency of lighters was enhanced by a Ce-based pyrophoric ignition source (mischmetal lighter flint). These devices were extensively used for manufacturing light burners until 1910, when the first electric lamps appeared. Nevertheless, today Auer gas mantles are still used in Coleman lanterns but with a thoria-free rare-earth oxide mixture.

3.6.4 Natural Occurrence, Minerals, and Ores

Rare-earth elements, by contrast with their historical name, are relatively abundant in the lithosphere (Earth's crust) and they occur in many economically viable ore deposits throughout the world with an estimated world reserve of 110 million tonnes. For instance, cerium (Ce) which is the most abundant rare earth has an abundance of 66.5 mg.kg^{-1} (ppm wt) similar to that of zinc, while thulium (Tm) which is the least abundant has an abundance of 0.52 mg.kg^{-1}, greater than that of cadmium and silver. Abundance of lanthanides in nature shows even–odd alteration with atomic number. As a general rule, owing to their extremely close chemical properties, especially valence and ionic radii, geochemical processes often concentrate these elements in the same minerals, where the elements are intimately mixed and therefore, they always occur in the same ore deposits. Nevertheless, owing to its smaller atomic and ionic size, scandium only occurs in rare-earth ores in minor amounts. Chief rare-earth ores are the mixed phosphate minerals **monazite** ((CeLaYTh)PO$_4$, monoclinic), and **xenotime** (YPO$_4$, tetragonal), and the fluorocarbonate **bastnaesite** ((Ce,La)(CO$_3$)F, hexagonal). However, bastnaesite and monazite are the principle ores of rare earths. The largest percentage of the world's rare-earth economic resources are bastnaesite ore deposits, however, significant quantities of rare earths are also recovered from the three other minerals: the thorium ore, **monazite**, in black sands, **xenotime**, and, to a lesser extent, from ion-adsorption clays. Major rare-earth ore world reserves are distributed in the following manner: China (43%), the Commonwealth of Independent States (19%), the United States (13%), Australia (5%), and finally India (1%). The United States is the largest rare earths producing country, followed by China, Australia, Brazil, India, South Africa and Malaysia. Except for one primary mine in the United States, essentially all rare earths are produced as by-products during processing of titanium and zirconium minerals (e.g., black sands containing ilmenite, and zircon), iron minerals, or tin ore (e.g., cassiterite, tin slag).

3.6.5 Processing and Industrial Preparation

Owing to their strong chemical reactivity with oxygen, highly pure grades of rare-earth metals are extremely difficult to prepare industrially. Fortunately, for many industrial applications, especially metallurgy, high-purity rare-earth metals are not required and hence there are no special industrial efforts to produce high-purity metals, alloys, and compounds. However, for special R&D or high-technology purposes, highly pure rare-earth metals and alloys can be required and hence particular industrial processes must be used to recover and purify the metals. Moreover, the chemical separation of rare-earth elements can only be industrially preformed by liquid–liquid extraction at a large scale or ion-exchange at the smaller scale. In addition, small quantities of rare earths are recovered from recycling of permanent magnet scraps.

Mining and mineral dressing – all world rare earths are extracted from monazite or bastnaesite or as a by-product during processing of black sands minerals for the recovery of titanium and zirconium. Actually, monazite is a minor constituent of black sands ore and when pure it contains on average 50 to 60wt% rare-earth oxides. Mining of black sands in placer ore deposits is usually carried out by bulldozers and front-end loaders, or suction dredging.

Ore concentration – concentration of monazite from other minerals (e.g., rutile, ilmenite, zircon, sillimanite, and garnet) is achieved by common ore beneficiation techniques. Firstly, the separation of heavier minerals from silica and other low-density minerals is done by specific gravity separation. Secondly, the heavier minerals are separated by strong electromagnets according to their respective magnetic susceptibility. During this step, ilmenite, magnetite, monazite (paramagnetism due to its rare-earth contents) and garnet are separated from nonmagnetic residue. The nonmagnetic residue is further treated by froth flotation to recover valuable minerals such as rutile, zircon, and sometimes gold. After a second specific gravity separation, rich monazite concentrate contains now more than 98wt%.

Hydrometallurgical concentration processes – because monazite is a relatively chemically inert mineral, only two hydrometallurgical dissolution processes can be efficiently used for recovering thorium and rare earths. Actually, the hydrometallurgical processing of monazite ore concentrate is carried out either by concentrated sulfuric acid or strong alkaline caustic hot digestion.

Alkali digestion of monazite – this process has been used on a large commercial scale in Brazil and India. The monazite concentrate is ground under water in a ball-mill to 325 mesh. The fine slurry is fed into a stainless steel reactor containing a strong caustic solution of sodium hydroxide with 65wt% NaOH. The mixture is then heated to 140°C for 3 h or until dissolution of the minerals is complete. After that, the mixture is diluted with a solution of sodium hydroxide and sodium triphosphate and digested again for 1 h at 105°C. The digested slurry, apart from trisodium phosphate, contains all the thorium, and rare earths as hydrous hydroxides (i.e., $Th(OH)_4$ and $Ln(OH)_3$). After filtration at 80°C through a sieve with Inconel[R] 625 wire, the filtration cake is washed with warm water. The filtrate which still contains 40wt% NaOH, is then evaporated in a steel kettle until the sodium hydroxide reaches 47wt% with a boiling point of 135°C and after precipitation of remaining trisodium phosphate is recycled for later digestion. The hydrated filtration cake is dissolved in 37wt% hydrochloric acid at 80°C for 1 h in a glass-lined steel reactor vessel. After dissolution, the acid solution and undissolved residue (i.e., monazite, and rutile) is then neutralized with recovered caustic soda from the evaporator and diluted with water. Thorium is separated from the rare earths by selective of thorium hydroxide $Th(OH)_4$ at pH 5.8. The filtered crude thorium hydroxide then undergoes several wash steps until 99.7% pure.

Sulfuric acid digestion process of monazite – the sulfuric acid digestion process has been extensively used in Europe, Australia, and the Untied States. Monazite concentrate is ground to less than 65 mesh, and is digested with 93wt% sulfuric acid at 210°C for 4 h in a stirred reactor. The temperature should be kept below 230°C in order to prevent formation of water-insoluble thorium pyrophosphate, ThP_2O_7. During digestion most of the thorium, rare earths, and uranium go into solution while insoluble silica and unreacted monazite form the sludge at the bottom of the reactor. The denser monazite is separated from the sludge and recycled. Radium is removed with the sludge by adding barium carbonate before decantation. The barium sulfate removes radium as insoluble radium sulfate. This process produces a solution of thorium, rare earths, and uranium cations with sulfate and phosphate anions. Because of the chemical similarity between cations in the solution, separation of thorium from lanthanides is a hard task and several routes have been developed. The precipitation of thorium oxalate has been developed at the Ames laboratory, while the solvent extraction with several organic solvents has been successfully developed for separation from rare earths.

Hydrochloric acid digestion process of bastnaesite – this second acid digestion process is only performed on the fluorocarbonate bastnaesite owing to the ability to the limestone gangue to be dissolved by diluted hydrochloric acid. The concentrate is ground to less than 200 mesh, and is digested with 10wt% hydrochloric acid in a stirred reactor. During digestion most of the carbonate minerals forming the gangue are dissolved. The slurry is then calcinated to oxidize Ce(III) to Ce(IV) and digested by concentrated sulfuric acid at 200°C. The solution is then filtered, producing the filtration cake which contains the insoluble ceria, CeO_2, while the filtrate contains light lanthanide sulfates and small amounts of thorium and heavier rare earths.

Purification or refining of rare earths – the separation of rare earths from thorium can be performed in different ways depending on the production scale. Small laboratory scale methods used originally the fractional crystallization of nitrates, followed by the fractional thermal decomposition of nitrates. Pilot scale separation can be achieved by ion-exchange, while the large commercial scale separation scale separation is based only on the solvent extraction process of an aqueous nitrate solution with n-tributyl phosphate (TBP) dissolved in kerosene.

Rare-earth metals preparation by metallothermic reduction (Ames laboratory process) – with this preparation process, 12 of the rare-earth metals are obtained pyrometallurgically by direct metallothermic reduction of the fluoride with molten calcium into a tantalum crucible under an inert atmosphere of argon. As a general rule, during this process, the highly pure fluoride, LnF_3, is obtained by reaction of the pure oxide, Ln_2O_3 or/and LnO_2, with pure anhydrous hydrogen fluoride gas, HF. The reaction takes place in platinum-lined Inconel tubular furnace at 650° under argon. After, the remaining impurities are removed by bubbling anhydrous hydrogen fluoride into the molten rare earth fluoride contained in a platinum crucible. Then, the fluoride, LnF_3, is mixed with a stoichiometric quantity of distilled calcium and the mixture is melted by induction heating under argon atmosphere in a tantalum crucible. After reduction, the floating slag separates from the molten rare-earth metal. After cooling, slag is mechanically removed and the metal is remelted under vacuum in order to remove the last volatile impurities (e.g., Ca, CaF_2, LnF_3). Nevertheless, according to their melting points and vapor pressures rare-earth metals must be split into separate groups. (i) Rare-earth metals (e.g., La, Ce, Pr, and Nd) with low melting points and high boiling points (i.e., largest melting range) are vacuum melted at 1800°C to remove volatile impurities and then cooled just above their melting point in order to permit the traces of tantalum dissolved at HT to precipitate out of the melt and settle to the bottom of the crucible. (ii) The rare-earth metals with both high melting and boiling points (e.g., Gd, Tb, Y, and Lu) dissolve too large an amount of tantalum at their melting point and hence they are purified by vacuum distillation leaving the tantalum in the crucible. Nevertheless, impurities such as carbon and oxygen are usually found in the distillate. (iii) The four remaining metals (i.e., Dy, Ho, Er, and Sc) have high melting points and low boiling points (i.e., narrow melting range) and hence they could be separated after reduction by a sublimation process. Moreover, the sublimation process purify the metal with respect to traces of O, N, C, and obviously Ta. (iv) The four remaining rare-earth metals (i.e., Sm, Eu, Tm, and Yb), owing to their low boiling points, are prepared directly by reduction of their oxides with lanthanum, cerium, or mischmetal.

Liquid–liquid extraction process – the trivalent rare-earth cations Ln(III), usually in a chloride or nitrate aqueous solution withstand several solvent extractions (e.g., often eight steps). During each solvent extraction stage, the lanthanide cations are gradually partitioned between the aqueous phase and an organic extracting solvent (e.g., tri-*n*-butyl phosphate, TBP in kerosene). These separations occupy large plant areas, with several rows of mixing tanks and settles. The downstream processing of the separated rare-earth streams is quite simple, i.e., the precipitation of the carbonate, and its calcination to produce the pure oxide. Nevertheless, the recovery of the pure rare-earth metal from its oxide is a hard task owing to the high thermodynamic stability of the rare-earth oxide. Actually, close examination of Ellingham's diagram (i.e., free energy of oxide formation versus temperature) indicates that only calcium metal can reduce the lanthanide oxide and that pyrometallurgical reduction with carbon is impossible. Moreover, owing to their strong electropositivity, they cannot be obtained by electrolysis in aqueous media, and only electrowinning in molten salts (e.g., molten chloride baths) can produce pure metal cathodic deposits.

3.6.6 Industrial Uses and Applications

Major industrial applications (Table 3.65) for the rare-earth compounds are, in order of importance: (i) metallurgical, such as mischmetal, or alloying elements in superalloys, magnesium, and aluminum alloys, (ii) magnetic materials, such as Sm–Co, and Nd–Fe–B for permanent magnets, (iii) chemical and petroleum engineering, such as fluid cracking catalysts or catalytic converter materials, (iv) glass industry, where cerium oxide compounds are used for glass polishing and glass additives, (v) hydrogen storage compounds for rechargeable secondary batteries such as Ni–MH, (vi) medical X-ray intensifying phosphors and phosphors for television screens, (vii) lighting and laser applications, (viii) nuclear, control rods for absorbing neutrons in nuclear power reactors (e.g., gadolinium), and finally (ix) pyrophoric materials.

3.6.7 World Rare-Earth Suppliers

See **Table 3.67** (page 225).

3.7 Uranides (Th and U)

See **Table 3.68** (page 226).

3.7.1 Uranium

3.7.1.1 General Properties and Description

Uranium [7440-61-1] with the chemical symbol U, the atomic number 92 and the relative atomic mass (atomic weight) 238.0289(1), is the fourth metallic element of the actinide family of Mendeleev's periodic chart. Uranium was named after the planet, Uranus, which astronomers had discovered several years before. When highly pure, uranium is a dense (18,950 kg.m^{-3}), silvery-white, lustrous metal which is malleable, ductile, slightly paramagnetic, and slightly radioactive. From a crystallographic point of view, uranium has three allotropic modifications: (i) the α-phase is orthorhombic (19,070 kg.m^{-3}), which transforms at 667.8°C to (ii) the tetragonal β-phase (18,369 kg.m^{-3}), and (iii) at 774.9°C to the γ-phase (18,070 kg.m^{-3}) which is body-centered cubic (bcc) which remains until the melting point at 1132.8°C. Mechanically it is important to remember that the large density changes of the metal with temperature, even at

Table 3.65 Rare-earth industrial applications and uses

Rare earth metal	Uses and applications
Cerium (Ce)	Cerium is the major component of mischmetal (about 50wt%). It is used as an alloying additive in ferrous alloys and as a scavenger of sulfur and oxygen in iron metallurgy. Alloying additions serve to strengthen magnesium alloys. Traces of cerium improve high-temperature corrosion resistance of superalloys. Cerium oxide, CeO_2, is used as abrasive for glass polishing. Cerium is also used in petroleum cracking catalysts, and in catalytic converters. Reactive cerium powder serves to prepare pyrophoric ordnance devices and in lighter flints. Finally, it is also employed as glass-decolorizing agents, in carbon arc lights, ceramic capacitors, and $CeCo_5$ as permanent magnets.
Dysprosium (Dy)	Owing to its important thermal neutron cross-section, dysprosium is used to produce control rods in nuclear reactors and also in neutron flux measurements. The alloy $Tb_{0.3}Dy_{0.7}Fe_2$ is used as magnetostrictive material. The alloys Nd–Fe–B are permanent magnets. Finally, dysprosium is also used as phosphors, catalysts, and garnet microwave devices.
Erbium (Er)	Erbium is used in lasers, phosphors, garnet microwave devices, ferrite bubble devices, and catalysts.
Europium (Eu)	Owing to its important thermal neutron cross-section, europium serves to produce control rods in nuclear reactors. Europium compounds are also used as phosphors for television screens.
Gadolinium (Gd)	Owing to its important thermal neutron cross-section, gadolinium is used as burnable poison in shields and in control rods in nuclear reactors. Its compounds are extensively used as phosphors, catalysts, and garnet microwave devices. Finally, Gd–Co alloys serve in magnetooptic storage devices and as magnetic refrigeration materials.
Holmium (Ho)	It is employed as phosphors, and ferrite bubble devices.
Lanthanum (La)	Lanthanum is the second major component of the mischmetal (about 25wt%) It is used as alloying additive in ferrous alloys and as a scavenger of sulfur and oxygen in iron metallurgy. Alloying additions serve to strengthen magnesium alloys. Traces of lanthanum improve high-temperature corrosion resistance of superalloys. Lanthanum is used in optical lenses, in petroleum cracking catalysts, and hydrogen storage alloys, catalytic converters, lighter flints, glass-decolorizing agents, carbon arc lights, and ceramic capacitors.
Lutetium (Lu)	It is employed as phosphors, ferrite and garnet bubble devices.
Neodymium (Nd)	Neodymium is the third component of mischmetal (about 20wt%). Nd–Fe–B serves to produce high-strength permanent magnets. It is used as an alloying additive in ferrous alloys and as a scavenger of sulfur and oxygen in iron metallurgy. Alloying additions serve to strengthen magnesium alloys. It also improves high-temperature corrosion resistance of superalloys. Compounds of neodymium are used in petroleum cracking catalysts and catalytic converters, as hydrogen storage alloys, lighter flints, glass-decolorizing agents, carbon arc lights, and finally ceramic capacitors.
Praseodymium (Pr)	Minor component of mischmetal (about 5wt%). It is used as an alloying additive in ferrous alloys and as a scavenger of sulfur and oxygen in iron metallurgy. Alloying additions serve to strengthen magnesium alloys. It also improves high-temperature corrosion resistance of superalloys. Compounds of praseodymium are used in petroleum cracking catalysts and catalytic converters, as hydrogen storage alloys, lighter flints, glass-decolorizing agents, carbon arc lights. The alloy $PrCo_5$ is a permanent magnet, and $PrNi_5$ is used in adiabatic demagnetization refrigeration for obtaining ultralow temperatures ($T < 1$ mK).
Promethium (Pm)	Promethium is used as luminous watch dials, and as lightly shielded radioisotope power sources.
Samarium (Sm)	The compound Sm_2Co_{17}–SmO_5 is a permanent magnet. Samarium is used also as a burnable poison in nuclear reactors. Finally, its compounds are used as phosphors for television screens, catalysts, and ceramic capacitors.
Scandium (Sc)	Scandium is used as neutron window or filter in nuclear reactors. Scandium compounds are used in high-intensity lamps.
Terbium (Tb)	The alloy $Tb_{0.3}Dy_{0.7}Fe_2$ is a magnetostrictive material, and amorphous Tb–Co alloys serve in magnetooptic devices. Other uses are in phosphors and catalysts.
Thulium (Tm)	Thulium is used as phosphors, in ferrite and garnet bubble devices, and as catalysts.
Ytterbium (Yb)	Ytterbium is used as phosphors, in ferrite and garnet bubble devices, and as catalysts, and finally in ceramic capacitors.

Table 3.65 (continued)

Rare earth metal	Uses and applications
Yttrium (Y)	Yttrium is used as an alloying component in magnesium alloys. Yttrium is used in ferrite and garnet bubble devices, in catalysts, in superalloys as dispersant, in ceramic capacitors, and for the stabilization of zirconia. Finally, $YBa_2Cu_3O_{7-x}$ is the precursor of superconductive oxides at room temperature.

Table 3.66 Rare-earth oxide prices (1996) (Rhône–Poulenc)

Rare-earth metal oxide	Purity (wt%)	Price ($US.kg^{-1})
Cerium (Ce)	99.50	23.00
Dysprosium (Dy)	95.00	69.00
Erbium (Er)	96.00	150.00
Europium (Eu)	99.99	700.00
Gadolinium (Gd)	99.99	115.00
Holmium (Ho)	99.90	485.00
Lanthanum (La)	99.99	23.00
Lutetium (Lu)	99.99	4500.00
Neodymium (Nd)	95.00	22.00
Praseodymium (Pr)	96.00	32.00
Samarium (Sm)	96.00	75.00
Terbium (Tb)	99.90	685.00
Thulium (Tm)	99.90	3600.00
Ytterbium (Yb)	99.00	230.00
Yttrium (Y)	99.99	85.00

Table 3.67 World rare-earth suppliers

Supplier	Address
Rhône-Poulenc SA	Division Terres Rares, 25, quai Paul Doumer, F-92408 Courbevoie Cedex, France Email: produit.info@rhone-poulenc.com Internet: http://www.rhone-poulenc.com
Molycorp Inc.	PO Box 469, Questa, NM 87556, USA Tel: +1 (505) 586 7606 Fax: +1 (505) 586 0811 Email: jrstark@questa.unocal.com

Table 3.68 Selected properties of thorium and uranium

Properties at 298.15 K, unless otherwise specified	Thorium	Uranium
Chemical symbol [IUPAC]	Th	U
Chemical abstract registered number [CAS RN]	[7440-29-1]	[7440-61-1]
Unified numbering system [UNS]	[n.a.]	[M08990]
Mineral occurrence and economics		
Earth's crust abundance (ppm wt)	12	2.4
Seawater abundance (ppm wt)	9.2×10^{-6}	3.13×10^{-3}
World reserves (R, 10^3 kg)	3.3×10^6	10×10^6
World production of ore (P, 10^3 kg.yr^{-1})	31,000 (ore)	35,000 (metal)
Cost of pure metal (1998)(C, $US.kg^{-1})(purity)	15,000 (99.8wt%)	200 (99.7wt%)
Atomic properties		
Atomic number (Z)	90	92
Relative atomic mass A_r (^{12}C = 12.000)[a]	232.0381(1)	238.0289(1)
Electronic configuration	[Rn]$6d^2 7s^2$	[Rn]$5f^3 6d^1 7s^2$
Fundamental ground state	$^3F_2$	$^5L_6$
Electronegativity χ_a (Pauling)	1.3	1.38
Electron work function (W_S, eV)	3.40	3.63
X-ray absorption coefficient CuK$_{a1,2}$((μ/ρ),cm^2.g^{-1})	327	352
Nuclear properties		
Thermal neutron cross-section (σ_n,10^{-28} m^2)	7.4	7.57
Isotopic mass range	212–236	226–242
Natural isotopes (including isomers)	25	17
Crystallographic properties at 293.15 K		
Crystal structure at room temperature (phase α or β)	fcc (α-Th)	orthorhombic
Stukturbericht designation	A1	A20
Space group (Hermann–Mauguin)	Fm3m	Cmcm
Pearson's symbol	cF4	oC4
Lattice parameters (pm)	$a = 508.42$	$a = 284.785$ $b = 585.801$ $c = 494.553$
Miller's indices of slip plane {hkl}	{111}	{010}
Phase transformation temperature α–β (T, K)	1673 (1400°C)	941 (668°C)
Mechanical properties (annealed)		
Density (293 K) (ρ, kg.m^{-3})	11,720	18,950
Young's modulus (300 K) (E, GPa)	72.4–78.3	201–176
Shear modulus (G, GPa)	47.8–30.8	75–73
Bulk modulus (K,GPa)	57.7–54	100.7–98
Vickers hardness (HV)	54–114	185–250
Yield strength proof 0.2% (σ_{YS}, MPa)	144	220
Ultimate tensile strength (σ_{UTS}, MPa)	219	650
Elongation (Z, %)	34	13–50
Longitudinal velocity of sound (V_S, m.s.$^{-1}$)	n.a.	3100
Poisson's ratio v (dimensionless)	0.270	0.210

Table 3.68 (continued)

Properties at 298.15 K, unless otherwise specified	Thorium	Uranium
Thermal properties (293.15 K)		
Melting point (mp, K)	2023 (1750°C)	1405.5 (1132°C)
Boiling point (bp, K)	5060 (4787°C)	4018 (3745°C)
Thermal conductivity (k, W.m^{-1}.K^{-1})	49.54	27.6
Thermal linear expansion coefficient (α, 10^{-6} K^{-1})	11.4–12.5	12.6
Molar heat capacity (C_p, J.mol^{-1}.K^{-1})	27.32	27.665
Specific heat capacity (c_p, J.kg^{-1}.K^{-1})	118	116
Enthalpy molar fusion (ΔH_{fus}, kJ.mol^{-1})(Δh_{fus}, kJ.kg^{-1})	19.2 (82.75)	15.5 (65.12)
Enthalpy molar vaporization (ΔH_{vap}, kJ.mol^{-1})(Δh_{vap}, kJ.kg^{-1})	543.9 (2344)	422.6 (1776)
Enthalpy molar sublimation (ΔH_{sub}, kJ.mol^{-1})	576.1	482.2
Enthalpy molar combustion (ΔH_{comb}, kJ.mol^{-1})	−1228	−1085
Electrical properties (293.15 K)		
Temperature superconductivity (T_c, K)	1.390	<0.5
Electrical resistivity (ρ, $\mu\Omega$cm)	15.7	30.8
Temperature coefficient of resistivity (0–100°C)(K^{-1})	0.003567	0.021
Hall coefficient (R_H, aΩm.T^1)	−0.088	+380
Thermoelectric potential versus platinum (emf, mV)	−130 (100°C)	n.a.
Magnetic properties		
Mass magnetic susceptibility (χ_m, 10^{-9} kg^{-1}.m^3)	+7.2	+21.6
Optical properties		
Spectral emissivity (650 nm)	0.380	0.265
Reflective index (650 mm)	n.a.	73.5

[a] Standard atomic masses from: Atomic weights of the elements 1995. Pure Appl Chem 68: (1996) 2339.

the lower temperature at which the orthorhombic α-phase is stable, combined with its unequal thermal expansion coefficient cause severe distortion and elongation of uranium metal during temperature cycling.

From a chemical point of view, uranium is a reactive metal similar in that sense to elements of group IVB(4), Ti, Zr, and Hf and another actinide, Th. Actually, at room temperature it rapidly tarnishes in moist air forming a thin protective passivating layer of dark-colored oxide, UO_2, which prevents further oxidation of base metal. Like other reactive and refractory metals such as titanium and zirconium, when finely divided, uranium is highly pyrophoric and may ignite spontaneously in air or oxygen and is attacked by cold water. Moreover, massive uranium rods burn steadily in air at temperatures above 700°C forming U_3O_8. In aqueous solution, uranium has several possible states of oxidation: trivalent (e.g., U^{3+}), tetravalent (e.g., U^{4+}), pentavalent (e.g., $U^V O_{2+}$), and hexavalent (e.g., $U^{VI}O_2^{2+}$). But trivalent cations, U^{3+}, are unstable, decomposing in water with evolution of hydrogen, while the pentavalent uranyl cation is also unstable, disproportionating into tetravalent and hexavalent uranyl cations. Hence, only the uranous, U^{4+}, and hexavalent uranyl cations, UO_2^{2+} are of practical importance in industrial processes involving uranium chemistry.

Uranium is readily dissolved by nitric and hydrochloric acid. Nevertheless, owing to the formation of a slightly protective oxide layer in oxidizing media, sulfuric, phosphoric, and nitric acid only dissolve uranium at a moderate rate. However, by contrast with Ti and Zr, uranium is only slightly soluble in hydrofluoric acid. Moreover, uranium metal is inert to alkaline media such as strong alkali hydroxides (e.g., NaOH, KOH) and ammoniac.

The natural uranium element is composed of three radionuclide isotopes: ^{238}U (99.2745at%), ^{235}U (0.7205at%), and ^{234}U (0.0054at%). However, slight variations of ^{235}U isotopic abundance in some geological specimens are known (e.g., Oklo ore deposit, Gabon, Africa) in which the element has an isotopic composition outside the limits for normal material. Moreover, modified isotopic compositions may also be found in commercially available material because it has been subject to an undisclosed or inadvertant isotopic fractionation. Hence, substantial deviations in atomic weight of the element can occur. Each natural uranium radionuclide isotope is an alpha-emitter, and is a member of one of the four possible radioactive decay series involving successive alpha and beta decay transitions. For instance, ^{238}U is the natural parent member and thus the longest living of the $4n + 2$ series, with a half-life of 4.46×10^9 years. ^{235}U is the natural parent member of the $4n + 3$ series, with a half-life of 710×10^6 years. Finally, ^{234}U with a decay half-life of 2.47×10^5 years is also a member of the $4n + 2$ series. Therefore, in natural uranium, assuming that the uranium radionuclides in the metal have not been undisturbed long enough to be in secular equilibrium with all their decay nuclides, the activity of all the radionuclides is the same. Hence, the specific radioactivity of pure uranium in ore is roughly 12.765 MBq.kg^{-1} of U (i.e., 0.345 Ci.tonne^{-1}). Actually, it is sufficiently radioactive to expose a photographic emulsion in 1 h. Moreover, owing to its strong toxicity, uranium is a chemical and a radiological hazard. For metallurgical and other non-nuclear applications where high density metals are needed, only depleted uranium (i.e., less than 0.3at% ^{235}U) is used and is referenced by the UNS as M08990.

3.7.1.2 History

In 1789, the German chemist Martin H. Klaproth (Berlin, Germany) isolated an oxide of uranium while analyzing pitchblende ore samples from the Joachimsthal silver mines in Bohemia (Czechoslovakia) and he suspected an unknown element in the mineral and attempted to isolate the new metal. However, the metal seems to be first isolated in the pure state in 1841 by the French chemist E.-M. Peligot (Paris, France),[96,97] who reduced the anhydrous chloride with molten potassium. For over 100 years uranium was mainly used as a colorant for ceramic enamels and glazes and for tinting in early photography. Uranium was produced in Bohemia, Cornwall, Portugal, and Colorado and total production amounted to about 300–400 tonnes. The discovery of radium in 1898 by the French chemist Marie Curie led to the construction of a number of radium extraction plants. Prized for its use in cancer radiotherapy, radium reached a price of 750,000 gold francs per gram in 1906 ($US10 million). It is estimated that 574 g were produced worldwide between 1898 and 1928. Uranium itself was simply dumped as waste. Later, the French physicist Henry Becquerel discovered its radioactivity and it was the starting point of an intense study and discovery of actinide elements. With the discovery of nuclear fission induced by thermal neutron in 1939 by the two German physicists Otto Hahn and Strahssmann the uranium industry entered a new era. On 2 December 1942, the first controlled nuclear chain reaction was achieved in Chicago by the Italo-American scientist Enrico Fermi, it was the starting point of the extensive use of uranium for nuclear industry. From a small beginning in 1951 when four lightbulbs were lit with nuclear electricity, the nuclear power industry now supplies some 17% of world electricity.

3.7.1.3 Natural Occurrence, Minerals, and Ores

Uranium, which has a natural elemental abundance in the Earth's crust of roughly 2.4 mg.kg^{-1} (i.e., ppm wt) is more abundant than mercury, cadmium, and silver, and hence, it is about 500 times more abundant than gold and about as common as tin. It is present in most rocks and soils as well as in many rivers and in seawater (3 μg.kg^{-1}). There are a number of areas around the world where the concentration of uranium in the ground is sufficiently high that extraction for use as nuclear fuel is economically feasible. Because of its chemical reactivity with oxygen, uranium never occurs in the free state in nature, but it occurs in numerous minerals in the tetravalent (i.e., uranous, U^{4+}) and hexavalent (i.e., uranyl, UO_2^{2+}) oxidation state. These minerals can be arbitrarily classified according to both the oxidation state of uranium cations and the treatment needed to recover the uranium. The first group are minerals in which uranium occurs as tetravalent cations with a high weight percentage of the element, such as **pitchblende** (U_3O_8, amorphous), **uraninite** (UO_2, cubic), **uranothorite** ($(U,Th)[SiO_4]$, tetragonal), and **coffinite** ($(U(SiO_4)_{1-x}(OH)_{4x}$, tetragonal). When found in massive form, these minerals can be concentrated by gravity methods, while acid or carbonate leaching is used for small particles. In the second group, uranium occurs as hexavalent uranyl cations in hydrated minerals such as **carnotite** ($K_2(UO_2)_2(VO_4)_2 \cdot 3H_2O$, rhombohedral), **autunite** ($Ca(UO_2)_2(PO_4)_2 \cdot 10H_2O$, tetragonal), **uranophane** ($Ca(UO_2)_2[Si_2O_7] \cdot 6H_2O$, orthorhombic), and **tobernite** ($Cu(UO_2)_2(PO_4)_2 \cdot 8H_2O$, tetragonal). In these ores, uranium must be recovered only by dilute sulfuric acid or alkali carbonate leaching. Finally, the last group concerns ubiquitous minerals found in combination with other minerals of refractory elements such as titanium, niobium, and tantalum. These minerals such as **davidite**, **brannerite**, and **pyrochlore** contain tetravalent uranium cations from which uranium can be only recovered by autoclave hot sulfuric acid leaching and hence, it is only a by-product of the recovery of other metals such as Ti, Nb, and Ta.

In addition to the previous well-characterized uranium minerals and ores, owing to its chemical properties, uranium is often found in low-grade sources in which it occurs as a minor element. These low-grade uranium sources can be classified in two main groups: (i) sources commercially used for recovering the metal – lignite in some coal deposits (South Dakota, USA) which contains uranium at concentrations greater than 100 mg.kg^{-1}, bituminous shales (Boliden, Sweden), gold (Witswatersrand, South Africa, and Chihuahua, Mexico), copper tailings (Katanga, Congo, Mount Mitaka in Japan, and Western United States), phosphate rocks (Florida, United States) in which uranium concentration can be as high as 400 mg.kg^{-1}, and (ii) non-economic sources such as shales, or granitic igneous rocks which make up 60wt% of the Earth's crust where uranium, found in concentrtions of about 4 mg.kg^{-1}, owing to similar ionic radii with Zr(IV), Hf(IV), and Th(IV), substitutes their cations, and hence is always found in zircon giving a metamicte (i.e., amorphization of the host lattice). However, as a general rule, commercial sources of uranium are mainly recovered from monazite black sands, pitchblende and phosphate rock ore deposits. The chief large ore deposits are found in Australia (e.g., Mary Kathleen, Rum Jungle, and Radium Hill), USA (e.g., Utah, New Mexico, and Colorado), Canada (e.g., Blind River, Elliot Lake), Russia (e.g., Ukraina, Baikalia, and the Urals), Africa (e.g., Congo, Gabon, and Rand Gold Fields in South Africa) and China. Uraninite and pitchblende are by far the most common primary uranium minerals which are sometimes associated with colorful (orange, yellow, green) secondary uranium minerals, called gummites ($UO_3 \cdot nH_2O$ with PbO), derived from weathering or/and hydrothermal hydration. Other ores of economic interest include coffinite and brannerite.

The minimum concentration of uranium in ore required to extract uranium on an economic basis from an ore deposit (i.e., Clarke index) varies widely depending on its geological setting and physical location. Average ore grades at operating uranium mines range from 300 mg.kg^{-1} to as high as 10wt% but are most frequently less than 1wt%. These figures do not apply to by-product operations. Uranium concentrations are sometimes expressed in terms of U_3O_8 content (U_3O_8 is an approximation of the chemical composition of typical naturally occurring oxides of uranium). For instance, a product that is said to be 60wt% U_3O_8 contains 51wt% uranium metal.

3.7.1.4 Mineral Dressing and Mining

The selection of a mining method to use for a particular uranium ore deposit is only governed by safety and economic considerations. Underground, open pit mining and in situ leaching techniques are used to recover uranium.

(i) The open pit process is used to mine shallow deposits. In general, open pit mining is used where ore deposits are close to the surface and underground mining is used for deep ore deposits, i.e., typically greater than 120 m deep. Open pit mines require large openings on the surface, larger than the size of the ore deposit, since the walls of the pit must be sloped to guard against collapse. As a result, the quantity of material that must be removed in order to access the ore is large. The economics depend on the ratio of ore to waste, higher grade ores being able to produce higher ratios. The largest open pit uranium mine in the world is at Rössing in Namibia.

(ii) Underground pit mining is used to mine deposits too deep for open pit mining. For mining to be viable, these deposits must be comparatively high grade. Underground mines have relatively small openings to the surface and the quantity of material that must be removed to access the ore is considerably less than in the case of an open pit mine. In the case of underground uranium mines special precautions, consisting primarily of increased ventilation, are required to protect against airborne radiation (i.e., radon and uranium ore dusts) exposure.

(iii) Finally, the in situ leaching method is only applicable to sandstone-hosted uranium ore deposits located below the water table in a confined aquifer. The uranium is dissolved in a mildly alkaline solution, containing sodium carbonate, which is injected into and recovered from the aquifer by means of wells. The bedrock remains undisturbed.

(iv) On the other hand, uranium, which often occurs in association with other minerals such as gold (e.g., Witswatersrand, South Africa), in phosphate rocks (e.g., United States and elsewhere), and in copper sulfide ores is recovered as by-product of the processing of these ores.

3.7.1.5 Processing and Industrial Preparation

Amongst all the possible preparation routes, which strongly depend on the types of crude uranium ores, the steps in producing refined uranium compounds from uranium ores may be always conveniently classified into: (i) size reduction and concentration, which consist of separating uranium ore from inert gangue materials and increasing uranium oxide content from a few tenths of one percent to 95wt% U_3O_8 in the concentrate, (ii) purification or refining, which consists of removing trace impurities and producing a pure uranium compound, and finally (iii) conversion, which consists of transforming uranium concentrate into pure uranium metal or uranium chemical compounds wanted for the required industrial application, mainly nuclear power reactors for electricity generation.

Crushing and grinding – relatively standard particle size reduction processes are always used in order to increase the specific surface area of the ore to facilitate separation of gangue material and later chemical dissolution during the concentration by leaching. The crude uranium ore (i.e., uranium minerals mixed with inert gangue minerals), such as pitchblende or uraninite, is crushed in a jaw crusher to particle sizes smaller than 1.25 cm and ground in a steel conical ball-mill in order to reduce the ore to sand/silt-sized particles under 200 mesh. The finely ground ore is ready for the concentration process unit.

Concentration by leaching – the powdered ore is concentrated in uranium by means of classical ore beneficiation processes. Although several powdered metallic ores are easily concentrated by a common froth flotation process, this operation unit is seldom applicable to uranium ores because relatively few uranium minerals can be selectively floated. Therefore, after particle size reduction, because of the unsuitability of operation units such as specific gravity or froth flotation to separate uranium ore from gangue materials, only leaching processes can be performed, using either diluted sulfuric acid or a alkali carbonate aqueous solutions. However, the leaching reagent strongly depends on the nature of gangue materials present in the ground ore. When the gangue materials are made of silica, or other insoluble minerals (e.g., fluorine, baryte), diluted sulfuric **acid leaching** is preferred, because it costs less than alkali carbonate and dissolves uranium minerals more rapidly. Acid leaching is performed in rubber-lined tanks stirred by turbines. The slurry of ground ore is mixed with sulfuric acid, water, and steam until the temperature rises to 45–55°C and the pH reaches 0.5. Gas evolution of H_2S and CO_2 obviously occurs during acid leaching owing to the dissolution of sulfide and carbonate minerals present in the gangue. Nevertheless, when insoluble tetravalent uranium cations are present in the ore, addition of an oxidant reagent such as natural pyrolusite or chemical manganese dioxide, MnO_2, or sodium chlorate, $NaClO_3$, is required to oxidize uranium to the soluble hexavalent uranyl cations, UO_2^{2+}. On the other hand, when the gangue materials comprise limestone, $CaCO_3$, or other carbonate minerals (e.g., dolomite) readily dissolved in acidic media, **alkaline leaching** is preferably achieved with sodium or ammonium carbonate allowing a cleaner solution to be produced, containing lower concentrations of impurities than when acid leaching is performed. The alkaline leaching solution consists of a mixture of sodium carbonate, Na_2CO_3, and sodium hydrogenocarbonate, $NaHCO_3$, to bring the pH to 10–11. The leaching is performed in a pressurized reactor (i.e., autoclave) at 4.5 atm and 95°C. However, in the two leaching processes, the uranium is always recovered under its soluble hexavalent uranyl cations, allowing the uranium-bearing solution to be separated from the leached solids by solid–liquid separation devices (e.g., clarifier, filter, cyclone), resulting in a filtered or clarified uranium-bearing mother solution containing between 1 and 4 g.l^{-1} U_3O_8 and washed fine and coarse tailings.

Recovery of uranium from leach liquors – processes for recovering uranium from leach liquors are different according to the nature of the leaching reagent. **Precipitation** – actually, this process is used to recover uranium from leach alkaline liquor. It consists of precipitating the insoluble yellow sodium diuranate, $Na_2U_2O_7$, known as "yellow cake", adding sodium hydroxide (i.e., caustic soda) to the mother liquor according to the reaction:

$$2Na_4UO_2(CO_3)_3 + 6NaOH \rightarrow Na_2U_2O_7\downarrow + 6Na_2CO_3 + 3H_2O.$$

Because vanadium pentoxide, V_2O_5, which coprecipitates simultaneously with sodium uranate is always present in the impure yellow cake, at levels ranging from 5 to 6wt%, it must be removed by roasting the impure yellow cake with sodium carbonate at 860°C for 30 min and after cooling the solid calcinated mass is leached with water to extract the soluble sodium vanadate, $NaVO_3$. The leached product is filtered and the washed yellow cake is dried, while the solution from which vanadium can be recovered is stored. **Solvent extraction and anion exchange** – by contrast, the processes for recovering uranium from acid leach liquor consist of using the Amex process or solvent extraction with amines (mixture of n-trioctyl and n-tridecylamine), the Dapex process or solvent extraction by organophosphorus compounds (di(2-ethylhexyl) phosphoric acid, DEHPA in kerosene), or finally, anion exchange on resins (e.g., amberlite, dowex, ionac). As a general rule, uranium concentrate now usually consists of uranium oxide (U_3O_8) if obtained from acid leaching liquor or sodium, ammonium diuranate ($M_2U_2O_7$ with $M = Na^+$, or NH_4^+) if obtained from alkaline leaching.

Purification or refining – The sodium diuranate is then dissolved in hot concentrated 40wt% nitric acid, HNO_3, to give the uranyle nitrate $UO_2(NO_3)_2$. The next step is to remove metallic impurities from the uranium by solvent extraction processes performed on the acidified solution with a mixture of organic solvents such as n-tributyl phosphate (n-C_4H_9)$_3PO_4$ dissolved at 30vol% in an inert hydrocarbon such as kerosene. Kerosene was chosen owing to its different density to that of water, low viscosity and high interfacial tension in order to prevent stable emulsions forming in the settler. The liquid–liquid extraction process is performed continuously in a large counterflow reactor. The optimum conditions are produced by increasing the distribution coefficient of uranium in the organic phase. Then, the pure uranyle nitrate is recovered by the reverse process by counterflowing dilute 0.01 M nitric acid while the organic solvent mixture is recycled in the loop. Dewatering and concentration of the aqueous solution by evaporation in a boil-down tank supplies a syrupy liquid with the approximate composition of $UO_2(NO_3)_2 \cdot 6H_2O$. The uranyle nitrate crystals are then denitrated by calcination in a fluidized bed reactor giving the oxide UO_3, evolving abundant nitric acid fumes and nitrogen oxides which complete the process. Another route consists of precipitating

$(NH_4)_2U_2O_7$ from uranyle nitrate solution with ammonia, filtering and drying the precipitate, and calcinating the crystals to drive off ammonia and UO_3.

At a conversion facility, uranium trioxide is converted to a specific uranium end product according to the nuclear power reactor class. Actually, there exist three uranium compounds used in nuclear power reactors: (i) the pure natural uranium metal, U_{nat}, was used at one time in graphite carbon dioxide gas nuclear power reactors (Magnox, or UNGG[4]), (ii) the natural uranium dioxide, UO_2 is used in heavy water nuclear power reactors (e.g., the Canadian CANDU type), and (iii) finally modern pressurized water nuclear power reactors (PWR) and boiling water reactors (BWR) uses enriched uranium dioxide nuclear fuel.

Reduction of UO_3 to UO_2 – UO_3 is reduced to UO_2 by direct reaction with hydrogen present in cracked ammonia (a stoichiometric mixture of nitrogen and hydrogen gases) at 590°C in a fluidized bed reactor through which solids and gases flow countercurrently according to the chemical reaction

$$UO_3 + H_2 \rightarrow UO_2 + H_2O.$$

Conditions must be carefully controlled to prevent sintering of the oxide particles during reaction, in order to keep a high specific surface area of the oxide needed in the fluorination reaction. Nevertheless, in particular cases, if the uranium dioxide is to be used directly as nuclear fuel in heavy-water nuclear reactors (CANDU), sintering can be allowed to produce a denser ceramic fuel.

Hydrofluorination of UO_2 to UF_4 – the hydrofluorination reaction is achieved in a series of stirred fluidized bed reactors with counterflow of solids and gases. During the exothermic reaction the conversion of the uranium dioxide to uranium tetrafluoride occurs with complete consumption of hydrogen fluoride gas according to the scheme

$$UO_2 + 4HF \rightarrow UF_4 + 2H_2O.$$

Fluorination of UF_4 to UF_6 – the fluorination is performed in a tower reactor, made of Inconel[®], in a slight excess of fluorine, F_2, with water-cooled walls at 500°C. Uranium hexafluoride produced in this way is exceptionally pure, with UF_6 content above 99.97wt%.

$$UF_4 + F_2 \rightarrow UF_6.$$

Enrichment – natural uranium consists, primarily, of a mixture of two isotopes of uranium. Only 0.717wt% of natural uranium, i.e., ^{235}U, is capable of undergoing fission, the process by which energy is produced in a nuclear reactor. The fissionable isotope of uranium is uranium ^{235}U. The remainder is uranium ^{238}U and, to a lesser extent, ^{234}U. In the most common types of nuclear power reactors (e.g., PWR and BWR), a higher than natural concentration of ^{235}U is required. The enrichment process produces this higher concentration, typically between 3.5% and 4.5% ^{235}U, by removing a large part of the ^{238}U (80% for enrichment to 3.5%). Today, there are two chief enrichment processes used on a large commercial scale, each of which uses uranium hexafluoride as feed: gaseous diffusion (e.g., program EURODIFF) and gas centrifugation (e.g., program URENCO). Other enrichment processes such as the obsolete electromagnetic process performed in the Calutron during the Manhattan project, and laser ionization, are under development. The two products of this stage of the nuclear fuel cycle are enriched and depleted uranium hexafluorides. But enriched and depleted uranium is commonly used as metallic uranium or as refractory ceramic dioxide, UO_2.

Preparation of uranium dioxide – to obtain UO_2, it is necessary to reconvert UF_6 to UF_4 or UO_2 to produce enriched uranium metal or uranium oxide. Three processes have been used for converting UF_6 to UO_2. (i) In the first process, uranium hexafluoride, UF_6, is reduced to UF_4 and then uranium tetrafluoride can be hydrolized with water according to the chemical reaction

$$UF_6 + H_2 \rightarrow UF_4 + 2HF$$
$$UF_4 + 2H_2O \rightarrow UO_2 + 4HF.$$

---

[4]French acronym of Uranium Naturel Graphite Gaz.

(ii) In a second process, UF_6 is hydrolyzed to produce uranyl fluoride, UO_2F_2, after which ammonium hydroxide is added to precipitate ammonium diuranate, $(NH_4)_2U_2O_7$ according to the reaction:

$$UF_6 + 2H_2O \rightarrow UO_2F_2 + 4HF$$

$$2UO_2F_2 + 6NH_4OH \rightarrow (NH_4)_2U_2O_7 + 4NH_4F + 3H_2O.$$

The ammonium diuranate is then reduced to uranium dioxide by hydrogen at 820°C:

$$(NH_4)_2U_2O_7 + 4H_2 \rightarrow 2UO_2 + 2NH_3 + 3H_2O.$$

(iii) Finally, in the last process, streams of gaseous UF_6, NH_3, and CO_2 are fed batchwise into deionized water, whereby ammonium uranocarbonate precipitates. The precipitate is then converted to uranium dioxide in a fluidized bed reactor by contacting it with hydrogen at 500°C, with recovery of carbon dioxide and ammonia gases. Subsequently, steam at 650°C is supplied to the fluidized bed to reduce by pyrohydrolysis the fluorine content below 50 ppm wt.

Preparation of uranium metal – as discussed previously, some nuclear power plant reactors such as the UNGG type, have required in the past a nonenriched uranium metal as nuclear fuel. Hence, it was the major consumer of pure uranium metal. Uranium metal can be prepared according to several reduction processes. First of all, it can be obtained by direct reduction of uranium halides (e.g., uranium tetrafluoride) by molten alkali metals (e.g., Na, K) or alkaline-earth metals (e.g., Mg, Ca). For instance, in the Ames process, uranium tetrafluoride, UF_4, is directly reduced by molten calcium or magnesium at 700°C in a steel bomb. Another process has reduced uranium oxides with Ca, Al (thermite process[R], or aluminothermy), or carbon. Thirdly, the pure metal can also be recovered by molten salt electrolysis of a fused bath made of a molten mixture of $CaCl_2$ and NaCl, with a solute of KUF_5 or UF_4. However, like hafnium or zirconium, high-purity uranium can be prepared according to the Van Arkel–DeBoer process, by the hot-wire process which consists of thermal decomposition of uranium halides on a hot tungsten filament (similar in that way to chemical vapor deposition, CVD).

The nuclear fuel cycle – the nuclear fuel cycle is the series of industrial processes which facilitate the production of electricity from crude uranium ores to uranium nuclear fuel in power reactors. The various activities associated with the production of electricity from nuclear reactions are referred to collectively as the nuclear fuel cycle. Actually, after crude uranium ore is mined, ore is processed (concentration, purification, and conversion) until nuclear fuel is obtained. Then the fuel is used for a nuclear power reactor. Electricity is created by using the heat generated in a nuclear reactor to produce steam and drive a turbine connected to a generator. Fuel removed from a reactor, after it has reached the end of its useful life (i.e., spent nuclear wastes), can be reprocessed to produce new fuel. Therefore, the nuclear fuel cycle starts with the mining of crude uranium ores and ends with the disposal of nuclear wastes. With the introduction of reprocessing as an option for nuclear fuel, the stages can now form a true cycle.

3.7.1.6 Industrial Uses and Applications

Uranium-235 is extensively used in nuclear power reactors and nuclear weapons; for instance it is used as an enriched form as nuclear fuel in PWR and BWR, or natural in CANDU. Depleted uranium (DU) is used as a fertile blanket material in fast neutron breeder reactors or is used in non-nuclear applications primarily owing to its high density. For instance, depleted uranium is extensively used in military applications to manufacture armor-piercing ammunitions, in initial guidance devices and gyro-compasses, as counterweights for missile re-entry vehicles, in spacecraft applications as radiation-shielding materials, and, owing to its high atomic number, in X-ray tubes as target cathodes in replacement of molybdenum, copper, or tungsten. Actually, depleted uranium is essentially selected over other dense metals such as iridium, platinum, gold, or tungsten owing to its ability to be easily cast and machined by contrast with refractory metals, and its lower cost than platinum group metals.

3

Less
Common
Nonferrou
Metals

3.7.2 Thorium

3.7.2.1 Description and General Properties

Thorium [7440-29-1], with the chemical symbol Th, the atomic number 90, and the atomic relative atomic mass (atomic weight) 232.0381(1) is the most abundant actinide of Mendeleev's periodic chart. Thorium is named after the Scandinavian god of war, Thor. Pure thorium is a soft, very ductile, silvery white, mildly dense (11,720 $kg.m^{-3}$) and refractory metal (mp 1750°C). By contrast with uranium metal, owing to its cubic lattice structure, thorium metal expands equally in all directions, and hence, it is not subjected to mechanical distortion on thermal cycling. From a crystallographic point of view thorium is a dimorphic element having two allotropes. Actually, the low-temperature alpha phase (α-Th) is face-centered cubic (fcc) and transforms at 1400°C to a beta phase which is body-centered cubic (bcc). Like zirconium, finely divided thorium, such as powder, is highly pyrophoric and may ignite spontaneously in dry air. Moreover, because the alpha–beta transition temperature in thorium is much higher than in uranium, thorium metal reactor fuel has much better dimensional stability than uranium metal. From a chemical point of view, thorium does not readily tarnish in air and keeps its silvery white luster several weeks when exposed to moist air owing to its impervious protective passivating oxide film. However, at temperatures above 200°C, progressive oxidation takes place. Thorium is slowly attacked by water but, however, at temperatures above 178°C, progressive oxidation takes place owing to the spalling off of ThO_2. For instance, the rate of weight loss in pressurized water reactor conditions at 315°C is 560 $g.m^{-2}.h^{-1}$. It is not dissolved by dilute mineral acids (e.g., HCl, HF, HNO_3, H_2SO_4) but readily dissolves in concentrated hydrochloric acid and aqua regia. Thorium is passivated by diluted nitric acid but passivation breakdown occurs if fluoride anions are added. Thorium is not attacked by molten sodium up to 500°C. Thorium reacts with hydrogen gas to form the two hydrides ThH_2 and Th_4H_{15} at temperatures above 250°C. Thorium reacts with nitrogen at temperatures above 670°C to form the nitride ThN. For these reasons, melting of pure thorium must be achieved only under a vacuum or in an inert atmosphere such as helium or argon.

3.7.2.2 History

Thorium was first discovered by Berzelius in 1828. Thorium, like uranium, is a nuclear fuel, but the use of thorium fuel, unlike the use of uranium, has nearly been forgotten. While uranium technology in pressurized water reactors (PWR) has been demonstrated to be dependable for over 30 years and is well understood today, the use of thorium technology has lagged behind uranium ever since the demise of the prototype high-temperature gas-cooled reactor (HTGR) in the US owing to certain operating difficulties. The sustained lack of interest in thorium as a nuclear fuel has resulted in limited research efforts; very few data have been compiled on the subject and even fewer have been published in recent years. Although natural thorium cannot be used to produce a nuclear chain reaction by itself, it can, under irradiation, be converted into the fissile fuel, uranium (^{233}U). Therefore, thorium (^{232}Th) is consequently of potential use in nuclear reactors. Use of thorium in addition to uranium would expand the nuclear fuel supply base. Further, advanced converter reactors using thorium would not generate plutonium. Plutonium produced during nuclear power generation and its recycling raises nuclear proliferation concerns. For these reasons, there were many studies in the 1960s and 1970s to determine the feasibility of using thorium in nuclear power reactors. Studies were focused toward potential applications on HTGR, light water breeder reactors (LWBR), and gas-cooled fast breeder reactors (GCFR). Also, the US Government considered a modified Canadian deuterium–uranium (CANDU) reactor capable of consuming thorium. The operation of the Fort St.-Vrain reactor, the full-scale commercial HTGR, however, became unsuccessful due to a combination of economic factors and lingering mechanical problems that resulted in over

2 years of delays in starting, followed by intermittent operations with a persistently low capacity factor.

3.7.2.3 Natural Occurrence, Minerals, and Ores

Thorium is widely distributed in nature with an Earth's crust abundance of 12 mg.kg^{-1} (i.e., ppm wt) which is about four times greater than that of uranium and abundant as lead and molybdenum. Owing to its chemical similarity with element of group IVB(4) such as zirconium, hafnium, and the other actinide uranium, thorium is usually associated in nature with uranium ores and other rare-earth containing minerals. Concentration of thorium occurs in the following three principal types of ore deposits: (i) vein deposits, (ii) beach or steam placer deposits such as black sands, and (iii) to a lesser extent, disseminated in massive carbonatites, and in igneous or metamorphic rocks. The chief thorium-bearing minerals are the oxide, **thorianite** (ThO$_2$, cubic), the nesosilicate, **thorite** (ThSiO$_4$, tetragonal), which is the highest grade thorium mineral, containing theoretically 63wt% thorium oxide (thoria, ThO$_2$), the orthophosphate, **monazite** ((Ce,La,Y,Th)PO$_4$, monoclinic), and finally **bastnaesite** (La,Ce,Th)CO$_3$F). However, monazite is the most commercially exploited thorium mineral, and contains up to 12wt% ThO$_2$. It is found mostly in the stream placer deposits in Brazil, India, Sri-Lanka, Australia, South Africa, and the United States (e.g., Northern Idaho, North and South Carolina, and beach sands in Green Cove Springs, Florida) along with heavy minerals like ilmenite, rutile, zircon, and sillimanite. Therefore, the coproduction of rare earths, titanium, and zirconium has defrayed much of the cost of extracting thorium. Bastnaesite ore is located in the Mountain Pass District of California. Rare-earth fluorocarbonate mineral and other carbonatite concentrate deposits can be found in the United States (e.g., South Platte District of Colorado, and the Barringer Hill District, Texas). This type of deposit consists of mostly low-grade ores. Nevertheless, although few thorite-containing vein-type deposits have been developed thus far in the United States, high-grade thorite is likely to be exploited in the event a large demand arises for domestic thorium in the near future. Most of the vein deposits are associated with quartz–feldspar iron oxide. Other vein-type thorium minerals include thorianite and bastnaesite. A substantial quantity of thorium also coexists with uranium in Precambrian conglomerates, such as in the Elliot Lake area of Ontario, Canada. The conglomerate deposit in this region is also a potentially important source for the long-term supply of thorium in North America.

3.7.2.4 Processing and Industrial Preparation

The principal steps in producing refined thorium compounds from crude thorium-bearing ores are: (i) the concentration of thorium ores, (ii) the extraction of thorium from concentrate, (iii) the purification or refining, and (iv) finally the conversion to the required metal or thorium compounds.

Mining and mineral dressing – all of the world's thorium is extracted from monazite or produced as a by-product from the processing of monazite for rare earths or as a by-product during processing of black sands minerals for the recovery of titanium and zirconium. Actually, monazite is a minor constituent of black sands ore and when pure it contains an average of 3 to 10wt% ThO$_2$ and 50 to 60wt% rare-earth oxides. Mining of black sands in placer ore deposits is usually carried out by bulldozers and front-end loaders, or suction dredging.

Ore concentration – concentration of monazite from other minerals (e.g., rutile, ilmenite, zircon, sillimanite, and garnet) is achieved by common ore benefication techniques. Firstly, the separation of heavier minerals from silica and other low-density minerals is done by specific gravity separation. Secondly, the heavier minerals are separated by strong electromagnets according to their magnetic susceptibility. During this step, ilmenite, magnetite, monazite (paramagnetism due to its rare-earth contents) and garnet are separated from nonmagnetic residue. The nonmagnetic residue is further treated by froth flotation to recover valuable minerals such as rutile, zircon, and sometimes gold. After a second specific gravity separation, rich monazite concentrate contains now more than 98wt% monazite, with ThO_2 content ranging from 3 to 9wt% according to the black sand location.

Hydrometallurgical concentration processes – because monazite is a relatively chemically inert mineral, only two hydrometallurgical dissolution processes can be efficiently used for recovering thorium. Actually, the hydrometallurgical processing of monazite ore concentrate is carried out either by concentrated sulfuric acid or strong alkaline caustic hot digestion.

Caustic soda digestion process – the caustic soda process has been used on a large commercial scale in Brazil and India. The monazite concentrate is ground under water in a ball-mill to 325 mesh. The fine slurry is fed into a stainless steel reactor containing a strong caustic solution of sodium hydroxide with 73wt% NaOH. The mixture is then heated to 140°C for 3 h or until the reaction of minerals is complete. After that, the mixture is diluted with a solution of sodium hydroxide and sodium triphosphate and digested again for 1 h at 105°C. The digested slurry, apart from trisodium phosphate, contains all the thorium, and rare earths as hydrous hydroxides (i.e., $Th(OH)_4$ and $Ln(OH)_3$). After filtration at 80°C through a sieve with Inconel wire, the filtration cake is washed with warm water. The filtrate, which still contains 40wt% NaOH, is then evaporated in a steel kettle until the sodium hydroxide reaches 47wt% with a boiling point of 135°C and after precipitation of the remaining trisodium phosphate is recycled for later digestion. The hydrated filtration cake is dissolved into 37wt% hydrochloric acid at 80°C for 1 h in a glass-lined vessel. After dissolution, the acid solution and undissolved residue (i.e., monazite, rutile) is then neutralized with recovered caustic soda from the evaporator and diluted with water. Thorium is separated from the rare earths by selective precipitation of thorium hydroxide $Th(OH)_4$ at pH 5.8. The filtered thorium hydroxide still undergoes several wash steps until 99.7wt% pure.

Sulfuric acid digestion process – the sulfuric acid digestion process has been extensively used in Europe, Australia, and the United States. Monazite concentrate is ground to less than 65 mesh, and is digested with 93wt% sulfuric acid at 210°C for 4 h in a stirred reactor. The temperature should be kept below 230°C in order to prevent formation of water-insoluble thorium pyrophosphate, ThP_2O_7. During digestion most of the thorium, rare earths, and uranium go into solution, while insoluble silica and unreacted monazite form the sludge at the bottom of the reactor. The denser monazite is separated from the sludge and recycled. Radium is removed with the sludge by adding barium carbonate before decantation. The barium sulfate removes radium as insoluble radium sulfate. This process produces a solution of thorium, rare earths, and uranium cations with sulfate and phosphate anions. Because of the chemical similarity between cations in the solution with phosphate anions, separation of thorium is a hard task and several routes have been developed. The precipitation of the thorium oxalate has been developed at the Ames laboratory, while the solvent extraction with several organic solvents has been successfully developed for separation from rare earths.

Purification or refining of thorium – thorium produced previously is too impure to be used as nuclear fuel. Actually, impurities such as rare earths and uranium, owing to their large thermal neutron cross-sections are objectionable. Hence, the objective of the thorium refining process is to remove these impurities down to concentrations below $\mu g.kg^{-1}$ (ppb wt). Solvent extraction of an aqueous thorium nitrate solution with n-tributyl phosphate (TBP) in kerosene is a common procedure to perform the refining of thorium. At the end of the purification process, the thorium is recovered in the form of an aqueous solution of thorium nitrate or crystals of hydrated thorium nitrate.

Conversion of thorium nitrate is performed to produce a thorium nuclear fuel required in power reactors. Five thorium compounds can be used as nuclear fuel, these compounds are: (i) thorium metal, (ii) thorium dioxide (thoria), ThO_2, (iii) thorium carbide, ThC_2, (iv) thorium tetrafluoride, ThF_4, and finally, (v) thorium tetrachloride, $ThCl_4$.

Conversion to thoria (ThO_2) is performed by three possible routes: (i) firstly, by thermal dissociation of the nitrate (i.e., denitration) according to the calcination reaction

$$Th(NO_3)_4 \cdot 4H_2O \rightarrow ThO_2 + 4HNO_3 + 2H_2O;$$

(ii) secondly, by precipitation of thorium hydroxide, $Th(OH)_4$, adding ammonia NH_3 to the nitrate solution followed by the calcination of the filtered hydroxide, to produce a colloidal sol of $Th(OH)_4$ suited for preparing thoria by a particular sol-gel process

$$Th(NO_3)_4 + 4NH_4OH \rightarrow Th(OH)_4\downarrow + 4NH_4NO_3$$
$$Th(OH)_4 \rightarrow ThO_2 + 2H_2O;$$

(iii) thirdly, preparation of thoria by precipitation of thorium oxalate dihydrate, $Th(C_2O_4)_2 \cdot 2H_2O$, adding oxalic acid to the thorium nitrate solution followed by the calcination of the filtered oxalate; this process has the advantage of separating thorium from other trace impurities (e.g., uranium, iron and titanium) that remain in the nitric acid solution:

$$Th(NO_3)_4 + 2H_2C_2O_4 + 2H_2O \rightarrow Th(C_2O_4)_2 \cdot 2H_2O\downarrow + 4HNO_3$$
$$Th(C_2O_4)_2 \cdot 2H_2O \rightarrow ThO_2 + 2H_2O + 4CO_2.$$

Conversion to thorium tetrafluoride – this conversion is performed at 566°C by a hydrofluorination exothermic reaction achieved in a series of screw-fed horizontal reactors made of AISI 309Nb stainless steel with a screw of Illium®R or Inconel® in which solid and gases flow countercurrently. During the exothermic reaction the conversion of the thorium dioxide to thorium tetrafluoride occurs with recovery of hydrofluoric acid as commercial by-product according to the scheme:

$$ThO_2 + 4HF \rightarrow ThF_4 + 2H_2O.$$

Conversion to thorium tetrachloride is performed by carbo-chlorination on a mixture of thoria and carbon at 600°C under a flow of chlorine gas. During the exothermic reaction the conversion of the thorium dioxide to thorium tetrachloride follows the scheme below. After production, the thorium tetrachloride must be purified by fractional distillation to liberate it from unreacted solids and other impurities. Because of its high boiling point (942°C) the operation is difficult and an alternative process in which thorium oxalate reacts with excess carbon tetrachloride with traces of chlorine as catalyst has been developed. The process is achieved batchwise in a vertical graphite reactor at 600°C, and produces a pure solid thorium tetrachloride in a single step:

$$ThO_2 + 2C + 2Cl_2 \rightarrow ThCl_4 + 2CO$$
$$\text{or } Th(C_2O_4)_2 + CCl_4 \rightarrow ThCl_4 + 2CO + 3CO_2.$$

Preparation of thorium metal – several industrial processes can produce thorium metal but winning of the metal is more difficult than the preparation of uranium owing to its higher melting point. The main processes developed on a commercial scale are: (i) the molten salt electrolysis of the fused mixture baths $KThF_5$–NaCl, ThF_4–NaCl–KCl, or $ThCl_4$–NaCl–KCl, (ii) the reduction of thorium oxide with calcium, (iii) the Kroll process in which the metallothermic reduction of thorium tetrachloride is performed by sodium, magnesium, or calcium as for titanium, (iv) decomposition of thorium tetraiodide by the Van Arkel–DeBoer or hot-wire process as for zirconium and hafnium.

3.7.2.5 Industrial Uses and Applications

Table 3.69 Thorium applications and uses

Applications	Description
Nuclear energy	Thorium cannot completely replace uranium in pressurized water reactors (PWR). It may only be used to replace ^{238}U as a fertile material. It must be used in conjunction with one of the fissionable materials such as ^{235}U, or ^{239}Pu.
Refractories	Because thoria has the highest melting point of all the oxides, $3300°C$, which is only surpassed by tungsten, or tantalum carbide, it is used as high-temperature refractories.
Lighting engineering	Thorium nitrate has long been used in the manufacture of thoriated Welsbach mantles for incandescent lanterns, including natural portable gas lights and oil lamps. The thoriated mantle consisted of thoria doped with 1wt% CeO_2 giving, when heated in the gas flame, a dazzling white light. Thoriated mantles, however, are no longer being produced due to the development of a suitable thorium-free substitute. Thorium fluoride is used in the manufacture of carbon arc lamps for movie projectors and searchlights to provide a high–intensity light.
Metallurgy	Thorium is an important alloying element in structural magnesium alloys, improving strength and creep resistance at high temperature. Thoria is also used to control grain size growth in tungsten used as filament in lamps or to stabilize the zirconia high-temperature cubic phase.
Electronic devices	Owing to its low work function, low vapor pressure and high melting point, thorium is extensively used in thermionic emitting devices or for coating onto tungsten for electron-emitting tubes or filaments in various electronic applications. Thorium nitrate also is used to produce thoriated tungsten welding electrodes.
Glass industry	At one time thorium oxide was used as coloring pigment in glasses. Because thoria largely increases the refractive index and allows a low dispersion of glass, it is extensively used for manufacturing high-quality optical lenses for scientific instruments.
Instrumentation	Thoria, owing to its high melting point, is used to manufacture high-temperature laboratory crucibles for melting refractory metals.
Chemical engineering	Thoria is an efficient catalyst in the conversion of ammonia to nitric acid, in petroleum cracking, and in the production of sulfuric acid.

References

[1]   Grady HR (1980) Lithium metal for the battery industry. J Power Sources 5: 127–135.
[2]   Saito E, Dirian G (1962) Process for the isotopic enrichment of lithium by chemical exchange. British Patent 902,755 9 Aug.
[3]   Ruedl, E, Coen V, Sasaki, T, Kolbe H (1982) Intergranular lithium penetration of low Ni–Cr–Mn austenitic stainless steels. J Nuclear Mater 110: 28–36.
[4]   Hoffmann EE, Mandly WD (1957) Corrosion resistance of the metal and alloys to sodium and lithium. US Atomic Energy Comm., ORNL-2271, Oak Ridge National Laboratory, 11 pp.
[5]   Beskorovainyi NM, Ivanov VK (1967) Mechanism underlying the corrosion of carbon steels in lithium. In: Emelíyanov VS, Evstyukin AI (eds) High purity metals and alloys. Consultants Bureau, pp 120–129.
[6]   Klueh RL (1974) Oxygen effects on the corrosion of niobium and tantalum by liquid lithium. Met Trans 5: 875–879.
[7]   Klueh RL (1973) Effect of oxygen on the corrosion of niobium and tantalum by liquid lithium. US Atomic Energy Comm. Report ORNL-TM-4069, Oak Ridge National Laboratory.
[8]   Smith DL, Natesan K (1974) Influence of nonmetallic impurity elements on the compatibility of liquid lithium with potential containment materials. Nucl Technol 22: 392–404.
[9]   Singh RN (1976) Compatibility of ceramics with liquid Na and Li. J Am Ceram Soc 59: 112–115.
[10] Weeks ME (1956) Discovery of the elements, 6th edn. J Chem Ed, ACS, Easton, PA, pp 484–490.
[11] Hajek J (1949) French Patent, 8 Oct.
[12] Herbert D, Ulam J (949) French Patent, 26 Nov.
[13] Haicang L, Wei Z (1995) Research, manufacture and applications of lithium metal materials in China. Rare Metals 14: 313–316.
[14] Averill WA, Olson D (1978) A review of extractive processes for lithium from ores and brines. Energy 3: 305–313.

[15] Warren (1896) Chem News 74: 6.

[16] Hanson (1936) US Pat. 2,028,390.

[17] Gunz (1893) Compt Rend Acad Sci 117: 732.

[18] Ruff, Johannsen (1906) Z Electrochem 12: 186.

[19] Muller J, Bauer R, Sermond B, Dolling E (Metallgesellschaft Aktiengesellschaft) (1988) Process and apparatus for producing high-purity lithium metal by fused-salt electrolysis. US Patent 4,740,279, 26 April.

[20] Weeks ME, Leicester HM (1968). Discovery of the elements, 7th edn. Published by the Journal of Chemical Education, Easton, Pennsylvania.

[21] Schwartz, US Pat. 3,170,812 (1970).

[22] Cardarelli F (1999) Scientific unit conversion: a practical guide to metrication, 2nd edn. Springer-Verlag, London.

[23] McIntyre DR, Dillon CP (1986) Pyrophoric behavior and combustion of reactive metals. MTI Publications/NACE.

[24] Kroll WJ (1940) Trans Electrochem Soc 78:35.

[25] Hunter MA (1910) J Am Chem Soc 32: 330.

[26] Standard specification for titanium and titanium alloy strip, sheet, and plate. ASTM B265–95a.

[27] Klaproth MH (1789) Ann Chim Phys 6: 1.

[28] Vauquelin NL (1797) Ann Chim Phys 22: 179.

[29] Berzelius JJ (1824) Ann Chim Phys 26: 43.

[30] Van Arkel AE, DeBoer JH (1925) Z Anorg Chem 148: 345.

[31] Yau TL, and Bird KW (1995) Manage corrosion with zirconium. Chem Eng Prog 91: 42–46.

[32] Marden JW and Rich MN (1927) Vanadium. Ind Eng Chem 19: 786–788.

[33] McKechnie RK and Seybolt AU (1950) Preparation of ductile vanadium by calcium reduction. J Electrochem Soc 97: 311–315.

[34] Boehni H (1967) Corrosion behavior of various rare metals in aqueous acid solutions with special consideration of niobium and tantalum. Schweiz Arch Angew Wiss Tech 33: 339–363.

[35] Bishop CR and Stern M (1961) Hydrogen embrittlement of tantalum in aqueous media. Corrosion 17: 379t–385t.

[36] Von Bolton W (1905) Z Elektrochem 11: 45.

[37] Boehni H (1967) Corrosion behavior of various rare metals in aqueous acid solutions with special consideration of niobium and tantalum. Schweiz Arch Angew Wiss Tech 33: 339–363.

[38] Bishop CR, Stern M (1961) Hydrogen embrittlement of tantalum in aqueous media. Corrosion 17: 379t–385t.

[39] Von Bolton W (1905) Z Elektrochem 11: 45.

[40] Driggs FH, Lilliendahl WC (1931) Preparation of metals powders by electrolysis of fused salts. III Tantalum. Ind Eng Chem 23: 634–637.

[41] Norton Grinding Wheel Co. (1958) Electrolytic production of Ti, Zr, Hf, V, Nb, Ta, Cr, Mo, or W. Brit Pat 792,716.

[42] Ervin G Jr, Ueltz HFG (1958) Apparatus for continuous production of refractory metal by electrolysis of fused salts. US Pat 2,837,478.

[43] Ervin G Jr, Ueltz HFG (1960) Electrolytic preparation of Th, U, Nb, Ta, V, W, Mo, and Cr. Ger Pat 1,078,776.

[44] Ueltz HFG (1960) Electrolytic extraction of refractory metals of groups IV, V, and VI from their carbide. US Pat 2,910,021.

[45] Sarla RM, Schneidersmann EO (1960) Fused salt electrolytic cell for producing high-melting reactive metals, such as tantalum. US Pat 2,957,816.

[46] Union Carbide Corporation (1966) Cell for plating heat resistant metals from molten salt mixtures. Neth Appl 6,516,263.

[47] Horizon Titanium Corp. (1957) Electrodeposition of Ti, Zr, Hf, V, Ta, and Nb. Brit Pat 778,218.

[48] Horizon Titanium Corp. (1958) Formation of hard intermetalic coatings from electrodeposited layers of refractory metals. Brit Pat 788,804.

[49] Horizon Titanium Corp. (1958) Electrodeposition of Ti, Zr, Hf, Ta, V, Nb, Cr, Mo, and W. Brit Pat 788,295.

[50] Horizon Titanium Corp. (1958) Fused-salt bath for electrodeposition of multivalent metals: Ti, Nb, Ta, and V. Brit Pat 791,151.

[51] Merlub-Sobel M, Arnoff MJ, Sorkin JL {1959) Chlorination and electrolysis of metal oxides in fused salts baths. US Pat 2,870,073.

[52] Wainer E (1959) Transition-metal halides for electrodeposition of transition metals. US Pat 2,894,886.

[53] Timax Corp. (1960) Electrolytic preparation of pure niobium and tantalum. Brit Pat 837,722.

[54] Hubert K, Fost E (1960) Preparation of niobium and tantalum by melt electrolysis. Ger Pat 1,092,217.

[55] Kern F (1961) Electrolytic production of niobium and tantalum. US Pat 2,981,666.

[56] Kern F (1962) Tantalum powders by electrolysis. Ger Pat 1,139,284.

[57] Scheller W, Blumer M (1962) Niobium and tantalum. Ger Pat 1,139,982.

[58] Pruvot E (1959) Electrolytic manufacture of tantalum. Fr Pat 1,199,033.

[59] Société Général du Vide (1964) Protective coating. Neth Appl 6,400,547.

[60] Nishio Y, Oka T, Ohmae T (1971) Steel plated with tantalum or a tantalum alloy. Ger Pat 2,010,785.

[61] Paschen P, Koeck W (1990) Fused salt electrolysis of tantalum. In: Refractory metals: Extraction, processing, and applications. The Minerals, Metals, and Minerals Society, TMS Publishing, Warrendale, Pennsylvania pp 221–230.

[62] Mellors GW, Senderoff S (1964) Electrolytic deposit of refractory metals. Belg Pat 640,801.

[63] Danzig IF, Dempsey RM, La Conti AB (1971) Characteristic of tantalided and hafnided samples in highly corrosive electrolyte solutions. Corrosion 27: 55–62.

[64] Christopher D (1961) Bimetallic pipe. Mech Eng 83: 68–71.

[65] Whiting KA (1964) Cladding copper articles with niobium or tantalum and platinum outside. US Pat 3,156,976.
[66] Cardarelli F, Taxil P, Savall A (1996) Tantalum protective thin coating techniques for the chemical process industry: molten salts electrocoating as a new alternative. Int J Refract Metals Hard Mater 14: 365–881.
[67] Grams WR (1968) Cladding the cleaned reactive refractory metals with lower-melting metals in the absence of a reactive atmosphere. US Pat 3,409,978.
[68] Krupin AV (1968) Rolling metal in vacuum. Mosk Inst Stali Splavov 52: 153–163.
[69] Chelius J (1968) Explosion-clad sheet metal for corrosion-resistant chemical equipment. Weikst Korros 19: 307–312.
[70] Bergmann OR, Cowan GR, Holtzman AH (1970) Metallic multilayered composites bonded by explosion detonation shock. US Pat 3,493,353.
[71] Glatz B (1970) Explosive cladding of metals. Huntn Listy 25: 398–406.
[72] Bouckaert GP, Hix HB, Chelius J (1974) Explosive-bonded tantalum-steel vessels. DECHEMA Monograph 76: 9–22.
[73] Umanshii YaS, Urazaliev US, Ivanov RD (1972) Formation of tantalum thin films prepared by cathodic sputtering. Fiz Metal Metalloved 33: 196–199.
[74] Fitzer E, Kehr D (1973) Processing studies of the chemical vapor deposition of niobium and tantalum. Proc 4th Int Conf CVD, pp 144–146.
[75] Spitz J, Chevallier J (1975) Comparative study of tantalum deposition by chemical vapor deposition and electron beam vacuum evaporation. Proc 5th Int Conf CVD, pp 204–216.
[76] Spitz J (1973) Proc 5th European congr corrosion, Paris.
[77] Beguin C, Horrath E, Perry AJ (1977) Tantalum coating of mild steel by CVD. Thin Solid Films 46: 209–212.
[78] Bobst J (1971) Niobium and tantalum electroplating. Ger Pat 2,064,586.
[79] Lantelme F, Inman D, Lovering DG (1984) Electrochemistry – I. In: Lovering DG, Gale RJ (eds.) Molten salt techniques, vol. 2. Plenum Press, New York, pp 138–220.
[80] Sadoway DR (1990) The synthesis of refractory-metal compounds by electrochemical processing. In: Non aqueous media in refractory metals: extraction, processing, and applications. The Minerals, Metals, and Materials Society, TMS Publishing, Warrendale, Pennsylvania, pp 213–220.
[81] Delimarskii IuK, Markov BF (1961) Electrochemistry of fused salts. Sigma Press Publishing, New York.
[82] Cook NC (1969) Metalliding. Sci Amer 221: 38–46.
[83] Cook NC (1962) Beryllide coatings on metals. US Pat 3,024,175.
[84] Cook NC (1962) Boride coatings on metal. US Pat 3,024,176.
[85] Cook NC (1962) Silicide coatings on metals. US Pat 3,024,177.
[86] Cook NC (1966) Corrosion-resistant chromide coating. US Pat 3,232,853.
[87] Ilyushchenko NG, Antinogenov AI, Belyaeva GI et al. (1968) Tr 4-ogo Vsesoyuzn Soveshch, 105.
[88] Mellors GW, Senderoff S (1964) Electrolytic deposit of refractory metals. Belg Pat 640,80!.
[89] Mellors GW, Senderoff S (1968) Novel compounds of tantalum and niobium (1968) US Pat 3,398,068.
[90] Mellors GW, Senderoff S (1969) Electrodeposition of Zr, Ta, Nb, Cr, Hf, W, Mo, V and their alloys. US Pat 3,444,058.
[91] Balikhin VS (1974) Electroplating of protective tantalum coating. Zasch Metal 10: 459–460.
[92] Balikhin VS, Sukhoverkov IN (1974) Tantalum electroplating. Tsvet Metal 3: 70–71.
[93] Noddack W, Tacke I (1925) Naturwiss 13: 57–574.
[94] Cardarelli F (1998) Scientific unit conversion: a practical guide to metrication, 1st edn. Springer-Verlag, London p 31.
[95] Lunge G, Naville J (1879) Traité de la grande industrie chimique, tome I. Acide sulfurique et oléum. Masson & Cie, Paris.
[96] Péligot EM (1841) CR Acad Sci 12: 735.
[97] Péligot EM (1842) Ann Chim Phys 5: 5.

Further Reading

Lithium

Addison CC (1984) The chemistry of the liquid alkali metals. Wiley, Chichester, UK.
Bach RO, and Wasson JR (1984) Lithium and lithium compounds. In: Kirk-Othmer encyclopedia of chemical technology, vol. 14. Wiley Interscience, New York, pp 448–476.
Chaudron G, Dimitrov C, and Dubois B (1977) Monographies sur les métaux de haute pureté. vol. 3 Groupes 1b, 4b, 5b. Masson & Cie, Paris.
Foltz GE (1993) Lithium metal. In: McKetta JJ (ed.) Inorganics chemical handbook, vol. 2. Marcel Dekker, New York.
Gabano JP (1983) Lithium batteries. Academic Press, London.
Greenwood NN, and Earnshaw A (eds) (1984) Chemistry of the elements. Pergamon Press, New York, pp 75–116.
Mahi P, Smeets AA, Fray DJ, and Charles JA (1986) Lithium: metal of the future. J Met 38: 20–26.
Meyer RJ, Pietsch E (eds) (1960) Lithium. In: Gmelin's Handbuch der Anorganischen Chemie (System Number 20) 8th edn. Springer-Verlag, Heidelberg.
Ober JA (1997) Lithium. In: The mineral yearbook, vol. 1, 1996 US Geological Survey (USGS).
Roskill Information Services Ltd (1994) The economics of lithium, 7th edn. Roskill Information Services Ltd, UK.
Whaley TP (1973) Sodium, potassium, rubidium, caesium, and francium. In: Trottman-Dickenson, AF (ed.) Comprehensive inorganic chemistry, vol. 1. Pergamon Press, Oxford, chap. 6, pp 369–529.

Sodium

Abramson R, Delisle J-P, Elie X, Salon G, and Peyrelongue J-P. (1981) Apparatus for the purification of a liquid metal for cooling in the core of a fast neutron reactor. US Patent 4,278,499, 14 Jul.

Chaudron G, Dimitrov C, Dubois B (1977) Monographies sur les métaux de haute pureté, vol. 3. Groupes 1b, 4b, 5b. Masson & Cie, Paris.

Dumay J-J, Malaval C (1987) Device for purifying liquid metal coolant for a fast neutron nuclear reactor. US Patent 4,713,214, 15 Dec.

Foust OJ (1972) Sodium-NaK engineering handbook, vol. 1. Sodium chemistry and physical properties. Gordon & Breach.

Foust OJ (1976) Sodium-NaK engineering handbook, vol. 2. Sodium flow, heat transfer, intermediate heat exchangers, and steam generators. Gordon & Breach.

Foust OJ (1978) Sodium-NaK engineering handbook, vol. 3. Sodium systems, safety, handling, and instrumentation. Gordon & Breach.

Foust OJ (1978) Sodium-NaK engineering handbook, vol. 4. Sodium pumps, valves, piping, and auxiliary equipment. Gordon & Breach.

Foust OJ (1979) Sodium-NaK engineering handbook, vol. 5. Sodium purification, materials, heaters, coolers, and radiators. Gordon & Breach.

Hundal R (US Energy Research and Development Administration) (1976) Liquid metal cold trap. US Patent 3,962,082, 8 Jun.

Pascal P, Chrétien A. (eds) (1966) Lithium, sodium. Nouveau traité de chimie minérale, tome 2/1, Masson & Cie, Paris.

Sittig M (1956). Sodium, its manufacture, properties and uses. American Chemical Society (ACS) Monograph Series (No. 133). Reinhold Publishing Corp., New York.

Whaley TP (1973) Sodium, potassium, rubidium, caesium, and francium. In: Trottman-Dickenson AF (ed.) Comprehensive inorganic chemistry, vol. 1. Pergamon Press, Oxford, chap. 6, pp 369–529.

Potassium

Greer JS et al. (1982). Potassium. In: The Kirk-Othmer encyclopedia of chemical technology, vol. 18. Wiley-Interscience, New York, pp 912–920.

Whaley TP (1973). Sodium, potassium, rubidium, caesium and francium. In: Trottman-Dickenson AF (ed.). Comprehensive inorganic chemistry, vol. 1. Pergamon Press, Oxford, chap. 6, pp 369–529.

Rubidium

Hampel CA (1961) In: Rare metals handbook, 2nd ed. Reinhold, New York, pp 434–440.

Cesium

Wessel FW (1959–62) Minor metals and minerals: cesium and rubidium. In: Mineral yearbook vol.1, US Geological Survey, Washington, DC.

Beryllium

Kjellgren (1954) Beryllium. In: Hampel, CA (ed.) Rare metals handbook. Reinhold Publishing Company, New York.

Pinto NP, Greenspan J (1968) Beryllium. In: Gonser BW (ed.) Modern materials, vol. 6. Academic Press, New York.

Stonehouse AJ, Marder JM (1995) Beryllium. In: ASM metals handbook, 10th edn. Vol. 2, Properties and selection: nonferrous alloys and special-purpose materials. ASM, Ohio Park, pp 683–687.

Magnesium and Magnesium Alloys

Avedesian M, Baker H (eds) (1998) ASM specialty handbook: magnesium and magnesium alloys. ASM International, Metal Park, OH.

Kipouros GJ, Sadoway, DR (1987) The chemistry and electrochemistry of magnesium production. In: Mammantov G, Mamantov CB, Braunstein J (eds) Advances in molten salts, vol. 6. Elsevier, Amsterdam, pp 127–209.

Strelets KL (1998) Electrolytic production of magnesium. International Magnesium Association, McLean, VA.

Calcium, Barium, and Strontium

Mantell CL (1968) Calcium. In: Hampel CA (ed.) Encyclopedia of chemical elements. Reinhold, New York, pp 94–103.
Mantell CL (1973) The alkaline earth metals: calcium, barium, and strontium. In: Hampel CA (ed.) Rare metals handbook, 2nd edn. Reinhold, New York, pp 15–25.
Mantell CL, Hardy C (1945) Calcium metallurgy and technology. Reinhold, New York.

Titanium and Titanium Alloys

Barksdale (1966) Titanium, its occurrence, chemistry and technology. Reinhold, New York.
Boyer R, Collings EW, Welsh G (1994) Materials properties handbook: titanium alloys. ASM Books, Ohio Park.
Bringas JE (1995) The metals red book, vol. 2: Nonferrous metals. CASTI Publishing, Edmonton, Canada.
Destefani JD (1995) Introduction to titanium and titanium alloys. In: ASM handbook of metals series, vol. 2: Properties and selection: Nonferrous alloys and special-purpose materials 10th edn. ASM, Ohio Park, pp 586–591.
Donachie MJ Jr (ed.) (1988) Titanium: a technical guide. ASM Books, Ohio Park.
Everhart (1954) Titanium and titanium alloys. Reinhold, New York.
Eylon D, Newman JR (1995) Titanium and titanium alloys castings. In: ASM handbook of metals series, vol. 2: Properties and selection: Nonferrous alloys and special-purpose materials 10th edn. ASM, Ohio Park, pp 634–646.
Lampman S (1995) Wrought titanium and titanium alloys. ASM Handbook of Metals Series, vol. 2: Properties and selection: nonferrous alloys and special-purpose materials 10th edn. ASM, Ohio Park, pp 592–633.
Schutz RW, Thomas DE (1995) Corrosion of titanium and titanium alloys castings. In: ASM handbook of metals series, vol. 2: Properties and selection: Nonferrous alloys and special-purpose materials 10th edn. ASM, Ohio Park, pp 669–706.
Smallwood RE (ed.) (1984) Refractory metals and their industrial applications. ASTM STP 849, ASTM, Philadelphia.
Timet Company (1997) Titanium and titanium alloys. Titanium Metals Corporation, Denver, CO, USA.
Titanium Industries Inc. (1998) Titanium: data and reference manual. Titanium Industries Incorporated, March.

Zirconium and Hafnium and Their Alloys

Blumenthal WB (1958) The chemical behavior of zirconium. Van Nostrand, Princeton.
Hedrick JB (1996) Zirconium. In: Mineral yearbook 1996. US Geological Survey.
Larsen (1970) Zirconium and hafnium chemistry. Advan Inorg Chem Radiochem 13: 1–333.
Lustman B, Kevse F Jr (eds) (1955) Metallurgy of zirconium. McGraw-Hill, New York.
Schemel JH (1977) Manual on zirconium and hafnium – STP 639. American Society for Testing and Materials (ASTM).
Thomas DE, Hayes ET (1960) The metallurgy of hafnium. Naval reactors, Division of Reactor Development, US Atomic Energy Commission.
Webster RT (1995) Zirconium and hafnium. In: ASM metals handbook, 9th edn, vol. 2: Properties and selection of nonferrous alloys and special purpose materials. ASM, Metal Park, OH, pp 661–721.
Webster RT (1995) Surface engineering of zirconium and hafnium. In: ASM metals handbook, 9th edn, vol. 2: Properties and selection of nonferrous alloys and special purpose materials. ASM, Metal Park, OH, pp 825–855.
Webster RT, Yau TL (1995) Corrosion of zirconium and hafnium. In: ASM metals handbook, 9th edn, vol. 2: Properties and selection of nonferrous alloys and special purpose materials. ASM, Metal Park, OH, pp 707–721.

Niobium and Niobium Alloys

Balliett RW (1986) Niobium and tantalum in material selection. J Metals September 25–27.
Bringas JE (ed.) (1993) The metals red book, vol. 2, nonferrous edition. CASTI Publishing.
Fairbrother, F (1967) Chemistry of niobium and tantalum. Elsevier, New York.
Hampel CA (ed.) (1967) Rare metals handbook, 2nd edn. Reinhold Publishing Corp., New York.
Kumar P (1988) High purity niobium for superconductor applications. J Less Common Metals 139: 149–158.
Lambert JB (1991) Refractory metals and alloys. In: ASM handbook of metals series, 9th ed., vol. 2: Properties and selection of nonferrous alloys and special-purpose materials. American Society of Metals (ASM), Ohio Park, pp 557–585.
Machlin I, Begley RT, Weisert ED (eds) (1968) Refractory metal alloys, metallurgy and technology. Plenum Press, New York.
Miller GL (1959) Tantalum and niobium. Academic Press, New York.

Niobium. Teledyne Wah Chang Albany Technical Note No. TWCA-9209NB (1992) Teledyne Wah Chang, Albany, OR.

Niobium and biobium alloys. Cabot Technical Note No. 506-95-2.5M (1996) Cabot Performance Material Inc., Boyertown, PA.

Payton PH (1984) Niobium and niobium compounds. In: Kirk-Othmer encyclopedia of chemical technology, 3rd edn, vol. 15. Wiley-Interscience, New York, pp 820–840.

Sisco FT, Epremian E (1963) Columbium and tantalum. John Wiley & Sons, New York.

Smallwood RE (1984) Use of refractory metals in chemical process industries. In: Smallwood RE (ed.) Refractory metals and their industrial applications. ASTM STP 849, ASTM, Philadelphia, pp 106–104.

Webster RT (1984) Niobium in industrial applications. In: Smallwood RE (ed.) Refractory metals and their industrial applications. ASTM STP 849, ASTM, Philadelphia pp 18–27.

Wojcik CC (1998) High-temperature niobium alloys. Adv Mater Processes 12: 22–31.

Tantalum and Tantalum Alloys

Bringas JE (eds) (1993) The metals red book, vol. 2, Nonferrous edition. CASTI Publishing.

Droegkamp RE, Schussler M, Lambert JB et al. (1984) Tantalum and tantalum compounds. In: Kirk-Othmer encyclopedia of chemical technology, 3rd edn., vol. 22. Wiley-Interscience, New York, pp 541–564.

Fairbrother F (1967) Chemistry of niobium and tantalum. Elsevier, New York.

Hampel CA (ed.) (1967) Rare metals handbook, 2nd edn. Reinhold Publishing Corp., New York.

Lambert JB (1991) Refractory metals and alloys. In: ASM Handbook of metals series, 9th edn, vol. 2: Properties and selection of nonferrous alloys and special-purpose materials. American Society of Metals (ASM), Ohio Park, pp 557–585.

Machlin I, Begley RT, Weisert ED (eds) (1968) Refractory metal alloys, metallurgy and technology. Plenum Press, New York.

Miller GL (1959) Tantalum and niobium. Academic Press, New York.

Sisco FT, Epremian E (1963) Columbium and tantalum. John Wiley & Sons, New York.

Smallwood RE (1984) Use of refractory metals in chemical process industries. In: Smallwood RE (ed.) Refractory metals and their industrial applications. ASTM STP 849, ASTM, Philadelphia, pp 106–104.

Tantalum and tantalum alloys. (1996) Cabot technical note no. 505-95-5M. Cabot Performance Material Inc., Boyertown, PA.

Tungsten and Tungsten Alloys

Yih SWH, Wang CT (1979) Tungsten: sources, metallurgy, properties, and applications. Plenum Press, New York.

Silver and Silver Alloys

Butts A, Coxe CD (1967) Silver: economics, metallurgy, and use. Van Nostrand, New York.

Platinum Group Metals

Duval C (1958) Platine. In: Pascal P (ed.) Nouveau Traité de chimie minérale, tome XIX: Ru-Os-Rh-Ir-Pd-Pt. Masson & Cie, Paris, pp 725–741.

Howe (ed.) (1949) Bibliography of the platinum metals. Baker and Co., Newark.

Rare-Earth Metals

Callow RJ (1968) The industrial chemistry of the lanthanons, yttrium, thorium, and uranium. Pergamon Press, New York.

Cotton SA (1991) Lanthanides and actinides. Macmillan, London.

Greenwood NN and Earnshaw, A (1997) Chemistry of the elements, 2nd edn. Butterworth-Heinman, London.

Uranium

Bellamy RG, Hill, NA (1963) Extraction and metallurgy of uranium, thorium, and beryllium. Macmillan, New York.

Benedict M, Pigford TH, Levi HW (1981) Nuclear chemical engineering, 2nd edn. McGraw-Hill, New York.
Callow RJ (1968) The industrial chemistry of the lanthanons, yttrium, thorium, and uranium. Pergamon Press, New York.
Cleg JW, Foley DD (1958) Uranium ore processing. Addison-Wesley, Reading Massachusetts.
Cordfunke EHP (1969) The chemistry of uranium. Elsevier, New York (1969).
Harrington CD, Ruehle AR (1959) Uranium production technology. Van Nostrand, Princeton.
Merritt RC (1971) The Extractive Metallurgy of Uranium. Colorado School of Mines Research Institute, Boulder, Colorado.
Roubeault M, Jurain G (1958) Geologie de l'uranium. Masson & Cie, Paris.

Thorium

Bellamy RG, Hill NA (1963) Extraction and metallurgy of uranium, thorium, and beryllium. Macmillan, New York.
Callow RJ (1968) The industrial chemistry of the lanthanons, yttrium, thorium, and uranium. Pergamon Press, New York.
Ross AM (1958) Thorium, production technology. Addison-Wesley, Reading MA.
Smith JF (ed.) (1975) Thorium: preparation and properties. Iowa State University Press.
Wilhelm HA (ed.) (1958) The metal thorium. American Society for Metals, Cleveland, OH.

4

Semiconductors

4.1 Band Theory of Bonding in Crystalline Solids

The theory of the chemical bonding in crystalline solids such as pure metals and alloys, insulators and semiconductor materials may be well understood by an expansion of the **linear combination of atomic orbitals** (LCAO). In this theory the atomic orbitals (AO) of two atoms can be combined together in order to form bonding and antibonding molecular orbitals (MO) symbolized by σ and σ^* respectively. In the case of three neighboring atoms, it creates a string of atoms with bonding that connects all three. Hence there appear a bonding orbital, an antibonding orbital, and a new orbital called a nonbonding orbital. Essentially a nonbonding orbital is an orbital that neither increases nor decreases the net bonding energy in the molecule. The important feature here is that three atomic orbitals must produce three molecular orbitals. Hence, the total number of orbitals must remain constant. If we apply this concept by considering combinations of four atoms, it will give four molecular orbitals, two bonding and two antibonding. Notice that the two bonding and two antibonding orbitals have not exactly the same energy. The lower bonding orbital is slightly more bonding than the other and symmetrically one antibonding orbital is slightly more antibonding than the other. As a general rule, if we consider a large number of atoms, N, where N could have an order of magnitude similar to that of Avogadro's number, it will lead to the combination of a large number of bonding and antibonding orbitals. These orbitals will be so close together in energy that they begin to overlap creating a definite band of bonding (i.e. highest occupied (HO) energy band or **valence band**) and a band of antibonding orbitals (i.e., lowest unoccupied (LU) or **conduction band**), the empty energy region between the valence and conduction bands is called the **energy band gap**. These definitions arise because electrons that enter the antibonding band are free to move about the crystal under an electric field strength (i.e., electrical conduction). It is this existence of valence and conduction bands that explains the electrical and optical properties of crystalline solids. The **Fermi level** with its energy E_F is a level at which the probability of an electron occupying it is 1/2. The Fermi level is the highest occupied state at absolute zero (i.e., -273.15 °C).

4.2 Electrical Classification of Solids

Atoms of a metal have many unoccupied levels with similar energies. A large number of mobile electric charge carriers are able to move across the material when an electrical potential difference (i.e., voltage) is applied. In a semiconductor or an insulator, the valence band is completely filled with electrons in bonding states, so that conduction cannot occur. There are no vacant levels of similar energy on neighboring atoms. At absolute zero, its antibonding states (i.e., the conduction band) are completely empty, with no electrons there able to conduct electricity. For this reason, insulators cannot conduct. In the case of semiconductors, as temperature increases, electrons in the valence band acquire, due to the Brownian thermal motion, sufficient kinetic energy to be promoted across the energy band gap into the conduction band. When this occurs, these promoted electrons can move and conduct electricity. Therefore, the narrower the energy band gap, the easier it is for electrons to move to the conduction band. Hence, according to the band theory, in crystalline solids, it is possible to classify solids into three distinct categories of materials: (i) if the solid presents a large gap in energy between valence and conduction band (i.e., above 3.0 eV or 290 kJ.mol^{-1}), the material is called an electrical **insulator**, (ii) if the energy gap is between 0.0 and 3.0 eV (i.e., 190 to 290 kJ.mol^{-1}), the material is said to be a **semiconductor**, and (iii), if there is effectively no gap (i.e., zero gap) or a gap below 0.01 eV between the bands, the materials are called **conductors** (i.e., pure metals and alloys). For instance, orders of magnitude of energy band gaps are: 5.3 eV (511 kJ.mol^{-1}) for diamond type I, showing its excellent electrical insulating properties, while it is 1.04 eV (i.e., 100 kJ.mol^{-1}) for pure silicon (Si) monocrystal used as semiconductors, and 0.69 eV (i.e., 67 kJ.mol^{-1}) for pure germanium (Ge) crystals; the two former examples are both intrinsic semiconductors. Moreover, electrical conductivity of materials is strongly temperature dependent. Actually, as the temperature increases, the conductivity of metals decreases, while the electrical conductivity of pure semiconductors and insulators increases as well.

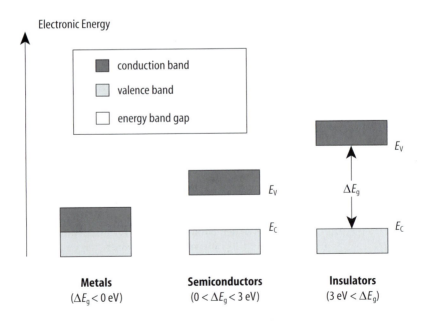

According to very general trends of properties in the periodic table, it can be shown that chemical elements increase in their **metallic character** when displacing to the left and bottom of the table. Thus, it would be expected that the most metallic of elements would be found at the lower left corner of the table and the least metallic at the upper right. However, there is a gradual transition of properties from metallic to nonmetallic when moving to the top right of the periodic table. This rule of the thumb can help to compare quickly the electrical properties of the elements. Actually, the two intermediate chemical elements of the group IVA (14) of the periodic table exhibit properties that are intermediate between metallic (e.g., Sn, Pb) and nonmetallic (e.g., C, diamond) and hence can be characterized as semiconductors (e.g., Si, Ge). Actually these elements have a cubic diamond type space lattice and both pure silicon and germanium behave as perfect insulators at absolute zero temperature ($-273.15°C$), but at moderate temperatures their resistance to the flow of electricity decreases measurably. Since they never become good conductors, they are classified as electrical semiconductors (sometimes called **semimetals** or **metalloids** in old textbooks).

4.3 Semiconductor Classes

Semiconductors can therefore be defined as crystalline or amorphous solid materials which carry an electric current by electromigration of both electrons and holes and have an energy band gap between 0.0 and 3.0 eV. The main characteristic of semiconductors, by contrast with metals, is the exponential rise of electrical conductivity with increase of temperature. Moreover, another important property of semiconductors is the ability to decrease their electrical resistivity at a given temperature by **doping**, i.e., introducing a definite amount of traces of electrically active impurities. As a general rule, semiconductors have electrical resistivities between 10 $\mu\Omega$.cm to 10 MΩ.cm. However, the semiconductors group can also be split into three main groups: (i) **intrinsic** or **elemental semiconductors**, (ii) extrinsic **doped** or **p-type semiconductors**, and extrinsic **doped** or **n-type semiconductors**, and finally (iii) **extrinsic** or **compound semiconductors**.

4.3.1 Intrinsic or Elemental Semiconductors

Intrinsic semiconductors are solids having an energy band gap between 0 and 3 eV and that are electrical insulators under normal conditions, but can become good electrical conductors in special circumstances such as a temperature increase or under electromagnetic irradiation. Intrinsic semiconductors are particular pure single elements of the group IVA(14) of Mendeleev's periodic chart: silicon (Si), germanium (Ge), and alpha-tin (α-Sn) with intentional doping and having the diamond crystal space lattice structure. Owing to their electronic configuration, the atoms in this class of semiconductors have exactly enough outershell electrons $ns^2 np^2$ to fill the valence or bonding band, while the conduction band remains totally empty. If an incident beam of electromagnetic radiation (UV, visible light) with the appropriate wavelength irradiates a silicon monocrystal, the light release is a sufficient quantum of energy that the electrons can jump from valence to conduction bands. Since electrons in the antibonding bands are free to move under an electric field this behavior leads to electrical conductivity of the crystal. For instance, this property has led to the use of silicon materials as photosensitive sensors in electronic circuits and to the manufacture of photovoltaic devices such as solar cells.

4.3.2 Extrinsic Doped Semiconductors

Over the last 30 years, inorganic chemists and materials scientists have recognized that it is possible to introduce trace levels of impurities (**dopants**) into silicon or germanium monocrystals that have little or no effect on their crystal lattice structure, but have beneficial effects on their electrical properties. In order to accomplish **doping**, the selected atoms must have an atomic radius similar to that of silicon atoms and satisfy the **Hume–Rothery rules** (atoms with two atomic radii differing no more than 15%), but have either one more or one less electron than does silicon. However, doping cannot be performed to the point where it disturbs the crystalline structure of the host semiconductor. Hence, doping can be achieved in the range of parts per million (i.e., mg.kg^{-1} or ppm wt) concentrations, but may be up to a few parts per thousand (‰). A semiconductor doped to several parts per thousand has a conductivity close to that of low-conductivity metal. According to the electronic configuration of the dopant, doping can produce two types of semiconductors or subgroups depending upon the element added.

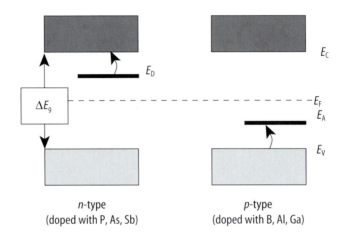

n-type
(doped with P, As, Sb)

p-type
(doped with B, Al, Ga)

Extrinsic or doped semiconductors

The **n-type semiconductors** are doped intrinsic semiconductors in which the dopant is a pentavalent element, for instance of the group VA(15) of the periodic chart such as arsenic (As), antimony (Sb), or phosphorus (P). The substitutional impurities will give a supplementary electron owing to its ns^2np^3 electronic configuration containing five rather than four outershell electrons. Therefore, the density of holes in the valence band is exceeded by the density of electrons in the conduction band. A hole is a mobile electron vacancy in a semiconductor that acts like a positive electron charge with a positive mass. Then, the n-type behavior is induced by doping with addition of pentavalent element impurities acting as **donor**, without distortion of the crystal structure of silicon or germanium. Owing to the fact that electrons are the **majority carriers** in this particular semiconductor, it is called **n-type**, while holes are **minority carriers**. The energy level of the donor electron is located just below the conduction band.

The **p-type semiconductors** are doped semiconductors in which the dopant is a trivalent element, for instance of the group IIIA(13) of the periodic chart such as boron (B), aluminum (Al), or gallium (Ga). The substitutional impurities will trap an electron owing to its ns^2np^1 electron configuration containing three rather than four outershell electrons. Therefore, the density of electrons in the conduction band is exceeded by the density of holes in the valence band. The p-type behavior is induced by trivalent element impurities acting as **acceptor**, without distortion of the crystal structure of silicon or germanium. Owing to the fact that holes are the majority carriers in this particular semiconductor, it is called **p-type**. The energy level of the acceptor electron is located just above the valence band.

4.3.3 Compound Semiconductors

These are chemical compounds for which the only required condition is to exhibit an average of four electrons in the valence band. The standard notation of these compounds is defined in the following way: it consists of the combination of the chemical group number in the periodic table of each element, with in some cases the stoichiometric coefficient in subscript, for instance, I–VII (e.g., CuCl), II–VI (e.g., CuS), III–V (e.g., AlAs, GaAs, GaP), I–III–VI$_2$ (e.g., CuGaTe$_2$), and I$_2$–VI (e.g., Ag$_2$S, Cu$_2$O).

4.3.4 Grimm–Sommerfeld Rule

In order to define quickly and exactly if a doped or compound semiconductor having a fraction y of impurity B, i.e., A_xB_y, is an n-type or a p-type, it is possible to use the following practical rule called the **Grimm–Sommerfeld rule**. For this purpose, it is important to introduce the dimensionless physical quantity called the average number of valence electrons, denoted n, and which is calculated by the following equation:

$$n = (xn_A + yn_B)/(x + y)$$

where x and y are the stoichiometric coefficients, n_A and n_B are the number of valence electrons of the constitutive atoms. The semiconductors are classified according to the numerical value of the average number of valence electrons:

$n = 4$, intrinsic semiconductor;
$n < 4$, extrinsic semiconductor p-type;
$n > 4$, extrinsic semiconductor n-type.

For instance, perfect stoichiometric semiconductor compounds such as GaAs and InAs exhibit an average number of valence electrons $n = 4$ and hence are pure intrinsic III–V semiconductors. In the case of GaAs doped with Zn, i.e., $Ga_{1-x}AsZn_x$ the average number of valence electrons is $n = (8 - x)/2$, and hence upon substitution of gallium atoms by zinc atoms with a fraction x slightly above zero the compound will be a p-type semiconductor, while GaAs doped with Te, i.e., $Ga_{1-x}AsTe_x$ exhibits $n = (8 + 3x)/2$, and hence the substitution of gallium atoms by tellurium atoms with a fraction x slightly above zero will mean that the compound will be an n-type semiconductor.

4.4 Semiconductor Physical Quantities

The **concentration of electric charge carriers** in the intrinsic semiconductors is represented by particle densities of the electrons, n, and holes, p, both expressed in m^{-3} (cm^{-3}). It can be calculated from the theory of energy bands in solids assuming that the energy difference between the Fermi level and the band edges is larger in comparison with the thermal motion kT. Therefore, the statistical distribution of charge carrier follows a classical Boltzmann distribution:

$$n = N_C \exp[-(E_C - E_F)/kT],$$
$$p = N_V \exp[-(E_F - E_V)/kT],$$

where N_C and N_V are respectively the electron and hole state densities in the conduction and valence band both expressed in m^{-3} (cm^{-3}). E_F, E_C, and E_V represent the energies in J (eV) of the Fermi level, the conduction-band edge and the valence-band edge respectively. The effective densities of states at the top edge of the valence band, N_V, and at the bottom edge of the conduction band, N_C can be defined as follows:

$$N_C = 2(2\pi m_n^* kT)^{3/2}/h^3,$$

$$N_V = 2(2\pi m_p^* kT)^{3/2}/h^3,$$

where T is the absolute thermodynamic temperature in K, h is Planck's constant in J.s, and k Boltzmann's constant in J.K^{-1}. The two new quantities m_n^* and m_p^* are the effective masses in kg (g), of the electrons and holes respectively.

The product of the electron and hole densities does not depend on the Fermi level and can be expressed as:

$$np = N_C N_V \exp[-(E_C - E_V)/kT] = N_C N_V \exp[-\Delta E_g/kT]$$

with $\Delta E_g = E_C - E_V$ the energy band gap.

In the case of intrinsic semiconductors $n = p = n_i$, thus the charge carrier density can be expressed as a single equation for both electrons and holes:

$$n_i = (N_C N_V)^{1/2} \exp[-\Delta E_g/2kT].$$

For instance, for silicon the charge carrier density is roughly 1.4×10^{10} cm^{-3}, while for germinium it is 1.8×10^6 cm^{-3}.

In the case of extrinsic semiconductors for which traces of impurities are added intentionally by doping in order to modify their electrical properties, **concentration of donors** (e.g., P, As, or Sb) is denoted N_D, while **concentration of acceptors** (e.g., B, Al, or Ga) is denoted N_A. In order to calculate the carrier concentration in this kind of semiconductor, it is necessary to use the electric charge neutrality equation:

$$p_n + N_D = n_n + N_A \ (n\text{-type}),$$

$$p_p + N_D = n_p + N_A \ (p\text{-type}),$$

where p_n and n_n are the hole and electron densities, m^{-3}(cm^{-3}), in the n-type semiconductors, while p_p and n_p are the hole and electron densities in the p-type semiconductors. Therefore combining the above equations leads to:

$$n_n = N_D - N_A \text{ and } p_n = n_i^2/(N_D - N_A) \ (n\text{-type}),$$

$$p_p = N_A - N_D \text{ and } n_p = n_i^2/(N_D - N_A)(p\text{-type}).$$

It can be seen that a reduction in charge carriers can be obtained in materials with low differential $N_D - N_A$. This lowering can be achieved, for instance, by adding donors to p-type semiconductors. Such a material is called compensated intrinsic.

4.5 Transport Properties

4.5.1 Electromigration

When an external electric field with strength E, in V.m^{-1}, is applied to a semiconductor, it induces a drift of the electric charge carriers (i.e. electrons and holes) denoted i. At a low electric field strength the drive velocity of the charge carrier i is directly proportional to the electric field strength according to the following equation:

$$v = u_i \times E.$$

The proportional coefficient u_i (sometimes denoted μ_i) is the intrinsic electric mobility of the charge carrier expressed in V^{-1}.m^2.s^{-1}. If we express the overall flux of all the electric charge carriers through the semiconductor, J, in $m^{-2}.s^{-1}$, which is the summation of the flux of the holes and of the electrons:

$$J = J_n + J_p = (n \times v_n + p \times v_p)$$

then the overall electric current density victor j, in $A.m^{-2}$, is the product of the overall flux and the elementary charge e, in C, and is expressed by the equation:

$$j = j_n + j_p = e(n \times v_n + p \times v_p) = e(n \times \mu_n + p \times \mu_p) \times E.$$

Introducing the generalized form of Ohm's law of electric conduction, $j = \sigma \times E$, where σ is the electrical conductivity, in $S.m^{-1}$, the electrical conductivity of the semiconductor is given by the equation:

$$\sigma = e(n\mu_n + p\mu_p).$$

For intrinsic semiconductors, because $n = p = n_i$ the equation can be simplified to:

$$\sigma = en_i(\mu_n + \mu_p)$$

and hence the Naperian logarithm of the electrical conductivity, plotted as a function of the absolute temperature, is a linear function with a negative slope equal to $-E_g/2k$:

$$\ln \sigma = A - E_g/2kT$$

while for extrinsic semiconductors the logarithm plot shows: (i) an intrinsic semiconductor behavior at low temperature with a negative slope $-E_g/2k$, (ii) a plateau for intermediate temperature with a constant conductivity, and (iii) an extrinsic behavior with a negative slope equal to $-(E_g - E_D)/2k$, for n-type doped semiconductors, and $-(E_A - E_g)/2k$ for p-type doped semiconductors.

The mobility of electrons and holes is affected by two main scattering mechanisms: (i) chemical impurities, (ii) and lattice scattering. The mobility temperature dependence due to impurity scattering is proportional to $T^{3/2}$, while the mobility due to lattice scattering is proportional to $T^{-3/2}$. The total mobility μ is related to the individual components as:

$$1/\mu = 1/\mu_C + 1/\mu_L.$$

However, lattice defects also scatter electrons. This scattering is small and has only a slight temperature dependence.

4.5.2 Diffusion

Existence of gradients of the electric charge carrier densities causes an electric current to flow from concentrated regions to depleted regions in addition to the normal flow caused by an electric field. In this case Fick's first law of diffusion applies and electron and hole diffusion fluxes can be expressed according to the following equations:

$$J_n = -D_n \times \nabla n \text{ and } J_p = -D_p \times \nabla p$$

where D_n and D_p are the diffusion coefficients of electrons and holes respectively, expressed in $m^{-2}.s^{-1}$. These diffusion coefficients are related to mobilities according to Einstein's equations:

$$D_n = \mu_n(kT/e) \text{ and } D_p = \mu_p(kT/e).$$

4.5.3 Hall Effect

When an external magnetic induction, B, expressed in tesla, T, is applied across a current-carrying material, it will force the moving electric charge carriers to crowd to one side of the conductor. An electric field with strength E_H, in V.m^{-1}, will develop as a result of this accumulation. For instance, the relation of the electric current density, j_x, in A.m^{-2}, along the x axis and the perpendicular applied magnetic induction is given by the following equation:

$$E_H = R_H j_x \times B_y$$

where the proportional quantity, R_H, is the Hall coefficient of the material, expressed in Ωm.T^{-1}. The n-type semiconductors have a negative Hall coefficient equal to $-1/en$, while p-type semiconductors have a positive Hall coefficient equal to $1/ep$.

4.6 Semiconductor Physical Properties

Semiconductor physical properties are shown in Table 4.1 (pages 254–257).

4.7 Applications of Semiconductors

Table 4.2 Selected applications of semiconductors

Effect	Application	Semiconductors
Electroluminescence	Light displays	GaAs, GaP, InAs, InSb, SiC
Gunn effect	High-frequency generation and amplification	GaAs, InP, CdTe
Laser effect	Laser diode	GaAs, InAs, InSb
Photovoltaic effect	Photocell, solar cells	Si, Ge
Piezoelectric effect	Electroacoustic amplifier	GaAs, CdS, CdSe
Piezoresistance	Pressure indicator	Si, Ge
Transistor effect	Amplifier	Si, Ge
Tunnel effect	High-frequency switch, oscillator, amplifier	Si, Ge, GaAs
Varactor effect	Parametric amplification, tunneling diode	Si, Ge, GaAs

4.8 Some Common Semiconductors

4.8.1 Silicon

4.8.1.1 Description and General Properties

Amongst the modern semiconductors used today, silicon is the most common. Silicon, with the chemical symbol Si, is the second element of the group IVA(14) of the periodic chart with the atomic number 14 and the relative atomic mass (i.e., atomic weight) 28.0855(3). Silicon is a crystalline gray brittle metalloid with a metallic luster and a cubic diamond space lattice structure ($a = 543.072$ pm). Owing to its larger energy-band gap (1.170 eV) than its chemical neighbor germanium, it can be used for higher-temperature operations. As a general rule pure silicon can be doped with donors such as elements of group VA(15): P, As, and Sb in order to give extrinsic n-type semiconductors, while it can give extrinsic p-type semiconductors when doped with elements of group IIIA(13): B, Al, and Ga. Doping is usually achieved by thermal diffusion, or ion implantation. Elemental silicon has a transmittance more than 95% for infrared radiations with wavelengths ranging from 1.3 to 6.0 μm. Moreover, pure silicon is sensitive to irradiation by nuclear radiations such as X-rays and gamma-rays, creating recombination centers and increasing surface state densities. From a chemical point of view, owing to its protective dioxide film, SiO_2, which spontaneously forms in oxidizing media containing oxygen, silicon is a relatively inert element. However, it is readily attacked by hydrogen fluoride, hydrofluoric acid, HF, ammonium dihydrogenofluoride, $NH_4H_2F_2$, gaseous halogens (i.e., F_2, Cl_2, Br_2, and I_2) and diluted alkaline solutions. Silicon is named after the Latin, silex, silicis, meaning flint.

4.8.1.2 History

4

Semi-
conductors

In 1800 the British scientist Sir Humphrey Davy thought silica (i.e., silicon dioxide, SiO_2) to be a chemical compound and not an element. Later in 1811, the two French chemists, Gay Lussac and Thenard probably prepared impure amorphous silicon by reducing silicon tetrafluoride, SiF_4 with potassium metal. In 1824, the Swedish chemist Jöns Berzelius, generally credited with the discovery, prepared amorphous silicon by the same general method but greatly improved it by purifying the product by removing the fluosilicates formed by repeated washings. In 1854, the French chemist and metallurgist Sainte-Claire Deville first prepared crystalline silicon by reducing pure acid-washed silica sand with carbon in an electric-arc furnace, using rods of carbon as electrodes.

4.8.1.3 Natural Occurrence

Silicon is present in the sun and stars and is a main component of a class of meteorites known as aerolites. It is also a component of tektite, a natural glass of uncertain origin. Silicon abundance in the Earth's crust is approximately 25.7wt%, and hence, it is the second most abundant element after oxygen. It was for this reason that the Austrian geologist Suess called the lithosphere "SiAl" for its high silicon and aluminum content. In nature, owing to its strong chemical reactivity for oxygen, silicon is not found free as a native element, but occurs chiefly as silicon dioxide, SiO_2 or silica, the basic element of the largest class of minerals: the silicates. Quartz, rock crystal, amethyst, agate, flint, jasper, and opal are some of various varieties of silica-bearing minerals. Quartz, feldspars both orthoclases and plagioclases, peridots, pyroxenes, amphiboles, micas, clays, and zeoliths, are few of the numerous rock-forming silicate minerals. Moreover, silicon is important to plant and animal life. Diatoms in both fresh and salt water extract silica from the water to build their cell walls. Silica is present in the ashes of plants and in the human skeleton.

4.8.1.4 Preparation

Several methods can be used for preparing the pure element. Polycrystalline silicon for electronics uses can be produced using three common methods: (i) the reduction of silicon tetrachloride with zinc, (ii) the reduction of trichlorosilane with hydrogen, and (iii) the thermal decomposition of monosilane. In method (i), silicon tetrachloride, $SiCl_4$, is firstly obtained from chlorination of pure silica sand by carbon and chlorine. Then a high-purity silicon tetrachloride is produced by rectification (i.e., fractional distillation). Simultaneously, zinc is purified by vaporization. Then both chemicals are added to the heated quartz reaction vessel. Needle-like polycrystalline silicon is grown inside the furnace as a result of the chemical reaction:

$$SiCl_4 + Zn = Si + ZnCl_2.$$

The zinc chloride is removed by water leaching.

Table 4.1 Semiconductor physical properties

Semiconductor	Relative atomic or molecular mass (M_r)	Crystal lattice type	Lattice parameters (pm)	Semiconductor type	Density (ρ, kg.m^{-3})	Specific heat capacity (c_p, J.kg^{-1}.K^{-1})	Melting point (mp, °C)	Linear coefficient of thermal expansion (α, $10^{-6}K^{-1}$)	Thermal conductivity (k, W.m^{-1}.K^{-1})	Min. energy band gap at 298.15 K (E_g, eV)	Mobility electrons (μ_e, cm^2.s^{-1}.V^{-1})	Mobility holes (μ_h, cm^2.s^{-1}.V^{-1})	Dielectric permittivity (ϵ_r)	Index of refraction (n_D at 589 nm)
α-Sn	118.710	Diamond	$a = 649.12$	IV–IV	5765	213	232	5.4	n.a.	0.80	2500	2400	24	2.750
AgBr	187.772	Sphalerite	$a = 577.45$	I–VII	6473	270	432	n.a.	n.a.	2.500	4000	n.a.	12.4	2.253
AgI	234.773	Sphalerite	$a = 650.2$	I–VII	5670	n.a.	558	n.a.	n.a.	2.220	30	n.a.	10.0	2.22
AgI	234.773	Wurtzite	$a = 458.0$ $c = 749.4$	I–VII	n.a.	n.a.	558	n.a.	n.a.	2.630	n.a.	n.a.	n.a.	n.a.
Al$_2$S$_3$	150.161	Wurtzite	$a = 357.90$ $c = 582.90$	III$_2$–VI$_3$	2550	n.a.	1127	n.a.	n.a.	4.100	n.a.	n.a.	n.a.	n.a.
Al$_2$Se$_3$	290.843	Wurtzite	$a = 389.90$ $c = 630.00$	III$_2$–VI$_3$	3910	n.a.	977	n.a.	n.a.	3.100	n.a.	n.a.	n.a.	n.a.
AlAs	101.903	Sphalerite	$a = 566.22$	III–V	3810	n.a.	1740	3.5	84	2.230	1200	420	8.5	n.a.
AlN	40.988	Wurtzite	$a = 311.10$ $c = 497.8$	III–V	3260	n.a.	n.a.	n.a.	n.a.	6.020	n.a.	n.a.	n.a.	n.a.
AlP	57.955	Sphalerite	$a = 545.10$	III–V	2420	n.a.	1827	n.a.	92	2.450	80	n.a.	11.6	n.a.
AlSb	148.742	Sphalerite	$a = 613.55$	III–V	4218	n.a.	1057	4.2	60	1.696	200	550	10.1	3.2
BAs	85.733	Sphalerite	$a = 477.70$	III–V	n.a.	n.a.	2027	n.a.	n.a.	1.500	n.a.	n.a.	n.a.	n.a.
BeS	41.078	Sphalerite	$a = 486.50$	II–VI	2360	n.a.	n.a.	n.a.	n.a.	4.170	n.a.	n.a.	n.a.	4.17
[BeS...]	87.0??	Sphalerite	$a = 513.90$	II–VI	4315	n.a.	n.a.	n.a.	n.a.	3.610	n.a.	n.a.	n.a.	3.61

4 — Semi-conductors

BN	24.818	Sphalerite	$a = 361.50$	III–V	3490	793	3030	n.a.	20	4.600	n.a.	n.a.	7.1	n.a.
BP	41.785	Sphalerite	$a = 453.80$	III–V	2970	n.a.	2530	n.a.	n.a.	2.200	500	70	11.6	n.a.
C	12.0107	Diamond	$a = 356.68$	IV–IV	3510	471.5	3577	1.18	990	5.480	1800	1400	5.7	2.419
CdO	128.410	Sphalerite	n.a.	II–VI	n.a.	n.a.	n.a.	n.a.	n.a.	2.500	n.a.	n.a.	n.a.	n.a.
CdS	144.477	Sphalerite	$a = 583.20$	II–VI	4838	330	1477	4.7	20	2.420	340	18	5.4	2.32
CdS	144.477	Wurtzite	$a = 413.48$ $c = 674.90$	II–VI	4820	n.a.	1750	n.a.	n.a.	2.420	350	40	9.12	2.32
CdSe	191.371	Sphalerite	$a = 605.00$	II–VI	5740	255	1239	3.8	9	1.740	650	50	10.0	n.a.
CdSe	191.371	Wurtzite	$a = 429.90$ $c = 701.10$	II–VI	5660	n.a.	1239	n.a.	31.6	1.740	900	50	n.a.	n.a.
CdTe	240.011	Sphalerite	$a = 647.70$	II–VI	5867	205	1092	4.9	5.85	1.440	1200	50	7.2	2.50
CdTe	240.011	Wurtzite	$a = 457.00$ $c = 747.00$	II–VI	n.a.	n.a.	n.a.	n.a.	n.a.	1.500	650	50	11.0	n.a.
CuBr	143.450	Sphalerite	$a = 569.06$	I–VII	4980	381	497	15.4	12.5	2.960	n.a.	n.a.	7.9	2.12
CuCl	98.999	Sphalerite	$a = 540.57$	I–VII	3530	490	422	12.1	0.84	3.202	n.a.	n.a.	7.9	1.93
CuI	190.450	Wurtzite	$a = 431.00$ $c = 709.00$	I–VII	n.a.	n.a.	n.a.	n.a.	n.a.	2.630	n.a.	n.a.	n.a.	n.a.
CuI	190.450	Sphalerite	$a = 660.43$	I–VII	5630	276	605	19.2	1.68	3.060	n.a.	n.a.	6.5	2.346
Ga$_2$Te$_3$	522.246	Sphalerite	$a = 589.90$	III$_2$–VI$_3$	5750	n.a.	790	n.a.	4.7	1.350	50	n.a.	n.a.	n.a.
GaAs	144.645	Sphalerite	$a = 565.32$	III–V	5318	n.a.	1237	5.4	56	1350	8800	500	10.4	3.30

Table 4.1 (continued)

Semiconductor	Relative atomic or molecular mass (M_r)	Crystal lattice type	Lattice parameters (pm)	Semiconductor type	Density (ρ, kg.m^{-3})	Specific heat capacity (c_p, J.kg^{-1}.K^{-1})	Melting point (mp, °C)	Linear coefficient of thermal expansion (α, $10^{-6}K^{-1}$)	Thermal conductivity (k, W.m^{-1}.K^{-1})	Min. energy band gap at 298.15 K (E_g, eV)	Mobility electrons (μ_e, cm^2.s^{-1}.V^{-1})	Mobility holes (μ_h, cm^2.s^{-1}.V^{-1})	Dielectric permittivity (ϵ_r)	Index of refraction (n_D at 589 nm)
GaN	83.730	Wurtzite	$a = 319.00$ $c = 518.90$	III–V	6100	n.a.	1230	n.a.	65.6	3.503	n.a.	n.a.	n.a.	n.a.
GaP	100.697	Sphalerite	$a = 545.05$	III–V	4131	n.a.	1477	5.3	75.2	2.350	300	150	8.5	3.20
GaSb	191.483	Sphalerite	$a = 609.54$	III–V	5619	320	712	6.1	27	0.670	4000	1400	14.0	3.8
Ge	72.612	Diamond	$a = 565.75$	IV–IV	5324	322	938	5.8	64	0.670	3900	1820	16	3.99
HgS	232.656	Sphalerite	$a = 585.17$	II–VI	7730	210	1547	n.a.	n.a.	2.275	250	n.a.	n.a.	2.85
HgSe	279.550	Sphalerite	$a = 608.40$	II–VI	8250	178	797	5.46	1.0	0.300	20,000	1.5	n.a.	n.a.
HgTe	328.190	Sphalerite	$a = 646.23$	II–VI	8170	164	670	4.6	2.0	0.150	25,000	350	n.a.	n.a.
In$_2$Te$_3$	612.436	Sphalerite	$a = 615.00$	III$_2$–VI$_3$	5800	n.a.	667	6.9	n.a.	1.040	50	n.a.	n.a.	n.a.
In$_2$Te$_3$	612.436	Wurtzite	n.a.	III$_2$–VI$_3$	n.a.	n.a.	n.a.	n.a	n.a.	1.100	n.a.	50	n.a.	n.a.
InAs	189.740	Sphalerite	$a = 605.84$	III–V	5667	268	942	4.7	29	0.360	33,000	460	11.7	3.5
InN	128.825	Wurtzite	$a = 353.3$ $c = 569.3$	III–V	6880	n.a.	927	n.a.	n.a	1.950	n.a.	n.a.	n.a	n.a.
InP	145.792	Sphalerite	$a = 586.88$	III–V	4791	n.a.	1057	4.6	80	1.270	4600	150	12.4	3.1
InSb	236.578	Sphalerite	$a = 647.88$	III–V	5778	144	525	4.7	16	0.165	78,000	750	15.7	3.96
Mg$_2$Ge	121.220	Antifluorite	$a = 638.00$	II$_2$–IV	3099	n.a.	1115	n.a.	n.a.	0.740	520	110	n.a.	n.a.

Compound														
Mg_2Sn	167.320	Antifluorite	a = 676.50	II_2–IV	3530	n.a.	778	n.a.	n.a.	0.360	320	260	n.a.	n.a.
MnTe	182.538	Wurtzite	a = 407.8, c = 670.1	II–VI	6701	n.a.	n.a.	n.a.	n.a.	1.000	n.a.	n.a.	n.a.	n.a.
PbS	239.262	Halite	a = 593.62	IV–VI	7597	n.a.	1177	n.a.	n.a.	0.500	600	200	17.0	n.a.
PbSe	286.160	Halite	a = 612.43	IV–VI	8275	n.a.	1067	n.a.	n.a.	0.370	1000	900	23.6	n.a.
PbTe	334.800	Halite	a = 645.40	IV–VI	8272	n.a.	907	n.a.	n.a.	0.260	1600	600	30.0	n.a.
Si	28.0855	Diamond	a = 543.07	IV–IV	2328	702	1414	2.49	124	1.170	1900	475	11.8	3.490
SiC (-b)	40.096	Sphalerite	a = 434.80	IV–IV	3210	n.a.	2797	n.a.	n.a.	2.300	400	50	10.0	2.697
Te	127.600	γ–Se	a = 445.65, c = 592.68	VI–VI	6237	201	450	18.2	5.93	0.330	1100	560	n.a.	n.a.
ZnO	81.389	Wurtzite	a = 324.95, c = 520.69	II–VI	5660	n.a.	1975	n.a.	n.a	3.200	180	n.a.	n.a.	n.a.
ZnO	81.389	Sphalerite	a = 463.00	II–VI	5675	494	1975	2.9	23.4	3.438	180	n.a.	n.a.	n.a.
ZnS	97.456	Sphalerite	a = 540.93	II–VI	4090	472	1850	6.36	25.1	3.540	180	5	5.2	2.356
ZnS	97.456	Wurtzite	a = 381.40, c = 625.76	II–VI	4100	n.a.	1850	n.a.	n.a.	3.670	n.a.	n.a.	n.a.	n.a.
ZnSe	144.350	Sphalerite	a = 566.76	II–VI	5266	339	1515	7.2	14	2.580	540	28	9.2	2.89
$ZnSnAs_2$	333.943	Sphalerite	a = 585.10	II–IV–V2	5530	n.a.	777	n.a.	7.6	0.700	n.a.	n.a.	n.a.	n.a
$ZnSnP_2$	246.048	Sphalerite	a = 565.00	II–IV–V2	n.a.	n.a.	927	n.a	n.a.	2.100	n.a.	n.a.	n.a.	n.a
ZnTe	192.990	Sphalerite	a = 610.10	II–VI	5645	264	1295	8.19	10.8	2.260	340	100	10.4	3.56

Method (ii) (also called thermal decomposition or chemical vapor deposition) is new. After distillation, the refined ultrapure trichlorosilane, $SiHCl_3$, is mixed with hydrogen gas and the gas mixture fed into a heated reaction vessel. A heating filament made of pure tantalum is heated by an electric current. The resulting purified polycrystalline silicon is deposited and grown on the surface of an electrically heated tantalum metal hollow wick according to the reaction:

$$SiHCl_3 + H_2 = Si + HCl + H_2.$$

Pure hydrogen can be easily refined in order to obtain high-purity hydrogen by diffusion through a palladium membrane, and this operation unit can also allow use of hydrogen gas produced by catalytic cracking of methanol. A separation method by deep cooling for waste hydrogen and trichlorosilane recovery was also introduced. The Czochralski or the vacuum float zone processes are commonly used to produce single monocrystals of silicon used for solid-state semiconductor devices.

4.8.1.5 Applications and Industrial Uses

Silicon is mainly used as base material processed by planar technology to produce Si-wafers fully described in the next section. Hyperpure silicon can be doped with boron, gallium, phosphorus, or arsenic to produce silicon for use in transistors, solar cells, rectifiers, and other solid-state devices which are used extensively in the electronics and space-age industries. Hydrogenated amorphous silicon has shown promise in producing economical cells for converting solar energy into electricity. Apart from electronics applications silicon and silicon compounds are extensively used in industry. In the form of silica sand and clays it is used to make concrete and brick; it is a useful refractory material for high-temperature work, and in the form of silicates it is used in making enamels and pottery. Silica, as sand, is a principal ingredient in glassmaking. Silicon is an important ingredient in steel; silicon carbide (well known under the common tradename Carborundum®) is one of the most important abrasives and has been used in lasers to produce coherent light of 456 nm. Silicones are important products of silicon. They may be prepared by hydrolyzing a silicon organic chloride, such as dimethyl silicon chloride. Hydrolysis and condensation of various substituted chlorosilanes can be used to produce a very great number of polymeric products, or silicones, ranging from liquids to hard, glasslike solids with many useful properties. Miners, stonecutters, and others engaged in work where siliceous dust is breathed in large quantities often develop a serious lung disease known as silicosis.

Cost (1998) – regular grade silicon (99wt% Si) is priced at about 500 $US.kg^{-1} (227 $US.lb^{-1}). Pure grade silicon (99.9wt% Si) costs about 1100 $US.kg^{-1} (i.e., 499 $US.lb^{-1}), while hyperpure silicon may cost as much as 3215 $US.kg^{-1} (1458 $US.lb^{-1}).

4.8.2 Germanium

4.8.2.1 Description and General Properties

Germanium, with the chemical symbol Ge, the atomic number 32 and the relative atomic mass (i.e., atomic weight) 72.61(2) is, like silicon, an element of the group IVA(14) of the periodic chart. Germanium is a gray-white metalloid with a cubic diamond space lattice structure ($a = 565.754$ pm), and in its pure state is crystalline and brittle, retaining its luster in air at room temperature. Germanium is, after silicon, the second most important semiconductor material which, owing to its energy band gap of 0.67 eV, can be used at temperatures higher than 80°C. Nevertheless, by contrast with silicon, germanium cannot be processed as wafers in planar technology owing to the chemical instability of germanium dioxide (GeO_2). Zone-refining techniques have led to production of crystalline germanium for semiconductor use with an impurity of only one part in 10^{10}. However, germanium, in which the transistor effect was first observed in 1948,[1] has limited uses in electronics compared with its homolog silicon. The two main reasons are: (i) its scarcity compared to silicon; actually its elemental abundance in the Earth's crust is very low (1.5 mg.kg^{-1}) compared to the widespread occurrence of silicon (25.7wt%) in the lithosphere, and moreover (ii) definite germanium ores are rare and only occur dispersed as trace impurities in ores such as sphalerite (ZnS). Therefore, almost all germanium is recovered from zinc smelters and to a lesser extent from copper smelters.

4.8.2.2 History

The Russian chemist Dimitri Mendeleev predicted the existence of the element in 1871. Owing to its similarity with silicon, he suggested the name ekasilicon, with the symbol Es. His predictions for the properties of germanium are remarkably close to reality. Finally the element was discovered in 1886 in a mineral called **argyrodite** by the German chemist Clemens Winkler. Therefore, it was named after the Latin name for Germany, Germania.

4.8.2.3 Natural Occurrence

Germanium element distribution is noted for: (i) its scarcity compared to silicon; actually its elemental abundance in the Earth's crust is very low (1.5 mg.kg^{-1}), and (ii) definite germanium ores are rare such as **argyrodite** (Ag_8GeS_6, orthorhombic), **briartite** ($Cu_2(Zn,Fe)GeS_4$, tetragonal) or **germanite** ($Cu_{26}Fe_4Ge_4S_{32}$, cubic) and usually it only occurs dispersed as trace impurities in lead or silver ores such as **sphalerite** or **zinc blende** (ZnS, cubic) and less commonly in many coals worldwide.[2] Therefore, almost all germanium is recovered from zinc smelters and to a lesser extent from copper smelters and from coal power-generating plants. Actually, when zinc or copper ores are fired and coal is burned, the germanium tends to concentrate in the fly ashes and flue dust produced.

4.8.2.4 Preparation

As described previously, the element is commercially obtained from the dusts of smelters processing zinc ores, as well as recovered from combustion by-products of certain coals. Actually, a large reserve of the elements for future uses is insured in coal sources. Germanium can be separated from other metals by fractional distillation of its volatile tetrachloride, $GeCl_4$. The techniques permit the production of germanium of ultrahigh purity. Actually, zone-refining techniques have led to production of crystalline germanium for semiconductor use with an impurity of only one part in 10^{10}.

4.8.2.5 Applications and Industrial Uses

Germanium doped with arsenic, gallium, or other elements, is used as a transistor element in thousands of electronic applications. The most common use of germanium is as a semiconductor. Germanium is also finding many other applications including use as an alloying agent, as a phosphor in fluorescent lamps, and as a catalyst. Germanium and germanium oxide are transparent in the infrared and are used in infrared spectroscopes and other optical equipment, including extremely sensitive infrared detectors. The high index of refraction and dispersion properties of its dioxide have made germanium useful as a component of wide-angle camera lenses and microscope objectives. The field of organo-germanium chemistry is becoming increasingly important. Certain germanium compounds have a low mammalian toxicity, but a marked activity against certain bacteria, which makes them useful as chemotherapeutic agents. **Cost** (1998) – The cost of germanium (purity 99.9999wt%) is about 3000 $US.kg^{-1} (1361 $US.lb^{-1}).

4.8.3 Other Semiconductors

Graphite is the hexagonal allotrope of carbon, which is the first element of the group IVA(14) of the periodic table. By contrast with carbon's other allotrope, diamond, which is a perfect insulator with an energy band gap of 5.3 eV, graphite can act as an *n*-type or *p*-type semiconductor by appropriate doping. Nevertheless, owing to its space lattice structure having an aromatic ring along the basal plane it exhibits a strong anisotropy, i.e., its electrical conductivity is greater along this plane and extremely low in the direction perpendicular to it. Other well-known semiconductors are: (i) **GaAs, InP,** and **CdTe** which have a wide direct energy band gap (1.35 eV, 1.27 eV, and 1.44 eV respectively), a complex valence band edge structure (warped surfaces) and finally a high electron mobility for such wide energy gap; (ii) **InSb, InAs,** with a narrow direct energy band gap (0.165 and 0.36 eV), a complex valence band edge structure (warped surfaces), a very high electron mobility with small effective mass; (iii) **GaSb,** with a moderately wide direct energy band gap (0.67 eV), low-lying subsidiary conduction band minima, and a complex valence band edge structure (warped surfaces); (iv) **GaP,** with a very wide energy band gap (2.35 eV) and a complex band edge structure similar to that of silicon; (v) **PbTe** with a narrow direct energy band gap (0.26 eV), a complex band edge structures with many valleys and high charge carrier mobilities.

4.9 Semiconductor Wafer Processing

The first step in semiconductor manufacturing begins with production of a **wafer,** i.e., a thin, round slice of a semiconductor material, usually silicon used in integrated circuit (IC, or chip) manufacturing. Generally, the process involves the creation of eight to 20 patterned layers on and into the substrate (i.e., the base semiconductor), ultimately forming the complete integrated circuit. This layering process creates electrically active regions in and on the semiconductor wafer surface. The overall process of manufacturing semiconductors or integrated circuits typically consists of more than a hundred steps, during which hundreds of copies of an integrated circuit or formed on a single wafer.

4.9.1 Monocrystal Growth

Because, in general, semiconductor grade devices cannot be fabricated directly from polycrystalline silicon, the first step in the wafer manufacturing process is the formation of a large, silicon single crystal (monocrystal) or ingot form. Silicon single crystal ingots are usually produced by either the Czochralski (CZ) method, or by the float zone (FZ) method. The material for these processes is purified polycrystalline silicon, produced from, for instance, the reduction of distilled silicon tetrachloride by hydrogen itself produced by chlorination of pure silica sand (see silicon preparation above).

The **Czochralski crystallization process** (CZ) or **pulling crystal growth technique** is the most frequently used method for producing large single monocrystals of pure silicon but also of germanium or gallium arsenide (GaAs). The CZ method begins by melting at 1400°C high-purity silicon with minute amount of a dopant (e.g., As, B, P, and Sb) in a highly pure quartz crucible under an inert gas atmosphere of argon. The crystal growth is initiated by dipping a small cylindrical piece of pure solid silicon (i.e., seed) in the molten silicon bath. At the same time, the crystal rod and the crucible are rotated in opposite directions. After thermal equilibrium is reached, the seed is slowly extracted or pulled upwards from the melt so that it grows, with a constant diameter and the liquid cools to form a single monocrystal ingot with the same crystallographic orientation as the seed. The temperature of the melt as well as the speed of extraction govern the final diameter of the ingot. The surface tension between the seed and molten silicon causes a small amount of the liquid to rise with the seed and cool. There are two heating methods: one is resistance heating using graphite resistors, and the other is induction heating using high-frequency waves. The main characteristics of the silicon monocrystal produced in this process is an uniform resistivity distribution and a dislocation-free monocrystal. Single crystals up to 200 mm diameter can be produced with this technique. More recently, the Sony Corp. has greatly improved the CZ crystal growth by applying a strong horizontal magnetic field to the molten silicon; this technique is called MCZ (Magnetic Field Applied CZ). The magnetic field serves to tighten the thermal convective current as each atom in the silicon melt is arrested and upward mobility prevented. The resulting single crystalline silicon has a lower oxygen concentration, more suitable for some device manufacturing, and a more homogeneous impurity gradient. After cooling, the monocrystal ingot is then ground to a uniform diameter, cut into thin wafers using a diamond saw blade, and processed.

The **float zone** (FZ) method for producing single crystal takes place like CZ in an inert gaseous atmosphere, keeping a polycrystalline rod and a seed crystal vertically face to face. Both are partially melted by a high ratio frequency ratio power inducted heating at the molten zone liquid phase. At the next step, this molten zone is rotated gradually upwards with the seed crystal until the entire polycrystalline rod has been converted to single crystal. Along with efforts to create a crystal rod with a larger diameter, research energy has also been devoted to achieving a narrower molten zone and attention also given to the coil shape. Thus, the FZ method compared to the CZ method is more difficult with larger diameters and proves to be extremely unstable and unpredictable. Even the slightest vibration often means that the whole process fails. Compared to the CZ method, the FZ method has a reduced yield and lower electric power efficiency.

4.9.2 Wafer Production

The wafer is processed through a series of machines, where it is ground smooth and chemically polished to a mirror-like finish (i.e., luster). The wafers are then ready to be sent to the wafer fabrication area where they are used as the starting material for manufacturing integrated circuits. The heart of semiconductor manufacturing is the wafer fabrication facility where the integrated circuit is formed in and on the wafer. The fabrication process, which takes place in a clean room, involves a series of principle steps described below. Typically it takes from 10 to 30 days to complete the fabrication process.

Shaping – the three respective slicing, lapping and etching operations are commonly referred to under the generic term of wafer shaping. However, before the monocrystal ingots are ready for slicing, the tops and tails must first be removed and the ingots ground to a uniform diameter with an orientation flat. This orientation flat is used by device makers to gage the crystal orientation.

(i) **Slicing** – wafer slicing can be done either with an outer or an inner diamond saw; however, in most cases an inner saw is used, allowing for less kerf on the blade, and fewer irregularities in the final shape.

(ii) **Lapping** – in order to increase symmetry and remove surface roughness from saw cuts and process damage, the sliced wafer (SW) is now mechanically lapped. A mixture of abrasive corundum (Al_2O_3) powder and water is fed through a chute onto two conversely rotating base plates. The wafers are simultaneously rotated separately and within a orbiting carrier which creates four-way rotation and makes for a very smooth finished product.

(iii) **Etching** – the surfaces of the lapped wafers (LW) are then etched to remove any remaining embedded abrasive particles, microcracks or surface damage introduced in the previous lapping stage giving etched wafers (HW). Etching is done chemically using a corrosive mixture of nitric acid (HNO_3) and glacial acetic acid (CH_3COOH) solution. This acid surface dissolution technique is preferred in Japan, while an alkaline etching method which uses a caustic aqueous solution of sodium hydroxide (NaOH) is used in the US.

Polishing – wafers are fixed to a hard ceramic base plate using a wax bonding method. These are then lowered against a synthetic-leather polishing pad, attached to a metal plate and buffed to a state of minimum roughness. This method improves parallelism and creates the surface mirror. The entire process is automated from beginning to end, and the possibility of inaccurate dimensions or damage through mishandling are hence eliminated.

Chemical mechanical polishing – this recent method is currently employed and involves both mechanical and chemical polishing mechanisms. Silica powder is mixed with deionized water and controlled at pH 10–11 with sodium hydroxide, and then fed onto the wafers which are simultaneously buffed by a peel and stick polishing pad of artificial leather. This method removes any remaining surface roughness and the combined effect of the mechanical and chemical approach also ensures that there is no additional damage during the process. The carrier plates, initially made of glass and resulting in a high incidence of thermal deformation, were replaced with a metal alloy, then eventually with ceramics. For newly developed highly integrated devices such as DRAM, the demand for surface flatness became even more stringent. For instance, the maximum permissible roughness for 16 MB of DRAM on a 4 cm^2 area is less than 0.5 μm. During these initial stages high-purity, low-particle chemicals (e.g., deionized water, etching acids) are mandatory for obtaining high-yield products.

Thermal oxidation or deposition – the silicon polished wafers (PW) are heated in the diffusion furnaces at high temperature ranging between 700°C and 1300°C and exposed to a reactive atmosphere of water and ultrapure oxygen under carefully controlled conditions forming an insulating layer of silicon dioxide, SiO_2, of uniform thickness on the wafer surface. The role of this layer is: (i) to form an impervious barrier against implant or diffusion of dopant into silicon, (ii) to passivate the surface and form a perfect dielectric layer, (iii) to form an active layer in particular in metal-oxide semiconductor (MOS) based devices. For thicker oxide passivation layers chemical vapor deposition (CVD) is preferred, while plasma oxidation, owing to the lower operating temperature (max. 600°C) induces almost no defect deformation.

Masking – masking is used to protect one area of the wafer while working on another. This process is referred to as photolithography or photomasking. A photoresist or light-sensitive polymer film is applied to the wafer, giving it characteristics similar to a piece of photographic paper. A photo-aligner aligns the wafer to a mask made of chromium coated onto glass with the circuit pattern. Then an intense UV light is focussed through the mask and through a series of reducing lenses, exposing the photoresist with the mask pattern. Sometimes ion or electron beams can replace classic UV radiation. Precise alignment of the wafer to the mask prior to exposure is critical. Most alignment tools are today fully automatic.

Etching – the wafer is then developed, i.e., the exposed photoresist is removed, and baked to harden the remaining photoresist pattern. It is then exposed to a chemical solution (e.g., hydrofluoric acid, HF, or ammonium dihydrogenofluoride, $NH_4H_2F_2$, solutions) or plasma so that areas not covered by the hardened photoresist are etched away. The photoresist is removed using additional chemicals or plasma and the wafer is inspected to ensure that the image transferred from the mask to the top layer is correct.

Doping – atoms, electron-acceptors such as boron, or electron-donors such as phosphorous, are introduced into the area exposed by the etch process to alter the electrical character of the pure silicon, which is an intrinsic semiconductor. These areas are called p-type (e.g., with boron) or n-type (e.g., with phosphorous) to reflect their particular charge carrier in the conduction process.

The previous steps, i.e., thermal oxidation, masking, etching and doping operations are repeated several times until the last front end layer is completed, i.e., all active devices have been formed.

Dielectric deposition and metallization – following completion of the front end, the individual devices are interconnected using a series of metal deposition and patterning steps of dielectric films (i.e., electrical insulators). Current semiconductor fabrication includes as many as three metal layers separated by dielectric layers.

Passivation – after the last metal layer is patterned, a final dielectric layer (i.e., passivation film) is deposited to protect the circuit from damage and contamination. Openings are etched in this film to allow access to the top layer of metal by electrical probes and wire bonds.

Cleaning and inspection – all the processing stages must be carried out with no handling or contamination. This means that each wafer must arrive in perfect condition ready for immediate use by the end user. In the 1970s, the cleaning method was developed and patented by the company RCA and instantly adopted by silicon wafer producers worldwide. The process has three steps, beginning with the SC1 solution which comprises a mixture of ammonia (NH_4OH), hydrogen peroxide (H_2O_2) and water (H_2O), in order to remove organic impurities and particles from the wafer surface. Next, natural oxides and metal impurities are removed with hydrofluoric acid solution (HF), then finally, the SC2 solution, a mixture of hydrochloric acid and hydrogen peroxide, is put onto the now bare surface. The surface is replaced by superclean new natural oxides which grow on the surface. Even now the RCA method is used although concentrations, temperatures, and time adjustments differ among companies, as do the chemical recycling processes and the option to use ultrasonic waves. In general, the number of particles on the surface of the final wafer does not exceed ten with diameters larger than 0.16 μm, with the density of surface metal ions at less than 10^{10} atoms cm^{-2}.

Electrical test – an automatic, computer-driven electrical test system then checks the functionality of each chip on the wafer. Chips that do not pass the test are marked with red ink for rejection.

Assembly – a diamond saw typically slices the wafer into single chips. The red inked chips are discarded, while the remaining chips are visually inspected under a stereo-microscope before packaging. The chip is then assembled into a package that provides the contact leads for the chip. A wire-bonding machine then attaches wires, a fraction of the width of a human hair, to the leads of the package. Encapsulated with a plastic coating for protection, the chip is tested again prior to delivery to the customer. Alternatively, the chip is assembled in a ceramic package for certain military applications.

4.10 The *P–N* Junction

The building block of most semiconductor devices involves the combination of *p*-type and *n*-type extrinsic semiconductors giving the so-called *p–n* junction. This will cause some electrons from the *n*-type to flow toward the *p*-type material. At the interface between the *p*-type and *n*-type regions, the electrons from the *n*-side fill the holes on the *p*-side. Then, a build-up of oppositely charged ions is generated, and, thus, an electric potential across the barrier appears. This build-up of electric charge is called the junction potential. With a net current of zero, the barrier prevents further migration of electrons. If a voltage is applied to the *p–n* junction with the negative terminal connected to the *n*-region and the *p*-region connected to the positive terminal, the electrons will flow toward the positive terminal, while the holes will flow toward the negative terminal. This is called forward bias and current flows. By contrast, if the positive terminal is connected to the *n*-type and the negative connected to the *p*-type, a reverse bias forms and no current flows due to the build-up of the potential barrier. In other words, these devices exhibit a so-called valve action property, that is, the system allows the current to flow in one direction but not the reverse. Therefore, the *p–n* junction must be correctly inserted in an electrical circuit with the right polarity, or the electric current will not flow across. In the formation of an electric potential at a *p–n* junction, the two semiconductor materials are placed close together, where electrons from the *n*-side combine with the holes on the *p*-side. This results in a positive charge on the *n*-side of the junction and a negative charge accumulation on the p-side. This separation of charge creates a junction electric potential. As a result, there are no electrons or holes at the junction because they have combined with each other. The applications of p–n junctions are essentially in electronics and electrical engineering. Actually, there are many polar electronic devices which are based on the p–n junction, such as diodes, photovoltaic cells, and solid-state rectifiers. Transistors are another application of the p–n junction. Transistors, by contrast with diodes, contain more than one p–n junction. Because of this, a transistor can be used in a circuit to amplify a small voltage or current into a larger one or function as an on–off switch.

References

[1]   Bardeen J, Brattain WH (1948) The transistor, a semiconductor triode. Phys Rev 74: 230.
[2]   Weber JN (ed.) (1973) Geochemistry of germanium. Dowden, Hutchinson, & Ross.

Further Reading

Braun E, MacDonald S (1982) The physics of solid state devices. Cambridge University Press, New York.
Coughlin R, Driscol F (1976) Semiconductor fundamentals. Prentice Hall, Englewood Cliffs, NJ.
Jaeger R (1993) Introduction to microelectronic fabrication. Addison-Wesley.
Madelung O (ed.) (1996) Semiconductors basic data, 2nd edn. Springer-Verlag, Berlin.
Sapoval B, Hermann C (1995) Physics of semiconductors. Springer-Verlag, New York.
Yu PY, Cardona M (1996) Fundamentals of semiconductors: physics and materials properties. Springer-Verlag, Heidelberg.

5

Superconductors

5.1 General Description

Superconducting materials (superconductors) are electrically conductive materials which under certain specific conditions; (i) show a decrease in their intrinsic electrical resistivity to a value near zero, and (ii) become perfect diamagnetic materials, i.e., there is a complete exclusion of an applied magnetic field from the bulk of the material. However, three particular conditions must be required for a material to exhibit the superconducting state with these unusual electric and magnetic properties: the temperature of the material must be lowered below a certain critical temperature (T_c, in K); the applied magnetic field strength must be less than the critical magnetic field strength (H_c, in $A.m^{-1}$); and the current density flowing in the superconductor must be less than the critical current density (j_c, in $A.m^{-2}$). If any of these values exceeds the respective critical value, the superconductor becomes quenched, and loses its superconductive properties. Hence, the two former parameters, i.e., the critical temperature and critical magnetic field strength are intrinsic properties for a given material or composition and they are not affected to any large extent by modification in processing or changes in microstructure. On the contrary, the critical current density may vary over several orders of magnitude within a single material, and it is strongly affected by: (i) the metallurgical processing, (ii) the presence of crystal lattice defects and traces of chemical impurities, and finally (iii) the microstructure. However, these three critical parameters are closely interdependent and can be defined by a three-dimensional thermo-dynamic phase diagram within which the superconducting state is stable. For all superconducting materials presently known, the critical temperature ranges between boiling point of liquefied gases to below room temperature and experimental measurements have clearly demonstrated that no measurable decay of the superconducting properties has been noted. The most common method to attain the low temperatures required by superconductors is to use low-temperature liquefied gases such as liquid helium (bp 4.22 K or $-268.93°C$) or liquid nitrogen (bp 77 K or $-195.79°C$). For understanding superconductivity, it is important to remember the meaning of the physical quantity called the electrical resistance. Actually, if we consider a common solid conductor (i.e., resistor senso-stricto) which undergoes a potential gradient on each side, the electric potential difference (i.e., voltage) causes the electric charge (e.g., electric carriers such as electrons, or ions) to flow through the material with a definite charge flow rate (i.e., electric current). In this microscopic description of electric charge migration in solid materials, the electric resistance can be easily understood as the friction encountered by the electric charge carriers during displacements. According to Ohm's law, $U = RI$, the electric current is directly proportional to the voltage, and inversely proportional to the electrical resistance. Higher resistance causes less current for a given voltage, and higher voltage causes more current for a given resistance. If the voltage drops to zero, no electric current will flow, on the contrary, for any resistance greater than zero. If the resistance is zero, there is theoretically no voltage needed for electric current to flow.

5.2 Superconductor Types

As predicted by the Ginzburg–Landau theory and demonstrated later by experimental studies, superconducting materials can be classified according to their magnetic behavior in two distinct classes: type I and type II superconductors. Only several hundred superconducting materials (e.g., pure metals, alloys, ceramics, and lately organic compounds) are known today and this leads us to consider superconductivity as a rare physical phenomenon.

5.2.1 Type I Superconductors

The original elemental superconductors are pure metals from group IIB(12): Zn, Cd, Hg; group IVB(4): Ti, Zr, Hf; VB(5): Ta; group VIB(6): Mo, W; group VIIB(7): Re group VIIIB(8): Ru, Os, Ir; group IIIA(13): Al, Ga, In, Tl; and group IVA(14): Sn, Pb of Mendeleev's periodic chart. They are classified as **type I** or **soft superconductors**. They have a critical temperature (T_c, in K) above which the superconducting state disappears, and by contrast with other superconductor types, they exhibit only one critical magnetic field (H_c, in A.m^{-1}) for a given temperature, i.e., perfect diamagnetic behavior, excluding a magnetic field up to this value. The temperature interval, ΔT_c, over which the transition between the normal and the superconductive states take place, may be narrow as 2×10^{-5} K (e.g., this narrow transition width was attained for ultrapure 99.9999wt% gallium monocrystals). Hence, the physical quantities T_c and H_c are intrinsic properties of type I superconducting materials. If they are in a magnetic field that is weaker than the critical magnetic field, they have zero resistance and exhibit perfect diamagnetism (i.e., $M = -\chi$, $\chi = -1$, and $B = 0$). If they are in a magnetic field that is stronger than the critical magnetic field, they have resistance greater than zero, and there is a magnetic flux penetration in the bulk material. It is also important to note that several elements become superconducting only when submitted to high pressure (e.g., Cs, Ba, Y, Ce, La, Lu, Rh, Al, Ga, Tl, Si, Ge, Sn, Pb, P, As, Sb, Bi, Se, Te Zr, and U).

5.2.2 Type II Superconductors

Type II or **high-field superconductors** enclose only four pure chemical elements, the two refractory metals of the group VB(5) of the periodic chart V and Nb, along with Gd and Tc, but are widely represented by binary and ternary alloys or intermetallics and other more complicated compounds. By contrast with previous superconducting materials, type II superconductors possess three critical magnetic fields ($H_{c1} < H_{c2} < H_{c3}$). For magnetic fields $H < H_{c1}$, type II superconductors behave like pure metals or type I superconductors and are perfectly diamagnetic. For $H > H_{c3}$, superconductivity vanishes and the material recovers its

Table 5.1 Type I superconductor properties

Element (low temp. phase)	Crystal lattice structure (Strukturbericht, Pearson, space group)	Critical temperature (T_c, K)	Critical magnetic field (H_c, A.m^{-1})	Debye temperature (T_D, K)	Electronic molar heat capacity (γ, mJ.mol^{-1}.K^{-1})
Al (α–)	A1, cF4, Fm$\bar{3}$m	1.175	8347.68	420	1.35
Be (α–)	A3, hP2, P6$_3$/mmc	0.026	n.a.	1000	0.21
Cd	A3, hP2, P6$_3$/mmc	0.517	2228.17	209	0.69
Ga (α–)	A11, oC8, Cmca	1.083	4639.37	n.a.	n.a.
Hf (α–)	A3, hP2, P6$_3$/mmc	0.128	1010.63	n.a.	2.21
Hg (α–)	A10, hR1, R$\bar{3}$m	4.154	32,706.34	87	1.81
Hg (β–)	Aa, tI2, I4/mmm	3.949	26,976.76	93	1.37
In	A6, tI2, I4/mmm	3.408	22,361.27	109	1.672
Ir	A1, cF4, Fm$\bar{3}$m	0.1125	1273.24	425	3.19
La (α–)	A3', hP4, P6$_3$/mmc	4.88	63,661.98	151	9.8
La (β–)	A1, cF4, Fm$\bar{3}$m	6.00	87,216.91	139	11.3
Lu	A3, hP2, P6$_3$/mmc	0.1	27,852.12	n.a.	n.a.
Mo	A2, cI2, Im$\bar{3}$m	0.915	7639.44	460	1.83
Os	A3, hP2, P6$_3$/mmc	0.66	5570.42	500	2.35
Pb (α–)	A1, cF4, Fm$\bar{3}$m	7.196	63,900.71	96	3.1
Re	A3, hP2, P6$_3$/mmc	1.697	15,915.49	4.5	2.35
Ru	A3, hP2, P6$_3$/mmc	0.49	5490.85	580	2.8
Sn (α–)	A4, cF8, Fd$\bar{3}$m	3.722	24,271.13	195	1.78
Ta	A2, cI2, Im$\bar{3}$m	4.47	65,969.72	258	6.15
Th (α–)	A1, cF4, Fm$\bar{3}$m	1.38	127.32	165	4.32
Ti (α–)	A3, hP2, P6$_3$/mmc	0.40	4456.34	415	3.3
Tl (α–)	A3, hP2, P6$_3$/mmc	2.38	14,164.79	78.5	1.47
U (α–)	A20, oC4, Cmcm	0.2	n.a.	n.a.	n.a.
W	A2, cI2, Im$\bar{3}$m	0.0154	91.51	383	0.90
Zn	A3, hP2, P6$_3$/mmc	0.85	4297.18	310	0.66
Zr (α–)	A3, hP2, P6$_3$/mmc	0.61	3740.14	290	2.77

normal resistive behavior. However, for a magnetic field between H_{c1} and H_{c2}, type II superconductors have a unique property which type I superconductors do not have: they exhibit a mixed superconducting state with zero electrical resistance, but allow partial magnetic flux penetration. In this particular region, the type II superconductor is said to be in the **vortex state**. This behavior was first discovered in 1957 by the Russian physicist Alexi Abriksov. In the vortex state, there are several cores of normal material, surrounded by material in the superconducting state. Quantized supercurrents surround each core, creating exactly one quantum of magnetic flux per core (i.e., $h/2e$, also called fluxons by solid-state physicists). Between H_{c2} and H_{c3}, the superconductor has a sheath of current-carrying superconductive materials at the body surface, and above, increasing the magnetic field increases the number of vortices, and when no more vortices can fit into the superconductor, the material becomes totally nonsuperconducting.

Table 5.2 Type II superconductor properties (pure metals)

Metals	Crystal lattice structure (Strukturbericht, Pearson, space group)	Critical temperature (T_c, K)	Critical magnetic field (H_c, A.m^{-1})	Debye temperature (T_D, K)	Electronic molar heat capacity (γ, mJ.mol^{-1}.K^{-1})
Gd (γ–)	hR3, R$\bar{3}$m	7.00	75,598.60	325	0.60
Nb	A2, cI2, Im$\bar{3}$m	9.25	163,929.59	277	7.80
Tc	A3, hP2, P6$_3$/mmc	7.80	112,204.23	411	6.28
V	A2, cI2, Im$\bar{3}$m	5.40	112,045.08	383	9.82

Table 5.3 Type II superconductors (alloys and compounds)

Materials	Crystal lattice structure (Strukturbericht, Pearson, space group)	Critical temperature (T_c, K)	Observed temperature (T_{obs}, K)	Critical magnetic fields (H_{c1}, A.m^{-1})	(H_{c2}, A.m^{-1})	(H_{c3}, A.m^{-1})
Bi$_x$Pb$_{1-x}$	n.a.	7.35–8.8	4.2	9709	2,387,324	n.a.
Hg$_{0.101}$Pb$_{0.899}$	n.a.	4.14–7.26	4.2	18,303	342,183	n.a.
In$_{0.94}$Pb$_{0.06}$	n.a.	3.90	3.12	7560	14,320	27,852
Nb$_{0.45}$Ti$_{0.55}$	n.a.	9.5	4.2	71,620	8,554,578	n.a.
Nb$_3$Al	A15, cP8, Pm$\bar{3}$n	18.9	4.2	29,842	n.a.	n.a.
Nb$_3$Ge	A15, cP8, Pm$\bar{3}$n	23.2	4.2	31,831	21,883,805	n.a.
Nb$_3$Sn	A15, cP8, Pm$\bar{3}$n	18.3	4.2	27,056	17,507,044	n.a.
NbN	B1, cF8, Fm3m	16.1	4.2	31,831	21,883,805	n.a.
V$_3$In	A15, cP8, Pm$\bar{3}$n	13.9	n.a.	n.a.	n.a.	n.a.
Nb$_3$Ga	A15, cP8, Pm$\bar{3}$n	20.3	n.a.	n.a.	n.a.	n.a.
Mo$_3$Os	A15, cP8, Pm$\bar{3}$n	11.68	n.a.	n.a.	n.a.	n.a.
Nb$_3$Pt	A15, cP8, Pm$\bar{3}$n	10	n.a.	n.a.	n.a.	n.a.
Cr$_3$Ru	A15, cP8, Pm$\bar{3}$n	3.43	n.a.	n.a.	n.a.	n.a.
Nb$_x$Zr$_{1-x}$	A3, hP2, P6$_3$/mmc	10.7	4.2	n.a.	7,480,282	n.a.
V$_3$Ga	A15, cP8, Pm$\bar{3}$n	15.4	4.2	31,831	18,302,819	n.a.
V$_3$Si	A15, cP8, Pm$\bar{3}$n	17.1	4.2	43,768	13,130,283	n.a.

5.2.3 High Critical Temperature Superconductors

Since 1986, Karl Alex Muller and Johannes Georg Bednorz at the IBM Research Laboratories in Zurich, Switzerland have prepared ceramic materials which are electrical insulators at room temperature but exhibit superconductivity above 77 K (the boiling point of nitrogen). These new superconductive materials have demonstrated much higher critical temperatures than many classic superconductors, and are therefore referred to as high critical temperature superconductors. Previously, the highest critical temperature supercoductor known was the compound Nb$_3$Ge with a critical temperature of 23 K. These superconductors are oxides, with a perovskite lattice structure type, for instance lanthanum barium copper oxide (i.e.,

Table 5.4 High-temperature oxide superconductors

Oxides	Space group (Hermann–Mauguin)	Critical temperature (T_c, K)
$Ba_{0.6}K_{0.4}BiO_3$	n.a.	30
$BaPb_{0.75}B_{0.25}O_3$	n.a.	13
$Bi_2Sr_2CaCu_2O_8$	A_2aa	84
$La_{1.85}Ba_{0.15}CuO_4$	n.a.	36
$La_{2-x}Sr_xCuO_4$	I4/mmm	35
$Nd_{2-x}Ce_xCuO_4$	I4/mmm	30
$Tl_2Ba_2Ca_2Cu_3O_{10}$	I4/mmm	25
$Tl_2Ba_2CaCu_2O_8$	P4/mmm	108
$Y_2Ba_4Cu_8O_{16}$	Ammm	81
$YBa_2Cu_3O_{7-x}$	Pmmm	92

$La_{1.85}Ba_{0.15}CuO_4$) was found to have a critical temperature of about 30 K. When yttrium was substituted for lanthanum in a ratio of 1:2:3 with barium and copper (e.g., $Yba_2Cu_3O_{7-x}$), the critical temperature was measured to be 93 K. Liquid nitrogen, which is easier to obtain, easier to work with, and much cheaper, could be used instead of the expensive liquid helium. Today, critical temperatures are approaching 200 K as scientists continue searching for room temperature superconductivity. However, one important drawback of high-temperature oxide superconductors is that they are ceramics. Hence, they are brittle and cannot easily be made into wires during processing such as the competitive Nb_3Sn.

5.2.4 Organic Superconductors

Although organic compounds rarely exhibit electrical conductivity either at low or high temperature, some organic compounds have demonstrated the ability to exhibit super-conductivity. Nevertheless, these compounds are restricted to (i) those containing fullerenes (i.e., spherical cluster C_{60}) with a critical temperature range between 1 and 12 K, and (ii) those made of fulvalenes (i.e., pentagonal rings containing S or Se) with critical temperature range between 8 and 32 K.

5.3 Basic Theory

Superconductivity still has not been totally explained. Most of the research done has been experimental rather than theoretical. For example, the formula for the critical magnetic field as a function of temperature is an empirical formula based on experimental data rather than theoretical prediction:

$$H_c(A.m^{-1}) = H_{c0}[1 - (T/T_c)^2]$$

The BCS theory, however, developed in 1957 by the three physicists John Bardeen, Leon Cooper, and Robert Schrieffer, does establish a model for the mechanism behind superconductivity.

Bardeen, Cooper, and Schrieffer received the Nobel Prize in Physics in 1972 for their theory. It was known that the flux quantum was inversely proportional to twice the charge of an electron, and it had also been observed that different isotopes of the same superconducting element had different critical temperatures. Actually, the heavier the isotope, the lower the critical temperature is. The critical temperature, T_c in K, of an isotope with an atomic mass, M, expressed in kg.mol^{-1} can be predicted by the following equation:

$$T_c = K/M^\alpha$$

where K is a constant expressed in K.kg$^\alpha$.mol$^{-\alpha}$, and α is the dimensionless **isotope-effect exponent**. Theoretically, according to BCS theory, the dimensionless isotopic exponent should have a value of roughly 1/2. For instance, practical values are: Cd (0.32), Tl (0.61), and Zr (0.00 ± 0.05). Close examination of these experimental values for the isotope-effect exponent shows that they are not exactly 1/2 because the forces between the electrons also affect it. It is interesting to note that zirconium has a value of 0.00 for α but the BCS theory cannot explain this. BCS theory states that in a superconductor, electrons form pairs. As the first electron moves through the crystal lattice, it pulls the nuclei of the atoms in the superconductor toward it. The second electron, rather than experiencing a backward force from the first electron, experiences a forward force from the positively charged nuclei in front of it. This is how the electrons stay together rather than being repelled as they move through the lattice. A pair of electrons moving through the lattice of a superconductor is referred to as a Cooper pair, after Leon Cooper. The forces on the nuclei of the atoms in the lattice by the electrons cause vibrations referred to as lattice vibrations, or phonons. The idea of electron–phonon interaction was actually first proposed in 1950 by Herbert Frohlich, but it took until 1957 for the necessary experiments to be completed and a formal theory to be written.

Although the BCS model is a brilliant theory describing the mechanism for super-conductivity, it cannot determine which materials are superconductors and which materials are not. The reason metals with low electrical resistivities at room temperature (e.g., metals of group IB(11) such as copper, silver, gold) are not superconductors and metals with higher resistivities at room temperature (e.g., mercury, tin, lead) are superconductors is due to the lattice quantum random vibrations (i.e., phonons) due to the transfer of thermal energy across the material. Actually, the more the lattice is vibrating, the more the electrons traveling through the lattice will be slowed own by the vibrating atoms in the lattice. Materials with lattices which vibrate easily generally have higher resistivities at room temperature, while materials with lattices which do not vibrate easily generally have lower resistivities at room temperature. However, at very low temperatures, the Cooper pairs can move more easily through the materials with lattices which are more susceptible to vibration.

5.4 The Meissner–Ochsenfeld Effect

According to Maxwell's laws, a perfect conductor would not allow any change in magnetic flux. Any change in magnetic flux would induce eddy-currents in the perfect conductor which would produce a magnetic flux opposite to the change in magnetic flux which was originally introduced. If a superconductor were a perfect conductor, it should behave the same way. A material becomes superconducting when its temperature drops below the critical temperature. So when the temperature is higher than the critical temperature, the material is not a perfect conductor, but when the temperature is lower than the critical temperature, the material superconducts. Consider a perfect conductor with zero magnetic flux inside it. If a permanent magnet were brought near the perfect conductor, the magnetic flux inside the conductor would remain zero, due to the induced currents. The magnetic field from the induced currents would oppose the magnetic field of the permanent magnet. If this permanent magnet had a magnetic field strong enough to support its own weight, it could be levitated above the perfect conductor. Now consider a material at a temperature above its critical temperature. It has a resistivity greater than zero, so when a permanent magnet is brought near it, the magnetic flux can penetrate the material. If the material were then cooled to a temperature below its critical temperature, and became perfectly conducting, the magnetic flux would be trapped inside it. No magnetic field opposite the magnetic field of the permanent magnet would be present, and it would not be possible to levitate the magnet.

5.5 History

The discovery of superconductivity was the indirect result of work performed during the second half of the nineteenth century to liquefy gases with low boiling points, and the study of the new properties of these cryogenic liquids. Actually, common gases such as carbon dioxide, and atmospheric nitrogen, oxygen, and hydrogen were liquefied before 1898. However, helium was the only gas which had not yet been liquefied until the work of the Dutch scientist Heike Kammerlingh Onnes in 1908 (Leiden, Netherlands), based on the theories of Johannes van der Waals. Afterwards, Onnes began to investigate the electrical resistivities of several metals maintained at the boiling point of helium (4 K). Among the first metals studied, gold and platinum showed that their resistivities did in fact stabilize at extremely low temperatures, due to amounts of impurities. Therefore, Onnes selected mercury because at that time it could be easily purified by distillation, providing samples exhibiting a higher purity than other metals. Hence, in 1911, when cooling highly pure mercury close to 4 K, Onnes observed that its electrical resistivity abruptly dropped to zero. This work was the starting point of the study of the rare physical phenomenon known today as superconductivity.

As a consequence of the discovery of superconductivity in mercury, other metals were found to be superconductive, and superconducting alloys were found in 1931. In the late 1930s, the Ukrainian physicists Shubnikov et al., studying effects of alloying elements on the superconductivity of lead, discovered the type II superconductor materials. In the 1950s, the two Russian physicists Ginzburg and Landau established the theoretical explanation of the superconducting state. The search for new superconducting materials led to the preparation in the 1960s of the intermetallic compound Nb_3Sn with a critical temperature of 18 K, and later in the mid-1970s of the intermetallic compound Nb_3Ge with 23 K. In 1986, Karl Alex Muller and Johannes Georg Bednorz (IBM Research Laboratories, Zurich, Switzerland) based on the work of French scientists C. Michel, L. Er-Rahko, and B. Raveau prepared a lanthanum barium copper oxide ($La_{1.85}Ba_{0.15}CuO_4$) exhibiting a critical temperature of about 30 K. This was a significant breakthrough, and a new class of superconductors had been found.

Soon after the discovery of superconductivity in lanthanum barium copper oxide, researchers around the world began investigating superconductivity in search of a higher temperature superconductor. In 1987, C.W. Paul Chu et al. (Huntsville, Alabama, USA), discovered a superconductive compound ($YBa_2Cu_3O_{7-x}$) with a critical temperature (95 K) significantly above the boiling point of nitrogen (77 K); this ceramic is now well known under the common acronym YBACUO. This discovery was extremely important from a technological point of view and opened a new frontier for superconducting applications because by contrast with liquid helium, liquid nitrogen is easier to obtain, easier to handle and to store, and is less costly, which could lead to the development of great applications of superconductivity.

However, although the science and engineering of superconductors had experienced an important growth since the discovery of high-temperature superconducting materials in the 1990s, today the major industrial and commercial applications of superconductor devices have mostly been confined to superconducting compounds, particularly alloys that can be easily processed and manufactured in the form of wires and thin films. This is the main reason for the large demand for niobium-based superconductors. The major applications of this type of superconductor include: (i) R&D, where powerful electromagnets are the essential component of nuclear magnetic resonance (NMR) spectrometers, extensively used in all analytical organic chemistry research laboratories; (ii) medicine, where electromagnets are used in magnetic resonance imaging (MRI) equipment which provides a powerful diagnostic technique widely used in hospitals worldwide for imaging the entire human body; (iii) in high-energy physics where enormous assemblies of large superconducting magnets are installed both in particle accelerators (CERN, Fermilab, etc.) and in nuclear fusion reactors. However, more recent articles have indicated that developments in the late 1990s allow rare-earth-based super-conductors to be obtained at the laboratory scale using new processing technologies, such as the melt process and powder metallurgy. These compounds exhibit both a high critical current density and magnetic field strength at 77 K with other suitable properties required for the commericalization of new devices. Hence, in the near future, these superconductors could open a new way to the manufacture of many bulk devices for various innovative applications: (i) bulk equipment for electric power such as flywheel energy storage devices, generator motors, transmission cables, fault current limiters, and transformers, and obviously (ii) high-performance magnets for high-energy physics.

In conclusion, today, critical temperatures of modern superconductors approach nearly 200 K and scientists and engineers continue intense research for room temperature super-conductivity and find new efficient superconductive compounds every year. However, the major critical issues to solve in order to make possible the commercialization of bulk superconducting devices on a larger scale (e.g., for magnetically levitated trains, power transmission cables) are (i) achievement of excellent processability and fabricability of these new superconductors to allow high-performance electromagnets having intricate geometrical shapes to be obtained, and to a lesser extent (ii) the selection of non-strategic (i.e., abundant and low cost) elements entering into the composition of superconductors.

5.6 Industrial Uses and Applications

The applications of superconducting materials can be summarized in two main categories: (i) high-magnetic-field, and (ii) low-magnetic-field applications.

High-magnetic-field applications: these applications, which include electromagnets and generators, require superconducting materials able to withstand high current densities flowing through the material without a catastrophic quenching effect,[1] and hence the selection of the best material is focused on obtaining superconductors with the highest critical current density. Amongst the high-magnetic-field applications, nuclear magnetic resonance imaging (NMRI), more commonly known as magnetic resonance imaging (MRI) is the first large-scale commercial application of superconductivity. For this purpose, a superconducting electro-magnet, made of a cylindrical wire of tin–niobium alloy, Nb_3Sn, surrounded by copper, is used. Tin–niobium was selected for several reasons. Actually, high-temperature superconductors would be more efficient, but they are ceramics, hence they are brittle and cannot easily be drawn into wires. Moreover, tin–niobium is a type II superconductor and has a high critical current density and high second critical magnetic field. The first critical magnetic field is unimportant, as the wire only needs to have zero resistivity. Diamagnetism is not important for this application. However, tin–niobium is not a high-temperature superconductor and liquid nitrogen cannot be used to cool the wire. Hence, liquid helium must be used to keep the tin-niobium well below its critical temperature and critical current density. The second major high-magnetic field application is the manufacture of electromagnets for the high energy accelerators and colliders. Finally, magnetic confinement for thermonuclear fusion is the third major high-magnetic-field application. In this application electromagnets serve to confine the high-temperature plasma produced by the thermonuclear fusion reaction.

Low-magnetic-field applications: these applications include Josephson-effect devices, magnetic flux shields, transmission lines, and resonant cavities required for superconducting materials having a high critical temperature and magnetic field.

Further Reading

ASTM Standard B713-82 (1997) Standard terminology relating to superconductors. In: The annual book of ASTM standards, pp 346–348.

Berger LI, Roberts BW (1997–1998) Properties of superconductors. In: Lide RD (ed) CRC handbook of chemistry and physics 78th edn. CRC Press, Boca Raton, FL, pp 12-60 to 12-91.

Doss JD (1989) Engineer's guide to high-temperature superconductivity. New York, John Wiley & Sons.

---

[1]The quench is the abrupt and uncontrolled loss of superconductivity produced by a disturbance.

Foner SS, Schwartz BB (1981) Superconductor materials science: metallurgy, fabrication and applications. Plenum Press, New York.

Ginsburg DM (ed.) (1989–1992) Physical properties of high-temperature superconductors, vol I–III. World Scientific, Singapore.

Ishigura T, Yamaji K (1990) Organic superconductors. Springer-Verlag, Berlin.

Ishigura T, Yamaji K, Saito G (1998) Organic superconductors, 2nd edn. Springer-Verlag, Berlin.

Larbaslestier DC, Shubnikov LV (1995) Superconducting materials, introduction. In: ASM metals handbook, 9th edn, vol 2. Properties and selection of nonferrous alloys and special purpose materials, ASM, Metal Park, OH, pp 1027–1029.

Lynton EA (1964) Superconductivity. London, Methuen.

Malik SK, Shah SS (ed) (1994) Physical and material properties of high temperature superconductors. Nova Science Publishing, Commack, New York.

Müller P, Ustinov AV (1997) The physics of superconductors: introduction to fundamentals and applications. Springer-Verlag, Heidelberg.

Phillips JC (1989) Physics of high-T_c superconductors. Academic Press, New York.

Rao CNR (ed.) (1990) Chemistry of high temperature superconductors. World Scientific, Singapore.

Rose-Innes AC, Rhoderick FH (1969) Introduction to supraconductivity. Pergamon Press, New York.

Smather DB (1995) A15 Superconductors. In: ASM metals handbook, 9th edn. vol 2: Properties and selection of nonferrous alloys and special purpose materials. ASM, Metal Park, OH, pp 1060–1076.

Warnes WH (1995) Principles of superconductivity. In: ASM metals handbook, 9th edn, vol 2: Properties and selection of nonferrous alloys and special purpose materials. ASM, Metal Park, OH, pp 1030–1042.

Williams J (1992) Organic superconductors. Prentice Hall, Englewood Cliffs, NJ.

6

Magnetic Materials

This chapter gives the description of the most common magnetic materials used in electrical engineering and other industrial applications. The first section details the most common physical quantities used to describe magnetic properties of materials, while the following sections describe the basic magnetic properties of the five classes of magnetic materials.

6.1 Magnetic Physical Quantities

The physical quantities essential to understand the most common magnetic properties of materials are briefly described in the following paragraphs.

6.1.1 Magnetic Field Strength

When an electric current, I, expressed in amperes (A), flows through a metallic wire, it generates a magnetic field strength perpendicular to the flow direction. The magnetic field strength or simply magnetic field, H, expressed in amperes per meter (A.m^{-1}), is an axial vectorial quantity produced by an electric current circulating in a finite long coil (i.e., solenoid) having a certain number of turns per unit length, N, expressed in reciprocal meters (m^{-1}), by the following equation:

$$H = N \times I.$$

6.1.2 Microscopic Magnetic Dipole Moment

A magnetic dipole can be theoretically described as an elementary pair of magnetic poles and can be represented by an infinitesimal current loop. Therefore, the magnetic moment is the product of the current and the loop area. The microscopic magnetic dipole moment, μ, expressed in ampere square meters (A.m^2), is a vectorial physical quantity which characterizes the magnetic properties of small entities such as elementary particles (e.g., proton, electron), nuclides, atoms or molecules having a spin angular momentum value different from zero. For nuclides having a nuclear spin angular momentum (I) different from zero, the magnetic dipole moment comes from protons and it has a value of the same order of magnitude as the nuclear magneton ($\mu_N = eh/4\pi m_p$, or β_N). The magnetic dipole moment of atoms is of electronic origin and it is several times the Bohr magneton ($\mu_B = eh/\pi m_e$, or β). Hence, the magnetic dipole moment of nuclides and atoms is given by the two equations:

$$\mu_{\text{nuclide}} = \gamma_N \times \hbar \times I = g_p \times \mu_N \times I,$$
$$\mu_{\text{atom}} = \gamma_e \times \hbar \times S = g_e \times \mu_B \times S,$$

where γ_N and γ_e are the nuclear and atomic gyromagnetic ratios, expressed in C.kg^{-1}, sometimes called the magnetomechanical ratio, $\hbar = h/2\pi$ the rationalized Planck constant in J.s, g_p and g_e the dimensionless proton and electron Landé's factors, and I, S, the nuclear and atomic spin angular momentum quantum number.

6.1.3 Macroscopic Magnetic Moment

The macroscopic magnetic moment, m, expressed in ampere square meters (A.m^2) is a vector summation of all the microscopic magnetic dipole moments present inside the material. However, the magnetism of materials, because the contribution of nuclide magnetism is negligible, is essential due to the atomic magnetism.

$$m = \Sigma_i \mu_i$$

6.1.4 Magnetization

The magnetization, M, expressed in amperes per meter (A.m^{-1}), is a macroscopic vectorial physical quantity defined as the density of magnetic dipole moments, n, expressed in reciprocal cubic meters (m^{-3}), i.e., the number of microscopic magnetic dipole moments per unit volume, V, of a selected magnetic material, according to the following equation:

$$M = n \times m = m/V.$$

For a material put in a magnetic field, H, the magnetization is proportional to the applied magnetic field and is given by:

$$M = \chi_m H = m/V$$

where the dimensionless factor χ_m is the **magnetic susceptibility** of the material, which is an intrinsic property of the material.

6.1.5 Magnetic Flux Density or Magnetic Induction

In empty space (i.e., vacuum) the vectorial quantity called the magnetic flux density, B, expressed in tesla (T), is collinear and directly proportional to the applied magnetic field, H according to the following equation:

$$B = \mu_0 H$$

where the physical quantity, μ_0, is the **magnetic permeability of a vacuum** expressed in henries per meter (H.m^{-1}). This is a fundamental physical constant equal by definition to exactly $4\pi \times 10^{-7}$ H.m^{-1}.

However, if a magnetic field is applied to a material, by reaction there appears a magnetic induction given by the equation:

$$B = \mu_0 \mu_r H$$

where the dimensionless physical quantity, μ_r, is the **relative magnetic permeability** of the material. It can also be defined as the **magnetic permeability of a material**, μ, expressed in henries per meter $(H.m^{-1})$ which is the product of the magnetic permeability of a vacuum and the relative magnetic permeability of the material.

$$\mu = \mu_0 \mu_r$$

For magnetic materials having magnetic losses the relative magnetic permeability can be written as a complex quantity as follows: $\mu_r = \mu_{real} - j\mu_{im}$. Moreover, the magnetic induction can also be written as the sum of two contributions (i) the magnetic moment and (ii) the applied magnetic field:

$$B = \mu_0 \mu_r H = \mu_0 H + \mu_0 M = \mu_0 H(1 + \chi_m).$$

Therefore, it appears clearly that the interdependence between the two dimensionless quantities, i.e., the relative magnetic permeability (μ_r) and the magnetic susceptibility (χ_m) of the material is:

$$\mu_r = 1 + \chi_m.$$

The **mass magnetic susceptibility** is the ratio of the susceptibility over the density of the materials (χ_m/ρ).

6.1.6 Magnetic Energy Density Stored

The magnetic energy density of a material, expressed in joules per cubic meter $(J.m^{-3})$, corresponds to the magnetic energy stored per unit volume of material. The magnetic energy density stored by increasing the magnetic induction from B_1 to B_2 is given by the following equation:

$$\frac{E}{V} = \int_{B_1}^{B_2} H dB$$

Table 6.1 Magnetic physical quantities and conversion factors

Magnetic quantity	IUPAC symbol	SI unit conversion factor for obsolete emu units[a]
Magnetic field strength	H	1 Oe = $(1000/4\pi)$ A.m^{-1}
Magnetic flux density	Φ	1 Mx = 10^{-8} Wb
Magnetic induction	B	1 G = 10^{-4} T
Magnetic moment	M	1 emu = $4\pi \times 10^{-10}$ A.m^2
Magnetic permeability	μ	1 emu = $4\pi \times 10^{-7}$ H.m^{-1}
Magnetization	J	1 emu = 1000 A.m^{-1}
Magnetomotive force	–	1 Gb = $(10/4\pi)$ A

[a]Cardarelli F (1999) Scientific unit conversion: a practical guide to metrication, 2nd edn. Springer-Verlag, London.

6

Magnetic Materials

6.2 Classification of Magnetic Materials

Magnetic materials are those materials that can be attracted or repelled by a magnet and can be magnetized themselves. The magnetic properties of materials are of atomic origin, arising from the orbital motion and spin angular momentum of electrons in the atoms. According to Maxwell's theory of magnetism, electric charges in motion form small magnetic dipole moments which react to an applied magnetic or electric field. Even if nuclear magnetism exists, it is seldom that the contribution of nuclear magnetism to the overall atomic magnetism is too low owing the several orders of magnitude between the Bohr magneton and nuclear magneton. There exist five classes of magnetic materials: (i) diamagnetic materials, (ii) paramagnetic materials, (iii) ferromagnetic materials, (iv) antiferromagnetic materials, and finally (v) ferrimagnetic materials.

6.2.1 Diamagnetic Materials

When a diamagnetic material is submitted to an external magnetic induction, the atomic electronic orbitals are strongly modified owing to the deviation of electron trajectory by the magnetic field according to Laplace's law. Therefore, a spontaneous induced magnetic field appears and it opposes the variations of the external magnetic field as predicted by Lenz's law. Actually, in spite of the weakness of the magnetic dipole moment of the atoms, they orientate along the field lines in order to compensate the external magnetic field. This behavior is totally reversible and the random magnetic moment orientation is restored when the application of external field has ceased. In conclusion, the diamagnetism originates from induced current opposing the external applied magnetic field. For this reason, diamagnetic materials have a small negative magnetic susceptibility ($\chi_m \cong -10^{-5}$), that is, they exhibit a relative magnetic permeability value slightly below unity ($\mu_r < 1$). As a general rule, because the diamagnetism originates from orbital deformation under an applied external magnetic field, all materials have obviously a basic diamagnetic component. Diamagnetic materials are for instance: (i) gases such as hydrogen, nitrogen, chlorine, bromine, and noble or inert gases such as He, Ne, Ar, Kr, Xe; (ii) the chemical elements from group IIA(2): Be; group IIIA(13): B, Ga, In, Tl; group IVA(14): C, Si, Ge, Pb; group VA(15): P, As, Sb, Bi; group VIA(16): S, Se, Te; group IA(11): Cu, Ag and Au; and finally group IIA(12): Zn, Cd, Hg, and (iii) crystalline solid materials such as MgO and diamond. On the other hand, perfect diamagnetic materials are particularly the superconductors of type I (see Chapter 5).

6.2.2 Paramagnetic Materials

For paramagnetic materials, the magnetism originates from the partial alignment of the existing magnetic dipole moments, which are randomly oriented by thermal agitation in the absence of an applied external magnetic field. When an external field is applied to the material, all the magnetic dipole moments orientate along the field lines and locally increase the magnetic field value. Paramagnetic materials have a positive value of magnetic susceptibility, commonly ranging from $+10^{-6}$ to 10^{-2}. Hence, their relative magnetic permeability is slightly above unity ($\mu_r > 1$). Examples of paramagnetic materials are: (i) gases such as oxygen, (ii) all the chemical elements not listed in the previous paragraph dealing with diamagnets, such as, Li, Na, Mg, Al, Ti, Zr, Sn, Mn, Cr, Mo, W, and all the platinum group metals: Ru, Rh, Pd, Os, Ir, Pt. On the other hand, the magnetic susceptibility of paramagnetic materials decreases with an increase in temperature. The temperature dependence of the magnetic susceptibility of paramagnetic materials is given by the Curie–Weiss law described by the following equation:

$$\chi_m = \frac{\mu_0 n m^2}{3k(T - T_C)} = \frac{C}{(T - T_C)}$$

where, μ_0 is the magnetic permeability of a vacuum in H.m^{-1}, n the atom density in m^{-3}, m the microscopic dipolar magnetic moment of an atom in A.m^2, k Boltzmann's constant in J.K^{-1}, T the absolute thermodynamic temperature in K, T_C the paramagnetic Curie temperature in K, at which the susceptibility reaches its maximum value, and C the paramagnetic Curie constant in K^{-1}.

6.2.3 Ferromagnetic Materials

Ferromagnetic materials have magnetic dipolar moments aligned parallel to each other even without an external applied magnetic field. Particular zones in the material, where all the magnetic dipole moments exhibit the same orientation are called **magnetic domains** or **Weiss domains**. Interfaces between the Weiss domains are called **Bloch's boundaries** or **walls**. For instance in a polycrystalline material, crystal borders which separate different lattice orientations are Bloch walls. Nevertheless, in the same crystal or monocrystal, several magnetic domains can coexist together. Therefore, the entire macroscopic material is divided into small magnetic domains, each domain having a net magnetization even without an external field. This magnetization is called spontaneous magnetization (M_S). However, a bulk sample will generally not have a net magnetization since the summation of all spontaneous magnetization vectors in the various domains is zero due to their random orientations. But application of a small external magnetic field will cause growth of favorable domains resulting in materials having a high magnetization and a high magnetic susceptibility (roughly 10^6). Therefore, their relative magnetic permeabilities are largely above unity. The main elements which exhibit ferromagnetism are the three transition metals of the group VIIIB(8) Fe, Co, Ni and some lanthanides such as Gd, Tb, Dy, Ho, and Tm, crystalline compounds such as MnAs, MnBi, MnSb, CrO_2, and Fe_3C, and alloys or intermetallic compounds containing Fe, Co, and Ni (e.g., steel, mumetal, alnico, permalloy). However, above a certain critical temperature, called the **Curie temperature**, T_c, these materials lose their spontaneous magnetization and become paramagnetic. There are two main requirements for an atom of an element to be ferromagnetic. First of all, the atom must have a total angular momentum different from zero ($J \neq 0$). This atomic condition is completed when both electronic nonspherical subshells 3-d or 4-f are not completely filled and the summation of all electrons' spin angular momentum is different from zero. The second condition is based on thermodynamics; it is dependent on the sign of the difference between electronic repulsion energy between Fermi gases of two adjacent atoms and the energy from the repulsion of electrons having the same spin. The total energy variation is positive for ferromagnetic materials, while it is negative for nonferromagnetic materials (e.g., Pt, Mn, and Cr). The physicist Slater has established a practical criterion to determine the ferromagnetic character of a material. This criterion is the ratio between the equilibrium radius between two adjacent atoms in the solid and the average orbital radius of electron in 3-d or 4-f subshells. If this ratio is above 3 the material is ferromagnetic while those for which the ratio is below 3 do not exhibit ferromagnetic properties. Properties of ferromagnetic elements are listed in Table 6.2, while properties of ferromagnetic compounds are reported in Table 6.3 and those of ferromagnetic ferrites are reported in Table 6.4.

Table 6.2 Properties of ferromagnetic elements

Element	Fe	Co	Ni	Gd	Tb	Dy	Ho	Er	Tm
Curie temperature (T_C, K)	1043.15	1394.15	631.15	292.15	222	87	20	32	25
Saturation magnetization (B_S, T) at 4 K	2.193	1.797	0.656	2.470	3.430	3.750	3.810	3.410	2.700
Atomic dipole magnetic moment (μ, μ_B)	2.22	1.72	0.62	7	9	10	10	9	7

Table 6.3 Properties of selected ferromagnetic compounds

Ferromagnetic compound	CrTe	EuO	Fe_3C	FeB	MnAs	MnB	MnBi	MnSb	MnSl
Curie temperature (T_C, K)	339	77	483	598	318	578	630	587	34
Saturation induction (B_S, T) at 293 K	0.0247	0.1910	n.a.	n.a.	0.0670	0.0152	0.0620	0.0710	n.a.

Table 6.4 Properties of selected ferromagnetic ferrites and garnets

Ferrite type	Chemical formula	Structure type, crystal system, lattice parameters, strukturbericht, Pearson's symbol, and space group	Magnetic induction saturation (B_S, T)	Curie temperature (T_C, °C)
Ba–Fe ferrite	$BaFe_{12}O_{19}$	Hexagonal	0.45	430
Cobalt ferrite	$CoFe_2O_4$	Spinel type, cubic H1$_1$, cF56 ($Z = 8$) Fd3m	0.53	520
Copper ferrite	$CuFe_2O_4$	Spinel type, cubic H1$_1$, cF56 ($Z = 8$) Fd3m	0.17	455
Eu–Fe garnet	$Eu_3Fe_5O_{12}$	Garnet type, cubic Fm3m ($Z = 8$)	0.116	293
Franklinite	$ZnFe_2O_4$	Spinel type, cubic ($a = 842.0$ pm) H1$_1$, cF56 ($Z = 8$) Fd3m	0.50	375
Gd–Fe garnet	$Gd_3Fe_5O_{12}$	Garnet type, cubic Fm3m ($Z = 8$)	0.017	291
Jacobsite	$MnFe_2O_4$	Spinel type, cubic ($a = 851.0$ pm) H1$_1$, cF56 ($Z = 8$) Fd3m	0.50	300
Lithium ferrite	$LiFe_{[5}O_8$	Spinel type, cubic H1$_1$, cF56 ($Z = 8$) Fd3m	0.39	670
Magnesioferrite	$MgFe_2O_4$	Spinel type, cubic ($a = 838.3$ pm) H1$_1$, cF56 ($Z = 8$) Fd3m	0.14	440
Magnetite	Fe_3O_4	Spinel type, cubic ($a = 839.4$ pm) H1$_1$, cF56 ($Z = 8$) Fd3m	0.60	585
Manghemite	γ-Fe_2O_3	Spinel type, cubic ($a = 834.0$ pm) H1$_1$, cF56 ($Z = 8$) Fd3m	0.52	575
Ni–Al ferrite	$NiAlFe_2O_4$	Spinel type, cubic H1$_1$, cF56 ($Z = 8$) Fd3m	0.05	1860
Nickel ferrite	$NiFe_2O_4$	Spinel type, cubic H1$_1$, cF56 ($Z = 8$) Fd3m	0.34	575
Sm–Fe garnet	$Sm_3Fe_5O_{12}$	Garnet type, cubic Fm3m ($Z = 8$)	0.170	305
Sr–Fe ferrite	$SrFe_{12}O_{19}$	Hexagonal	0.40	450
Y–Fe garnet	$Y_3 Fe_5O_{12}$	Garnet type, cubic Fm3m ($Z = 8$)	0.178	292

6.2.4 Antiferromagnetic Materials

Antiferromagnetic materials have an antiparallel arrangement of equal spins resulting in very low magnetic susceptibility similar to that of paramagnetic materials. The spin arrangement of antiferromagnetic materials is not stable above a critical temperature, called the Néel

Table 6.5 Néel temperature of antiferromagnetic elements

Element	Ce	Nd	Tm	Er	Eu	Mn	Sm	Ho	Dy	Tb	Cr
Néel temperature (T_N, K)	12.5	19.2	56	80–84	90.5	100	106	131–133	176–179	229–230	311–475

temperature,[1] T_N. For instance, antiferromagnetic materials are transition metals oxides such as MnO, FeO, and NiO, and other solids such as MnS, CrSb, $FeCO_3$, and MnF_2. The Néel temperatures of some antiferromagnetic elements and ferrite compounds are listed in Table 6.5 and Table 6.6.

6.2.5 Ferrimagnetic Materials

Ferrimagnetic materials have two kinds of magnetic ions with unequal spins, oriented in an anti-parallel fashion. The spontaneous magnetization can be regarded as the two opposing and unequal magnetizations of the ions on the two sublattices. Ferrimagnetic materials become paramagnetic above a Curie temperature.

6.3 Ferromagnetic Materials

6.3.1 *B–H* Magnetization Curve and Hysteresis Loop

The magnetic induction, B, of a ferromagnetic material is depicted in detail on page 283 as a function of the applied external magnetic field, H (i.e., $B–H$ diagram or hysteresis loop). At the beginning, the magnetic induction starts from zero at zero magnetic field (i.e., $B = 0$ and $H = 0$), increasing gradually the magnetic field. During this stage, the Weiss domains, in which spontaneous magnetization has the same orientation of the applied magnetic field, grow despite the other magnetic domains with displacement of the Bloch walls inside the material. Reversible motion of Bloch walls for very low magnetic fields explains the initial linear slope of the curve. However, when the external magnetic field reaches a maximum value, H_S, all the material forms a single Weiss domain having a net maximum magnetic induction value called **saturation magnetic induction**, B_S which is collinear to the applied magnetic field. When the external magnetic field is decreased, the magnetic induction follows a curve with higher values than the original curve owing to the irreversibility of wall motion.

At zero magnetic field ($H = 0$), magnetic domains tend to reappear slowly, and remain inside the material a residual magnetic induction, called **remanence or remanent magnetic induction**, B_R. The maximum residual magnetic induction at which the materials were fully magnetized is called **retentivity**. In order to remove completely the retentivity of the materials, it is necessary to apply an opposite magnetic field called **coercivity or coercitive force or coercive magnetic field** of strength H_c. This process, which cancels the magnetic induction, is called demagnetization. Application of a higher magnetic field causes the reversal behavior previously described saturating the material. Reversing the magnetic field leads to completion of the $B–H$ curve. The entire curve is called the **hysteresis curve or loop**. A condition of zero magnetization at zero field can only be again achieved by heating the materials past the Curie temperature to generate a new system of random magnetic domains. Therefore, magnetic properties of ferromagnetic materials are entirely described by the four parameters B_S, B_R, H_C and H_S. On the other hand, the surface area below the curve represents the stored magnetic energy by unit volume of material, expressed in $J.m^{-3}$. Therefore, the particular surface area between the magnetization and the demagnetization curve, i.e., inside the hysteresis loop, represents the **magnetic energy losses** per unit volume of materials. It is important to note that the magnetic permeability of ferromagnetic materials is not a constant physical quantity and depends on a

Table 6.6 Néel temperature of selected antiferromagnetic compounds

Ferrite type	Chemical formula	Structure type, crystal system, lattice paramaters, strukturbericht, Pearson's symbol, and space group	Néel temp. $(T_N, °C)$
Ca–Mn oxide	$CaMnO_3$	Orthorhombic, E2$_1$, cP5 ($Z = 4$), Pm3m, perowskite type	−163.15
Chromium arsenide	CrAs	Hexagonal, B8$_1$, hP4 ($Z = 2$), P63/mmc, niccolite type	+26.85
Chromium (III) oxide	Cr_2O_3	Trigonal (rhombohedral), D5$_1$, hR10 ($Z = 6$), R-3c, corundum type	+44.85
Co–Ti oxide	$CoTiO_3$	Trigonal (rhombohedral), D5$_1$, hR10 ($Z = 6$), R-3c, corundum type	−235.15
Cobalt (II) fluoride	CoF_2	Tetragonal, C4, tP6 ($Z = 2$), P4$_2$/mnm, rutile type	−235.15
Cobalt (II) oxide	CoO	Face centered cubic	+17.85
Cobalt oxide	Co_3O_4	Cubic, H1$_1$, cF56, Fd3m, spinel type	−233.15
Copper (I) oxide	CuO	Monoclinic	−43.15
Erbia	Er_2O_3	Cubic, D5$_3$, c180, Ia-3, α-Mn$_2$O$_3$ type	−269.75
Franklinite	$ZnFe_2O_4$	Cubic ($a = 842.0$ pm), H11, cF56 ($Z = 8$), Fd3m (Spinel type)	−264.15
Gadolinia	Gd_2O_3	Cubic, D5$_3$, c180, Ia-3, α-Mn$_2$O$_3$ type	−271.55
Hematite	α-Fe$_2$O$_3$	Trigonal (rhombohedral) ($a = 503.29$ pm 13.749°), D5$_1$, hR10 ($Z = 6$), R-3c, corundum type	+674.85
Illmenite	$FeTiO_3$	Trigonal (rhombohedral) ($a = 509.3$ pm $c = 1405.5$ pm), R$\bar{3}$m	−205.15
Iron (II) fluoride	FeF_2	Tetragonal, C4, tP6 ($Z = 2$), P42/mnm, rutile type	−194.15
Iron (II) oxide	FeO	Face centered cubic	−75.15
La–Cr oxide	$LaCrO_3$	Orthorhombic, E2$_1$, cP5 ($Z = 4$), Pm3m, perowskite type	+8.85
La–Mn oxide	$LaMnO_3$	Orthorhombic, E2$_1$, cP5 ($Z = 4$), Pm3m, perowskite type	−173.15
Manganese (II) oxide	MnO	Face centered cubic	−151.15
Manganese telluride	MnTe	Hexagonal, B8$_1$, hP4 ($Z = 2$), P63/mmc, niccolite type	+49.85
Manganese (III) oxide	Mn_2O_3	Cubic, D5$_3$, c180, Ia-3, α-Mn$_2$O$_3$ type	−183.15
Mn–Ti oxide	$MnTiO_3$	Trigonal (rhombohedral), D5$_1$, hR10 ($Z = 6$), R-3c, Corundum type	−232.15
Nd–Fe oxide	$NdFeO_3$	Orthorhombic, E2$_1$, cP5 ($Z = 4$), Pm3m, perowskite type	+486.85
Niccolite	NiAs	Hexagonal ($a = 360.9$ pm, $c = 501.9$ pm), B8$_1$, hP4 ($Z = 2$), P6$_3$/mmc, niccolite type	−10.15
Nickel fluoride	NiF_2	Tetragonal, C4, tP6 ($Z = 2$), P42/mnm, rutile type	−190.15
Nickel oxide	NiO	Face centered cubic	+251.85
Pyrolusite	MnO_2	Tetragonal, ($a = 438.8$ pm, $c = 286.5$ pm) C4 tP6 ($Z = 2$), P4$_2$/mnm, rutile type	−189.15
Uranium (IV) oxide	UO_2	Cubic ($a = 546.82$ pm), C1, cF12 ($Z = 4$), Fm3m (fluorite type)	−242.35

particular region of the **B–H** diagram. Actually, the initial slope of the **B–H** curve is called the **initial magnetic permeability** (μ_{in}), and the maximum slope measured from the origin is called the **maximum magnetic permeability** (μ_{max}), while magnetic permeability measured for an applied alternating magnetic field is termed **ac magnetic permeability**.

From the close examination of hysteresis curves of numerous ferromagnetic materials, it is possible to classify ferromagnetic materials into two distinct chief categories: (i) the **retentive** or **hard** (ferro)**magnetic materials** having high coercivity, high losses and a large hysteresis loop, and (ii) **nonretentive** or **soft** (ferro)**magnetic materials** with a high permeability having low coercivity, low magnetic losses and a narrow hysteresis loop. From a practical point of view, hard magnetic materials such as cobalt steel retain their magnetism for long periods of time when the magnetizing field is removed, i.e., they can act as permanent magnets.

6.3.2 Eddy Current Losses

When an alternating magnetic field is applied to a ferromagnetic material, it induces an electromotive force, denoted emf (in V), which generates eddy currents. The Joule heating due to the current is termed eddy current losses. For a low frequency, f (in Hz), the flux penetration into the material is complete and proportional to f^2 and to the reciprocal of the electrical resistivity. At high frequency and for a constant flux amplitude, the eddy current losses remain chiefly in a thin layer at the surface of the material (the skin effect) and are proportional to $f^{3/2}$. The former effect is extensively used in induction heating.

6.3.3 Soft Ferromagnetic Materials

Soft or **nonretentive** ferromagnetic materials exhibit a high magnetic permeability combined with a high saturation induction, and a low coercivity. Moreover, their **B–H** diagram curve shows a narrow hysteresis loop, low magnetic losses and low eddy current losses when an alternating magnetic flux is applied. This particular class of ferromagnetic materials are able to be magnetized with a small magnetic field and they are used in electrical applications requiring the magnetic characteristics described previously and involving changing magnetic induction such as solenoids, motors, relays, transformers, and magnetic shielding. Magnetically soft materials are for instance: pure iron (e.g., Armco®), highly pure iron (e.g., carbonyl iron), low-carbon steels, gray and ductile cast irons, silicon steels, Fe–Ni alloys (e.g., MuMetal®, Permalloy®, Perminvar®), Fe–Co alloys (e.g., Permindur®, Hyperco®), Fe–Cr alloys, Fe–Al–Si alloys (e.g., Alfer®, Sendust®), spinels (e.g., Ni–Zn, Mn–Fe, and Ni–Co ferrites), hexagonal ferrites, and synthetic garnets. On the other hand, properties of magnetically soft materials can be easily split into two main groups, the first being properties which are strongly sensitive, influenced by the space lattice structure, the content of trace impurities, and the macroscopic grain size. These properties are the relative magnetic permeability (μ_r), coercive magnetic field (H_c), hysteresis losses, and remanent induction (B_r). By contrast, other magnetic properties such as magnetic induction at saturation (B_S), electrical resistivity (ρ), and the Curie temperature (T_C) are sensitive to the material structure. Therefore, trace impurities, alloying elements and heat treatment (e.g., annealing, work hardening) are main factors that strongly affect the structure-sensitive magnetic properties of the final magnetically soft material, and they must be carefully controlled during their processing in order to prepare the appropriate material. Actually, the impurities of elements such as C, N, O, and S even in minute amounts tend to locate in the interstitial sites in the material space lattice and hence, they avoid the reversible displacement of Weiss magnetic domains increasing the hysteresis losses of the materials. By contrast the effect of alloying ferromagnetic elements such as Fe, Co, and Ni is most favorable because they contribute to promote high magnetic permeability, low coercivity, and low hysteresis losses. Moreover, these additions also increase the electrical resistivity which reduce eddy current losses when the material is submitted to an alternating current. Finally, heat treatment is often required for magnetically soft materials prepared by metalworking operations such as cold working, rolling or stamping. As a general rule, the heat treatments (e.g., annealing, tempering) are essential: (i) to restore the grain size, (ii) to reduce or eliminate residual stresses, and to a lesser extent (iii) to improve the formability. Properties of selected magnetically soft materials are listed in Table 6.7.

6.3.4 Hard Magnetic Materials

Hard or retentive ferromagnetic materials are characterized by retaining a high remanence or residual magnetization even after exposure to a strong magnetic field, a high coercivity, and an elevated energy content. These materials exhibit a high coercivity, commonly between 5000 to 900,000 A.m^{-1} and their B–H diagram shows a wide hysteresis loop. From a practical point of view, hard magnetic materials are alloys that when magnetized retain their magnetization for long periods of time when the magnetizing field is removed, i.e., they can act as **permanent magnets**. Actually, the term permanent magnet is used to describe magnetic materials having a high coercivity, and exhibiting a high output magnetic flux. The main property useful for selecting permanent magnet material is the maximum value of the BH product, $(BH)_{max}$. Magnetically hard materials are composed of a fine crystallographic structure, and they can be classified in four major classes: (i) martensitic lattice transformation alloys (e.g., high-carbon steels containing alloying elements such as Al, V, Co, Cr, Mo, W), (ii) precipitation hardened alloys (e.g., Alnico®, Cunife®, Cunico®, Vicalloy®) obtained by quenching or work hardening, (iii) ordered alloys exhibiting ordered superlattice structure on aging (e.g., Fe–Pt and Pt–Co alloys), (iv) compacted fine-particle magnets obtained from sintered or bonded particles such as hard ferrites, and rare-earth-containing alloys. Therefore, the main magnetic characteristics of permanent magnets are: the intrinsic induction (B_i), the remanent or residual induction (B_r), the coercive magnetic field or force (H_c), the intrinsic coercive magnetic field or force (H_{ci}), the maximum magnetic energy stored (BH_{max}). Properties of selected magnetically hard materials are listed in Table 6.8.

Table 6.7 Properties of soft magnetic metals and alloys

Material	Average chemical composition (x, wt%)	Curie temperature T_C, °C	Maximum magnetic permeability (μ_{max}, μ_0)	Remanence magnetic induction (B_R, T)	Coercive magnetic field (H_C, A.m^{-1})	Saturation magnetic induction (B_S, T)	Electrical resistivity (ρ, $\mu\Omega$cm)
Alfenol® 16 (Alperm)	84Fe–16Al	450	55,000–116,000	0.38	1.98–3.20	0.78–0.80	150
Alfer®	87Fe–13Al	400–426	3700	n.a.	53	1.20	n.a.
Armco® Iron	Fe	770	6000–8000	0.11–0.58	32–70	2.158	10
Ferrosilicon	99Fe–1Si	740	7700	0.80–1.10	44	2.10	25
Ferrosilicon	96Fe–4Si	735	18,500	1.08	24	1.95	58
Ferroxcube® 101 or B	(Ni,Zn)Fe$_2$O$_4$	n.a.	n.a.	0.11	14.3	0.23	> 10^5
Ferroxcube® 3 or A	(Mn,Zn)Fe$_2$O$_4$	n.a.	1500	0.10	7.9	0.30	> 10^6
HyMu® 80	80Ni–20Fe	n.a.	100,000	n.a.	n.a.	0.87	57
Hyperco®	64.5Fe–35Co–0.5Cr	970	10,000	n.a.	80	2.42	n.a.
Hyperco® 27	72.4Fe–27Co–0.6Cr	925	n.a.	n.a.	n.a.	2.40	19
Hypernik® V	51Fe–49Ni	480	180,000	0.90	4.8	1.55	47
MuMetal®	77Ni–18Fe–5Cu–2Cr	405	100,000–375,000	0.30–0.34	0.4–0.6	0.77	56
Permalloy 45	55Fe–45Ni	480	90,000	0.68–0.87	4	1.58	50
Permalloy® 78	78.5Ni–21.5Fe	378	100,000–300,000	0.50	4.0	1.05	16
Permendur®	50Fe–50Co	980	6000	n.a.	160	2.46	26
Permendur® 2V	49Fe–49Co–2V	980	4500	1.40	159	2.40	43
Perminvar®	43Ni–34Fe–23Co	n.a.	400,000	n.a.	2.4	1.50	n.a.
Perminvar® 25	45Ni–30Fe–25Co	n.a.	2000	n.a.	100	1.55	n.a.
Perminvar® 7	70Ni–23Fe–7Co	n.a.	4000	n.a.	50	1.25	n.a.
Rhometal®	64Fe–36Ni	n.a.	5000	0.36	39.79	1.00	90
Sendust®	85Fe–10Si–5Al	480	120,000	0.50	3.980	1.00	60–80
Silicon iron alloys	97Fe–3Si	757	30,000	n.a.	12	2.00	50
Steel AISI 1010	Fe (0.1wt% C, 0.33wt% Si, and 0.67wt% Mn)	n.a.	2420–3800	0.80–0.90	136–160	2.15	13
Supermalloy®, Magnifer® 7904	79Ni–15Fe–5Mo–0.5Mn	443–455	400,000–1,000,000	0.35–0.70	0.3–0.4	0.79	59
Supermendur®	49 Fe–49Co–2V	980	70,000	2.14	16	2.40	n.a.

Conversion factors: 1 T = 10^4 G (E); 1 Oe = (1000/4π) A.m^{-1} (E) ∼ 79.57747155 A.m^{-1}; 1 $\mu\Omega$cm = 10^{-8} Ωm(E)

6

Magnetic Materials

Table 6.8 Properties of selected hard magnetic materials

Hard magnetic alloy	Average chemical composition (wt%)	Density (ρ, kg.m^{-3})	Curie temperature (T_C, °C)	Maximum operating temperature (T_{max}, °C)	Remanence magnetic induction (B_r, T)	Coercive magnetic field (H_C, A.m^{-1})	Intrinsic coercive magnetic field (H_C, A.m^{-1})	Maximum magnetic energy (BH)$_{max}$, kJ.m^{-3}	Demagnetization induction (B_d, T)	Electrical resistivity (ρ, $\mu\Omega$cm)
Alnico® 1 (cast)	Fe–12Al–21Ni–5Co–3Cu	6900	780	540	0.710	35,000	36,000	11	0.45	75
Alnico® 12 (cast)	Fe–6Al–18Ni–35Co–8Ti	7400	n.a.	480	0.600	64,000	76,000	14	0.315	62
Alnico® 2 (cast)	Fe–10Al–19Ni–13Co–3Cu	7100	810	540	0.725	44,000	46,000	13	0.45	65
Alnico® 2 (sintered)	Fe–10Al–17Ni–12.5Co–6Cu	6800	610	480	0.670	42,000	44,000	12	0.43	68
Alnico® 3 (cast)	Fe–12Al–25Ni–3Cu	6900	760	480	0.700	38,000	39,000	11	0.43	60
Alnico® 4 (cast)	Fe–12Al–27Ni–5Co	7000	800	590	0.535	58,000	62,000	10	0.30	75
Alnico® 4 (sintered)	Fe–12Al–28Ni–5Co	6900	800	590	0.520	56,000	61,000	10	0.30	68
Alnico® 5 (cast)	Fe–8.5Al–14.5Ni–24Co–3Cu	7300	860–900	540	1.250–1.280	50,930	50,930	42–44	1.02	47
Alnico® 5 (sintered)	Fe–8.5Al–14.5Ni–24Co–3Cu	6900	860	525	1.040–1.090	49,338	50,134	29–31	0.785	50
Alnico® 5-7 (cast)	Fe–8.5Al–14.5Ni–24Co–3Cu	7300	900	540	1.320	58,000	59,000	59	1.15	47
Alnico® 5DG (cast)	Fe–8.5Al–14.5Ni–24Co–3Cu	7300	900	540	1.290	52,000	52,000	49	1.05	47
Alnico® 6 (cast)	Fe–8Al–16Ni–	7400	860	540	1.050	60,000	75,000	30	0.71	50

Material	Composition									
Alnico®7 (cast)	24Co–3Cu–2Ti	7300	840	540	0.857	84,000	n.a.	30	n.a.	58
Alnico®8 (cast)	Fe–8Al–18Ni–24Co–5Cu–5Ti	7300	860	550	0.820–0.830	131,303	148,014	40–42	0.506	50
Alnico®8 (sintered)	Fe–7Al–15Ni–35Co–4Cu–5Ti	7000	860	n.a.	0.760	125,000	134,000	36	0.46	n.a.
Alnico®9 (cast)	Fe–7Al–15Ni–35Co–4Cu–5Ti	7300	n.a.	520	1.050	115,000	145,000	68	n.a.	n.a.
Chromindur® II	Fe–28Cr–10.5Co	n.a.	630	500	0.980	32,000	n.a.	16	n.a.	n.a.
Chromium steel	Fe–3.5Cr–1Cr	7770	745	n.a.	0.950	5300	n.a.	2.3	n.a.	29
Cobalt samarium 1	$SmCo_5$	8200	725	250	0.920	720,000	1,600,000	170	n.a.	50
Cobalt samarium 2	$SmCo_5$	8200	725	500	0.860	640,000	2,000,000	145	0.44	50
Cobalt samarium 3	$SmCo_5$	8200	725	500	0.800	535,000	1,200,000	120	0.40	50
Cobalt samarium 4	Sm_2Co_{17}	n.a.	800	500	1.130	640,000	640,000	240	0.60	50
Cunico®	Cu–21Ni–29Fe	n.a.	n.a.	n.a.	0.340	500	n.a.	8	n.a.	n.a.
Cunife®	20Fe–20Ni–60Cu	8600	410	350	0.540	44,000	44,000	12	0.40	18
Ferrite 1 (sintered)	BaO–$6Fe_2O_3$	4800	450	400	0.220	145,000	276,000	8	0.11	10^{12}
Ferrite 2 (sintered)	BaO–$6Fe_2O_3$	5000	450	400	0.380	175,000	185,000	27	0.185	10^{12}
Ferrite 3 (sintered)	BaO–$6Fe_2O_3$	4500	450	400	0.320	240,000	292,000	20	0.16	10^{12}
Ferrite 4 (sintered)	SrO–$6Fe_2O_3$	4800	460	400	0.400	175,000	185,000	30	0.215	10^{12}
Ferrite 5 (sintered)	SrO–$6Fe_2O_3$	4500	460	400	0.355	250,000	287,000	24	0.173	10^{12}

Table 6.8 (continued)

Hard magnetic alloy	Average chemical composition (wt%)	Density (ρ, kg.m^{-3})	Curie temperature (T_C, °C)	Maximum operating temperature (T_{max}, °C)	Remanence magnetic induction (B_r, T)	Coercive magnetic field (H_C, A.m^{-1})	Intrinsic coercive magnetic field (H_C, A.m^{-1})	Maximum magnetic energy (BH)$_{max}$, kJ.m^{-3}	Demagnetization induction (B_d, T)	Electrical resistivity (ρ, $\mu\Omega$cm)
Ferrite A (bonded)	BaO-6Fe$_2$O$_3$+ org. binder	3700	450	95	0.214	155,000	n.a.	8	0.116	10^{12}
Ferrite B (bonded)	BaO-6Fe$_2$O$_3$+ org. binder	3700	450	n.a.	0.140	92,000	n.a.	3	n.a.	10^{12}
Ferroxdur®	BaFe$_{12}$O$_{19}$	n.a.	450	400	0.400	160,000	192,000	29	n.a.	n.a.
High-cobalt steel	Fe–36Co–3.75W–5.75Cr–0.8C	8180	890	n.a.	0.975	19,000	n.a.	7.4	n.a.	27
Low-cobalt steel	Fe–17Co–8.5W–2.5Cr–0.7C	8.350	n.a.	n.a.	0.950	14,000	n.a.	5.2	n.a.	28
Neodymium® 27	NdFeB	7400	280	80	1.080	740,070	875,350	215	n.a.	10^{12}
Neodymium® 27H	NdFeb	7400	300	100	1.080	779,860	1,352,282	215	n.a.	10^{12}
Neodymium® 30	NdFeB	7400	280	80	1.100	795,780	1,432,400	238	n.a.	10^{12}
Neodymium® 30H	NdFeB	7400	300	100	1.100	835,563	1,352,820	223–238	n.a.	10^{12}
Neodymium® 35	NdFeB	7400	280	80	1.180–1.230	835,563	954,930	269–279	n.a.	10^{12}
Neodymium® 35H	NdFeB	7300	280–315	80	1.160–1.200	899,000	1,474,930	236–262	n.a.	10^{12}
Platinum cobalt	76.7Pt–23.3Co	15,500	480	350	0.645	355,000	430,000	74	0.35	28
Tungsten steel	Fe–6W–0.5Cr–0.7C	8120	760	n.a.	0.950	5900	n.a.	2.6	n.a.	30
Vicaloy® II	Fe–52Co–14V	n.a.	700	500	1.000	42,000	n.a.	28	n.a.	n.a.

Conversion factors: 1 T = 10^4 G (E); 1 Oe = (1000/4π) A.m^{-1}(E) ~ 79.57747155 A.m^{-1}, 1 MGOe = (25/4π)kJ.m^{-3} ~ 7.957747155 kJ.m^{-3}

6.3.5 Magnetic Shielding and Materials Selection

Magnetic shielding protects sensitive electronic circuitry from electromagnetic interference (EMI). Usually, the common sources of EMI are: (i) permanent magnets, transformers, motors, solenoids, and electric cables generating a strong magnetic field; (ii) magnetic fields present in the environment or emanating from other emitting sources. As a general rule, the appropriate magnetic shielding materials deflect magnetic flux by providing a path around the sensitive volume to protect from EMI. In addition shielding may be used to contain magnetic flux around a component generating a magnetic flux. Actually, a suitable shielding material is a magnetically soft material having a high relative magnetic permeability ranging between 200 and 350,000 giving it the ability to conduct magnetic lines of force; the most common is the alloy designated commercially as MuMetal®, nevertheless selected properties of other suitable soft magnetic materials can be found in Table 6.7. The design of an efficient magnetic shield for a definite external magnetic field neads the careful selection of several parameters either magnetic properties of the shielding material (e.g., permeability, magnetic induction at saturation) or pure geometric considerations (e.g., shield thickness).

6.3.5.1 Maximum Allowed Magnetic Induction in the Shield

The maximum allowed magnetic induction, B_{max} in T, generated by the applied magnetic field inside the shielding materials is given by the following equation:

$$B_{max} = \frac{\mu_0 \mu_r dH}{t}$$

where H is the applied external magnetic field, in $A.m^{-1}$, d the diameter or diagonal of the shield in m, t the thickness of the shield in m, and μ_r the relative permeability of the shielding material. However, the maximum induction field B_{max} must be always maintained below the saturation magnetic induction at saturation B_S for the selected shielding material. For withstanding a higher external magnetic field, two main solutions can be used together or separately: (i) a thicker shield can be used, either as one layer or multiple layers of thinner shields, and (ii) another appropriate magnetically soft material having a higher B_S value can be selected. The attenuation or **shielding efficiency** of a magnetic shield is called the **attenuation ratio**; it is a dimensionless physical quantity denoted a, which corresponds to the ratio of the measured magnetic field before and after shielding. As a general rule, the attenuation decreases in shields of large volume, or having a wide opening and unusual configurations. The attenuation ratio is hence given by the following equation:

$$a = \frac{t}{\mu_r D}.$$

Reference

[1] Néel L (1948) Ann Phys 3: 137.

Further Reading

Ball R (1979) Soft magnetic materials. Heyden & Sons Ltd., London.
Bozorth RM (1951) Ferromagnetism. Van Nostrand Company, Princeton.
Bozorth RM, McGuire TR, Hudson RP (1972) Magnetic properties of materials. In: Gray DE (ed.) American Institute of physics handbook, 3rd edn. McGraw-Hill, New York, pp 5–139 to 5–145.
Burke HE (1986) Handbook of magnetic phenomena. Van Nostrand Reinhold, New York.

Campbell P (1994) Permanent magnetic materials and their applications. Cambridge University Press, New York.

Cullity RD (1972) Introduction to magnetic materials. Addison-Wesley, New York.

Douglas WD (1995) Magnetically soft materials. In: ASM metals handbook, 9th edn, vol. 2: Properties and selection of nonferrous alloys and special purpose materials. ASM, Metal Park OH, pp 761–781.

Fiepke JW (1995) Permanent magnet materials. In: ASM metals handbook, 9th edn, vol. 2: Properties and selection of nonferrous alloys and special purpose materials. ASM, Metal Park OH, pp 782–803.

Frederikse HPR (1997–1998) Properties of magnetic materials. In: Lide DR Handbook of chemistry and physics, 78th edn. CRC Press, Boca Raton, FL, pp 12–117 to 12–118.

Heck C (1974) Magnetic materials and their applications. Butterworths, London.

Hubert A, Shäfer R (1998) Magnetic domains: the analysis of magnetic microstructures. Springer-Verlag, Heidelberg.

Ishiguro T, Yamaji K, Saito G (1998) Organic superconductors, 2nd edn. Springer-Verlag, Heidelberg.

Jiles D (1991) Introduction to magnetism and magnetic materials. Chapman and Hall, London.

Kittel C (1987) Introduction to solid state physics, 6th edn. John Wiley & Sons, New York.

McCaig M, Clegg AE (1986) Permanent magnetics in theory and practice, 2nd edn. Pentech and Wiley, New York.

McCurrie RA (1994) Ferromagnetic materials. Academic Press, London.

Parker RJ (1962) Permanent magnets and their applications. John Wiley & Sons, New York.

Snelling EC (1987) Soft ferrites, properties, and applications, 2nd edn. Butterworths, London.

Wijn HPJ (1991) Magnetic properties of metals: d-elements, alloys, and compounds. Springer-Verlag, Heidelberg.

7

Insulators and Dielectrics

7.1 Physical Quantities of Dielectrics

The set of physical quantities essential to understand the most common electrical properties of dielectric materials are briefly described in the following paragraphs.

7.1.1 Permittivity of a Vacuum

The permittivity of a vacuum (sometimes called permittivity of empty space by electrical engineers), is denoted ε_0, and is expressed in F.m^{-1}. It is defined by Coulomb's law for a vacuum. Actually, the modulus of the electrostatic force, F_{12}, expressed in newtons (N), between two point electric charges in a vacuum, q_1, and q_2, expressed in coulombs (C) separated by a distance r_{12} in meters (m) is given by the following equation:

$$F_{12} = \frac{1}{4\pi\varepsilon_0} \frac{q_1 q_2}{r_{12}} e_r$$

with $\varepsilon_0 = 1/\mu_0 \cdot c^2 = 8.8541187817 \times 10^{-12}$ F.m^{-1}.

7.1.2 Dielectric Permittivity

The dielectric permittivity of a medium, denoted ε, is expressed in farads per meter (F.m^{-1}), and is defined by Coulomb's law for a medium. Actually, the modulus of the electrostatic force, F_{12}, in newtons (N), between two point electric charges in a medium q_1, and q_2, in coulombs (C) separated by a distance r_{12} in meters (m) is given by the following equation:

$$F_{12} = \frac{1}{4\pi\varepsilon} \frac{q_1 q_2}{r_{12}} e_r.$$

7.1.3 Relative Dielectric Permittivity

The relative dielectric permittivity, denoted ε_r, is a dimensionless physical quantity equal to the ratio of the dielectric permittivity over the permittivity of vacuum; it was also called the dielectric constant of the medium. Hence, it is defined as follows:

$$\varepsilon = \varepsilon_0 \varepsilon_r.$$

Nevertheless, the dielectric constant of an insulating material strongly depends on the frequency, v, in hertz (Hz) of the electric field applied. Common dielectric constants listed in handbooks and databases are measured at a frequency of 1 MHz. Hence, the total permittivity can be entirely described as a complex physical quantity, where the imaginary part is related to dielectric losses:

$$\varepsilon = \varepsilon'_r + j\varepsilon''_r.$$

7.1.4 Capacitance

Two conducting bodies or **electrodes**, separated by a dielectric constitute a **capacitor** (formerly **condenser**). If a positive charge is placed on one electrode, an equal negative charge in simultaneously induced in the other electrode in order to maintain electrical neutrality. Therefore, the principal characteristic of a capacitor is that it can store an electric charge Q, expressed in coulombs (C), which is directly porportional to the voltage applied expressed in volts (V) according to the following equation:

$$Q = CV$$

where C is termed the capacitance expressed in farads (F). Hence, the value of capacitance is defined as one farad when the electric potential difference (i.e., voltage) across the capacitor is one volt, and a charging current of one ampere flows for one second. The required charging current i, in amperes (A) is therefore defined as:

$$i = dQ/dt = CdV/dt.$$

Because the farad is a very large unit of measurement, and is not encountered in practical applications, SI submultiples of the farad are commonly used; in increasing order the most used are: the picofarad (pF), the nanofarad (nF), and the microfarad (μF). Another important point, is that the dielectric properties of a medium relate to its ability to conduct dielectric lines. This must be clearly distinguished from its insulating properties which relate to its ability not to conduct an electric current. For instance, an excellent electrical insulator can rupture dielectrically at low breakdown voltages.

7.1.5 Temperature Coefficient of Capacitance

The temperature coefficient of capacitance, denoted a or TCC, and expressed in K^{-1}, is determined accurately by measurement of the capacitance change at various temperatures from a reference point set usually at room temperature (T_1) up to a required higher temperature (T_2), by means of an environmental chamber:

$$a = \frac{1}{C}\frac{\partial C}{\partial T}.$$

Note, however, that in the electrical industry, the temperature coefficient of capacitance is usually expressed as the percent change in capacitance, or in parts per million per degree

Celsius (ppm.$°C^{-1}$). Moreover, for industrial dielectrics, it is usually plotted in the temperature range $-55°C$ to $+125°C$.

7.1.6 Capacitance of a Parallel Electrode Capacitor

The capacitance of a parallel electrode capacitor is directly proportional to the active electrode area, and inversely proportional to the dielectric thickness as described by the following equation:

$$C = \varepsilon_0 \varepsilon_r \frac{A}{d}.$$

7.1.7 Capacitance of Other Capacitor Geometries

For more complicated capacitor geometries the capacitance is given in Table 7.1.

7.1.8 Electrostatic Energy Stored in a Capacitor

The electrostatic energy, W, expressed in joules (J) stored in a capacitor is given by the following equation:

$$W = \frac{1}{2}QV = \frac{1}{2}CV^2 = \frac{1}{2}\frac{Q^2}{C}.$$

Table 7.1 Capacitance of capacitors of different geometries

Capacitor Geometry	Description	Theoretical formula
	Parallel plates	$C = \varepsilon_0 \varepsilon_r \dfrac{A}{d}$
	Coaxial cylinders	$C = 2\pi\varepsilon_0\varepsilon_r \dfrac{l}{\left[\ln\left(\dfrac{R_2}{R_1}\right)\right]}$
	Concentric spheres	$C = 4\pi\varepsilon_0\varepsilon_r \dfrac{R_2 R_1}{(R_2 - R_1)}$
	Two parallel wires	$C = \varepsilon_0\varepsilon_r \dfrac{l}{\left[\ln\left(\dfrac{D}{r}\right)\right]}$

7.1.9 Electric Field Strength

The electric field strength, symbolized by E, is a vector quantity directed from negative charge regions to positive charge regions. Its modulus is expressed in volts per meter ($V.m^{-1}$). It is clearly defined by a vectorial equation as follows:

$$E = -\nabla V.$$

For insulating materials, it is common to define the dielectric field strength, symbolized by E_d, which is the maximum electric field that the material can withstand before the sparking begins (i.e., dielectric breakdown). The common non-SI units are the volt per micrometer ($V.\mu m^{-1}$) or in the US or UK systems the volt per mil ($V.mil^{-1}$). The dielectric field strength is a measure of the ability of the material to withstand a large electric field strength without occurrence of an electrical breakdown.

7.1.10 Electric Flux Density

The electric flux density or electric displacement, symbolized by D, is a vector quantity defined in vacuum as the product of the electric field strength by the permittivity of vacuum. Its modulus is expressed in coulombs per square meter ($C.m^{-2}$). It is defined according to the following equation:

$$D = \varepsilon_0 E.$$

In a medium, the electric flux density is defined as the product of the electric field strength and permittivity of the medium as follows:

$$D = \varepsilon E = \varepsilon_0 \varepsilon_r E.$$

7.1.11 Atomic and Molecular Polarizability

The polarizability of an atom or a molecule is a tensor physical quantity, denoted α, expressed in $C.m^2.V^{-1}$, which describes the response of the electron cloud (i.e., Fermi gas) to an external electric field strength E. The atomic or molecular energy shift dW is proportional to E^2 for an external field strength weaker than the internal electric field according to the equation:

$$dW = -(\alpha E^2/2).$$

The polarization of a dielectric material, containing n atoms per unit volume (m^{-3}), and having a relative dielectric permittivity ε_r is given by the Clausius–Mosotti equation:

$$\alpha = \frac{3}{4\pi n} \left[\frac{\varepsilon_r - 1}{\varepsilon_r + 2} \right].$$

7.1.12 Microscopic Electric Dipole Moment

Molecules having a asymmetric electron cloud distribution exhibit a permanent electric dipole moment. Hence, the electric dipole moment is a vector physical quantity denoted μ_i or p_i, with a modulus expressed in C.m (in some textbooks, it was expressed in the obsolete unit the Debye (D), with $1 D = 3.58 \times 10^{-29}$ C.m). The electric dipole moment between two identical electric charges q in coulombs (C) separated by a distance d_i, in meters (m) is given by the equation:

$$\mu_i = q_i d_i$$

Note that when a dipole is composed of two point charges $+q$ and $-q$, separated by a distance r, the direction of the dipole moment vector is taken to be from the negative to the positive electric charge. However, the opposite convention was adopted in physical chemistry but is to be discouraged. Moreover, the dipole moment of an ion depends on the choice of the origin.

7.1.13 Macroscopic Electric Dipole Moment

The macroscopic electric dipole moment, μ, expressed in C.m, is the summation of all the contributions of individual microscopic electric dipole moments:

$$\mu = \sum \mu_i = \sum q_i r_i.$$

7.1.14 Dielectric Polarization

The dielectric polarization within dielectric materials is a vector physical quantity, denoted P, and its modulus is expressed in coulombs per square meter (C.m^{-2}). Electric polarization arises due to the existence of atomic and molecular forces, and appears whenever electric charges in a material are displaced with respect to one another under the influence of an applied external electric field of strength E. On the other hand, the electric polarization represents the total electric dipole moment contained per unit volume of the material averaged over the volume of a crystal cell lattice, V, expressed in m^3:

$$P = \frac{\mu}{V} = n\mu.$$

The negative charges within the dielectric are displaced toward the positive region, while the positive charges shift in the opposite direction. Because electric charges are not free to move in an insulator owing to atomic forces involving them, restoring forces are activated which either do work, or cause work to be done on the circuit, i.e. energy is transferred. On charging a dielectric, the polarization effect opposing the applied field draws charges onto the electrodes, storing energy. By contrast, on discharge, this energy is released. The result of the above microscopic interaction is that materials which possess easily polarizable charges will greatly influence the degree of charge which can be stored in the material. The proportional increase in storage ability of a dielectric with respect to a vacuum is defined as the relative dielectric permittivity (sometimes called dielectric constant) of the material. The degree of polarization P is related to the relative dielectric permittivity and to the electric field strength E as follows:

$$P = D - \varepsilon_0 E = (\varepsilon_r - 1)\varepsilon_0 E.$$

7.1.15 Electrical Susceptibility

The electrical susceptibility of a material, denoted χ_e, is a dimensionless physical quantity defined as the ratio of the electric polarization over the electric flux density, according to the following equation:

$$P = \varepsilon_0 \chi_e E.$$

Therefore, the relation between electrical susceptibility and relative dielectric permittivity is given by:

$$\chi_e = (\varepsilon_r - 1).$$

In fact, when electric field strength is greater than the value of interatomic electric field (i.e., 100 to 10,000 MV.m^{-1}), the relationship between the polarization and the electric field strength becomes nonlinear and the relation can be expanded in a Taylor series:

$$P = \varepsilon_0 \left[\chi_e^{(1)} E + \left(\frac{1}{2}\right)\chi_e^{(2)} E^2 + \left(\frac{1}{6}\right)\chi_e^{(3)} E^3 + \cdots \right]$$

where $\chi_e^{(1)}$ is the linear dielectric susceptibility, and $\chi_e^{(2)}$, $\chi_e^{(3)}$ are respectively the first, and second hypersusceptibilities (sometimes called optical susceptibilities) expressed in C^3.m^3.J^{-2}, and C^4.m^4.J^{-3}. In a medium that is anisotropic these electric susceptibilities are tensor quantities of rank 2, 3, and 4, while for an isotropic medium (e.g., gases, liquids, amorphous solids, and cubic crystals) or for a crystal with a centrosymmetric unit cell, the second hypersusceptibility is zero by symmetry.

7.1.16 Dielectric Breakdown Voltage

The dielectric breakdown voltage of a dielectric material which is sometimes shortened by breakdown voltage is the maximum value of the potential difference that the material can sustain without losing its insulating properties and before sparking appears (i.e., dielectric failure). It is commonly symbolized by U_b, and expressed in volts (V). However, the breakdown voltage its a quantity which depends on the thickness of the insulator and hence it is not an intrinsic property which describes the material. Therefore, it is preferable to use the dielectric field strength.

7.1.17 Dielectric Absorption

Dielectric absorption is the measurement of a residual electric charge on a capacitor after it is discharged, and is expressed as the percent ratio of the residual voltage to the initial charge voltage. The residual voltage, or charge, is attributed to the relaxation phenomena of polarization. Actually, the polarization mechanisms can relax the applied electric field. The inverse situation, whereby there is a relaxation on depolarization, or discharge, also applies. A small fraction of the polarization may in fact persist after discharge for long time periods, and can be measured in the device with a high impedance voltmeter. Dielectrics with higher dielectric permittivity, and therefore more polarizing mechanisms, typically display more dielectric absorption than lower dielectric permittivity materials.

7.1.18 Dielectric Losses

In an AC circuit, the voltage and current across an ideal capacitor are $\pi/2$ radians out of phase. This is evident from the following relationship:

$$Q = CV.$$

In an alternating applied electric field, $V = V_o \sin(\omega t)$ where V_o is the amplitude of the sinusoidal signal, ω is the angular frequency in rad.s^{-1}, and t is time in seconds (s). Therefore:

$$Q = CV_o \sin(\omega t)$$
$$i = dQ/dt = d/dt\, CV_o \sin(\omega t)$$
$$i = CV_o \omega \cos(\omega t).$$

Because $\cos(\omega t) = \sin(\omega t + \pi/2)$, the current flow is therefore $\pi/2$ radians out of phase with the voltage in an ideal capacitor. Real dielectrics, however, are not ideal devices, as the resistivity of the material is not infinite, and the relaxation time of the polarization mechanisms with frequency generates losses. The above model for an ideal capacitor in practical applications must be modified. A practical model for a real capacitor can be considered to be an ideal capacitor in parallel with an ideal resistor. For the real capacitor, the voltage $V = V_o \sin(\omega t)$ in an alternating electric field, and the electric current, i_C, flowing through the capacitor is given by $i_C = CV_o \omega \cos(\omega t)$. For the resistor, the electric current, i_R, flowing through is: $i_R = V/R = V_o \sin(\omega t)/R$. The net electric current flow is therefore the summation of the two contributions $i_C + i_R$ or $i = CV_o \omega \cos(\omega t) + V_o \sin(\omega t)/R$. The two parts of the net electric current equation indicate that a fraction of the electric current (contributed by the resistive portion of the capacitor) will not be $\pi/2$ radians out of phase with the voltage.

7.1.19 Loss Tangent or Dissipation Factor

In an ideal lossless dielectric material, there is a phase difference between the current and the voltage of 90°. On the contrary, in the presence of dielectric losses, the phase difference is then equal to $90° - \delta$, and the power losses are proportional to $\tan\delta$, which is called the loss tangent or dissipation factor. The angle by which the current is out of phase from the ideal can be determined, and the tangent of this angle is defined as the loss tangent or dissipation factor of the capacitor. The loss tangent, $\tan\delta$, is an intrinsic material property, and is not dependent on the geometry of a capacitor. The loss tangent factor greatly influences the usefulness of a dielectric in electronic applications. In practice it is found that a lower dissipation factor is associated with materials of lower dielectric constant. Higher ε_r materials, which develop this property by virtue of high polarization mechanisms, display higher dissipation factors.

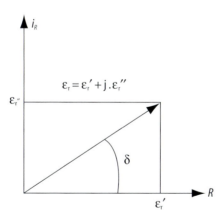

7.1.20 Dielectric Heating

Dielectric losses are usually dissipated as heat inside the dielectric material, and in the surroundings (e.g., air). The dissipated power (i.e., power input for the material), expressed in watts (W), is given by the following equation, where f is the frequency in hertz (Hz), with $2\pi f = \omega$:

$$P = 2\pi f E^2 C \tan \delta.$$

The dissipation of heat strongly depends on the applied parameters such as: (i) the frequency, f, of the electric field and its amplitude, E, and intrinsic properties of the material such as (ii) the loss factor and (iii) thermophysical properties of the material, geometrical factors such as (iv) the shape and (v) dimensions of the material, and finally (vi) the type and nature of the surroundings. On the other hand, this behavior is extensively used to heat industrial materials in the plastics industry, in woodworking, and for drying many insulating materials.

7.2 Insulator Physical Properties

7.2.1 Insulation Resistance

The insulation resistance, R, expressed in ohms (Ω), is a measure of the capability of a material to resist current flow under a potential difference (i.e., voltage). As a general rule, insulators are materials that have no free electrons in their conduction band that can move under an applied electric field. The insulation resistance of insulators is dependent on the chemical composition, the processing, and temperature at measurement. In all dielectrics, the resistivity decreases with temperature, and a considerable drop is observed from low temperatures ($-55°C$) to high temperatures ($+150°C$). As for conductors, Ohm's law states that the current i, in amperes (A), circulating in a conductor is related to the applied voltage, V in volts (V) and the resistance R according to the following equation:

$$U = Ri.$$

The resistance R is a dimensionally dependent (i.e., extensive) property, and is related to the intrinsic resistivity of the material described in the next section.

7.2.2 Volume Electrical Resistivity

The volume electrical resistivity is the proportional physical quantity which relates the electric current density, j, in $A.m^{-2}$, and the electric field strength, E, in $V.m^{-1}$, according to the generalized Ohm's law:

$$j = \sigma E = \frac{E}{\rho}.$$

Therefore, the electrical resistance, R, expressed in Ω of a homogeneous material having a regular cross-section area, A, expressed in m^2, and a length, L in m, is given by the following equation, where the proportional quantity ρ, is the electrical volume resistivity of the material, expressed in $\Omega.m$:

$$R = \rho \frac{L}{A}.$$

The electrical resistivity is the summation of two resistive contributions: (i) the lattice contribution or the thermal resistivity, i.e., the thermal scattering of conduction electrons due to atomic vibrations of the material crystal lattice (i.e., phonons) and (ii) the residual resistivity, which cames from the scattering of electrons by crystal lattice defects (e.g., vacancies, dislocations, and voids), solid solutes, and chemical impurities (i.e., interstitials). Therefore, the overall resistivity can be described by Matthiessen's equation as follows:

$$\rho = \rho_L + \rho_R.$$

7.2.3 Temperature Coefficient of Electrical Resistivity

Over a narrow range of temperature, the electrical resistivity ρ and hence electrical resistance R, vary linearly with the temperature according to the equation below, where α is the temperature coefficient of electrical resistivity expressed in $\Omega.m.K^{-1}$:

$$\rho(T_2) = \rho(T_1)[1 + \alpha(T_2 - T_1)].$$

The temperature coefficient of electrical resistivity is an algebraic physical quantity

$$\alpha = \frac{1}{\rho}\frac{\partial\rho}{\partial T}$$

(i.e., negative for semiconductors and dielectrics and positive for metals and alloys). Moreover, in the previous equation the dimensional change of the material with temperature is not taken into account. Actually, the dimensional change of the material is directly described by the coefficient of linear thermal expansion, which is generally smaller than the temperature coefficient of electrical resistivity. Therefore, if dimensional variations are negligible, values of the coefficient of electrical resistivity and that of electrical resistance are identical.

7.2.4 Surface Electrical Resistivity

At high frequencies, the surface of the insulator may possess different resistivity than the bulk of the material due to absorbed impurities on the surface, external contamination, or water moisture, hence, the electric current is chiefly conducted near the surface of the conductor (**skin effect**). The depth δ at which the current density falls to $1/e$ of its value at the surface is called the **skin depth**. The skin depth and the surface resistance are dependent upon the AC frequency. The surface resistivity R_S, expressed in Ω is the DC sheet resistivity of a conductor having a thickness of one skin depth:

$$R_S = \frac{\rho}{\delta}.$$

7.2.5 Leakage Electric Current

When considering a ceramic capacitor with parallel electrodes, the leakage electric current, denoted i_{leak} in amperes (A), through the dielectric insulator can be expressed as:

$$i_{leak} = \frac{VA_{active}}{\rho t}$$

where V is the applied voltage, in volts (V), A the active surface area in m^2, ρ the electrical volume resistivity of the dielectric material in $\Omega.m$, and t the dielectric thickness in m. From the above relationship, for any given applied voltage, the leakage electric current is directly proportional to the active electrode area of the capacitor, and is inversely proportional to the thickness of the dielectric layer. Similarly, the capacitance is directly proportional to the active electrode area, and inversely proportional to the dielectric thickness (see Section 7.1.6). Therefore, the leakage electric current depends on the dielectric material's intrinsic properties (i.e., resistivity and permittivity), and on capacitance and applied voltage as follows:

$$i_{leak} = \frac{CV}{\rho\varepsilon_0\varepsilon_r} = \left(\frac{C}{\rho\varepsilon_0\varepsilon_r}\right)V = \frac{V}{R_{ins}}.$$

Expressing the leakage electric current according to Ohm's law, it is possible to identify another physical quantity denoted insulation resistance R_{ins}. For any given capacitor, the insulation resistance is largely dependent on the resistivity of the dielectric, which is a property of the material, dependent on formulation and temperature of measurement.

The measured insulation resistance is inversely proportional to the capacitance value of the unit under test. Insulation resistance is a function of capacitance and hence minimum standards for insulation resistance in the industry are established as the product of the resistance and the capacitance, RC. For instance, the minimum RC product (as it is commonly known in the industry) for ceramic capacitors must be 1 kΩ.F when measured at 25°C, and is 0.1 kΩ.F at 125°C.

7.3 Dielectric Behavior

The origin of the dielectric behavior occurring in insulators is due to the atomic and molecular structure of the materials. The electric polarization is determined by the motion of electric charges and dipoles present in the material under an external electric field. Therefore, the overall polarization of a dielectric material is the summation of the contribution of individual polarizations, i.e., the summation of four types of electric charge displacement that occur in the dielectric material. These charge displacements, are in order of importance: (i) the **electronic polarization** due to the displacement of atomic and molecular electon clouds, (ii) the **ionic polarization** due to the displacements of ions (i.e., both cations and anions), (iii) the **dipole polarization** due to the reorientation of permanent molecular electric dipoles, and finally (iv) the **space charge polarization** due to the macroscopic displacement of electric charges submitted to an electric field. The total polarization is defined according to the following equation:

$$P = P_e + P_i + P_d + P_c$$

7.3.1 Electronic Polarization

This effect is common to all materials, as it involves distortion of the center of charge symmetry of the atomic electron cloud around the nucleus under the action of the electric field. Under the influence of an applied field, the nucleus of an atom and the negative charge center of the electrons shift, creating a small electric dipole. This polarization effect is small, despite the large number of atoms within the material, because the angular moment of the dipoles is very short, perhaps only a small fraction of an angstrom. The electronic polarization resulting from this behavior is given by the relation below, where n is the atom density, in m^{-3}, α_e the electronic polarizability in C.m^2.V^{-1}, and E_0 the local electric field in the atom in V.m^{-1}:

$$P_e = n\alpha_e E_0.$$

7.3.2 Ionic Polarization

Ionic displacement is common in ionic solids such as ceramic materials, which consist of crystal lattice structures occupied by cations and anions. Under the influence of an electric field, dipole moments are created by the shifting of these ions towards the opposing polarity. The displacement of the dipoles can be relatively large in comparison to the electronic displacement and therefore can give rise to high relative dielectric permittivity in some ceramics. The ionic polarization resulting from this behavior is given by the relation below, where n is the atom density, in m^{-3} α_i the ionic polarizability in C.m^2.V^{-1}, and E_0 the local electric field in the atom in V.m^{-1}:

$$P_i = n\alpha_i E_0.$$

7.3.3 Dipole Orientation

This is a phenomenon involving rotation of permanent electric dipoles under an applied electric field. Although permanent dipoles exist within many ceramic compounds such as SiO_2

which has no center of symmetry for positive and negative charges, dipole orientation is not found to occur in most cases, as the dipole is restricted from shifting by the rigid crystal lattice of ceramic materials; reorientation of the dipole is precluded as destruction of the lattice would ensue. Dipole orientation is more common in polymers which by virtue of their atomic structure permit reorientation. Note that this mechanism of permanent dipoles is not the same as that of induced dipoles of ionic polarization. The dipole polarizability is a function of the permanent dipole moment of the molecule as described by the equation:

$$\alpha_d = \frac{\mu^2}{3kT}.$$

7.3.4 Space Charge Polarization

This mechanism is extrinsic to any crystal lattice. The phenomenon arises due to charges which exist from contaminants or irregular geometry of the interfaces of polycrystalline ceramics, and is therefore an extraneous contribution. These charges are partially mobile and migrate under an applied field.

7.3.5 Effect of Frequency on Polarization

Because the four mechanisms of polarization have different varying time responses, dielectric solid properties strongly depend on the frequency of the applied electric field. Actually, electrons respond rapidly to electric potential reversals, and hence no relaxation of the electronic displacement polarization contribution occurs up to frequencies of 10^6 GHz. Secondly, ions (which are larger and heavier than electrons and must shift within the crystal structure) are less mobile, and have a less rapid response to field changes. The polarization effect of ionic displacement decreases at 10^4 GHz. At this frequency, the ionic displacement begins to lag the field reversals, increasing the loss factor and contributing less to the dielectric constant. At higher frequencies the field reversals are such that the ions no longer see the field (i.e., the ions do not have time to respond) and no polarization or loss factor contribution is made by ionic displacement. Dipole orientation and space charge polarization have even slower frequency responses. The net effect of the four polarization mechanisms is illustrated below. The peaks which occur near the limiting frequency for ionic and electronic polarization are due to resonance points, where the applied frequency equals the natural frequency of the material. The variation with frequency of the polarization mechanisms is reflected when measuring the dielectric constant of a capacitor. As expected, capacitance value, i.e. dielectric constant, always decreases with increased frequency for all ceramic materials, although with varying degrees depending upon which type of polarization mechanism is dominant in any particular dielectric formulation.

7.3.6 Frequency Dependence of the Dielectric Losses

As was illustrated previously, the frequency at which a dielectric is used has an important effect on the polarization mechanisms, notably the relaxation time (sometimes called lag time) displayed by the material in keeping up with field reversals in an alternating circuit. Short relaxation time is associated with instantaneous polarization processes, long relaxation time with slower polarization processes. The loss contribution is maximized at a frequency where the applied electric field has the same frequency as the polarization process. Hence, dielectric losses are low when the relaxation time and period $T = 1/f$ of the applied electric field differ greatly. When the relaxation frequency is greater than the electric field frequency, the dielectric losses are small because the polarization mechanism is much slower than the field reversals and the ions cannot follow the field at all, hence creating no heat loss. When the relaxation frequency is lower than the electric field frequency, the dielectric losses are small because the polarizing processes can easily follow the field frequency with no lag. Finally, for a relaxation frequency similar to the field frequency, the ions can follow the field reversals but are limited by their lag time and thus generate the highest loss. The variation of dielectric loss with frequency coincides with the change in dielectric constant, as both are related to the polarizing mechanisms. In high-frequency applications, a measurement known as the **Q factor** is often used. The Q factor is the reciprocal of the loss tangent: $Q = 1/\tan\delta$.

7.4 Dielectric Breakdown Mechanisms

Electrical breakdown is a sequence of often rapid processes leading to a change from an insulating to a conducting state. Actually, dielectric failure occurs in insulators when the applied electric field strength reaches a threshold point (e.g., 10^5 To 10^9 V.m^{-1}) where the restoring forces within the crystal lattice are overcome and a field emission of electrons occurs, generating sufficient numbers of free electrons which, on collision, create an avalanche effect, resulting in a sudden burst of current which punctures the dielectric irreversibly. In addition to this electrical type of failure, high voltage stresses create heat, which lowers the resistivity of the material to the point where with sufficient time, a leakage path may develop through the weakest portion of the dielectric. This type of thermal failure is of course temperature dependent, and the dielectric strength decreases with temperature. There are three postulated mechanisms to explain the above dielectric breakdown of insulating materials. These chief mechanisms are in order of importance: (1) the electronic discharge or corona mechanism, (2) the thermal discharge mechanism, and finally (3) the internal discharge or intrinsic mechanism. When dielectric breakdown occurs in an insulator, all the three mechanisms operates with a probability depending on the material type and conditions.

7.4.1 Electronic Breakdown or Corona Mechanism

The most common dielectric breakdown mechanism occurring in insulating materials is the breakdown caused by several processes such as impact ionization, field emission, double injection, and insulator-to-metal transition, leading to an electric discharge, which produces locally a high electric field. In the particular case of solid materials, the corona discharge is located in the surrounding medium. Suitable regions inside the materials in which electric discharge can occur preferably are macroscopic defects such as voids, cracks, or bubbles that are present after material processing or may develop on aging. This mechanism leads to local erosion or chemical decomposition producing a failure path for electric carriers. Electronic breakdown has been commonly observed for electric field strength ranging between 10^5 to 10^9 V.m^{-1}.

7.4.2 Thermal Discharge or Thermal Mechanism

After electrical discharges have produced an electrical pathway, in which the displacement of electric charge carriers (i.e., electrons, holes, and ions) causes a current to flow across the material and generate heat by the Joule effect, the cumulative heating increases the vibrations of the lattice structure (i.e., phonons) which induces dielectric and ionic conduction losses by electric carrier–phonon interactions. The heat generated is greater than the heat dissipated, causing thermal instability of the insulator. Thermal instability has been commonly observed for electric field strengths ranging between 10^5 and 10^9 V.m^{-1}.

7.4.3 Internal Discharge or Intrinsic Mechanism

This mechanism occurs when the electric field strength applied to the material is sufficient to accelerate electrons through the material and cause sparking in the cavities of the solid. Discharges in cavities can cause erosion by sputtering, chemical reactions, local melting, and evaporation. These local destructive processes can be rapid for high electric fields and high frequencies but relatively slow at lower voltages and with DC current.

7.5 Electro-mechanical Coupling

Dielectric materials always display an elastic deformation when stressed by an electric field, due to displacements of ions within the crystal lattice. The mechanism of polarization, i.e. the shifting of ions in the direction of an applied field, results in a constriction of surrounding ions in the atomic lattice, as restoring forces between atoms seek to balance the system. This behavior is called electro-striction, and is common to all crystals endowed with a center of symmetry. Electro-striction is a one-sided relationship, in that an electric field causes deformation, but an applied mechanical stress does not induce an electric field, as charge centers are not displaced.

7.6 Piezoelectricity

7
Insulators
and
Dielectrics

Piezoelectric materials are those which display a two-sided relationship of mechanical stress and polarization, which is attributed to crystal lattice configurations which lack a center of symmetry. Upon compression, the centers of charge shift and produce a dipole moment, resulting in polarization. This effect is a true linear coupling, as the elastic strain observed is directly proportional to the applied field intensity, and the polarization obtained is directly proportional to the applied mechanical stress. For instance, barium titanate, the major constituent of ferroelectric dielectrics, lacks a center of symmetry in the crystal lattice at temperatures below the Curie point. The material is therefore piezoelectric in nature. When heated past the Curie temperature, the crystal lattice changes from the tetragonal to the cubic configuration, which possesses a center of symmetry, and piezoelectric effects are no longer observed.

7.7 Ferroelectrics

Ferroelectric materials (i.e., paraelectrics) have electric dipole moments aligned parallel to each other even without an external applied electric field. Particular zones in the material, where all the electric dipole moments exhibit the same orientation are called **ferroelectric domains** or **Weiss domains**. Interfaces between the Weiss domains are called **Bloch's boundaries** or **walls**. For instance in a polycrystalline material, crystal borders within the which separate different lattice orientations are Bloch walls. Nevertheless, within the same crystal or monocrystal, several ferroelectric domains can coexist together. Therefore, the entire macroscopic material is divided into small ferroelectric domains, each domain having a net polarization even without an external electric field. This polarization is called **spontaneous polarization** (P_S). However, a bulk sample will generally not have a net polarization since the summation of all spontaneous polarization vectors in the various domains is zero due to their random orientations. But application of a small external electric field will cause growth of favorable domains resulting in materials having a high polarization and a high electrical susceptibility. Hence ferroelectrics can be summarized as materials which are characterized by a net spontaneous polarization that can

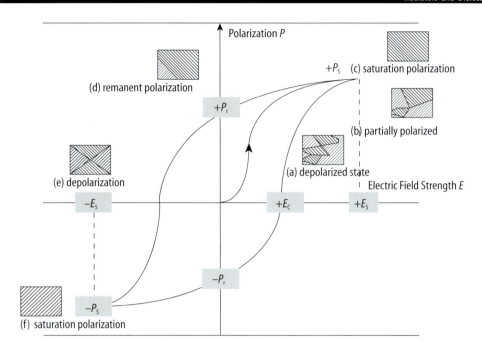

be reversed or reoriented along certain crystallographic directions of the crystal when the applied electric field direction is changed. The spontaneous polarization has its origin in a noncentrosymmetric arrangement of ions in the space lattice structure, which leads to an electric dipole moment associated with the unit cell. Ferroelectric materials differ from dielectrics in that a nonlinear response of charge versus voltage occurs due to the crystal structure of the material. Actually, the polarization as a function of the applied electric field is nonlinear, and the resulting plot is called the ferroelectric hysteresis loop or P–E diagram. The hysteresis behavior (i.e., irreversibility) is caused by the existence of permanent electric dipoles in the ferroelectric material (e.g., barium titanate, $BaTiO_3$), which develop spontaneously below a certain critical temperature called the Curie temperature, T_C, in K. For instance, in the case of $BaTiO_3$ these dipoles arise due to the fact that in its tetragonal unit cell, the single Ti^{4+} cation is surrounded by six O^{2-} anions in a tetragonal arrangement, and can occupy one of two asymmetric sites. In either position, the Ti^{4+} cation is not concentric with the negative barycenter of charge of the oxygen anions by a small fraction of an atomic distance, creating an electric dipole. The energy barrier between the two possible Ti cation positions is sufficiently low to permit motion of the atom between sites by coercion of an electric field, and the material can thus be polarized with ease. The interaction between adjacent unit cells, in fact, is sufficient to create domains of parallel polarity the instant the material assumes its ferroelectric state on cooling through the Curie point.

Upon creation, the ferroelectric domains are random in orientation, and the material has no macroscopic polarization. As shown in the ferroelectric hysteresis diagram above, this state is equivalent to a point of origin ($E = 0$, and $P = 0$). If an external field is now applied, Ti^{4+} cations become displaced in the direction of the electric field such that domains more favorably aligned with the field will grow, at the expense of those that are not favorably aligned, creating a rapid and major polarizing effect until a maximum orientation with the field is achieved until saturation polarization point (E_S, P_S). Removal of the field at this point will eliminate any normal ionic polarization, the Ti^{4+} cations remain in their oriented sites so that a **remanent polarization** (P_r) is observed. In order to remove this polarization, it becomes necessary to apply an opposite or **coercive electric field strength** (E_C) which reverts half the volume of the domains to favor the new electric field direction. Continuation of the field cycle inverts the polarization to a maximum ($-E_S$) and removal of the negative field leaves a **net polarization** ($-P_S$). Further cycles of the electric field retrace the original path, creating a continuous hysteresis effect. A condition of zero polarization at zero field can only be again achieved by short-circuiting the capacitor and heating past the Curie temperature to generate a new system of random ferroelectric domains. Ferroelectric domains can actually be observed in polycrystalline barium titanate, where etched polished surfaces reveal differences in orientation of the grain structure of the material. The ferroelectric hysteresis loop varies in shape with temperature. At lower temperatures there is less thermal motion of atoms and a greater field is required to orient the domains. Measurements at higher temperatures show that the coercive electric field required for polarization decreases, until at the Curie temperature the hysteresis disappears and linearity is approximated. It should be noted that barium titanate undergoes other phase transformations below the Curie point, which are accompanied by changes in the dielectric constant of the ceramic. In addition to the cubic to tetragonal transformation on cooling through the Curie point, a change from tetragonal to orthorhombic occurs at approximately 0°C, which then changes to the rhombohedral crystal structure at –90°C. The variations of electrical properties of $BaTiO_3$, in addition to the changes with temperature, present some obvious problems. The polarization obtained is a function of the electric field intensity, due to the energy required for domain orientation, i.e. is a function of the applied field. In addition, the dielectric constant is highly temperature dependent and practical applications specify stability over –55°C to +125°C temperature range. Moreover, ferroelectric ceramics display aging and piezoelectric effects. There are fortunately other elements which can be incorporated into the $BaTiO_3$ crystal structure to modify its properties. Lead titanate for example, readily forms solid solutions with barium titanate, in which the Pb^{2+} cation substitutes for the Ba^{2+} cation. Other partial substitutions for the barium cation such as strontium, cadmium, and calcium, as well as replacement for the titanium cation by tin, zircon, and hafnium, are also used to modify the dielectric behavior and temperature dependence of barium titanate. These additives greatly enhance the range of compositions and possible dielectric characteristics, and much effort has been expended in recent years to optimize these materials for practical purposes.

7.8 Aging of Ferroelectrics

The ferroelectric group of dielectrics is based on the lattice structure of barium titanate ($BaTiO_3$) as the main constituent, an oxide which undergoes changes in crystal lattice structure that give rise to ferroelectric domains. At its Curie temperature (120°C), $BaTiO_3$ transforms from a tetragonal to a cubic configuration. On cooling through and below the Curie point, the material again transforms from a cubic to a tetragonal crystal structure. When in the tetragonal configuration, the Ti^{4+} cation can occupy one of two off-center sites within the cell. The fact that the Ti^{4+} cation is off-center gives rise to a permanent electric dipole within the cell. These dipoles are somewhat ordered because adjacent cells within the lattice structure influence each other. Weiss domains are formed in which the Ti^{4+} cation occupies the same location within many neighboring cells. These domains are random in orientation without the influence of an electric field and impart a certain strain energy to the system. The relaxation of this strain energy is attributed to be the mechanism of aging of the dielectric constant. Aging is found to have the following relationship with time:

$$\varepsilon = \varepsilon_{t=0} - r\ln(t)$$

where ε is the dielectric permittivity at the time t, $\varepsilon_{t=0}$ the dielectric permittivity at time t_0, and r the rate of decay. The microstructural details which affect polarization (material purity, grain size, sintering, grain boundaries, porosity, internal stresses) also determine freedom of domain

Table 7.2 Properties of selected ferroelectric materials

Name (acronym)	Chemical Formula	Curie temperature (T_C, K)	Maximum relative dielectric permittivity (ε_r)	Application
Ammonium cadmium sulfate (AMCS)	$(NH_4)_2Cd_2(SO_4)_3$	95	n.a.	n.a.
Ammonium fluoroborate (AFB)	$(NH_4)_2BeF_4$	176	n.a.	n.a.
Ammonium hydrogenodichloroacetate (AHDCA)	$NH_4H(ClCH_2COOH)_2$	128	n.a.	n.a.
Ammonium hydrogenosulfate (AMHS)	$(NH_4)HSO_4$	271	n.a.	n.a.
Ammonium nitrate (AMN)	NH_4NO_3	398	n.a.	n.a.
Ammonium Rochelle's salt	$NH_4KC_4H_4O_6 \cdot 4H_2O$	297	n.a.	n.a.
Barium titanate (BT)	$BaTiO_3$	406	12,000	Capacitors, sensors, phase shifter
Barium titanium niobiate (BATIN)	$Ba_6Ti_2Nb_8O_{30}$	521	n.a.	n.a.
Barium sodium niobiate (BANAN)	$Ba_2NaNb_5O_{15}$	833	86	n.a.
Cadmium pyroniobiate (CDPN)	$Cd_2Nb_2O_7$	185	n.a.	n.a.
Cesium dihydrogen arsenate (CDA)	CsH_2AsO_4	143	n.a.	n.a.
Cesium dihydrogen phosphate (CDP)	CsH_2PO_4	159	n.a.	n.a.
Cesium hydrogenoselenite (CHSE)	$CsH_3(SeO_3)_2$	143	n.a.	n.a.
Gadolynium molybdate (GDMO)	$Gd_2(MoO_4)_3$	432	n.a.	n.a.
Lead iron niobiate (PFN)	PbFeNbO	385	24,000	Multilayered capacitors
Lead iron tungstate (PFW)	PbFeWO	183	20,000	Multilayered capacitors
Lead magnesium niobiate (PMN)	PbMgNbO	272	18,000	Multilayered capacitors
Lead titanate (PT)	$PbTiO_3$	765	8000	Pyrodetector, acoustic transducer
Lead zinc niobiate (PZN)	PbZnNbO	413	22,000	Multilayered capacitors
Lithium ammonium tartrate	$NH_4LiC_4H_4O_6 \cdot H_2O$	106	n.a.	n.a.
Lithium iodate (LII)	$LiIO_3$	529	n.a.	n.a.
Lithium niobate (LN)	$LiNbO_3$	1483	37	Pyrodetector
Lithium tantalate (LT)	$LiTaO_3$	891	n.a.	Waveguide device, optical modulator, surface acoustic wave filter, SHG
Methyl ammonium alum (MASD)	$NH_3(CH_3)Al(SO_4)_2 \cdot 12H_2O$	177	n.a.	n.a.
Potassium dihydrogen arsenate (KDA)	KH_2AsO_4	97	n.a.	n.a.
Potassium dihydrogen phosphate (KDP)	KH_2PO_4	123	n.a.	n.a.
Potassium iodate (KI)	KIO_3	485	n.a.	n.a.
Potassium niobate (KN)	$KNbO_3$	712	30–137	Waveguide device, frequency doubler, holographic storage

Table 7.2 *(continued)*

Name (acronym)	Chemical Formula	Curie temperature (T_C, K)	Maximum relative dielectric permittivity (ε_r)	Application
Potassium nitrate (KN)	KNO_3	397	n.a.	n.a.
Potassium selenate (KSE)	K_2SeO_4	93	n.a.	n.a.
Potassium tantalate niobiate (KTN)	$K_3Ta_2NbO_9$	271	n.a.	Pyrodetector
Rochelle's salt	$NaKC_4H_4O_6$ $\cdot 4H_2O$	255–297	n.a.	n.a.
Rubidium dihydrogen arsenate (KDA)	RbH_2AsO_4	111	n.a.	n.a.
Rubidium dihydrogen phosphate (RDP)	RbH_2PO_4	146	n.a.	n.a.
Rubidium nitrate (KN)	$RbNO_3$	437–487	n.a.	n.a.
Sodium nitrate (KN)	$NaNO_3$	548	n.a.	n.a.
Triglycine fluoroborate (TGFB)	$(NH_2CH_2CO OH)_3 \cdot H_2BeF_4$	346	n.a.	n.a.
Triglycine selenate (TGSE)	$(NH_2CH_2CO OH)_3 \cdot H_2SeO_4$	295	n.a.	n.a.
Triglycine sulfate (TGS)	$(NH_2CH_2CO OH)_3 \cdot H_2SO_4$	322	n.a.	n.a.

7
Insulators and Dielectrics

wall movement and reorientation. It is found that the aging rate is composition and process dependent and insensitive to variables which also influence the dielectric constant of the material. The loss of capacitance with time is unavoidable with ferroelectric formulations, although it can be reversed by heating the dielectric above the Curie point and reverting the material back to its cubic state. On cooling, however, spontaneous polarization will again occur as the material transforms to the tetragonal crystal structure, and new domains recommence the aging process. As is expected, no aging is observed in paraelectric formulations which do not possess the mechanism of spontaneous polarization. The rate at which aging may occur can be influenced by voltage conditioning. It is found that units stressed by a DC voltage at elevated temperature below the Curie point will experience a loss in capacitance, but at a lower aging rate. It is theorized that the voltage stress at the elevated temperature accelerates the domain relaxation process. This voltage conditioning effect is eliminated when the capacitor experiences temperatures above the Curie point.

7.9 Industrial Dielectrics Classification

In electrical engineering, it is common to classify dielectrics in three main classes. Actually, dielectric materials are identified and classified in the electrical industry according to their capacitance temperature coefficient. Two basic groups (Class I and Class II) are used in the manufacture of ceramic chip capacitors, while to a lesser extent a third group (Class III) identifies the barium titanate solid structure type barrier-layer formulations utilized in the production of disk capacitors.

7.9.1 Class I Dielectrics or Linear Dielectrics

This group identifies the linear dielectrics. These materials display the most stable characteristics, as they are non-ferroelectric (i.e., paraelectric) formulations and hence show a linear relationship of polarization to voltage, and are formulated to have a linear temperature coefficient of capacitance. These materials are primarily of rutile type (TiO_2), and therefore exhibit lower relative dielectric permittivity (< 150) but more importantly, also lower dielectric loss and no reduction of capacitance with time. These properties, along with negligible dependence of capacitance on voltage or frequency, make these dielectrics useful in capacitor applications where close tolerance and stability are required. Linear dielectrics are also referred to as temperature compensating, as the temperature coefficient can be modified to give predictable slopes of the temperature coefficient over the standard $-55°C$ to $+125°C$ range. These slopes vary from approximately $+100$ ppm.$°C^{-1}$ to -750 ppm.$°C^{-1}$. These slopes are called P100 and N750 respectively. A flat slope, neither positive nor negative, is denoted NPO and is one of the most common of all dielectric characteristics. The extended temperature-compensating ceramics are a sub-group of formulations which utilize small additions of other (ferroelectric) oxides such as $CaTiO_3$ or $SrTiO_3$ and which display near-linear and predictable temperature characteristics with relative dielectric permittivity ranging up to 500. Both categories are used in circuitry requiring stability of the capacitor (i.e., negligible aging rate, low loss, no change in capacitance with frequency or voltage, and predictable linear capacitance changes with temperature).

7.9.2 Class II Dielectrics or Ferroelectrics

Class II dielectrics comprise the ferroelectrics. These materials offer much higher dielectric constants than Class I dielectrics, but with less stable properties with temperature, voltage, frequency, and time. The diverse range of properties of the ferroelectric ceramics requires a sub-classification into two categories, defined by temperature characteristics. (1) Stable class II: these types of materials display a maximum capacitance temperature coefficient of $+/-15\%$ from $25°C$ reference over the temperature range from $-55°C$ to $+125°C$. These materials typically have relative dielectric permittivity ranging from 600 to 4000. (2) High relative dielectric permittivity Class II dielectrics: these capacitors have steep temperature coefficients, due to the fact that the Curie point is shifted towards room temperature for maximization of the dielectric constant. These bodies have dielectric constants from 4000 to 28,000, typically. It is important to note that ceramic capacitors made with ferroelectric formulations display a decay of capacitance and dielectric loss with time. This phenomenon, called aging, is reversible and occurs due to the crystal structure and to its changes with temperature.

7.10 Selected Properties of Insulators and Dielectric Materials

The insulating and dielectric materials include numerous and various classes of materials amongst which are liquid organic solvents, ceramics, glasses, polymers and elastomers, woods, minerals, rocks, and paper. Dielectric properties of selected dielectrics are reported in Table 7.3. For more physical properties on the selected materials, it is recommended to refer to the appropriate section in the book dealing with the desired class of material (e.g., ceramics, polymers, etc.).

Table 7.3 Insulator and dielectric electrical properties

Dielectric material	Class	Density (ρ, kg.m^{-3})	Relative dielectric permittivity (1 MHz)	Dielectric field strength (MV.m^{-1})	Loss factor at 1 MHz (tan δ)	Electrical resistivity (ρ, Ω.cm)
Alumina (90wt% Al_2O_3)	Ceramics	3600	8.8	92.5	0.0004	10^{14}
Alumina (96wt% Al_2O_3)	Ceramics	3720	9.0	82.6	0.0001	10^{14}
Alumina (99.5wt% Al_2O_3)	Ceramics	3890	7.45–9.7	33.1	0.0002	10^{15}
Alumina (99.5wt% Al_2O_3)	Ceramics	3870	9.7	86.6	0.0003	10^{14}
Alumina (99.99wt% Al_2O_3)	Ceramics	3990	10.1	90.5	0.0004	10^{15}
Alumina porcelain	Ceramics	3100–3900	8–9	9.8–15.7	0.001–0.002	10^{14}
Barite ($BaSO_4$)	Minerals	4490	11.4	n.a.	n.a.	n.a.
Barium zirconate ($BaZrO_3$)	Ceramics	5520	43	n.a.	n.a.	n.a.
Barium titanate ($BaTiO_3$)	Ceramics	6020	12,000	n.a.	n.a.	n.a.
Beryllia (99.5wt% BeO)	Ceramics	2850	6.8	11.81	0.0004	n.a.
Boron nitride (BN)	Ceramics	2250	4.15	n.a.	0.0002	n.a.
Calcium oxide (CaO, calcia)	Ceramics	3320	11.8	n.a.	n.a.	n.a.
Calomel (Hg_2Cl_2)	Minerals	6480	14	n.a.	n.a.	n.a.
Celestite ($SrSO_4$)	Minerals	3970	11.5	n.a.	n.a.	n.a.
Cellulose acetate (CA)	Thermoplastic	1310	3.2	197	0.100	10^{13}
Cellulose triacetate (CTA)	Thermoplastic	1310	3.3	146	0.100	10^{15}
Cinnabar (HgS)	Minerals	8170	20	n.a.	n.a.	n.a.
Cordierite ($Mg_2Al_3[Si_5AlO_{18}]$)	Minerals	2800	5.0	1.6–9.9	n.a.	10^{14}
Corning® 0010	Glasses	2860	6.32	n.a.	0.015	10^{17}
Corning® 0080	Glasses	2470	6.75	n.a.	0.058	10^{12}
Corning® 0120	Glasses	3050	6.65	n.a.	0.012	10^{17}
Cuprite (Cu_2O)	Minerals	6106	7.6	n.a.	n.a.	n.a.
Fluorite (CaF_2)	Minerals	3180	6.81	n.a.	n.a.	n.a.
Fluoroelastomer	Elastomers	2200	2.3	118	0.037	10^{17}
Forsterite ($MgSiO_4$)	Minerals	3220	6.2	7.9–11.8	0.0003	10^{13}
Hard rubber	Elastomers	1050	2.95–4.80	n.a.	0.007–0.0028	n.a.
Hematite (Fe_2O_3)	Minerals	5256	12	n.a.	n.a.	n.a.
Hexagonal boron nitride (96wt% BN)	Ceramics	2110	4.2	39.4	0.0002	n.a.
Lead zirconate ($PbZrO_3$)	Ceramics	n.a.	200	n.a.	n.a.	n.a.
Macor®	Glasses	2520	5.92–6.03	118	0.003	10^{16}
Magnetite (Fe_3O_4)	Minerals	5201	20	n.a.	n.a.	n.a.
Mica muscovite ($KAl[Si_3O_{10}(OH)_2]$)	Minerals	2880	6.5–8.7	10–20	0.0001	10^{14}
Mica phlogopite ($KMg_3Al[Si_3O_{10}(OH)_2]$)	Minerals	2800	5–6	n.a.	0.004–0.070	n.a.

7

Insulators and Dielectrics

Table 7.3 *(continued)*

Dielectric material	Class	Density (ρ, kg.m^{-3})	Relative dielectric permittivity (1 MHz)	Dielectric field strength (MV.m^{-1})	Loss factor at 1 MHz (tan δ)	Electrical resistivity (ρ, Ω.cm)
Mullite ($Al_6Si_2O_{13}$)	Ceramics	3260	6.3	9.8	n.a.	10^{13}
Niobium pentaoxide (Nb_2O_5)	Ceramics	4470	67	n.a.	n.a.	n.a.
Paper (Kraft, dry)	Paper	930	4–6	6–12	0.001	n.a.
Periclase (MgO)	Minerals	3560	9.65	n.a.	n.a.	n.a.
Polyamide (PA)	Thermoplastic	1130	3.7	66.9	0.016	n.a.
Polyester	Thermoplastic	1400	3.0	276	0.016	10^{18}
Polyethylene (PE)	Thermoplastic	940	2.2	19.7	0.0003	10^{16}
Polyimide (PI)	Thermoplastic	1420	3.4	276	0.010	10^{18}
Polypropylene (PP)	Thermoplastic	925	2.1	118	0.0003	10^{16}
Polytetrafluoroethylene (PTFE)	Thermoplastic	2200	2.1	16.9	0.0002	10^{18}
Polytrifluorochloroethylene (PTFCE)	Thermoplastic	2150	2.5	n.a.	0.017	10^{18}
Polyurethane (PU)	Themoset	1260	7.1	19.7	0.060	10^{11}
Polyvinyl chloride (PVC)	Thermoplastic	1500	4.0	23.6	0.14	10^{14}
Polyvinyl fluoride (PVF)	Thermoplastic	1380	7.4	n.a.	0.009	10^{13}
Polyvinylidene chloride (PVDC)	Thermoplastic	1640	5.0	197	0.075	10^{16}
Polyvinylidene fluoride (PVDF)	Thermoplastic	1760	6.43	50.4	0.159	10^{14}
Pryex® 1710	Glasses	2520	6.00	n.a.	0.025	n.a.
Pyrex® 3320	Glasses	2230	4.71	n.a.	0.019	n.a.
Pyrex® 7040	Glasses	2240	4.65	n.a.	0.013	n.a.
Pyrex® 7050	Glasses	2240	4.77	n.a.	0.017	n.a.
Pyroceram®	Ceramics	2600	5.58	n.a.	0.0015	n.a.
Pyrolussite (MnO_2)	Minerals	5234	12.8	n.a.	n.a.	n.a.
Rutile (TiO_2)	Minerals	4230–4250	85–170	3.9–8.3	n.a.	10^{14}
Silicium oxide (SiO_2, fused silica)	Ceramics	2202	3.79	35	0.00002	n.a.
Silicon nitride (Si_3N_4)	Ceramics	3200–3300	7.9–8.1	11.8	0.0006	10^{14}
Sodalime glass	Glasses	2530	7.2	n.a.	0.009	n.a.
Spinel ($MgAl_2O_4$)	Minerals	3583	8.6	n.a.	n.a.	n.a.
Steatite	Ceramics	2500–2700	5.5–7.5	7.9–13.8	0.0008	10^{13}
Strontianite ($SrCO_3$)	Minerals	3720	8.85	n.a.	n.a.	n.a.
Strontium oxide (SrO, strontia)	Ceramics	5100	13.3	n.a.	n.a.	n.a.
Strontium titanate ($SrTiO_3$)	Ceramics	5100	2080	n.a.	n.a.	n.a.
Sulfur (sublimed)	Minerals	2000	3.69	n.a.	n.a.	n.a.
Tantalum pentoxide (Ta_2O_5)	Ceramics	8200	27.6	n.a.	n.a.	n.a.
Thorium dioxide (ThO_2, thoria)	Ceramics	9860	18.9	n.a.	n.a.	n.a.
Uranium dioxide (UO_2)	Ceramics	10,960	24	n.a.	n.a.	n.a.
Vycor® (96wt% SiO_2, 3wt% B_2O_3)	Glasses	2180	6.0	40	n.a.	10^{16}

Table 7.3 (continued)

Dielectric material	Class	Density (ρ, kg.m^{-3})	Relative dielectric permittivity (1 MHz)	Dielectric field strength (MV.m^{-1})	Loss factor at 1 MHz (tan δ)	Electrical resistivity (ρ, Ω.cm)
Witherite (BaCO$_3$)	Minerals	4290	8.53	n.a.	n.a.	n.a.
Yttrium sesquioxide (Y$_2$O$_3$, yttria)	Ceramics	6030	10	n.a.	n.a.	n.a.
Zinc oxide (ZnO)	Ceramics		8.15	n.a.	n.a.	n.a.
Zirconium oxide (ZrO$_2$, zirconia)	Ceramics	5700	12.5–24.7	n.a.	n.a.	10^{15}

Conversion factors: 1 MV.m^{-1} = 25.4 V.mil^{-1} (E); 1 Ω.cm = 10^{-2} Ω.m (E)

Further Reading

Sessler GM (ed.) (1987) Electrets. Springer-Verlag, New York.
Strukov BA, Levanyuk AP (1998) Ferroelectric phenomena in crystals: physical foundations. Springer-Verlag, Heidelberg.
Whitehead S (1951) Dielectric breakdown of solids. Oxford University Press, New York.

7
Insulators
and
Dielectrics

8 Miscellaneous Electrical Materials

8.1 Thermocouple Materials

8.1.1 The Seebeck Effect

In 1821, Thomas Seebeck observed that if two different electrically conductive materials (i.e., metals, alloys, or semiconductors), made of two dissimilar materials, A and B, are joined together at both ends and the two junctions kept at two different temperatures (cold junction denoted T_c and hot junction T_h), this thermal differential induces an electric current to flow continuously through the circuit. In the open circuit, an electric potential difference, called Seebeck electromotive force (emf or e_{AB}) in honor of its discoverer, appears and this voltage is a complex function of the temperature difference and of material type (i.e., $e_{AB} = f(\Delta T, A, B)$). As a general rule, the Seebeck electromotive force is related to the temperature difference by a polynomial equation, where the respective polynomial coefficients (c_0, c_1, c_2, c_3, etc.) are intrinsic properties of the thermocouple selected:

$$\text{emf} = c_0 \Delta T + c_1 \Delta T^2 + c_2 \Delta T^3 + c_3 \Delta T^4 + \ldots = e_{AB}.$$

However, for small temperature differences the Seebeck emf is directly proportional to the temperature difference:

$$\text{emf} = \Delta \alpha \Delta T = \Delta \alpha (T_h - T_c)$$

where the quantity $\Delta \alpha$, the relative Seebeck coefficient, is the difference between the absolute Seebeck coefficients of each metal:

$$\Delta \alpha = \alpha_B - \alpha_A.$$

The absolute Seebeck coefficient is an algebraic physical quantity, which represents an intrinsic property of the material, and is expressed in $V.K^{-1}$. The difference between absolute Seebeck coefficients (i.e., global thermoelectromotive force per unit of temperature) is called the thermoelectric power (denoted Q) and its order of magnitude is commonly expressed in $\mu V.K^{-1}$ for semiconductors and $mV.K^{-1}$ for metals and alloys. For instance, for semiconducting materials, the thermoelectric power can be defined by the following equation:

$$Q_{12} = \frac{\pi^2 k^2 T}{3e} \left(\frac{\partial}{\partial E} \ln \sigma(E) \right)$$

where $\sigma(E)$ is the electrical conductivity due to displacement of charge carriers having a kinetic energy E.

8.1.2 Thermocouple

A thermocouple is a particular temperature sensing device consisting, in its simplest design, of two wires made of two dissimilar metals A and B, which are joined by two junctions (1 and 2). When the two junctions are maintained at two different temperatures (cold junction T_c and hot junction T_h, this thermal differential induces an electric current to flow continuously through the circuit. When n identical thermocouples are connected electrically in series, the total electromotive force ΔV is the summation of all the individual emf of each thermocouple.

$$\Delta V = \Delta \alpha \sum_{i=1}^{n} \Delta T_i$$
$$= \Delta \alpha (T_{h,1} - T_{c,1}) + \Delta \alpha (T_{h,2} - T_{c,2}) + \Delta \alpha (T_{h,3} - T_{c,3}) + \ldots + \Delta \alpha (T_{h,n} - T_{c,n}).$$

8.1.3 Properties of Common Thermocouple Materials

Table 8.1 Standard thermocouple types and common uses

Type	Description
Type J	Suitable in reducing, vacuum, or inert atmospheres but limited use in oxidizing atmosphere at high temperature. Not recommended for low temperatures
Type K	Clean and oxidizing atmosphere, limited use in vacuum or reducing atmosphere. Wide temperature range
Type S	Oxidizing or inert atmospheres. Beware of contamination. For high temperatures
Type R	Oxidizing or inert atmospheres. Beware of contamination. For high temperatures
Type N	More stable than type K at high temperature
Type B	Oxidizing or inert atmospheres. Beware of contamination. For high temperatures
Type E	Oxidizing or inert atmosphere, limited use in vacuum or reducing atmosphere. Highest thermoelectric power
Type C	Suitable in reducing, vacuum, or inert atmospheres. Beware of embrittlement. Not suitable for oxidizing atmosphere and not practical below 400°C
Type T	Suitable in mid-oxidizing, reducing, vacuum, or inert atmospheres. Good for cryogenic applications

8.2 Electron-Emitting Materials

To extract an electron of an atom with a kinetic energy K, it is required to provide an **ionizing energy** E (thermal, mechanical, chemical, electrical, or optical) superior to the **binding energy** of the electron, B. The kinetic energy released to the ionized electron is given by the equation $K = E - B$. In a solid, extraction of electrons requires provision of sufficient energy to the electrons to reach the difference between the Fermi level (i.e., electrochemical potential of an electron inside the solid crystal lattice) and the surface potential energy level in a vacuum at absolute zero; this energy difference is called the electron work function, denoted W_S, and it is expressed in J (eV). The thermal emission of electrons, **thermoelectronic** or **thermionic emission**, consists of electrons leaving the surface of the material due to thermal activation. Actually, electrons having sufficient kinetic energy due to thermal motion escape from the material surface. Therefore, increasing the temperature at the surface of a material will increase the flow of electrons (i.e., electric current). The electric current density, expressed in $A.m^{-2}$, as a function of the absolute temperature of the material surface is given by the **Richardson–Dushman** equation as follows:

$$J_S = AT^2(1 - r)\exp\left[-\frac{W_S}{kT}\right]$$

where A is the Richardson constant expressed in $A.m^{-2}K^{-2}$, and r the dimensionless reflection coefficient of the surface for zero applied electric field (usually negligible). The Richardson constant would be equal in theory to $1.2\ MA.m^{-2}.K^{-2}$. However, in practice because the work function is also a function of temperature, A varies over a wide range of magnitude. The theoretical value of A is given by the quantum theory and described below:

$$A = \frac{4\pi mk^2 e}{h^3}.$$

Electron-emitting materials or commonly, thermionic emitters can be classified as: (i) pure-metal emitters (e.g., W, Ta), (ii) monolayer-type emitters, (iii) oxide emitters, (iv) chemical compound emitters and finally (v) alloy emitters. Thermionic properties of selected materials are listed in Table 8.3 (page 318).

8.3 Photocathode Materials

When a monochromatic electromagnetic radiation with a frequency, ν, expressed in Hz, illuminates the surface of a solid, some electrons (i.e., photoelectrons) can be emitted if the incident photon energy, $h\nu$, is equal or superior to the binding energy of the electron in the atom of the solid. Because the energy transfer occurs between photons and electrons, this behavior is called the **photoelectric effect**. More precisely, Einstein demonstrated at the beginning of the century that the **maximum kinetic energy**, K_{max} released by the electromagnetic radiation to the photoelectrons is given by the energy difference between the incident photon energy and electron binding energy in the atoms of the solid: $K_{max} = h\nu - h\nu_0$, where $B = h\nu_0$ is the **binding energy** of the electron inside the solid which corresponds to the **electron work function** in the emitting material, i.e., $e\Phi$. For a given incident radiation energy, the ratio between the number of photoelectrons and the number of incident photons is called the **photoelectric quantum yield** or **efficiency**. Owing to the order of magnitude of binding electronic energies, the photoelectric effect occurs in metals for electromagnetic radiations having a frequency higher than that of the near UV. Even if all solid materials can exhibit a photoelectric effect under irradiation by an appropriate electromagnetic radiation (e.g., UV, X-rays, gamma-rays), the common metals exhibiting a photoelectric effect for low-energy photons and currently used as photocathodes are the alkali- and alkaline-earth metals and some of their alloys deposited onto an antimony coating. For instance, rhenium metal, with an electronic work function of roughly 5.0 eV, requires at least a UV radiation with a wavelength of 248 nm for emitting photoelectrons, while cesium requires only irradiation by visible light with wavelengths of 652 nm or lower. Selected properties of some common photocathode materials are listed in Table 8.4 (page 319). As a general rule, photocathode materials are extensively used in photocells and photomultiplier tubes.

Table 8.2 Physical properties of selected thermocouple materials

Metal and alloy trade names	Iron	Constantan®	Chromel®	Alumel®	Nicrosil®	Nisil®	Copper OFHC	Constantan®	Tophel®	Constantan®	Platinum-13 rhodium	Platinum	Platinum-10 rhodium
Temperature coefficient of resistance (0–100°C)	0.0065	0.00002	0.0032	0.00188	0.00011	0.00078	0.0043	0.00002	0.00032	0.00002	0.0016	0.00393	0.0017
Electrical resistivity (ρ, $\mu\Omega.cm$)	9.67	48.9	70	32	93	37	1.74	48.9	70	48.9	19.6	10.4	18.9
Thermal conductivity (k, $W.m^{-1}.K^{-1}$)	67.78	22.18	19.25	29.71	130	230	376.8	22.18	19.25	22.18	36.81	71.54	37.66
Specific heat capacity (c_p, $J.kg^{-1}.K^{-1}$)	447.7	397.5	447.7	523	15.06	26.61	384.9	397.5	447.7	397.5	n.a.	133.9	n.a.
Coefficient of linear thermal expansion (α, $10^{-6}\ K^{-1}$)	11.7	14.9	13.1	12.0	13.3	12.1	16.6	14.9	13.1	14.9	9.0	9.1	10.0
Melting range (°C)	1539	1270	1350	1400	1410	1400	1083	1270	1430	1270	1860	1769	1830
Thermoelectric power ($\mu V.K^{-1}$)	50.2 (0°C)		39.4 (0°C)		26.2 (0°C)		38 (0°C)		58.5 (0°C)		11.5 (600°C)		10.3 (600°C)
Temperature range (°C)	0 to 760		−270 to 1372		−270 to 1260		−200 to 370		−200 to 870		−50 to 1768		−50 to 1768
Elongation (Z, %)	40	32	27	32	30	35	46	32	n.a.	32	32	38	32
Ultimate tensile strength (σ_{UTS}, MPa) (annealed)	234	552	655	586	760	655	221	552	670	552	331	166	317
Yield strength 0.2% proof (σ_{YS}, MPa) (annealed)	n.a.	n.a.	n.a.	n.a.	415	380	69	n.a.	n.a.	n.a.	190	70	180
Density (ρ, $kg.m^{-3}$)	7860	8890	8730	8600	8520	8700	8930	8890	8730	8890	19,550	21,450	19,950
Junction polarity	JP	JN	KP	KN	NP	NN	TP	TN	EP	EN	RP	RN	SP
Average chemical composition (wt%)	Fe	45Ni–55Cu	90Ni–9Cr	95Ni–2Mn–2Al	84.3Ni–14Cr–1.4Si–0.1Mg	95.5Ni–4.4Si–0.1Mg	Cu (99.9wt%)	45Ni–55Cu	90Ni–10Cr	45Ni–55Cu	87Pt–13Rh	Pt	90Pt–10Rh
Thermocouple type (ANSi)	Type J	Type J	Type K	Type K	Type N	Type N	Type T	Type T	Type E	Type E	Type R	Type R	Type S

		BP						(600°C)						
rhodium								to 1820		1810	n.a.	n.a.	n.a.	17.5	0.0014
Platinum-6 rhodium	94Pt-6Rh	BP	20,510	n.a.	255	34	to 1820		1810	n.a.	n.a.	n.a.	17.5	0.0014
Alloy 19®	Ni-1Co	P	8900	170	415	35	0 to 1260	n.a.	1450	13.6	n.a.	50	8	n.a.
Alloy 20®	Ni-18Mo	N	9100	515	895	35			1425	11.9	n.a.	15	165	n.a.
Platinum-5 molybdenum	Pt5-Mo	P	n.a.	n.a.	n.a.	n.a.	1100 to 1500	29	1788	n.a.	n.a.	n.a.	n.a.	n.a.
Platinum-molybdenum	Pt-0.1Mo	N	n.a.	n.a.	n.a.	n.a.			1770	n.a.	n.a.	n.a.	n.a.	n.a.
Palladium alloy	83Pd-14Pt-3Au	P	n.a.	n.a.	n.a.	n.a.	n.a.	41.9	1580	n.a.	n.a.	n.a.	n.a.	n.a.
Gold-palladium	65Au-35Pd	N	n.a.	n.a.	n.a.	n.a.			1426	n.a.	n.a.	n.a.	n.a.	n.a.
Palladium alloy	55Pd-31Pt-14Au	P	n.a.	n.a.	n.a.	n.a.	n.a.	42.4	1500	n.a.	n.a.	n.a.	n.a.	n.a.
Gold-palladium	65Au-35Pd	N	n.a.	n.a.	n.a.	n.a.			1426	n.a.	n.a.	n.a.	n.a.	n.a.
Tungsten	W	P	19,900	n.a.	552	3	0 to 2760	16.7	3410	n.a.	n.a.	n.a.	n.a.	0.0048
Tungsten-Re	W-26Re	N	19,700	n.a.	1517	10	0 to 2760		3120	n.a.	n.a.	n.a.	n.a.	n.a.
Tungsten-Re	W-3Re	P	19,400	n.a.	1241	10	0 to 2760	17.1	3360	n.a.	n.a.	n.a.	9.14	0.0003
Tungsten-Re	W-25Re	N	19,700	n.a.	1448	10			3130	n.a.	n.a.	n.a.	27.43	0.0012
Tungsten-Re	W-5Re	P	19,400	n.a.	1379	10	0 to 2760	19.5	3350	n.a.	n.a.	n.a.	11.63	n.a.
Tungsten-Re	W-26Re	N	19,700	n.a.	1517	10	0 to 2760	(600°C)	3120	n.a.	n.a.	n.a.	28.3	n.a.
Type B	Alloy 19/20	Pt-Mo		Platinel I		Platinel II		W-Re		W-Re		Type C		

8
Misc.
Electrical
Materials

Table 8.3 Thermionic properties of selected materials

Material	Electron work function (W_S, eV)	Richardson constant (A, kA.m^{-2}.K^{-1})	Material	Electron work function (W_S, eV)	Richardson constant (A, kA.m^{-2}.K^{-1})
Ferrous metals			**Refractory carbides**		
Iron (ferrite)	4.5	260	Carbon	5.0	150
Cobalt	5.0	410	TaC	3.14	3
Nickel	4.61	300	TiC	3.35	250
Common nonferrous metals			ZrC	2.18	3
Copper	4.65	1200	SiC	3.5	640
Other metals			ThC$_2$	3.5	5500
Beryllium	4.98	3000	**Refractory borides**		
Barium	2.52	600	CeB$_6$	2.6	36
Cesium	2.14	1600	LaB$_6$	2.7	290
Platinum group metals (PGMs)			ThB$_6$	2.9	5
Osmium	5.93	1,100,000	CaB$_6$	2.9	26
Rhodium	4.98	330	BaB$_6$	3.5	160
Iridium	5.27	1200	**Refractory oxides**		
Platinum	5.65	320	ThO$_2$	2.6	50
Refractory metals (group IVB, VB, and VIB)			CeO$_2$	2.3	10
Titanium	4.53	n.a.	La$_2$O$_3$	2.5	9
Zirconium	4.05	3300	Y$_2$O$_3$	2.4	10
Hafnium	3.60	220	BaO–SrO	1.0	10
Niobium	4.19	1200	**Uranides**		
Tantalum	4.25	1200	Uranium	3.27	60
Molybdenum	4.15	550	Thorium	3.38	700
Tungsten	4.55	600			

Important note: it is important to make a clear distinction between the photoelectric effect, which occurs during the extraction of electrons of an atom from a crystal lattice in a solid by an incident electromagnetic radiation, and **photoemission**, which consists of the extraction of electrons (i.e., ionization) of a free atom in a vapor by an incident electromagnetic radiation.

8.4 Secondary Emission

When a flux of electrons is incident upon a surface of a solid, secondary electrons are produced and emitted in a vacuum. These secondary electrons can be grouped into several types according to their origin: (i) true secondary electrons with a kinetic energy about 10 eV independent of that of the primary energy, (ii) primary electrons scattered both elastically (coherent scattering) or inelastically (incoherent scattering). The dimensionless ratio of secondary electrons to other primary electrons is called the **secondary emission coefficient**, denoted δ. The secondary emission coefficient reaches a maximum value, δ_{max}, for a definite maximum energy of incident electrons E_{max}, and after decreases slowly for higher kinetic energies.

Table 8.4 Photocathode metals and alloys

Photocathode materials	Electron work function (W_s, eV)	Wavelength (λ, nm)	Photoelectric quantum yield (η)
Lithium	2.4	517	10^{-4}
Sodium	2.2	564	10^{-4}
Potassium	2.2	564	10^{-4}
Rubidium	2.1	591	10^{-4}
Cesium	1.9	653	10^{-4}
Rb_3Sb	2.2	564	0.10
Cs_3Sb	2.05	605	0.25
NaK_3Sb	2.0	620	0.30
$CsNaK_3Sb$	1.55	800	0.40
Calcium	2.9	428	10^{-4}
Strontium	2.7	459	10^{-4}
Barium	2.5	496	10^{-4}
Na_3Sb	3.1	400	0.02
K_3Sb	2.6	478	0.07
Rb_3Sb	2.2	564	0.10
Cs_3Sb	2.05	605	0.25
NaK_3Sb	2.0	620	0.30
$CsNaK_3Sb$	1.55	800	0.40

Note: the correspondence between the energy of the incident photon expressed in electron–volts and the wavelength expressed in nanometers of the associated electromagnetic radiation is given by the Duane and Hunt relation: $\lambda(nm) = 1239.85207/E(eV)$.

8.5 Electrode Materials

8.5.1 Electrode Materials for Batteries and Fuel Cells

In power sources, i.e., primary and secondary batteries, and fuel cells, the electrode material both cathode and anode must exhibit a high standard electrode potential expressed in volts (V). Actually, an anode material must be highly electropositive, i.e., reducing, while a cathode material highly electronegative, i.e., oxidizing. The second important physical quantity required to select the most appropriate electrode material is its electrochemical equivalence. The **electrochemical equivalence** denoted E_q of an electrode material is defined as the available electric charge stored per unit mass of material, hence it is expressed in $C.kg^{-1}$ ($A.h.g^{-1}$) and calculated with the following equation:

$$E_q = \frac{nF}{vM}$$

where n is the number of electrons required to oxidize or reduce the electrode material, F is Faraday's constant 96,485.309 $C.mol^{-1}$, v the dimensionless stoichiometric coefficient of the electrochemical reduction or oxidation, M the atomic or molecular mass of the electrode material in $kg.mol^{-1}$ ($g.mol^{-1}$). Sometimes the electrochemical equivalence per unit volume is used, and it is expressed as the electric charge stored per unit volume of material ($A.h.m^{-3}$); in

Table 8.5 Secondary emission characteristics of selected materials

Material	Maximum incident energy (E_{max}, eV)	Maximum secondary emission coefficient (δ_{max})	Material	Maximum incident energy (E_{max}, eV)	Maximum secondary emission coefficient (δ_{max})
Ferrous metals			**Halides**		
Iron	200	1.30	CsCl	n.a.	6.50
Cobalt	500	1.35	LiF	n.a.	5.60
Nickel	450	1.35	NaF	n.a.	5.70
Common nonferrous metals			NaBr	n.a.	6.30
Copper	600	1.28	NaCl	600	6.80
Other metals			KCl	1500	8.00
Beryllium	200	0.50	**Oxides and sulfides**		
Barium	300	0.85	BeO	440	8.00
Cesium	400	0.72	MgO	1600	15
Platinum group metals (PGMs)			Al_2O_3	1300	3.00
Palladium	550	1.65	Cu_2O	440	1.20
Ruthenium	570	1.40	SiO_2	300	2.20
Iridium	700	1.50	ZnS	350	1.80
Platinum	720	1.60	MoS_2	n.a.	1.10
Reactive and refractory metals (group IVB, VB, and VIB)					
Titanium	280	0.90	Niobium	350	1.20
Zirconium	350	1.10	Tantalum	600	1.25
Chromium	400	1.10	Molybdenum	350	1.20
Tungsten	650	1.35	Thorium	800	1.10

this particular case, it can be calculated by multiplying the specific electrochemical equivalence by the density of the electrode material.

In addition, in primary and rechargeable batteries, apart from the two previous scientific parameters, several technological requirements must also be considered when selecting the most appropriate electrode. These requirements are as follows: (i) a high electrical conductivity, (ii) a chemical inertness, (iii) ease of fabrication, (iv) involvement of nonstrategic materials, (v) low cost, and finally (vi) commercial availability. As a general rule, metals and alloys represent the major anode materials in batteries, except for the particular case of hydrogen in fuel cells, while metallic oxides, hydroxides, chlorides, and sulfides represent the major anodic materials except oxygen in fuel cells.

8.5.2 Electrode Materials for Electrolytic Cells

Today, in the modern chemical process industries, electrochemistry occupies an important place. The electrochemical processes are widely used in inorganic synthesis.[1] Actually, it is the only method for preparing and recovering several pure elements (e.g., aluminum, alkali and alkaline-earth metals, chlorine and fluorine).[2] Furthermore, it takes an important place in hydrometallurgy for electrowinning and electrorefining metals of groups IB (e.g., Cu, Ag, Au), IIB (e.g., Zn, Cd) and IVA (e.g., Sn, Pb).[3,4] In addition, its development also concerns organic

Table 8.6 Common anode materials used in batteries and fuel cells

Anode material	Relative molecular mass ($^{12}C = 12.000$)	Standard electrode potential (V/ENH)	A.h.kg^{-1}	A.h.dm^{-3}
Al	26.9	−1.66	2980	8100
Ca	40.1	−2.84	1340	2060
Cd	112.4	−0.40	480	4100
Fe	55.8	−0.44	960	7500
H$_2$	2.01	0.00	26,590	–
Li	6.94	−3.01	3860	2060
Mg	24.3	−2.38	2200	3800
Na	23.0	−2.71	1160	1140
Pb	207.2	−0.13	260	2900
Zn	65.4	−0.76	820	5800

Table 8.7 Common cathode materials used in batteries and fuel cells

Cathode material	Relative molecular mass ($^{12}C = 12.000$)	Standard electrode potential (V/SHE)	A.h.kg^{-1}	A.h.dm^{-3}
Ag$_2$O	231.7	+0.35	231	1640
AgO	123.8	+0.57	432	3200
Cl$_2$	71.0	+1.36	755	–
CuCl	99.0	+0.14	270	950
HgO	216.6	+0.10	247	2740
MnO$_2$	86.9	+1.23	308	1540
NiOOH	91.7	+0.49	292	2160
O$_2$	32.0	+1.23	3350	n.a.
PbO$_2$	239.2	+1.69	224	2110
SO$_2$	64.0	+1.37	419	n.a.
SOCl$_2$	119	+1.63	450	n.a.
V$_2$O$_5$	181.9	+3.6	150	530

synthesis, where some processes have reached the industrial scale (e.g., Monsanto, Nalco, and Philips processes).[5] Apart from the electrochemical processes for preparing inorganic and organic compounds, other electrolytic processes are also used in different fields: (i) in hydrometallurgy (e.g., the electrolytic recovery of zinc),[6] (ii) in electroplating (e.g., the high-speed zinc electroplating of steel plates), (iii) in electrodialysis (e.g., the salt-splitting regeneration of sulfuric acid and sodium hydroxide from sodium sulfate waste brines,[7,8] the regeneration of the leaching solutions of the uranium ores, the electrolytic regeneration of the spent pickling solutions,[9] and (iv) in processes for a cleaner environment, where electrochemistry is used to achieve the electrooxidation of organic pollutants (i.e., electrolytic mineralization), and in the removal of metal cations from liquid waste effluents.[10]

The electrochemical processes are performed in an electrolytic cell[11] (**electrolyzer**). The electrolyzer is a reactor vessel, filled with an electrolytic bath or **electrolyte**, in which the electrodes are immersed and electrically connected via busbars to a power supply. When the electrolyzer is split into two compartments by a **separator** (e.g., diaphragm, membrane), the electrolyte has two different compositions (anolyte and catholyte). The electrodes are the main parts of the electrolyzer, and consist of the anode (positive, +) where the oxidation reaction occurs, while at the cathode (negative, –) a reduction takes place. Among the several issues encountered by engineers for designing an industrial electrochemical reactor, one of them consists in reducing the specific electric energy consumption (i.e., electric energy per unit mass of product). The specific energy consumption can be minimized in two ways: increasing the current efficiency, and lowering the operating cell voltage. Other issues for designing the electrochemical cells are discussed in more detail elsewhere in the literature.[12–14] The overall cell voltage can be classically described as the following algebraic sum:

$$\Delta U_{cell} = (E_{a,th} - E_{c,th}) + \sum_{k=1}^{n}(\eta_{a,k} - \eta_{c,k}) + i\sum_{k=1}^{n} R_k = \Delta U_{th} + \Delta\eta + iR_{total}$$

where the first term corresponds to the theoretical or thermodynamic cell voltage, and consists of the algebraic difference between the thermodynamic potentials of the anode and cathode respectively (Nernst electrode potentials), the second term is the summation of all the electrode overpotentials (e.g. activation, concentration, passivation, etc.), and the third term the summation of all the ohmic drops (e.g., electrolyte both anolyte and catholyte, separators, connectors, and busbars). Hence, the operating cell voltage could be reduced in several ways.[15] Firstly, an appropriate counter electrode reaction minimizes the reversible cell voltage. Secondly, a narrow inter-electrode gap and electrode–membrane gap in association with a highly conductive electrolyte and separator and highly conductive metals for busbars, feeders, and connectors diminish the overall ohmic drop. Thirdly, the turbulent promoters should be used to enhance convection and hence the mass transfer coefficient in order to reduce the concentration overpotential. Finally, the activation overpotential could be reduced using an efficient and appropriate electrocatalyst. The selection of a catalyst is an important problem to solve, particularly in the case of the oxygen or chlorine anodes. The theoretical aspects of the electrocatalysis are extensively reviewed in more detail by Trasatti.[16] Indeed, because of the complex behavior of an electrode, the selection of an electrocatalyst for a definite process could not be simply made from only electrochemical kinetics considerations (i.e., exchange current density, Tafel slopes). An experimental approach is compulsory. Actually, the prediction of the electrode service life needs real standardized tests (i.e., accelerated service life test). For the practitioner engineers, several scientific and technical criteria must be considered when selecting an appropriate electrode material. Therefore the electrode material must exhibit the following requirements as follows:

(1) high exchange current density (j_o), and good electronic transfer coefficient (α or β) for the selected electrochemical reaction, in order to decrease activation overpotential,
(2) good electronic conductivity in order to decrease the ohmic drop and the Joule heating,
(3) good corrosion resistance to both chemical and electrochemical reactions, combined with no passivating and blistering behavior leading to electrode degradation and consumption,
(4) a good set of mechanical properties suited for industrial use (i.e., low density, high tensile strength, stiffness),
(5) easy fabrication (i.e., machining, joining, and descaling) allowing clean and intricate shapes to be obtained,
(6) low cost combined with commercial availability and a wide diversity of mill products (e.g., rod, sheet, expanded metal),
(7) nonhazardous, nontoxic and environmentally friendly.

It is important to note that the combination of criteria (3), (4) is essentially for the dimensional stability of the electrode and of its service life. Generally speaking, the selection of the cathode

Table 8.8 Cathode materials for hydrogen evolution in acid media

Overvoltage range	Cathode material	Electrolyte composition	Molarity (C, mol.dm^{-3})	Temperature (T, °C)	Cathodic Tafel slope (b_c, mV.log$_{10}$ j_0^{-1})	Exchange current density decimal logarithm (log$_{10}$ j_0, A.cm^{-2})	Overvoltage at 200 A.m^{-2} (|η|, mV) (absolute value)
Extra low hydrogen overvoltage	Iridium (Ir)	H$_2$SO$_4$	0.5	25	30	−2.699	30
	Palladium (Pd)	HCl	1	25	30	−2.500	24
		H$_2$SO$_4$	1	25	29	−3.200	44
	Platinum (Pt)	HCl	1	25	29	−3.161	43
		H$_2$SO$_4$	2	25	25	−3.200	38
	Rhodium (Rh)	H$_2$SO$_4$	4	25	28	−3.200	42
	Ruthenium	HCl	1	25	30	−4.200	75
Low hydrogen overvoltage	Molybdenum (Mo)	HCl	0.1	25	104	−6.400	343
	Tungsten (W)	HCl	5	25	110	−5.000	363
	Nickel (Ni)	HCl	1	25	109	−5.222	384
		H$_2$SO$_4$	1	25	124	−5.200	434
	Silver (Ag)	HCl	5	25	120	−5.301	432
		H$_2$SO$_4$	1	25	120	−5.400	444
	Iron (Fe)	HCl	0.5	25	133	−5.180	425
		H$_2$SO$_4$	0.5	25	118	−5.650	466
	Gold (Au)	HCl	0.1	25	123	−5.500	468
		H$_2$SO$_4$	1	25	116	−5.400	430
			4	25	130	−6.500	624
	Copper (Cu)	HCl	0.1	25	120	−6.823	615
		H$_2$SO$_4$	0.5	25	120	−7.700	720
High hydrogen overvoltage	Niobium (Nb)	HCl	1	25	110	−9.000	803
		H$_2$SO$_4$	2	25	120	−8.400	804
	Titanium (Ti)	HCl	1	25	130	−7.500	754
		H$_2$SO$_4$	0.5	25	135	−8.200	877
			1	25	119	−8.150	767
	Tin (Sn)	H$_2$SO$_4$	4	25	120	−9.00	877
	Zinc (Zn)	HCl	1	25	120	−10.800	1092
		H$_2$SO$_4$	2	25	120	−10.800	1092
	Cadmium (Cd)	H$_2$SO$_4$	0.25	25	135	−10.769	1225
	Lead (Pb)	HCl	1	25	117	−12.900	1311
		H$_2$SO$_4$	0.5	25	120	−12.700	1320
	Mercury (Hg)	HCl	1	25	118	−12.500	1475
		H$_2$SO$_4$	2	25	119	−12.107	1239

$$\eta_c = E_{c,j} - E_{th} = b_c(\log_{10} j_{eq} - \log_{10} j) = \frac{\ln 10 RT}{\beta_n F}\log_{10} j_{eq} - \frac{\ln 10 RT}{\beta_n F}\log j$$

8

Misc. Electrical Materials

Table 8.9 Anode materials for oxygen evolution in acid media

Overvoltage range	Anode material	Electrolyte composition	Molarity (C, mol.dm^{-3})	Temperature (T, °C)	Anodic Tafel slope (b_a, mV.log$_{10}$ j_0^{-1})	Exchange current density decimal logarithm (log$_{10}$ j_0, A.cm^{-2})	Overvoltage at 200 A.m^{-2} (η_a, mV)
Low oxygen overvoltage	Ta/Ta$_2$O$_5$–IrO$_2$	H$_2$SO$_4$ 30wt%	3.73	80	52, 133	−3.630, −10.21	101
	Ti–Pd (Gr.7)/Ta$_2$O$_5$–IrO$_2$	H$_2$SO$_4$ 30wt%	3.73	80	54, 164	−4.53, −8.21	153
	Ti/TiO$_2$–IrO$_2$	H$_2$SO$_4$ 30wt%	3.73	80	60	−4.886	191
	Ti (Gr.2)/Ta$_2$O$_5$–IrO$_2$	H$_2$SO$_4$ 30wt%	3.73	80	51, 158	−5.82, −7.69	210
Medium oxygen overvoltage	Ti/TiO$_2$–RuO$_2$ (DSA®–Cl$_2$)	H$_2$SO$_4$	1	80	66	−7.900	409
		CF$_3$SO$_3$H	1	80	65	−8.000	410
	Ruthenium–iridium	H$_2$SO$_4$	1	80	74	−7.020	400
		CF$_3$SO$_3$H	1	80	86	−6.630	419
	Iridium (Ir)	H$_2$SO$_4$	1	80	85	−6.800	433
		CF$_3$SO$_3$H	1	80	84	−6.800	428
High oxygen overvoltage	alpha-PbO$_2$	H$_2$SO$_4$	4	30	45	−15.700	630
	Platinum–ruthenium	H$_2$SO$_4$	1	80	120	−7.700	710
		CF$_3$SO$_3$H	1	80	120	−7.500	670
	Platinum–rhodium	H$_2$SO$_4$	1	25	115	−7.600	679
	Rhodium (Rh)	HClO$_4$	1	25	125	−7.520	727
	Platinum (Pt)	H$_2$SO$_4$	1	80	90	−10.900	828
		CF$_3$SO$_3$H	1	80	94	−9.800	762
		HClO$_4$	1	25	110	−9.000	803
	Pt/MnO$_2$	H$_2$SO$_4$	0.5	25	110	n.a.	n.a.
	beta-PbO$_2$	H$_2$SO$_4$	4	30	120	−9.200	900
	PbO$_2$	H$_2$SO$_4$	4	30	120	−10.000	996

$$\eta_a = E_{a,j} - E_{th} = b_a(\log_{10} j - \log_{10} j_{eq}) = \frac{\ln 10 RT}{anF}\log_{10} j - \frac{\ln 10 RT}{anF}\log_{10} j_{eq}$$

materials is easier for the electrochemical engineer than the selection of the anode materials. Actually, there exists a wide diversity of electronic conductive materials according to the needed hydrogen evolution potential for both acid or alkaline electrolytes. For instance, some materials exhibit a high hydrogen evolution overpotential (e.g., Pb, Cd, Hg), or by contrast a low overvoltage (e.g., Cu, Ag, platinized C, Ni). Nevertheless, some metals (e.g., Ti,[17] Nb, Ta, iron and steels) are susceptible to hydrogen pick-up and hence extremely sensitive to embrittlement by nascent electrolytic hydrogen gas evolved during cathodic polarization of the metal.[18] This phenomenon leads to electrode blistering. Therefore, these metals are not suited to cathode manufacture in aqueous media.

8.5.3 Industrial Anode Materials

By contrast with the cathode, the selection of the anode material is often a critical issue which leads to the abandonment of some industrial processes. Actually, the anode is defined as the electrode where the electro-oxidation occurs, hence, the anode material must withstand harsh conditions due to both the elevated positive potential and the corrosiveness of the electrolyte. Several electrode materials were used since the beginning of the industrial processes, and they can be grouped into the following categories.

8.5.3.1 Precious and Noble Metal Anodes

Electrochemists observed early on that noble and precious metals were stiff materials, with both good tensile properties and high electronic conductivity, easy to fabricate and have a high chemical and electrochemical inertness in the corrosive media. Consequently, the first industrial anodes used in the electrochemical processes requiring excellent dimensional stability were made of: (i) the noble and precious metals (e.g., Au, Ag), (ii) the six platinum group metals (Ru, Rh, Pd, Os, Ir, Pt) or (iii) their alloys (e.g., Pt90–Ir10, Pt90–Rh10).[19] The noble and platinum group metals (PGMs) took a particular place owing to both their exceptional chemical and electrochemical corrosion resistance in several harsh environments,[20] and their intrinsic electrocatalytic properties.[21] Moreover, PGMs, such as platinum are the most appropriate anode material for the preparation of persulfates, perchlorates, periodates, and cerium (IV). However, the extremely high cost of these solid metals combined with their high density has restricted their industrial uses. However, early in the century there was an attempt to develop noble anodes made of a base metal coated with a thin layer of platinum or rhodium. Actually, these composite electrodes were first described in 1913 by Stevens in three US patents.[22] The platinum and the rhodium layers were electroplated onto a refractory metal such as tungsten or tantalum. Despite its novelty, this idea was not industrially developed at this time because it was impossible to obtain industrial production of these refractory metals, especially of the mill products (e.g., plates, rods, sheet, strips) needed for manufacturing industrial anodes with large surface areas. It was not until the 1960s that the first commercial platinized anodes appeared (see Section 8.5.3.4). Accordingly, in replacement the lead and the carbon anodes underwent increasing development owing to their low cost in comparison with the platinum group metals.

8.5.3.2 Lead and Lead Alloy Anodes

Besides the precious metal anodes, electrochemists early on used anodes made of two inexpensive materials such as lead and carbon. Lead and graphite were actually the only cheap anode materials that were industrially used up to the 1960s. The use of lead anodes results from its behavior studies in the lead–acid battery invented by Planté in 1859.[23] Lead is inexpensive (0.8 $US.kg^{-1}) and has also several attractive features, such as good electronic conductivity (20.64 $\mu\Omega$.cm) and good chemical and electrochemical corrosion resistance in numerous corrosive and oxidizing environments (e.g., chromates, sulfates, carbonates, and phosphates).[24] This chemical and electrochemical inertness is due to the self-formation of a protective passivating layer. Owing to this passivation film, it has been widely used in the electrochemical processes using sulfuric acid as electrolyte.[25] The lead anode was typically made of: (i) commercially pure metal, (ii) silver alloy (Pb95–Ag5)[26] where the small amount of silver increases oxygen overvoltage, (iii) copper alloy (i.e., chemical lead), or (iv) the other lead alloys such as antimonial and tellurial lead.[27] These alloying elements (e.g., Sn, Bi, Sb, Te) increase its stiffness and corrosion resistance. For instance, the corrosion rate of the pure metal in sulfuric acid 50wt% is 130 μm per year at 25°C. When the metal undergoes an anodic current density of 1 kA.m^{-2} in sulfuric acid 60wt%, the corrosion rate reaches only 9 mm per year.[28] Nevertheless, its high density (11,350 kg.m^{-3}), combined with a low melting point (327.5°C), and a high thermal linear expansion coefficient (30 μm.m^{-1}.K^{-1}) leads to severe creep phenomena when the electrolysis is conducted above the ambient temperature. Nevertheless, more recently, reinforced lead anodes were developed for the electrolytic production of zinc from sulfate baths. Indeed, this anode comprises a skin portion formed by conventional anode metal, lead containing from 0.25 to 1.0wt% Ag, and a stiffening reinforcing member of titanium or zirconium. The reduction in thickness of the anodes, which is made possible by the provision of the reinforcing member, results in a substantial saving in the amount of silver-bearing lead which is immobilized, and a substantial reduction in the unit weight of the anode.

8.5.3.3 Carbon Anodes

Carbon is an attractive material because it is a cheap material with an excellent chemical inertness, is easy to machine and has a low bulk density (2260 kg.m^{-3}). Furthermore, there is a great diversity of commercially available products (e.g., graphite, pyrolytic, impervious, or glassy) and in several forms (e.g., cloth, blacks, powders, or reticulated). The graphite variety, despite its high electrical resistivity (1375 $\mu\Omega$.cm) and brittleness, was widely used for the electrolysis of brines, and especially for the electrolysis of hydrochloric acid for production of chlorine.[29] This last process, initially developed in Germany during World War II by Holemann and Messner at IG Farben Industrie[30,31] was continued during the 1950s by De Nora–Monsanto[32,33] and Hoechst–Udhe.[34,35] However, in the 1960s, the improvement of the chlor-alkali processes (e.g., mercury cathode cell and diaphragm cell) required a great effort in R&D. This research was essentially focussed on improving graphite anodes which had serious drawbacks: (i) the non-dimensional stability of the carbon anodes during the electrolysis leads to a continual increase in the inter-electrode gap enhancing ohmic drop, (ii) a high chlorine evolution overpotential, and (iii) a very short service life (6 to 24 months) due to the corrosion by the chlorine and the inescapable traces of oxygen forming chlorinated hydrocarbons, and carbon dioxide. These efforts which started in the 1950s led to the birth of the third generation of the industrial anodes. Carbon anodes are also the only appropriate anode material in some particular processes where no other materials exhibit both low cost and a satisfactory corrosion resistance. Actually, several industrial electrolytic processes performed in molten salt electrolyte continue to use carbon anodes. These processes are: (i) the electrowinning of aluminum by the Heroult–Hall process, (ii) the electrolytic production of alkali metals (e.g., Na, Li), and alkaline-earth metals (e.g., Be, Mg), and finally (iii) the electrolytic production of fluorine. Actually, the use of carbon anodes in the chlor-alkali process for the production of chlorine is now discontinued due to the replacement by modern dimensionally stable anodes.

8.5.3.4 Titanium and Tantalum Platinized Anodes

In the 1950s–60s, under the full expansion of the American and Russian aircraft and space programs, and the development of nuclear power reactors, several processes for the production of the main reactive and refractory metals (e.g., Ti, Zr, Hf, Nb, Ta) reached the industrial production. These processes (e.g., the Kroll process for titanium[36]) allowed the reactive and refractory metals to be obtained under a great diversity of shapes and alloy compositions. This development brought several advantages: (i) the reduction of the cost of these metals, (ii) the standardization of the alloy composition and properties (e.g., standard ASTM B265 for titanium grades), (iii) a great effort in R&D in using these metals outside the aircraft and nuclear applications. At this stage, all the difficulties in preparing anodes of refractory metals coated with precious metals vanished and the idea invented 40 years ago by Stevens reappeared (see Section 8.5.3.1). Hence, the niobium and the tantalum platinized anodes became known in the 1950s with the works of Rhoda[37] and Rosenblatt[38] in 1955. In this last patent, a layer of platinum was deposited onto a tantalum base metal by thermal decomposition of hydrogen hexachloroplatinate (IV) under an inert atmosphere. This thermal treatment, which was conducted between 800 and 1000°C, gave a thin layer of few micrometers of an interdiffusion alloy between the tantalum and platinum. Furthermore, the titanium, commercially available at this time because of the increasing demand of this construction material in turbine blades for aircraft engines, has been widely studied as an anode material. Its valve action properties (VA), which allow the metal to be protected by an insulating rutile (TiO_2) layer under anodic polarization, were discovered by Cotton.[39,40] It was only in 1957 that Beer[41] at Magnetochemie in Netherlands, and Cotton,[42,43] at ICI in the UK, showed independently but simultaneously that the addition of a small amount of precious metal deposited onto a titanium base metal, by electrolysis, or spot welding, gave a sufficiently conductive electrode under anodic polarization in spite of the passivation of the metal. These anodes were named **noble metal coated titanium** (NMCT). During the following decade (1960s–70s), several companies registered patents about this type of anode. They were often prepared by electrodeposition because this method allows a smooth and nonporous deposit to be obtained with a complete coverage of complex anode shapes and does not require the high surface mass density of platinum.[44] These firms were for example: Heraeus Gmbh,[45] Metallgeselschaft Gmbh,[46] Texas Instrument, Inc.,[47] Engelhard,[48,49] and IMI–Kynock.[50] The electroplating baths commonly encountered in these works contained platinum salts such as the "P-salt", hexachloroplatinic acid or sodium hexachloroplatinate (IV).[51] These anodes, which were initially supplied by IMI Marston Ltd under the trade name K-type® Pt/Ir 70/30[52] were made from a titanium base metal coated by the PGMs alloy Pt70–Ir30. Niobium and tantalum could also be used as a substrate but they were only considered where the titanium showed deficiencies, owing to their greater cost. The small iridium content served to improve the corrosion resistance. These anodes were rapidly used in all the numerous processes requiring a long service life under severe conditions. For example, they were employed for the cathodic protection of immersed plants such as oil-rigs, storage tanks, and subterranean pipelines,[53,54] in the electrolytic processes for the production of sodium hypochlorite,[55] electrodialysis, for generation of Ce(IV) in perchloric or nitric acid[56] and for oxidation of sulfuric acid in peroxodisulfuric acid.[57] These anodes are coated with platinum or rhodium. Nevertheless, these two metals have a high chlorine overvoltage (e.g. 300 mV for rhodium and 486 mV in the case of platinum under an anodic current density of 10 kA.m^{-2}) and exhibit a slight corrosion with the rate depending on the electrolyte and the nature of impurities (e.g. chlorides, fluorides, organic pollutants). The improvement of this type of electrode is the starting point of the study and the preparation of platinized anodes by the thermal decomposition of a precursor. The precursor consists of a salt of a precious metal dissolved in an appropriate organic solvent (e.g. linalool, isopropanol, ethyl acetoacetate). It was deposited by painting of the base metal plate. The plate also underwent a long thermal treatment at high temperature. The pioneers in this work were Angell and Deriaz from ICI.[58,59] This protocol for the deposition of the catalyst was inspired by Taylor's methods[60] used in the 1930s to obtain a perfect platinum coating onto glasses for the manufacture of optical mirrors. The study of the thermal decomposition of these particular painting solutions was conducted by Hopper[61] in 1923 and more recently by Kuo[62] in 1974. Other patents were registered on the preparation of the titanium platinized anodes by thermal decomposition (Engelhard,[63] Ionic[64]). It is interesting to note that De Nora registered a patent on a Pt-coated anode in which the base metal was a ferro-silicon with some amounts of chromium.[65] A few years later, Millington[66] observed that the use of tantalum in place of the titanium base metal for the same catalyst, increased the service life of the anode. This also appeared in some technical specifications of electrode suppliers.[67] This continuous research effort, always developed under industrial pressure, resulted in the 1960s in a new generation of anodes widely used in all the electrochemical industries.

8.5.3.5 DSA® Type Anodes

In the 1960s, Henri B. Beer who worked at Permelec Corp.,[68] and the Italian team of Bianchi, De Nora, Gallone, and Nidola (Diamond Shamrock Division Research) started studying the

Table 8.10 Definition of Dimensionally Stable Anodes[®]

A Dimensionally Stable Anode is a composite electrode made of:	Base metal or substrate	A base metal with a valve action property, such as the refractory metals (e.g. Ti, Zr, Hf, Nb, Ta, Mo, W) or their alloys (e.g. Ti–0.2Pd, Ti–Ru). This base metal acts as a current collector.[a] Sometimes, it is possible to find in the claims of some particular patents, unusual base materials (e.g. Al, Si–cast iron, Bi, C, Fe_3O_4).
	Protective passivating layer	A thin and impervious layer (some micrometers thickness) of a protective passivating oxide (e.g. TiO_2, ZrO_2, HfO_2, Nb_2O_5, Ta_2O_5. NbO_2 and TaO_2).
	Electrocatalyst	An electrocatalytic oxide of a noble metal or more often an oxide of the PGMs. This oxide allows an increase in the electrical conductivity of the previous passivating film (e.g. RuO_2, PtO_x, IrO_2). Sometimes, other oxides are added (e.g. SnO_2, Sb_2O_5, Bi_2O_3) and also carbides (e.g. B_4C) or nitrides.

[a]De Nora O, Nidola A, Trisoglio G, Bianchi G. Brit. Pat. 1,399,576 (1973).

electrocatalytic behavior of the mixed oxide and nitride coatings for the evolution of chlorine and oxygen.[69,70] These oxides were obtained by the calcination of precursors but under an oxidizing atmosphere (i.e., air, or pure oxygen). These RuO_2-based anodes or so-called "ruthenized titanium anodes", composed of a mixed oxide layer (TiO_2–RuO_2) coated onto a titanium base metal, have been developed with great success by Beer since 1966.[71] At this stage, the selection of the ruthenium was only made according to the low cost of the metal and its commercial availability. These electrodes were initially protected by South African[72,73] and American patents.[74–77] It was the birth of the **activated titanium anode** (ATA) also named **oxide coated titanium anode** (OCTA). These designations are now obsolete. These anodes are characterized by a geometrical stability and a constant potential over a long time (above 2–3 years). It is this dimensional stability in comparison with the graphite anodes which gives its actual trade name: Dimensionally Stable Anodes (the acronym DSA[®] is a trademark of Electronor Corp.). The classical composition of the composite anodes is defined in Table 8.10.[78]

As a general rule, these anodes were made from a titanium base metal covered by a rutile layer, TiO_2, doped with RuO_2 (30% mol).[79] They had an important industrial development (e.g. De Nora, Permelec, Eltech, Heraeus, and Magnetochemie) and today they are used in all the chlor-alkali processes and for the production of chlorates.[80] The Dimensionally Stable Anodes for chlorine evolution were described in the technical literature by the common acronyms DSA[®] (RuO_2) or DSA[®]–Cl_2 and they had a great industrial success for the following two reasons: (i) ruthenium has the lowest price of all the PGMs, and (ii) its density is about half of its neighbors. Moreover, its electrocatalytic characteristics for the evolution of chlorine are satisfactory. Under industrial conditions (2 to 4 $kA.m^{-2}$) the service life of these electrodes is over 5 years. Therefore, today, titanium is the only base metal used for manufacturing Dimensionally Stable Anodes for chlorine evolution. The contribution of Beer's discovery to the development of industrial electrochemistry is very important. The reader can also find a complete story of the invention of the DSA[®] by the inventor himself, and written on the occasion of his award of an Electrochemical Society Medal.[81]

8.5.3.6 Anodes for Oxygen Evolution in Acidic Media

Several processes need anodes for oxygen evolution in an acidic medium. In comparison with chlorine evolution, the evolution of oxygen in an acidic medium leads to a higher anodic potential with a high acidity which represents harsh conditions for the electrode material. In these conditions, most of the materials are put in their anodic dissolution or transpassive zone. These conditions greatly restrict the choice of the material. The suitable materials which resist in these conditions are gold and PGMs. However, it is obvious that their high density and very high cost prohibit their use. Today they are only used cladded onto common base metal, in industrial electrolytic cells under low anodic current densities (1 kA.m^{-2}); when high-valued chemicals are produced a hydrogen gas anode is used. As has been previously discussed, their high corrosion resistance associated with a high electrocatalytic activity are responsible for their exceptional properties. As a general rule, their decreasing electrochemical activity could be classified as follows: Ir > Ru > Pd > Rh > Pt > Au.[82] The carbon anodes, sometimes impregnated with a dispersion of PGMs, are now today obsolete because of their high oxygen overvoltage and rapid material damage during electrolysis. This last drawback is due to the swelling of the anodic material submitted to the gas evolution. Indeed, owing to its high porosity an intercalation phenomenon occurs: the anions penetrate in the lattice and they expand the structure leading rapidly to disintegration of the electrode.[83]

Lead dioxide – lead dioxide exhibits two polymorphic crystal phases: (i) orthorhombic crystals (α-PbO$_2$), and (ii) tetragonal crystals with the rutile type structure (β-PbO$_2$). It is used as supported anodes, such as Pb/PbO$_2$ or Ti/PbO$_2$.[84] These anodes are commonly prepared by in situ anodization in a bath containing small amounts of Cu(II) and Ni(II) in order to prevent the lead depositing at the cathode. However, reproductible and high-performance PbO$_2$ anodes require the use of carefully selected and controlled conditions. The typical preparation is discussed by Thangappan et al.[85] Lead dioxide has actually several attractive features such as a low electrical resistivity (40 to 50 $\mu\Omega$cm), a good chemical and electrochemical corrosion resistance in sulfate media even at low pH, and a high overvoltage for the evolution of oxygen. However, its brittleness, even when supported by lead, exhibits some issues for industrial applications. Nevertheless, PbO$_2$ deposited onto an inert base material such as graphite, Pt, Ti, or Ta, was developed and is now commercially available.[86,87] Sometimes a thin intermediate Pt layer is inserted between the base metal and the PbO$_2$ coating to enhance the service life. The lead anodes coated with PbO$_2$ replaced the platinum anodes before the 1950s and the graphite anodes for the regeneration of potassium dichromate, and in the production of chlorates and perchlorates.[88] These anodes were also used in: (i) hydrometallurgy for oxygen anodes for electroplating copper and zinc in sulfate baths,[89–91] and (ii) organic electrosynthesis in the process for the production of glyoxilic acid from oxalic acid using sulfuric acid as supporting electrolyte.[92] In fact, PbO$_2$-based anodes are used for their inertness, low cost, and when oxidation should be achieved without an oxygen-evolving competitive reaction.

Manganese dioxide – MnO$_2$ was used for a long time in hydrometallurgy for the electrowinning or electroplating of Zn,[93] Cu, and Ni in sulfate baths. They are prepared by solution impregnation–calcination[94] or by the anodization protocol such as Pb/PbO$_2$ of a sulfuric solution containing manganese ions. However, they never have the same full expansion of lead dioxide owing to their high corrosion rate under extreme conditions (e.g. high temperature, high pH, and high anodic current density). Nevertheless, some improvements have been made to increase their stability. Feige prepared a Ti/MnO$_2$ anode made by sintering titanium and lead particles with MnO$_2$.[95] De Nora et al. obtained a Ti/MnO$_2$ type anode by the application of the classical procedure employed for the preparation of DSA®.[96]

Spinel – the oxides with the spinel type structure (A^{II}B$^{III}_2$O$_4$) are sometimes used supported by titanium.[97] There are three classes: ferrites, cobaltites, and chromites. Since 1870, the pure magnetite (Fe$_3$O$_4$) or its doped form[98] has been used as anode material but today the cobaltites are preferred and only developed (e.g. MCo$_2$O$_4$ with M = Mg, Cu, Zn). They are produced by calcination in a moderately oxidizing atmosphere (e.g. steam, or argon–carbon dioxide mixture) at high temperature (700–900°C). Sometimes, as in the PbO$_2$-based anodes an intermediate thin Pt layer is deposited between the base metal and the magnetite to enhance the service life. These anodes, in spite of their good stability under high positive potential,[99,100] have two drawbacks: (i) they are brittle which involves supporting the ceramic with a substrate, and (ii) they have a very high electrical resistivity (27,000 $\mu\Omega$.cm) with respect to the other electrode materials.

Ebonex® – since the mid-1980s, substoichiometric titanium oxides were developed by the UK company Atraverda Ltd and commercialized under the common trade name Ebonex®.[101]

Ebonex$^{®}$ consists of substoichiometric titanium oxides with the chemical formula TiO_{2-x} (e.g., Ti_5O_9 and Ti_4O_7). Ebonex$^{®}$ material exhibits both a good corrosion resistance in acidic media, an electrical conductivity similar to that of graphite (around 1200 $\mu\Omega$.cm), and a high over-potential for oxygen evolution. As a general rule, the use of these materials is limited, especially under severe conditions. These anodes were suggested for treatment of waste effluents, but their high cost combined with mechanical fragility has restricted their industrial uses.

DSA$^{®}$-O$_2$ – however, for the high anodic current densities involved in some industrial processes (e.g., 2 to 15 kA.m^{-2}) such as in the high-speed electroplating of steel plates, or in the zinc electrowinning, the previously described materials are rapidly corroded and therefore not satisfactory. The success of DSA$^{®}$ in the chlor-alkali industry inspired the use of these anodes for the oxygen evolution reaction. According to the good results obtained with DSA$^{®}$-Cl$_2$, several compositions of mixed oxides such as TiO_2-RuO_2, Ta_2O_5-RuO_2, TiO_2-IrO_2, and base metals were studied as anode materials for oxygen evolution in acidic media. Only the oxide mixtures with iridium dioxide (IrO_2) were selected. Indeed, ruthenium dioxide (RuO_2), despite being a very efficient catalyst for DSA$^{®}$-Cl$_2$, is not suited for oxygen anodes because it reacts with oxygen to give off the volatile ruthenium tetroxide (RuO_4), and RuO_2 is too sensitive to the electrochemical disolution.[102,103] The service life of this anode is only several dozen hours, due to its corrosion behavior. In this case, the IrO_2-based anodes such as Ti/TiO_2-IrO_2 are effectively used. However, it was in 1975 that De Nora et al.[104] registered an important patent disclosing the oxide mixture Ta_2O_5-IrO_2. Later, Comninellis et al. optimized the composition of this mixture and showed that an electrocatalyst with 70mol% Ta_2O_5-IrO_2 content was the best choice for an IrO_2-based anode for oxygen evolution in acidic media.[105] This product has been developed commercially by the Eltech System Corp. under the common trade name TIR2000$^{®}$. Nevertheless, during the electrolysis, these anodes are quickly deactivated. Actually, a complex corrosion–passivation mechanism occurs throughout the crystals of the catalysts. Indeed, during the calcination, the thermal retraction stresses give to the catalytic film $Ta_2O_5IrO_2$ a typical dried mud crack structure.[106] These assemblies of catalyst grains separated by the cracks lead to the electrolyte penetrating through to the base metal: this is called the undermining process.[107-109] According to Hine et al.,[110] by analogy with the anodic deactivation behavior of the tantalum and titanium PbO_2- and MnO_2-coated anodes, this behavior involves damage to the interface between the iridium dioxide crystal catalyst and the base metal by the formation of a thin rutile layer. This insulating film decreases the conductive area and increases the local anodic current density. This behavior ends up with the occurrence of a mechanical failure with desquamation of the coating. As a general rule, this phenomenon is observed industrially because the operating cell potential increases continuously up to the limiting potential delivered by the power supply. At this stage the anode is considered to be deactivated and it is returned to the supplier in order to be treated and refurbished. The costly electrocatalytic noble coating is then removed from the base metal by chemical stripping; the etching operation is performed in a molten mixture of alkali metal hydroxides containing small amounts of an oxidizing salt.[111] The precious catalyst is then recovered in the slurries at the bottom of the reactor, while the pickled base metal is treated and reactivated by the classical procedure. Furthermore, the service life of the anode depends on the following parameters: (i) the anodic current density, (ii) the coating preparation, and (iii) the impurities. These inorganic and organic pollutants can lead to the dissolution of titanium (e.g., fluoride anions[112]), scaling (e.g., Mn(II) cations) and the loss of coating (e.g., organic acids, nitroalcohols). For example, in organic electrosynthesis, the service life of these electrodes ranges from 500 to 1000 h in molar sulfuric acid at 60°C.[113] But their high cost per unit area (20,000 to 30,000 $US.m^{-2}) forbids their industrial use. At the end of the 1970s, as the consequence of the work on the cathodically modified alloys initially conducted in 1940s by the Tomashov group, followed in the 1960s by Stern and Cotton,[114] appeared the titanium–palladium alloy (ASTM grade 7) in which the small amount of palladium (0.12–0.25wt%) increases the corrosion resistance in the reducing acids. Hence, several patents included this alloy for preparing oxygen anodes. Recent work of Cardarelli et al. indicated that the service life of Ti/Ta_2O_5-IrO_2 type anodes depends both on the impurities and the type of alloying elements in titanium and titanium alloys.[115]

Nevertheless, these service lives are not long enough to allow these anodes to be satisfactory, and their short lifetime is the origin of the abandonment of several industrial projects. The formation of a passivating layer of rutile under a high anodic potential is always the source of the deactivation behavior. Therefore, several other refractory metals were tested. Vercesi et al.[116] showed that mixtures of catalytic oxides deposited onto a solid tantalum base metal provided anodes with a extended service life. This good behavior was due to the exceptional valve action property of the tantalum pentoxide in the sulfuric acid. However, in spite of its remarkable corrosion resistance, the development of a tantalum base metal is rarely used in the industrial plants because of its two drawbacks: high cost (461 $US.kg^{-1}) combined with high density (16,654 kg.m^{-3}), in comparison with titanium (4540 kg.m^{-3}) which has a medium cost (49 $US.kg^{-1}). As a consequence, an anode of tantalum is 35 times more expensive than a titanium anode. Furthermore, owing to its high reactivity versus oxygen above 350°C, the preparation of the tantalum anodes involves great difficulties during the thermal treatment required during the electrode manufacture. For these reasons, the tantalum anode does not reach a full industrial level. In order to decrease the cost, a thin tantalum layer deposited onto a common base metal is very attractive alternative. This idea appeared for the first time in 1968 in a patent[117] registered by the German company Farbenfabriken Bayer AG and also in 1974 proposed by Jeffes in a patent of Allbright & Wilson.[118] In this last patent a composite DSA$^{®}$ – Cl$_2$ was made from a steel plate coated by chemical vapor deposition (CVD) with a 0.5 mm tantalum layer. Then the tantalum is coated with RuO$_2$ (steel/Ta/RuO$_2$). Twenty years after, the anodes made according to this idea were still not industrially developed. However, in 1990, in a European patent[119] Registered by ICI, Denton and Hayfield described the preparation of oxygen anodes made of a thin tantalum coating deposited onto common base metal by several techniques. Finally in 1993, Kumagai et al.[120] Prepared an anode made of a thin tantalum layer deposited onto a titanium base metal by sputtering (Ti/Ta/Ta$_2$O$_5$–IrO$_2$). In order to select the most optimized method for depositing tantalum onto a common substrate, a comprehensive comparison of tantalum coating techniques used in the chemical process industries was recently reviewed by Cardarelli et al.[121] Moreover, the same authors developed anodes made from a thin tantalum layer deposited onto a common base metal (e.g., copper, nickel, or stainless steel) coated with the electrocatalytic mixture of oxides Ta$_2$O$_5$–IrO$_2$ produced by calcination (stainless steel/Ta/Ta$_2$O$_5$–IrO$_2$). The performance of these anodes (stainless steel/Ta/IrO$_2$) is identical to that obtained with solid tantalum base metal (Ta/IrO$_2$).[122]

8.5.4 Electrodes for Corrosion Protection and Control

Apart from batteries, fuel cells, and industrial electrolyzers, corrosion protection and control is another field in which the electrode materials occupy an important place.

8.5.4.1 Cathodes for Anodic Protection

Anodic protection[123] is a modern electrochemical technique used for protecting against corrosion of metallic equipment used in the chemical process industries, and in handling highly corrosive chemicals (e.g. concentrated sulfuric and orthophosphoric acids). The technique consists of applying a very low anodic current (usually 10 μA.m^{-2}) to metallic equipment (e.g., tank, thermowells, column) to protect against corrosion. This anodic polarization puts the electrochemical potential of the metal in the passivity region of its Pourbaix diagram, i.e., where the dissolution reaction does not occur, and hence this leads to a negligible corrosion rate (less than 25 μm.yr^{-1}). The anodic protection method can only be used to protect against corrosion of metals and alloys exhibiting a passive state (e.g., reactive and refractory metals, stainless steels, etc.). Usually, the equipment it is required to protect is a cathode, a reference electrode, and a power supply. The various cathode materials used in anodic protection are listed in Table 8.11.

Table 8.11 Cathode materials for anodic protection[a]

Cathode	Corrosive chemicals
Hastelloy[®] C	Nitrate aqueous solutions, sulfuric acid
Illium[®] G	Sulfuric acid (78–100wt%), oleum
Nickel-plated steel	Electroless nickel plating solutions
Platinized copper or brass	Acids
Silicon-cast iron (Duriron[®])	Sulfuric acid (89–100wt%), oleum
Stainless steels (AISI 304, 316L)	Nitrates aqueous solutions
Steel	Kraft digester liquid

[a]From Locke CE (1992) Anodic protection. In: ASM metal handbook, 10th edn. Corrosion, vol. 9. ASM, Ohio Park, OH, pp 463–465.

8.5.4.2 Anodes for Cathodic Protection

Cathodic protection consists of polarizing cathodically a metal in order to maintain its immunity in a corrosive environment. There exist two ways for achieving efficiently cathodic polarization: (i) passive protection which consists of connecting electrically the metal to a less noble material that will result in a galvanic coupling of the two materials, leading to the anodic dissolution of the **sacrificial anode**, and (ii) active protection which consists of applying a current in order to polarize cathodically the workpiece versus a nonconsumable anode.

Table 8.12 Sacrificial anode materials

Sacrificial anode material	Oxidation reaction	Electrode potential at 298.15 K (E_0,mV vs SHE)	Capacity (A.h.kg^{-1})	Consumption rate (kg.A^{-1}.yr^{-1})	Notes
Magnesium	Mg^0/Mg^{2+}	−2.360	1100	7.9	Buried soils, suitable for high-resistivity environments. Not suited to marine applications due to the high corrosion rate of magnesium in seawater.
Zinc	Zn^0/Zn^{2+}	−760	810	10.7	Fresh water, brackish and marine water
Aluminum–Zinc–Mercury	Al^0/Al^{3+}	−1660	920–2600	3.0–3.2	Seawater, brines. Offshore and oil rigs, marine technology. Addition of In, Hg, and Sn avoids passivation
Aluminum–Zn–Indium			1670–2400	3.6–5.2	
Aluminum–Zinc–Tin			2750–2840	3.4–9.4	

References

[1]   Gopa R, Gibbons DW (1994) J Electrochem Soc 14: 2918.
[2]   Pletcher D, Walsh FC (1990) Industrial electrochemistry, 2nd edn. Chapman and Hall, London.
[3]   Kuhn AT (ed.) (1977) Electrochemistry of lead. Academic Press, London.
[4]   Gonzalez-Dominguez JA, Peters E, Dreisinger DB (1991) J Appl Electrochem 21: 189.
[5]   Baizer MM, Lund H (1983) Organic electrochemistry: an introduction and a guide, 2nd edn. Marcel Dekker, New York.
[6]   Karavasteva M, Karaivanov S (1993) J Appl Electrochem 23: 763.
[7]   Thompson J, Genders D (1992) US Pat 5,098,532.

Table 8.13 Impressed current anode materials

Sacrificial anode material	Composition	Typical anodic current density (A.m^{-2})	Consumption rate (g.A^{-1}.yr^{-1})	Notes
Dimensionally Stable Anodes (DSA®)	Ti/IrO$_2$, Nb/IrO$_2$	700 to 2000	less than 1	Cathodic protection of water tanks and buried steel structures
Duriron®	Fe–14.5Si–0.5C–0.75Mn	10 to 40	200 to 500	Both good corrosion and abrasion/wear resistance. Extensively used in offshore, oil rigs and other marine technology applications
Ebonex®	Ti$_5$O$_9$, Ti$_4$O$_7$	50 (uncoated) 2000 (IrO$_2$ coated)	n.a.	Corrosion resistant to both alkaline and acid media
Graphite and carbon	Carbon	10 to 40	225 to 450	Brittle and shock-sensitive materials. Extensively used buried for cathodic protection of ground pipelines
Lead alloy anodes	Pb-6Sb-1Ag/PbO$_2$	160 to 220	45 to 90	Cathodic protection for equipment immersed in seawater
Platinized titanium or niobium anodes	Ti/Pt–Ir, Nb/Pt–Ir, Ta/Pt–Ir	500 to 1000	1 to 6	Low consumption rate, high anodic current but expensive
Polymeric	Conductive polymer	n.a.	n.a.	Cathodic protection of reinforced steel bars in salt contaminated concrete

8

Misc. Electrical Materials

[8]   Pletcher D, Genders D, Weinberg N, Spiege E (1993) US Pat 5,246,551.
[9]   Schneider L (1995) US Pat 5,478,448.
[10]   Genders D, Weinberg N (eds) (1992) Electrochemistry for a cleaner environment. The Electrosynthesis Co., Lancaster, NY.
[11]   Wendt S (1998) Electrochemical engineering. Springer-Verlag, Heidelberg.
[12]   Pickett DJ (1979) Electrochemical reactor design. Elsevier, Amsterdam.
[13]   Rousar I, Micka K, Kimla A (1985) Electrochemical engineering, vols 1 and 2. Elsevier, Amsterdam.
[14]   Hine F (1985) Electrode processes and electrochemical engineering. Plenum Press, New York.
[15]   Couper AM, Pletcher D, Walsh FC (1990) Chem Rev 90: 937.
[16]   Trasatti S (1994) In Lipkowski J, Ross PN (eds) The electrochemistry of novel materials. VCH, New York, Chap. 5, pp 207–295.
[17]   La Conti AB, Fragala AR, Boyack JR (1977) ECS meeting, Philadelphia, May.
[18]   Bishop CR, Stern M (1961) Hydrogen embrittlement of tantalum in aqueous media. Corrosion 17: 379t–385t.
[19]   Howe (ed.) (1949) Bibliography of the platinum metals. Baker & Co., Newark.
[20]   Dreyman EW (1972) Mater Prot Perform 11: 17.
[21]   Cailleret L, Collardeau E (1894) CR Acad Sci 830.
[22]   Stevens RH (1913) US Pat 1,077,827, 1,077,894 and 1,077,920.
[23]   Bode H (1977) Lead-acid batteries. John Wiley, New York.
[24]   Greenwood NN, Earnshaw N (1984) Chemistry of the elements. Pergamon Press, Oxford, Chap. 10, p 435.
[25]   De Nora O (1962) Brit Pat 902,023 (1962).
[26]   Hoffmann W (1960) Lead and lead alloys. Springer-Verlag, Berlin.
[27]   Mao GW, Larson J, Rao GP (1969) J Inst Metal 97: 343.
[28]   Beck F (1972) Electrochim Acta 17: 2317.
[29]   Isfort H (1985) DECHEMA 98 144.
[30]   Gardiner WC (1947) Chem Eng 28 January: 100.
[31]   Holemann H (1962) Chem Ing Techn 34: 371.
[32]   De Nora O (1960) Chem Eng 25 July.
[33]   Messner G (1966) US Pat 3,236,760.
[34]   Grosselfinger FB (1964) Chemical Eng 71: 172.
[35]   Donges E, Janson HG (1966) Chem Ing Techn 38: 443.

[36] Kroll WJ (1940) Trans Electrochem Soc 78: 35.
[37] Rhoda RN (1952) Trans Inst Met Finish 36: 82.
[38] Rosenblatt EF, Cohn JF (1955) US Patent 2,719,797.
[39] Cotton JB (1958) Chem Ind 3: 492.
[40] Cotton JB (1958) Chem Ind 3: 640.
[41] Beer HB (1958) Brit Pat 855,107.
[42] Cotton JB (1958) Brit Pat 877,901.
[43] Cotton JB (1958) Platinum Met Rev 2: 45.
[44] Balko EN (1991) Electrochemical applications of the platinum group: metal coated anodes. In Hartley FR (ed.) Chemistry of the platinum group metals: recent developments, chap. 10. Elsevier, New York.
[45] Muller P, Spiedel H (1960) Metall 14: 695.
[46] Schleicher HW (1963) Brit Pat 941,177.
[47] Whiting KA (1964) US Pat 3,156,976.
[48] Haley AJ, Keith CD, May JE (1969) US Pat 3,461,058.
[49] May JE, Haley AJ (1970) US Pat 3,505,178.
[50] Cotton JB, Hayfield PCS (1965) Brit Pat 1,113,421.
[51] Lowenheim FA (1965) Modern Electroplating, 3rd edn. Wiley, New York.
[52] Hayfield PCS, Jacob WR (1980) In: Coulter MO (ed.) Modern chlor-alkali technology. Ellis Horwood, London, chap. 9, pp 103–120.
[53] Cotton JB, Williams EC, Barber, AH (1958) Brit Pat 877,901.
[54] Anderson EP (1961) US Pat 2,998,359.
[55] Adamson AF, Lever BG, Stones WF (1963) J Appl Chem 13, 483.
[56] Ibl N, Kramer R, Ponto L, Robertson PM (1979) AIChE Symp Ser 75: 45.
[57] Rakov AA, Veselovskii VI, Kasatkin EV, Potapova GF, Sviridon VV (1977) Zh Prikl Khim 50: 334.
[58] Angell CH, Deriaz MG (1960) Brit Pat 885,819.
[59] Angell CH, Deriaz MG (1963) Brit Pat 984,973.
[60] Taylor JF (1929) J Opt Soc Amer 18: 138.
[61] Hopper RT (1923) Ceram Indust June.
[62] Kuo CY (1974) Solid State Technol 17: 49.
[63] Anderson EP (1961) US Pat 2,998,359.
[64] Tirrel CE (1964) US Pat 3,117,023.
[65] Bianchi G, Gallone P, Nidola AE (1970) US Pat 3,491,014.
[66] Millington JP (1974) Brit Pat 1,373,611.
[67] Haley AJ (1967) Engelhardt Indust Techn Bull 7: 157.
[68] Beer HB (1963) US Pat 3,096,272.
[69] Bianchi G, De Nora V, Gallone P, Nidola A (1971) US Pat 3,616,445.
[70] Bianchi G, De Nora V, Gallone P, Nidola A (1976) US Pat 3,948,751.
[71] Trasatti S, O'Grady WE (1981). In: Gerisher H, Tobias CW (eds) Advances in electrochemistry and electrochemical engineering, vol. 12. Wiley Interscience, New York, pp 177–261.
[72] Beer HB (1966) Z Afrik Pat ZA 662,667.
[73] Beer HB (1968) Z Afrik Pat ZA 680,034.
[74] Beer HB (1966) US Pat 3,214,110.
[75] Beer HB (1972) US Pat 3,632,498.
[76] Beer HB (1973) US Pat 3,711,385.
[77] Beer HB (1973) US Pat 3,751,291.
[78] Nidola A (1981) In Trasatti S (ed.) Electrodes of conductive metallic oxides, Part B. Elsevier, Amsterdam, chap. 11, pp 627–659.
[79] Comninellis Ch, Vercesi GP (1991) J Appl Electrochem 21: 335.
[80] Gorodtskii VV, Tomashpol'skii Yu Ya, Gorbacheva LB et al. (1984) Elektrokhimiya 20: 1045.
[81] Beer HB (1980) J Electrochem Soc 127: 303C.
[82] Miles MH, Thomason J (1976) J Electrochem Soc 123: 1459.
[83] Jasinski R, Brilmyer, G, Helland L (1983) J Electrochem Soc 130: 1634.
[84] Pohl JP, Richert H (1980) In: Trasatti S (ed.) Electrodes of conductive metallic oxides, part A. Elsevier, Amsterdam, chap. 4, pp 183–220.
[85] Thangappan R, Nachippan S, Sampathi S (1970) Ind Eng Chem Prod Res Dev 9: 563.
[86] De Nora O (1962) Brit Pat 902,023.
[87] Clarke JS, Ehigamusoe RE, Kuhn AT (1976). J Electroanal Chem 70: 333.
[88] Grigger JC, Miller HC, Loomis FD (1958) J Electrochem Soc 105: 100.
[89] Engelhardt V, Huth M (1909) US Pat 935,250.
[90] Gaunce FS (1964) Fr Pat 1,419,356.
[91] Higley LW, Dressel WM, Cole ER (1976) US Bureau of Mines, Report No. 8111.
[92] Goodridge F, Lister K, Plimley R et al. (1980) J Appl Electrochem 10: 55.
[93] Bennett JE, O'Leary KJ (1973) US Pat 3,775,284.
[94] Huth M (1919) US Pat 3,616,302.
[95] Feige NG (1974) US Pat 3,855,084.
[96] De Nora O, Nidola O, Spaziante PM (1978) US Pat 4,072,586.

[97]   Hayes M, Kuhn AT (1978) J Appl Electrochem 8: 327.
[98]   Kuhn AT, Wright PM (1971) In: Kuhn AT (ed.) Industrial electrochemical processes, chap. 14. Elsevier, New York.
[99]   Itai R, Shibuya M, Matsumura T, Ishi G (1968) Electrochem Technol 6: 402.
[100] Itai R, Shibuya M, Matsumura T, Ishi G (1971) J Electrochem Soc 118: 1709.
[101] Clarke R, Pardoe R (1992) Applications of Ebonex conductive ceramics in effluent treatment. In: Genders D, Weinberg N (eds) Electrochemistry for a cleaner environment. Electrosynthesis, New York, pp 349–363.
[102] Manoharan R, Goodenough JB (1991) Electrochim Acta 36: 19.
[103] Yeo RS, Orehotsky J, Visscher W, Srinivasan S (1981) J Electrochem Soc 128: 1900.
[104] De Nora O, Bianchi G, Nidola A, Trisoglio G (1975) US Pat 3,878,083.
[105] Comninellis Ch, Vercesi GP (1991). J Appl Electrochem 21: 335.
[106] Kuznetzova EG, Borisova TI, Veselovskii VI (1968). Elektrokhimiya 10: 167.
[107] Warren HI, Wemsley D, Seto K (1975) Inst mining met branch meeting, 11 February, 53.
[108] Seko K (1976) Am Chem Soc Centennial Meeting, New York.
[109] Antler M, Butler CA (1967) J Electrochem Technol 5: 126.
[110] Hine F, Yasuda M, Yoshida T, Okuda J (1978) ECS Meeting, Seattle, 15 May, Abstract 447.
[111] Colo ZJ, Hardee KL, Carlson RC (1992) US Pat 5,141,563.
[112] Fukuda K, Iwakura C, Tamura H (1980) Electrochim Acta 25: 1523.
[113] Savall A (1992) In: Électrochimie 92, L'Actualité Chimique, special issue, Janvier.
[114] Potgieter JH, Heyns AM, Skinner W (1990) J Appl Electrochem 20: 711.
[115] Cardarelli F, Comninellis Ch, Savall A, Taxil P, Manoli G, Leclerc O (1998) J Appl Electrochem 28: 245.
[116] Vercesi GP, J, Rolewicz J, Comninellis Ch (1991) Thermochim Acta 176: 31.
[117] Farbenfabriken Bayer Aktiengesellschaft (1968) Fr Pat 1,516,524.
[118] Jeffes JHE (1974) Brit Pat 1,355,797.
[119] Denton DA, Hayfield PCS (1990) Eur Pat A0 383,412.
[120] Kumagai N, Jikihara S, Samata Y, Asami K, Hashimoto AM (1993) Proc Hawaii meet ISE, Honolulu, Hawaii, pp 16–21.
[121] Cardarelli F, Taxil P, Savall A (1996) Int J Refract Metals Hard Mater 14: 365.
[122] Cardarelli F, Comninellis Ch, Leclerc O, Saval A, Taxil P, Manoli G (1996) WO 9743465 Al, FR 96-5916, 28 May.
[123] Riggs OL Jr, Locke CE (1981) Anodic protection: theory and practice in the prevention of corrosion. Plenum Press, New York.

8
Misc.
Electrical
Materials

9 Ceramics and Glasses

9.1 Ceramics and Refractories

9.1.1 Definitions

Ceramics were initially named from the Greek, keramos, meaning solids obtained from firing of clays. But according to a wider modern definition, ceramics are either crystalline or amorphous solid materials involving only ionic, covalent, or ionocovalent chemical bonds between metallic and nonmetallic elements. Basic examples are silica, alumina, magnesia, or zirconia. Although oxides, and silicates, are the common well-known compounds for ceramic materials, ceramics also include for instance borides, carbides, silicides, nitrides, phosphides, and sulfides.

9.1.2 Basic Classification

As a general rule ceramic materials can be grouped into two main categories: (i) traditional ceramics, and (ii) advanced or engineered ceramics.

9.1.2.1 Traditional Ceramics

Traditional ceramics are those obtained only from the firing of clay-based materials. The common initial composition before firing consists usually of: (i) a clay mineral (i.e., phyllosilicate minerals such as **kaolinite**, **montmorillonite**, and **illite**, (ii) fluxing agents or fluxes (e.g., feldspars: orthoclases and plagioclases), and (iii) filler materials (e.g., silica, alumina, magnesia). The traditional ceramics can be prepared using two main groups of clays: (i) kaolin or china clays made from the phyllosilicate kaolinite, and to a lesser extent micas, but free of quartz, and (ii) ball clays containing a mixture of kaolinite, montmorillonite, illite, and micas. The classical procedure for preparing traditional ceramics consists of the following operations sequence: raw material selection and preparation (i.e., grinding, mixing), forming (e.g., molding, extrusion, slip casting, and die pressing), drying, prefiring operations (i.e., glazing), firing, and postfiring operations (e.g., enameling, cleaning, and machining). The common classes of traditional ceramics are: whiteware (e.g., stoneware, china, porcelain), glaze, porcelain enamels, high-temperature refractories, mortars, cements, and concretes (see Chapter 14).

Silica brick – the earliest silica bricks were composed of crushed minerals of 90wt% silica, with as much as 3.5wt% flux materials (usually CaO) and fired at about 1010°C. These bricks found use primarily in steel mills and coke by-products operations, and from the second quarter of this century in chemical service, primarily in exposure to strong phosphoric acid where shale and fireclay brick do not survive long. They can serve at higher temperatures to about 1093°C and are more resistant to thermal shock due to their greater porosity, as high as 16%. For chemical service, those of the highest silica content (not below 98wt% SiO_2) should be used. The purity of the silica and its percentage of alkali, along with the manufacturing techniques, determine the uniformity or the wideness of ranges of the physical properties. Ranges of chemical composition of silica brick are: 98.6–99.6wt% SiO_2, 0.2–0.5wt% Al_2O_3, 0.02–0.3wt% Fe_2O_3, 0.02–0.1wt% MgO, 0.02–0.03wt% CaO, and 0.01–0.2wt% (Na_2O, K_2O, Li_2O). Silica bricks serve well and for long periods in acid environments (except hydrofluoric), without noticeable damage, showing greater resistance, especially to strong hot mineral acids, and particularly phosphoric, than acid brick, and in exposure to halogens (except fluorine), solvents and organic chemicals. Silica bricks are not recommended for service in strong alkali environments. They also exhibit better shock resistance than shale or fireclay acid bricks, but they have lower strength and abrasion resistance.

Porcelain brick – porcelain bricks are made from high-fired clays, the temperature of firing depending on the amount of alumina in the clay, 15% to 38% usually at approximately 1200–1300°C, 85% at 1500–1550°C, and 95–98% at 1600–1700°C. The bodies of these bricks are extremely dense and nonporous, with zero absorption, and Mohs hardness ranging from 6 to 9 for 99wt% alumina. As alumina content increases Mohs hardness, maximum service temperature and chemical resistance increases. Major uses of porcelain include particularly the lining of ball mills, where they will outlast almost all other abrasion-resistant linings, and (glazed) as pole line hardware for the power industry where, exposed to abrasion, weathering and cycling temperature changes, they outlast all other materials in similar service. All chemists and laboratory personnel are familiar with glass and porcelain equipment, and so are aware of the fact that they give excellent service in hot chemicals except hydrofluoric acid and acid fluorides and strong sodium or potassium hydroxides, especially in the molten state. Due to the high cost of porcelain bricks, they are used sparingly in the process industries, chiefly in dye manufacture, due to their density preventing inter-batch contamination, and ease of cleaning. The use of porcelain brick is primarily limited by cost.

Table 9.1 Examples of traditional ceramics

Type	Properties	Applications
Fired bricks	Porosity: 15–30% Firing temperature: 950–1050°C Enameled or not	Bricks, pipes, ducts, walls, ground floors
China	Porosity: 10–15% Firing temperature: 950–1200°C Enamel, opacity	Sanitary, tiles
Stoneware	Porosity: 0.5–3% Firing temperature: 1100–1300°C Glassy surface	Crucibles, labware, pipes
Porcelain	Porosity: 0–2% Firing temperature: 1100–1400°C Glassy, translucent	Insulators, labware, cook-ware

Table 9.2 Selected common advanced ceramics

Ceramics	Properties and characteristics	Typical uses
Alumina (Al_2O_3)	Polymorphic varieties of alumina are generally more corrosion resistant, harder, stronger, and more refractory than mullite. Alumina offers good electrical insulating characteristics and high-temperature capabilities. High-purity alumina (99.8wt%) can be used at operating temperatures to 1950°C in both oxidizing and reducing atmospheres. Alumina is chemically inert to hydrogen, carbon, and refractory metal in many severe situations. Moreover, alumina is exceptionally tough, designed especially for wear-resistant applications.	Noble metal thermocouple protection, furnace tubes, high-temperature vacuum furnaces, heat-treating rollers, radiant tubes, laser tubes, corona treater tubes and ozone generators. Abrasive in sand-blasting and polishing operations, and as a container materials in corrosive environments.
Graphite (impervious)	Impervious graphite is manufactured by processing graphite at temperatures above 2000°C, evacuating the pores, and impregnating with a phenolic resin. The impregnation seals the porosity. Thermal conductivity in impervious graphite is close to that of copper alloys. An important limitation of this material is its low tensile strength, and all components manufactured from carbon or graphite are highly susceptible to brittle fracture by mechanical shock or vibration. Impervious graphite is almost completely inert to all but the most severe oxidizing conditions. Actually, impervious graphite is recommended for use in 60% HF 20% HNO_3, 96%H_2SO_4, bromine, fluorine, or iodine.	The excellent heat transfer property of impervious graphite has made it very popular in heat exchangers handling corrosive media, but also for a number of other devices used in the chemical process industries such as piping, pumps, valves, brick lining for process or storage vessels, anode material in electrochemical processes, ring packing for columns. Impervious graphite is also used for liquid metals handling devices.
Mullite ($3Al_2O_3$ $\cdot 2SiO_2$)	Mullite is a refractory silicate material combining a low thermal expansion coefficient, a good mechanical strength, and resilience at elevated temperatures. Mullites are formulated to produce dense shapes, some in a glass matrix in order to yield maximum thermal shock resistance and good strength. Dense mullite in a glassy matrix formulated to offer a high-quality economical insulating tubing for thermocouple applications is an extremely versatile and economical material. Its workability allows an extensive range and flexibility in fabrication. It is well suited to the casting of special shapes.	Its typical applications are: insulators in oxidizing conditions for base and noble metal thermocouples used in conditions up to 1450°C, protection tubes, target and sight tubes, furnace muffles, diffusion liners, combustion tubes, radiant furnace tubes and kiln rollers.
Sialon (SiAlON)	Sialon is an alloy of silicon nitride and aluminum oxide. This superior refractory material has the combined properties of silicon nitride, i.e., high strength, hardness, fracture toughness, and low thermal expansion, and aluminum oxide, i.e., corrosion resistance, chemical inertness, high-temperature capabilities, and oxidation resistance. Most refractory products are capable of surviving one or two specific environments that typically involve high temperature, mechanical abuse, corrosion, wear, or electrical resistance.	Good for molten metal applications. high-wear or high-impact environments up to 1250°C.
Silica (SiO_2)	High-purity amorphous or fused silica is a high-performance ceramic. It has a very low thermal expansion coefficient, a remarkable thermal shock resistance, a low thermal conductivity, excellent electrical insulation up to 1000°C, and excellent resistance to corrosion from molten metal and glass.	Common industrial uses for fused silica are: steel making, coke making, metallurgy, glass production, nonferrous foundries, precision foundries, ceramics, the chemical industry, the nuclear Industry, and aeronautics.
Silicon carbide (SiC)	Silicon carbide (carborundum ®) is nonwetting for most nonferrous metals, gas tight, chemically inert, resistant to corrosion from metals and features high strength, hardness, and thermal shock resistance. It is resistant to most organics, inorganic acids, alkalis and salts in a variety of concentrations except to HF and fluorides.	Silicon carbide is a premium-priced material which is employed in lining work for its uniformity, abrasion resistance, and dimensional stability.

9
Ceramics
and
Glasses

Table 9.2 (continued)

Ceramics	Properties and characteristics	Typical uses
Zircon ($ZrSiO_4$)	Because of its chemical inertness and high melting point, zircon is wetted less easily by molten metal, producing smoother surfaces on iron, high-alloy steel, aluminum, and bronze casting.	The largest use of zircon sand is foundry sand, where zircon is used as the basic mold material, as facing material on mold cores, and in ram mixes. Zircon-sand molds have greater thermal shock resistance and better dimensional stability than quartz-sand molds. The zircon grains are usually bonded with sodium silicate.
Zirconia (ZrO_2)	Polymorphic varieties of zirconias (ZrO_2) are extremely refractory materials. They possess excellent chemical inertness and corrosion resistance at temperatures well above the melting point of alumina. Zirconia products should not be used in contact with alumina above 1600°C. Zirconias are stabilized with yttria (10.5%) in the fine-grained, high-density, cubic crystal structure to avoid cracking and mechanical weakening during heating and cooling. In addition to its high melting point, it exhibits a low thermal conductivity, and is an electrical conductor above 800°C. It also possesses the unique ability to allow oxygen anions to move freely through the crystal structure above 600°C.	Zirconia has a high index of refraction which allows it to be used to increase the refractive index of some optical glasses. Highly pure and transparency zirconia crystals are good synthetic gem materials. Industrial zirconia is used for laboratory crucibles that will withstand heat shock, for linings of metallurgical furnaces, and by the glass and ceramic industries as a refractory material deodorants. Zirconium oxide is fused with alumina in electric-arc furnace to make alumina–zirconia abrasive grains for use in grinding wheels, coated abrasive disks, and belts. Although fine zirconia powder has been used for polishing glass, ceria seems to be preferred.

9.1.2.2 Advanced Ceramics

Advanced and engineered ceramics are various chemical compounds, not only oxides and silicates, and exhibit excellent physical properties. They can be grouped according to their application fields: electrical (e.g., semiconductors, insulators and dielectrics, piezo- and pyro-electrics, and superconductors), optical (e.g., phosphors, lasing crystals, mirrors, and reflectors), magnetic (e.g., permanent magnets), and structural ceramics. The properties of these ceramic materials are extensively described in the chapters in the book relating to their properties (e.g., insulators in Chapter 7, and superconducting ceramics in Chapter 5).

9.1.3 Properties of Pure Ceramics and Refractories

Physical properties of advanced and high-temperature refractories, ordered by groups (i.e., borides, carbides, etc.), are reported in Tables 9.3–9.5. (Table 9.3 on pages 342–365.)

9.2 Glasses

9.2.1 Definitions

Glass is from a thermodynamic point of view, a **supercooled liquid**, i.e., a molten liquid cooled at a rate sufficiently rapid to fix the random microscopic organization of a liquid and avoid the crystallization process. Therefore, by contrast with crystallized solids, glasses do not exhibit a

Table 9.4 Carbon product properties[a]

Carbon derivate	Density (ρ, kg.m^{-3})	Young's modulus (E, GPa)	Compressive strength (MPa)	Flexural strength (MPa)	Thermal conductivity (k, W.m^{-1}.K^{-1})	Specific heat capacity (c_p, J.kg^{-1}.K^{-1})	Coeff. linear thermal expansion (α, 10^{-6} K^{-1})	Electrical resistivity (ρ, $\mu\Omega$.cm)	Vickers hardness (HV)
Graphite (industrial)	1400–2266	3–12	14–42	5–21	85–350	709	1.3–3.8	1385	HM 1
Pyrolitic carbon (impervious graphite)	1400–2210	16–30	72	32	480–1950	707	4.5	1500	145
Diamond	3514	900	7000	n.a.	900	506.3	n.a.	10^{16}	8000
Vitreous carbon (treated at 1000°C)	1500–1550	28	300	100	4	710	3.2	5500	225
Vitreous carbon (treated at 2500°C)	1500–1550	22	150–200	60–80	8	710	3.2	4500	150–175

[a]Sources: Technical data from Le Carbone Lorraine, Sigradur, and Tokkai.

Table 9.5 Fired refractory bricks

Brick (major chemical components)	Density (ρ, kg.m^{-3})	Melting temperature (°C)	Thermal conductivity (k, W.m^{-1}.K^{-1})
Alumina brick (6–65wt% Al$_2$O$_3$)	1842	1650–2030	4.67
Building brick	1842	1600	0.72
Carbon brick (99wt% graphite)	1682	3500	3.6
Chrome brick (100wt% Cr$_2$O$_3$)	2900–3100	1900	2.3
Chroma–magnesite brick (52wt% MgO, 23wt% Cr$_2$O$_3$)	3100	3045	3.5
Fire clay brick (54wt% SiO$_2$, 40wt% Al$_2$O$_3$)	2146–2243	1740	0.3–1.0
Fired dolomite (55wt% CaO, 37wt% MgO)	2700	2000	n.a.
High-alumina brick (90–99wt% Al$_2$O$_3$)	2810–2970	1760–2030	3.12
Magnesite brick (95.5wt% MgO)	2531–2900	2150	3.7–4.4
Mullite brick (71wt% Al$_2$O$_3$)	2450	1810	7.1
Silica brick (95–99wt% SiO$_2$)	1842	1765	1.5
Silicon carbide brick (80–90wt% SiC)	2595	2305	20.5
Zircon (99wt% ZrSiO$_4$)	3204	1700	2.6
Zirconia (stabilized) brick	3925	2650	2.0

9
Ceramics
and
Glasses

Table 9.3 Physical properties of pure ceramics and high-temperature refractories

IUPAC name (synonyms, common trade names)	Theoretical chemical formula, [CASRN], relative molecular mass ($^{12}C = 12.000$)	Crystal system, lattice parameters, structure type, Strukturbericht, Pearson, space group, structure type (Z)	Density (ρ, kg.m^{-3})	Electrical resistivity (ρ, $\mu\Omega$cm)	Dielectric permittivity at 1 MHz	Dielectric field strength (E, MV.m^{-1})	Dissipation or loss factor (tan ε)	Melting point (mp, °C)	Thermal conductivity (k, W.m^{-1}.K^{-1})	Specific heat capacity (cp, J.kg^{-1}.K^{-1})	Coeff. linear thermal expansion (α, 10^{-6}K^{-1})	Young's modulus (E, GPa)	Shear modulus (G, GPa)
Borides													
Aluminum diboride	AlB$_2$ [12041-50-8] 48.604	Hexagonal a = 300.50 pm c = 325.30 pm C32, $hP3$, P6/mmm, AlB$_2$ type (Z = 1)	3190	n.a.	n.a.	n.a.	n.a.	1654	n.a.	897.87	n.a.	n.a.	n.
Aluminum dodecaboride	AlB$_{12}$ [12041-54-2] 156.714	Tetragonal a = 1016 pm c = 1428 pm	2580	n.a.	n.a.	n.a.	n.a.	2421	n.a.	954.48	n.a.	n.a.	n.
Beryllium boride	Be$_4$B [12536-52-6] 46.589	n.a.	n.a.	n.a.	n.a.	n.a.	n.a.	1160	n.a.	n.a.	n.a	n.a.	n.
Beryllium diboride	BeB$_2$ [12228-40-9] 30.634	Hexagonal a = 979 pm c = 955 pm	2420	10,000	n.a.	n.a.	n.a.	1970	n.a.	n.a.	n.a.	n.a.	n.
Beryllium hemiboride	Be$_2$B [12536-51-5]	Cubic a = 467.00 pm Cl, $cF12$, Fm3m, CaF$_2$ type (Z = 4)	1890	1000	n.a.	n.a.	n.a.	1520	n.a.	n.a.	n.a	n.a.	n.
Beryllium hexaboride	BeB$_6$ [12429-94-6]	Tetragonal a = 1016 pm c = 1428 pm	2330	10^{13}	n.a.	n.a.	n.a.	2070	n.a.	n.a.	n.a	n.a.	n.
Beryllium monoboride	BeB [12228-40-9]	n.a.	n.a.	n.a.	n.a.	n.a.	n.a.	1970	n.a.	n.a.	n.a.	n.a.	n.
Boron	β-B [7440-42-8] 10.811	Trigonal (Rhombohedral) a = 1017 pm α = 65° 12' $hR105$, R3m, β-B type	2460	18,000	n.a.	n.a.	n.a.	2190	n.a.	n.a.	n.a.	320	n.
Chromium boride	Cr$_5$B$_3$ [12007-38-4] 292.414	Orthorhombic a = 302.6 pm b = 1811.5 pm c = 295.4 pm D8$_1$, $tI32$, I4/mcm, Cr$_5$B$_3$ type (Z = 4)	6100	n.a.	n.a.	n.a.	n.a.	1900	15.8	n.a	13.7	n.a.	n.
Chromium diboride	CrB$_2$ [12007-16-8] 73.618	Hexagonal a = 292.9 pm c = 306.6 pm C32, $hP3$, P6/mmm, AlB$_2$ type (Z = 1)	5160–5200	21	n.a.	n.a.	n.a.	1850–2100	20–32	712	6.2–7.5	211	n.
Chromium monoboride	CrB [12006-79-0] 62.807	Tetragonal a = 294.00 pm c = 1572.00 pm B$_f$, $oC8$, Cmcm, CrB type (Z = 4)	6200	64.0	n.a.	n.a.	n.a.	2000	20.1	n.a.	12.3	n.a.	n.
Hafnium diboride	HfB$_2$ [12007-23-7] 200.112	Hexagonal a = 314.20 pm c = 347.60 pm C32, $hP3$, P6/mmm, AlB$_2$ type (Z = 1)	11,190	8.8–11	n.a.	n.a.	n.a	3250–3380	51.6	247.11	6.3–7.6	500	n.
Lanthanum hexaboride	LaB$_6$ [12008-21-8] 203.772	Cubic a = 415.7 pm D2$_1$, $cP7$, Pm3m, CaB$_6$ type (Z = 1)	4760	17.4	n.a.	n.a.	n.a.	2715	47.7	n.a.	6.4	479	n.

Poisson's ratio (ν)	Ultimate tensile strength (σ_{UTS}, MPa)	Flexural strength (τ, MPa)	Compressive strength (α, MPa)	Fracture toughness (K_{IC}, MPa.m$^{1/2}$)	Vickers hardness (HV) (HM)	Other physico-chemical properties, corrosion resistance^a, and uses	IUPAC name (synonyms, common trade names)
							Borides
n.a.	n.a.	n.a.	n.a	n.a.	n.a.	Temperature transition to A1B$_{12}$ at 920°C. Soluble in dilute HCl. Nuclear shielding material.	Aluminum diboride
n.a.	n.a.	n.a.	n.a	n.a.	n.a.	Soluble in hot HNO$_3$, insoluble in other acids and alkalis. Neutron shielding material.	Aluminum dodecaboride
n.a.	n.a.	n.a.	n.a	n.a.	n.a.		Beryllium boride
n.a.	n.a.	n.a.	n.a	n.a.	n.a.		Beryllium diboride
n.a.	n.a.	n.a.	n.a	n.a.	870		Beryllium hemiboride
n.a.	n.a.	n.a.	n.a	n.a.	n.a.		Beryllium hexaboride
n.a.	n.a.	n.a.	n.a	n.a.	n.a.		Beryllium monoboride
n.a.	n.a.	n.a.	n.a	n.a.	2055 (HM) (11)	Brown or dark powder, unreactive to oxygen, water, acids and alkalis. DH$_{vap.}$ = 480 kJ mol^{-1}.	Boron
n.a.	n.a.	n.a.	n.a	n.a.	n.a.		Chromium boride
n.a.	n.a.	607	1300	n.a.	1800	Strongly corroded by molten metals such as Mg, Al, Na, Cu, Si, V, Cr, Mn, Fe, Ni. It is corrosion resistant to the following liquid metals: Cu, Zn, Sn, Rb, and Bi.	Chromium diboride
n.a.	n.a.	n.a.	n.a	n.a.	n.a.		Chromium monoboride
0.12	n.a.	350	n.a.	n.a.	2900	Gray crystals, attacked by HF else highly resistant.	Hafnium diboride
n.a.	n.a.	126	n.a	n.a.	n.a.	Wear-resistant, semiconducting, thermoionic conductor film.	Lanthanum hexaboride

Table 9.3 *(continued)*

IUPAC name (synonyms, common trade names)	Theoretical chemical formula, [CASRN], relative molecular mass ($^{12}C = 12.000$)	Crystal system, lattice parameters, structure type, Strukturbericht, Pearson, space group, structure type (Z)	Density (ρ, kg.m^{-3})	Electrical resistivity (ρ, $\mu\Omega$cm)	Dielectric permittivity at 1 MHz	Dielectric field strength (E, MV.m^{-1})	Dissipation or loss factor (tan δ)	Melting point (mp, °C)	Thermal conductivity (k, W.m^{-1}.K^{-1})	Specific heat capacity (cp, J.kg^{-1}.K^{-1})	Coeff. linear thermal expansion (α, 10^{-6}K^{-1})	Young's modulus (E, GPa)	Shear modulus (G, GPa)
Molybdenum boride	Mo_2B_5 [12007-97-5] 245.935	Trigonal $a = 301.2$ pm $c = 2093.7$ pm D8$_1$, $hR7$, R$\bar{3}$m, Mo$_2$B$_5$ type (Z = 1)	7480	22–55	n.a.	n.a.	n.a.	1600	50	n.a.	8.6	672	n.a
Molybdenum diboride	MoB_2 117.59	Hexagonal $a = 305.00$ pm $c = 311.30$ pm C32, $hP3$, P6/mmm, A1B$_2$ type (Z = 1)	7780	45	n.a.	n.a.	n.a.	2100	n.a.	527	7.7	n.a.	n.a
Molybdenum hemiboride	Mo_2B [12006-99-4] 202.691	Tetragonal $a = 554.3$ pm $c = 473.5$ pm C16, $tI2$, I4/mcm, CuAl$_2$ type (Z = 4)	9260	40	n.a.	n.a.	n.a.	2280	n.a.	377	5.0	n.a.	n.a
Molybdenum monoboride	MoB 106.77	Tetragonal $a = 311.0$ $c = 169.5$ B$_g$, $tI4$, I4$_1$/amd, MoB type (Z = 2)	8770 n.a. n.a.	α–MoB 45, β–MoB 25	n.a.	n.a.	n.a.	2180	n.a.	368	n.a	n.a.	n.a
Niobium diboride	ε-NbB$_2$ [12007-29-3] 114.528	Hexagonal $a = 308.90$ pm $c = 330.03$ pm C32, hP3, P6/mmm, A1B$_2$ type (Z = 1)	6970	26–65	n.a.	n.a.	n.a.	2900	17–23.5	418	8.0–8.6	637	n.a
Niobium monoboride	δ-NbB [12045-19-1] 103.717	Orthorhombic $a = 329.8$ pm $b = 316.6$ pm $c = 87.23$ pm B$_f$, $oC8$, Cmcm, CrB type (Z = 4)	7570	40–64.5	n.a.	n.a.	n.a.	2270–2917	15.6	n.a.	12.9	n.a.	n.a
Silicon hexaboride	SiB_6 [12008-29-6]	Trigonal (Rhombohedral)	2430	200,000	n.a.	n.a.	n.a.	1950	n.a.	n.a	n.a.	n.a.	n.a
Silicon tetraboride	SiB_4 [12007-81-7] 71.330	n.a.	2400	n.a.	n.a.	n.a.	1870	n.a. (dec.)	n.a	n.a.	n.a.	n.a.	n.a
Tantalum diboride	TaB_2 [12077-35-1] 202.570	Hexagonal $a = 309.80$ pm $c = 324.10$ pm C32, $hP3$, P6/mmm, AlB$_2$ type (Z = 1)	12,540	33–	n.a.	n.a.	n.a.	3037–3200	10.9–16.0	237.55	8.2–8.8	257	n.a
Tantalum monoboride	TaB [12007-07-7] 191.759	Orthorhombic $a = 327.6$ pm $b = 866.9$ pm $c = 315.7$ pm B$_f$, $oC8$, Cmcm, CrB type (Z = 4)	14,190	100	n.a.	n.a.	n.a.	2340–3090	n.a.	246.85	n.a.	n.a.	n.a
Thorium hexaboride	ThB_6 [12229-63-9] 296.904	Cubic $a = 411.2$ pm D2$_1$, $cP7$, Pm3m, CaB$_6$ type (Z = 1)	6800	n.a.	n.a	n.a.	n.a.	2149	44.8	n.a.	7.8	n.a.	n.a
Thorium tetraboride	ThB_4 [12007-83-9] 275.53	Tetragonal $a = 725.6$ pm $c = 411.3$ pm D1$_e$, $tP20$, P4/mbm, ThB$_4$type (Z = 4)	8450	n.a.	n.a.	n.a.	n.a.	2500	25	510	7.9	148	n.a

Poisson's ratio (ν)	Ultimate tensile strength (σ_{UTS}, MPa)	Flexural strength (τ, MPa)	Compressive strength (σ, MPa)	Fracture toughness (K_{IC}, MPa.m$^{1/2}$)	Vickers hardness (HV) (HM)	Other physico-chemical properties, corrosion resistance[a], and uses	IUPAC name (synonyms, common trade names)
n.a.	n.a.	345	n.a	n.a.	n.a.	Corroded by the following molten metals: Al, Mg, V, Cr, Mn, Fe, Ni, Cu, Nb, Mo, and Ta. It is corrosion resistant to molten Cd, Sn, Bi, and Rb.	Molybdenum boride
n.a.	n.a.	n.a.	n.a	n.a.	1280		Molybdenum diboride
n.a.	n.a.	n.a.	n.a	n.a.	n.a. (HM 8–9)	Corrosion resistant film.	Molybdenum hemiboride
n.a.	n.a.	n.a.	n.a	n.a.	1570		Molybdenum monoboride
n.a.	n.a.	n.a.	n.a	n.a.	3130 (HM > 8)	Corrosion resistant to molten Ta; corroded by molten rhenium.	Niobium diboride
n.a.	n.a.	n.a.	n.a.	n.a.	n.a.	Wear resistant and semiconductive films, neutron absorbing layer on nuclear fuel pellets.	Niobium monoboride
n.a.	n.a.	n.a.	n.a	n.a.	n.a.		Silicon hexaboride
n.a.	n.a.	n.a.	n.a	n.a.	n.a.		Silicon tetraboride
n.a.	n.a.	n.a.	n.a	n.a.	(HM > 8)	Gray metallic powder. Severe oxidation in air above 800°C. Corroded by the following molten metals: Nb, Mo, Ta, and Re.	Tantalum diboride
n.a.	n.a.	n.a.	n.a	n.a.	2200 (HM > 8)	Severe oxidation above 1100–1400°C in air.	Tantalum monobride
n.a.	n.a.	n.a.	n.a	n.a.	n.a.		Thorium hexaboride
n.a.	n.a.	137	n.a	n.a.	n.a.		Thorium tetraboride

9
Ceramics
and
Glasses

Table 9.3 (continued)

IUPAC name (synonyms, common trade names)	Theoretical chemical formula, [CASRN], relative molecular mass ($^{12}C = 12.000$)	Crystal system, lattice parameters, structure type, Strukturbericht, Pearson, space group, structure type (Z)	Density (ρ, kg.m⁻³)	Electrical resistivity (ρ, $\mu\Omega$cm)	Dielectric permittivity at 1 MHz	Dielectric field strength (E, MV.m⁻¹)	Dissipation or loss factor (tan δ)	Melting point (mp, °C)	Thermal conductivity (k, W.m⁻¹.K⁻¹)	Specific heat capacity (cp, J.kg⁻¹.K⁻¹)	Coeff. linear thermal expansion (α, 10⁻⁶K⁻¹)	Young's modulus (E, GPa)	Shear modulus (G, GPa)
Titanium diboride	TiB₂ [12045-63-5] 69.489	Hexagonal a = 302.8 pm c = 322.8 pm C32, hP3, P6/mmm, AlB₂ type (Z = 1)	4520	16–28.4	n.a.	n.a.	n.a.	2980–3225	64.4	637.22	7.6–8.6	372–551	n.a.
Tungsten hemiboride	W₂B [12007-10-2] 378.491	Tetragonal a = 556.4 pm c = 474.0 pm C16, tI12, I4/mcm, CuAl₂ type (Z = 4)	16,720	n.a.	n.a.	n.a.	n.a.	2670	n.a.	168	6.7	n.a.	n.a.
Tungsten monoboride	WB [12007-09-9] 194.651	Tetragonal a = 311.5 pm c = 1692 pm	15,200 16,000	4.1	n.a.	n.a.	n.a.	2660	n.a.	n.a.	6.9	n.a.	n.a.
Uranium diboride	UB₂ [12007-36-2] 259.651	Hexagonal a = 313.10 pm c = 398.70 pm C32, hP3, P6/mmm, AlB₂ type	12,710	n.a.	n.a	n.a.	n.a.	2385	51.9	n.a.	9	n.a.	n.a.
Uranium dodecaboride	UB₁₂ 367.91	Cubic a = 747.3 pm D2f, cF52, Fm3m, UB₁₂ type (Z = 4)	5820	n.a.	n.a	n.a.	n.a.	1500	n.a.	n.a	4.6	n.a.	n.
Uranium tetraboride	UB₄ [12007-84-0] 281.273	Tetragonal a = 707.5 pm c = 397.9 pm D1e, tP20, P4/mbm, ThB₄ type (Z = 4)	5350	n.a.	n.a	n.a.	n.a.	2495	4.0	n.a.	7.0	440	n.
Vanadium diboride	VB₂ [12007-37-3] 72.564	Hexagonal a = 299.8 pm c = 305.7 pm C32, hP3, P6/mmm, AlB₂ type (Z = 1)	5070	23	n.a.	n.a.	n.a.	2450–2747	42.3	647.43	7.6–8.3	268	n.
Zirconium diboride	ZrB₂ [12045-64-6] 112.846	Hexagonal a = 316.9 pm c = 353.0 pm C32, hP3, P6/mmm, A1B₂ type (Z = 1)	6085	9.2	n.a.	n.a.	n.a.	3060–3245	57.9	392.54	5.5–8.3	343–506	22
Zirconium dodecaboride	ZrB₁₂ 283.217	Cubic a = 740.8 pm D2f, cF52, Fm3m, UB₁₂ type (Z = 4)	3630	60–80	n.a.	n.a.	n.a.	2680	n.a.	523	n.a.	n.a.	n.
Carbides													
Aluminum carbide	Al₄C₃ [1299-86-1] 143.959	Trigonal (Rhombohedral) a = 333 pm c = 2494 pm D71	2360	n.a.	n.a.	n.a.	n.a.	2798	n.a.	n.a.	n.a.	n.a.	n.
Beryllium carbide	Be₂C [506-66-1] 30.035	Cubic a = 433 pm C1, cF12, Fm3m, CaF₂ type (Z = 4)	1900	n.a.	n.a.	n.a.	n.a.	2100	21.0	1397	10.5	314.4	n.

	Poisson's ratio (v)	Ultimate tensile strength (σ_{UTS}, MPa)	Flexural strength (τ, MPa)	Compressive strength (σ_c, MPa)	Fracture toughness (K_{IC}, MPa.m$^{1/2}$)	Vickers hardness (HV) (HM)	Other physico-chemical properties, corrosion resistance^a, and uses	IUPAC name (synonyms, common trade names)
a.	0.11	131	240	669	6.7	3370 (HM) > 9)	Gray crystals, superconducting at 1.26 K. High-temperature electrical conductor, cermet, used as crucible material for handling molten metals such as Al, Zn, Cd, Bi, Sn, and Rb. It is strongly corroded by liquid metals such as T1, Zr, V, Nb, Ta, Cr, Mn, Fe, Co, Ni, Cu. Begins to be oxidized in air above 1100–1400°C. Corrosion resistance in hot concentrated brines. Maximum operating temperature: 1000°C (reducing), and 800°C (oxidizing).	Titanium diboride
.	n.a.	n.a.	n.a.	n.a	n.a.	(HM 9) 2420	Black powder.	Tungsten hemiboride
.	n.a.	n.a.	n.a.	n.a	n.a.	(HM 9)	Black powder.	Tungsten monoboride
.	n.a.	n.a.	n.a.	n.a	n.a.	1390		Uranium diboride
.	n.a.	n.a.	n.a.	n.a	n.a.	n.a.		Uranium dodecaboride
a.	n.a.	n.a.	413	n.a	n.a.	2500		Uranium tetraboride
.	n.a.	n.a.	n.a.	n.a	n.a.	(HM 8–9)	Wear resistant and semiconductive films.	Vanadium diboride
.	0.15	n.a.	305	n.a.	n.a.	1900–3400 (HM 8)	Gray metallic crystals, excellent thermal shock resistance, greatest oxidation inertness of all refractory hard metals. Hot-pressed crucible for handling molten metals such as Zn, Mg, Fe, Cu, Zn, Cd, Sn, Pb, Rb, Bi, Cr, brass, carbon steel, cast irons, and molten cryolithe, yttria, zirconia, and alumina. It is readily corroded by liquid metals such as Si, Cr, Mn, Co, Ni, Nb, Mo, Ta, and attacked by molten salts such as Na$_2$O, alkali carbonates, and NaOH. Severe oxidation in air occurs above 1100–1400°C. Stable above 2000°C under inert or reducing atmosphere.	Zirconium diboride
.	n.a.	n.a.	n.a.	n.a.	n.a.	n.a.		Zirconium dodecaboride
								Carbides
	n.a.	n.a.	n.a.	n.a.	n.a.	n.a.	Decomposed in water with evolution of CH$_4$.	Aluminum carbide
	0.100	155	n.a.	723	n.a.	n.a.	Brick red or yellowish red octahedra. Nuclear reactor cores.	Beryllium carbide

9

Ceramics and Glasses

Table 9.3 *(continued)*

IUPAC name (synonyms, common trade names)	Theoretical chemical formula, [CASRN], relative molecular mass ($^{12}C = 12.000$)	Crystal system, lattice parameters, structure type, Strukturbericht, Pearson, space group, structure type (Z)	Density (ρ, kg·m^{-3})	Electrical resistivity (ρ, $\mu\Omega$cm)	Dielectric permittivity at 1 MHz	Dielectric field strength (E, MV·m^{-1})	Dissipation or loss factor (tan δ)	Melting point (mp, °C)	Thermal conductivity (k, W·m^{-1}·K^{-1})	Specific heat capacity (cp, J·kg^{-1}·K^{-1})	Coeff. linear thermal expansion (α, 10^{-6} K^{-1})	Young's modulus (E, GPa)	Shear modulus (G, GPa)
Boron carbide (Norbide®)	B$_4$C [12069-32-8] 55.255	Hexagonal $a = 560$ pm $c = 1212$ pm D1$_g$, $hR15$, R$\bar{3}$m, B$_4$C type	2512	4500	n.a.	n.a.	n.a.	2350–2427	27	1854	2.63 –5.6	440–470	18
Chromium carbide	Cr$_7$C$_3$ 400.005	Hexagonal $a = 1398.02$ pm $c = 453.20$ pm	6992	109.0	n.a.	n.a.	n.a.	1665	n.a.	n.a.	11.7	n.a.	n.a.
Chromium carbide	Cr$_3$C$_2$ [12012-35-0] 180.010	Orthorombic $a = 282$ pm $b = 553$ pm $c = 1147$ pm D5$_{10}$, $oP20$, Pbnm, Cr$_3$C$_2$ type ($Z = 4$)	6680	75.0	n.a.	n.a.	n.a.	1895	19.2	n.a.	10.3	386	n.
Diamond	C [7782-40-3] 12.011	Cubic $a = 356.683$ pm A4, $cF8$, Fd3m, diamond type ($Z = 8$)	3515 .24	>10^{16} (I, IIa) >10^3 (IIb)	n.a.	n.a.	n.a.	3550	900 (I) 2400 (IIa)	n.a.	n.a.	930	n.
Graphite	C [7782-42-5] 12.011	Hexagonal $a = 246$ pm $b = 428$ pm $c = 671$ pm A9, $hP4$, P6$_3$/mmc, graphite type ($Z = 4$)	2250	1385	n.a.	n.a.	n.a.	3650	n.a.	n.a.	0.6–4.3	6.9	n.
Hafnium monocarbide	HfC [12069-85-1] 190.501	Cubic $a = 446.0$ pm B1, $cF8$, Fm3m, rock salt type ($Z = 4$)	12,670	45.0	n.a.	n.a.	n.a.	3890–3950	22.15	n.a.	6.3	424	1
Lanthanum dicarbide	LaC$_2$ [12071-15-7] 162.928	Tetragonal $a = 394.00$ pm $c = 657.20$ pm C11a, $tI6$, I4/mmm CaC$_2$ type ($Z = 2$)	5290	68.0	n.a.	n.a.	n.a.	2360–2438	n.a.	n.a.	12.1	n.a.	n
Molybdenum hemicarbide	β-Mo$_2$C [12069-89-5] 203.891	Hexagonal $a = 300.20$ pm $c = 427.40$ pm L'3, $hP3$, P6$_3$/mmc, Fe$_2$N type ($Z = 1$)	9180	71.0	n.a.	n.a.	n.a.	2687	n.a.	29.4	7.8	221	n
Molybdenum monocarbide	MoC [12011-97-1] 107.951	Hexagonal $a = 290$ pm $c = 281$ pm B$_k$, P6$_3$/mmc, BN type ($Z = 4$)	9159	50.0	n.a.	n.a.	n.a.	2577	n.a.	n.a.	5.76	197	n
Niobium hemicarbide	Nb$_2$C [12011-99-3] 197.824	Hexagonal $a = 312.70$ pm $c = 497.20$ pm L'3, $hP3$, P6$_3$/mmc, Fe$_2$N type ($Z = 1$)	7800	n.a.	n.a.	n.a.	n.a.	3090	n.a.	n.a.	n.a.	n.a.	n

Poisson's ratio (ν)	Ultimate tensile strength (σ_UTS, MPa)	Flexural strength (τ, MPa)	Compressive strength (α, MPa)	Fracture toughness (K_Ic, MPa.m^1/2)	Vickers hardness (HV) (HM)	Other physico-chemical properties, corrosion resistance[a], and uses	IUPAC name (synonyms, common trade names)
0.207	310–350	n.a.	2900	n.a.	3200 3500 HK (HM 9.32)	Hard black shiny crystals, fourth most hard material known after diamond, cubic boron nitride, boron oxide. It does not burn in an O_2 flame if temperature is maintained below 983°C. Maximum operating temperature: 2000°C (inert, reducing), or 600°C oxidizing). It is not attacked by hot HF or chromic acid. Used as abrasive, crucible container for molten salts except molten alkalis hydroxides. In form of molded shapes, it is used for pressure blast nozzles, wire-drawing dies, and bearing surfaces for gages. For grinding and lapping applications available mesh sizes cover the range 240 to 800.	Boron carbide (Norbide®)
n.a.	n.a.	n.a.	n.a.	n.a.	1336	Resists oxidation in the range of 800–1000°C. Corroded by the following molten metals: Ni, Zn.	Chromium carbide
0.280	n.a.	n.a.	1041	1.350	2650	Corroded by the following molten metals: Ni, Zn, Cu, Cd, Al, Mn, Fe. Corrosion resistant in molten Sn and Bi.	Chromium carbide
n.a.	n.a.	n.a.	7000	n.a.	8000 HK (HM 10)	Type I: containing 0.1T to 0.2% N, Type IIa: N-free, Type IIb: very pure, generally blue in color. Electric insulator(E_g = 7 eV.) Burns in oxygen.	Diamond
n.a.	28	n.a.	n.a.	n.a.	(HM 2)	High-temperature lubricant, crucible container for handling molten metals such as Mg, Al, Zn, Ga, Sb, and Bi.	Graphite
0.17	n.a.	n.a.	n.a.	n.a.	1870 – 2900	Dark gray brittle solid, most refractory binary material known. Control rods in nuclear reactors, crucible containers for melting HfO_2 and other oxides. Corrosion resistant to liquid metals such as Nb, Ta, Mo, and W. Severe oxidation in air above 1100–1400°C, and stable up to 2000°C in helium.	Hafnium monocarbide
n.a.	n.a.	n.a.	n.a.	n.a.	n.a.	Decomposed by H_2O.	Lanthanum dicarbide
n.a.	n.a.	n.a.	n.a.	n.a.	1499 (HM >7)	Gray powder. Wear-resistant film. Oxidized in air at 700-800°C. Corroded in the following molten metals: Al, Mg, V, Cr, Mn, Fe, Ni, Cu, Zn, Nb. Corrosion resistant in molten Cd, Sn, Ta.	Molybdenum hemicarbide
0.204	n.a.	n.a.	n.a.	n.a.	1800 (HM >9)	Oxidized in air at 700-800°C.	Molybdenum monocarbide
n.a.	n.a.	n.a.	n.a.	n.a.	2123		Niobium hemicarbide

9
Ceramics and Glasses

Table 9.3 *(continued)*

IUPAC name (synonyms, common trade names)	Theoretical chemical formula, [CASRN], relative molecular mass (^{12}C = 12.000)	Crystal system, lattice parameters, structure type, Strukturbericht, Pearson, space group, structure type (Z)	Density (ρ, kg.m^{-3})	Electrical resistivity (ρ_{el},$\mu\Omega$cm)	Dielectric permittivity at 1 MHz	Dielectric field strength (E, MV.m^{-1})	Dissipation or loss factor (tan δ)	Melting point (mp, °C)	Thermal conductivity (k, W.m^{-1}.K^{-1})	Specific heat capacity (cp, J.kg^{-1}.K^{-1})	Coeff. linear thermal expansion (α, 10^{-6}K^{-1})	Young's modulus (E, GPa)	Shear modulus (G, GPa)
Niobium monocarbide	NbC [12069-94-2] 104.917	Cubic a = 447.71 pm B1, $cF8$, Fm3m, rock salt type (Z = 4)	7820	51.1–74.0	n.a.	n.a.	n.a.	3760	14.2	n.a.	6.84	340	n.a.
Silicon monocarbide (Moissanite, (Carbolon®, Crystolon®, Carborundum®)	α-SiC [409-21-2] 40.097	Hexagonal a = 308.10 pm c = 503.94 pm B4, $hP4$, P6$_3$/mmc, wurtzite type (Z = 2)	3160	4.1×10^5	10.2	n.a.	n.a.	2093 trans.	42.5	690	4.3–4.6	386–414	n.a.
Silicon monocarbide (Carbolon®, Crystolon® Carborundum®)	β-SiC [409-21-2] 40.097	Cubic a = 435.90 pm B3, $cF8$, F43m, ZnS type (Z = 4)	3160	107–200	n.a.	n.a.	n.a.	2093–2400	135	1205	4.5	262–468	168
Tantalum hemicarbide	Ta$_2$C [12070-07-4] 373.907	Hexagonal a = 310.60 pm c = 493.00 pm L'3, $hP3$, P6$_3$/mmc, Fe$_2$N type (Z = 2)	15,100	80.0	n.a.	n.a.	n.a.	3327	n.a.	n.a.	n.a.	n.a.	n.a.
Tantalum monocarbide	TaC [12070-06-3] 194.955	Cubic a = 445.55 pm B1, $cF8$, Fm3m, rock salt type (Z = 4)	14,800	30–42.1	n.a.	n.a.	n.a.	3880–3920	22.2	190	6.64–8.4	364	n.a.
Thorium dicarbide	α-ThC$_2$ [12071-31-7] 256.060	Tetragonal a = 585 pm c = 528 pm C11a, $tI6$, I4 mmm CaC$_2$ type (Z = 2)	8960–9600	30.0	n.a.	n.a.	n.a.	2655	23.9	n.a.	8.46	n.a.	n.a.
Thorium monocarbide	ThC [12012-16-7] 244.089	Cubic a = 534.60 pm B1, $cF8$, Fm3m, rock salt type (Z = 4)	10670	25.0	n.a.	n.a.	n.a.	2621	28.9	n.a.	6.48	n.a.	n.a.
Titanium monocarbide	TiC [12070-08-5] 59.878	Cubic a = 432.8 pm B1, $cF8$, Fm3m, rock salt type (Z = 4)	4938	52.5	n.a.	n.a.	n.a.	2940–3160	17–21	841	7.5–7.7	310–462	17
Tungsten hemicarbide	W$_2$C [12070-13-2] 379.691	Hexagonal a = 299.82 pm c = 472.20 pm L'3, $hP3$, P6$_3$/mmc, Fe$_2$N type (Z = 1)	17,340	81.0	n.a.	n.a.	n.a.	2730	n.a.	n.a	3.84	421	n.a.
Tungsten monocarbide (Widia®)	WC [12070-12-1] 195.851	Hexagonal a = 290.63 pm c = 283.86 pm L'3, $hP3$, P6$_3$/mmc, Fe$_2$N type (Z = 1)	15,630	19.2	n.a.	n.a.	n.a.	2870	121	n.a.	6.9	710	n.a.
Uranium carbide	U$_2$C$_3$ [12076-62-9] 512.091	Cubic a = 808.89 pm D5c, $cI40$, I43d, Pu$_2$C$_3$ type (Z = 8)	12,880	n.a.	n.a.	n.a.	n.a.	1777	n.a.	n.a.	11.4	179–221	n.a.

	Poisson's ratio (ν)	Ultimate tensile strength (σ_{UTS}, MPa)	Flexural strength (τ, MPa)	Compressive strength (α, MPa)	Fracture toughness (K_{Ic}, MPa.m$^{1/2}$)	Vickers hardness (HV) (HM)	Other physico-chemical properties, corrosion resistance[a], and uses	IUPAC name (synonyms, common trade names)
1.	n.a.	n.a.	n.a.	n.a.	n.a.	2470 (HM >9)	Lavender gray powder, soluble in the H–HNO$_3$ mixture. Wear-resistant film, coating graphite in nuclear reactors. Oxidation in air becomes severe only above 1000°C.	Niobium monocarbide
a. a.	n.a. n.a.	450–520	n.a.	500	n.a.	2400 – 2500 (HM 9.2)	Semiconductor ($E_g = 3.03$ eV) soluble in fused alkalis hydroxides.	Silicon monocarbide (Moissanite, (Carbolon®, Crystolon®, Carborundum®)
.14	0.192	550	n.a.	1000	n.a.	2700 – 3350 (HM 9.5)	Green to bluish black, irridescent crystals. Soluble in fused alkalis hydroxides. Abrasives best suited to the grinding of low-tensile strength materials such as cast iron, brass, bronze, marble, concrete, stone and glass, optical structural and wear resistant components. Corroded by molten metals such as Na, Mg, Al, Zn, Fe, Sn, Rb, Bi. Resistant to oxidation in air up to 1650°C. Maximum operating temperature of 2000°C in reducing or inert atmosphere.	Silicon monocarbide (Carbolon®, Crystolon®, Carborundum®)
.	n.a.	n.a.	n.a.	n.a.	n.a.	1714 – 2000		Tantalum hemicarbide
.	0.172	n.a.	n.a.	n.a.	n.a.	1599 – 1800 (HM 9–10)	Golden brown crystals, soluble in HF–HNO$_3$ mixture. Crucible container for melting ZrO$_2$ and similar oxides with high melting point. Corrosion resistant to molten metals such as Ta, Re. Readily corroded by liquid metals such as Nb, Mo, Sn. Burning occurs in pure oxygen above 800°C. Severe oxidation in air above 1100–1400°C. Maximum operating temperature of 3760°C under helium.	Tantalum monocarbide
.	n.a.	n.a.	n.a.	n.a.	n.a.	600	α–β transition at 1427°C and β–γ at 1497°C. Decomposed by H$_2$O with evolution of C$_2$H$_6$.	Thorium dicarbide
.	n.a.	n.a.	n.a.	n.a.	n.a.	1000	Readily hydrolyzes in water evolving C$_2$H$_6$.	Thorium monocarbide
56	0.182	275–450	n.a.	1310	n.a.	2620 – 3200 (HM 9–10)	Gray crystals. Superconducting at 1.1 K. Sol. in HNO$_3$ and aqua regia. Resistant to oxidation in air up to 450°C. Maximum operating temperature 3000°C under helium. Crucible containers for handling molten metals such as Na, Bi, Zn, Pb, Sn, Bi, Rb, and Cd. Corroded by the following liquid metals: Mg, Al, Si, Ti, Zr, V, Nb, Ta, Cr, Mo, Mn, Fe, Co, and Ni. Attacked by molten NaOH.	Titanium monocarbide
.	n.a.	n.a.	n.a.	n.a.	n.a.	3000	Black. Resistant to oxidation in air up to 700°C. Corrosion resistant to Mo.	Tungsten hemicarbide
3	0.26	n.a.	n.a.	530	n.a.	2700 (HM > 9)	Gray powder, dissolve by the HF–HNO$_3$ mixture. Cutting tools, wear-resistant semi-conductor film. Corroded by the following molten metals: Mg, Al, V, Cr, Mn, Ni, Cu, Zn, Nb, Mo. Corrosion resistant to molten Sn and Tn.	Tungsten monocarbide (Widia®)
.	n.a.	n.a.	n.a.	434	n.a.	n.a.		Uranium carbide

Table 9.3 (continued)

IUPAC name (synonyms, common trade names)	Theoretical chemical formula, [CASRN], relative molecular mass (^{12}C = 12.000)	Crystal system, lattice parameters, structure type, Struktürbericht, Pearson, space group, structure type (Z)	Density (ρ, kg.m^{-3})	Electrical resistivity (ρ_{el}, $\mu\Omega$cm)	Dielectric permittivity at 1 MHz	Dielectric field strength (E, MV.m^{-1})	Dissipation or loss factor (tan δ)	Melting point (mp, °C)	Thermal conductivity (k, W.m^{-1}.K^{-1})	Specific heat capacity (c_p, J.kg^{-1}.K^{-1})	Coeff. linear thermal expansion (α, 10^{-6}K^{-1})	Young's modulus (E, GPa)	Shear modulus (G, GPa)
Uranium dicarbide	UC$_2$ [12071-33-9] 262.051	Tetragonal a = 352.24 pm c = 599.62 pm C11a, $tI6$, I4/mmm, CaC$_2$ type (Z = 2)	11280	n.a.	n.a.	n.a.	n.a.	2350–2398	32.7	147	14.6	n.a.	n.a.
Uranium monocarbide	UC [12070-09-6] 250.040	Cubic a = 496.05 pm B1, $cF8$, Fm3m, rock salt type (Z = 4)	13,630	50.0	n.a.	n.a.	n.a.	2370–2790	23.0	n.a.	11.4	172.4	66.
Vanadium hemicarbide	V$_2$C [2012-17-8] 113.89	Hexagonal a = 286 pm c = 454 pm L'3, $hP3$, P6$_3$ mmc, Fe$_2$N type (Z = 2)	5750	n.a.	n.a.	n.a.	n.a.	2166	n.a.	n.a.	n.a.	n.a.	n.a
Vanadium monocarbide	VC [12070-10-9] 62.953	Cubic a = 413.55 pm B1, $cF8$, Fm3m, rock salt type (Z = 4)	5770	65.0–98.0	n.a.	n.a.	n.a.	2810	24.8	n.a.	4.9	614	43
Zirconium monocarbide	ZrC [12020-14-3] 103.235	Cubic a = 469.83 pm B1, $cF8$, Fm3m, rock salt type (Z = 4)	6730	68.0	n.a.	n.a.	n.a.	3540–3560	20.61	205	6.82	345	12
Nitrides													
Aluminum mononitride	AlN [24304-00-5] 40.989	Hexagonal a = 311.0 pm c = 497.5 pm B4, $hP4$, P6$_3$mc, wurtzite type (Z = 2)	3050	10^{17}	n.a.	n.a.	n.a.	2230	29.96	820	5.3	346	n.a
Beryllium nitride	α-Be$_3$N$_2$ [1304-54-7] 55.050	Cubic a = 814 pm D5$_3$, $cI80$, Ia3, Mn$_2$O$_3$ type (Z = 16)	2710	n.a.	n.a.	n.a.	n.a.	2200	n.a.	1221	n.a.	n.a.	n.a
Boron mononitride	BN [10043-11-5] 24.818	Hexagonal a = 250.4 pm c = 666.1 pm B$_k$, $hP8$, P6$_3$/mmc, BN type (Z = 4)	2250	10^{19}	n.a.	n.a.	0.0002	2730 (dec.)	15.41	711	7.54	85.5	n.
Boron mononitride (Borazon®, CBN)	BN 24.818	Cubic a = 361.5 pm	3430	1900 (200° C)	2.54	n.a.	0.0002	1540	n.a.	n.a.	n.a.	n.a.	n.
Chromium heminitride	Cr$_2$N [12053-27-9] 117.999	Hexagonal a = 274 pm c = 445 pm L'3, $hP3$, P6$_3$/mmc, Fe$_2$N type (Z = 1)	6800	76	n.a.	n.a.	n.a.	1661	22.5	630	9.36	n.a.	n.
Chromium mononitride	CrN [24094-93-7] 66.003	Cubic a = 415.0 pm B1, $cF8$, Fm3m, rock salt type (Z = 4)	6140	640	n.a.	n.a.	n.a.	1499 (dec.)	12.1	795	2.34	n.a.	n

	Poisson's ratio (ν)	Ultimate tensile strength (σ_{uts}, MPa)	Flexural strength (τ, MPa)	Compressive strength (α, MPa)	Fracture toughness (K_{Ic}, MPa.m$^{1/2}$)	Vickers hardness (HV, (HM))	Other physico-chemical properties, corrosion resistance^a, and uses	IUPAC name (synonyms, common trade names)
a.	n.a.	n.a.	n.a.	n.a.	n.a.	600	Transition tetragonal to cubic at 1765°C. Decomposed in H_2O, slightly soluble in alcohol. Used in microsphere pellets to fuel nuclear reactors.	Uranium dicarbide
a.	0.29	n.a.	n.a.	351.6	n.a.	750–935 (HM >7)	Gray crystals with metallic appearance, reacts with oxygen. Corroded by the following molten metals: Be, Si, Ni, Zr.	Uranium monocarbide
a.	n.a.	n.a.	n.a.	n.a.	n.a.	3000	Corroded by molten Nb, Mo, and Ta.	Vanadium hemicarbide
a.	n.a.	790–825	n.a.	613	n.a.	2090	Black crystals soluble in HNO_3 with decomposition. Wear-resistant film, cutting tools. Resistant to oxidation in air up to 300°C.	Vanadium monocarbide
38	0.257	110	n.a.	1641	n.a.	1830–2930 (HM >8)	Dark gray brittle solid, soluble in HF solns containing nitrates or peroxide ions. UC-nuclear power reactors, crucible containers for handling molten metals such as Bi, Cd, Pb, Sn, Rb, and molten zirconia ZrO_2. Corroded by the following liquid metals: Mg, Al, Si,V, Nb, Ta, Cr, Mo, Mn, Fe, Co, Ni, and Zn. In air oxidized rapidly above 500°C. Maximum operating temperature of 2350°C under helium.	Zirconium monocarbide
								Nitrides
8	n.a.	270	n.a.	2068	n.a.	(HM 9–10) 1200	Insulator ($E_g = 4.26$ eV). Decomposed by water, acids and alkalis to $Al(OH)_3$ and NH_3. Crucible container for GaAs crystal growth.	Aluminum mononitride
.	n.a.	n.a.	n.a.	n.a.	n.a.	n.a.	Hard white or grayish crystal. Oxidized in air above 600°C. Slowly decomposed in water, quickly in acids and alkalis with evolution of NH_3.	Beryllium nitride
1	n.a.	41–62	n.a.	310	n.a.	(HM 2.0) 230	Insulator ($E_g = 7.5$ eV). Crucibles for melting molten metals such as Na, B, Fe, Ni, Al, Si, Cu, Mg, Zn, In, Bi, Rb, Cd, Ge, and Sn. Corroded by these molten metals: U, Pt, V, Ce, Be, Mo, Mn,Cr, V, and Al. Attacked by the following molten salts: PbO_2, Sb_2O_3, AsO_3, Bi_2O_3, KOH, K_2CO_3. Used in furnace insulation diffusion masks, and passivation layers.	Boron mononitride
.	n.a.	n.a.	n.a.	7000	n.a.	4700–5000 (HM 10)	Tiny reddish to black grains. Used as abrasive for grinding tool and die steels and high-alloy steels when chemical reactivity of diamonds is a problem.	Boron mononitride (Borazon®, CBN)
.	n.a.	n.a.	n.a.	n.a.	n.a.	1200–1571		Chromium heminitride
.	n.a.	n.a.	n.a.	n.a.	n.a.	1090		Chromium mononitride

9

Ceramics and Glasses

Table 9.3 (continued)

IUPAC name (synonyms, common trade names)	Theoretical chemical formula, [CASRN], relative molecular mass ($^{12}C = 12.000$)	Crystal system, lattice parameters, structure type, Struktürbericht, Pearson, space group, structure type (Z)	Density (ρ, kg·m^{-3})	Electrical resistivity ($\rho,\mu\Omega$cm)	Dielectric permittivity at 1 MHz	Dielectric field strength (E, MV·m^{-1})	Dissipation or loss factor (tan δ)	Melting point (mp, °C)	Thermal conductivity (k, W·m^{-1}·K^{-1})	Specific heat capacity (cp, J·kg^{-1}·K^{-1})	Coeff. linear thermal expansion (α, 10^{-6}K^{-1})	Young's modulus (E, GPa)	Shear modulus (G, GPa)
Hafnium mononitride	HfN [25817-87-2] 192.497	Cubic $a = 451.8$ pm B1, cF8, Fm3m, rock salt type (Z = 4)	13,840	33	n.a.	n.a.	n.a.	3310	21.6	210	6.5	n.a.	n.a.
Molybdenum heminitride	Mo$_2$N [12033-31-7] 205.887	Cubic $a = 416$ pm L'1, cP5, Pm3m, Fe$_4$N type (Z = 2)	9460	19.8	n.a.	n.a.	n.a.	760–899	17.9	293	6.12	n.a.	n.a.
Molybdenum mononitride	MoN [12033-19-1] 109.947	Hexagonal $a = 572.5$ pm $c = 560.8$ pm B$_h$, hP2, P6/mmm, WC type (Z = 1)	9180	n.a.	n.a.	n.a.	n.a.	1749	n.a.	n.a.	n.a.	n.a.	n.a.
Niobium mononitride	NbN [24621-21-4] 106.913	Cubic $a = 438.8$ pm B1, cF8, Fm3m, rock salt type (Z = 4)	8470	78	n.a.	n.a.	n.a.	2575	3.63	n.a.	10.1	n.a.	n.a.
Silicon nitride	β-Si$_3$N$_4$ [12033-89-5] 140.284	Hexagonal $a = 760.8$ pm $c = 291.1$ pm P6/3m	3170	10^6	n.a.	n.a.	n.a.	1850	28	713	2.25	55	n.a.
Silicon nitride (Nitrasil®)	α-Si$_3$N$_4$ [12033-89-5] 140.284	Hexagonal $a = 775.88$ pm $c = 561.30$ pm P31c	3184	10^{19}	9.4	n.a.	n.a.	1900 (sub.)	17	700	2.5–3.3	304	n.a.
Tantalum heminitride	Ta$_2$N 375.901	Hexagonal $a = 306$ pm $c = 496$ pm L'3, hP3, P6$_3$/mmc, Fe$_2$N type (Z = 1)	15,600	263	n.a.	n.a.	n.a.	2980	10.04	126	5.2	n.a.	n.a.
Tantalum mononitride (ϵ)	TaN [12033-62-4] 194.955	Hexagonal $a = 519.1$ pm $c = 290.6$ pm	13,800	128–135	n.a.	n.a.	n.a.	3093	8.31	210	3.2	n.a.	n.a.
Thorium mononitride	ThN [12033-65-7] 246.045	Cubic $a = 515.9$ B1, cF8, Fm3m, rock salt type (Z = 4)	11,560	20	n.a.	n.a.	n.a.	2820	n.a.	n.a.	7.38	n.a.	n.a.
Thorium nitride	Th$_2$N$_3$ [12033-90-8]	Hexagonal $a = 388$ pm $c = 618$ pm D5$_2$, hP5, $\bar{3}$m1, La$_2$O$_3$ type (Z = 1)	10400	n.a.	n.a.	n.a.	n.a.	1750	n.a.	n.a.	n.a.	n.a.	n.a.
Titanium mononitride	TiN [25583-20-4] 61.874	Cubic $a = 424.6$ pm B1, cF8, Fm3m, rock salt type (Z = 4)	5430	21.7	n.a.	n.a.	n.a.	2930 (dec.)	29.1	586	9.35	248	n.
Tungsten dinitride	WN$_2$ [60922-26-1] 211.853	Hexagonal $a = 289.3$ pm $c = 282.6$ pm	7700	n.a.	n.a.	n.a.	n.a.	600 (dec.)	n.a.	n.a.	n.a.	n.a.	n.
Tungsten heminitride	W$_2$N [12033-72-6] 381.687	Cubic $a = 412$ pm L'1, cP5, Pm3m, Fe$_4$N type (Z = 2)	17,700	n.a.	n.a.	n.a.	n.a.	982	n.a.	n.a.	n.a.	n.a.	n.
Tungsten mononitride	WN [12058-38-7]	Hexagonal	15,940	n.a.	n.a.	n.a.	n.a.	593	n.a.	n.a.	n.a.	n.a.	n.a.

Poisson's ratio (ν)	Ultimate tensile strength (σ_{UTS}, MPa)	Flexural strength (τ, MPa)	Compressive strength (α, MPa)	Fracture toughness (K_{IC}, MPa.m$^{1/2}$)	Vickers hardness (HV) (HM)	Other physico-chemical properties, corrosion resistance^a, and uses	IUPAC name (synonyms, common trade names)
n.a.	n.a.	n.a.	n.a.	n.a.	1640 (HM > 8–9)	Most refractory of all nitrides.	Hafnium mononitride
n.a.	n.a.	n.a.	n.a.	n.a.	1700	Temperature transition at 5.0 K.	Molybdenum heminitride
n.a.	n.a.	n.a.	n.a.	n.a.	650		Molybdenum mononitride
n.a.	n.a.	n.a.	n.a.	n.a.	1400 (HM >8)	Dark gray crystals. Transition temperature 15.2 K. Insoluble in HCL, HNO$_3$, H$_2$SO$_4$, but attacked by hot caustic, lime or strong alkalis evolving NH$_3$.	Niobium mononitride
n.a.	n.a.	n.a.	n.a.	n.a.	(HM >9)		Silicon nitride
n.a.	400–580	n.a.	n.a.	n.a.	(HM >9)	Gray amorphous powder or crystals. Corrosion resistant to molten metals such as: Al, Pb, Zn, Cd, Bi, Rb, and Sn, and molten salts NaCl–KCl, NaF, and silicate glasses. Corroded by molten Mg, Ti, V, Cr, Fe, and Co, cryolite, KOH, Na$_2$O.	Silicon nitride (Nitrasil®)
n.a.	n.a.	n.a.	n.a.	n.a.	3200	Decomposed by KOH with evolution of NH$_3$.	Tantalum heminitride
n.a.	n.a.	n.a.	n.a.	n.a.	1110 (HM >8)	Bronze or black crystals. Transition temperature 1.8 K. Insoluble in water, slowly attacked by aqua regia, HF, and HNO$_3$.	Tantalum mononitride (ϵ)
n.a.	n.a.	n.a.	n.a.	n.a.	600	Gray solid. Slowly hydrolyzed by water.	Thorium mononitride
n.a.	n.a.	n.a.	n.a.	n.a.	n.a.		Thorium nitride
n.a.	n.a.	n.a.	972	n.a.	1900 (HM 8–9)	Bronze powder. Transition temperature 4.2 K. Corrosion resistant to molten metals such as Al, Pb, Mg, Zn, Cd, Bi. Corroded by molten Na, Rb, Ti, V, Cr, Mn, Sn, Ni, Cu, Fe, and Co. dissolved by boiling acqua regia, decomposed by boiling alkalis evolving NH$_3$.	Titanium mononitride
n.a.	n.a.	n.a.	n.a.	n.a.	n.a.	Brown crystals.	Tungsten dinitride
n.a.	n.a.	n.a.	n.a.	n.a.	n.a.	Gray crystals.	Tungsten heminitride
n.a.	n.a.	n.a.	n.a.	n.a.	n.a.	Gray solid. Slowly hydrolyzed by water.	Tungsten mononitride

Table 9.3 (continued)

IUPAC name (synonyms, common trade names)	Theoretical chemical formula, [CASRN], relative molecular mass ($^{12}C = 12.000$)	Crystal system, lattice parameters, structure type, Strukturbericht, Pearson, space group, structure type (Z)	Density (ρ, kg.m^{-3})	Electrical resistivity (ρ, $\mu\Omega$cm)	Dielectric permittivity at 1 MHz	Dielectric field strength (E, MV.m^{-1})	Dissipation or loss factor (tan δ)	Melting point (mp, °C)	Thermal conductivity (k, W.m^{-1}.K^{-1})	Specific heat capacity (cp, J.kg^{-1}.K^{-1})	Coeff. linear thermal expansion (α, 10^{-6}°K^{-1})	Young's modulus (E, GPa)	Shear modulus (G, GPa)
Uranium nitride	U$_2$N$_3$ [12033-83-9] 518.259	Cubic $a = 1070$ pm D5$_3$, $c180$, Ia3, Mn$_2$O$_3$ type (Z = 16)	11,240	n.a.	n.a.	n.a.	n.a.	n.a.	n.a.	n.a.	n.a	n.a.	n.a
Uranium mononitride	UN [25658-43-9] 252.096	Cubic $a = 489.0$ pm B1, $cF8$, Fm3m, rock salt type (Z = 4)	14320	208	n.a.	n.a.	n.a.	2900	12.5	188	9.72	149	60
Vanadium mononitride	VN [24646-85-3] 64.949	Cubic $a = 414.0$ pm B1, $cF8$, Fm3m, rock salt type (Z = 4)	6102	86	n.a.	n.a.	n.a.	2360	11.25	586	8.1	n.a.	n.a
Zirconium mononitride	ZrN [25658-42-8] 105.231	Cubic $a = 457.7$ pm B1, $cF8$, Fm3m, rock salt type (Z = 4)	7349	13.6	n.a.	n.a.	n.a.	2980	20.90	377	7.24	n.a.	n.a
Silicides													
Chromium disilicide	CrSi$_2$ [12018-09-6] 108.167	Hexagonal $a = 442$ pm $c = 635$ pm C40, $hP9$, P6$_2$22, CrSi$_2$ type (Z = 3)	4910	1400	n.a.	n.a.	n.a.	1490	106	n.a.	13.0	n.a.	n.a
Chromium silicide	Cr$_3$Si [12018-36-9] 184.074	Cubic $a = 456$ pm A15, $cP8$, Pm3n, Cr$_3$Si type (Z = 2)	6430	45.5	n.a.	n.a.	n.a.	1770	n.a.	n.a.	10.5	n.a.	n.a
Hafnium disilicide	HfSi$_2$ [12401-56-8] 234.66	Orthorhombic $a = 369$ pm $b = 1446$ pm $c = 364$ pm C49, $oC12$, Cmcm, ZrSi$_2$ type (Z = 4)	8030	n.a.	n.a.	n.a.	n.a.	1699	n.a.	n.a.	n.a.	n.a.	n.a
Molybdenum disilicide	MoSi$_2$ [12136-78-6] 152.11	Tetragonal $a = 319$ pm $c = 783$ pm C11b, $tI6$, I4/mmm, MoSi$_2$ type (Z = 2)	6260	21.5	n.a.	n.a.	n.a.	1870	58.9	n.a.	8.12	407	16.
Niobium disilicide	NbSi$_2$ [12034-80-9] 149.77	Hexagonal $a = 479$ pm $c = 658$ pm C40, $hP9$, P6$_2$22, CrSi$_2$ type (Z = 3)	5290	50.4	n.a.	n.a.	n.a.	2160	n.a.	n.a.	n.a.	n.a.	n.a
Tantalum disilicide	TaSi$_2$ [12039-79-1] 237.119	Hexagonal $a = 477$ pm $c = 655$ pm C40, $hP9$, P6$_2$22, CrSi$_2$ type (Z = 3)	9140	8.5	n.a.	n.a.	n.a.	2299	n.a.	n.a.	9.54–8.8	n.a.	n.a
Tantalum silicide	Ta$_5$Si$_3$ [12067-56-0] 988.992	Hexagonal	13,060	n.a.	n.a.	n.a.	n.a.	2499	n.a.	n.a.	n.a.	n.a.	n.a
Thorium disilicide	ThSi$_2$ [12067-54-8] 288.209	Tetragonal $a = 413$ pm $c = 1435$ pm Cc, $tI12$, I4 amd, ThSi$_2$ type (Z = 4)	7790	n.a.	n.a.	n.a.	n.a.	1850	n.a.	n.a.	n.a.	n.a.	n.a

Bulk modulus (K, GPa)	Poisson's ratio (v)	Ultimate tensile strength (σ_{UTS}, MPa)	Flexural strength (τ, MPa)	Compressive strength (α, MPa)	Fracture toughness (K_{IC}, MPa.m$^{1/2}$)	Vickers hardness (HV) (HM)	Other physico-chemical properties, corrosion resistance^a, and uses	IUPAC name (synonyms, common trade names)
n.a.	n.a.	n.a.	n.a.	n.a.	n.a.	n.a.		Uranium nitride
n.a.	0.24	n.a.	n.a.	n.a.	n.a.	455		Uranium mononitride
n.a.	n.a.	n.a.	n.a.	n.a.	n.a.	1520 (HM 9–10)	Black powder. Transition temperature 7.5 K. Soluble in aqua regia.	Vanadium mononitride
n.a.	n.a.	n.a.	n.a.	979	n.a.	1480 (HM >8)	Yellow solid. Transition temperature 9 K. Corrosion resistant to steel, basic slag and cryolithe and molten metals such as Al, Pb, Mg, Zn, Cd, Bi. Corroded by molten Be, Na, Rb, Ti, V, Cr, Mn, Sn, Ni, Cu, Fe, and Co. Soluble in concentrated HF, slowly soluble in hot H_2SO_4.	Zirconium mononitride
								Silicides
n.a.	n.a.	n.a.	n.a.	n.a.	n.a.	1000 – 1130		Chromium disilicide
n.a.	n.a.	n.a.	n.a.	n.a.	n.a.	1005		Chromium silicide
n.a.	n.a.	n.a.	n.a.	n.a.	n.a.	865 – 930		Hafnium disilicide
0.344	0.165	276	n.a.	2068– 2415	n.a.	1260	The compound is thermally stable in air up to 1000°C. Corrosion resistant to molten metals such as: Zn, Pd, Ag, Bi, and Rb. It is corroded by the following liquid metals: Mg, Al, Si, V, Cr, Mn, Fe, Ni, Cu, Mo, and Ce.	Molybdenum disilicide
n.a.	n.a.	n.a.	n.a.	n.a.	n.a.	1050		Niobium disilicide
n.a.	n.a.	n.a.	n.a.	n.a.	n.a.	1200 – 1600	Corroded by molten Ni.	Tantalum disilicide
n.a.	n.a.	n.a.	n.a.	n.a.	n.a.	1200 – 1500	The compound is thermally stable in air up to 400°C.	Tantalum silicide
n.a.	n.a.	n.a.	n.a.	n.a.	n.a.	1120	Corrosion resistant to molten Cu, while corroded by molten Ni.	Thorium disilicide

9
Ceramics and Glasses

Table 9.3 (continued)

IUPAC name (synonyms, common trade names)	Theoretical chemical formula, [CASRN], relative molecular mass ($^{12}C = 12.000$)	Crystal system, lattice parameters, structure type, Strukturbericht, Pearson, space group, structure type (Z)	Density (ρ, kg.m^{-3})	Electrical resistivity (ρ_e, $\mu\Omega$cm)	Dielectric permittivity at 1 MHz	Dielectric field strength (E, MV.m^{-1})	Dissipation or loss factor (tan δ)	Melting point (mp, °C)	Thermal conductivity (k, W.m^{-1}.K^{-1})	Specific heat capacity (c_p, J.kg^{-1}.K^{-1})	Coeff. linear thermal expansion (α, 10^{-6}K^{-1})	Young's modulus (E, GPa)	Shear modulus (G, GPa)
Titanium disilicide	TiSi$_2$ [12039-83-7] 104.051	Orthorhombic $a = 360$ pm $b = 1376$ pm $c = 360$ pm C49, oC12, Cmcm, ZrSi$_2$ type (Z = 4)	4150	123	n.a.	n.a.	n.a.	1499	n.a.	n.a.	10.4	n.a.	n.a.
Titanium trisilicide	Ti$_5$Si$_3$ [12067-57-1] 323.657	Hexagonal $a = 747$ pm $c = 516$ pm D8$_8$, hP16, P6$_3$mcm, Mn$_5$Si$_3$ type (Z = 2)	4320	55	n.a.	n.a.	n.a.	2120	n.a.	n.a.	110	n.a.	n.a.
Tungsten disilicide	WSi$_2$ [12039-88-2] 240.01	Tetragonal $a = 320$ pm $c = 781$ pm C11b, tI6, I4/mmm, MoSi$_2$ type (Z = 2)	9870	33.4	n.a.	n.a.	n.a.	2165	n.a.	n.a.	8.28	n.a.	n.a.
Tungsten silicide	W$_5$Si$_3$ [12039-95-1] 1003.46		12,210	n.a.	n.a.	n.a.	n.a.	2320	n.a.	n.a.	n.a.	n.a.	n.a.
Uranium disilicide	USi$_2$ 294.200	Tetragonal $a = 397$ pm $c = 1371$ pm Cc, tI12, I4/amd, ThSi$_2$ type (Z = 4)	9250	n.a.	n.a.	n.a.	n.a.	1700	n.a.	n.a.	n.a.	n.a.	n.a.
Uranium silicide	β-U$_3$Si$_2$ 770.258	Tetragonal $a = 733$ pm $c = 390$ pm D5a, tP10, P4/mbm, U$_3$Si$_2$ type (Z = 2)	12,200	150	n.a.	n.a.	n.a.	1666	14.7	n.a.	14.8	77.9	33.1
Vanadium disilicide	VSi$_2$ [12039-87-1] 107.112	Hexagonal $a = 456$ pm $c = 636$ pm C40, hP9, P6$_2$22, CrSi$_2$ type (Z = 3)	5100	9.5	n.a.	n.a.	n.a.	1699	n.a.	n.a.	11.2	n.a.	n.a.
Vanadium silicide	V$_3$Si [12039-76-8] 147.9085	Cubic $a = 471$ pm A15, cP8, Pm3n, Cr$_3$Si type (Z = 2)	5740	203	n.a.	n.a.	n.a.	1732	n.a.	n.a.	8.0	n.a.	n.a.
Zirconium disilicide	ZrSi$_2$ [12039-90-6] 147.395	Orthorhombic $a = 372$ pm $b = 1469$ pm $c = 366$ pm C49, oC12, Cmcm, ZrSi$_2$ type (Z = 4)	4880	161	n.a.	n.a.	n.a.	1604	n.a.	n.a.	8.6	n.a.	n.a.
Oxides													
Aluminum sesquioxide (Alumina, corundum, sapphire)	α-Al$_2$O$_3$ [1344-28-1] [1302-74-5] 101.961	Trigonal (Rhomboedral) $a = 475.91$ pm $c = 1298.4$ pm D5$_1$, hR10, R$\bar{3}$c, corundum type (Z = 2)	3987	2×10^{23}	9.1–9.8	28–47	0.0005	2054	35.6 39	795.5–880	7.1–8.3	365–393	162–184

Bulk modulus (K, GPa)	Poisson's ratio (ν)	Ultimate tensile strength (σ_{UTS}, MPa)	Flexural strength (τ, MPa)	Compressive strength (σ, MPa)	Fracture toughness (K_c, MPa·m$^{1/2}$)	Vickers hardness (HV) (HM)	Other physico-chemical properties, corrosion resistance[a], and uses	IUPAC name (synonyms, common trade names)
n.a.	n.a.	n.a.	n.a.	n.a.	n.a.	890 – 1039		Titanium disilicide
n.a.	n.a.	n.a.	n.a.	n.a.	n.a.	986		Titanium trisilicide
n.a.	n.a.	n.a.	n.a.	n.a.	n.a.	1090		Tungsten disilicide
n.a.	n.a.	n.a.	n.a.	n.a.	n.a.	770	Corroded by molten Ni.	Tungsten silicide
n.a.	n.a.	n.a.	n.a.	n.a.	n.a.	700		Uranium disilicide
n.a.	0.170	n.a.	n.a.	n.a.	n.a.	796		Uranium silicide
n.a.	n.a.	n.a.	n.a.	n.a.	n.a.	1400		Vanadium disilicide
n.a.	n.a.	n.a.	n.a.	n.a.	n.a.	1500		Vanadium silicide
n.a.	n.a.	n.a.	n.a.	n.a.	n.a.	1030 – 1060		Zirconium disilicide
								Oxides
234– 496	0.254	206– 255	282	2549 – 3103	4.0	2100 – 3000 (HM 9}	White and translucent, hard material used as abrasive for grinding. Excellent electric insulator and also wear resistant. Insoluble in water, insoluble in strong mineral acids, readily soluble in strong alkali hydroxides, attacked by HF, or NH_4HF_2. Owing to its corrosion resistance, under inert atmosphere, in molten metals such as Mg, Ca, Sr, Ba, Mn, Sn, Pb, Ga, Bi, As, Sb, Hg, Mo, W, Co, Ni, Pd, Pt, and U it is used as crucible containers for these liquid metals. Alumina is readily attacked under inert atmosphere by molten metals such as: Li, Na, Be, Al, Si, Ti, Zr, Nb, Ta, and Cu. Maximum service temperature 1950°C.	Aluminum sesquioxide (Alumina, corundum, sapphire)

Table 9.3 (continued)

IUPAC name (synonyms, common trade names)	Theoretical chemical formula, [CASRN], relative molecular mass ($^{12}C = 12.000$)	Crystal system, lattice parameters, structure type, Strukturbericht, Pearson, space group, structure type (Z)	Density (ρ, kg.m^{-3})	Electrical resistivity ($\rho,\mu\Omega$cm)	Dielectric permittivity at 1 MHz	Dielectric field strength (E, MV.m^{-1})	Dissipation or loss factor (tan δ)	Melting point (mp, °C)	Thermal conductivity (k, W.m^{-1}.K^{-1})	Specific heat capacity (cp, J.kg^{-1}.K^{-1})	Coeff. linear thermal expansion (α, 10^{-6}K^{-1})	Young's modulus (E, GPa)	Shear modulus (G, GPa)
Beryllium monoxide (Beryllia)	BeO [1304-56-9] 25.011	Trigonal (Hexagonal) $a = 270$ pm $c = 439$ pm B4, $hP4$, P6$_3$mc, wurtzite type (Z = 2)	3008–3030	1.0×10^{22}	6.8–7.66	11.8	0.0004	2550–2565	245–250	996.5	7.5–9.7	296.5–345	n.a.
Calcium monoxide (Calcia, lime)	CaO [1305-78-8] 56.077	Cubic $a = 481.08$ pm B2, $cP2$, Pm3m, CsCl type (Z = 1)	3320	1.0×10^{14}	11.1	n.a.	n.a.	2927	8–16	753.1	3.88	n.a.	n.a.
Cerium dioxide (Ceria, Cerianite)	CeO$_2$ [1306-38-3] 172.114	Cubic $a = 541.1$ pm C1, $cF12$, Fm3m, fluorite type (Z = 4)	7650	10^{10}	n.a.	n.a.	n.a.	2340	n.a.	389	10.6	181	70.3
Chromium oxide (Eskolaite)	Cr$_2$O$_3$ [1308-38-9] 151.990	Trigonal (Rhombohedral) $a = 538$ pm, 54° 50' D5$_1$, $hR10$, R$\bar{3}$c, corundum type (Z = 2)	5220	1.3×10^9 (346°C)	n.a.	n.a.	n.a.	2330	n.a.	921.1	10.90	n.a.	n.a.
Dysprosium oxide (Dysprosia)	Dy$_2$O$_3$ [1308-87-8] 373.00	Cubic D5$_3$, $cI80$, Ia$\bar{3}$, Mn$_2$O$_3$ type (Z = 16)	8300	n.a.	n.a.	n.a.	n.a.	2408	n.a.	n.a.	7.74	n.a.	n.a.
Europium oxide (Europia)	Eu$_2$O$_3$ [1308-96-9] 351.928	Cubic D5$_3$, $cI80$, Ia$\bar{3}$, Mn$_2$O$_3$ type (Z = 16)	7422	n.a.	n.a.	n.a.	n.a.	2350	n.a.	n.a.	7.02.	n.a.	n.a.
Hafnium dioxide (Hafnia)	HfO$_2$ [12055-23-1] 210.489	Monoclinic [1790°C] $a = 511.56$ pm $b = 517.22$ pm $c = 529.48$ pm C43, $mP12$, P2$_1$ c, baddeleyite type (Z = 4)	9680	5×10^{15}	n.a.	n.a.	n.a.	2900	1.14	121	5.85	57	n.a.
Gadolinium oxide (Gadolinia)	Gd$_2$O$_3$ [12064-62-9] 362.50	Cubic D5$_3$, $cI80$, Ia$\bar{3}$, Mn$_2$O$_3$ type (Z = 16)	7630	n.a.	n.a.	n.a.	n.a.	2420	n.a.	276	10.44	124	n.a.
Lanthanum dioxide (Lanthania)	La$_2$O$_3$ [1312-81-8] 325.809	Trigonal (Hexagonal) D5$_2$, $hP5$, P$\bar{3}$m1, lanthania type (Z = 1)	6510	1×10^{14} (550°C)	n.a.	n.a.	n.a.	2315	n.a.	288.89	11.9	n.a.	n.a.
Magnesium monoxide (Magnesia, periclase)	MgO [1309-48-4] 40.304	Cubic $a = 420$ pm B1, $cF8$, Fm3m, rock salt type (Z = 4)	3581	1.3×10^{15}	9.65–9.8	n.a.	n.a.	2852	50–75	962.3	11.52	303.4	117–130
Niobium pentoxide (Columbite, niobia)	Nb$_2$O$_5$ [1313-96-8] 265.810	Trigonal (Rhombohedral) $a = 211.6$ pm $b = 382.2$ pm $c = 193.5$ pm Columbite type	4470	5.5×10^{12}	n.a.	n.a.	n.a.	1520	n.a.	502.41	n.a.	n.a.	n.a.

Bulk modulus (K, GPa)	Poisson's ratio (ν)	Ultimate tensile strength (σ_{UTS}, MPa)	Flexural strength (τ, MPa)	Compressive strength (α, MPa)	Fracture toughness (K_{Ic}, $MPa.m^{1/2}$)	Vickers hardness (HV) (HM)	Other physico-chemical properties, corrosion resistance[a], and uses	IUPAC name (synonyms, common trade names)
n.a.	0.340	103.4	241–250	1551	n.a.	1500 (HM 9)	It is the only material apart from diamond which combines both excellent thermal shock resistance, high electrical resistivity and high thermal conductivity, and hence, it is used for heat sinks in electronics. Beryllia is very soluble in water, but dissolves slowly in conc. acids and alkalis. It is highly toxic. It exhibits outstanding corrosion resistance in the following liquid metals: Li, Na, Al, Ga, Pb, Ni, and Ir. It is readily attacked by molten metals such as Be, Si, Ti, Zr, Nb, Ta, Mo, and W. Maximum service temperature: 2400°C.	Beryllium monoxide (Beryllia)
n.a.	n.a.	n.a.	n.a.	n.a.	n.a.	560 (HM 4.5)	White or grayish ceramics. It readily absorbs CO_2 and water from air to form spent lime and calcium carbonate. It reacts readily with water to give $Ca(OH)_2$. Volumic expansion coefficient $0.225 \times 10^{-9}.K^{-1}$. It exhibits outstanding corrosion resistance in the following liquid metals: Li, Na.	Calcium monoxide (Calcia, lime)
n.a.	0.311	n.a.	n.a.	589	n.a.	(HM 6)	Pale yellow cubic crystals. Abrasive for polishing glass, interference filters, anti-reflection coating. Insoluble in water, soluble in H_2SO_4 and HNO_3 but insoluble in dilute acids.	Cerium dioxide (Ceria, Cerianite)
n.a.	n.a.	n.a.	n.a.	n.a.	n.a.	(HM >8)		Chromium oxide (Eskolaite)
n.a.	n.a.	n.a.	n.a.	n.a.	n.a.	n.a.		Dysprosium oxide (Dysprosia)
n.a.	n.a.	n.a.	n.a.	n.a.	n.a.	n.a.		Europium oxide (Europia)
n.a.	n.a.	n.a.	n.a.	n.a.	n.a.	780–1050		Hafnium dioxide (Hafnia)
n.a.	n.a.	n.a.	n.a.	n.a.	n.a.	480		Gadolinium oxide (Gadolinia)
n.a.	n.a.	n.a.	n.a.	n.a.	n.a.	n.a.	Insoluble in water, soluble in diluted strong mineral acids.	Lanthanum dioxide (Lanthania)
n.a.	0.33–0.36	200–300	441	1300–1379	n.a.	750 (HM 5.5–6)	White ceramics, with a high reflective index in the visible and near UV regions. Used as linings in steel furnace. Crucible containers for fluoride melts. Very slowly soluble in pure water but soluble in diluted strong mineral acids. It exhibits outstanding corrosion resistance in the following liquid metals: Mg, Li, and Na. It is readily attacked by molten metals such as Be, Si, Ti, Zr, Nb and Ta. MgO reacts with water, CO_2, and diluted acids. Maximum service temperature 2400°C. Transmittance of 80%, and $n = 1.75$ in the IR region 7 to 300 μm.	Magnesium monoxide (Magnesia, periclase)
n.a.	n.a.	n.a.	n.a.	n.a.	n.a.	1500 HV	Dielectric used in film supercapacitors. Insoluble in water, soluble in HF, and in hot conc. H_2SO_4.	Niobium pentoxide (Columbite, niobia)

Table 9.3 (continued)

IUPAC name (synonyms, common trade names)	Theoretical chemical formula, [CASRN], relative molecular mass ($^{12}C = 12.000$)	Crystal system, lattice parameters, structure type, Strukturbericht, Pearson, space group, structure type (Z)	Density (ρ, kg.m^{-3})	Electrical resistivity (ρ_{el},$\mu\Omega$cm)	Dielectric permittivity at 1 MHz	Dielectric field strength (E, MV.m^{-1})	Dissipation or loss factor (tan δ)	Melting point (mp, °C)	Thermal conductivity (k, W.m^{-1}.K^{-1})	Specific heat capacity (cp, J.kg^{-1}.K^{-1})	Coeff. linear thermal expansion (α, 10^{-6}K^{-1})	Young's modulus (E, GPa)	Shear modulus (G, GPa)
Samarium oxide (Samaria)	Sm_2O_3 [12060-58-1] 348.72	Cubic D5$_3$, cI80, Ia$\bar{3}$, Mn$_2$O$_3$ type (Z = 16)	7620	n.a.	n.a.	n.a.	n.a.	2350	2.07	331	10.3	183	n.a.
Silicium dioxide (Silica, α-quartz)	α-SiO$_2$ [7631-86-9] [14808-60-7] 60.085	Trigonal (Rhombohedral) a = 491.27 pm c = 540.46 pm C8, hP9, R-3c, α-quartz type (Z = 3)	2202–2650	1×10^{20}	3.79	50	0.0002	1710	1.38	787	0.55	72.95	29.9
Tantalum pentoxide (Tantalite, tantala)	Ta_2O_5 [1314-61-0] 441.893	Trigonal (Rhomboedral) Columbite type	8200	1.0×10^{12}	n.a.	n.a.	n.a.	1882	n.a.	301.5	n.a.	n.a.	n.a.
Thorium dioxide (Thoria, thorianite)	ThO_2 [1314-20-1] 264.037	Cubic a = 559 pm Cl, cF12, Fm3m, fluorite type (Z = 4)	9860 n.a.	4×10^{19}	n.a.	n.a.	n.a.	3390	14.19	272.14	9.54	144.8	94.2
Titanium dioxide (Anatase)	TiO_2 [13463-67-7] [1317-70-0] 79.866	Tetragonal a = 378.5 pm c = 951.4 pm C5, tI12, I4$_1$ amd, anatase type (Z = 4)	3900 [3890]	n.a.	n.a.	n.a.	n.a.	700°C (Rutile)	n.a.	n.a.	n.a.	n.a.	n.a.
Titanium dioxide (Brookite)	TiO_2 [13463-67-7] 79.866	Orthorhombic a = 916.6 pm b = 543.6 pm c = 513.5 pm C21, oP24, Pbca, brookite type (Z = 8)	4140 [4000]	n.a.	n.a.	n.a.	n.a.	1750	n.a.	n.a.	n.a.	n.a.	n.a.
Titanium dioxide (Rutile, titania)	TiO_2 [13463-67-7] [1317-80-2] 79.866	Tetragonal a = 459.37 pm c = 296.18 pm C4, tP6, P4/mnm rutile type (Z = 2)	4240 [4250]	10^{19}	110–117	769	n.a.	1855	10.4 (// c) 7.4 ($\perp$ c)	711	7.14	248–282	111
Uranium dioxide (Uraninite)	UO_2 [1344-57-6] 270.028	Cubic a = 546.82 pm Cl, cF12, Fm3m, fluorite type (Z = 4)	10,960	3.8×10^{10}	n.a.	n.a.	n.a.	2880	10.04	234.31	11.2	145	74.2
Yttrium oxide (Yttria)	Y_2O_3 [1314-36-9] 225.81	Trigonal (Hexagonal) D5$_2$, hP5, P$\bar{3}$m1), lanthania type (Z = 1)	5030	n.a.	n.a.	n.a.	n.a.	2439	n.a.	439.62	8.10	114.5	48.3
Zirconium dioxide (Baddeleyite)	ZrO_2 [1314-23-4] [12036-23-6] 123.223	Monoclinic a = 514.54 pm b = 520.75 pm c = 531.07 pm 99.23° C43, mP12, P2$_1$ c, baddeleyite type (Z = 4)	5850	n.a.	n.a.	n.a.	n.a.	2710	n.a.	711	7.56	241	97

Poisson's ratio (ν)	Ultimate tensile strength (σ_{UTS}, MPa)	Flexural strength (τ, MPa)	Compressive strength (α, MPa)	Fracture toughness (K_{Ic}, MPa.m$^{1/2}$)	Vickers hardness (HV) (HM)	Other physico-chemical properties, corrosion resistance^a, and uses	IUPAC name (synonyms, common trade names)
n.a.	n.a.	n.a.	n.a.	n.a.	n.a.	438		Samarium oxide (Samaria)
0.170	69–276	310	680–1380	n.a.	550–1000 (HM 7)	Colorless amorphous (fused silica) or crystalline (quartz) material having a low thermal expansion coefficient and excellent optical transmittance in the far UV, Silica is insoluble in strong mineral acids and alkalis except HF, conc. H_3PO_4, NH_4HF_2, concentrated alkali metal hydroxides. Owing to its good corrosion resistance to liquid metals such as Si, Ge, Sn, Pb, Ga, In, Tl, Rb, Bi, and Cd, it is used as crucible containers for melting these metals. Silica is readily attacked under inert atmosphere by molten metals such as Li, Na, K Mg, and Al. Quartz crystals are piezoelectric and pyroelectric. Maximum service temperature 1090°C.	Silicium dioxide (Silica, α-quartz)
n.a.	n.a.	n.a.	n.a.	n.a.	n.a.	Dielectric used in film supercapacitors. Tantalum oxide is a high refractive index, and low absorption material usable for making optical coating in the near-UV (350 nm) to IR (8 μm). Insoluble in most chemicals except HF, HF–HNO_3 mixtures, oleum, fused alkali hydroxides (e.g., NaOH, KOH) and molten pyrosulfates.	Tantalum pentoxide (Tantalite, tantala)
0.280	96.5	n.a.	1475	n.a.	(HM 6.5) 945	Corrosion resistant container material for the following molten metals: Na, Hf, Ir, Ni, Mo, Mn, Th, U. Corroded by the following liquid metals: Be, Si, Ti, Zr, Nb, Bi. Radioactive.	Thorium dioxide (Thoria, thorianite)
n.a.	n.a.	n.a.	n.a.	n.a.	(HM 5.5–6)		Titanium dioxide (Anatase)
n.a.	n.a.	n.a.	n.a.	n.a.	(HM 5.5–6)		Titanium dioxide (Brookite)
.06	0.278	69–103	340	800–940	n.a.	(HM 7–7.5)	White and translucent hard ceramic material. Readily soluble in HF and inconcentrated H_2SO_4 and reacts rapidly in molten alkali hydroxides, and fused alkali carbonates. Owing to its good corrosion resistance to liquid metals such as Ni, and Mo, it is used as crucible containers for melting these metals. Titania is readily attacked under inert atmosphere by molten metals such as Be, Si, Ti, Zr, Nb, and Ta.	Titanium dioxide (Rutile, titania)
n.a.	0.302	n.a.	n.a.	n.a.	n.a.	600 (HM 6–7)	Used in nuclear power reactors as nuclear fuel sintered elements containing either natural or enriched uranium.	Uranium dioxide (Uraninite)
n.a.	0.186	n.a.	n.a.	393	n.a.	700	Yttria is a medium refractive index, low absorption material usable for optical coating in the near-UV (300 nm) to IR (12 μm) regions. Hence, used to protect Al and Ag mirrors. Used for crucibles containing molten lithium.	Yttrium oxide (Yttria)
.a.	0.337	n.a.	n.a.	2068	n.a.	(HM 6.5) 1200	Zirconia is highly corrosion resistant to molten metals such as Bi, Hf, Ir, Pt, Fe, Ni, Mo, Pu, and V, while is strongly attacked by the following liquid metals: Be, Li, Na, K, Si, Ti,Zr, and Nb. Insoluble in water, but slowly soluble in HCl, HNO_3, soluble in boiling concentrated H_2SO_4 and alkali hydroxides but readily attacked by HF. Monoclinic (baddeleyite) below 1100°C, tetragonal between 1100 and 2300°C, cubic (fluorine type) above 2300°C. Maximum service temperature 2400°C.	Zirconium dioxide (Baddeleyite)

9

Ceramics and Glasses

Table 9.3 (continued)

IUPAC name (synonyms, common trade names)	Theoretical chemical formula, [CASRN], relative molecular mass ($^{12}C = 12.000$)	Crystal system, lattice parameters, structure type, Strukturbericht, Pearson, space group, structure type (Z)	Density (ρ, kg.m^{-3})	Electrical resistivity (ρ_{el},$\mu\Omega$cm)	Dielectric permittivity at 1 MHz	Dielectric field strength (E, MV.m^{-1})	Dissipation or loss factor (tan ε)	Melting point (mp, °C)	Thermal conductivity (k, W.m^{-1}.K^{-1})	Specific heat capacity (cp, J.kg^{-1}.K^{-1})	Coeff. linear thermal expansion (α, 10^{-6}K^{-1})	Young's modulus (E, GPa)	Shear modulus (G, GPa)
Zirconium dioxide PSZ (stabilized with MgO) (Zirconia >2300°C)	ZrO$_2$ [1314-23-4] [64417-98-7] 123.223	Cubic Cl, cF12, Fm3m, fluorite type (Z = 4)	5800–6045	n.a.	24.7	400–480	n.a.	2710	1.8	400	10.1	200	n.a.
Zirconium dioxide TTZ (stabilized Y$_2$O$_3$) (Zirconia >2300°C)	ZrO$_2$ [1314-23-4] [64417-98-7] 123.223	Cubic Cl, cF12, Fm3m, fluorite type (Z = 4)	6045	n.a.	n.a.	n.a.	n.a.	2710	n.a.	n.a.	n.a.	n.a.	n.a.
Zirconium dioxide, TZP (Zirconia >1170°C)	ZrO$_2$ [1314-23-4] 123.223	Tetragonal C4, tP6 P4$_2$/mnm, rutile type (Z = 2)	5680–6050	7.7 × 10^7	n.a.	n.a.	n.a.	2710	n.a.	n.a.	10–11	200–210	n.a.
Zirconium dioxide (stabilized 10–15% Y$_2$O$_3$) (Zirconia >2300°C)	ZrO$_2$ [1314-23-4] [64417-98-7] 123.223	Cubic Cl, cF12, Fm3m, fluorite type (Z = 4)	6045	n.a.	n.a.	n.a.	n.a.	2710	n.a.	n.a.	n.a.	n.a.	n.a.

^{a}Corrosion data on molten salts from Geirnaert G (1970) Céramiques et mètaux liquides: Compatibilités et angles de mouillages. Bull Soc Fr Ceram 106: 7–50.

clear melting temperature and the structural change is only indicated by an inflection in the temperature–time curve. This change occurs at what is known as the **glass transition temperature**. As a general rule, glasses are amorphous inorganic solids usually made of silicates, but other inorganic or organic compounds can exhibit a vitreous structure (e.g., sulfides, polymers). As a general rule, commercial glasses are hard but both brittle and thermal-shock-sensitive materials, excellent electrical insulators, optically transparent media, and exhibit for certain particular chemical compositions (e.g., Vycor® and borosilicated glasses, such as Pyrex®) an excellent corrosion resistance to a wide range of chemicals, except hydrofluoric acid and strong alkali hydroxides. Owing to their good transmission in the visible range, glasses are extensively used for industrial and scientific lenses, spectacle lenses, and windows. Corrosion resistant glasses are widely used for cookware and laboratory glassware. The basic components of silicate glasses are silica, SiO$_2$ (e.g., from siliceous sand), lime, CaO (from fired limestone, CaCO$_3$), and soda, Na$_2$O (from soda ash, Na$_2$CO$_3$). Other oxides are used for special purposes, such as B$_2$O$_3$, K$_2$O, BaO, Li$_2$O. Colored glasses require minute additions of transition metal oxides (e.g., FeO, Co$_2$O$_3$).

Silicate glasses can be grouped into: (i) **A-glass** (high alkali or soda-lime), (ii) **C-glass** (chemical resistant), (iii) **E-glass** (calcium alumino borosilicate or borosilicated glasses), (iv) **S-glass** (high-strength magnesium alumino silicate).

9.2.2 Physical Properties of Glasses

Physical properties of selected common commercial glasses are shown in Table 9.6.

(…)	Poisson's ratio (ν)	Ultimate tensile strength (σ_{UTS}, MPa)	Flexural strength (τ, MPa)	Compressive strength (α, MPa)	Fracture toughness (K_{Ic}, MPa·m$^{1/2}$)	Vickers hardness (HV) (HM)	Other physico-chemical properties, corrosion resistance[a], and uses	IUPAC name (synonyms, common trade names)
n.a.	0.230	700	690	1850	9.5	1600		Zirconium dioxide PSZ (stabilized with MgO) (Zirconia >2300°C)
n.a.	n.a.	n.a.	n.a.	n.a.	n.a.	n.a.		Zirconium dioxide, TTZ (stabilized Y_2O_3) (Zirconia >2300°C)
n.a.	0.310	n.a.	800–	>2.900	7–12	n.a.		Zirconium dioxide, TZP (Zirconia >1170°C)
n.a.	n.a.	n.a.	n.a.	n.a.	n.a.	n.a.		Zirconium dioxide (stabilized 10–15% Y_2O_3) (Zirconia >2300°C)

9
Ceramics and Glasses

Further Reading

Ceramics and Refractories

Alper AM (1970) High temperature oxides. Material Science and Technology.

Aronsson B, Lundstrom T, Rundquist S (1965) Borides, silicides, and phosphides. Methuen, London.

Billups WE, Ciufolini MA (1993) Buckminsterfullerenes. VCH, Weinheim.

Blesa MA, Morando PJ, Regazzoni, AE (1994) Chemical dissolution of metal oxides. CRC Press, Boca Raton, FL.

Bradshaw WG, Matthews CO (1958) Properties of refractory materials: collected data and references. Lockheed Aircraft Corporation, Sunnyvale CA, US Government Report AD 205 452.

Brixner LH (1967) High temperature materials and technology. John Wiley, New York.

Freer R et al. (1989) The physics and chemistry of carbides, nitrides and borides. Kluwer, Boston.

Goodenough JB, Longo JM (1970) Crystallographic and magnetic properties of perowskite and perowskite related compounds. Springer-Verlag, Berlin.

Kosolapova TA (1971) Carbides, properties, productions and applications. Plenum Press, New York.

Levin EM, McMurdie HF, Hall FP (1956) Phase diagrams for ceramists, Part I. American Ceramic Society.

Levin EM, McMurdie HF, Hall FP (1956) Phase diagrams for ceramists, Part II. American Ceramic Society.

Matkovich, VI (ed.) (1977) Boron and refractory borides. Springer-Verlag, New York.

Samsonov, GV (1974) The oxides handbook. Plenum Press, New York.

Storms EK (1967) The refractory carbides. Academic Press, New York.

Toropov (ed.) (1974) Phase diagrams of silicates systems handbooks. Document NTIS AD 787517.

Toth LE (1971) Transition metals, carbides and nitrides. Academic Press, New York.

Glasses

Bach H, Neuroth N (1998) The properties of optical glass. Springer-Verlag, Heidelberg.

Table 9.6 Physical properties of selected commercial glasses

Glass trade name	Chemical composition (wt%)	Density (kg.m^{-3})	Young's modulus (GPa)	Poisson's ratio (ν)	Knoop hardness (HK)[a]	Thermal conductivity (k, W.m^{-1}.K^{-1})	Specific heat capacity (c$_p$, J.kg^{-1}.K^{-1})	Coefficient of linear thermal expansion (0–300°C) (10^{-6} K)
Corning® 0080	n.a.	2470	71	0.22	465	n.a.	n.a.	93.5
Corning® 7570	High-leaded glass	5420	56	0.28	n.a.	n.a.	n.a.	84
Float glass (soda lime glass)	74SiO$_2$–1Al$_2$O$_3$–CaC–15Na$_2$O–MgO	2530	72	0.23	n.a.	0.937	n.a.	89
Kimble® EG11	Leaded glass	2850	n.a.	n.a.	n.a.	n.a.	n.a.	108
Macor®	n.a.	2520	n.a.	n.a.	n.a.	1.46	n.a.	93
Pyrex® 0211	64SiO$_2$–3Al$_2$O$_3$–5.9CaO–7ZnO–9B$_2$O$_3$–7Na$_2$O–7K$_2$O–3TiO$_2$	2570	76	0.22	593	0.98	n.a.	73.8
Pyrex® 7059	n.a.	2760	n.a.	n.a.	n.a.	n.a.	n.a.	46
Pyrex® 7070	n.a.	2130	52	0.22	n.a.	n.a.	n.a.	32.0
Pyrex® 7740	80.6SiO$_2$–13B$_2$O$_3$–4Na$_2$O–2.3Al$_2$O$_3$–0.1K$_2$O	2230	76	0.20	418	1.13	n.a.	32.5
Pyrex® 7789	81SiO$_2$–13B$_2$O$_2$–3Na$_2$O–	2220	64.3	n.a.	n.a.	n.a.	n.a.	32.5
Pyrex® 7799	70SiO$_2$–10B$_2$O$_3$–9Na$_2$O–6Al$_2$O$_3$–2BaO–1K$_2$O–1CaO–0.5MgO–0.5ZnO	2470	n.a.	n.a.	n.a.	n.a.	n.a.	62
Pyrex® 7800	n.a.	2340	n.a.	n.a.	n.a.	n.a.	n.a.	n.a.
Pyrex® 7913 (Vycor®)	96.4SiO$_2$–3B$_2$O$_3$–0.5Al$_2$O$_3$	2180	89	0.19	487	0.19	n.a.	5.52–7.5
Pyrex® plus	n.a.	2302	n.a.	n.a.	438	n.a.	n.a.	57
Robax®	n.a.	2580	92	0.25	n.a.	1.6	n.a.	0.3
Sapphire glass	Fused Al$_2$O$_3$	3980	379	0.29	1500	16–23	n.a.	n.a.
Schott® BaK1	n.a.	3190	73	0.252	530	0.795	687	8.60
Schott® Bkl	n.a.	2460	74	0.210	560	1.069	825	8.80
Schott® BK7	n.a.	2510	82	0.206	610	1.114	858	8.30
Schott® FK3	n.a.	2270	46	0.243	380	0.90	840	9.40
Schott® FK5	n.a.	2450	62	0.232	520	0.925	808	10.00
Schott® FK51	n.a.	3730	81	0.293	430	0.911	636	15.30
Schott® FK52	n.a.	3640	78	0.291	400	0.861	716	16.00
Schott® FK54	n.a.	3180	76	0.286	390	n.a.	n.a.	16.50
Schott® K5	n.a.	2590	71	0.224	530	0.950	783	9.60
Schott® KF9	n.a.	2710	67	0.202	490	1.160	490	7.90
Schott® LaK9	n.a.	3510	110	0.285	700	0.908	649	7.50
Schott® LF5	n.a.	3220	59	0.223	450	0.866	657	10.60
Schott® PK3	n.a.	2590	84	0.209	640	1193	779	8.30
Schott® PK50	n.a.	2590	66	0.235	430	0.772	812	10.30
Schott® PSK3	n.a.	2910	84	0.226	630	0.990	682	7.30

	Annealing point (°C)[c]	Strain point (°C)[d]	Working point (°C)[e]	Continuous operating temperature (°C)	Refractive index (at 589.3 nm) (n_D)	Relative dielectric permittivity 1 MHz (ϵ^n)	Dielectric field strength (E_d, MV.m^{-1})	Loss factor (tanδ)	Electrical volume resistivity (Ωm)	Glass trade name
6	514	473	1005	110	1.512	7.2	n.a.	0.009	10^{12}	Corning® 0080
0	363	342	558	100	1860	15	n.a.	0.0022	10^{17}	Corning® 7570
6	546	514	n.a.	230	1523	n.a.	n.a.	n.a.	n.a.	Float glass (soda lime glass)
6	434	394	980	n.a.	1540	n.a.	n.a.	n.a.	10^9	Kimble® EG11
a.	n.a.	n.a.	n.a.	1000	n.a.	6.30	40	n.a.	10^{16}	Macor®
0	550	508	1008	n.a.	1523	6.7	n.a.	0.005	n.a.	Pyrex® 0211
4	639	593	n.a.	n.a.	1533	n.a.	n.a.	n.a.	n.a.	Pyrex® 7059
a	496	456	n.a.	230	1469	4.1	n.a.	0.006	10^{17}	Pyrex® 7070
1	560	510	1252	230	1474	4.6	n.a.	0.005	10^{17}	Pyrex® 7740
5	560	510	n.a.	n.a.	1474	n.a.	n.a.	n.a.	n.a.	Pyrex® 7789
0	560	525	n.a.	n.a.	n.a.	n.a.	n.a.	n.a.	n.a.	Pyrex® 7799
a.	n.a.	n.a.	n.a.	n.a.	n.a.	n.a.	n.a.	n.a.	n.a.	Pyrex® 7800
30	1020	890	n.a.	900	1458	3.8	n.a.	0.0015	10^{17}	Pyrex® 7913 (Vycor®)
6	502	467	n.a.	n.a.	1492	n.a.	n.a.	n.a.	n.a.	Pyrex® plus
a.	n.a.	n.a.	n.a.	680	n.a.	n.a.	n.a.	n.a.	n.a.	Robax®
a	n.a.	n.a.	n.a.	n.a.	n.a.	9–11	n.a.	48	10^{18}	Sapphire glass
a.	n.a.	n.a.	n.a.	n.a.	1.5725	n.a.	n.a.	n.a.	n.a.	Schott® BaK1
a.	n.a.	n.a.	n.a.	n.a.	15101	n.a.	n.a.	n.a.	n.a.	Schott® Bkl
a.	n.a.	n.a.	n.a.	n.a.	1.5168	n.a.	n.a.	n.a.	n.a.	Schott® BK7
a.	n.a.	n.a.	n.a.	n.a.	1.4650	n.a.	n.a.	n.a.	n.a.	Schott® FK3
a.	n.a.	n.a.	n.a.	n.a.	1.4875	n.a.	n.a.	n.a.	n.a.	Schott® FK5
a.	n.a.	n.a.	n.a.	n.a.	1.4866	n.a.	n.a.	n.a.	n.a.	Schott® FK51
a.	n.a.	n.a.	n.a.	n.a.	1.4861	n.a.	n.a.	n.a.	n.a.	Schott® FK52
a.	n.a.	n.a.	n.a.	n.a.	1.4370	n.a.	n.a.	n.a.	n.a.	Schott® FK54
a.	n.a.	n.a.	n.a.	n.a.	1.5225	n.a.	n.a.	n.a.	n.a.	Schott® K5
a.	n.a.	n.a.	n.a.	n.a.	1.5474	n.a.	n.a.	n.a.	n.a.	Schott® KF9
a.	n.a.	n.a	n.a.	n.a.	1.6910	n.a.	n.a.	n.a.	n.a.	Schott® LaK9
a.	n.a.	n.a.	n.a.	n.a.	1.5814	n.a.	n.a.	n.a.	n.a.	Schott® LF5
a.	n.a.	n.a.	n.a.	n.a.	1.5254	n.a.	n.a.	n.a.	n.a.	Schott® PK3
a.	n.a.	n.a.	n.a.	n.a.	1.5205	n.a.	n.a.	n.a.	n.a.	Schott® PK50
a.	n.a.	n.a.	n.a.	n.a.	1.5523	n.a.	n.a.	n.a.	n.a.	Schott® PSK3

9 Ceramics and Glasses

Table 9.6 (continued)

Glass trade name	Chemical composition (wt%)	Density (kg.m^{-3})	Young's modulus (GPa)	Poisson's ratio (ν)	Knoop hardness (HK)[a]	Thermal conductivity (k, W.m^{-1}.K^{-1})	Specific heat capacity (c$_p$, J.kg^{-1}.K^{-1})	Coefficient of linear thermal expansion (0–300°C) (10^{-6} K)
Schott® SF63	n.a.	4620	58	0.235	390	0.744	431	9.00
Schott® SK2	n.a.	3550	78	0.263	550	0.776	595	7.10
Schott® ZKN7	n.a.	2490	70	0.214	530	1042	770	5.20

[a]Microhardness 100 g load.
[b]For a dynamic viscosity of 10$^{8.6}$ Pa.s.
[c]For a dynamic viscosity of 10^{13} Pa.s.
[d]For a dynamic viscosity of 10$^{15.5}$ Pa.s.
[e]For a dynamic viscosity of 10^4 Pa.s.

Eitel W (ed.) (1964–1973) Silicate science, 6 volumes. Academic Press, New York.
Feltz A (1993) Amorphous inorganic materials and glasses. VCH, Weinheim.
Jones GO (1956) Glass. John Wiley, New York.
Morey, GW (1954) The properties of glasses, 2nd edn. Reinhold-Van Nostrand, New York.
Shand EB (1958) Glass engineering handbook. McGraw-Hill, New York.
Stanworth JE (1950) The physical properties of glasses. Clarendon Press, Oxford.
Zarzycky J (1981) Les verres et l'état vitreux. Masson, Paris.

Annealing point (°C)[c]	Strain point (°C)[d]	Working point (°C)[e]	Continuous operating temperature (°C)	Refractive index (at 589.3 nm) (n_D)	Relative dielectric permittivity 1 MHz (ϵ^n)	Dielectric field strength (E_d, MV·m^{-1})	Loss factor (tanδ)	Electrical volume resistivity (Ωm)	Glass trade name
n.a.	n.a.	n.a.	n.a	1.7484	n.a.	n.a.	n.a.	n.a.	Schott[R] SF63
n.a.	n.a.	n.a.	n.a.	1.6074	n.a.	n.a.	n.a.	n.a.	Schott[R] SK2
n.a.	n.a.	n.a.	n.a.	1.5085	n.a.	n.a.	n.a.	n.a.	Schott[R] ZKN7

9

Ceramics
and
Glasses

10 Polymers and Elastomers

Polymers are readily divided into three chief classes: (i) thermoplastics, (ii) thermosetting plastics or thermosets, and (iii) elastomers. The **thermoplastics** soften with increasing temperature and return to their original hardness when cooled. Most are meltable, e.g. nylon is extruded into fibers or filaments from the molten state. The **thermosets** harden when heated and retain hardness when cooled. They set into permanent shapes by catalysis or when heated under pressure. Generally, they cannot be reworked by scrap. **Elastomers** such as rubbers can withstand high deformation without rupture.

10.1 Thermoplastics

10.1.1 Polyolefins or Ethenic Polymers

These thermoplastics all have in common the same basic monomer structure of ethylene ($H_2C=CH_2$).

Polyethylene (PE) was first produced on a commercial basis in 1940 and it is prepared from petroleum or natural gas. Polyethylene is a thermoplastic material which varies from type to type according to the particular molecular structure of each type. Actually, several products can be made by varying the molecular weight (i.e., the chain length), the crystallinity (i.e., the chain orientation), and the branching characteristics (i.e., chemical bonds between adjacent chains). Polyethylene can be prepared in four commercial grades: (i) **low-density**, (ii) **medium-density**, (iii) **high-density** and (iv) **ultrahigh-molecular-weight** polyethylene.

Low-density polyethylene (**LDPE**) exhibits a melting point of 105°C, toughness, stress-cracking resistance, clarity, flexibility, and elongation. Hence, it is used extensively for piping and packaging, because of its ease of handling and fabrication. The chemical resistance of the product is outstanding, although not as good as high-density polyethylene or polypropylene but it is resistant to many strong mineral acids (e.g., HCl, HF) and alkalis (e.g., NaOH, KOH, NH_4OH), and it be used for handling most organic chemicals but alkanes, aromatic hydrocarbons, chlorinated hydrocarbons, and strong oxidants (e.g., HNO_3) must be avoided. Assembly of parts made of PE can be achieved by fusion welding of the material which is readily accomplished with appropriate equipment. For instance, installations of piping made in this manner are the least expensive and most durable of any material available for waste lines, water lines, and other miscellaneous services not subjected to high pressures or temperatures. Nevertheless several limitations prevent its use in some applications. These limitations are: a low modulus, low strength, and low heat resistance (the upper temperature limit for the material is 60°C) combined with degradation under UV irradiation (e.g., sunlight exposure). However, polyethylene can be compounded with a wide variety of materials to increase strength, rigidity, and other suitable mechanical properties. It is now available in a fiber-reinforced product to further increase the mechanical properties. Stress cracking can be a problem without careful selection of the basic resin used in the product or proper compounding to reduce this effect. Compounding of the product is also recommended to reduce the effect of atmospheric exposures over long periods.

High-density polyethylene (**HDPE**) has considerably improved mechanical properties, has better permeation barrier properties, and its chemical resistance is also greatly increased compared to the low-density grade. Only strong oxidants will attack the material appreciably within the appropriate temperature range. Stress cracking of HDPE can again be a problem if proper selection of the resin is not made. The better mechanical properties of these products extend their use into larger shapes, the application of the sheet materials on the interior of appropriately designed vessels, as packing in columns, and as solid containers to compete with glass and steel. Fusion welding can be achieved with a hot nitrogen gun.

Ultrahigh-molecular-weight polyethylene (**UHMWPE**) is a linear polyethylene with an average relative molecular mass ranging from 3×10^6 to 5×10^6. Its long linear chains provide great impact strength, wear resistance, toughness, and freedom from stress cracking in addition to the common properties of PE such as chemical inertness, self-lubrication, and low coefficient of friction. Therefore, this thermoplastic is suitable for applications requiring high wear/abrasion resistance for components used in machinery.

As a general rule, polyethylenes are highly sensitive to UV irradiation, especially sunlight exposure. Nevertheless, it is possible to avoid UV-light sensitivity by adding particular UV stabilizers.

Polypropylene (PP) – with the basic methyl-substituted ethylene (propylene) structure as monomer has considerably improved mechanical properties compared to polyethylene, actually it has a low density (900 to 915 $kg.m^{-3}$), it is stiffer, harder, and has a higher strength than many grades. Moreover, it can be used at higher temperatures than PE. Its chemical resistance is also greatly increased, and it is only attacked by strong oxidants. Stress cracking of PP can be a problem if proper selection of the resin is not made. In comparison with PE there exist few commercial grades but the plastic is stereospecific and can be isotactic and atactic. The better mechanical properties of these products extend their use into larger shapes, the application of the sheet materials on the interior of appropriately designed vessels, as packing in columns, and as solid containers to compete with glass and steel. The modulus of PP is somewhat higher than PE, which is beneficial in certain instances. The coefficient of thermal expansion is less for PP than for HDPE. Fusion welding with a hot nitrogen gun is practical in the field for both materials when the technique is learned. The two main applications of PP are injected molded parts and fibers and filaments.

Polybutylene (PB) is made from polyisobutylene, a distillation product of crude oil. Its monomer is ethylene with two methyl groups replacing two hydrogen atoms.

Polyvinyl chloride (PVC) was the first thermoplastic to be used in any quantity in industrial applications. It is prepared by reacting acetylene gas with hydrochloric acid in the presence of a suitable catalyst. PVC has grown steadily in favor over the years, primarily because of the ease of fabrication. It is easily worked and can be solvent welded or machined to accommodate fittings. It is very resistant to strong mineral acids and bases, and as a consequence, the material has been extensively used for over 40 years as piping for cold water and chemicals. However, in the design of a piping structure, the thermal coefficient of linear expansion must be taken into consideration, and the poor elastic modulus of the material must be considered. With these limitations, the product as a piping material can accommodate a wide range of products found in the chemical process industries. Two classes of the primary PVC material are available: normal grade and high-impact grade. The latter is normally used.

Chlorinated poly vinyl chloride (CPVC) – polyvinyl chloride can be modified through chlorination to obtain a vinyl chloride plastic with improved corrosion resistance and the ability to withstand operating temperatures that are 20 to 30°C higher. Hence, CPVC, which has about the same range of chemical resistance as rigid PVC, is extensively used as piping, fittings, ducts, tanks, and pumps for handling highly corrosive liquids and for hot water. For instance, it has been determined that its chemical resistance is satisfactory in comparison with PVC on exposure for 30 days in such environments as 20wt% acetic acid, 40–50wt% chromic acid, 60–70wt% nitric acid at 300°C, and 80wt% sulfuric acid, hexane at 50°C and 80wt% sodium hydroxide up to 80°C.

Polyvinylidene chloride (PVDC) or polyvinyl dichloride is based on a dichloroethylene monomer. It has superior chemical resistance and mechanical properties. It has better strength than common PVC. The material has an upper temperature limitation of 65°C for the normal (Type I) and 60°C for the high-impact (Type II) products. The chemical resistance is good in inorganic corrosive media with an outstanding resistance to oxidizing agents. However, contamination by solvents of almost all types must be avoided. This material has had a great significance in chemical industry applications over the years. The product has been made into a number of specific items designed to serve the chemical process industry. Among these are valves, pumps, piping, and liners, particularly on the inside of the pipe. PVDC was the first thermoplastic to be used for this purpose and found extensive and useful service as its trade name **Saran**[R]. Also, the material is available as a rigid or pliable sheet liner for application on the interior of vessels. It must be recognized that many modifications to the PVDC can be made. Fiber-reinforced products are also available.

Polyvinyl Acetate (PVA) is based on a monomer where an acetate group substitutes a hydrogen atom in the ethylene monomer. It is not used as structural polymer because it is a relatively soft thermoplastic and hence it is only used for coatings and adhesives.

Polystyrene (PS) is based on the styrene monomer –$C_6H_5CH=CH_2$– (phenylbenzene). Polystyrene is essentially a light amorphous and atactic thermoplastic. The aromatic ring contributes to the plastic stiffness and avoids chain displacement which renders the plastic brittle. The material is not recommended for applications involving corrosive chemicals because its chemical resistance by comparison with other available thermoplastics is poor and the material will stress crack in certain specific media. However, it has a high light transmission in the visible region, it has an excellent moldability rendering ease of fabrication is of low cost, meaning that it can always be considered if the properties are adequate for the use. Nevertheless, polystyrene is sensitive to UV irradiation (e.g., sunlight exposure) which gives a yellowish color to the material and the heat resistance of the material is only 65°C. The plastic will be encountered as casings for equipment and in various electrical applications. Fittings for piping have been made from the plastic, and many containers may be found made of modified polystyrene. Joining can be achieved by solvent welding of the product to fabricate devices but restricts its use to water and services not containing organic and inorganic chemicals. Polystyrene is the third most widely used thermoplastic today after PE and PP with 20% of the market.

Polymethyl penthene (PMP) is a plastic with good transparency, good electrical properties similar to those of fluorocarbons, and it can be used up to 150°C.

Acrylonitrile butadiene styrene (ABS) is a terpolymer with one monomer made from butadiene; the second acrylonitrile monomer consists of an ethylene molecule in which a hydrogen atom is replaced by a nitrile group (CN); the third monomer consists of an ethylene molecule with a phenyl group replacing a hydrogen atom (styrene). The material can be varied considerably in properties by changing the ratio of acrylonitrile to the other two components of the terpolymer. This offshoot of the original styrene resins has achieved a place in industrial work of considerable importance. Actually, the strength, toughness, dimensional stability, and

10
Polymers
and
Elastomers

other mechanical properties are improved at the expense of other properties. Although the material has poor heat resistance (90°C), a relatively low strength, and a restricted chemical resistance, the low price, ease of joining, and ease of fabrication make the material most attractive for distribution piping for gas, water, waste and vent lines, automotive parts, and numerous consumer service items ranging from the telephone to automobile parts. The plastic withstands attack by very few organic compounds, but is readily attacked by oxidizing agents and strong mineral acids. Moreover, stress cracking can occur in the presence of certain organic products.

Polymethyl metacrylate (PMMA) – the basic monomer has a methyl group (–CH$_3$) replacing one hydrogen atom on the ethylene molecule, while the second hydrogen atom on the same carbon location is replaced by an acetyl group (–CH$_3$COO). The polymer is obtained by esterification by reacting methylacrylic acid (CH$_2$=C(CH$_3$)COOCH$_3$) with an alcohol. The commercial product is well known under the common trade name **Plexiglas**$^®$. PMMA is a clear and rigid thermoplastic and it is readily formed by injection moulding. Main applications are guards, and covers.

Fluorinated polyolefins (fluorocarbons) represent certainly the most versatile and important group of thermoplastics for use in the chemical process industries (CPI). Most of these fluorinated polymers are able to handle without any corrosion extremely harsh environments and highly corrosive chemicals that only refractory, noble or precious metals or, in particular, ceramics can tolerate. Nevertheless, owing to the presence of the highly electronegative fluorine in the chain, the fluorocarbons are occasionally attacked by molten alkali metals such as sodium and lithium, forming graphite. As with other products, when such inertness is obtained in the material, certain other properties must be sacrificed. In this case, the fluorocarbon materials are more difficult to work in any manner and are much more limited in design and application than are other thermoplastic materials. The materials are porous and the permeation of specific chemicals must be considered to insure the proper selection of the proper fluorocarbon for the intended service. There are currently six types of commercial fluorocarbon thermoplastics. All have exceedingly good chemical stability, but there are differences which should be noted. These materials have been designed into a number of solid items for chemical service, such as impellers, mixers, spargers, packing, smaller containers, and a few more intricate shapes. However, the major use of fluorocarbons in the chemical process industry is as linings in steel or ductile iron. All shapes of lined pipe can be obtained. In addition, certain fluorocarbons can be cut and jointed in the field using appropriate tools. Lined pumps and valves are available.

Polytetrafluoroethylene (PTFE) – the basic monomer is a totally fluorinated ethylene molecule. It is well known under its common trade name **Teflon**$^®$. It remains the most difficult of the fluorocarbons to work. Actually, shapes and parts can be only made from sintering in order to produce it in usable form. However, owing to its high melting point (327°C) it has the highest temperature stability with a heat resistance up to 280°C, and it is one of the most chemically inert materials known apart from glass, refractory metals such as tantalum, and platinum group metals such as iridium or platinum for service in various severe corrosive chemicals even at high temperatures. Nevertheless, permeation is an issue depending on the specific exposure but sometimes no greater than many of the newer materials. Some problems associated with thermal cycling which can cause fatigue due to repeated expansion and contraction over a period of time when going through high temperatures were reported. Nevertheless, owing to their porosity, one particular mode of deterioration for fluorocarbons is the adsorption of a chemical, followed either by reaction with another component inside the thermoplastic or by polymerization of the product within the plastic. When this phenomenon occurs, it leads to surface degradation such as blistering. The material has also a definite heat limitation and overheating should be avoided. Cold flow of the resins is well known and implies that the design and use of the fluorocarbon should be such that excessive compressive stresses are not imposed to create a cold flow condition.

Polytrifluorochloroethylene (PTFCE) – this chlorotrifluoropolymer has heat stability up to 175°C, and a slightly lower chemical resistance than the totally fluorinated PTFE. It is well known under its common trade name **Kel-F**®. In addition, the working properties of the plastic are relatively good as it can be formed by injection molding, and consequently the material is found as a coating as well as a prefabricated liner for severe chemical applications.

Fluorinated ethylene propylene (FEP) – this fluorocarbon has a lower heat resistance of 200°C than PTFE but is more workable and is a clearer resinous product than the PTFE. Certain carefully prepared films of the FEP can be used as windows in equipment when necessary. The product has found extensive use as a pipe-fitting liner as well as a liner in smaller vessels.

Perfluorinated alkoxy (PFA) – this fluorocarbon has a good heat stability of 260°C near to that of PTFE but in addition it has a better creep resistance.

Polyvinylidene fluoride (PVDF) – this product has both less heat resistance (150°C) and chemical stability than other fluorocarbons. However, the material is much more workable and has been made into essentially any shape necessary to the chemical process industry. Complete pumps, valves, piping, smaller vessels, and other accoutrements have been made and have served successfully. The material may also be applied as a coating or as a liner.

10.1.2 Polyamides (PA)

Polyamide thermoplastics are prepared by condensation by reacting carboxylic acid (RCOOH) and an amine (R'NH$_2$) giving off water. These resins are well known under the common name **Nylon**®, one of the first resinous products to be used as an engineering material. Their excellent mechanical properties combined with their ease of fabrication has assured their continued growth for mechanical applications. Excellent strength, toughness, abrasion/wear resistance, and a high Young's modulus are the chief valuable properties of nylons and explained the important applications as mechanical parts in various operating equipment such as gears, electrical fittings, valves, fasteners, tubing, and wire coatings. Actually, some nylon grades have tensile properties comparable to that of the softer aluminum alloys. In addition, coatings and structural items can be obtained. The heat resistance of nylon can be varied, but must be considered in the range of 100°C. The chemical resistance is remarkably good for a thermoplastic, the most notable exception being the poor resistance to strong mineral acids. Moreover, stress corrosion cracking of nylon parts can occur, particularly when in contact with acids and alkaline solutions. Owing to the wide diversity of different additives or copolymers as starting materials, there are several commercial grades of nylon resins available, each of them with particular properties. The main grades are Nylon® and Nylon® 66, these two grades having the highest strength. More recently, new commercial grades of nylon resins were developed in order to overcome limitations of the previous common nylon grades. These products consist of polyamides that contain an aromatic functional group in their monomer, and are hence called aramid resins (**ar**omatic **amid**es) They are well known under the trade names **Kevlar**® and **Nomex**®.

10.1.3 Polyacetals

Polyacetals under the common trade name **Delrin**® differ from other polymers due to the presence of an oxygen heteroatom in their monomer giving a heterochain polymer. The basic polymer unit is usually formaldehyde. The excellent dimensional stability and toughness of the acetal resins recommends their use for gears, pump impellers, other types of threaded connections such as plugs, and mechanical parts. The material has an upper useful temperature limitation of ~105°C. The chemical resistance indicated in the literature shows a wide range of tolerance for various inorganic and organic products. As with many other resins, this formaldehyde polymer will not withstand strong acids, strong alkalis, or oxidizing media.

10.1.4 Cellulosics

The major cellulosic derivatives of interest in polymers are the acetate, butyrate, and propionate thermoplastics. These do not constitute any major use but are encountered daily in a number of smaller items such as name plates, electrical component cases, high-impact lenses, and other applications requiring a transparent plastic with good impact resistance. Weathering properties of the materials are good, particularly that of propionate, but overall chemical resistance is not comparable to other thermoplastics. Water and salt solutions are readily handled, but any appreciable quantity of acid, alkali, or solvent can have an adverse effect on the plastic. The upper temperature of usefulness is 60°C.

10.1.5 Polycarbonates (PC)

Polycarbonate is prepared by reacting bisphenol A and phosgene or by reacting a polyphenol with dichloromethane and phosgene. The basic monomer unit is $-OC_6H_4C(CH_3)_2C_6H_4COO-$. Polycarbonate is a linear, low-crystalline, transparent, high-molecular-mass thermoplastic commonly known under the commercial trade name **Lexan**®. It exhibits a good chemical resistance to greases and oils but has a poor organic solvent resistance. Moreover, it is greatly restricted in its resistance by a severe propensity to stress crack. This property can be modified greatly by proper compounding but remains the most serious problem when considering polycarbonate for chemical exposure. The exceedingly high impact resistance of this thermoplastic (30 times that of safety glass) combined with high electrical resistivity, ease of fabrication, fire resistance, and good light transmission (90%) has promoted its use into a wide range of industrial applications. The most notable of these for industrial applications is the use of the sheet material as a glazing product. Where a high-impact, durable, transparent shield is required, polycarbonate materials are used extensively. In addition, many smaller mechanical parts for machinery, particularly those with very intricate molding requirements, impellers in pumps, safety helmets, and other applications requiring light weight and high impact resistance have been satisfied by the use of polycarbonate plastics. The material can be used from –170°C up to a temperature of 121°C.

10.1.6 Polyimides (PI)

These plastics offer the most unique combinations of properties available for use in industrial service. The plastics are usable from –190°C to +370°C. These excellent low-temperature properties are often overlooked where a plastic is required to retain some ductility and toughness at such low temperatures. Some combinations of the resins can be taken to 510°C for short periods without destroying the parts. The plastic has excellent creep and abrasion resistance, excellent elastic modulus for a thermoplastic, and good tensile strength that does not drop off rapidly with temperature. The chemical resistance must be rated as good.

10.1.7 Polysulfone (PSU)

Polysulfone plastics have added another dimension to the thermoplastic field in heat resistance and strength at high temperature. The ease of molding the material, and its retention of properties as temperatures increase, has made it one of the faster growing resins in the market place. Use of the product for chemical equipment applications has not been noted to date but is anticipated. Polysulfone has specific resistance to a large variety of acids, alkalis, and oxidants. However, certain solvents do attack the resin. In addition, stress cracking can be a problem and should be considered in any application of this thermoplastic.

10.1.8 Polyphenylene Oxide (PPO)

This thermoplastic has excellent heat and dimensional stability, and satisfactory chemical resistance. The material may be found primarily in pump parts and certain other applications where impact strength, good modulus, and reasonable abrasion resistance are required. The chemical resistance of the material is good and the allowable temperature limit of 120°C, under appropriate conditions, extends the attractiveness for use of the product. The cost requires specific need for an identifiable application before choosing the product over many less costly thermoplastics.

10.1.9 Polyphenylene Sulfide (PPS)

Chemical resistance is outstanding and the temperature usefulness ranges from −170°C to +190°C. Coatings prepared from the resin are available. Considerable strength with high elastic modulus can be obtained by the addition of glass or other fillers to the material.

10.1.10 Acrylic

The good atmospheric stability and clarity of these resins have made their use as high-impact window panes and other seethrough barriers important in industry. Various modifications are being made to alter the properties of the basic acrylic resins for specific services. However, these are not found in any great use in the industrial area to date. Upper temperature of usefulness is approximately 90°C. The loss in light transmission is only 1% after five years exposure in locations subject to extensive sunlight exposure.

10.2 Thermosets

10.2.1 Aminoplastics

These thermosetting polymers are represented by urea formaldehyde and melamine formaldehyde.

10.2.2 Phenolics

Amongst the important thermosetting materials are phenolic plastics filled with asbestos, carbon or graphite, and silica. Relatively low cost, good mechanical properties, and chemical resistance except to strong alkalis make phenolics popular for chemical equipment.

10.2.3 Polyurethanes (PUR)

Polyurethanes are thermosets prepared by a condensation reaction involving diisocyanate (e.g., toluene diisocyanate, polymethylene diphenylene diisocyanate) with an appropriate polyol. The polymers can be used in several forms such as flexible and rigid foams, elastomers, and liquid resin. The polyurethanes exhibit low corrosion resistance to strong acids and alkalis, and to organic solvents. Flexible foams are extensively used for domestic applications (e.g., bedding and packaging), while rigid foams are used as thermal insulation material for transportation of cryogenic fluids and cold food products.

10.2.4 Furane Plastics

They are more expensive than the phenolics but also offer somewhat higher tensile strengths. Furane plastics, filled with asbestos, have much better alkali resistance than phenolic asbestos. Some special materials in this class, based on bisphenol, are more alkali resistant. Temperature limit for polyesters is about 80°C.

10.2.5 Epoxy Resins (EP)

Glycidal ether-based epoxies represent perhaps the best combination of corrosion resistance and mechanical properties. Epoxies reinforced with fiberglass have very high strengths and resistance to heat. Chemical resistance of the epoxy resin is excellent in non-oxidizing and weak acids but not good against strong acids. Alkali resistance is excellent in weak solutions. Chemical resistance of epoxy–glass laminates may be affected by any exposed glass in the laminate. Epoxies are available as castings, extrusions, sheet, adhesives, and coatings. They are used as pipes, valves, pumps, small tanks, containers, sinks, bench tops, linings, protective coatings, insulation, adhesives, and dies for forming metal.

10.3 Rubbers and Elastomers

Rubbers and elastomers are widely used as lining materials for columns, vessels, tanks, and piping. The chemical resistance depends on the type of rubber and its compounding. A number of synthetic rubbers have been developed to meet the demands of the chemical industry. Although none of these has all the properties of natural rubber, they are superior in one or more ways. *Trans*-polyisoprene and *cis*-polybutadiene synthetic rubbers are close duplicates of natural rubber. A variety of rubbers and elastomers have been developed for specific uses.

10.3.1 Natural Rubber (NR)

Natural rubber or *cis*-1,4-polyisoprene has as basic monomer unit a *cis*-1,4-isoprene (it is sometimes called caoutchouc). Natural rubber is made by processing the sap of the rubber tree (*Hevea brasiliensis*) with steam, and compounding it with vulcanizing agents, antioxidants, and fillers. Desired colors can be obtained by incorporation of suitable pigments (e.g., red: iron oxide, Fe_2O_3, black: carbon black, and white: zinc oxide, ZnO). Natural rubber has good dielectric properties, an excellent resilience, a high damping capacity and a good tear resistance. As a general rule, natural rubbers are chemically resistant to non-oxidizing dilute mineral acids, alkalis, and salts. However, they are readily attacked by oxidizing chemicals, atmospheric oxygen, ozone, oils, benzene, and ketones and mostly they have also poor chemical resistance to petroleum and its derivatives and many organic chemicals, in which the material softens. Moreover, natural rubbers are highly sensitive to UV-irradiation (e.g., sunlight exposure). Hence, natural rubber is a general-purpose material for applications requiring abrasion/wear resistance, electrical resistance, and damping or shock absorbing properties. Nevertheless, due to its mechanical limitation natural rubber, as well as many of the synthetic rubbers, is converted into a more hard and stable product by vulcanization and compounding with additives. The **vulcanization process** consists of mixing crude natural or synthetic rubber with 25wt% sulfur and heating the blend at 150°C in a steel mold. The resulting rubber material is harder and stronger than the previous raw material due to the cross-linking reaction between adjacent carbon chains. Therefore, industrial applications of vulcanized natural rubber include components such as: internal lining for pumps, valves, piping, hoses, and for machined components. However, because natural rubber has a low chemical resistance and is sensitive to sunlight, unsuitable properties in many industrial applications, it is today replaced by newer improved elastomers.

10.3.2 *Trans*-Polyisoprene Rubber (PIR)

Trans-1,4-polyisoprene rubber (sometimes called gutta percha in the past) is a synthetic rubber with similar properties to its natural counterpart. It was first industrially prepared during World War II due to the lack of supply of natural rubber. Although it contains fewer impurities than natural rubbers and the preparation process is more simple, it is not widely used because it has a higher cost. Mechanical properties and chemical resistance are identical to that of natural rubber. Like many other rubbers its mechanical properties can be also improved by the vulcanization process.

10.3.3 Polybutadiene Rubber (BR)

Polybutadiene rubber is similar to natural rubber in its properties but it is more costly to process into intricate shapes than rubbers such as styrene butadiene rubber. Hence, it is essentially used as additive in order to increase tear resistance of other rubbers.

10.3.4 Styrene Butadiene Rubber (SBR)

Styrene butadiene rubber is a copolymer of styrene and butadiene. It is well known under the common trade name **Buna**[R]**S**. Its chemical resistance is similar to that of natural rubber; that is a poor resistance to oxidizing media, hydrocarbons and mineral oils. Hence, it offers no particular advantages in chemical service in comparison with other rubbers. SBR is a general-purpose material, used in automobile tires, belts, gaskets, hoses, and other miscellaneous products.

10.3.5 Nitrile Rubber (NR)

Nitrile rubber is a copolymer of butadiene and acrylonitrile. It is produced in different ratios varying from 25:75 to 75:25. The manufacturer's designation should identify the percentage of acrylonitrile. Nitrile rubber under the common trade name **Buna**[R]**N** is well known for its excellent resistance to oils and solvents owing to its resistance to swelling when immersed in mineral oils. Moreover, its chemical resistance to oils is proportional to the acrylonitrile content. However, it is not resistant to strong oxidizing chemicals such as nitric acid, and it exhibits fair resistance to ozone and to UV-irradiation, which severely embrittles it at low temperatures. Nitrile rubber is used for gasoline hoses, fuel pump diaphragms, gaskets, seals and packings (e.g. O-rings), and finally oil-resistant soles for safety work shoes.

10.3.6 Butyl Rubber (IIR)

Butyl rubber is a copolymer of isobutylene and isoprene. Butyl rubber is chemically resistant to non-oxidizing dilute mineral acids, salts, and alkalis, and has a good chemical resistance to concentrated acids, except sulfuric and nitric acids. Moreover, it has a low permeability to air and an excellent resistance to aging and ozone. However, it is readily attacked by oxidizing chemicals, oils, benzene, and ketones, and as a general rule it has also poor chemical resistance to petroleum and its derivatives and many organic chemicals. Moreover, butyl rubber is sensitive to UV-irradiation (e.g., sunlight exposure). Like other rubbers, its mechanical properties can be largely improved by the vulcanization process. Industrial applications are the same as for natural rubber. Butyl rubber is used for tire inner tubes and hoses.

10.3.7 Chloroprene Rubber (CPR)

Polychloroprene is a chlorinated rubber material well known under its common trade name **Neoprene**[R] or grade M. This elastomer is an extremely versatile synthetic rubber with nearly 70 years of proven performance in a broad industry spectrum. It was the first commercial synthetic rubber originally developed in 1930s as an oil-resistant substitute for natural rubber. The polymer structure can be modified by copolymerizing chloroprene with sulfur and/or 2,3-dichloro-1,3-butadiene to yield a family of materials with a broad range of chemical and physical properties. By proper raw material selection and formulation of these polymers, the compounder can achieve optimum performance for a given end-use. Initially developed for resistance to oils and solvents, it may resist various organic chemicals including mineral oils, gasoline, and some aromatic or halogenated solvents. It also exhibits good chemical resistance to aging and attack by ozone, and good resistance to UV-irradiation (e.g., exposure to sunlight), up to moderately elevated temperatures. Moreover, it has outstanding resistance to damage caused by flexing and twisting, high toughness, and it resists burning, but its electrical properties are inferior to that of natural rubber. Therefore, neoprene is noted for a unique combination of properties which has led to its use in thousands of applications throughout industry. It is extensively used as wire and cable jacketing, hoses, tubes, and covers. In the automotive industry, neoprene serves as gaskets, seals, boots, air springs, and power transmission belts, molded and extruded goods, cellular products, adhesives and sealants, both solvent- and water-based, foamed wet suits, latex-dipped goods (e.g., gloves, balloons), paper, and industrial binders (e.g., shoe boards). In civil engineering and construction applications, neoprene is used for bridge pads/seals, soil pipe gaskets, waterproof membranes, and asphalt modification.

10.3.8 Chlorosulfonated Polyethylene (CSM)

Chlorosufonated polyethylene (CSM) is well known under its common trade name **Hypalon**[R]. It is prepared by reacting polyethylene with sulfur dioxide and chlorine. This elastomer has outstanding chemical resistance to oxidizing environments including ozone, but it is readily attacked by fuming nitric and sulfuric acids. It is oil resistant but it has poor resistance to aromatic solvents and most fuels. Except for its excellent resistance to oxidizing media, its physical and chemical properties are similar to that of neoprene with, however, improved resistance to abrasion, heat, and weathering.

10.3.9 Polysulfide Rubber (PSR)

Usually known under the trade name **Thiokol**[R].

10.3.10 Ethylene Propylene Rubber

Ethylene propylene rubber (EPR or EPDM) is a copolymer of ethylene and propylene. It has much of the chemical resistance of the related plastics, excellent resistance to heat and oxidation, and good resistance to steam and hot water. It is used as standard lining material for steam hoses and is widely used in chemical services as well, having a broad spectrum of resistance.

10.3.11 Silicone Rubber

Polysiloxanes are inorganic polymeric materials well known under the common name of silicone rubbers. Instead of the classic carbon chain skeleton, this particular class of polymers is based on a structure of silicon to oxygen bonds (Si–O) similar to that found in silicates. They have outstanding temperature resistance over an unusually wide temperature range (e.g., −75 to +200°C). Silicones have a relatively poor chemical resistance to aromatic oils, fuels, high-pressure steam, but withstand aging and ozone, as well as aliphatic solvents, oils and greases. They have poor resistance to abrasion.

10.3.12 Fluorelastomers

Fluoroelastomers combine excellent chemical resistance (e.g. oxidizing acids and alkalis) and high-temperature resistance (up to 275–300°C for short periods of time); excellent oxidation resistance; good resistance to fuels containing up to 30% aromatics; mostly poor resistance in solvents or organic media by contrast with fluorinated plastics.

Viton[R] **fluoroelastomers** – there are three major general-use families of Viton[R] fluoroelastomers: A, B, and F. They differ primarily in their resistance to fluids, and in particular aggressive lubricating oils and oxygenated fuels, such as methanol and ethanol automotive fuel blends. There is a full range of Viton[R] grades that accommodate various manufacturing processes including transfer and injection molding, extrusion, compression molding, and calendering. There is also a class of high-performance Viton[R] grades such as GB, GBL, GF, GLT, and GFLT.

Viton[R] A is a family of fluoroelastomer dipolymers, that is they are polymerized from two monomers, vinylidene fluoride (VF2) and hexafluoropropylene (HFP). Viton[R] A fluoroelastomers are general-purpose types that are suitable for general molded goods such as o-rings and v-rings, gaskets, and other simple and complex shapes.

Viton[R] B is a family of fluoroelastomer terpolymers, that is they are polymerized from three monomers, vinylidene fluoride (VF2), hexafluoropropylene (HFP), and tetrafluoroethylene (TFE). Viton[R] B fluoroelastomers offer better fluid resistance than A type fluoroelastomer.

Viton[R] F is a family of fluoroelastomer terpolymers, that is they are polymerized from three monomers, vinylidene fluoride (VF2), hexafluoropropylene (HFP), and tetrafluoroethylene (TFE). Viton[R] F fluoroelastomers offer the best fluid resistance of all Viton types. F types are particularly useful in applications requiring resistance to fuel permeation.

Viton[R] GBL is a family of fluoroelastomer terpolymers, that is they are polymerized from three monomers, vinylidene fluoride (VF2), hexafluoropropylene (HFP), and tetrafluoroethylene (TFE). Viton[R] GBL uses peroxide cure chemistry that results in superior resistance to steam, acid, and aggressive engine oils.

Viton[R] GLT is a fluoroelastomer designed to retain the high heat and the chemical resistance of general-use grades of Viton fluoroelastomer, while improving the low-temperature flexibility of the material. Viton[R] GLT shows a glass transition temperature 8 to 12°C lower than general-use Viton grades.

Viton[R] GFLT is a fluoroelastomer designed to retain the high heat and the superior chemical resistance of the GF high-performance types, while improving the low-temperature performance of the material. Viton[R] GFLT shows a glass transition temperature 6 to 10°C lower than general-use Viton grades.

10.4 Polymer Physical Properties

Physical properties of common polymers and elastomers are reported in Table 10.1, while physical quantities commonly used in the table to describe polymers characteristics are listed in

Table 10.1 Properties of thermoplastics (TP), thermosets (TS), and elastomers (E)

Usual chemical name	Trade name	Acronym, abbreviation or symbol	Category	Density (ρ, kg.m^{-3})	Tensile modulus (E, GPa)	Flexural modulus (G, GPa)	Compressive modulus (K, GPa)	Poisson's ratio (ν)	Yield tensile strength (σ_{YS}, MPa)	Ultimate tensile Strength (σ_{UTS}, MPa)	Elongation at break (Z, %)	Ultimate compressive strength (σ_{UCS}, MPa)	Flexural yield strength (MPa)	Notched izod impact Energy per unit width (J.m^{-1})	Rockwell hardness (or Shore SHD)	Static friction Coefficient (μ/nil)	Wear resistance (i.e., weight loss per 1000 cycles) (mg)	Minimum operating temperature range (°C)	Maximum operating temperature range (°C)
Acrylonitrile butadiene styrene	Cycolac®	ABS	TP	1040–1180	1.7–2.6	0.92–3.03	1.03–2.90	n.a.	32–45	41–62	20–100	36–69	28–97	105–440	R75–115	0.5	n.a.	n.a.	70–110
Butyl rubber	Kalar®, GR-1	IIR	E	917	0.3–3.4	n.a.	n.a.	n.a.	n.a.	17–21	700–950	n.a.	n.a.	n.a.	SHA30–100	n.a.	n.a.	–45	150
Cellulose acetate		CA	TP	1270–1340	1.0–4.0	8.3–27.6	n.a.	n.a.	17–43	12–110	6–70	20–55	14–110	100–450	R34–125	n.a.	65	–20	55–95
Cellulose acetate-butyrate		CAB	TP	1150–1220	0.3–2.0	0.62–4.14	n.a.	n.a.	10.3–48.3	20–60	38–74	14.5–52	12.4–110	260	R31–99	n.a.	n.a.	–40	60–100
Cellulose acetate-propionate		CAP	TP	1150–1220	0.34–1.38	0.69–1.93	n.a.	n.a.	10.3–48.3	13.8–51.7	38–74	21–70	21–75	182	R20–120	n.a.	n.a.	n.a.	60–105
Chlorinated polyvinyl chloride		CPVC	TP	1490–1500	2.48–3.0	2.6–3.15	2.3–4.14	n.a.	n.a.	n.a.	n.a.	42–75	53–299	n.a.	R117	n.a.	n.a.	n.a.	110
Chlorosulfinated polyethylene	Hypalon®	CSM	E	n.a.	n.a.	n.a.	n.a.	n.a.	n.a.	21	600	n.a	n.a.	n.a.	SHA 40–90	n.a.	n.a.	n.a.	n.a.
Epichloridrin rubber		ECO	E	1270	n.a.	n.a.	n.a.	n.a.	n.a.	17	400	n.a.	n.a.	n.a.	SHA 60–90	n.a.	n.a.	–46	121
Epoxy resin	Novalac®	n.a.	TS	1120–1180	1.5–3.6	n.a.	n.a.	n.a.	69–121	n.a.	n.a.	n.a.	0.4–0.9	n.a.	n.a.	n.a.	n.a.	n.a	200–260
Ethylene propylene diene rubber	Dutral®, Nordel®	EPDM	E	850	n.a.	n.a.	n.a.	n.a.	n.a.	21	100–300	n.a.	n.a.	n.a.	SHA 30–90	n.a.	n.a.	–51	150
Ethylene tetrafluoroethylene	Tefzel®, Halon®	ETFE	TP	1700	1.4	1.2–1.4	n.a.	n.a.	45	44.85	300	n.a.	38	1000	SHD 67–75	0.4	n.a.	0	150
Ethylene-propylene rubber		EBR	E	n.a.	n.a.	n.a.	n.a.	n.a.	n.a.	n.a.	n.a.	n.a.	n.a.	n.a.	n.a.	n.a.	n.a.	n.a.	n.a.
Ethylene chlorotrifluoroethylene	Halar®	ECTFE	TP	1680	1.7	1.7	1.7	n.a.	n.a.	31–48	200–300	n.a.	48	nil	R93	n.a.	n.a.	n.a.	150
Fluorinated ethylene propylene	Neoflon®	FEP	TP	2150	n.a.	0.62	n.a.	n.a.	n.a.	23	325	21	18	nil	SHD 50–65	0.27	n.a.	n.a.	204
Melamine formaldehyde		MF	TS	1500	7.6–10	n.a.	n.a.	n.a.	n.a.	36–90	n.a.	n.a.	n.a.	11–21	M115–125	n.a.	n.a.	n.a.	120–200
Natural rubber (cis-1, 4-polyisoprene)	Caoutchouc	NR	E	920–1037	3.3–5.9	n.a.	2.2	0.4995	17.1–31.7	29	660–850	n.a.	n.a.	n.a.	SHA 30–95	n.a.	n.a.	–56	82

Table 10.1 (continued)

Usual chemical name	Vicat softening temperature (°C)	Glass transition temperature (T_g, °C)	Melting point or range (mp, °C)	Specific heat capacity (c_p, J.kg^{-1}.K^{-1})	Thermal conductivity (k, W.m^{-1}.K^{-1})	Coefficient of linear thermal expansion (α, 10^{-6}K^{-1})	Deflection temperature under 0.455 MPa flexural load (T, °C)	Heat deflection temperature under 1.82 MPa flexural load (T, °C)	Electrical resistivity (ρ, Ω.cm)	Relative electrical permittivity (@ 1MHz) ($\varepsilon\rho$)	Dielectric field strength (E_d, kV.cm^{-1})	Loss factor	Refractive index (n_D)	Transmittance (T, %)	Water absorption per 24 hours (wt% day^{-1})	Water absorption at saturation (wt%)	Flame rating ASTM UL94
Acrylonitrile butadiene styrene	n.a.	n.a.	88–120	1506	0.17–0.34	53–110	77–98	99–112	1.E+15	2.4–3.3	140–250	0.0200	1.49	92	0.3–0.7	n.a.	HB
Butyl rubber	n.a.	−75 to −67	n.a.	1950	0.13–0.23	n.a.	n.a.	n.a.	n.a.	n.a.	n.a.	n.a.	1.5081	n.a.	n.a	n.a.	n.a.
Cellulose acetate	n.a.	n.a.	230	1200–1900	0.16–0.36	80–180	52–105	73	1.E+12	5	110	0.0600	1.49	n.a.	1.9–7.0	n.a.	comb
Cellulose acetate-butyrate	n.a.	n.a.	140	1464	0.16–0.32	140	73	62	1.E+11	2.5–6.2	100	0.0400	1.478	n.a.	0.9–2.2	n.a.	comb
Cellulose acetate-propionate	n.a.	n.a.	n.a.	1200–1600	0.16–0.33	120–160	n.a.	n.a.	1.E+11	n.a.	n.a.	n.a.	1.478	n.a.	1.0–3.0	n.a.	comb
Chlorinated polyvinyl chloride	n.a.	n.a.	110	n.a.	0.14	68–78	n.a	n.a.	1.E+15	3.3–3.8	480–590	0.0019	n.a.	n.a.	0.1	n.a.	n.a.
Chlorosulfinated polyethylene	n.a.	n.a.	n.a.	n.a.	n.a.	n.a.	n.a.	n.a.	n.a.	n.a.	n.a.	n.a.	n.a.	n.a.	n.a.	n.a.	n.a.
Epichloridrin rubber	n.a.	n.a.	n.a.	n.a.	n.a.	n.a.	n.a.	n.a.	n.a.	n.a.	n.a.	n.a.	n.a.	n.a.	n.a.	n.a.	n.a.
Epoxy resin	n.a.	n.a.	n.a.	n.a.	n.a.	n.a.	n.a.	230–260	n.a.	n.a.	n.a.	n.a.	n.a.	n.a.	n.a.	n.a.	n.a.
Ethylene propylene diene rubber	n.a.	n.a.	n.a.	n.a.	2.22	n.a.	n.a.	n.a.	n.a.	2.5	n.a.	n.a.	1.474	n.a.	n.a.	n.a.	n.a.
Ethylene tetrafluoroethylene	n.a.	n.a.	270	n.a.	n.a.	90	105	70	1.E+17	2.6	800	0.0050	1.4028	n.a.	0.03	n.a.	V0
Ethylene-propylene rubber	n.a.	n.a.	n.a.	n.a.	n.a.	n.a.	n.a.	n.a.	n.a.	n.a.	n.a.	n.a.	n.a.	n.a.	n.a.	n.a.	n.a.
Ethylene chlorotrifluoroethylene	n.a.	n.a.	245	n.a.	0.16	80	115	77	n.a.	2.6	190	n.a.	n.a.	n.a.	0.01	n.a.	V0
Fluorinated ethylene propylene	n.a.	n.a.	260	n.a.	n.a.	1135	n.a.	n.a.	1.E+18	2.1	110–160	0.0007	n.a.	n.a.	0.01	n.a.	n.a.
Melamine formaldehyde	n.a.	n.a.	n.a.	1674	0.42	22	n.a.	183	n.a.	n.a.	n.a.	n.a.	n.a.	n.a.	0.1	n.a.	n.a.
Natural rubber (cis-1, 4-polyisoprene)	n.a.	−70	n.a.	1830	0.15	n.a.	n.a.	n.a.	n.a.	2.6	n.a.	n.a.	n.a.	n.a.	n.a.	n.a.	n.a.

10

Polymers and Elastomers

Table 10.1 (continued)

Usual chemical name	Trade name	Acronym, abbreviation or symbol	Category	Density (ρ, kg·m^{-3})	Tensile modulus (E, GPa)	Flexural modulus (G, GPa)	Compressive modulus (K, GPa)	Poisson's ratio (ν)	Yield tensile strength (σ_{YS}, MPa)	Ultimate tensile Strength (σ_{UTS}, MPa)	Elongation at break (Z, %)	Ultimate compressive strength (σ_{UCS}, MPa)	Flexural yield strength (MPa)	Notched izod impact Energy per unit width (J·m^{-1})	Rockwell hardness (or Shore SHD)	Static friction Coefficient (μ/nil)	Wear resistance (i.e., weight loss per 1000 cycles) (mg)	Minimum operating temperature range (°C)	Maximum operating temperature range (°C)
Butadiene acrylonitrile rubber	Buna® N, Nyrek®	NBR	E	1000	n.a.	n.a.	n.a.	n.a.	n.a.	21	510	100–600	n.a.	n.a.	SHA 30–90	n.a.	n.a.	−40	121
Perfluorinated alkoxy		PFA	TP	2140–2150	0.66	0.66	n.a.	n.a.	n.a.	21–29	300	n.a.	n.a.	nil	SHD 60	0.2	n.a.	n.a.	260
Phenol formaldehyde		PF	TS	1360	n.a.	n.a.	n.a.	n.a.	n.a.	n.a.	n.a.	n.a.	n.a.	n.a.	n.a.	n.a.	n.a.	n.a.	n.a.
Polyacrylic butadiene rubber		ABR	E	n.a.	n.a.	n.a.	n.a.	n.a.	n.a.	n.a.	n.a.	n.a.	n.a.	n.a.	n.a.	n.a.	n.a.	n.a.	n.a.
Polyamide-imide	Torlon®, Ultem®	PAI	TP	1420–1460	4.5–6.8	2.9–3.3	n.a.	0.38	n.a.	110–190	7–15	170–220	76–200	60–140	E72–86	n.a.	n.a.	−200	200–260
Polyamide nylon 11	Nylon® 11	PA	TP	1040	1.5	n.a.	n.a.	n.a.	38	54	320	n.a.	n.a.	96	M60	n.a.	n.a.	−50	70–130
Polyamide nylon 12	Nylon® 12	PA	TP	1010	2.0	n.a.	n.a.	n.a.	45	50–55	290–300	n.a.	n.a.	n.a.	R84–107	n.a.	n.a.	n.a.	n.a.
Polyamide nylon 4,6	Nylon® 46	PA	TP	1180	3.1–3.3	3.1	n.a.	n.a.	95	55–100	50	n.a.	n.a.	80	M92	n.a.	n.a.	−40	100–200
Polyamide nylon 6	Nylon® 6	PA	TP	1130	2.6–3.0	0.97	n.a.	n.a.	44	78	300	n.a.	n.a.	30–250	M82	0.2–0.3	5	−40	80–160
Polyamide nylon 6,10	Nylon® 610	PA	TP	n.a.	n.a.	n.a.	n.a.	n.a.	55	n.a.	100–250	n.a.	n.a.	n.a.	R107	n.a.	n.a.	n.a.	n.a.
Polyamide nylon 6,12	Nylon® 612	PA	TP	1060	2.1	1.241	n.a.	n.a.	57	52–61	100–250	n.a.	n.a.	5–70	R95–120	n.a.	n.a.	n.a.	n.a.
Polyamide nylon 6,6	Nylon® 66	PA	TP	1140	3.3	1.207	n.a.	n.a.	59	82	300	n.a.	n.a.	40–110	M89	0.2–0.3	3–5	−30	80–180
Polyaramide	Kevlar®	PAR	TP	1440	59–124	n.a	na.	n.a.	n.a.	2760	n.a.	n.a.	n.a.	n.a.	n.a.	n.a.	n.a.	−200	180–245
Polyaramide	Nomex®	PAR	TP	n.a.	n.a.	n.a.	n.a.	n.a.	n.a.	n.a.	n.a.	n.a.	n.a.	n.a.	n.a.	na.	n.a.	n.a.	n.a.
Polyarylate resins	Durel®	PAR	TP	1210	16.60	13.80	n.a.	n.a.	58.9–75.8	138	50	n.a.	130.90	117–294	n.a.	n.a.	n.a.	−150	100
Polybenzene-imidazole		PBI	TP	n.a.	n.a.	n.a.	n.a.	n.a.	n.a.	n.a.	n.a.	n.a.	n.a.	n.a.	n.a.	n.a.	n.a.	n.a.	n.a.
Polybutadiene rubber		BR	E	910	2.1–10.3	n.a.	na.	n.a.	13.8–17.2	n.a.	450	n.a.	n.a.	n.a.	SHA 45–80	n.a.	n.a.	−100	95
Polybutadiene terephtalate		PBT	TP	1310	2.6	2.5	n.a.	n.a.	52	50	250	n.a.	n.a.	60	M70	n.a.	n.a.	n.a.	120
Polybutylene		PB	TP	935	0.3	0.75	n.a.	n.a.	16–18	n.a.	n.a.	n.a.	n.a.	640–800	SHD 60	n.a.	n.a.	−150	n.a.
Polycarbonate	Lexan®, Macrolon®	PC	TP	1200	2.3–2.4	2.2	n.a.	n.a.	62	55–75	100–150	80	n.a.	600–850	M70	0.31	10–15	−135	115–130
Polychloroprene rubber	Neoprene®	CPR	E	1230–1250	0.7–20.1	n.a.	n.a.	n.a.	3.4–24.1	n.a.	100–800	n.a.	n.a.	n.a.	SHA 30–95	n.a.	n.a.	−43	107

Table 10.1 (continued)

Usual chemical name	Flame rating ASTM UL94	Water absorption at saturation (wt%)	Water absorption per 24 hours (wt% day^{-1})	Transmittance (T, %)	Refractive index (n_D)	Loss factor	Dielectric field strength (E_d, kV.cm^{-1})	Relative electrical permittivity (@ 1MHz) ($\varepsilon\rho$)	Electrical resistivity (ρ, Ω.cm)	Heat deflection temperature under 1.82 MPa flexural load (T, °C)	Deflection temperature under 0.455 MPa flexural load (T, °C)	Coefficient of linear thermal expansion (α, 10^{-6}K^{-1})	Thermal conductivity (k, W.m^{-1}.K^{-1})	Specific heat capacity (c_p, J.kg^{-1}.K^{-1})	Melting point or range (mp, °C)	Glass transition temperature (T_g, °C)	Vicat softening temperature (°C)
Butadiene acrylonitrile rubber	n.a.	n.a.	n.a.	n.a.	n.a.	n.a.	n.a.	n.a.	n.a.	n.a.	n.a.	n.a.	n.a.	n.a.	n.a.	n.a.	n.a.
Perfluorinated alkoxy	V0	n.a.	0.03	n.a.	n.a.	0.0001	198	2.1	1.E+18	n.a.	n.a.	n.a.	0.25	n.a.	305	n.a.	n.a.
Phenol formaldehyde	n.a.	n.a.	0.2	n.a.	n.a.	0.0060	120–160	5.0–6.5	1.E+12	163	n.a.	16	0.25	1226	n.a.	n.a.	n.a.
Polyacrylic butadiene rubber	n.a.	n.a.	n.a.	n.a.	n.a.	n.a.	n.a.	n.a.	n.a.	n.a.	n.a.	n.a.	n.a.	n.a.	317	104	n.a.
Polyamide-imide	V0	n.a.	0.3	n.a.	1.42–1.46	0.0420	230	3.9–5.4	1.E+17	278	n.a.	n.a.	0.26–0.54	1000	n.a.	280	n.a.
Polyamide nylon 11	V2	1.1	n.a.	n.a.	n.a.	0.0500	160–200	3	1.E+13	55	150	125	0.3	1226	n.a.	n.a.	n.a.
Polyamide nylon 12	V2	1.1	n.a.	n.a.	n.a.	0.0600	260–300	3.5	1.E+13	48–55	130–135	100–120	0.19	1226	n.a.	n.a.	n.a.
Polyamide nylon 4,6	V2	n.a.	1.3	n.a.	n.a.	0.3500	200	3.8–4.3	1.E+13	160	220	25–50	0.3	n.a.	290	42.85	n.a.
Polyamide nylon 6	Self-E	1.1	2.7	n.a.	1.53	0.2000	250	3.6	5.E+12	65–80	200	45	0.23	1600	223	53	n.a.
Polyamide nylon 6,10	V2	n.a.	n.a.	n.a.	n.a.	n.a.	n.a.	n.a.	n.a.	55–90	130–180	120–130	0.22	n.a.	n.a.	49.85	n.a.
Polyamide nylon 6,12	HB-V2	n.a.	3.0	n.a.	n.a.	0.0280	270	3.1–3.8	n.a.	n.a.	n.a.	n.a.	n.a.	1670	212	45.85	n.a.
Polyamide nylon 6,6	Self-E	1.1	2.3	n.a.	n.a.	0.2000	250	3.4	1.E+13	90–100	233	40	0.25	1670	255	49.85	n.a.
Polyaramide	n.a.	n.a.	n.a.	n.a.	n.a.	n.a.	n.a.	n.a.	n.a.	n.a.	n.a.	2(‖)	0.04	1400	640	375	n.a.
Polyaramide	n.a.	n.a.	n.a.	n.a.	n.a.	n.a.	n.a.	n.a.	n.a.	n.a.	n.a.	n.a.	n.a.	n.a.	640	375	n.a.
Polyarylate resins	V0	n.a.	n.a.	n.a.	n.a.	0.0220	134–183.1	2.93–3.30	1.E+12	174	355	50.4–72.0	0.178	n.a.	n.a.	n.a.	n.a.
Polybenzene-imidazole	nil	n.a.	n.a.	n.a.	n.a.	n.a.	n.a.	n.a.	n.a.	n.a.	n.a.	n.a.	n.a.	n.a.	n.a.	n.a.	n.a.
Polybutadiene rubber	n.a.	n.a.	0.01	n.a.	n.a.	n.a.	n.a.	2.5	1.E+15	n.a.	n.a.	n.a.	n.a.	n.a.	n.a.	n.a.	n.a.
Polybutadiene terephtalate	HB	n.a.	n.a.	n.a.	n.a.	0.0020	200	3.2	1.E+15	60	150	45	0.21	1350	223	50	210
Polybutylene	HB	n.a.	0.003	n.a.	n.a.	0.0005	n.a.	2.53	n.a.	54–60	102–113	13	0.22	n.a.	127	n.a.	113
Polycarbonate	V0-V2	n.a.	0.01	n.a.	1.585	0.0010	150–670	2.92	1.E+16	128–138	140	38–70	0.19–0.22	1200	n.a.	n.a.	n.a.
Polychloroprene rubber	n.a.	n.a.	n.a.	n.a.	n.a.	n.a.	n.a.	2.0–6.3	1.E+11	n.a.	n.a.	n.a.	0.192	2170	80	–50	n.a.

10

Polymers and Elastomers

Table 10.1 (continued)

Usual chemical name	Trade name	Acronym, abbreviation or symbol	Category	Density (ρ, kg.m^{-3})	Tensile modulus (E, GPa)	Flexural modulus (G, GPa)	Compressive modulus (K, GPa)	Poisson's ratio (ν)	Yield tensile strength (σ_{YS}, MPa)	Ultimate tensile Strength (σ_{UTS}, MPa)	Elongation at break (Z, %)	Ultimate compressive strength (σ_{UCS}, MPa)	Flexural yield strength (MPa)	Notched izod impact Energy per unit width (J.m^{-1})	Rockwell hardness (or Shore SHD)	Static friction Coefficient (μ/nil)	Wear resistance (i.e., weight loss per 1000 cycles) (mg)	Minimum operating temperature range (°C)	Maximum operating temperature range (°C)
Polyether ether ketone	Victrex®	PEEK	TP	1320	3.7–4.0	n.a.	n.a.	0.4	n.a.	70–100	50	n.a.	n.a.	85	M99	0.18	n.a.	n.a.	250
Polyether imide	Ultem®	PEI	TP	1270	2.9	n.a.	2.9	n.a.	n.a.	85	60	140	n.a.	50	R125	n.a.	10	n.a.	n.a.
Polyether sulfone		PESV	TP	1370	2.4–2.6	n.a.	n.a.	n.a.	25–40	70–95	40–80	n.a.	n.a.	85	M88	n.a.	6	–110	180–200
Polyethylene (high density)		HDPE	TP	950–968	0.414–1.24	0.062–0.105	n.a.	n.a.	25–40	16–40	5–12	n.a.	20–38	20–210	SHD 60–73	0.29	n.a.	n.a.	55–120
Polyethylene (low density)		LDPE	TP	912–925	0.14–1.86	0.069–0.207	n.a.	n.a.	6.2–11.5	5–26	20–40	n.a.	n.a.	1000	SHD 41–46	n.a.	n.a.	–60	50–70
Polyethylene (medium density)		MDPE	TP	926–940	0.17–0.38	0.240–0.790	n.a.	n.a.	10–19	10–19	10–20	n.a.	n.a.	1000	SHD 45–60	n.a.	n.a.	n.a.	n.a.
Polyethylene (ultrahigh molecular weight)	Lennite®	UHMW	TP	940	0.135–6.90	0.520–0.970	n.a.	n.a.	n.a.	20–40	500	n.a.	n.a.	1000	R50–70	0.1–0.2	n.a.	n.a.	55–95
Polyethylene naphtalate		PEN	TP	1360	5.0–5.5	n.a.	5.5	n.a.	n.a.	200	60	n.a.	n.a.	n.a.	n.a.	0.27	n.a.	n.a.	155
Polyethylene oxide		PEO	TP	n.a.	n.a.	n.a.	n.a.	n.a.	n.a.	n.a.	n.a.	n.a.	n.a.	n.a.	n.a.	n.a.	n.a.	n.a.	n.a.
Polyethylene terephtalate	Mylar®	PET	TP	1560	2.0–4.0	3.0	n.a.	n.a.	81	80	70	n.a.	n.a.	13–35	M94–101	0.2–0.4	n.a.	–40 to –60	115–170
Polyhydroxybutyrate (biopolymer)		PHB	TP	1250	3.5	n.a.	n.a.	n.a.	n.a.	40	n.a.	n.a.	n.a.	35–60	n.a.	n.a.	n.a.	n.a.	95
Polyimide	Vespel®	PI	TP	1420	2.0–3.0	3.1–3.45	n.a.	n.a.	n.a.	70–150	8–70	n.a.	n.a.	80	E52–99	0.42	n.a.	–270	250–320
Polyisoprene (trans-1,4-polyisoprene)	Gutta Percha	PIP	E	n.a.	n.a.	n.a.	n.a.	n.a.	n.a.	n.a.	n.a.	n.a.	n.a.	n.a.	n.a.	n.a.	n.a.	n.a.	n.a.
Polymethyl methacrylate	Plexiglas®	PMMA	TP	1180–1190	3.036	2.24–3.17	2.55–3.17	n.a.	54–73	72.4	2.5–4	72–124	72–131	16–32	M92–100	n.a.	n.a.	–40	50–90
Polymethyl pentene	TPX®	PMP	TP	835	1.5	n.a.	n.a.	n.a.	n.a.	25.5	15–75	n.a.	n.a.	49	R85	n.a.	n.a.	–20 to –40	75–115
Polyoxymethylene (heteropolymer)	Acetal®	POM	TP	1400	2.9–3.2	2.41–3.10	4.62	n.a.	65–69	69–83	15	110	90	53–80	R120–M78	n.a.	n.a.	n.a.	105
Polyoxymethylene (homopolymer)	Delrin® 500	POMH	TP	1420	3.60	2.62–3.585	3.11	n.a.	57–70	72	40–75	107–124	94–110	75–130	M94–101	0.20–0.35	n.a.	n.a.	80–120
Polyphenylene oxide	Noryl®	PPO	TP	1090	2.5	2.59	n.a.	n.a.	n.a.	55–65	50	110	n.a.	200	M78–R115	0.35	20	–40	80–120

Table 10.1 (continued)

Usual chemical name	Flame rating ASTM UL94	Water absorption at saturation (wt%)	Water absorption per 24 hours (wt% day⁻¹)	Transmittance (T, %)	Refractive index (n_D)	Loss factor	Dielectric field strength (E_d, kV.cm⁻¹)	Relative electrical permittivity (@ 1MHz) ($\varepsilon\rho$)	Electrical resistivity (ρ, Ω.cm)	Heat deflection temperature under 1.82 MPa flexural load (T, °C)	Deflection temperature under 0.455 MPa flexural load (T, °C)	Coefficient of linear thermal expansion (α, $10^{-6}K^{-1}$)	Thermal conductivity (k, W.m⁻¹.K⁻¹)	Specific heat capacity (c_p, J.kg⁻¹.K⁻¹)	Melting point or range (mp, °C)	Glass transition temperature (T_g, °C)	Vicat softening temperature (°C)
Polyether ether ketone	V0	n.a.	0.3	n.a.	n.a.	0.0030	190	3.1	1.E+16	142	147	26–108	0.25	320	334	143	n.a.
Polyether imide	V0	n.a.	0.25	n.a.	n.a.	0.0013	280–330	3.1	1.E+15	190–210	210	31–56	0.22	n.a.	n.a.	215	n.a.
Polyether sulfone	V0	n.a.	1	n.a.	1.65	0.0030	160	3.7	1.E+17	203	260	55	0.13–0.18	n.a.	n.a.	225	n.a.
Polyethylene (high density)	Comb	n.a.	0.01	n.a.	1.53–1.54	0.0010	420–520	2.30–2.40	1.E+15	46	75	100–200	0.42–0.52	1900	125–137	−90 to −200	n.a.
Polyethylene (low density)	Comb	n.a.	0.015	n.a.	1.51–1.52	0.0005	270–390	2.25–2.35	1.E+15	35	50	100–200	0.33	1900	102–112	−110 to −20	n.a.
Polyethylene (medium density)	Comb	n.a.	0.01	n.a.	1.52–1.53	0.0010	180–390	2.25–2.35	1.E+17	n.a.	n.a.	100–200	n.a.	1900	110–120	−118	n.a.
Polyethylene (ultrahigh molecular weight)	Comb	n.a.	0.01	n.a.	n.a.	0.0010	190–280	2.3	1.E+18	42	69	130–200	0.45–0.52	1900	125–135	n.a.	n.a.
Polyethylene naphtalate	n.a.	n.a.	n.a.	84	n.a.	0.0048	1600	3.2	1.E+15	n.a.	n.a.	20–21	n.a.	n.a.	n.a.	n.a.	n.a.
Polyethylene oxide	n.a.	n.a.	n.a.	n.a.	n.a.	n.a.	n.a.	n.a.	n.a.	n.a.	n.a.	n.a.	n.a.	n.a.	n.a.	n.a.	n.a.
Polyethylene terephtalate	HB	n.a.	0.1	n.a.	1.58–164	0.0020	170	3.0	1.E+14	80	115	15–65	0.17–0.40	1200	255	68.85	235
Polyhydroxybutyrate (biopolymer)	n.a.	n.a.	n.a.	n.a.	n.a.	n.a.	n.a.	3.0	1.E+16	n.a.	n.a.	n.a.	n.a.	n.a.	n.a.	n.a.	n.a.
Polyimide	V0	n.a.	0.2–2.9	n.a.	1.42	0.0018	220	3.4	1.E+18	360	n.a.	30–60	0.10–0.36	1130	365	280–330	n.a.
Polyisoprene (trans-1,4-polyisoprene)	n.a.	n.a.	n.a.	n.a.	n.a.	n.a.	n.a.	2.5	1.E+15	n.a.	n.a.	n.a.	n.a.	n.a.	45	n.a.	n.a.
Polymethyl methacrylate	HB	n.a.	0.2–0.3	92	1.49	0.0140	150	2.76	1.E+15	74–95	105	34–77	0.17–0.19	1450	n.a.	104.85	n.a.
Polymethyl pentene	Comb	n.a.	0.01	n.a.	n.a.	0.0020	n.a.	2.12	1.E+16	40	100	117	0.17	2000	250	29	n.a.
Polyoxymethylene (heteropolymer)	HB	n.a.	0.2	n.a.	n.a.	0.0048	200	3.8	1.E+15	136	n.a.	85	0.37	1464	173	n.a.	n.a.
Polyoxymethylene (homopolymer)	HB	n.a.	0.25	n.a.	n.a.	0.0050	200	3.7	1.E+15	136	170	122	0.22–0.24	1464	173	n.a.	n.a.
Polyphenylene oxide	V0	n.a.	0.1–0.5	n.a.	n.a.	0.0040	160–200	2.59	2.E+17	100–125	137–179	38–60	0.22	n.a.	267	84.9	n.a.

10

Polymers and Elastomers

Table 10.1 (continued)

Usual chemical name	Trade name	Acronym, abreviation or symbol	Category	Density (ρ, kg.m^{-3})	Tensile modulus (E, GPa)	Flexural modulus (G, GPa)	Compressive modulus (K, GPa)	Poisson's ratio (v)	Yield tensile strength (σ_{YS}, MPa)	Ultimate tensile Strength (σ_{UTS}, MPa)	Elongation at break (Z, %)	Ultimate compressive strength (σ_{UCS}, MPa)	Flexural yield strength (MPa)	Notched izod impact Energy per unit width (J.m^{-1})	Rockwell hardness (or Shore SHD)	Static friction Coefficient (μ/nil)	Wear resistance (i.e., weight loss per 1000 cycles) (mg)	Minimum operating temperature range (°C)	Maximum operating temperature range (°C)
Polyphenylene sulfide	Milkon®, Ryton®	PPS	TP	1350	1350	3.8	n.a.	n.a.	n.a.	65.5	1.6	110	96	16	R120	n.a.	n.a.	−170	190
Polypropylene (atactic)	Propylux®	PP	TP	850–900	0.689–1.520	0.9	n.a.	n.a.	n.a.	21.4	300	n.a.	n.a.	763	R95	0.10–0.30	13–16	n.a.	240
Polypropylene (isotactic)	Propylux®	PP	TP	920–940	0.689–1.520	1.2–1.7	n.a.	n.a.	n.a.	31–41	100–600	n.a.	n.a.	20–53	R95	n.a.	n.a.	n.a.	160
Polypropylene (syndiotactic)	Propylux®	PP	TP	890–915	0.689–1.520	n.a.	n.a.	n.a.	n.a.	35	n.a.	n.a.	n.a.	n.a.	R95	n.a.	n.a.	n.a.	140
Polystyrene (high-impact)	Propylux®	HIPS	TP	1040	1.6	2.07	n.a.	0.34	24.8	35–100	36–50	n.a.	35–39.3	133	L73	n.a.	n.a.	n.a.	n.a.
Polystyrene (normal)	Crystal®	PS	TP	1054–1070	2.3–4.1	3.17	n.a.	n.a.	n.a.	27–69	1.6–3	83–117	90	19–24	M60–90	n.a.	n.a.	n.a.	50–95
Polysulfide rubber	Thiokol®	PSR	E	1340	n.a.	n.a.	n.a.	n.a.	n.a.	4.83–8.63	100–400	n.a.	n.a.	n.a	n.a.	n.a.	n.a.	−54	n.a.
Polysulfone	Udel®, Thermalux®	PSU	TP	1240	2.48	2.69	2.58	0.37	n.a.	70.3	75	96	106	69	M69	n.a.	n.a.	n.a.	82–100
Polytetrafluoroethylene	Teflon®	PTFE	TP	2130–2220	0.48–0.76	0.19–0.55	0.41	n.a.	17–27	10–40	200–400	11.7	nil	160	R45 (SHD50)	0.05–0.20	−260	180–260	n.a.
Polytrifluorochloroethylene	Kel-F®	PTFCE	TP	2100	1.3	1.17–1.38	1.03–2.07	n.a.	37	32–35	80–250	32–51	51–76	65	R75–112	n.a.	n.a.	n.a.	280
Polyurethane		PUR	TS	1050–1250	0.6	n.a.	n.a.	n.a.	n.a.	n.a.	n.a.	n.a.	n.a.	n.a.	n.a.	n.a.	n.a.	n.a.	175
Polyvinyl acetate		PVA	TP	1191	n.a.	n.a.	n.a.	n.a.	n.a.	29–49	10–21	n.a.	n.a.	102	n.a.	n.a.	n.a.	n.a.	120
Polyvinyl alcohol		PVAL	TP	n.a.	n.a.	n.a.	n.a.	n.a.	n.a.	n.a.	n.a.	n.a.	n.a.	n.a.	n.a.	n.a.	n.a.	n.a.	120
Polyvinylidene chloride	Saran®	PVDC	TP	1630	0.3–0.55	1.17–8.3	n.a.	n.a.	20–57	48	200	n.a.	n.a.	16–53	R98–106	0.24	n.a.	n.a.	80–100
Polyvinylidene fluoride	Kynar®, Foraflon®	PVDF	TP	1760	1.0–3.0	n.a.	2.09–2.90	n.a.	33–41	5–25	50–300	55–110	67–94	120–320	R77–83	0.20–0.40	24	−40	135–150
Polyvinyl fluoride		PVF	TP	1380–1720	0.44–1.10	n.a.	n.a.	n.a.	n.a.	n.a.	n.a.	n.a.	n.a.	n.a.	n.a.	n.a.	n.a.	−70	175
Polyvinyl chloride	Vinyl®	PVC	TP	1160–1550	2.1–2.7	1.0	n.a.	0.4	55	7–27	4.5–65	n.a.	n.a.	21	SHD 65–85	n.a.	n.a.	n.a.	60–105

Table 10.1 (continued)

Usual chemical name	Flame rating ASTM UL94	Water absorption at saturation (wt%)	Water absorption per 24 hours (wt% day⁻¹)	Transmittance (T, %)	Refractive index (n_D)	Loss factor	Dielectric field strength (E_d, kV.cm⁻¹)	Relative electrical permittivity (@ 1MHz) ($\varepsilon\rho$)	Electrical resistivity (ρ, Ω.cm)	Heat deflection temperature under 1.82 MPa flexural load (T, °C)	Deflection temperature under 0.455 MPa flexural load (T, °C)	Coefficient of linear thermal expansion (α, 10⁻⁶K⁻¹)	Thermal conductivity (k, W.m⁻¹.K⁻¹)	Specific heat capacity (c_p, J.kg⁻¹.K⁻¹)	Melting point or range (mp, °C)	Glass transition temperature (T_g, °C)	Vicat softening temperature (°C)
Polyphenylene sulfide	V0	n.a.	0.05.	n.a.	n.a.	0.0014	177–240	3.8	1.E+14	135	260	30–49	0.17–0.28	1090	285	85	n.a.
Polypropylene (atactic)	Comb	n.a.	0.01	n.a.	n.a.	0.0005	200–260	2.2–2.3	1.E+16	43	85	68–95	0.12	1966	176	–18	n.a.
Polypropylene (isotactic)	Comb	n.a.	0.01	n.a.	n.a.	0.0005	200–260	2.2–2.3	1.E+16	50–60	110–120	81–100	0.154	1966	165	–15 to –10	n.a.
Polypropylene (syndiotactic)	Comb	n.a.	0.01	n.a.	n.a.	0.0005	200–260	2.2–2.3	1.E+16	n.a.	n.a	60–90	0.154	1966	n.a	–10 to –8.2	n.a.
Polystyrene (high-impact)	V0	n.a.	0.1	n.a.	1.59–1.60	0.0004	177–240	2.3–2.5	1.E+16	82	91	90	0.124	1250	n.a	100	98
Polystyrene (normal)	V0	n.a.	0.4	n.a.	1.59–1.60	0.0002	180–240	2.4–3.1	1.E+16	80	90	30–210	0.10–0.13	1250	115	100	109
Polysulfide rubber	n.a.	n.a.	n.a.	n.a.	n.a.	n.a.	n.a.	1.3	1.E+08	n.a.	n.a.	n.a.	n.a.	n.a.	n.a.	n.a.	n.a.
Polysulfone	V0	n.a.	0.22	99	1.63	0.0050	166	3.14	1.E+16	174	n.a.	31–51	0.259	1255	n.a.	193	n.a.
Polytetrafluoroethylene	V0	n.a.	0.01	nil	1.38	0.0001	400–800	2.0–2.1	1.E+18	54	120	100–160	0.25	1000	327	–97	n.a.
Polytrifluorochloroethylene	V0	n.a.	nil	n.a.	n.a.	n.a.	197–230	2.46	n.a.	75	130	126–216	0.19–0.22	9.20	215	45	n.a.
Polyurethane	n.a.	n.a.	1	n.a.	n.a.	n.a.	n.a.	n.a.	1.E+12	n.a.	n.a.	100–200	0.21	1800	141–178	n.a.	n.a.
Polyvinyl acetate	n.a.	n.a.	3.6	n.a.	1.4669	n.a.	394	3.5	n.a.	n.a.	n.a.	100–200	0.159	n.a.	n.a.	29	n.a.
Polyvinyl alcohol	Self-E	n.a.	n.a.	n.a.	n.a.	n.a.	n.a.	n.a.	n.a.	n.a.	n.a.	n.a.	0.795	1.255	n.a.	84.85	n.a.
Polyvinylidene chloride	Self-E	n.a.	0.1	n.a.	1.63	0.0450	160–240	3.2–6.0	1.E+12	n.a.	n.a.	190	0.13	1339	258	–18.15	n.a.
Polyvinylidene fluoride	V0	n.a.	0.04–0.06	n.a.	1.42	0.0490	100–130	6.4–8.9	1.E+14	80–115	120–150	80–140	0.10–0.25	1381	198	–40 to –35	n.a.
Polyvinyl fluoride	Self-E	n.a.	n.a.	n.a.	1.460	n.a.	80–130	6.2–7.7	10⁹	n.a.	n.a.	50	0.17	n.a.	200	–20 to 41	n.a.
Polyvinyl chloride	Self-E	n.a.	0.15–1	n.a.	1.540	0.0070	160–590	2.9–3.6	10¹⁶	n.a.	n.a.	60–70	0.167	1674	212	81 to 87	n.a.

10

Polymers and Elastomers

Table 10.1 (continued)

Usual chemical name	Trade name	Acronym, abreviation or symbol	Category	Density (ρ, kg.m^{-3})	Tensile modulus (E, GPa)	Flexural modulus (G, GPa)	Compressive modulus (K, GPa)	Poisson's ratio (v)	Yield tensile strength (σ_{YS}, MPa)	Ultimate tensile Strength (σ_{UTS}, MPa)	Elongation at break (Z, %)	Ultimate compressive strength (σ_{UCS}, MPa)	Flexural yield strength (MPa)	Notched izod impact Energy per unit width (J.m^{-1})	Rockwell hardness (or Shore SHD)	Static friction Coefficient (μ/nil)	Wear resistance (i.e., weight loss per 1000 cycles) (mg)	Minimum operating temperature range (°C)	Maximum operating temperature range (°C)
Propylene-vinylidene hexafluoride	Viton®, Fluorel®	PVHF	E	1800–1860	2.07–15.17	n.a.	n.a.	n.a.	8.96–18.62	4.8–11.0	100–700	n.a.	n.a.	n.a.	SHA50–95	n.a.	n.a.	−29	204
Silicone rubber (polysiloxane)	Rhodorsil®	SR	E	n.a.	n.a.	n.a.	n.a.	n.a.	n.a.	6.5	n.a.	n.a.	n.a.	n.a.	SHD60A	n.a.	n.a.	−60	232
Styrene-butadiene styrene rubber	Kraton-D®	SBS	E	n.a.	n.a.	n.a.	n.a.	n.a.	n.a.	n.a.	n.a.	n.a.	n.a.	n.a.	n.a.	n.a.	n.a.	n.a.	n.a.
Styrene-butadiene rubber	Buna®S, GR-S	SBR	E	940	2.1–10.3	n.a.	n.a.	n.a.	12.4–20.7	21	450–500	n.a.	n.a.	n.a.	SHA30–90	n.a.	n.a.	−54	82
Synthetic isoprene rubber		IR	E	940	n.a.	n.a.	n.a.	n.a.	n.a.	25	750	n.a.	n.a.	n.a.	SHA30–95	n.a.	n.a.	n.a.	n.a.
Unplasticized polyvinyl chloride		UPVC	TP	1300–1450	24–40	n.a.	n.a.	n.a.	n.a.	38–62	2–40	n.a.	n.a.	21	n.a.	n.a.	n.a.	n.a.	n.a.
Unsaturated polyester		UP	TS	1780	5.5	n.a.	n.a.	n.a.	n.a.	41	n.a.	n.a.	n.a.	32	M88	n.a.	n.a.	n.a.	n.a.
Urea formaldehyde		UF	TS	1470–1520	n.a.	n.a.	n.a.	n.a.	n.a.	n.a.	n.a.	n.a.	n.a.	n.a.	n.a.	n.a.	n.a.	n.a.	77

Table 10.1 (continued)

Usual chemical name	Vicat softening temperature (°C)	Glass transition temperature (T_g, °C)	Melting point or range (mp, °C)	Specific heat capacity (c_p, J.kg^{-1}.K^{-1})	Thermal conductivity (k, W.m^{-1}.K^{-1})	Coefficient of linear thermal expansion (α, 10^{-6}K^{-1})	Deflection temperature under 0.455 MPa flexural load (T, °C)	Heat deflection temperature under 1.82 MPa flexural load (T, °C)	Electrical resistivity (ρ, Ω.cm)	Relative electrical permittivity (@ 1MHz) ($\varepsilon\rho$)	Dielectric field strength (E_d, kV.cm^{-1})	Loss factor	Refractive index (n_D)	Transmittance (T, %)	Water absorption per 24 hours (wt% day^{-1})	Water absorption at saturation (wt%)	Flame rating ASTM UL94
Propylene-vinylidene hexafluoride	n.a.	n.a.	n.a.	n.a.	n.a.	n.a.	n.a.	n.a.	n.a.	n.a.	n.a.	n.a.	n.a.	n.a.	n.a.	n.a.	Self-E
Silicone rubber (polysiloxane)	n.a.	n.a.	n.a.	n.a.	n.a.	n.a.	n.a.	n.a.	n.a.	n.a.	n.a.	n.a.	n.a.	n.a.	0.1	n.a.	n.a.
Styrene-butadiene styrene rubber	n.a.	n.a.	n.a.	n.a.	n.a.	n.a.	n.a.	n.a.	n.a.	n.a.	n.a.	n.a.	n.a.	n.a.	n.a.	n.a.	n.a.
Styrene-butadiene rubber	n.a.	n.a.	n.a.	n.a.	n.a.	n.a.	n.a.	n.a.	1.E+14	2.4	n.a.	n.a.	n.a.	n.a.	n.a.	n.a.	n.a.
Synthetic isoprene rubber	n.a.	n.a.	n.a.	n.a.	n.a.	n.a.	n.a.	n.a.	1.E+15	2.4	n.a.	n.a.	n.a.	n.a.	n.a.	n.a.	n.a.
Unplasticized polyvinyl chloride	n.a.	n.a.	n.a.	n.a.	n.a.	n.a.	n.a.	n.a.	n.a.	n.a.	n.a.	n.a.	n.a.	n.a.	n.a.	n.a.	Self-E
Unsaturated polyester	n.a.	n.a.	n.a.	n.a.	n.a.	16	n.a.	n.a.	n.a.	n.a.	n.a.	n.a.	n.a.	n.a.	n.a.	n.a.	HB
Urea formaldehyde	n.a.	n.a.	n.a.	250	0.30–0.42	22–90	120–160	n.a.	n.a.	n.a.	n.a.	n.a.	n.a.	n.a.	0.5	n.a.	n.a.

10

Polymers and Elastomers

Table 10.2 Polymer physical quantities and ASTM standards

Physical quantities		ASTM standard	SI unit	US customary unit
Processing	Processing temperature range	n.a.	°C	°F
	Molding pressure range	n.a.	Pa	psi
	Compression ratio	n.a.	nil	nil
	Melt mass flow rate	D1238	$kg.s^{-1}$	$lb.h^{-1}$
	Mold linear shrinkage	D955	nil	$in.in^{-1}$
Mechanical	Density (ρ)	D792	$kg.m^{-3}$	$lb.ft^{-3}$
	Specific gravity (d)	D792	nil	nil
	Poisson's coefficient (v)	n.a.	nil	nil
	Yield tensile strength (σ_{YS})	D638	Pa	psi
	Ultimate tensile strength (σ_{UTS})	D638	Pa	psi
	Elongation at yield (Z)	D638	nil	%
	Elongation at break (Z)	D638	nil	%
	Compressive strength	D695	Pa	psi
	Flexural yield strength	D790	Pa	psi
	Tensile or elastic modulus (E)	D638	Pa	psi
	Compressive or bulk modulus (K)	D695	Pa	psi
	Flexural or shear modulus (G)	D790	Pa	psi
	Unnotched Izod impact strength (i.e., impact energy per unit width)	D256 A	$J.m^{-1}$	$ft.lb.in^{-1}$
	Abrasion resistance per 1000 cycles	D1044	kg.Hz	$lb.cycle\ s^{-1}$
	Rockwell hardness (HR) scale M and R	D785	nil	nil
	Shore durometer hardness (SH) scale A and D	D2240	nil	nil
	Barcol durometer hardness	D2583	nil	nil
Thermal properties	Minimum operating temperature (T_{min})	n.a.	°C	°F
	Maximum operating temperature (T_{max})	n.a.	°C	°F
	Brittle temperature (T_{brit})	D746	°C	°F
	Glass transition temperature (T_g)	n.a.	°C	°F
	Vicat softening point (T_{vicat})	D1525	°C	°F
	Melting point (mp)	n.a.	°C	°F
	Thermal conductivity (k)	C177	$W.m^{-1}.K^{-1}$	$Btu.ft^{-1}.h^{-1}.°F^{-1}$
	Specific heat capacity (c_P)	n.a.	$J.kg^{-1}.K^{-1}$	$Btu.lb^{-1}.°F^{-1}$
	Coefficient of linear thermal expansion (a)	D696	K^{-1}	$°F^{-1}$
	Deflection temperature under flexural load (0.455 MPa)	D648	°C	°F
	Deflection temperature under flexural load (1.82 MPa)	D648	°C	°F
Electrical	Dielectric permittivity (e_r) (1 MHz)	D150	nil	nil
	Dielectric field strength (E_d)	D149	$V.m^{-1}$	$V.mil^{-1}$
	Dissipation or loss factor (δ)	D149	nil	nil
	Electrical volume resistivity (ρ)	D257	ohm.m	$ohm.cir\ ft.in^{-1}$
Miscellaneous	Refractive index (n_D) (589 nm)	n.a.	nil	nil
	Optical transmission (T) (visible light)	n.a.	nil	%
	Nuclear radiation resistance (e.g., α, β, γ, and X-rays)	n.a.	nil	nil

Table 10.2 (*continued*)

Physical quantities		ASTM standard	SI unit	US customary unit
Chemical	Water absorption in 24 hours	D570	s^{-1}	wt% day^{-1}
	Water absorption at saturation	D570	nil	wt%
	Chemical resistance	n.a.	nil	nil
	Flammability rating index	ANSI/ UL-94	nil	nil

Table 10.2 with the corresponding ASTM standards. Particular mechanical properties are briefly described below.

Shore hardness (durometer) is a property that, as applied to elastomers measures resistance to indentation. **Shore A** scale is used for soft elastomers, while **shore D** scale is for harder materials.

Compression modulus is the stress required to achieve a specific deflection, typically 50% deflection. This test measures the polymer rigidity or toughness.

Flexural or tear strength measures the resistance to growth of a nick or cut when tension is applied to a test specimen. Tear strength is critical in predicting an elastomer's work life in demanding and aggressive applications.

Tensile strength describes the ultimate strength of a material when enough stress is applied to cause it to break. In combination with elongation and modulus, tensile strength can predict a material's toughness.

Elongation at break relates to the ability of an elastomer to stretch without breaking. Ultimate elongation is the percentage of the original length of the sample and is measured at the point of rupture. This property is useful in identifying the appropriate elastomer for stress or stretching applications.

Further Reading

Ash MB, Ash IA (eds) (1992) Handbook of plastic compounds, elastomers, and resins. An international guide by category, trade name, composition, and suppliers. VCH, Weinheim.

Bost J (1982) Matières plastiques, vol. 2. Techniques & Documentation, Paris.

Bost J (1985) Matières plastiques, vol. 1, 2nd edn. Techniques & Documentation, Paris.

Elias HG (1993) An introduction to plastics. VCH, Weinheim.

FIZ Chemie (ed.) (1992) Part-index of polymer trade names, 2nd edn. VCH, Weinheim.

10
Polymers
and
Elastomers

11 Minerals, Ores, and Gemstones

11.1 Definitions

In this section are detailed and explained the main definitions, properties, and physical quantities used in the mineral properties table.

Crystal – a crystal is a homogeneous solid with an ordered atomic space lattice which has developed a crystalline morphology when external crystallographic planes have had the possibility to grow freely without external constraints and under favorable conditions. Moreover, it is a chemical substance with a definite theoretical chemical formula. Nevertheless, the theoretical chemical composition is usually variable within a limited range owing to the isomorphic substitutions (diadochy), and/or occasionally presence of traces of impurities.

Minerals – a mineral is defined as a naturally occurring, inorganic, and homogeneous crystal that has been formed as a result of geological processes with a definite but generally not fixed chemical composition. Therefore, minerals are the basic building entities of Earth's crust i.e., rocks and soils. On the other hand, amongst the 4000 mineral species, the most abundant minerals found in common rocks (igneous, sedimentary, metamorphic, and meteorites) are called by petrologists the **rock-forming minerals**.

Mineraloids – the mineraloids are naturally occurring substances having a structure which can be partially crystalline or noncrystalline, i.e., solids with an irregular atomic arrangement within the solid. For instance, compounds such as obsidian, opal, amber, or succinite are defined as mineraloids.

Ores – an ore is a naturally occurring mineral or association of minerals containing a high percentage of a metallic element, which form deposits from which this metal can be mined, extracted, and processed at a profit under favorable conditions. Therefore, it is economically defined. However, a distinction must be made between ore and ore minerals. A deposit of **ore minerals** in geological terms is not always an ore deposit, while an ore mineral is a mineral from which a metal can feasibly be extracted, and an **ore deposit** (or an **orebody**) is a mass of rock from which a metal or mineral can be profitably produced. What is, or is not, becomes dependent upon economic, technological, and political factors as

Table 11.1 Common gangue minerals

Class	Mineral
Oxides	Quartz
	Limonite
Carbonates	Calcite
	Dolomite
	Rhodocrosite
Sulfates	Baryte
	Gypsum
Halides	Fluorspar
Phosphates	Apatite
Silicates	Feldspars
	Clays
	Chlorites

well as geological criteria. A **protore** is a low-grade metalliferrous material which is not itself valuable but from which ore may be formed by superficial enrichment.

Gangue – the gangue is an earthy or nonmetallic mineral associated with the ore minerals of a deposit, i.e., a worthless material in which the ore mineral is disseminated and must be concentrated by classical ore beneficiation techniques (e.g., gravity separation, flotation, leaching). The most common gangue minerals are listed in Table 11.1.

Vein deposits – a **vein** is a mineral mass, more or less tabular, deposited by solutions in or along a fracture or group of fractures. The **country rock** is the rock that encloses a metalliferous deposit. Vein **walls** are the rock surfaces on the borders of the veins. The **footwall** is the rock below an inclined vein, a bed, or a fault. The **hanging wall** is the rock above an inclined vein, bed or fault. A **druse** or **vug** is an unfilled portion of a vein usually lined with crystals. **Gouge** (salbandes in French) is a soft claylike material that occurs at some places as a selvage between a vein and country rock or in a vein.

Industrial minerals or nonmetallics – this designation includes all the minerals with economic importance, except those defined as ore, which are processed industrially. In fact, the industrial minerals class also includes (i) sedimentary rocks such as limestone, dolomite, clays, sand, gravel, diatomite, and phosphates, (ii) metamorphic rocks such as slate, and (iii) igneous rocks such as granite and basalt. However, in order to be rigorous from a mineralogical and petrological point of view it is preferable to split the previous group into two distinct subgroups: (i) industrial minerals, sensu stricto, and (ii) industrial rocks, sensu stricto. A conventional listing of the more important nonmetallics is presented in Table 11.2.

Gemstones – a gemstone is a semi-precious or precious natural mineral with exceptional physical properties which, when cut and polished, can be used in jewelry. Only four minerals are considered as precious gemstones sensu stricto: **diamond**, one gem variety of beryl (**emerald**: green), and the two gem varieties of corundum (**ruby**: deep red, and **sapphire**: deep blue). Besides, natural minerals, synthetic gemstones and their simulants are also found in jewelry.

11.2 Mineralogical, Physical, and Chemical Properties

Amongst the approximately 4000 mineral species found in nature, only the major rock-forming minerals, chief metal ores, and gemstones are listed in Table 11.4 presented in Section 11.4 with their common physical and chemical properties useful for mineralogical identification. These

Table 11.2 Industrial minerals and rocks

Nonmetallics	Material	Uses and applications
Industrial minerals	Asbestos (chrysoytile, crocidolite, anthophyllite)	(i) Spinning fibers: woven brake lining, clutch facing, fireproof and safety clothing, and blankets. (ii) Nonspinning fibers: roofing shingles, millboard, and corrugated panels for thermal insulation
	Barite (baryte, heavyspar)	Oil-well drilling muds, filler in rubbers, paint extender, aggregate in speciality concretes, and barium chemicals
	Beryl	Beryllia, beryllium chemicals
	Borax and borates	Fluxing agents in the manufacture of glass and vitreous enamel, borosilicated glasses (i.e., Pyrex[R]), synthetic cubic boron nitride (i.e., Borazon[R]) for industrial abrasives, boron-doped semiconductors
	Chalk	Aggregate
	Chromite	Refractories
	Cryolithe (cryolite)	Fluxing agent in the Héroult–Hall process in the aluminum industry
	Diamond (bort varieties)	Abrasives, diamond drills in the mining industry, wire-drawing dies
	Emery (corundum, magnetite, spinel)	Abrasive for paper grit
	Feldspars (microcline, orthose, plagioclases)	Glass industry for porcelain, enamels, and glazes
	Fluorspar (fluorite)	Foundry fluxes in steelmaking (metallurgical grade), preparation of hydrofluoric acid (acid grade), glass industry (ceramic grade)
	Garnet	Abrasives
	Graphite	Foundry mold facing (70%), crucibles, and lubricant
	Kyanite	Refractories
	Magnesite	After calcination gives periclase (MgO) used for refractories
	Micas (muscovite, phlogopite)	Electrical sheet insulators, furnace windows, roofing materials
	Nitrates (saltpeter, niter, ammonium nitrate)	Fertilizers in agriculture, raw material for the chemical industry (i.e., pyrotechnics and explosives)
	Potash (sylvite, carnallite)	Fertilizer in agriculture and potassium chemicals (e.g., soaps, detergents, dyes, explosives)
	Quartz	Piezoelectric crystals, optical lenses, speciality glassware, optical fibers, silicon for semiconductors
	Sillimanite	Refractories
	Sulfur (native)	Chemical industry for the manufacture of sulfuric acid
	Talc (steatite)	Manufacture of whiteware and porcelain, inert extender in paint, lubricant in paper-making, absorbant in pharmaceutical and chemical industry
Industrial rocks		
Igneous rocks	Basalt and diabase (crushed)	Concrete aggregate, railroad ballast, and roofing granules
	Granite and granodiorite	Monuments and memorials, building foundation blocks, steps curbstones and paving blocks
	Perlite (rhyolitic obsidian)	Aggregate in plasters, loose-fill insulation, filtration medium, paint filler, oil-well drilling muds, inert packing materials
	Pumice	Abrasives, polishing metals and woodworking
Sedimentary rocks	Clays	Filler material, oil-well drilling mud, wax, fat, and oil adsorbents, Portland cement, enamels and ceramics, pottery

11

Minerals, Ores and Gemstones

Table 11.2 Industrial minerals and rocks

Nonmetallics	Material	Uses and applications
	Diatomite (kieselguhr)	Filter aid, filler material
	Dolomite	(i) Crushed as aggregate in concrete, railroad ballast, sewage filter beds, (ii) fluxing agent in smelting and refining of steel, (iii) soil conditioner, (iv) source of lime and magnesia (dolime), (v) chemical raw material, (vi) high-grade refractories, and (vii) dimension stone
	Gypsum and anhydrite	Gypsum plasters for building purposes, fertilizers, sulfates and sulphuric acid
	Limestone	(i) Crushed as aggregate in concrete, railroad ballast, sewage filter beds, (ii) fluxing agent in smelting and refining of steel, (iii) soil conditioner, (iv) source of lime, (v) raw material for Portland cement, (vi) chemical raw material, and (vii) dimension stone
	Phosphate rocks (phosphorites)	Fertilizer in agriculture, raw material for the chemical industry for the manufacture of orthophosphoric acid, steelmaking and pyrotechnics
	Quartzite	Ferrosilicon, refractories, abrasives, pottery and enamels
	Rocksalt (halite)	Raw material for the chemical industry (e.g., chlor-alkali process)
	Sand and gravel (silica sand)	Aggregate in Portland-cement concrete, foundry sands, glass sands
Metamorphic rocks	Marble	Architectural and statuary, dimension stone
	Slate	Roofing slates and flagstones

properties are sufficient to identify common rock-forming and ore minerals occurring in common geological materials (e.g., rock and soils) with common field laboratory equipment (magnification lenses, polarizing microscope, pycnometer, microchemical analysis spot tests). The selected properties of minerals shown in the table are explained in detail in the following paragraphs.

11.2.1 Mineral Names

Minerals are most commonly classified on the basis of the presence of a major chemical component (anion or anionic complex) into several mineral classes such as native elements, sulfides and sulfosalts, oxides, carbonates, sulfates, phosphates, silicates, etc. Today, there exist two main mineralogical classifications of minerals according to either: (i) the modernized **Dana's classes** or (ii) **Strunz's classes**. Strunz's classes adopted in the mineral properties table are presented in detail in Section 11.3. However, the naming of minerals is not based on such a logical scheme. The careful description and identification of minerals often requires highly specialized physical or/and chemical techniques such as inorganic spectrochemical analysis (AAS, AES, XRF) and measurement of common physical properties (e.g., density, microhardness, optical properties, X-ray lattice parameters, etc.). However, for historical reasons, the names of minerals were not arrived at in an analogous scientific manner. Minerals may be given names on the basis of some physical property (e.g., barite from the Greek, baryos, meaning heavy due to its density) or chemical composition (e.g., germanite from its germanium content), or they may be named after the locality of discovery (e.g., aragonite from the Spanish region of Aragon), a public figure (e.g., perovskite from the Russian Count Perowski), a mineralogist (e.g., haüyne from the French mineralogist René-Just d'Haüy), or almost any other subject considered appropriate. An international committee, the Commission on New Minerals and New Mineral Names of the International Mineralogical Association (IMA), now reviews all new mineral descriptions and judges the appropriateness of new mineral names as well as the scientific characterization of newly discovered mineral species. As for all the chemical compounds, each mineral can also be identified by its chemical abstract registered number [CAS RN].

11.2.2 Chemical Formula and Theoretical Chemical Composition

The theoretical chemical formula of a mineral is unique and identifies only one species. Nevertheless, the actual chemical composition is usually variable within a limited range owing to the isomorphic substitutions (diadochy), and/or occasionally the presence of traces of impurities. The relative atomic or molecular mass (based on $^{12}C = 12.000$) of minerals is calculated from the theoretical formula using the last value of atomic masses adopted by the International Union of Pure and Applied Chemistry (IUPAC) in 1995, and the theoretical chemical composition is commonly expressed in percentage by weight (wt%) of elements and sometimes oxides for oxygenated minerals.

11.2.3 Crystallographic Properties

Minerals, with few exceptions (i.e., amorphous species), possess the ordered internal arrangement that is characteristic of crystalline solids. When conditions are favorable, they may be bounded by smooth plane surfaces and assume regular geometric forms known as crystals. The study of crystalline solids and the principles that govern their growth, external shape, and external structure is called crystallography. Morphological crystallography refers to the study of the external form, or morphology of crystals. Crystals are formed from solutions, melts, and vapors. The atoms in these disordered states have a random distribution but with changing temperature and pressure (T, P), and concentration they may join in an ordered arrangement characteristic of the crystalline state. Most well-formed mineral crystals are the result of chemical deposition from a solid (or a melt) into an open space, such as a vug, or a cavity in a rock formation. The main crystallographic properties are the **crystal system**, the **space lattice parameters** expressed in picometers (1 pm $= 10^{-12}$ m) and plane angle in degrees (°), the **strukturbericht designation, Pearson's notation**, and the number of atoms or molecules per unit space lattice are listed. Finally, the **space group** and **point group** according to the international Hermann–Mauguin notation and the crystal **space lattice structure type** are also given when known.

11.2.4 Habitus or Crystal Form

Some crystals grow in a characteristic morphological form called their **crystal habit** or **habitus**. For example, quartz may grow to form crystals with a hexagonal outline and pyramid-like ends. Habitus is generally well-developed only if a mineral is allowed to grow in an environment without space limitations and in this case it is called **euhedral** (idiomorph or automorph). On the other hand, it is called **anhedral** (xenomorph, or allotriomorph) if no external form can be identified. If the habitus is partially developed, the mineral is called **subhedral** (subautomorph). The habitus or appearance of single crystals as well as the manner in which crystals grow together in aggregates are of considerable aid in mineral recognition. Terms used to express habit and state of aggregation are given below. Single crystals, i.e., minerals in isolation or distinct crystals may be described as: (i) **acicular** (needlelike), (ii) **capillary** or **filiform** (hairlike or threadlike), (iii) **bladed** (lamellar, tabular), and (iv) **columnar** (prismatic). For aggregates, i.e., groups of distinct crystals the following terms are used: (v) **dendritic** (branching); (vi) **reticulated** (lattice-like); (vii) **divergent** or **radiated** (radiating), (viii) **drusy** (layer of small crystals on a surface). Parallel or radiating groups of individual crystals are described as: (ix) **columnar**, (x) **bladed**, (xi) **fibrous**, (xii) **stellated** (starlike), (xiii) **globular**, (xiv) **botryroidal** (bunch of grapes), (xv) **reniform** (kidney-shaped masses), (xvi) **mammillary,** (xvii) **colloform**. A mineral aggregate composed of scales or lamellae is described as: (xviii) **foliated**, (xix) **micaceous**, (xx) lamellar or tabular, and (xxi) **plumose**. Miscellaneous terms are: (xxii) **stalactitic**, (xxiii) **concentric**, (xxiv) **pisolitic**, (xxv) **oolitic**, (xxvi) **banded**, (xxvii) **massive**, (xxviii) **amygdaloidal**, (xxix) **geode**, and (xxx) **concentric**.

11.2.5 Color

The color variations of a nonmetallic mineral are often the result of ionic trace impurities in the crystal space lattice structure. Since the impurities vary from sample to sample, the color may vary. Some nonmetallic minerals have no color and are referred to as colorless. Because of variability in color, which sometimes can be extreme, color is one of the least useful properties for identifying nonmetallic minerals even though it is probably the most obvious one. Therefore, the origin of the color of a mineral can be explained in by three types of electronic transitions in the crystalline solids. (i) According to the **crystal field theory (CFT)** the color of nonmetallic minerals is often due to traces of transition element cations inside the crystal lattice of minerals. Actually, all the first group of first transition elements (from Ti to Cu) have partially filled 3d electron shell orbitals. The electrostatic interactions between the 3d electrons with the electric field imposed by the lattice of surrounding coordinating anions is responsible for the degeneration of the electron energy levels found in the free atom. (ii) Another possible origin of the color of nonmetallic minerals is due to the **charge transfer transitions (CTT)**. The charge transfer electronic transitions occur when valence electrons transfer back and forth between adjacent cations. Several important charge transfer transitions have energies within the visible region and therefore cause selective absorption. A characteristic of the absorption spectrum is that the intensity of absorption depends of particular orientations. This phenomenon gives rise to the important property of pleochroism discussed in Section 11.2.14. (iii) Finally, other important electronic transitions within minerals which cause color are **electron color centers** and **hole color centers**. Actually, in some ionic solids having lattice defects such as anion vacancies, electrons occupy vacancies in order to preserve the overall electric neutrality. For instance, fluorite and rock salt are common minerals exhibiting F-centers (from German, Farben, meaning color). In contrast, a hole color center arises when an electron is missing from a location normally occupied by an electron pair. Smoky quartz and amethyst are common examples of minerals exhibiting a hole color center.

Color is much more useful in identifying metallic minerals. Actually the origin of color in metallic minerals depends on the energy involved in the electronic transition between the conduction band and valence band described in the theory of bands (see Chapter 4). Therefore, it directly depends on the energy gap of the minerals.

11.2.6 Diaphaneity or Transmission of Light

The interaction of electromagnetic radiation with minerals only depends on the particular region of the spectra considered. As a general rule, the visible region (i.e., light wavelength between 380 nm and 700 nm) is considered in optical mineralogy. Two main categories of mineral can be clearly identified: (i) transparent and translucent minerals which may transmit light to varying degrees and (ii) opaque minerals which do not transmit visible light at all. Actually, minerals which are transparent transmit light much like glass. These minerals are essentially solids with ionic or covalent bonds such as oxides, carbonates, silicates (e.g., calcite, quartz), or native elements (e.g., diamond). Minerals which are translucent transmit light on thin edges or in thin section. By contrast, opaque minerals do not transmit light even in thin section and comprise solids with metallic or partially metallic bonds characterized by a free electron cloud (Fermi gas) such as native elements (e.g, Cu, Ag, Au), most iron- and copper-bearing sulfides (e.g., CuS, FeS_2), and several transition metal oxides (e.g., Fe_3O_4 $FeTiO_3$, $FeCr_2O_4$). As a general rule, all minerals with a metallic luster are commonly opaque.

11.2.7 Luster

The term luster refers to the external appearance of the mineral owing to the reflection of light by its surface. The most important distinction to be made is between minerals with a metallic luster and a nonmetallic luster. Minerals with a **metallic** luster (e.g., pyrite) reflect visible light like polished metals and alloys, and are often very shiny. Nevertheless, some minerals with a metallic luster may tarnish on exposure to moist air and become less shiny, taking on a darker color. Minerals with a **nonmetallic** luster do not reflect light like metals. There are a variety of nonmetallic lusters, each is descriptive of its appearance. A luster resembling light reflected from the surface of broken window glass is termed **glassy** or **vitreous** (e.g., quartz). A mineral which reflects light as if it were coated by a thin film of oil has a **greasy** luster (e.g., calcite). A dull luster resembling the appearance of dry soil is termed **earthy** (e.g., limonite). Other nonmetallic lusters include **pearly** (e.g., moonstone), **resinous** (e.g., garnets, and realgar), **waxy** (e.g., turquoise), **silky** (e.g., tiger's eye quartz), and **adamantine** such as diamond luster.

11.2.8 Cleavage and Parting

Some minerals tend to break repeatedly along certain planes parallel to atomic planes (i.e., flat crystallographic surfaces) owing to the weakness in their atomic structure due to the lowest binding energy between adjacent atoms. These planes are referred to as cleavage surfaces defined by miller indices (*hkl*) with cleavage directions perpendicular to them [*hkl*]. The result is flat regular surfaces. Parallel cleavage surfaces represent a single cleavage direction. A break may be very well developed (i.e., perfect) in some crystals (e.g., micas, calcite), or it may be fairly obscure (e.g., beryl). Cleavage surfaces which are not parallel represent different cleavage directions. While cleavage surfaces tend to reflect light all in the same direction, rougher fracture surfaces scatter light reflected off them. As a result, cleavage surfaces are generally shinier than fracture surfaces. Sometimes cleavage may appear as a series of surfaces on one side of a sample which are parallel to each other but at different heights somewhat like steps. Such parallel surfaces can be recognized as a cleavage direction by the fact that they will reflect light all at once. Many minerals may be identified by their number of cleavage directions and plane angle(s) between cleavage directions. **Parting** is like cleavage, but only occurs along planes of structural weakness in twinned crystals.

11.2.9 Fracture

The way in which a mineral breaks is determined by the arrangement of atoms in its crystal structure and the strength of the different types of chemical bonds between atoms. All minerals may break somewhat randomly in any direction across a crystal. This type of breakage is called fracture and this word refers to the way minerals break along an uneven surface when they do not yield along cleavage or parting surfaces. Different kinds of fracture are designated as: (i) **conchoidal** fracture, a curving shell-like fracture similar to the way glass breaks with concentric rings, named after the smooth curving surface of a conch shell, (ii) **fibrous or splintery** fracture, which produces long splintery fibers (e.g., nephrite), (iii) **hackly**, and (iv) **uneven or irregular** fracture, which simply produces a rough, broken, and irregular surface.

11.2.10 Streak

Streak is the color of a mineral when finely powdered, found by rubbing the mineral against an unglazed, typically white, porcelain plate called a streak plate. Minerals with a Mohs hardness much greater than 6 do not give a streak because their hardness is higher than that of the silicate solids found in porcelain. Instead, they scratch the streak plate. Most soft colorless or pale-colored minerals have a white streak which is only visible against a dark-colored streak plate or if the mineral is rubbed against a hard, dark-colored mineral such as pyroxene or amphibole. Although the color of a mineral may vary, the streak is usually constant and is thus useful in mineral identification. Actually, the streak color of a mineral is usually the same regardless of the color of the whole mineral (and may or may not be the same as the color of the whole mineral). Thus streak is a more reliable property than the color of the mineral. Streak is particularly useful for identifying metallic minerals. For instance, the mineral pyrite, which is often referred to as "fools' gold" because of its resemblance to true gold, has a black streak, while the streak of true gold is yellow.

11.2.11 Tenacity

The cohesiveness of a mineral is known as tenacity. The following terms are used to describe tenacity in minerals: (i) **brittle** (i.e., breaks and powders easily); (ii) **malleable** (i.e., may be hammered into thin sheets); (iii) **sectile** (i.e., can be cut into thin shavings with a knife); (iv) **ductile** (i.e., can be drawn into a wire); (v) **flexible** (i.e., can be bent without breaking); and finally (vi) **elastic** (i.e., will spring back after being bent, e.g. mica).

11.2.12 Density and Specific Gravity

Density (symbol d or ρ) is a physical quantity equal to the ratio of mass to volume expressed in the SI in $kg.m^{-3}$ while specific gravity or relative density (SG, D) is a dimensionless physical quantity equal to the ratio of the density of a mineral at a given temperature to the density of water at a referenced temperature, usually defined as the temperature of its maximum density (3.98°C). Qualitatively, mineral specific gravities can be classified in petrology as: **barylites** (heavy or dense with a density above that of quartz 2650 $kg.m^{-3}$), and **coupholites** (light with a density below that of quartz). The specific gravity of a mineral is frequently an important aid in its identification, particularly in working with fine crystals or gemstones, when other tests would injure the specimens. The specific gravity of a crystalline substance depends on: (i) its chemical composition, and (ii) its crystal space lattice structure type. For instance, the two allotropic forms of carbon, diamond and graphite, owing to their different space lattice structures exhibit different densities; diamond, owing to its closely packed cubic structure has a specific gravity of 3.512, while graphite, with its loosely packed hexagonal lamellar structure has specific gravity 2.230.

11.2.13 Mohs Hardness

The resistance of a mineral to scratching and abrasion is called its hardness. Hardness is a direct measure of the binding energy of atoms in the solid. In mineralogy, a series of 10 common standard minerals was chosen arbitrarily by the German mineralogist Friedrich Mohs [1,2] in 1824 as a relative scale, by which the relative hardness of any mineral can be told. Hence, the following minerals arranged in order of increasing hardness comprise what is known as the Mohs scale of hardness:

1. Talc
2. Gypsum
3. Calcite
4. Fluorite
5. Apatite
6. Orthoclase
7. Quartz
8. Topaz
9. Corundum
10. Diamond

However, the numbers of Mohs scale do not have a linear relationship to hardness. Diamond is actually much more than 10 times the hardness of talc. The numbers only represent a simple qualitative ordering of minerals by hardness. A mineral's hardness may be determined by attempting to scratch an object of known hardness such as glass or a coin with the mineral. Alternately, one may attempt to scratch a mineral sample with an object of known hardness. A harder mineral can scratch a softer mineral, but a softer mineral cannot scratch a harder mineral. Often a powder is produced when attempting to make a scratch. This powder may be mistaken for a scratch. Remember that while a powder can be wiped away, a scratch must remain after the removal of powder. Usually, the groove of a scratch in glass can be felt with a fingernail. The following common materials serve in addition to the above scale: the hardness of the fingernail is roughly 2.5, a US copper coin about 3, the stainless steel (AISI 440C grade) of a pocket knife blade a little over 6, window glass 5.5, and the quenched carbon steel of a knife blade file 6.5, and grit paper made of Carborundum® (silicon carbide) 9.25.

For more accurate and quantitative measurements, hardness of minerals can be measured, such as for metal and alloys by micro-indentation testing, e.g., microhardness Vickers and Knoop tests (see hardness tests definitions and scales in the appendices, Chapter 15). Nevertheless, since 1824 several other scales for reporting hardness of minerals have been established in order to improve the reliability and accuracy of hardness measurements. The Rosival scale is an improved version of the original Mohs scale using corundum in place of diamond as reference mineral, with a hardness number defined as 1000. Later in 1933, Ridgeway et al.[3] suggested an extended Mohs hardness. For this purpose, they introduced the hardness of fused silica between those of feldspar and quartz and the hardness of garnet between those of quartz and topaz respectively. However, more precise scientific studies were performed, in the 1960s, in the former Soviet Union, by Povarennikh.[4,5] This rational scale of hardness was established from accurate measurements of the hardness of minerals by the micro-Vickers diamond indenter. Hence, the original Mohs scale was increased by five additional synthetic minerals in order to decrease the gap existing between the hardness of corundum and that of diamond. Moreover, he also reported the crystallographic plane used in the measurement in order to take into account the anisotropy of mechanical properties of crystals. A brief comparison of these scales is shown in Table 11.3.

11.2.14 Optical Properties

In optical mineralogy, transparent and translucent minerals are classified according to five possible classes (i.e., with different indicatrices) to which a crystal can belong: isotropic, uniaxial (+/−), or biaxial (+/−). The main physical quantity is the **refractive index** (n). Refractive index at a given wavelength is the dimensionless ratio of velocity of the light in the crystal to the velocity of the light in a vacuum, $n = c/v$. In most tables and databases, this index is measured, unless otherwise specified, for a standardized wavelength taken equal to that of the D-line (589 nm) of the atomic transition resonance in the sodium vapor. Therefore, the symbol used is n_D. The three-dimensional surface describing the variation in refractive index with

11
Minerals,
Ores and
Gemstones

Table 11.3 Comparison of hardness scales of minerals

Mineral or material (original Mohs minerals in **bold**)	Mohs scale (1822)	Rosival scale	Ridge-way scale (1933)	Povarennikh scale (1962)	Micro Vickers hardness	Knopp hardness
Talc	1	0.033	1	1 (001)	n.a.	65
Graphite	1.5	n.a.	n.a.	n.a.	25	n.a.
Gypsum	2	1.25	2	n.a.	68	32–25
Halite	n.a.	n.a.	n.a.	2	n.a.	n.a.
Fingernail	2.5	n.a.	n.a.	n.a.	n.a.	150
Galena	2.5	n.a.	n.a.	3 (100)	n.a.	n.a.
Calcite	3	4.5	3	n.a.	110	135–190
Fluorite	4	5	4	4 (111)	n.a.	163–310
Scheelite	4.5–5	n.a.	n.a.	5 (111)	n.a.	n.a.
Apatite	5	6.5	5	n.a.	n.a.	430–435
Knife blade	5.5	n.a.	n.a.	n.a.	n.a.	n.a.
Feldspar (orthoclase)	6	37	6	n.a.	n.a.	625
Magnetite	5.5–6	n.a.	n.a.	6 (111)	n.a.	n.a.
Pyrex glass	6.5	n.a.	n.a.	n.a.	n.a.	n.a.
Silica (fused)	n.a.	n.a.	7	n.a.	n.a.	n.a.
Quartz	7	120	8	7 (10$\bar{1}$1)	n.a.	820–875
Garnet	6	n.a.	9	n.a.	n.a.	n.a.
Stellite®	n.a.	n.a.	8	n.a.	n.a.	n.a.
Zircon	7.5	n.a.	n.a.	n.a.	n.a.	n.a.
Porcelain (hard)	8	n.a.	n.a.	n.a.	n.a.	n.a.
Topaz	8	175	10	8 (001)	n.a.	1340
Zirconia (fused)	9	n.a.	11	n.a.	n.a.	n.a.
Tantalum carbide	n.a.	n.a.	11	n.a.	n.a.	n.a.
Alumina (fused)	n.a.	n.a.	12	n.a.	n.a.	n.a.
Tungsten carbide (WC + Co cermet)	n.a.	n.a.	12	n.a.	n.a.	n.a.
Corundum	9	1000	n.a.	9 (11$\bar{2}$0)	2100	2100
Carborundum®	n.a.	n.a.	13	n.a.	n.a.	2400
Titanium carbide	n.a.	n.a.	n.a.	10	n.a.	n.a.
Aluminum boride	n.a.	n.a.	n.a.	11	n.a.	n.a.
Sialon®	9	n.a.	n.a.	n.a.	n.a.	n.a.
Boron carbide	n.a.	n.a.	n.a.	12	n.a.	n.a.
Boron carbide	n.a.	n.a.	n.a.	13	n.a.	n.a.
Borazon®	n.a.	n.a.	14	14	n.a.	4700
Diamond	10	140,000	15	15	8000	7000

relationship to vibration direction of incident light is called the **indicatrix. Isotropic** materials have the same refractive index regardless of vibration directions, and the indicatrix is a sphere. Isotropic materials are: (i) crystals with a cubic crystal lattice, (ii) amorphous materials (vitreous or glassy), or fluids (e.g., liquids and gases). In contrast, a solid material with more than one principal refractive index is called anisotropic. **Anisotropic** materials are divided into two subgroups: (i) solid materials having a tetragonal, hexagonal, and rhombohedral crystal space lattice structure are called **uniaxial**, while (ii) solid materials having a orthorhombic, monoclinic, and triclinic crystal space lattice structure are called **biaxial**. Uniaxial crystals belong to either the rhombohedral, the hexagonal or tetragonal crystal systems and possess two mutually perpendicular refractive indices, ε and ω, which are called the principal refractive indices. Intermediate values occur and are called $\varepsilon\prime$, a non-principal refractive index. The uniaxial indicatrix is an ellipsoid, either **prolate** ($\varepsilon > \omega$), termed positive $(+)$, or **oblate** ($\varepsilon < \omega$), termed negative $(-)$. In either case, ε coincides with the single optic axis of the crystal, yielding the name uniaxial. The optic axis also coincides with the axis of highest symmetry of the crystal, either the four-fold for tetragonal minerals or the three- or six-fold of the hexagonal class. Because of the symmetry imposed by the three-, four- or six-fold axis, the indicatrix contains a circle of radius ω perpendicular to ε (i.e., perpendicular to the optic axis). Light vibrating parallel to any of the vectors would exhibit the refractive ω. Light vibrating parallel to the optic axis would exhibit ε. Light that does not vibrate parallel to one of these special directions within the uniaxial indicatrix would exhibit a refractive index intermediate to ϵ and ω and is termed $\varepsilon\prime$. Biaxial crystals belong to either the orthorhombic, monoclinic, or triclinic crystal systems and possess three mutually perpendicular refractive indices (α, β, and γ), which are the principal refractive indices. Intermediate values also occur and are labeled α' and γ'. The relationship between these values is $\alpha < \alpha' < \beta < \gamma' < \gamma$. The three principal refractive indices coincide with three mutually perpendicular lattice vector directions, **a**, **b**, and **c**, which form the framework for the biaxial indicatrix. The point group symmetry of the biaxial indicatrix is 2/m 2/m 2/m. In orthorhombic minerals the **a**, **b**, and **c** vectors coincide with either the two-fold axes or normals to mirror planes. In monoclinic minerals, either **a**, **b**, or **c** vectors coincide with the single symmetry element. In triclinic minerals, no symmetry elements necessarily coincide with the axes of the indicatrix.

Birefringence (double refraction), δ, is the physical quantity equal to the mathematical difference between the largest and smallest refractive index for an anisotropic mineral. **Pleochroism** is the property of exhibiting different colors as a function of the vibration direction. **Dichroism** refers to uniaxial minerals, while **trichroism** refers to biaxial minerals. **Dispersion** is the variation of the refractive index with the wavelength of incident light. Opaque minerals are more commonly studied in reflected light and that study is generally called ore microscopy or ore metallography. The main parameter is the reflective index (R_λ for a given wavelength, λ (generally taken as 650 nm), expressed in percentage of intensity of light reflected to intensity of incident light.

11.2.15 Magnetism

Some minerals can be strongly ferromagnetic, i.e., they are readily attracted by a permanent magnet. For instance, lodestone or magnetite, ilmenite and pyrrhotite are the most common ferromagnetic minerals found both in igneous and sedimentary rocks. Sometimes, hematite may be contaminated by magnetite and appear to be ferromagnetic.

11.2.16 Luminescence

This effect is noticed when some minerals when submitted to long (366 nm, Wood's light) or short (256 nm) wavlength UV-irradiation can simultaneously emit visible light (**fluorescence**) or emit light after the irradiation has stopped (**phosphorescence**). In particular cases, minerals owing to the relaxation of point defects (e.g., Schottky, Frenkel, or F-center) in their crystal lattices can also emit light when submitted to heating (**thermoluminescence**), or when scratched or rubbed to a rough surface (**triboluminescence**). **Cathodoluminescence** is displayed by some particular minerals when they are irradiated by a beam of high energy charged particles (electrons, protons, etc.). For instance, several uranium ores are fluorescents, while sphalerite bombarded by an electron beam is cathodoluminsecent and was the first compound used as screen-phosphor in spynthariscopes, while fluorspar (fluorite) is thermoluminescent and was used in dating of archeological stoneware.

11.2.17 Piezoelectricity and Pyroelectricity

The property of piezoelectricity refers to the development of a momentary electric current when crystals are squeezed suddenly in certain crystallographic directions. The strain caused by squeezing is very small and purely elastic. Some common rock-forming minerals exhibit piezoelectricity such as: low-temperature quartz, tourmaline, sphalerite, boracite, and topaz. The property of pyroelectricity refers to the development of a momentary electric charge displacement when crystals are submitted to a sudden change in temperature. The effect is proportional to the magnitude of the temperature change. Like piezoelectricity, pyroelectricity is strongly dependent on the crystal symmetry. Tourmaline is the most common pyroelectric mineral.

11.2.18 Play of Colors and Chatoyancy

Interference of light either at the surface or in the interior of a mineral may produce a series of colors as the angle of incident light changes. **Iridescence** refers to the color effects occurring when an incident ray of light falls upon a thin transparent layer; some fraction of the incident light is reflected, whilst the remainder is refracted and is subsequently reflected back along a different path parallel to the first. Owing to the path difference between the two rays, interference occurs with either cancelation when in phase opposition or intensification when in phase. **Opalescence** consists of the reflection of incident light by small lamellar or spherical inclusions in the mineral giving a milky or pearly aspect (e.g. precious opal). Some specimens of labradorite show colors ranging from blue to green or yellow with changing angle of incident light. This iridescence, also called **schiller** and **labradorescence**, is the result of light scattered by extremely fine exsolution lamellae. **Chatoyancy** consists of wavy band of light that is seen to pass accross the mineral at right angles to the direction of the fibres. **Asterism** is a star-like effect of minerals cut in cabochon, caused by the reflection of light from fibers or fibrous cavities crossing at $60°$ (six-rayed star) or $90°$ (four-rayed star).

11.2.19 Radioactivity

Several uranium- and thorium-containing minerals and ores are obviously radioactive owing to the decay of the actinides and particularly uranides elements that they contain, while some minerals such as zircon, which should not be radioactive, owing to the isomorph (diadochy) substitution of cations Zr(IV) by U(IV) and Th(IV) are often radioactive. The metamicte (amorphization) habitus is due to the destruction of the crystal lattice structure by self-irradiation.

11.2.20 Miscellaneous Properties

Halite tastes salty, while epsomite exhibits a bitter taste. Some sulfides when rubbed exhibit the odor of sulfur. Talc feels slippery like soap. Plagioclase may have tiny parallel grooves called striations on cleavage surfaces. Striations are best seen when a cleavage surface is oriented to reflect light. Micas break into thin sheets which are elastic. The sheets may be bent and will spring back. Transparent varieties of gypsum with obvious cleavage may be flexible. They can be bent but will not spring back. Some varieties of alkali feldspar may have an irregular pattern of veins.

11.2.21 Chemical Reactivity

Reaction to common strong mineral acids (e.g., HCl, HNO_3, H_2SO_4, HF, or aqua regia) either diluted or concentrated is another important and rapid identification test. A few minerals effervesce, i.e., they produce bubbles, when a few drops of dilute hydrochloric acid are placed on a sample. Calcite evolves carbon dioxide vigorously and can easily be detected using this test, while dolomite must be powdered (streak powder) or the acid must be concentrated and heated to produce the same strong effervescence. When fluorite is heated in concentrated sulfuric acid it evolves the hazardous hydrogen fluoride gas which strongly corrodes the test tube glassware.

11.3 Strunz Classification of Minerals

1/Class I: Native Elements
  1/A1 Metals and intermetallic alloys
  1/A2 Semimetals and nonmetals
2/Class II: Sulfides and Sulfosalts
  2/A1 Alloys and alloylike compounds
  2/B1 Sulfides, selenides, and tellurides (M:S, Se, Te > 1:1)
  2/C1 Sulfides, selenides, and tellurides (M:S, Se, Te = 1:1)
  2/D1 Sulfides, selenides, and tellurides (M:S, Se, Te < 1:1)
  2/E1 Sulfosalts ($M_mX_nY_p$ with X = As, Sb, Bi and Y = S, Se, Te)
  2/F1 Arseno-sulfides
3/Class III: Halides (i.e., Fluorides, Chlorides, Bromides, and Iodides)
  3/A1 Simple halides $[M_mX_n]$
  3/B1 Anhydrous double halides $[M_mX_nY_p]$
  3/C1 Hydrous double halides $[M_mX_n \cdot pH_2O]$
  3/D1 Oxyhalides and hydroxyhalides $[M_mX_nO_p(OH)_q]$
4/Class IV: Oxides and Hydroxides
  4/A1 Oxides $[M_2O]$ and $[MO]$
  4/B1 Oxides $[M_3O_4]$, spinel type
  4/C1 Oxides $[M_2O_3]$
  4/D1 Oxides $[MO_2]$
  4/E1 Oxides $[M_2O_5]$ and $[MO_3]$
  4/F1 Hydroxides and hydrated oxides $[M_m(OH)_n$ and $M_mO_n \cdot pH_2O]$
  4/G1 Vanadium oxides (V^{4+}/V^{5+})
  4/H1 Uranyl hydroxides and hydrates, with $[UO_2]^{2+}$
  4/J1 Arsenites
  4/K1 Sulfites, selenites, and tellurites
  4/L1 Iodates
5/Class V: Nitrates, Carbonates, and Borates
  5/A1 Nitrates with $[NO_3]$
  5/B1 Anhydrous carbonates $[CO_3]$, without additional anion
  5/C1 Anhydrous carbonates $[CO_3]$, with additional anions
  5/D1 Hydrous carbonates $[CO_{[3}]$, without additional anion
  5/E1 Hydrous carbonates $[CO_3]$, with additional anions
  5/F1 Uranylcarbonates with $[UO_2]^{2+}$ and $[CO_3]$

5/G1   Nesoborates with $[BO_3]$ with $[BO_4]$
5/H1   Soroborates with $[B_2O_5]$ to $[B_2O_7]$ with $[B_3O_5]$ to $[B_6O_{10}]$
5/J1   Inoborates $[B_2O_4]$ to $[B_6O_{10}]$
5/K1   Phylloborates with complex $[B_x(O,OH)_y]$ groups
5/L1   Tectoborates with $[BO_2]$ to $[B_6O_{10}]$
6/Class VI: Sulfates, Chromates, Molybdates, and Tungstates
6/A1   Anhydrous sulfates $[SO_4]$, without additional anion
6/B1   Anhydrous sulfates $[SO_4]$, with additional anions (e.g., F, Cl, O, OH)
6/C1   Hydrous sulfates $[SO_4]$, without additional anion
6/D1   Hydrous sulfates $[SO_4]$, with additional anions
6/F1   Chromates $[CrO_4]$
6/G1   Molybdates $[MoO_4]$, wolframates (tungstates) $[WO_4]$
7/Class VII: Phosphates, Arsenates, and Vanadates
7/A1   Anhydrous phosphates $[PO_4]$, without additional anion
7/B1   Anhydrous phosphates $[PO_4]$, with additional anions (F, C1, O, OH)
7/C1   Hydrous phosphates without additional anion
7/D1   Hydrous phosphates with additional anions
7/E1   Uranylphosphates and vanadates $[UO_2]$ and $[PO_4]$, $[AsO_4]$, $[UO_2]$ and $[V_2O_8]$
8/Class VIII: Silicates
8/A1   Nesosilicates (i.e., isolated tetrahedrons) $[SiO_4]$
8/B1   Nesosubsilicates with tetrahedron additional anion
8/C1   Sorosilicates (i.e., paired tetrahedrons) $[Si_2O_7]$
8/D1   Unclassified silicates
8/E1   Cyclosilicates (i.e., ring tetrahedrons) $[SiO_3]$, $[Si_3O_9]$, $[Si_4O_{12}]$, $[Si_6O_{18}]$
8/F1   Inosilicates (i.e., layers, chains, bands and ribbons of tetrahedrons) $[SiO_3]$,$[Si_4O_{11}]$
8/J1   Tectosilicates (i.e., framework or network tetrahedrons) $[(Si,A1)O_2]$, $[SiO_2]$
8/H1   Phyllosilicates (i.e., foliated)
9/Class IX: Organic Compounds and Mineraloids
9/A1   Salts of organic acids
9/B1   Hydrocarbon without nitrogen
9/C1   Resins and wax compounds
9/D1   Hydrocarbons with nitrogen

11.4 Mineral Properties Table

Table 11.4 (pages 410–466) contains descriptions of the most common minerals, with their main physical, chemical, and mineralogical properties.

Column 1: mineral name according to the International Mineralogical Association (IMA); its **Chemical Abstract Registered Number** in brackets [CAS RN], the **synonyms** (syn.) and its **etymology** are also indicated when available.

Column 2: (1) the theoretical **chemical formula**, (2) the relative **atomic or molecular mass** (^{12}C = 12,000) of minerals based on the previous theoretical formula using the last value of atomic masses adopted by the IUPAC in 1995, (3) the theoretical **chemical composition** expressed in percentage by weight (wt%), (4) the common **trace impurities**, (5) the **coordinance number** of major cations, and finally (6) the Strunz's **mineralogical class**.

Column 3: main crystallographic properties: (1) the **crystal system**, (2) the **space lattice parameters** expressed in picometers (1 pm = 10^{-12} m) and plane angle in degrees (°), (3) the **strukturbericht designation**, (4) **Pearson's notation**, (5) the number of atoms or molecules per unit space lattice (Z). Finally, (6) the **space** and (7) **point group** according to the international Hermann–Mauguin notation and the crystal **space lattice structure type** are also listed when known.

Column 4: mineral optical properties either in transmitted light with refractive index for isotropic (n_D), uniaxial (ε, ω), or biaxial (α, β, γ, $2V$), with the birefringence (δ), at 589 nm or in reflected light the reflective index (R) at 650 nm.

Column 5: Mohs hardness and Vickers hardness in brackets when available.

Column 6: density or range of density in kg.m^{-3}.

Column 7: the common habitus, the color, the diaphaneity, the luster, the luminescence, the streak, the cleavage planes, the fracture, the twinning planes, the chemical reactivity, the deposits and other miscellaneous properties of the mineral.

11.5 Mineral Synonyms

Absite = brannerite
Acerdese = manganite
Achrematite = mimetite + wulfenite
Achroite = elbaite
Acmite = aegirine
Adularia = orthoclase
Agalmatolite = talc or pyrophyllite or pinite
Agate = layered chalcedony (quartz)
Agricolite = eulytite
Alabaster = gypsum
Alexandrite = chrysoberyl
Allcharite = goethite
Allenite = pentahydrite
Allopalladium = stibiopalladinite
Altmarkite = lead amalgam
Alum = hydrous alkali aluminum sulfates
Alurgite = Mg, Fe, Mn muscovite
Alvite = Hf-zircon
Amazonite = microline
Amethyst = purple quartz
Amianthus = tremolite, actinolite, chrysotile
Amosite = grunerite
Ampangabeite = samarskite
Amphigene = leucite
Anarakite = paratacamite
Andrewsite = hentschelite + rockbridgeite + chalcosiderite
Annivite = Bi-tetrahedrite
Antimonite = stibnite
Aplome = andradite
Applelite = calcite
Apyrite = rubellite (elbaite)
Aquamarine = beryl
Arduinite = mordenite
Argentite = acanthite
Argyrose = argentite-acanthite
Argyrythrose = pyrargyrite
Arizonite = pseudorutile
Arkansite = brookite
Asbestos = tremolite, actinolite, chrysotile
Ashtonite = mordenite
Asphaltum = mineral pitch

(Continues page 467)

Table 11.4 Mineral properties

Mineral name (IMA) [CAS RN] (Synonyms) (Etymology)	Theoretical chemical formula, relative molecular mass, coordinance number, major impurities, mineral class (Strunz)	Crystal system, lattice parameters, Strukturbericht, Pearson symbol, Z, point group, space group, structure type	Optical properties	Mohs hardness (HM)	Density (ρ, kg.m^{-3})	Other mineralogical, physical, and chemical properties
Acanthite [21548-73-2] (syn, silver glance, argentite) (from the Greek, akanta, arrow, and after the Latin, argentum, silver)	Ag_2S MM = 247.8024 87.06wt% Ag 12.94wt% S Coordinance Ag(4) (Sulfides and sulfosalts)	Monoclinic a = 422.9 pm b = 692.8 pm c = 786.2 pm β = 99.58° C34, mC6 (Z = 4) S.G. P2$_1$/m AuTe$_2$ type	Biaxial (n.a.) R = 33%	2–2.5	7300	**Habitus:** acicular, octahedral, blocky, skeletal, arborescent. **Color:** black or lead gray. **Streak:** shining black. **Diaphaneity:** opaque. **Luster:** metallic. **Fracture:** subconchoidal, sectile. **Cleavage:** (001) poor, (110) poor. Argentite (cubic) is stable above 179°C while acanthite is stable below 179°C, melt at 825°C. Electrical resistivity 1.5–2.0 x 10^5 $\mu\Omega$cm.
Actinolite (syn, byssolite, nephrite, smaragdite; emerald green, asbestos) (from the Greek, aktinos, meaning gray, an allusion to actinolite fibrous nature)	$Ca_2(Mg,Fe)_5[Si_8O_{22}(OH)_2]$ 51–55wt% SiO_2 11–13wt% CaO 13–20wt% MgO 6–15wt% FeO Traces Mn and Al. (Inosilicates, double-width unbranched chains and band)	Monoclinic a = 984.00 pm b = 180.52 pm c = 527.5 pm 104°70 (Z = 2) Tremolite type	Biaxial (–) α = 1.613–1.688 β = 1.627–1.697 γ = 1.638–1.705 δ = 0.017–0.027 2V = 65–86°	5–6	3020–3440	**Habitus:** acicular, bladed, radial, fibrous. **Color:** pale yellow, green, or dark green. **Diaphaneity:** translucent to transparent. **Streak:** white. **Cleavage:** (110) perfect. **Fracture:** splintery, brittle. **Chemical:** insoluble in strong mineral acids. Other: attracted by an electromagnet, dielectric constant 6.6 to 6.82. **Deposits:** metamorphic rocks.
Aegirine (syn, acmite) (after the Teutonic god of the sea. Acmite is from the Greek, point, an allusion to the pointed crystals)	$NaFe[Si_2O_6]$ MM = 231.00416 9.95wt% Na 24.18wt% Fe 24.32wt% Si 41.56wt% O (Inosilicates, double-width unbranched chains and band)	Monoclinic a = 965.8 pm b = 879.5 pm c = 529.4 pm 107.42° (Z = 4) Diopside type	Biaxial (–) α = 1.72–1.778 β = 1.74–1.819 γ = 1.757–1.839 δ = 0.037–0.061 2V = 60–90° OAP (010)	6–6.5	3550	**Habitus:** acicular. **Color:** blackish green or reddish brown. **Diaphaneity:** transparent to opaque. **Luster:** vitreous, resinous. **Streak:** yellowish gray. **Fracture:** brittle. **Deposits:** sodium-rich igneous nepheline syenites.
Akermanite	$Ca_2Mg[Si_2O_7]$ (Sorosilicates, pair)	Tetragonal a = 784.35 gm c = 501.0 pm (Z = 2) Melilite type	Uniaxial (+)	n.a.	n.a.	
Albite (syn, clevelandite) (from the Latin, alba, an allusion to the common white color)	$Na[Si_3AlO_8]$ An0–Ab100 MM = 263.02222 8.30wt% Na 0.76wt% Ca 10.77wt% Al 31.50wt% Si	Triclinic a = 814.4 pm b = 1278.7 pm c = 716 pm α = 93.17° β = 115.85° γ = 87.65°	Biaxial (+) α = 1.527–1.533 β = 1.531–1.536 γ = 1.538–1.542 δ = 0.009–0.010 2V = 76–82°	6–6.5	2620	**Habitus:** blocky, striated, granular. **Color:** white, gray, greenish gray, or bluish green. **Diaphaneity:** transparent to translucent. **Luster:** vitreous (ie, glassy). **Cleavage:** (001) perfect, (010) good. **Twinning:** albite {010}, pericline {010}. **Fracture:** uneven. **Streak:** white. **Deposits:** magmatic and pegmatitic rocks.

Mineral	...osilicates, framework)	Chemistry	S.G., PI	Optical	Hardness	S.G.	Properties
Allanite	(Ca,Ce,Th)₂(Al,Fe,Mn,Mg)(Al,Fe)₂ O·OH[Si₂O₇][SiO₄] (Neso-sorosilicates)		Monoclinic a = 898 pm b = 575 pm c = 1023 pm 115° Epidote group	Biaxial (−/+) α = 1.690-1.791 β = 1.700-1.815 γ = 1.706-1.828 $2V$ = 40-123° OAP (010)	5-6.5	3400–4200	**Habitus:** massive, lamellar, granular. **Color:** reddish black or brownish red. **Streak:** white. **Diaphaneity:** transparent to translucent. **Luster:** vitreous (i.e., glassy), resinous. **Fracture:** brittle, conchoidal. **Cleavage:** (110). **Deposits:** metamorphic and pegmatitic rocks.
Almandine (syn., almandite) (after the locality, Alabanda, in Asia Minor)	Fe₃Al₂(SiO₄)₃ MM = 497.75338 10.84wt% Al 33.66wt% Fe 16.93wt% Si 38.57wt% O Coordination Fe(8), Al(6), Si(4) (Nesosilicates)		Cubic a = 1152.6 pm (Z = 8) P.G. 432 S.G. Ia3d Garnet group (Pyralspites series)	Isotropic n_D = 1.830	7-8	4318	
Alunite	KAl₃(SO₄)₂(OH)₆ MM = 414.214334 9.44wt% K 19.54wt% Al 15.48wt% S 54.08wt% O 1.46wt% H Coordination K(12), Cu(6), S(4) (Sulfates, chromates, molybdates, and tungstates)		Trigonal (Rhombohedral) a = 697 pm c = 1738 pm (Z = 3) P.G. 3m S.G. P3m	Uniaxial (+) ε = 1.592 ω = 1.572 δ = 0.020	3.5-4	2600–2900	**Habitus:** rhombohedral. **Color:** white, gray. **Streak:** white, gray. **Diaphaneity:** transparent to translucent. **Luster:** vitreous. **Fracture:** conchoidal. **Cleavage:** good (001), poor (101).
Amber (syn., succinite, bernstein)	C₄H₆O₂ MM = 180.2902 11.18wt% H 79.94wt% C 8.87wt% O (Mineraloids)		Amorphous	Isotropic	n.a.	n.a.	
Amblygonite	(Li, Na)PO₄(F,OH) Coordination Li(6), Na(6), P(4) (Phosphates, arsenates, and vanadates)		Triclinic a = 519 pm b = 712 pm c = 504 pm α = 112.02° β = 97.82° γ = 68.12° (Z = 2) P.G. $\bar{1}$ S.G. P$\bar{1}$	Biaxial (−) α = 1.590 β = 1.600 γ = 1.620 δ = 0.03 $2V$ = 52-90°	6	3000	**Habitus:** equant. **Color:** white, green. **Streak:** white. **Diaphaneity:** transparent to translucent. **Luster:** vitreous, greasy. **Fracture:** subconchoidal. **Cleavage:** (100) perfect, (110) good, (011) perfect. **Twinning:** {111}.
Analcime (syn, analcite, analcidite) (from the Greek, analcis, weak,	NaAlSi₂O₆·(H₂O) MM = 220-15398 10.44wt% Na		Cubic a = 1373.8 pm (Z = 16)	Isotropic n_D = 1.479-1.493	5.5	2260	**Habitus:** euhedral crystals, granular, massive. **Color:** white, grayish white, greenish white, yellowish white, or reddish white. **Diaphaneity:** transparent to translucent. **Luster:** vitreous (i.e., glassy). **Luminescence:** fluorescent. **Streak:** white.

11

Minerals, Ores and Gemstones

Table 11.4 (continued)

Mineral name (IMA) [CAS RN] (Synonyms) (Etymology)	Theoretical chemical formula, relative molecular mass, coordination number, major impurities, mineral class (Strunz)	Crystal system, lattice parameters, Strukturbericht, Pearson symbol, Z, point group, space group, structure type	Optical properties	Mohs hardness (HM)	Density (ρ, kg.m^{-3})	Other mineralogical, physical, and chemical properties
Analcime (continued) referring to a weak electrical charge developed on rubbing)	12.26wt% Al 25.51wt% Si 0.92wt% H 50.87wt% O Coordination Na(6), Si(4), Al(4) (Tectosilicates)	P.G. 432 S.G. I4;ad				Cleavage: (100) poor. Fracture: subconchoidal, uneven. Twinning: {100}, {110}. Deposits: occurs frequently in basalts and other basic igneous rocks associated with other zeolites.
Anatase (from the Greek, anatasis, tall direction owing to the great vertical space lattice parameter)	TiO_2 $MM = 79.8788$ 59.94wt% Ti 40.06wt% O Traces Fe, Sn Coordination Ti(6) (Oxides and hydroxides)	Tetragonal $a = 379.3$ pm $c = 951.2$ pm $C4$, $tP6$ ($Z = 2$) P.G. 422 S.G. I4$_1$/amd Anatase type	Uniaxial (+) $\epsilon = 2.488$ $\omega = 2.561$ $\delta = 0.073$ Dispersion strong	5.5–6	3877	Habitus: acicular, prismatic, massive. Color: reddish brown, yellowish brown, black or bluish violet. Diaphaneity: transparent, translucent, to opaque. Luster: adamantine. Streak: grayish black. Fracture: uneven. Cleavage: (111). Chemical: insoluble in water, slightly soluble in HCl, HNO_3, sol. HF and in hot H_2SO_4, or $KHSO_4$. Attacked by molten Na_2CO_3.
Andalusite [12183-80-1] (syn., chiastolite: carbonaceous inclusions) (after the province of Andalucia (Spain))	$Al_2O[SiO_4] = Al_2SiO_5$ $MM = 162.04558$ 33.30wt% Al 17.33wt% Si 49.37wt% O Traces of Fe, Mn Coordination Al(5), Al(6), Si(4) (Nesosubsilicates)	Orthorhombic $a = 779.59$ pm $b = 789.83$ pm $c = 555.83$ pm ($Z = 4$) P.G. mmm S.G. Pnnm	Biaxial (−) $\alpha = 1.629–1.640$ $\beta = 1.633–1.644$ $\gamma = 1.638–1.650$ $\delta = 0.009–0.010$ $2V = 73–86°$	6.5–7.5	3130–3160	Habitus: acicular, blocky, prismatic, euhedral crystals. Color: usually pink, white, rose, dark green, gray, brown, red, or green or with clouded inclusions. Luster: vitreous (i.e, glassy). Diaphaneity: transparent to translucent. Cleavage: (110) distinct, (100) indistinct, (010) poor. Fracture: uneven, splintery, brittle. Streak: white. Chemical: insoluble in strong mineral acids but attacked by molten alkali hydroxides (e.g., NaOH) and carbonates (e.g., Na_2CO_3). Heating in $Co(NO_3)_2$ gives the Thénard blue color. Nonfusible but transforms to sillimanite on heating. Other: dielectric constant: 8.28. Deposits: metamorphosed peri-aluminous sedimentary rocks.
Andesine (syn., acmite) (after Andes Mountains, South America)	$(Na,Ca)(Si,Al)_4O_8$ An40–Ab60 $MM = 268.61671$ 5.14wt% Na 5.97wt% Ca 14.06wt% Al 27.18wt% Si 47.65wt% O (Tectosilicates, framework)	Triclinic $a = 815.5$ pm $b = 129$ pm $c = 916$ pm $Z = 6$	Biaxial (+/−) $\alpha = 1.543–1.554$ $\beta = 1.547–1.559$ $\gamma = 1.552–1.562$ $\delta = 0.008–0.009$ $2V = 78–84°$	7	2670	Habitus: granular, crystalline. Color: white or gray. Diaphaneity: transparent to translucent. Luster: vitreous (i.e., glassy). Streak: white. Cleavage: (001) perfect, (010) good. Fracture: uneven. Deposits: magmatic and metamorphic rocks.
Andradite (syn., demantoid: green, melanite: black, topazolite: yellow)	$Ca_3Fe_2(SiO_4)_3$ $MM = 508.1773$ 23.66wt% Ca	Cubic $a = 1204.8$ pm ($Z = 8$) P.G. 432	Isotropic $n_D = 1.887$	6.5–7	3859	Habitus: dodecahedral crystals, massive. Color: black, yellowish brown, red, greenish yellow, or gray. Diaphaneity: transparent to translucent. Luster: vitreous (i.e, glassy). Streak: white. Fracture: subconchoidal. Deposits: igneous and metamorphic

Top-of-table fragments (continued from previous row):
adamantine luster)
Coordinance Ca(6), Fe(6), Si(4) (Nesosilicates)
(U-grandite series)

Mineral	Composition	Crystallography	Optical	Hardness	S.G.	Physical properties
Anglesite [7446-14-2] (syn. lead spar, lead vitriol) (after the island of Anglesey, Wales, UK)	$PbSO_4$, MM = 303.2636, 68.32wt% Pb, 10.57wt% S, 21.10wt% O, Coordinance Pb(6), S(4) (Sulfates, chromates, molybdates, and tungstates)	Orthorhombic, $a = 848.0$ pm, $b = 539.8$ pm, $c = 695.8$ pm, $(Z = 4)$, P.G. mmm, S.G. P2₁nma, Barite type	Biaxial (+), $\alpha = 1.878$, $\beta = 1.883$, $\gamma = 1.894$, $\delta = 0.017$, $2V = 68\text{–}73°$, Dispersion strong	2.5–3	6380	**Habitus:** tabular, prismatic, granular, stalactitic. **Color:** white, gray, or yellow. **Diaphaneity:** transparent to translucent. **Luster:** adamantine. **Streak:** white. **Cleavage:** perfect (001), good (210). **Twinning:** {011}. **Fracture:** conchoidal, brittle. **Chemical:** melts at 1087°C. **Deposits:** secondary, weathered deposits of lead ore.
Anhydrite [7778-18-9] (From the Greek, anhydros, meaning dry, in contrast to gypsum, which is hydrated)	$CaSO_4$, MM = 136.1416, 29.44wt% Ca, 23.55wt% S, 47.01wt% O, Coordinance Ca(6), S(4) (Sulfates, chromates, molybdates, and tungstates)	Orthorhombic, $a = 699.1$ pm, $b = 699.6$ pm, $c = 623.8$ pm, $(Z = 4)$, P.G. mmm, S.G. Ccmm, Anhydrite type	Biaxial (+), $\alpha = 1.569\text{–}1.574$, $\beta = 1.574\text{–}1.579$, $\gamma = 1.609\text{–}1.618$, $\delta = 0.040\text{–}0.045$, $2V = 36\text{–}45°$	3–3.5	2980	**Habitus:** massive, granular, fibrous, plumose. **Color:** colorless, white, bluish white, violet white, or dark gray. **Luster:** vitreous, pearly. **Diaphaneity:** transparent to translucent. **Streak:** white. **Cleavage:** (010) perfect, (001) good. **Fracture:** conchoidal, brittle. **Chemical:** melts at 1460°C. **Deposits:** sedimentary beds, gangue in ore veins, and in traprock zeolite occurrences.
Ankerite	$Ca(Mg,Fe,Mn)(CO_3)_2$, Coordinance Ca(6), Fe(6), C(3) (Nitrates, carbonates, and borates)	Trigonal (Rhombohedral), $a = 482$ pm, $c = 1614$ pm, $a_{Rh} = 605.0$ pm, $47°00'$, $(Z = 3)$, P.G. 3, S.G. R$\bar{3}$, Dolomite type	Uniaxial (−), $\epsilon = 1.500\text{–}1.548$, $\omega = 1.690\text{–}1.750$, $\delta = 0.182\text{–}0.202$, Dispersion strong	3.5–4	2930–3100	**Habitus:** crystalline, massive. **Color:** white, reddish white, brownish white, or gray. **Luster:** vitreous (i.e, glassy). **Diaphaneity:** transparent to translucent. **Streak:** white. **Fracture:** brittle, subconchoidal. **Cleavage:** (1011) perfect. **Twinning:** {0001} and {1010}.
Annabergite (syn. nickel bloom) (after the locality Annaberg, Germany)	$Ni_3(AsO_4)_2 \cdot 8(H_2O)$, MM = 598.03064, 29.44wt% Ni, 25.06wt% As, 2.70wt% H, 42.81wt% O (Phosphates, arsenates and vanadates)	Monoclinic, P.G. 2/m	Biaxial (−), $\alpha = 1.622$, $\beta = 1.658$, $\gamma = 1.687$, $\delta = 0.065$, $2V = 78°$, Dispersion weak.	2	3000–3100	**Habitus:** earthy, encrustations, massive. **Color:** green, green white, or apple green. **Luster:** pearly. **Diaphaneity:** transparent to translucent. **Streak:** light green. **Cleavage:** (010) perfect. **Fracture:** brittle.
Anorthite (syn. indianite) (from the Greek, an, and orthos, not upright, an allusion to the oblique crystals)	$Ca_2[Si_2Al_2O_8]$, An₁₀₀ Ab₀, MM = 277.40806, 0.41wt% Na, 13.72wt% Ca, 18.97wt% Al, 20.75wt% Si, 46.14wt% O, Coordinance Ca(7), Si(4), Al(4) (Tectosilicates, framework)	Triclinic, $a = 817.7$ pm, $b = 1287.7$ pm, $c = 1416.9$ pm, $\alpha = 93.33°$, $\beta = 115.60°$, $\gamma = 91.22°$, $(Z = 8)$, P.G. $\bar{1}$, S.G. P$\bar{1}$	Biaxial (−), $\alpha = 1.577$, $\beta = 1.585$, $\gamma = 1.590$, $\delta = 0.013$, $2V = 78°$, Dispersion weak.	6	2760	**Habitus:** granular, euhedral crystals, striated. **Color:** white gray, or reddish white. **Diaphaneity:** transparent to translucent. **Luster:** pearly vitreous (i.e, glassy). **Cleavage:** (001) perfect, (010) good. **Twinning:** albite {010}, pericline {010}. **Fracture:** uneven. **Streak:** white. **Deposits:** magmatic and metamorphic rocks.

11
Minerals, Ores and Gemstones

Table 11.4 (continued)

Mineral name (IMA) [CAS RN] (Synonyms) (Etymology)	Theoretical chemical formula, relative molecular mass, coordinance number, major impurities, mineral class (Strunz)	Crystal system, lattice parameters, Strukturbericht, Pearson symbol, Z, point group, space group, structure type	Optical properties	Mohs hardness (HM)	Density (ρ, kg.m^{-3})	Other mineralogical, physical, and chemical properties
Antigorite (2M$_1$) (Fibrous: chrysotile)	$Mg_3(OH)_4Si_4O_{10}$ Coordinance Mg(6), Si(4) (Phyllosilicates, layered)	Monoclinic $a = 532$ pm $b = 950$ pm $c = 1490$ pm $101.9°$ $(Z = 2)$ P.G. 2/m S.G. C2/m	Biaxial (−) $\alpha = 1.560$ $\beta = 1.570$ $\gamma = 1.570$ $\delta = 0.007$ $2V = 20$–$60°$ Dispersion weak	3–4	2600	**Habitus:** platy, massive. **Diaphaneity:** transparent to translucent. **Luster:** resinous, silky. **Cleavage:** (001) perfect. **Fracture:** uneven, flexible. **Streak:** white. Color: green, yellow.
Antimony (syn. stibium) (from the Arabic, aluthmud, the medieval Latin, antimonium, originally applied to stibnite)	Sb $MM = 121.75$ Coordinance Sb(3) (Native elements)	Trigonal (Rhombohedral) $a = 429.96$ pm $c = 1125.16$ pm A7, hR2 $(Z = 2)$ P.G. $\bar{3}$m S.G. $\bar{3}$m α-Arsenic type	Uniaxial (n.a.) $n_D = 1.70$–1.80 $R = 75\%$	3–3.5	6600	**Habitus:** massive, lamellar, massive, reticulate. **Color:** tin white. **Diaphaneity:** opaque. **Luster:** metallic. **Streak:** lead gray. **Cleavage:** (0001) perfect. **Fracture:** brittle.
Antlerite	$Cu_3SO_4(OH)_4$ $MM = 354.73108$ 53.74wt% Cu 9.04wt% S 36.08wt% O 1.14wt% H Coordinance Cu(6), S(4) (Sulfates, chromates, molybdates, and tungstates)	Orthorhombic $a = 824$ pm $b = 1199$ pm $c = 603$ pm $(Z = 4)$ P.G. mmm S.G. P2$_1$/aam	Biaxial(+) $\alpha = 1.726$ $\beta = 1.738$ $\gamma = 1.789$ $\delta = 0.063$ $2V = 53°$	3.5–4	3900 (010).	**Habitus:** prismatic, tabular. **Color:** white, gray. **Streak:** green, gray. **Diaphaneity:** transparent to translucent. **Luster:** vitreous. **Fracture:** uneven. **Cleavage:** perfect
Apatite (from the Greek, apatos, misleading, owing to the confusion with beryl, tourmaline, and olivine)	$Ca_5(PO_4)_3(OH, F, Cl)$ $MM = 504.30248$ 39.74wt% Ca 18.43wt% P 38.07wt% O 3.77wt% F Coordinance Ca(6), P(4) (Phosphates, arsenates, and vanadates)	Hexagonal $a = 938$ pm $c = 686$ pm $(Z = 2)$ P.G. 6/m S.G. P6$_3$/m Apatite type	Uniaxial (−) $\epsilon = 1.624$–1.666 $\omega = 1.629$–1.667 $\delta = 0.001$–0.007 Dispersion moderate	5	3100–3350	**Habitus:** prismatic, colloform, massive, granular, earthy. **Color:** white, yellow green, red, or blue. **Diaphaneity:** transparent to translucent. **Luster:** subresinous. **Streak:** white. **Cleavage:** (0001) indistinct, (1010) indistinct. **Fracture:** conchoidal. **Chemical:** soluble in HCl, and HNO$_3$.
Aphthitalite (syn. glaserite)	$(K,Na)_3Na(SO_4)_2$ (Sulfates, chromates,	Trigonal	Biaxial (+)	3	2700	**Habitus:** massive, encrustations. **Color:** white.

Mineral (name / etymology)	Chemical	Crystallography	Optical	H	ρ	Properties
[471-34-1] (after the Spanish locality of Aragon where the mineral was first discovered)	$MM = 100.0872$ 40.04wt% Ca 12.00wt% C 47.96wt% O Traces of Sr (up to 5.6wt%), Mg, Fe and Zn Coordination Ca(6), C(3) (Carbonates, aragonite group)	$a = 574.1$ pm $b = 796.8$ pm $c = 459.9$ pm $(Z = 4)$ S.G. Pmcn P.G. mmm Aragonite type	$\alpha = 1.530\text{-}1.531$ $\beta = 1.680\text{-}1.681$ $\gamma = 1.685\text{-}1.686$ $\delta = 0.155\text{-}0.156$ $2V = 18\text{-}19°$ Dispersion weak	(280 HV)	2950	less, white, gray, yellowish white, or reddish white. **Luster:** vitreous (i.e., glassy). **Diaphaneity:** transparent to translucent. **Streak:** white. **Cleavage:** (010) distinct. **Twinning:** [110]. **Fracture:** subconchoidal. **Chemical:** readily dissolved by cold diluted strong mineral acids (e.g., HCl) with evolution of CO_2. Decomposed at 825°C giving off CO_2 and CaO. Meigen's spot test: it exhibits a pink to violet color after immersion in boiling $Co(NO_3)_2$ solution. **Other:** dielectric constant 7.4. **Deposits:** fossil skeletons, with gypsum and celestine in marl and clays, near geysers and stalactites in caverns.
Arsenic (syn, arsenicum) (from the Greek, arsenikon, a name originally applied to the mineral orpiment)	As $MM = 74.9216$ Coordination As(3) (Native elements)	Trigonal (rhombohedral) $a = 413.19$ pm $54.12°$ A7, $hR2$ $(Z = 2)$ P.G. $\bar{3}m$ S.G. R3m α-Arsenic type	Uniaxial (n.a.) R = 62%	3.5	5700	**Habitus:** nodular, reniform, lamellar. **Color:** tin white or gray. **Diaphaneity:** opaque. **Luster:** metallic. **Streak:** black. **Cleavage:** (0001) perfect. **Fracture:** uneven. **Deposits:** in ore veins in igneous crystalline rocks.
Arsenopyrite (syn, arsenical pyrite, mispickel) (after the mineral's chemical composition)	FeAsS $MM = 162.8346$ 34.30wt% Fe 46.01wt% As 19.69wt% S (Sulfides and sulfosalts) Coordination Fe(6)	Monoclinic $a = 576.0$ pm $b = 569.0$ pm $c = 578.5$ pm $112.23°$ EO_7, $mP24$ $(Z = 8)$ S.G. $B2_1/d$ P.G. 2/m Arsenopyrite type	Biaxial R = 52%	5.5-6	6100	**Habitus:** faces striated, euhedral crystals, prismatic. **Color:** tin white or light steel gray. **Luster:** metallic. **Diaphaneity:** opaque. **Streak:** black. **Cleavage:** (110) distinct. **Twinning:** {100}, {101}, {012}. **Fracture:** uneven, brittle Electrical resistivity 20 to 300 $\mu\Omega$.cm.
Atacamite (after the Atacama desert province in Northern Chile)	$Cu_2Cl(OH)_3$ $MM = 213.56672$ 59.51wt% Cu 1.42wt% H 16.60wt% Cl 22.47wt% O (Halides)	Orthorhombic $a = 602$ pm $b = 915$ pm $c = 685$ pm P.G. 222 S.G. $P2_1/nam$ $(Z = 4)$	Biaxial (−) $\alpha = 1.831$ $\beta = 1.861$ $\gamma = 1.880$ $\delta = 0.049$ $2V = 75°$ Dispersion strong	3-3.5	3760-3780	**Habitus:** acicular, striated prisms, euhedral crystals, fibrous, granular. **Color:** green, dark green, or blackish green. **Diaphaneity:** transparent to translucent. **Luster:** adamantine. **Streak:** apple green. **Cleavage:** (010) perfect. **Fracture:** conchoidal. **Deposits:** arid climates with oxidizable copper minerals.
Augelite	$Al_2PO_4(OH)_3$ $MM = 199.956547$ 26.99wt% Al 15.49wt% P 56.01wt% O 1.51wt% H Coordination Al(5, and 6), P(4) (Phosphates, arsenates, and vanadates)	Monoclinic $a = 1312$ pm $b = 799$ pm $c = 507$ pm $\beta = 112.25°$ $(Z = 4)$ P.G. 2/m S.G. C2/m	Biaxial(+) $\alpha = 1.574$ $\beta = 1.588$ $\gamma = 1.576$ $\delta = 0.014$ $2V = 51°$	5	2700	**Habitus:** tabular, massive. **Color:** colorless, white. **Streak:** white. **Diaphaneity:** transparent to translucent. **Luster:** vitreous. **Fracture:** uneven. **Cleavage:** (101) good.
Augite (syn, fassaite) (from the Greek, auge, luster)	(Ca, Na)(Mg, Fe, Al, Ti)(Si, Al)$_2$O$_6$ (Inosilicates, double chains) $\beta = 1.684\text{-}1.711$ $\gamma = 1.706\text{-}1.729$	Orthorhombic	Biaxial (+) $\alpha = 1.68\text{-}1.703$	5-6.5	3400	**Habitus:** massive, fibrous, columnar. **Color:** white, green, or black. **Diaphaneity:** translucent to opaque. **Luster:** vitreous, resinous. **Cleavage:** (110) perfect, (010) indistinct. **Fracture:** brittle, conchoidal. **Streak:** greenish gray. **Deposits:** basic igneous and metamorphic rocks.

Table 11.4 (continued)

Mineral name (IMA) [CAS RN] (Synonyms) (Etymology)	Theoretical chemical formula, relative molecular mass, coordination number, major impurities, mineral class (Strunz)	Crystal system, lattice parameters, Strukturbericht, Pearson symbol, Z, point group, space group, structure type	Optical properties	Mohs hardness (HM)	Density (ρ, kg.m⁻³)	Other mineralogical, physical, and chemical properties
Augite (continued)			$\delta = 0.026$ $2V = 40\text{-}52°$ Dispersion weak			
Autunite	$Ca(UO_2)_2(PO_4)_2 \cdot 10H_2O$ $MM =$ Coordination Ca(6), U(2), P(4) (Urano-phosphates)	Tetragonal $a = 700$ pm $c = 2.067$ pm ($Z = 4$) P.G. 422 S.G. I4 mmm	Biaxial (−) $\epsilon = 1.553$ $\omega = 1.577$ $\delta = 0.024$	2-2.5	3.150	**Habitus:** tabular, foliated. **Color:** yellow. **Diaphaneity:** transparent to translucent. **Luster:** adamantine. **Streak:** yellow. **Cleavage:** (001) perfect, (100) good, (010) good. **Fracture:** uneven. Radioactive.
Azurite (syn. chessylite) (from the Persian, lazhward, blue)	$Cu_3(CO_3)_2(OH)_2$ $MM = 344.67108$ 55.31wt% Cu 0.58wt% H 6.97wt% C 37.14wt% O Coordination Cu(5), C(3) (Nitrates, carbonates, and borates)	Monoclinic $a = 500.8$ pm $b = 584.4$ pm $c = 1033.6$ pm $\beta = 92.45°$ ($Z = 2$)	Biaxial (+) $\alpha = 1.730$ $\beta = 1.756$ $\gamma = 1.836$ $\delta = 0.108$ $2V = 68°$ Dispersion weak	3.5-4	3770	**Habitus:** tabular, massive, prismatic, stalactitic. **Color:** azure blue or very dark blue. **Diaphaneity:** transparent to translucent. **Luster:** vitreous (i.e., glassy). **Streak:** light blue. **Cleavage:** (011) perfect, (100) good. **Fracture:** conchoidal, brittle. **Deposits:** secondary mineral in the oxidized zone of copper ore deposits in association with malachite.
Baddeleyite [1314-23-4] (from J. Baddeley)	ZrO_2 $MM = 123.2228$ 74.03wt% Zr 25.97wt% O (Oxides and hydroxides)	Monoclinic $a = 514.54$ pm $b = 520.75$ pm $c = 531.07$ pm $99.23°$ ($Z = 4$) Baddeleyite type	Biaxial (−) $\alpha = 2.13$ $\beta = 2.19$ $\gamma = 2.2$ $\delta = 0.070$ $2V = 30°$	6.5	5500-6000	**Habitus:** tabular, crystalline. **Color:** brown, colorless, black. **Diaphaneity:** transparent, translucent, opaque. **Luster:** adamantine. **Streak:** white. **Fracture:** uneven. **Cleavage:** (001). **Chemical:** slightly sol. HCl, HNO₃, and dil. H₂SO₄. sol. hot. conc. H₂SO₄, and HF. Attacked by molten KHSO₄, NaOH, and Na₂CO₃. Melting point 2710°C.
Barite [7727-43-7] (syn. heavy spar, barytine, baryte) (from the Greek, baryos, meaning heavy)	$BaSO_4$ $MM = 233.3906$ 58.84wt% Ba 13.74wt% S 27.42wt% O Coordination Ba(6), S(4) (Sulfates, chromates, molybdates, and tungstates)	Orthorhombic $a = 887.8$ pm $b = 545.00$ pm $c = 715.2$ pm ($Z = 4$) P.G. mmm S.G. P2₁ nma Barite type	Biaxial (+) $\alpha = 1.634\text{-}1.637$ $\beta = 1.636\text{-}1.639$ $\gamma = 1.647\text{-}1.649$ $\delta = 0.011\text{-}0.012$ $2V = 37\text{-}40°$ Dispersion weak	3-3.5	4490	**Habitus:** tabular, prismatic, lamellar, massive, fibrous, cockscomb aggregates. **Color:** white, yellowish white, grayish white, brownish white, or dark brown. **Luster:** vitreous (i.e., glassy), pearly. **Diaphaneity:** transparent, translucent, opaque. **Fracture:** uneven. **Streak:** white. **Cleavage:** (001) perfect, (210) perfect. **Twinning:** [011]. **Luminescence:** phosphorescent. **Chemical:** decomposes at 1580°C, insoluble in HCl. **Deposits:** sedimentary rocks and late gangue mineral in ore veins.
Bastnaesite (syn. hamartite) after the Swedish locality Bastnas	$(Ce,La,Th)(CO_3)F$	Hexagonal P.G. 6m² (Ditrigonal	Uniaxial (+) $\epsilon = 1.717$ $\omega = 1.818$	4-5	4970	**Habitus:** prismatic, granular. **Color:** yellow or reddish brown. **Luster:** vitreous, greasy. **Cleavage:** (1011) imperfect, (0001) poor. **Fracture:** uneven. **Streak:** white.

Name	Formula / Composition	Crystal system	Optical	Hardness	Density	Properties
(after the French mineralogist E. Bertrand)			$\alpha = 1.589$ $\beta = 1.602$ $\gamma = 1.614$ $\delta = 0.023$ $2V = 76°$ Dispersion none			translucent to transparent. **Color:** colorless or pale yellow. **Streak:** white. **Cleavage:** (001) perfect, (110) distinct, (101) distinct. **Fracture:** brittle. **Deposits:** commonly found in Be-bearing pegmatites and may be derived from the alteration of beryl.
Beryl [1302-52-9] (syn. emerald: green, aquamarine: blue, morganite: pink, goshenite: colorless, heliodor: yellow; from the Greek, beryllos, signifying a blue-green color of a gemstone)	$Be_3Al_2[Si_6O_{18}]$ $MM = 537.50182$ 5.03wt% Be 10.04wt% Al 31.35wt% Si 53.58wt% O Coordinance Al(6), Si(4), and Be(4) Traces of Fe, Cr, Mg, Li, Na, K, Cs. (Cyclosilicates, ring)	Hexagonal $a = 921.5$ pm $c = 919.2$ pm $(Z = 2)$ P.G. 6/mmm S.G. P6/mcc Beryl type	Uniaxial (−) $\epsilon = 1.564\text{-}1.598$ $\omega = 1.565\text{-}1.602$ $\delta = 0.003\text{-}0.008$	7.5-8.0	2640	**Habitus:** crystalline, prismatic, columnar. **Color:** green, blue, yellow, colorless, or pink. **Luster:** vitreous, resinous. **Diaphaneity:** transparent to subtranslucent. **Streak:** white. **Cleavage:** (0001) imperfect. **Twinning:** {311}, {110}. **Fracture:** brittle, conchoidal. **Chemical:** insoluble in strong mineral acids. **Others:** dielectric constant 3.9 to 7.7. **Deposits:** mainly in granitic pegmatites.
Bindheimite (after the German chemist, J.J. Bindheim)	$Pb_2\text{~}Sb_2O_6(O,OH)$ (Oxides and hydroxides)	Cubic P.G. 432	Isotropic $n_D = 1.84\text{-}1.87$	4-5	4600-7300	**Habitus:** encrustations, earthy, encrustations. **Color:** yellow, greenish yellow, green, brownish white, or grayish white. **Luster:** greasy. **Streak:** light greenish brown. **Fracture:** conchoidal.
Biotite (1M) (syn. manganophyllite: Mn, lepidomelane: Fe) (after the French physicist, J.B. Biot)	$K(Mg,Fe)_3Si_3(Al,Fe)O_{10}(OH,F)_2$ Coordinance K(6), Fe(6), Mg(6), Si(4), Al(4) (Phyllosilicates, layered)	Monoclinic $a = 533$ pm $b = 931$ pm $c = 1016$ pm $\beta = 99.3°$ $(Z = 2)$ P.G. 2/m S.G. C2/m	Biaxial (−) $\alpha = 1.565\text{-}1.625$ $\beta = 1.605\text{-}1.696$ $\gamma = 1.605\text{-}1.696$ $\delta = 0.040\text{-}0.080$ $2V = 0.32°$ Dispersion weak	2.5-3	2700-3300	**Habitus:** micaceous, foliated, lamellar, pseudo hexagonal. **Color:** dark brown, greenish brown, blackish brown, yellow, or white. **Diaphaneity:** transparent to opaque. **Luster:** vitreous, pearly. **Streak:** white gray. **Cleavage:** (001) perfect. **Twinning:** [310]. **Fracture:** uneven, elastic. **Deposits:** granitic rocks. Forms a series with phlogopite.
Bischofite [7791-18-6]	$MgCl_2 \cdot 6(H_2O)$ $MM = 203.301$ (Halides)	Monoclinic	Biaxial (+)			**Habitus:** fibrous, deliquescent, massive, granular. **Color:** colorless or white. **Luster:** vitreous, greasy. **Diaphaneity:** translucent to transparent. **Streak:** white. **Deposits:** marine evaporites. Dehydration occurs at 120°C. Soluble in water.
Bismuth (from the Arabic, biismid, having the properties of antimony)	Bi $MM = 208.9804$ Coordinance Bi(3) (Native elements)	Trigonal (Rhombohedral) $a = 474.60$ pm $57.23°$ A7, hR2 $(Z = 2)$ P.G. $\bar{3}$m S.G. R$\bar{3}$m α-Arsenic type	Uniaxial (n.a.) $n_D = 2.26$ $R = 68\%$	2-2.5	9750	**Habitus:** platy, lamellar, granular, reticulated. **Color:** silver white, pinkish white, or red. **Diaphaneity:** opaque. **Luster:** metallic. **Streak:** lead gray. **Cleavage:** (0001) perfect. **Fracture:** uneven.
Blodite (syn. bloedite)	$Na_2Mg(SO_4)_2 \cdot 4H_2O$	Monoclinic	Biaxial (−) $\alpha = 1.48$ $\beta = 1.48$ $\gamma = 1.48$ $\delta = 0.00$ $2V = 71°$ Dispersion strong	3	2230	**Habitus:** massive, granular. **Color:** colorless, green, yellow, or red. **Diaphaneity:** transparent to translucent.

11

Minerals, Ores and Gemstones

Table 11.4 (*continued*)

Mineral name (IMA) [CAS RN] (Synonyms) (Etymology)	Theoretical chemical formula, relative molecular mass, coordination number, major impurities, mineral class (Strunz)	Crystal system, lattice parameters, Strukturbericht, Pearson symbol, Z, point group, space group, structure type	Optical properties	Mohs hardness (HM)	Density (ρ, kg.m^{-3})	Other mineralogical, physical, and chemical properties
Boehmite [14457-84-2] (after the German geologist and paleontologist, J. Böhm)	γ-AlO(OH) $MM = 59.98828$ 44.98wt% Al 1.68wt% H 53.34wt% O Coordination Al(6) (Oxides and hydroxides)	Orthorhombic $a = 286.8$ pm $b = 1222.7$ pm $c = 370$ pm ($Z = 4$) P.G. mmm S.G. A2/mam Lepidocrite type	Biaxial (+) $\alpha = 1.646\text{-}1.650$ $\beta = 1.652\text{-}1.660$ $\gamma = 1.650\text{-}1.670$ $\delta = 0.015$ $2V = 80°$	3.5–4	3440	**Habitus:** flaky, nodular, pisolitic, massive. **Color:** white, light yellow, or yellowish green. **Diaphaneity:** transparent to translucent. **Luster:** vitreous, pearly. **Streak:** white. **Cleavage:** perfect (010). **Fracture:** brittle. **Deposits:** subtropical areas, lateritic soils develop on Al-bearing igneous rocks, major constituent of most bauxite ore.
Boracite	$Mg_3B_7O_{13}Cl$ $MM = 768.0744$ 18.99wt% Mg 19.71wt% B 9.22wt% Cl 52.08wt% O Coordination Mg(6), B(3 and 4) (Nitrates, carbonates, and borates)	Orthorhombic $a = 854$ pm $b = 854$ pm $c = 1270$ pm ($Z = 2$) P.G. m2m S.G. Pc2a	Biaxial (+) $\alpha = 1.662$ $\beta = 1.647$ $\gamma = 1.673$ $\delta = 0.011$ $2V = 82°$	7	2950	**Habitus:** pseudocubic. **Color:** white, yellow. **Diaphaneity:** transparent to translucent. **Luster:** vitreous. **Streak:** white. **Cleavage:** (111). **Fracture:** conchoidal, pyroelectric.
Borax [1303-96-4] (syn, tincal) (from the Arabic, buraq, white)	$Na_2B_4O_5(OH)_4 \cdot 10(H_2O)$ $MM = 381.36813$ 12.06wt% Na 11.34wt% B 5.29wt% H 71.32wt% O Coordination Na(6), B(3 and 4) (Nitrates, carbonates, and borates)	Monoclinic $a = 1185.8$ pm $b = 1067.4$ pm $c = 1267.4$ pm $106.58°$ ($Z = 4$) P.G. 2/m S.G. C2/c	Biaxial (−) $\alpha = 1.447$ $\beta = 1.469$ $\gamma = 1.472$ $\delta = 0.025$ $2V = 39\text{-}40°$	2–2.5	1730–1900	**Habitus:** prismatic, tabular, massive. **Color:** colorless, white, gray, or greenish white. **Diaphaneity:** translucent to opaque. **Luster:** resinous, greasy. **Streak:** white. **Cleavage:** (100) perfect, (110) perfect. **Fracture:** conchoidal, brittle. Sweet alkaline taste. Easily fusible acting as flux for several metal oxides (mp 75°C).
Bornite	Cu_5FeS_4 $MM = 501.823$ 63.31wt% Cu 11.13wt% Fe 25.56wt% S (Sulfides and sulfosalts) Coordination Cu(4), Fe(4)	Cubic $a = 1094$ pm ($Z = 8$) S.G. Fm-3m P.G. 4-32	Isotropic $R = 10\%$	3	6000	**Habitus:** cubic euhedral crystals, tarnishes to purple. **Color:** bronze. **Luster:** metallic. **Diaphaneity:** opaque. **Cleavage:** (111). **Twinning:** [111]. **Fracture:** conchoidal. Electrical resistivity 3 to 570 Ω.m.
Boulangerite	$Pb_5Cu_5Sb_4S_{11}$ $MM = 501.823$ (Sulfides and sulfosalts) Coordination Pb(7), Sb(3)	Monoclinic $a = 215.6$ pm $b = 235.1$ pm $c = 80.9$ pm $100.8°$ ($Z = 8$) S.G. P2/a	Biaxial (?) $R = 35\%$	2.5–3	6000–6200	**Habitus:** prismatic, tabular. **Color:** purple gray. **Luster:** metallic. **Diaphaneity:** opaque. **Streak:** gray. **Fracture:** conchoidal. **Cleavage:** (100). Electrical resistivity 2000 to 40,000 Ω.m.

Name	Formula / Composition	Crystal system	Optical	Hardness	Density	Properties
(syn. wheel ore endellionite) (after the French mineralogist, J.L. de Bournon)	$MM = 974.348$ 13.04wt% Cu 12.18wt% Sn 12.50wt% Sb 42.53wt% Pb 19.75wt% S (Sulfides and sulfosalts)	$a = 816$ pm $b = 870$ pm $c = 780$ pm $(Z = 4)$		n.a.	n.a.	opaque. **Luster:** metallic. **Cleavage:** (010) imperfect. **Fracture:** subconchoidal. **Streak:** gray.
Bradleyite	$Na_3Mg(PO_4)(CO_3)$ (Nitrates, carbonates, and borates)	Monoclinic	Biaxial	n.a.	3.970	
Brochantite (syn. blanchardite) (after the French geologist and mineralogist, A.J.M. Brochant de Villiers)	$Cu_4(SO_4)(OH)_6$ $MM = 452.29164$ 56.20wt% Cu 1.34wt% H 7.09wt% S 35.37wt% O (Sulfates, chromates, molybdates, and tungstates)	Monoclinic Prismatic (2/m) $a = 1306.7$ pm $b = 985.0$ pm $c = 602.2$ pm $(Z = 4)$	Biaxial (−) $\alpha = 1.728$ $\beta = 1.771$ $\gamma = 1.8$ $\delta = 0.072$ $2V = 72°$ Dispersion weak	3.5–4		**Habitus:** acicular, prismatic, druse. **Color:** emerald green or blackish green. **Luster:** vitreous, pearly. **Diaphaneity:** transparent to translucent. **Streak:** pale green. **Cleavage:** (100) perfect. **Fracture:** conchoidal, brittle. **Deposits:** secondary, formed in arid climates or in rapidly oxidizing copper sulfide deposits.
Bromargyrite (syn. bromyrite) (from Greek, bromos, stench and Latin, argentum, silver)	$AgBr$ $MM = 187.7722$ 57.45wt% Ag 42.55wt% Br (Halides)	Cubic $a = 577.45$ pm $(Z = 4)$ Rock salt type	Isotropic $n_D = 2.25$	1.5–2	5800	**Color:** bright yellow or amber yellow. **Luster:** adamantine-greasy. **Diaphaneity:** transparent to translucent. **Deposits:** oxidized portions of silver deposits.
Brookite (after the English mineralogist, H.J. Brucke)	TiO_2 $MM = 79.8788$ 59.94wt% Ti 40.06wt% O Coordinance Ti(6) (Oxides and hydroxides)	Orthorhombic $a = 545.6$ pm $b = 918.2$ pm $c = 514.3$ pm C2/, oP24 (Z = 8) P.G. mmm S.G. Pbca Brookite type	Biaxal (+) $\alpha = 2.583–2.584$ $\beta = 2.584–2.586$ $\gamma = 2.700–2.741$ $\delta = 0.117–0.158$ $2V = 0–30°$ Dispersion strong	5.5–6.0	4100–4140	**Habitus:** tabular. **Color:** reddish brown, yellowish brown, dark brown, black. **Diaphaneity:** transparent, translucent, opaque. **Luster:** submetallic, adamantine. **Streak:** yellowish white. **Fracture:** subconchoidal. **Cleavage:** (120). **Chemical:** insoluble in water, slightly soluble in HCl, HNO_3, soluble in HF and in hot H_2SO_4, or $KHSO_4$. Attacked by molten Na_2CO_3.
Brucite [1309-42-8]	$Mg(OH)_2$ $MM = 58.31974$ Coordinance Mg(6) (Oxides and hydroxides)	Trigonal (hexagonal) $a = 314.7$ pm $c = 476.9$ pm C6, hP3 (Z = 1) P.G. 32 S.G. P-3ml CdI_2 type	Uniaxial (+) $\epsilon = 1.560–1.590$ $\omega = 1.580–1.600$ $\delta = 0.012–0.020$ Dispersion strong	2.5	2390	**Habitus:** tabulated, foliated. **Color:** white, green. **Diaphaneity:** transparent to translucent. **Luster:** pearly, vitreous. **Streak:** white. **Fracture:** sectile. **Cleavage:** (001). Melting point: 350° C.
Bytownite (after Bytown old name of Ottawa, Ontario, Canada)	$(Ca,Na)(Si,Al)_4O_8$ An80–Ab20 $MM = 275.01042$ 1.67wt% Na 11.66wt% Ca 17.66wt% Al	Triclinic $a = 817$ pm $b = 1285$ pm $c = 1316$ pm $(Z = 7)$ P.G.$\bar{1}$	Biaxial (+/−) $\alpha = 1.563–1.572$ $\beta = 1.568–1.5784$ $\gamma = 1.573–1.583$ $\delta = 0.010–0.011$	7	2710	**Habitus:** granular, euhedral, striated. **Color:** white or gray. **Diaphaneity:** translucent to transparent. **Luster:** vitreous (i.e, glassy). **Cleavage:** (001) perfect, (010) good. **Fracture:** uneven. **Streak:** white. **Deposits:** magmatic and metamorphic rocks.

11

Minerals, Ores and Gemstones

Table 11.4 (continued)

Mineral name (IMA) [CAS RN] (Synonyms) (Etymology)	Theoretical chemical formula, relative molecular mass, coordinance number, major impurities, mineral class (Strunz)	Crystal system, lattice parameters, Strukturbericht, Pearson symbol, Z, point group, space group, structure type	Optical properties	Mohs hardness (HM)	Density (ρ, kg.m^{-3})	Other mineralogical, physical, and chemical properties
Bytownite (continued)	22.47wt% Si 46.54wt% O (Tectosilicates, framework)		2V = 80–88			
Calaverite (after Staislaus mine, Carson Hill, Calaveras Co. California)	AuTe$_2$ (Sulfides and sulfosalts)	Monoclinic a = 718 pm b = 440 pm c = 507 pm 90°13′ C34.mC6 (Z = 2) P.G. 2/m S.G. C2/m	Biaxial	2.5	9040	**Habitus:** striated, massive, crystalline, fine. **Color:** white. **Luster:** metallic. **Diaphaneity:** opaque.
Calcite [471-34-1] (syn. travertine, nicols, calcareous spar) (from the Latin, *calx*, meaning quicklime)	CaCO$_3$ MM = 100.0872 40.04wt% Ca 12.00wt% C 47.96wt% O Traces of Mg, Fe, Mn, and Zn Coordinance Ca(6), C(3) (Nitrates, carbonates, and borates)	Trigonal (Rhombohedral) a = 498.9 pm c = 1706.2 pm (Z = 6) P.G. –32/m S.G. R-3c Calcite type	Uniaxial (–) ϵ = 1.486–1.550 ω = 1.658–1.740 δ = 0.172–0.190 Dispersion strong	3.0 (110 HV)	2715–2940	**Habitus:** crystalline, coarse, stalactitic, massive. **Color:** colorless, white, pink, yellow, or brown. **Luster:** vitreous (i.e., glassy). **Diaphaneity:** transparent, translucent, to opaque. **Streak:** white. **Cleavage:** (1011) perfect. **Twinning:** (0001), [1014], {0118}. **Fracture:** brittle, conchoidal. **Luminescence:** fluorescent. **Chemical:** decomposes at 1330°C giving CaO and readily dissolves in diluted acids with evolution of carbon dioxide. Alizarine's spot test: a soln. of 0.5wt%. Alizarine S in dil. HCl colors the calcite crystal deep pink, while dolomite, ankerite, and magnesite remain colorless. **Deposits:** sedimentary rocks.
Calomel [10112-91-1] (syn. horn quicksilver) (from the Greek, *kalos*, beautiful, and *melas*, black)	Hg$_2$Cl$_2$ MM = 472.0854 84.98wt% Hg 15.02wt% Cl Coordinance Hg(5) (Halides)	Tetragonal a = 447.8 pm c = 1091.0 pm (Z = 4) S.G. I4/mmm P.G. 4/mmm	Uniaxial (+) ω = 1.973 ϵ = 2.656 δ = 0.683	1.5–2	6480	**Habitus:** tabular, pyramidal, prismatic, earthy. **Color:** white, yellowish gray, gray, yellowish white, or brown. **Luster:** adamantine, resinous. **Luminescence:** fluorescent. **Diaphaneity:** translucent to subtranslucent. **Streak:** pale yellowish white. **Cleavage:** (100), (011). **Twinning:** {110}. **Fracture:** conchoidal, sectile. **Deposits:** oxidized mercury deposits. Easy fusible (mp 525°C). Insoluble in water.
Carnallite (after the German mining engineer, R. von Carnall)	KMgCl$_3$·6H$_2$O MM = 277.85308 14.07wt% K 8.75wt% Mg 4.35wt% H 38.28wt% Cl 34.55wt% O Coordinance K(6), Mg(6) (Halides)	Orthorhombic a = 956 pm b = 1605 pm c = 2256 pm (Z = 12) P.G. mmm S.G. Pban	Biaxial (+) α = 1.467 β = 1.474 γ = 1.496 δ = 0.029 2V = 70°	2.5	1602	**Habitus:** massive, granular, pseudo hexagonal, fibrous. **Color:** colorless, milky white, reddish white or yellowish white. **Luster:** greasy (oily), vitreous. **Luminescence:** fluorescent. **Streak:** white, red. **Cleavage:** none. **Fracture:** conchoidal. **Deposits:** marine evaporites.
Carnotite (after the French chemist, M.A.	K$_2$(UO$_2$)$_2$(VO$_4$)$_2$·3(H$_2$O) MM = 902.1760	Orthorhombic a = 1047 pm	Biaxial (–) α ~ 1.75	1.5–2	4200	**Habitus:** earthy, encrustations, platy. **Color:** canary yellow or greenish yellow. **Luster:** dull, earthy. **Diaphaneity:** translucent. **Streak:** light yellow. **Cleavage:**

Name	Formula / Composition	Crystal system	Optical properties	Hardness	Density	Properties
	0.67wt% H 26.60wt% O Coordination V(5), V(4), K(9), U(2) (Uranylphosphates and uranylvanadates)	P.G. 2/m S.G. P2₁/a	2V = 38–44°			
Cassiterite [18282-10-5] (syn. tin ore, wood tin) (from the Greek kassiteros, tin)	SnO₂ MM = 150.7088 78.77wt% Sn 21.23wt% O Coordination Sn(6) (Oxides and hydroxides)	Tetragonal a = 473.8 pm c = 318.8 pm C4, tP6 (Z = 2) P.G. 422 S.G. P4/mmm Rutile type	Uniaxial (+) ε = 1.990–2.010 ω = 2.093–2.100 δ = 0.096–0.098 Dispersion strong	6–7	6994	**Habitus:** acicular, prismatic, massive, botryoidal, fibrous 'wood tin'. **Color:** yellow, reddish, brownish black, or white. **Diaphaneity:** transparent to opaque. **Luster:** adamantine. **Streak:** brownish white. **Fracture:** subconchoidal, irregular. **Cleavage:** (100) perfect, (110) indistinct. **Deposits:** granite pegmatites and alluvial placer deposits. Melting point: 1630° C.
Celestite [7759-02-6] (syn. celestine) (from the Latin, coelestis, meaning celestial)	SrSO₄ MM = 183.6836 47.70wt% Sr 17.46wt% S 34.8wt% O Coordination Sr(6), S(4) (Sulfates, chromates, molybdates, and tungstates)	Orthorhombic a = 835.9 pm b = 535.2 pm c = 686.6 pm (Z = 4) P.G. mmm S.G. P2₁nma Barite type	Biaxial (+) α = 1.622 β = 1.624 γ = 1.631 δ = 0.009 2V = 51° Dispersion moderate	3–3.5	3970	**Habitus:** tabular, radiated fibrous, crystalline, massive, granular. **Color:** colorless, bluish white, yellow white, or reddish white. **Diaphaneity:** transparent to translucent. **Luster:** vitreous (i.e., glassy). **Streak:** white. **Cleavage:** (001) perfect, (210) good. **Fracture:** uneven to conchoidal, brittle **Chemical:** decomposes at 1607° C. **Deposits:** sedimentary rocks.
Celsian (after the Swedish astronomer and natural scientist, A. Celsius)	Ba[Si₂Al₂O₈] (Tectosilicates, framework)	Monoclinic	Biaxial (+) α = 1.58–1.584 β = 1.585–1.587 γ = 1.594–1.596 δ = 0.012–0.014 2V = 86–90°	6–6.5	3250	**Habitus:** massive, granular, euhedral. **Color:** white or yellow. **Diaphaneity:** transparent. **Luster:** vitreous (i.e., glassy). **Cleavage:** (001) perfect, (010) good. **Fracture:** brittle, uneven. **Streak:** white. **Deposits:** contact metamorphic rocks with significant barium.
Cerussite [598-63-01] (syn. white lead ore) (from the Latin, cerussa, meaning white lead)	PbCO₃ MM = 267.2092 77.54wt% Pb 4.49wt% C 17.96wt% O Coordination Pb(6), C(3) (Nitrates, carbonates, and borates)	Orthorhombic a = 615.2 pm b = 843.6 pm c = 519.5 pm (Z = 4) P.G. mmm S.G. Pmcn Aragonite type	Biaxial (–) α = 1.804 β = 2.076 γ = 2.079 δ = 0.274 2V = 8–14° Dispersion strong	3–3.5	6580	**Habitus:** reticulate, tabular, massive, granular, crystalline, clustered. **Color:** colorless, gray, smoky gray, or grayish white. **Diaphaneity:** transparent to translucent. **Luster:** adamantine. **Streak:** white. **Cleavage:** (110) distinct, (021) distinct. **Twinning:** {110}, {130}. **Fracture:** conchoidal, brittle. **Chemical:** decomposes at 315° C giving off PbO and CO₂. Soluble in strong mineral acids with evolution of CO₂.
Cervantite (after the locality Cervantes, Spain)	Sb₂O₄ MM = 307.4976 79.19wt% Sb 20.81wt% O (Oxides and hydroxides)	Orthorhombic Pyramidal mm2	Biaxial	4–5	4000–6600	**Habitus:** reniform, earthy, acicular. **Color:** yellow, yellowish orange, white, or cream. **Luster:** vitreous–pearly. **Diaphaneity:** transparent to translucent. **Streak:** light yellow. **Cleavage:** (001) perfect. **Fracture:** conchoidal. **Deposit:** alteration product of stibnite.
Chabasite	CaSi₄Al₂O₁₂·6H₂O Coordination Ca(7), Si(4), and Al(4) (Tectosilicates, framework)	Trigonal (Rhombohedral) a = 1317 pm c = 1506 pm	Uniaxial (–) ε = 1.481 ω = 1.484 δ = 0.003	4–5	2100	**Habitus:** euhedral. **Luster:** vitreous. **Diaphaneity:** transparent to translucent. **Streak:** white. **Cleavage:** (101) perfect. **Twinning:** {100}, and {001}. **Fracture:** uneven.

11

Minerals, Ores and Gemstones

Table 11.4 (continued)

Mineral name (IMA) [CAS RN] (Synonyms) (Etymology)	Theoretical chemical formula, relative molecular mass, coordination number, major impurities, mineral class (Strunz)	Crystal system, lattice parameters, Strukturbericht, Pearson symbol, Z, point group, space group, structure type	Optical properties	Mohs hardness (HM)	Density (ρ, kg.m^{-3})	Other mineralogical, physical, and chemical properties
Chabasite (continued)		P.G. 32/m S.G. R32/m Zeolite group				
Chalcanthite [7758-99-8] (syn., copper vitriol, blue vitriol) (from the Greek, chalkos, copper, and, anthos, flower)	$CuSO_4 \cdot 5(H_2O)$ $MM = 249.686$ 25.45wt% Cu 4.04wt% H 12.84wt% S 57.67wt% O Coordinance Cu(6), S(4) (Sulfates, chromates, molybdates, and tungstates)	Triclinic $a = 610.45$ pm $b = 1072.0$ pm $c = 594.9$ pm $\alpha = 97.57°$ $\beta = 107.28°$ $\gamma = 77.43°$ $(Z = 2)$ P.G. $\bar{1}$ S.G. P$\bar{1}$	Biaxial (−) $\alpha = 1.514$ $\beta = 1.537$ $\gamma = 1.543$ $\delta = 0.029$ $2V = 56°$ Dispersion none	2.5	2120–2300	**Habitus:** tabular, encrustations, stalactitic, reniform. **Color:** berlin blue, sky blue, or greenish blue. **Luster:** vitreous (i.e., glassy). **Diaphaneity:** transparent to translucent. **Streak:** white. **Fracture:** conchoidal. **Cleavage:** (110) good, (111) indistinct. **Deposits:** secondary, formed in arid climates or in rapidly oxidizing copper deposits. Decomposes at 110°C.
Chalcocite [22205-45-4] (from the Greek, chalkos, copper)	Cu_2S $MM = 159.158$ 79.85wt% Cu 20.15wt% S (Sulfides and sulfosalts)	Orthorhombic $a = 1188.1$ pm $b = 273.23$ pm $c = 1349.1$ pm $(Z = 96)$	Biaxial $R = 32\%$	2.5–3	5800	**Habitus:** massive, granular, euhedral crystals. **Color:** black or iron black. **Luster:** metallic. **Diaphaneity:** opaque. **Streak:** grayish black. **Cleavage:** (110) indistinct. **Twinning:** {110}, {032}, {112}. **Fracture:** conchoidal. **Deposits:** secondary mineral in/near the oxidized zone of copper sulfide ore deposits. Electrical resistivity 80 to 100 $\mu\Omega$.cm, melting point of 1100°C.
Chalcopyrite [1308-56-1] (from the Greek, chalkos, copper, hence copper pyrite)	$CuFeS_2$ $MM = 183.525$ 30.43wt% Fe 34.63wt% Cu 34.94wt% S Traces of Ag, Au, Pt, Co, Ni, Pb, Sn, Zn, As, and Se (Sulfides and sulfosalts) Coordinance Cu(4), Fe(4)	Tetragonal $a = 529.88$ pm $c = 1043.4$ pm E1$_1$, tII6 $(Z = 4)$ S.G. 142d P.G. 42m Chalcopyrite type	Uniaxial $R = 41–46\%$	3.5–4	4190	**Habitus:** rare tetrahedral disphenoidal crystals, botryoidal, striated, druse, usually zoned. **Color:** brass yellow or honey yellow. **Luster:** metallic. **Diaphaneity:** opaque. **Streak:** greenish black. **Cleavage:** (011), (111). **Fracture:** uneven, brittle. **Twinning:** {112}. **Chemical:** attacked by HNO_3, corroded by a mixture of KOH + KMnO4. **Deposits:** veins and disseminated in metamorphic and igneous rocks (e.g., gabbros, norites). Electrical resistivity 150 to 9000 $\mu\Omega$.cm, melting point 950°C.
Chloanthite (syn., white nickel) (from Greek, chloantos, greenish)	$(Ni,Co)As_3$	Cubic 432	Isotropic	5.5	6400–6600	**Habitus:** massive, granular, euhedral crystals. **Color:** tin white or dark gray. **Luster:** metallic. **Diaphaneity:** opaque. **Streak:** grayish black. **Fracture:** uneven.
Chlorite (1M) (from the Greek, chloros, green)	$(Mg,Fe,Al)_6(Si,Al)_4O_{10}(OH)_8$ Coordinance Mg(6), Fe(6), Si(4) Al(4) (Phyllosilicates, layered)	Monoclinic $a = 537$ pm $b = 930$ pm $c = 1425$ pm $\beta = 101.77°$ $(Z = 2)$	Biaxial (−) $\alpha = 1.56–1.60$ $\beta = 1.57–1.61$ $\gamma = 1.58–1.61$ $\delta = 0.006–0.020$ $2V = 0–40°$ Dispersion strong	2–3	3000	**Habitus:** foliated, scaly, lamellar. **Color:** green. **Luster:** vitreous (i.e., glassy). **Diaphaneity:** transparent to translucent. **Twinning:** {001} simple, lamellar, common. **Fracture:** uneven in steps, brittle. **Cleavage:** (001) perfect. **Twinning:** {310}. **Deposits:** metamorphic rocks, weathering of Al-rich sedimentary rocks.

Mineral	Composition	Crystallography	Hardness	Optical	Density (kg/m³)	Properties
Chloritoid(2M) (syn., ottrelite; contains MnO) (from its similarity with chlorite)	$FeAl_2O[SiO_4]_{12}(OH)_4$ 26–28wt% FeO, 2–4wt% MgO, 39–41wt% Al_2O_3, 24–26wt% SiO_2, 2–7wt% H_2O Coordinance Fe(6), Mg(6), Al(6), Si(4) (Phyllosilicates, layered)	Monoclinic $a = 948$ pm $b = 548$ pm $c = 1818$ pm $\beta = 101.74°$ $(Z = 8)$ P.G. 2/m S.G. C2/c	6.5 (178–218H V)	Biaxial (+) $\alpha = 1.713{-}1.730$ $\beta = 1.719{-}1.734$ $\gamma = 1.723{-}1.740$ $2V = 45{-}68°$ O.A.P. ⊥(010) Dispersion strong	3510–3800	**Habitus:** similar to that of micas, lamellar. **Color:** dark green, colorless to green in thin section. **Luster:** vitreous (i.e., glassy) and submetallic for dark varieties. **Diaphaneity:** transparent to translucent. **Twinning:** [001] simple, lamellar [221], common. **Fracture:** uneven in steps, brittle. **Cleavage:** (001) perfect. **Chemical:** insol. in HCl but attacked by H_2SO_4. Slightly fusible on thin edges. Dielectric constant: 6.9. **Deposits:** metamorphic rocks, weathering of Al-rich sedimentary rocks.
Chlorargyrite [7783-90-6] (Halides)	$AgCl$ $MM = 143.321$ 75.26wt% Ag 24.74wt% Cl. Coordinance Ag(6)	Cubic $a = 554.91$ pm B1, cF8 $(Z = 4)$ S.G. Fm3m P.G. 4-32 Rock salt type	2.5	Isotropic $n_D = 2.071$	5550	**Habitus:** massive, cubic euhedral crystals. **Color:** colorless gray, violet tarnish. **Luster:** resinous. **Diaphaneity:** transparent to translucent. **Streak:** white. **Cleavage:** (100). **Fracture:** subconchoidal, sectile.
Chondrodite	$Mg(OH,F)_2 \cdot 2Mg_2[SiO_4]$ Coordinance Mg(6), Si(4) (Nesosilicates)	Monoclinic $a = 789$ pm $b = 474.3$ pm $c = 1029$ pm $109.03°$ $(Z = 2)$ P.G. 2/m S.G. P2₁/b (Humite group)	6.5	Biaxial $\alpha = 1.592{-}1.615$ $\beta = 1.602{-}1.627$ $\gamma = 1.621{-}1.646$ $\delta = 0.028{-}0.038$ $2V = 71{-}85°$ O.A.P. (010)	3150–3180	**Habitus:** massive. **Color:** brown, yellow, orange, red, colorless. **Luster:** vitreous (i.e., glassy). **Diaphaneity:** translucent to transparent. **Cleavage:** (100) poor. **Twinning:** {001}. **Fracture:** uneven brittle. **Chemical:** attacked by strong mineral acids giving a silica gel. **Deposits:** dolomites, limestones, skarns.
Chromite [1308-31-2] (syn., chromic iron, chrome iron ore) (after its chemical composition)	$FeCr_2O_4$ $MM = 223.8348$ 46.46wt% Cr 24.95wt% Fe 28.59wt% O Coordinance Fe(4), Cr(6) (Oxides and hydroxides)	Cubic $a = 839.40$ pm H1, cF56 $(Z = 8)$ P.G. m3m S.G. Fd3m Spinel type (Chromite Series)	5.5	Isotropic $n_D = 2.16$ $R = 13\%$	5090	**Habitus:** massive, granular. **Color:** black or brownish black. **Luster:** metallic. **Diaphaneity:** opaque. **Streak:** brown. **Cleavage:** none. **Twinning:** {111}. **Fracture:** uneven to conchoidal.
Chrysoberyl [12004-06-7] (syn., alexandrite; green) (from the Greek, chrysos, golden and the mineral beryl)	$BeAl_2O_4$ $MM = 126.97286$ 7.10wt% Be 42.50wt% Al 50.40wt% O Coordinance Al(6), Be(4) (Oxides and hydroxides)	Orthorhombic $a = 547.56$ pm $b = 940.41$ pm $c = 442.67$ pm $(Z = 4)$ P.G. mmm S.G. P2₁ bnm Olivine type	8.5	Biaxial (+) $\alpha = 1.747$ $\beta = 1.748$ $\gamma = 1.757$ $\delta = 0.010$ $2V = 45°$ Dispersion none	3699	**Habitus:** twinning, prismatic, tabular, striated (001). **Color:** green, white, brown, or yellow. **Streak:** white. **Luster:** vitreous (i.e., glassy). **Diaphaneity:** transparent to translucent. **Cleavage:** (110) distinct, (010) imperfect. **Twinning:** {031}. **Fracture:** brittle. **Deposits:** granitic pegmatite dikes.
Chrysocolla (syn., Bisbeeite) (from the Greek, chrysos, gold, and kolla, glue in allusion to the name of the material used to solder gold)	$(Cu,Al)_2H_2Si_2O_5(OH)_4 \cdot n(H_2O)$ (Cyclosilicates, ring)	Monoclinic	2.5–3.5	Uniaxial(+) $\varepsilon = 1.46$ $\omega = 1.57$ $\delta = 0.110$	2200–2400	**Habitus:** botryoidal, earthy, stalactitic. **Color:** green, bluish green, blue, blackish blue, or brown. **Diaphaneity:** translucent to opaque. **Luster:** vitreous, dull. **Cleavage:** none. **Fracture:** sectile. **Streak:** light green. **Deposits:** mineral of secondary origin commonly associated with other secondary copper minerals.

11
Minerals, Ores and Gemstones

Table 11.4 (continued)

Mineral name (IMA) [CAS RN] (Synonyms) (Etymology)	Theoretical chemical formula, relative molecular mass, coordinance number, major impurities, mineral class (Strunz)	Crystal system, lattice parameters, Strukturbericht, Pearson symbol, Z, point group, space group, structure type	Optical properties	Mohs hardness (HM)	Density (ρ, kg.m^{-3})	Other mineralogical, physical, and chemical properties
Cinnabar [1344-45-5] (from the Latin, cianabaris)	HgS $MM = 232.656$ 86.22wt% Hg 13.78wt% S (Sulfides and sulfosalts) Coordinance Hg(6)	Trigonal (Hexagonal) $a = 414.9$ pm $c = 949.5$ pm B9, hP6 ($Z = 3$) S.G. P3$_1$21 P.G. 32 Cinnabar type	Uniaxial (+) $\omega = 2.814$ $\varepsilon = 3.143$ $\delta = 0.351$ $R = 25\%$	2–2.5	8170	**Habitus:** druse, disseminated, tabular, massive. **Color:** intense red, brownish red, or gray. **Diaphaneity:** transparent, translucent to opaque. **Luster:** adamantine. **Streak:** scarlet, bright red. **Cleavage:** (1010) perfect. **Twinning:** {0001}. **Fracture:** subconchoidal, brittle, sectile. Decomposes at 386°C in HgO.
Clinohumite (from the British mineralogist Sir Abraham Hume, and clinos, for its monoclinic crystal system)	Mg(OH,F)$_2$·2Mg$_2$[SiO$_4$] Coordinance Mg(6), Si(4) (Nesosilicates)	Monoclinic $a = 475$ pm $b = 1027$ pm $c = 1368$ pm $100.83°$ ($Z = 2$) P.G. 2/m S.G. P2$_1$/b (Humite group)	Biaxial (+) $\alpha = 1.629–1.638$ $\beta = 1.641–1.643$ $\gamma = 1.662–1.674$ $\delta = 0.028–0.041$ $2V = 65–84°$ O.A.P. (100)	6	3210–3350	**Habitus:** massive. **Color:** brown, yellow, orange, red, colorless. **Luster:** vitreous (i.e, glassy). **Diaphaneity:** translucent to transparent. **Cleavage:** (100) poor. **Twinning:** {010}. **Fracture:** uneven, brittle. **Chemical:** attacked by strong mineral acids giving a silica gel. **Deposits:** dolomites, limestones, skarns.
Clinozoisite (after the Austrian natural scientist, S. Von Zois, and clinos for its monoclinic lattice structure)	Ca$_2$·Al·Al$_2$O(OH)[SiO$_4$][Si$_2$O$_7$] $MM = 427.37572$ 18.76wt% Ca 12.63wt% Al 19.71wt% Si 0.24wt% H 48.67wt% O Traces of Fe(III) Coordinance Ca(7), Al(6), Si(4) (Sorosilicates and nesosilicates)	Monoclinic $a = 888.7$ pm $b = 558.1$ pm $c = 1014.0$ pm $115°93$ ($Z = 2$) P.G. 2/m S.G. P2$_1$/m (Epidote group)	Biaxial (+) $\alpha = 1.670–1.715$ $\beta = 1.674–1.725$ $\gamma = 1.690–1.734$ $\delta = 0.0005–0.015$ $2V = 14–90°$ Dispersion strong	6–6.5 (680 HV)	3120–3380	**Habitus:** prismatic according to b, striated, columnar. **Color:** pale yellow, cream yellow, pink, greenish, colorless. **Diaphaneity:** transparent to opaque. **Luster:** vitreous (i.e., glassy). **Cleavage:** (001) perfect. **Twinning:** {100}. **Fracture:** uneven. **Chemical:** insoluble in HCl. Other: dielectric constant 8.51. **Deposits:** regional metamorphic and pegmatite rocks.
Cobaltite [12254-52-9] (from the German, Kobold, underground spirit, or goblin, an allusion to the refusal of cobaltiferous ores to smelt properly, hence bewitched)	CoAsS $MM = 197.9868$ 29.77wt% Co 37.84wt% As 32.39wt% S (Sulfides and sulfosalts) Coordinance Co(6)	Cubic $a = 557$ pm F01, cP12 ($Z = 4$) S.G. P213 P.G. 23 NiSbS type	Isotropic $R = 53\%$	5.5	6330	**Habitus:** massive, granular, faces striated. **Color:** reddish silver white, violet steel gray, or black. **Diaphaneity:** opaque. **Luster:** metallic. **Streak:** grayish black. **Cleavage:** (100) good, (010) good, (001) good. **Fracture:** uneven, brittle. Electrical resistivity 6.5 to 130 mΩ.m.
Coesite	α-SiO$_2$ $MM = 60.0843$	Monoclinic $a = 715.2$ pm	Biaxial (+) $\alpha = 1.590$	7–8	2930	**Habitus:** tubular, high pressure. **Color:** colorless. **Luster:** vitreous (i.e, glassy). **Streak:** white. **Fracture:** conchoidal. **Twinning:** {011}, {012}. **Diaphaneity:**

Mineral (name, synonyms, etymology)	Chemistry / Coordination	Crystal system	Optical	Hardness	Density	Properties
Coffinite	(Tectosilicates, framework) Coordinance Si(4); $U[SiO_{4(1-x)}(OH)_{4x}]$	Tetragonal; $(Z = 16)$ P.G. 2/m S.G. C2/c	Uniaxial; $\delta = 0.010$; $2V = 64°$			
Colemanite (syn., neocolmanite) (after W.T. Coleman, owner of the Death Valley, California mine where this species was first found)	$Ca_2B_6O_{11}\cdot 5(H_2O)$; $MM = 205.5429$; 19.50 wt% Ca; 15.78 wt% B; 2.45 wt% H; 62.27 wt% O; Coordinance Ca(7), B(3, and 4) (Carbonates and borates)	Monoclinic; $a = 874.3$ pm, $b = 1126.4$ pm, $c = 610.2$ pm, $\beta = 110.12°$; $(Z = 4)$; P.G. 2/m; S.G. $P2_1/a$	Biaxial (+); $\alpha = 1.586$, $\beta = 1.592$, $\gamma = 1.614$; $\delta = 0.028$; $2V = 55–56°$; Dispersion weak	4–4.5	2420	**Habitus:** blocky, crystalline, coarse, massive, granular. **Color:** white, yellowish white, gray, or yellowish white. **Streak:** white. **Luster:** vitreous (i.e., glassy). **Cleavage:** (010) perfect, (001) distinct. **Fracture:** subconchoidal, brittle. **Diaphaneity:** transparent to translucent. **Deposits:** lacusterine limestone hosted borate deposits.
Coloradoite (after its occurrence in Boulder, Colorado)	$HgTe$; $MM = 328.19$; 61.12 wt% Hg; 38.88 wt% Te (Sulfides and sulfosalts)	Cubic; Sphalerite type	Isotropic	2.5	8070	**Habitus:** massive, granular. **Color:** iron black. **Luster:** metallic. **Diaphaneity:** opaque. **Streak:** black. **Fracture:** conchoidal.
Columbite (syn., niobite)	$(Fe,Mn)(Nb,Ta)_2O_6$; Coordinance Fe(6), Mn(6), Nb(6), Ta(6) (Oxides and hydroxides)	Orthorhombic; $a = 510$ pm, $b = 1427$ pm, $c = 574$ pm; $(Z = 2)$; P.G. mmm; S.G. $P2_1bcn$	Biaxial (+); $\alpha = 2.44$, $\beta = 2.32$, $\gamma = 2.38$; $2V = 75°$	5	6000	**Habitus:** prismatic. **Color:** black, brown. **Luster:** submetallic. **Diaphaneity:** translucent to opaque. **Fracture:** subconchoidal. **Cleavage:** (010).
Copper (syn., cuprum) (from the Greek, kyprios, the name of the island of Cyprus, once producing this metal)	Cu; $MM = 63.546$; Coordinance Cu(12) (Native elements)	Cubic; $a = 361.5$ pm; A1, cF4 $(Z = 4)$; P.G. m3m; S.G. Fm3m; Copper type	Isotropic; $n_D = 0.641$; $R = 81\%$	2.5–3	8935	**Habitus:** nodular, dendritic, arborescent. **Color:** light rose, copper red, or brown. **Diaphaneity:** opaque. **Luster:** metallic. **Streak:** rose. **Cleavage:** none. **Fracture:** hackly. **Chemical:** readily dissolved in nitric acid, HNO_3, giving a blue solution of $Cu(NO_3)_2$, which deposits copper onto a pure zinc rod immersed in the solution. **Deposits:** cap rock of copper sulfide veins and in some types of volcanic rocks.
Cordierite (after the French mining engineer and geologist P.L.A. Cordier)	$Mg_2Si_5Al_4O_{18}$; Coordinance Mg(6), Si(4), Al(4) (Cyclosilicate, ring)	Orthorhombic; $a = 1713$ pm, $b = 980$ pm, $c = 935$ pm; $(Z = 4)$; P.G. 2/mmm; S.G. C2/ccm	Biaxial (+/−); $\alpha = 1.54$, $\beta = 1.55$, $\gamma = 1.56$; $2V = 65–105°$	7	2500–2800	**Habitus:** prismatic. **Color:** white. **Luster:** vitreous. **Diaphaneity:** transparent to translucent. **Fracture:** even. **Cleavage:** (010) good, (100) poor. **Twinning:** (110), [130]. **Deposits:** alumina rich metamorphic rocks.
Corundum [1344-25-1] (syn., alumina, sapphire: blue, ruby: red) (from India, Karund, name of the mineral)	$\alpha\text{-}Al_2O_3$; $MM = 101.961$; 52.93 wt% Al; 47.07 wt% O (Oxides, and hydroxides); Coordinance Al(6)	Trigonal (Rhombohedral); $a = 513.29$ pm; 55° 17′; D5₁, hR10 $(Z = 2)$; S.G. R3c; P.G. 32 m; Corundum type	Uniaxial (−); $\varepsilon = 1.768$, $\omega = 1.760$; $\delta = 0.008$; Dispersion moderate	9 (2000 HV)	3980–4020	**Habitus:** prismatic, tabular. **Color:** colorless, yellow, red, blue green, violet, black. **Luster:** vitreous to adamantine. **Diaphaneity:** transparent to opaque. **Streak:** none. **Twinning:** [101]. **Parting:** (101). **Fracture:** uneven. **Chemical:** soluble in strong mineral acids only after fusion with $KHSO_4$ or $CaSO_4$. **Melting point:** 2054°C. **Deposits:** metamorphic rocks, sedimentary rocks.

11
Minerals, Ores and Gemstones

Table 11.4 (continued)

Mineral name (IMA) [CAS RN] (Synonyms) (Etymology)	Theoretical chemical formula, relative molecular mass, coordination number, major impurities, mineral class (Strunz)	Crystal system, lattice parameters, Strukturbericht, Pearson symbol, Z, point group, space group, structure type	Optical properties	Mohs hardness (HM)	Density (ρ, kg.m^{-3})	Other mineralogical, physical, and chemical properties
Cotunnite	$PbCl_2$ (Halides)	Orthorhombic	Uniaxial (+)	1.5–2	5300–5800	**Habitus:** acicular, massive, granular. **Color:** white or yellowish white. **Luster:** adamantine. **Diaphaneity:** transparent to translucent. **Streak:** white. **Cleavage:** (001) perfect. **Deposits:** oxidized lead deposits.
Covellite [1317-40-4] (syn., coveline)	CuS (Sulfides and sulfosalts) Coordinance Cu(3), Cu(4)	Trigonal (Hexagonal) a = 380 pm c = 1636 pm B18, hP12 (Z = 6) S.G. P6$_3$/mmc P.G. 622 Covellite type	Uniaxial (+) ε = 1.450 ω = 1.600 R = 15–24%	1.5–2	4600	**Habitus:** tabular, tarnishes purple. **Color:** blue. **Luster:** submetallic. **Diaphaneity:** opaque. **Streak:** dark gray. **Fracture:** conchoidal. **Cleavage:** (0001) perfect. Electrical resistivity 3 to 83 $\mu\Omega$.cm. Melting point: 507°C.
Cristobalite (alpha) [14464-46-1]	SiO_2 MM = 60.0843 46.74wt% Si 53.26wt% O (Tectosilicates, framework) Coordinance Si (4)	Tetragonal a = 497.1 pm c = 691.8 pm (Z = 4) P.G. 422 S.G. P4$_3$2$_1$2	Uniaxial (–) ε = 1.482 ω = 1.489 δ = 0.007	6–7	2330	**Habitus:** coarse aggregate. **Color:** colorless. **Streak:** white. **Diaphaneity:** transparent to translucent. **Luster:** vitreous (i.e., glassy). **Fracture:** conchoidal. **Twinning:** {111}. **Chemical:** resistant to strong mineral acids, attacked by HF and molten alkali hydroxides. Melting point: 1713°C.
Crocoite [7758-97-6]	$PbCrO_4$ MM = 327.1937 63.33wt% Pb 15.89wt% Cr 19.56wt% O Coordinance Pb(6), Cr(4) (Sulfates, chromates, molybdates, and tungstates)	Monoclinic a = 711 pm b = 741 pm c = 681 pm β = 102.55° (Z = 4) P.G. 2/m S.G. P2$_1$/n Crocoite type	Biaxial (+) α = 2.31 β = 2.37 γ = 2.66 δ = 0.35 2V = 54°	2.5–3	6000	**Habitus:** acicular, striated. **Color:** red orange. **Streak:** orange. **Diaphaneity:** translucent. **Luster:** adamantine. **Fracture:** subconchoidal. **Cleavage:** (010). **Twinning:** [011]. Melting point: 844°C.
Cryolite [13775-53-6] (from the Greek, kryos, frost, and lithos, stone)	Na_3AlF_6 MM = 209.94126 32.85wt% Na 12.85wt% Al 54.30wt% F (Halides) Coordinance Na(12), Na(6), Al(6)	Monoclinic a = 546 pm b = 560 pm c = 778 pm 90.18° (Z = 2) S.G. P2$_1$/n P.G. 2/m Prismatic	Biaxial (+) α = 1.3385 β = 1.3389 γ = 1.3396 δ = 0.0011 2V = 43°	2.5–3	2970	**Habitus:** massive, granular, pseudocubic euhedral crystals. **Color:** snow white, gray, reddish white, or brownish white. **Diaphaneity:** transparent to translucent. **Luster:** vitreous, greasy. **Streak:** white. **Cleavage:** (001), (110), (101). **Fracture:** uneven. **Deposits:** large bed in a granitic vein in gray gneiss. Melting point: 1009°C.
Cubanite	$CuFe_2S_3$	Orthorhombic	Biaxial	3.5	4100	**Habitus:** lamellar. **Color:** gray, black. **Luster:** metallic. **Diaphaneity:** opaque.

Name	Formula / Class	Crystallography	Optical	Hardness	Density	Properties
		c = 623 pm, E9e, oP24 (Z = 4), S.G. Pmma, P.G. mmm				
Cuprite [1317-39-1] (syn, chalcotrichite) (from the Latin. cuprum, copper; chalcotrichite from the Greek, khalkos, hairy copper)	Cu_2O, MM = 143.0914, 88.82wt% Cu, 11.18wt% O, Coordinance Cu(2) (Oxides and hydroxides)	Cubic, a = 426.96 pm (Z = 2), C3, cP6 (Z = 2), P.G. 432, S.G. Pn3m, Cuprite type	Isotropic, n_D = 2.849, R = 23%	3.5–4	6106	Habitus: cubic, octahedral, massive, granular, capillary. Color: brownish red or dark red. Diaphanity: transparent to translucent. Luster: submetallic to adamantine. Streak: brownish red. Cleavage: (111) imperfect. Fracture: brittle, conchoidal. Other: electrical resistivity 10 to 50 Ω.m. Deposits: oxidized zone of copper deposits.
Datolite	$CaB[SiO_4](OH)$ (Nesosilicates)	Monoclinic, a = 962 pm, b = 760 pm, c = 484 pm, 90.15° (Z = 4)	Biaxial (−), α = 1.622–1.626, β = 1.649–1.654, γ = 1.666–1.670, δ = 0.044–0.046, 2V = 72–75°, O.A.P. (010), Dispersion weak	5–5.5	2960–3000	Color: colorless or white, yellow, green, pink. Cleavage: none. Twinning: none. Chemical: insoluble in HCl but gelatinizes. Gives intense yellowish color to bunsen flame when moistened with H_2SO_4. Deposits: in cavities and veins in hyapbyssal and volcanic igneous rocks. Skarns. Serpentine and hornblende schists. Secondary mineral associated with calcite, prehnite, zeolites and axinite. Melting point: 1235°C.
Diamond [7782-40-3] (syn, boart, carbonado) (from the Greek, adamas, invincible, or hardest)	C, MM = 12.0107, Coordinance C(4) (Native elements)	Cubic, a = 356.68 pm (Z = 8), A4, cF8 (Z = 8), P.G. m3m, S.G. Fd-3m, Diamond type	Isotropic, n_D = 2.4175–2.4178	10	3510	Habitus: eubedral crystals, granular. Color: colorless, white, gray, black, or blue. Diaphanity: transparent to translucent. Streak: colorless. Cleavage: (111) perfect. Fracture: conchoidal. Luminescence: fluorescent. Deposits: gas rich, ultra-basic diatremes from mantle depths (>30 km) and alluvial placer deposits derived from kimberlite rocks. Melting point: 4400°C under 12 GPa.
Diaspore [14457-84-2] (from the Greek, diaspores, to scatter, referring to its easy disintegration in the bunsen flame)	$AlO(OH)$, MM = 59.98828, 44.98wt% Al, 1.68wt% H, 53.34wt% O, Coordinance Al(6) (Oxides and hydroxides)	Orthorhombic, a = 440.1 pm, b = 942.1 pm, c = 284.5 pm (Z = 4), P.G. mmm, S.G. P2₁/bnm, Geothite type structure	Biaxial (+), α = 1.682–1.706, β = 1.705–1.725, γ = 1.730–1.752, δ = 0.047–0.050, 2V = 85–88°, Dispersion weak	6.5–7	3440	Habitus: platy, massive, tabular, disseminated. Color: white, greenish gray, grayish brown, colorless, or yellow. Luster: vitreous, pearly. Diaphanity: transparent to translucent. Streak: white, yellow. Cleavage: perfect (010), poor (210). Fracture: brittle, conchoidal. Deposits: metamorphic and sedimentary bauxite ores.
Diopside (syn, diallage) (from the Greek, dis, two kinds, and opsis, opinion)	$CaMg[Si_2O_6]$, MM = 216.5504, 18.51wt% Ca, 11.22wt% Mg, 25.94wt% Si, 44.33wt% O, Coordinance Ca(8), Mg(6), Si(4) (Inosilicates, double chains)	Monoclinic, a = 970 pm, b = 890 pm, c = 525 pm, β = 105.83° (Z = 4), P.G. 2/m, S.G. C2/c (Z = 4)	Biaxial (+), α = 1.665, β = 1.672, γ = 1.695, δ = 0.030, 2V = 56–63°, Dispersion weak	6	3400	Habitus: prismatic, blocky, granular. Color: white, yellowish green, black, or grayish blue. Diaphanity: transparent to translucent. Luster: vitreous (i.e., glassy). Streak: white, green. Cleavage: (110) good. Twinning: [001], [100]. Fracture: brittle, conchoidal. Deposits: basic and ultrabasic igneous and metamorphic rocks.
Dioptase (from the Greek, dia, through, and optomai, vision)	$CuSiO_2(OH)_2$ (Cyclosilicates, ring)	Trigonal	Uniaxial (+), ε = 1.644–1.658, ω = 1.697–1.709, δ = 0.051–0.053	5	3331	Habitus: massive, cryptocrystalline. Color: dark green or emerald green. Diaphanity: transparent to translucent. Luster: vitreous (i.e., glassy). Cleavage: (10ll) good. Fracture: conchoidal. Streak: green. Deposits: secondary mineral in oxidized zones of copper deposits.

11
Minerals, Ores and Gemstones

Table 11.4 (continued)

Mineral name (IMA) [CAS RN] (Synonyms) (Etymology)	Theoretical chemical formula, relative molecular mass, coordinance number, major impurities, mineral class (Strunz)	Crystal system, lattice parameters, Strukturbericht, Pearson symbol, Z, point group, space group, structure type	Optical properties	Mohs hardness (HM)	Density (ρ, kg.m^{-3})	Other mineralogical, physical, and chemical properties
Dolomite (named after the French mineralogist and geologist, D. de Dolomieu)	$CaMg(CO_3)_2$ MM = 184.4014 21.73wt% Ca 13.18wt% Mg 13.03wt% C 52.06wt% O Coordinance Mg(6), Ca(6), C(3) (Nitrates, carbonates, and borates)	Trigonal (Rhombohedral) a = 480.79 pm c = 1601.00 pm a_{rh} = 601.5 pm, 47° 07' (Z = 3) P.G. $\overline{3}$ S.G. R$\overline{3}$ Dolomite type	Uniaxial (−) ε = 1.500–1.520 ω = 1.679–1.703 δ = 0.179–0.185	3.5–4	2860–2930	**Habitus:** rhombohedral, crystalline, massive, botryoidal, globular, stalactitic. **Color:** white, gray, reddish white, brownish white. **Luster:** vitreous (i.e. glassy). **Diaphaneity:** transparent to translucent. **Streak:** white. **Cleavage:** {1011} perfect. **Twinning:** {0001}, {10$\overline{1}$0}, {1110}, {01$\overline{1}$2}. **Fracture** brittle, subconchoidal. **Chemical:** readily dissolved by strong mineral acids with evolution of carbon dioxide. **Deposits:** sedimentary rocks.
Dravite	$NaMg_3Al_6(OH)_4B_3O_3[Si_6O_{18}]$ Coordinance Na(6), Mg(6), Al(6), B[3], Si(4) (Cyclosilicates, ring)	Trigonal a = 1594 pm c = 722 pm (Z = 3) P.G. 3m S.G. R3m Tourmaline group	Uniaxial (−) ε = 1.650 ω = 1.628 δ = 0.022	7–7.5	3020	**Habitus:** prismatic. **Color:** brown, yellow. **Luster:** resinous. **Diaphaneity:** transparent to translucent. **Streak:** white. **Cleavage:** (101) poor, (110) poor. **Twinning:** {101}. **Fracture:** subconchoidal.
Eglestonite	$Hg_4Cl_3O(OH)$ (Halides)	Cubic	Isotropic n_D = 2.49	2–3	8300	**Color:** yellow or brownish yellow. **Luster:** adamantine, resinous. **Diaphaneity:** transparent to translucent.
Elbaite	$NaLi_3Al_6(OH)_4B_3O_3[Si_6O_{18}]$ Coordinance Na(6), Li(6), Al(6), B(3), Si(4) (Cyclosilicates, ring)	Trigonal a = 1646 pm c = 710 pm (Z = 3) P.G. 3m S.G. R3m Tourmaline group	Uniaxial(−) ε = 1.650 ω = 1.628 δ = 0.022	7–7.5	2900	**Habitus:** prismatic. **Color:** white. **Luster:** resinous. **Diaphaneity:** transparent to translucent. **Streak:** white. **Cleavage:** (101) poor, (110) poor. **Twinning:** {101}. **Fracture:** subconchoidal.
Electrum	(Au, Ag) (Native elements)	Cubic	Isotropic	n.a.	n.a.	
Embolite	Ag/Br, Cl) (Halides)	Cubic	Isotropic n_D = 2.15	n.a.	n.a.	**Habitus:** massive. **Color:** yellowish green or grayish yellow. **Luster:** adamantine, greasy. **Diaphaneity:** transparent to translucent. **Streak:** white. **Deposits:** oxidized portions of silver deposits.
Enargite	Cu_3AsS_4 MM = 398.806 47.80wt% Cu 20.04wt% As 32.16wt% S (Sulfides and sulfosalts) Coordinance Cu(4), As(4)	Orthorhombic a = 642.6 pm b = 742.2 pm c = 614.4 pm (Z = 2) S.G. Pn2m P.G. m2m	Biaxial R = 22–25%	3	4500	**Habitus:** striated tabular crystals. **Color:** bronze. **Luster:** metallic. **Diaphaneity:** opaque. **Streak:** dark gray. **Fracture:** uneven. **Electrical resistivity** 0.2 to 40 mΩ.m.

Mineral	Formula / Composition	Crystallography	Optical	Hardness / Density	Description
Enstatite [13776-74-4] (syn., bronzite) (from the Greek, enstates, opponent or resistant owing to the unfusibility of the mineral and bronze, brown)	$Mg[Si_2O_6]$ $MM = 200.778$ 60wt% SiO_2 40wt% MgO Traces of Ca, Fe, Mn, Ni, Cr, Al and Ti Coordinance Mg(6, 8), Si(4) (Inosilicates, single chain)	Orthorhombic $a = 1822.0$ pm $b = 882.9$ pm $c = 519.2$ pm P.G. mmm S.G. P2₁bca Enstatite type	$\alpha = 1.650\text{-}1.668$ $\beta = 1.652\text{-}1.673$ $\gamma = 1.658\text{-}1.680$ $\delta = 0.008\text{-}0.011$ $2V = 54\text{-}90°$ Dispersion weak	5-6 / 3500	Habitus: massive, tabular, radiated, prismatic. Color: white, yellowish green, brown, greenish white, or gray. Luster: vitreous (i.e., glassy), pearly. Diaphaneity: translucent to opaque. Streak: gray. Cleavage: (100) distinct, (010) distinct. Twinning: [101]. Fracture: uneven, brittle. Chemical: insoluble in HCl but sol, in HF. Decomposed at 1550°C. Other: dielectric constant 8.23. Deposits: magmatic and mafic rocks.
Epidote (from the Greek, epidosis, increase, owing to the high lattice parameter)	$Ca_2Fe(Al,Fe)_2O(OH)[SiO_4][Si_2O_7]$ (Sorosilicates and nesosilicates)	Monoclinic $a = 889.0$ pm $b = 563.0$ pm $c = 1019.0$ pm $115°40$ (Z = 2) Epidote group	Biaxial (−) $\alpha = 1.715\text{-}1.751$ $\beta = 1.725\text{-}1.784$ $\gamma = 1.734\text{-}1.797$ $\delta = 0.015\text{-}0.049$ $2V = 90\text{-}116°$ Dispersion strong	6-6.5 (680 HV) / 3380-3490	Habitus: prismatic according to b, striated, columnar. Color: gray, apple green, brown, blue, or rose red. Diaphaneity: transparent to translucent. Luster: vitreous (i.e., glassy), resinous, pearly. Streak: white. Cleavage: (001), perfect, (100) imperfect. Fracture: uneven. Chemical: insoluble in strong mineral acids, but attacked by HCl after calcination giving a gel of silica, fusible giving a dark green globule. Deposits: regional metamorphic and pegmatite rocks.
Epsomite	$MgSO_4 \cdot 7H_2O$ $MM = 246.47598$ 9.86wt% Mg 13.01wt% S 71.40wt% O 5.73wt% H Coordinance Mg(6), S(4) (Sulfates, chromates, molybdates, and tungstates)	Orthorhombic $a = 1196$ pm $b = 1199$ pm $c = 685.8$ pm (Z = 4) P.G. 222 S.G. P2₁2₁2₁ Epsomite type	Biaxial (+) $\alpha = 1.433$ $\beta = 1.455$ $\gamma = 1.461$ $\delta = 0.028$ $2V = 52$	2-2.5 / 1680	Habitus: botryoidal, prismatic. Color: colorless. Streak: white. Diaphaneity: transparent to translucent. Luster: vitreous. Fracture: conchoidal.
Erythrite (syn., cobalt bloom)	$Co_3(AsO_4)_2 \cdot 8H_2O$ $MM = 598.76072$ 29.53wt% Co 25.03wt% As 42.75wt% O 2.69wt% H Coordinance Co(6), As(4) (Phosphates, arsenates, and vanadates)	Monoclinic $a = 1026$ pm $b = 1337$ pm $c = 474$ pm $\beta = 105.1°$ (Z = 2) P.G. 2/m S.G. C2/m Vivianite type	Biaxial (−) $\alpha = 1626$ $\beta = 1.661$ $\gamma = 1.699$ $\delta = 0.073$ $2V = 90°$	1-2 / 3060	Habitus: reniform, fibrous. Color: purple red. Streak: pale purple. Diaphaneity: translucent. Luster: adamantine. Fracture: sectile. Cleavage: (010) perfect.
Eudialyte (syn., eucolite) (from the Greek, eu, well, and dialytos, decomposable)	$Na_4(Ca,Ce)_2(Fe,Mn,Y)Zr[Si_6O_{22}](OH,Cl)_2$ (Cyclosilicates, ring)	Trigonal (Rhombohedral)	Uniaxial (+) $\omega = 1.593\text{-}1.643$ $\varepsilon = 1.597\text{-}1.634$ $\delta = 0.001\text{-}0.010$	5-5.5 / 2900	Habitus: massive granular, tabular. Luster: vitreous (i.e., glassy). Cleavage: (0001) imperfect. Fracture: uneven. Color: pinkish red, red, yellow, yellowish brown, or violet. Streak: white. Deposits: hepheline-syenite rocks.
Falcondoite (syn., garnierite, genthite)	$(Ni,Mg)_4[Si_6O_{15}](OH)_2 \cdot 6(H_2O)$ (Phyllosilicates, layered)	Orthorhombic	Biaxial	3-4 / 2410	Habitus: stalactitic. Color: green or yellowish green. Luster: resinous.
Faujasite (after Barthélemy Faujas de Saint Fond, French geologist and writer on the origin of volcanoes)	$(Na_2,Ca,Mg)_{3.5}[Al_7Si_{17}O_{48}]/32 \cdot (H_2O)$ (Tectosilicates, network)	Cubic $a = 247$ pm (Z = 8)	Isotropic $n_D = 1.47\text{-}1.48$	5 / 1923-1940	Habitus: euhedral. Color: colorless, white, or pale brown. Diaphaneity: transparent to translucent. Luster: vitreous (i.e., glassy). Cleavage: (111) perfect, (111) perfect, (111) perfect. Fracture: brittle, uneven. Deposits: usually occurs in basaltic volcanics, metapyroxenite, also with augite in limburgite.

11
Minerals, Ores and Gemstones

Table 11.4 (continued)

Mineral name (IMA) [CAS RN] (Synonyms) (Etymology)	Theoretical chemical formula, relative molecular mass, coordination number, major impurities, mineral class (Strunz)	Crystal system, lattice parameters, Strukturbericht, Pearson symbol, Z, point group, space group, structure type	Optical properties	Mohs hardness (HM)	Density (ρ, kg.m^{-3})	Other mineralogical, physical, and chemical properties
Fayalite [10179-73-4] (syn., hortonolite: Mn,Mg, knebelite: Mn) (named after the locality, Fayal, one of the islands of the Acores archipelago)	$Fe_2[SiO_4]$ Fe-pole Fo10-0 $MM = 203.7771$ 54.81wt% Fe 13.78wt% Si 31.41wt% O Traces of Mg, Ca, and Mn Coordination Fe(6), Si(4) (Nesosilicates)	Orthorhombic $a = 481.7$ pm $b = 1047.7$ pm $c = 610.5$ pm $(Z = 4)$ P.G. mmm S.G. Pbnm Olivine group	Biaxial (–) $\alpha = 1.827$ $\beta = 1.869$ $\gamma = 1.879$ $\delta = 0.052$ $2V = 47$–$54°$ Dispersion weak O.A.P (001)	6.5–7 (820 HV)	4390	**Habitus:** massive, granular. **Color:** brownish black, or black. **Luster:** vitreous (i.e., glassy). **Diaphaneity:** translucent to transparent. **Streak:** white. **Cleavage:** (010) distinct, (001) poor. **Twinning:** [100], [011], [012]. **Fracture:** conchoidal. **Chemical:** attacked by strong mineral acids giving a gel of silica. Fusible giving a magnetic globule. **Deposits:** ultramafic silica-poor igneous rocks such as gabbros, basalts, peridotites.
Ferberite	$FeWO_4$ $MM = 303.6826$ 18.39wt% Fe 60.54wt% W 21.07wt% O Coordination Fe(6), W(4) (Sulfates, chromates, molybdates, and tungstates)	Monoclinic $a = 473.2$ pm $b = 570.8$ pm $c = 496.5$ pm $\beta = 90.00°$ $(Z = 2)$ P.G. 2/m S.G. P2/c Wolframite type	Biaxial (+) $\alpha = 2.31$ $\beta = 2.40$ $\gamma = 2.46$ $\delta = 0.15$ $2V = 79°$	4–4.5	7600	**Habitus:** short prismatic. **Color:** brown, black. **Luster:** submetallic. **Diaphaneity:** opaque. **Streak:** black. **Cleavage:** (010) poor. **Twinning:** [100], [023]. **Fracture:** uneven.
Fergusonite	$YNbO_4$ $MM = 245.80983$ 36.17wt% Y 37.80wt% Nb 26.04wt% O (Oxides and hydroxides)	Tetragonal	Isotropic 5.5–6 $n_D = 2.190$	5050		**Color:** brown or brownish black. **Diaphaneity:** translucent to opaque. **Luster:** submetallic. **Cleavage:** indistinct. **Fracture:** subconchoidal. **Streak:** brown.
Fluorite [7789-75-5] (syn., fluor spar) (after Latin, fluere, to flow)	CaF_2 $MM = 78.0748$ 51.33wt% Ca 48.67wt% F Traces of Y and Ce (Halides) Coordination Ca(8)	Cubic $a = 546.36$ pm Cl, cF12 $(Z = 4)$ S.G. Fm3m P.G. 4-32 Fluorite type	Isotropic $n_D = 1.434$	4	3180	**Habitus:** crystalline, massive, granular, octahedral crystals. **Color:** white, yellow, green, red, or blue. **Luster:** vitreous (i.e., glassy). **Luminescence:** fluorescent blue under short and long UV light. **Diaphaneity:** transparent to translucent. **Streak:** white. **Cleavage:** (111). **Fracture:** conchoidal, splintery. **Chemical:** Insoluble in water, dissolves in hot conc. H_2SO_4, evolving gaseous HF, slightly soluble in HCl and HNO_3. Decrepitation when fired. Fusible (mp 1418°C) giving a white enamel. Sometimes radioactive owing to traces of Th. **Deposits:** low-temperature vein deposits, guangue materials, sedimentary rocks.
Forsterite [26686-77-1] (syn., olivine, chrysolite, peridot)	$Mg_2[SiO_4]$ Mg-pole: Fo90–100 57.1wt% MgO	Orthorhombic $a = 475.8$ pm $b = 1021.4$ pm	Biaxial (+) $\alpha = 1.635$ $\beta = 1.651$	6.5–7 (820 HV)	3220	**Habitus:** prismatic, tabular. **Color:** white, yellow, or greenish yellow. **Diaphaneity:** transparent to translucent. **Luster:** vitreous (i.e., glassy). **Streak:** white. **Cleavage:** (010) good, (001) poor. **Fracture:** conchoidal. **Chemical:** attacked by concentrated

Mineral	Formula / Composition	Crystallography	Optical	Hardness	Density (kg/m³)	Properties
	Coordinance Mg(6), Si[4] (Nesosilicates)	P.G. mmm S.G. Pbnm Olivine group	$2V = 82°$ Dispersion weak O.A.P. (001)			Habitus: octahedral. Color: black, red brown. Diaphaneity: opaque. Luster: metallic. Cleavage: none. Parting: {111}.
Franklinite	$ZnFe_2O_4$ MM = 241.0776 Coordinance Zn(4), Fe(6) (Oxides and hydroxides)	Cubic a = 842.0 pm H1₁, cF56 (Z = 8) P.G. m3m S.G. Fd3m Spinel type	Isotropic n_D = 2.36 R = 15%	5.5	5320	
Gahnite	$ZnAl_2O_4$ (Oxides and hydroxides)	Cubic a = 808 pm H1₁, cF56 (Z = 8) P.G. m3m S.G. Fd3m Spinel type	Isotropic n_D = 1.805	7.5	4620	
Galaxite	$MnAl_2O_4$ (Oxides and hydroxides)	Cubic a = 828 pm H1₁, cF56 (Z = 8) P.G. m3m S.G. Fd3m Spinel type	Isotropic n_D = 1.920	7.5	4040	
Galena [1314-87-0] (syn. lead glance, blue lead) (the Roman naturalist, Pliny, used the Latin name, galena, to describe lead ore)	PbS MM = 239.266 86.60wt% Pb 13.40wt% S Traces Ag, As Sb, Zn, Cd, Cu. (Sulfides and sulfosalts) Coordinance Pb(6)	Cubic a = 593.6 pm B1, cF8 (Z =4) S.G. Fm3m P.G. 4-32 Rock salt type	Isotropic n_D = 3.921 R = 38–42.5%	2.5–3 (76 HV)	7597	Habitus: octahedral crystals, massive, granular. Color: light lead gray or dark lead gray. Luster: metallic. Diaphaneity: opaque. Streak: grayish black. Cleavage: (001), (010), (100). Twinning: {111}, {114}. Fracture: subconchoidal, brittle. Chemical: attacked by HNO₃, with evolution of H₂S, with sulfur and precipitate of PbSO₄, soluble in hot HCl. Yellow precipitate of lead iodide with KI. Electrical resistivity 6.8 to 90,000 µΩ.cm, melting point: 1118°C. Deposits: veins, and seminated in igneous and sedimentary rocks.
Gaylussite (syn. natrocalcite) (after the French chemist and physicist, J.-L. Gay-Lussac)	Na₂Ca(CO₃)₂·5(H₂O) MM = 296.15233 15.53wt% Na 13.53wt% Ca 3.40wt% H 8.11wt% C 59.43wt% O (Nitrates, carbonates, and borates)	Monoclinic P.G. 2/m	Biaxial (–) α = 1.444 β = 1.5155 γ = 1.523 δ = 0.0790 $2V = 34°$ Dispersion strong	2.5	1930–1990	Habitus: disseminated, tabular. Color: colorless, white, or yellowish white. Luster: vitreous (i.e., glassy). Diaphaneity: translucent. Streak: white. Cleavage: (110) perfect, (001) indistinct. Fracture: conchoidal.
Gehlenite	Ca₂Al[SiAlO₇] (Sorosilicates, pair)	Tetragonal a = 769.0 pm c = 506.75 pm (Z = 2) Melilite type	Uniaxial (–)	n.a.	n.a.	
Gersdorffite (syn. gray nickel pyrite, nickel glance)	NiAsS MM = 165.6776 35.42wt% Ni	Cubic Cobaltite group	Isotropic	5.5	5900–6330	Habitus: massive, granular, euhedral crystals, tabular. Color: tin white or white. Luster: metallic. Diaphaneity: opaque. Streak: grayish black. Cleavage: (100) good, (010) good, (001) good. Fracture: brittle. Antiferromagnetic.

Table 11.4 (continued)

Mineral name (IMA) [CAS RN] (Synonyms) (Etymology)	Theoretical chemical formula, relative molecular mass, coordination number, major impurities, mineral class (Strunz)	Crystal system, lattice parameters, Strukturbericht, Pearson symbol, Z, point group, space group, structure type	Optical properties	Mohs hardness (HM)	Density (ρ, kg.m^{-3})	Other mineralogical, physical, and chemical properties
Gersdorffite (*continued*) (after Herr von Gersdorff, owner of Schladming Mine, Austria)	45.22wt% As 19.35wt% S (Sulfides and sulfosalts)					
Gibbsite [21645-51-2]	$Al(OH)_3$ $MM = 78.991618$ Coordination Al(6) (Oxides and hydroxides)	Monoclinic $a = 971.9$ pm $b = 507.05$ pm $c = 864.12$ pm $\beta = 94.57°$ ($Z = 8$) P.G. 2/m S.G. P2$_1$/n	Biaxial (+) $\alpha = 1.560$–1.580 $\beta = 1.560$–1.580 $\gamma = 1.580$–1.600 $\delta = 0.02$ $2V = 0$–$40°$ Dispersion strong	2.5–3.5	2400	**Habitus:** tabular, foliated. **Color:** white, gray. **Luster:** pearly, vitreous. **Diaphaneity:** transparent to translucent. **Streak:** white. **Cleavage:** (001). **Twinning:** [310], and [001]. **Fracture:** uneven, tough.
Glauberite (syn. Glauber's salt) (after Glauber's salt, of alchemist origin)	$Na_2Ca(SO_4)_2$ (Sulfates, chromates, molybdates, and tungstates)	Monoclinic	Biaxial (−) $\alpha = 1.507$–1.515 $\beta = 1.527$–1.535 $\gamma = 1.529$–1.536 $\delta = 0.021$–0.022 $2V = 24$–$34°$ Dispersion strong	2.5–3	2700–2850	**Habitus:** tabular, prismatic. **Color:** yellow, reddish gray, or red. **Luster:** vitreous (glassy). **Diaphaneity:** transparent to translucent. **Streak:** white. **Fracture:** brittle-conchoidal. **Cleavage:** (001) perfect. **Deposits:** dry salt-lake beds in desert climates.
Glauconite	$(K,Na)(Fe,Al,Mg)_2(Si,Al)_4O_{10}(OH)_2$ (Phyllosilicates, layered)	Monoclinic P.G. 2/m Mica type	Biaxial (−) $\alpha = 1.590$–1.612 $\beta = 1.609$–1.643 $\gamma = 1.610$–1.644 $\delta = 0.020$–0.032 $2V = 0$–$20°$	n.a.	n.a.	**Habitus:** micaceous. **Color:** green.
Glaucophane (from the Greek, glaukos, blue, and fanos, appearing)	$Na_2(Mg,Fe)_3Al_2Si_8O_{22}(OH)_2$ (Inosilicates, chain, ribbon) Coordination Na (8), Mg(6),Al(6), and Si(4)	Monoclinic $a = 974.80$ pm $b = 1791.50$ pm $c = 527.70$ pm 102°78' ($Z = 2$) Tremolite type P.G. 2/m S.G. 2/m	Biaxial (−) $\alpha = 1.606$–1.661 $\beta = 1.622$–1.667 $\gamma = 1.627$–1.670 $\delta = 0.009$–0.021 $2V = 0.50°$ Dispersion strong	6–6.5	3080–3300	**Habitus:** massive, fibrous, columnar, granular. **Color:** colorless, grayish blue, bluish black, or lavender blue. **Luster:** vitreous (i.e., glassy), pearly. **Diaphaneity:** translucent. **Streak:** grayish blue. **Cleavage:** (110) good, (110) good. **Fracture:** conchoidal, uneven, brittle. **Chemical:** insoluble in strong mineral acids. Easy fusible giving a green enamel. The powdered mineral colors a bunsen flame yellow (Na). Slightly attracted by electromagnet. **Deposits:** metamorphic blue schists.

Name	Chemistry	Crystallography	Optical	Hardness	Density	Description
iron ore, limonite, groutite: Mn-varieties) (after the German poet, J.W. Goethe)	36.01wt%. O 1.13wt%.H Coordinance Fe(6) (Oxides and hydroxides)	c = 302.1 pm (Z = 4) P.G. mmm S.G. P2$_1$/bnm	γ = 2.398–2.515 δ = 0.138–0.140 2V = 0–27° Dispersion strong			hackly. **Deposits:** iron ore deposits.
Gold [7440-57-5] (syn. aurum)	Au MM = 196.96655 Coordinance Au(12) (Native elements)	Cubic a = 407.86 pm A1, $cF4$ (Z = 4) P.G. m3m S.G. Fm3m Copper type	Isotropic n_D = 0.368 R = 85%	2.5–3	19,287	**Habitus:** octahedral, dendritic, arborescent, platy, granular. **Color:** yellow, pale yellow, orange, yellow white, or reddish white. **Diaphaneity:** opaque. **Luster:** metallic **Streak:** yellow. **Cleavage:** none. **Twinning:** {111}. **Fracture:** hackly. **Deposits:** quartz veins and alluvial placers deposits. **Chemical:** inert to most strong mineral acids (eg, HCl, H_2SO_4, HF) but readily dissolved by aqua regia (3 vol. HCl + 1 vol. HNO_3).
Goslarite (after the locality of Goslar, Germany)	$ZnSO_4 \cdot 7H_2O$ MM = 287.56056 22.74wt%.Zn 4.91wt%.H 11.15wt%.S 61.20wt%. O (Sulfates, chromates, molybdates and fungstates)	Orthorhombic P.G. 222	Biaxial (–)	2–2.5	2000	**Habitus:** stalactitic, massive, acicular. **Color:** white, yellowish white, or reddish white. **Luster:** vitreous (i.e, glassy). **Streak:** white. **Cleavage:** (010) perfect **Fracture:** conchoidal
Graphite [7440-44-0] (syn., plumbago, black lead) from the Greek, graphein, to write	C MM = 12.0107 Coordinance C(3) (Native elements)	Hexagonal a = 246.4 pm c = 673.6 pm A9, $hP4$ (Z = 4) P.G. 6/mmm S.G. P6$_3$/mmc Graphite type	Uniaxial (n.a.) n_D = 1.93–2.07	1.5–2	2230	**Habitus:** foliated, tabular, earthy. **Color:** dark gray, black, or steel gray. **Diaphaneity:** opaque. **Luster:** submetallic. **Streak:** black. **Cleavage:** (0001) perfect. **Fracture:** sectile. **Deposits:** metamorphosed limestones, organic-rich shales, and coal beds. Melting point: 4492°C (10 MPa).
Greenalite	$Fe_2Fe_3Si_2O_5(OH)_4$ (Phyllosilicates, layered)	Monoclinic Sphenoidal 2	Istropic n_D = 1.65–1.675	n.a.	n.a.	
Grossular (syn. grossularite, hessonite brownish orange) from the Latin, grossularia, gooseberry; hessonite is from the Greek, hesson, slight, in reference to the smaller specific gravity	$Ca_3Al_2(SiO_4)_3$ MM = 450.44638 26.69wt%.Ca 11.98wt%.Al 18.71wt%.Si 42.62wt%.O Coordinance Ca(8), Al(6), Si(4) (Nesosilicates)	Cubic a = 1185.1 pm (Z = 8) P.G. 432 S.G. Ia3d Garnet group Ugrandite series)	Isotropic n_D = 1.734	6.5–7.5	3594	**Habitus:** dodecahedral, massive, granular, euhedral crystals, crystalline. **Color:** colorless, white, green, or yellow. **Streak:** brownish white. **Diaphaneity:** transparent to translucent. **Luster:** vitreous (i.e, glassy), resinous. **Cleavage:** none. **Fracture:** conchoidal. **Chemical:** soluble in HCl, and HNO_3. **Deposits:** contact metasomatic deposits.
Gummite [12326-21-5]	$UO_3 \cdot nH_2O$ MM = 304.043	Amorphous	Isotropic	n.a.	n.a.	
Gypsum [10101-41-4] (syn. selenite, alabaster, satin spar) (from the Greek, gypsos, meaning burned mineral; selenite from the	$CaSO_4 \cdot 2H_2O$ MM = 172.17216 23.28wt% Ca 2.34wt% H 18.62wt% S 55.76wt% O	Monoclinic a = 568.0 pm b = 1551.8 pm c = 629.0 pm β = 113°83' (Z = 4)	Biaxial (+) α = 1.510 β = 1.523 γ = 1.529 δ = 0.009 2V = 58°	2.0	2320	**Habitus:** tabular, crystalline, massive, reniform, fibrous cockscomb aggregates: "desert roses". **Color:** white, colorless, yellowish white, greenish white, or brown. **Luster:** vitreous, pearly. **Diaphaneity:** transparent to translucent. **Streak:** white. **Cleavage:** (010), (100). **Twinning:** {100}. **Fracture:** conchoidal, fibrous. **Chemical:** decomposes at 150°C giving off water. **Deposits:** sedimentary evaporites.

11 Minerals, Ores and Gemstones

Table 11.4 (continued)

Mineral name (IMA) [CAS RN] (Synonyms) (Etymology)	Theoretical chemical formula, relative molecular mass, coordinance number, major impurities, mineral class (Strunz)	Crystal system, lattice parameters, Strukturbericht, Pearson symbol, Z, point group, space group, structure type	Optical properties	Mohs hardness (HM)	Density (ρ, kg.m^{-3})	Other mineralogical, physical, and chemical properties
Greek, selenos, an allusion to its pearly luster (moonlight) on cleavage fragments	Coordinance Cat(8), S(4) (Sulfates, chromates, molybdates, and tungstates)	P.G. 2/m, S.G. A2/n	Dispersion strong			
Halite [7647-14-5] (syn, rock salt) (from the Greek, halos, salt, and lithos, rock)	NaCl $MM = 58.44246$ 39.34wt% Na 60.66wt% Cl (Halides) Coordinance Na(6)	Cubic $a = 564.02$ pm B1, cF8 ($Z = 4$) S.G. Fm3m P.G. 4-32 Rock salt type	Isotropic $n_D = 1.5446$	2.5	2160–2170	**Habitus:** euhedral crystals, granular, crystalline. **Color:** white, clear, light blue, dark blue, or pink. **Luster:** vitreous (i.e., glassy). **Diaphaneity:** transparent. **Streak:** white. **Cleavage:** (100), (010), (001). **Fracture:** conchoidal, brittle. **Deposits:** marine or continental evaporite. Soluble in water, the solution colors the flame of a bunsen yellow. Fusible (mp 801°C).
Hanksite	$KNa_{22}(SO_4)_9(CO_3)_2Cl$ (Sulfates, chromates, molybdates, and tungstates)	Hexagonal	Uniaxial (−) $\epsilon = 1.46$ $\omega = 1.48$ $\delta = 0.02$	3	2500	**Habitus:** crystalline, coarse. **Color:** white or yellow. **Diaphaneity:** transparent. **Luster:** vitreous, or greasy. **Streak:** white. **Fracture:** conchoidal. **Deposits:** continental evaporitic deposits under desert climates.
Haussmannite	$Mn_3O_4 = Mn^{II}Mn^{III}_2O_4$ $MM = 228.81175$ Coordinance Mn(4, and 6) (Oxides and hydroxides)	Tetragonal $a = 813.6$ pm $c = 944.2$ pm ($Z = 8$) P.G. 4/mmm S.G. I4$_1$/amd Distorted spinel type	Uniaxial (−) $\epsilon = 2.15$ $\omega = 2.46$ $\delta = 0.31$ $R = 17\%$	5–5.5	4863	**Habitus:** pseudo-octahedral. **Color:** brown, black. **Diaphaneity:** translucent to opaque. **Luster:** submetallic. **Streak:** light brown. **Fracture:** uneven. **Cleavage:** perfect (001). **Twinning:** {112}, {101}.
Haüyne (syn, hauynite) (after the French crystallographer, R.J. Haüy)	$(Na,Ca)_{4-8}[Al_6Si_6O_{24}](SO_4,S,Cl)_{1-2}$ (Tectosilicates, network)	Cubic $a = 913$ pm ($Z = 1$) Sodalite type	Isotropic $n_D = 1.496$–1.505	5.5–6	2440–2500	**Habitus:** euhedral crystals. **Color:** white, gray, blue, green, red, yellow. **Luster:** vitreous (i.e., glassy), greasy. **Diaphaneity:** transparent to translucent. **Streak:** bluish white. **Cleavage:** (110) perfect, (011) perfect, (101) perfect. **Twinning:** {111} common. **Fracture:** conchoidal. **Deposits:** igneous rocks low in silica and rich in alkalis.
Hedenbergite (after the Swedish mineralogist, M.A.L. Hedenberg)	$CaFe[Si_2O_6]$ Coordinance Ca(8), Fe(6), Si(4) Traces Al, Mn, Ti (Inosilicates, simple chain)	Monoclinic $a = 985.4$ pm $b = 902.4$ pm $c = 526.3$ pm $104.33°$ ($Z = 4$) P.G. C2/m S.G. C2/c Diopside type	Biaxial (+) $\alpha = 1.699$–1.739 $\beta = 1.705$–1.745 $\gamma = 1.728$–1.757 $\delta = 0.018$–0.029 $2V = 52$–63° Dispersion strong	5.5–6.5	3550	**Habitus:** granular, crystalline, lamellar. **Color:** grayish green, brownish green, dark green, black. **Luster:** vitreous (i.e., glassy), pearly. **Diaphaneity:** translucent to opaque. **Streak:** white green. **Cleavage:** (110) perfect, (110) indistinct. **Twinning:** {100}, {001} simple, multiple, common. **Fracture:** conchoidal, brittle. **Chemical:** insoluble in strong mineral acids, fusible giving a magnetic globule. **Other:** dielectric constant 8.99, attracted by a strong permanent magnet. **Deposits:** contact metamorphic rocks.

Name	Chemistry	Crystallography	Optical	Hardness	Density	Properties
[1309-37-1] (syn., kidney ore, specularite, martite) (from the Greek, haematites, bloodlike, in allusion to vivid red color of the powder)	$MM = 159.6922$ 69.94wt% Fe 30.06wt% O (Oxides and hydroxides) Traces Ti, Mg, Coordinance Fe(6)	(Rhombohedral) $a = 503.29$ pm $13.749°$ D5, hR10 ($Z = 2$) S.G. R-3c P.G. -32/m Corundum type	$\varepsilon = 2.96$ $\omega = 3.22$ $\delta = 0.28$ Dispersion strong $R = 28\%$	6.5		red. Luster: metallic, greasy. Diaphaneity: translucent to opaque. Twinning: none. Cleavage: reddish brown. Cleavage: none. Twinning: {001}, {101}. Fracture: subconchoidal. Chemical: soluble in hot HCl. Becomes magnetic after firing in reducing flame. Antiferromagnetic. Deposits: magmatic, hydrothermal, metamorphic and sedimentary rocks.
Hercynite (from Latin, hercynia silva, forested mountains)	$FeAl_2O_4$ $MM = 173.80768$ 31.05wt% Al 32.13wt% Fe 36.82wt% O (Oxides and hydroxides)	Cubic $a = 813.5$ pm H1, cF56 ($Z = 8$) S.G. Fd3m Spinel type (Spinel Series)	Isotropic $n_D = 1.835$	7.5	3950–4400	Habitus: euhedral crystals, massive, granular. Color: black. Luster: vitreous (i.e, glassy). Diaphaneity: opaque. Streak: dark green. Cleavage: (111). Fracture: uneven. Deposits: magnetic rocks.
Hessite (syn., telluric silver) (after the Swiss chemist, GH Hesse)	Ag_2Te $MM = 235.4682$ 45.81wt% Ag 54.19wt% Te (Sulfides and sulfosalts)	Monoclinic P.G. 2	Biaxial	1.5–2	7200–7900	Habitus: euhedral crystals, granular. Color: lead gray or steel gray. Luster: metallic. Diaphaneity: opaque. Streak: light gray. Cleavage: (100) indistinct. Fracture: uneven.
Heulandite	$(Ca,Na)Si_7Al_2O_{18}·6H_2O$ Coordinance Ca(6), Na(6), Si(4), Al(4) (Tectosilicates, framework)	Monoclinic $a = 1773$ pm $b = 1782$ pm $c = 743$ pm $\beta = 116.3°$ ($Z = 4$) P.G. m S.G. Cm Zeolite group	Biaxial (+) $\alpha = 1.490$ $\beta = 1.500$ $\gamma = 1.500$ $\delta = 0.005$ $2V = 35°$	3.5–4	2150	Habitus: platy. Color: white. Luster: vitreous. Diaphaneity: transparent to translucent. Cleavage: (010) perfect. Fracture: subconchoidal. Deposits: volcanic tuffs and volcano-clastic sediments.
Huebnerite	$MnWO_4$ $MM = 302.775649$ Coordinance Mn(6), W(4) (Sulfates, chromates, molybdates, and tungstates)	Monoclinic $a = 483.4$ pm $b = 575.8$ pm $c = 449.9$ pm $\beta = 91.18°$ ($Z = 2$) P.G. 2/m S.G. P2/c Wolframite type	Biaxial (+) $\alpha = 2.17$ $\beta = 2.22$ $\gamma = 2.30$ $\delta = 0.13$ $2V = 73°$	4–4.5	7250	Habitus: long prismatic. Color: red brown. Luster: submetallic. Diaphaneity: opaque. Cleavage: (010) perfect. Fracture: uneven, brittle.
Humite (the British mineralogist Sir Abraham Hume)	$Mg(OH,F)_2·3Mg_2[SiO_4]$ (Nesosilicates)	Orthorhombic ($Z = 2$) (Humite group)	Biaxial $\alpha = 1.607$–1.643 $\beta = 1.619$–1.653 $\gamma = 1.639$–1.675 $\delta = 0.029$–0.031 $2V = 65$–84° O.A.P (001)	6	3200–3320	Habitus: massive. Color: brown, yellow, orange, red, colorless. Luster: vitreous (i.e, glassy). Diaphaneity: translucent to transparent. Cleavage: (100) poor. Fracture: uneven, brittle. Chemical: attacked by strong mineral acids giving a silica gel. Deposits: dolomites, limestones, skarns.

11

Minerals, Ores and Gemstones

Table 11.4 (continued)

Mineral name (IMA) [CAS RN] (Synonyms) (Etymology)	Theoretical chemical formula, relative molecular mass, coordinance number, major impurities, mineral class (Strunz)	Crystal system, lattice parameters, Strukturbericht, Pearson symbol, Z, point group, space group, structure type	Optical properties	Mohs hardness (HM)	Density (ρ, kg.m^{-3})	Other mineralogical, physical, and chemical properties
Illite (syn, hydromuscovite, hydromica, gumbelite)	(K,H)(Al,Mg,Fe)$_2$(Si,Al)$_4$O$_{10}$ [(OH)$_2$·(H$_2$O)] (Phyllosilicates, layered)	Monoclinic	Biaxial (−) $\alpha = 1.535-1.57$ $\beta = 1.555-1.6$ $\gamma = 1.565-1.605$ $\delta = 0.030-0.035$ $2V = 5-25°$ Dispersion none	n.a.	n.a.	
Ilmenite [12168-52-4] (after the lake Ilmen, Russia)	FeTiO$_3$ $MM = 151.7252$ 31.56wt% Ti 36.81wt% Fe 31.63wt% O (Oxides and hydroxides) Coordinance Fe(6), Ti(6)	Trigonal (Rhombohedral) $a = 509.3$ pm $c = 1405.5$ pm $(Z = 6)$ S.G.R3m P.G. $\bar{3}$ Ilmenite type	Uniaxial (−) $\omega = 2.700$ $\varepsilon = 2.700$ Dispersion strong $R = 20\%$	5.0-5.5 (519-703 HV)	4720-4780	**Habitus:** tabullar, lamellar, massive. **Color:** iron black. **Diaphaneity:** opaque. **Luster:** submetallic. **Streak:** brownish black. **Fracture:** subconchoidal. **Cleavage:** none. **Chemical:** insol. in HCl, or HNO$_3$ but attacked by boiling H$_2$SO$_4$. Slightly magnetic. **Nonfusible:** mp 1470°C. Electrical resistivity 0.001 to 4 Ω.m. Antiferromagnetic. **Deposits:** igneous rocks (e.g., peridotites, gabbros, diorites, syenites), sedimentary rocks.
Iodargyrite (syn., iodyrite) (from the Greek, iodos, violet, and Latin, argentum, silver)	AgI $MM = 234.77267$ 45.95wt%Ag 54.05wt%I (Halides)	Hexagonal	Uniaxial (+)	1.5-2	5500-5700	**Habitus:** platy. **Color:** pale yellow or green. **Luster:** adamantine-greasy. **Diaphaneity:** transparent to translucent.
Iridium (from the Latin, iris, rainbow, an allusion to the colored salts derived from its compounds)	(Ir,Os,Ru) (Native elements)	Cubic $a = 383.92$pm A1, cF4 ($Z = 4$) Fm3m Copper type	Isotropic	6-7	22,600-22,800	**Color:** white. **Diaphaneity:** opaque. **Luster:** metallic. **Cleavage:** none. **Streak:** white.
Jacobsite	MnFe$_2$O$_4$ (Oxides and hydroxides)	Cubic $a = 851$ pm Hl1, cF56 ($Z = 8$) Fd3m Spinel type	Isotropic $n_D = 2.30$	5.5	4870	Ferromagnetic
Jadeite (from the Spanish, piedra de ijada, stone of the side, because its supposed to cure kidney ailments if applied to the side of the body)	Na(Al,Fe)[Si$_2$O$_6$] $MM = 202.1387$ 11.37wt% Na 13.35wt% Al 27.79wt% Si 47.49wt% O Coordinance Na(8), Al(6), Si(4) Traces of Mg, Ca, and Fe	Monoclinic $a = 940.9$ pm $b = 856.4$ pm $c = 522.0$ pm $107.43°$ $(Z = 4)$ P.G. 2/m S.G. C2/c	Biaxial (+) $\alpha = 1.640-1.658$ $\beta = 1.645-1.663$ $\gamma = 1.652-1.673$ $\delta = 0.012-0.013$ $2V = 67-75°$ Dispersion moderate	6-6.5	3240-3430	**Habitus:** granular, fibrous, massive. **Color:** green, white, pale bluish gray, grayish green, or pale purple. **Luster:** vitreous (i.e, glassy), pearly. **Diaphaneity:** translucent. **Streak:** white. **Cleavage:** (110) good. **Twinning:** {100}, {001}. **Fracture:** tough. **Chemical:** insoluble in strong mineral acids. **Deposits:** strongly metamorphosed sodium-rich serpentinous rocks.

Name / Etymology	Composition	Crystallography	Optical	Hardness	Density	Description
(from the Greek, kaolos, contemporary)	$MM = 248.96944$ 15.70wt% K 9.76wt% Mg 2.43wt% H 12.88wt% S 14.24wt% Cl 44.98wt% O (Sulfates, chromates, molybdates, and tungstates)	P.G.2/m	$\alpha = 1.494$ $\beta = 1.505$ $\gamma = 1.516$ $\delta = 0.022$ $2V = 88°$			dark flesh red. **Luster:** vitreous, greasy. **Streak:** white. **Fracture:** conchoidal, **Cleavage:** (001) good.
Kaolinite (1Tc) (after the locality, Kao-Ling, China, where clay was extracted, from the Chinese, kao high, ling, mountain)	$Al_4[Si_4O_{10}(OH)_8]$ $MM = 516.32088$ 20.90wt% Al 21.76wt% Si 1.56wt% H 55.78wt% O Coordination Al(6), Si(4) (Phyllosilicates, layered)	Triclinic $a = 515$ pm $b = 892$ pm $c = 738$ pm $\alpha = 91.8°$ $\beta = 104.8°$ $\gamma = 90.0°$ $(Z = 1)$ P.G. $\bar{1}$ S.G. P$\bar{1}$	Biaxial (−) $\alpha = 1.553–1.563$ $\beta = 1.559–1.569$ $\gamma = 1.565–1.570$ $\delta = 0.007$ $2V = 40–44°$ Dispersion none	2–2.5	2600	**Habitus:** platy, massive. **Color:** white, brownish white, grayish white, yellowish white, or grayish green. **Diaphaneity:** transparent to translucent. **Luster:** earthy (i.e. dull). **Streak:** white. **Cleavage:** (001) perfect. **Fracture:** flexible.
Kermite (syn. rasorite) (after Kern County, State of California, USA)	$Na_2B_4O_6(OH)_2 \cdot 3(H_2O)$ $MM = 290.28379$ 15.84wt% Na 14.90wt% B 3.13wt% H 66.14wt% O Coordinance Na(5), B(3 and 4) (Carbonates and borates)	Monoclinic $a = 702.2$ pm $b = 915.1$ pm $c = 1567.6$ pm $(Z = 4)$ P.G 2/m S.G. P2/a	Biaxial(−) $\alpha = 1.454$ $\beta = 1.472$ $\gamma = 1.488$ $\delta = 0.034$ $2V = 80°$ Dispersion distinct	2.5–3	1900–1920	**Habitus:** crystalline, coarse, acicular, massive. **Color:** colorless or white. **Luster:** vitreous, pearly. **Diaphaneity:** transparent to translucent. **Cleavage:** (100) perfect, (001) perfect, (201) good. **Fracture:** uneven, brittle. **Deposits:** continental sedimentary basins.
Krennerite	(Au, Ag)Te$_2$ (Sulfides and sulfosalts)	Orthorhombic $a = 1654$ pm $b = 882$ pm $c = 446$ pm C46, oP24 $(Z = 8)$ S.G. Pma	Biaxial	2.5	8530	**Habitus:** striated. **Color:** white or blackish yellow. **Luster:** metallic. **Diaphaneity:** opaque.
Kyanite (syn. disthene) (from the Greek, kyanos, blue, and dis, two-fold, and sthenos, force, owing to the variation of hardness according to crystal planes)	$Al_2O[SiO_4] = Al_2SiO_5$ $MM = 162.04558$ 33.30wt% Al 17.33wt% Si 49.37wt% O Traces of Fe, Ca, Cr Coordinance Al(6), Si(4) (Nesosubsilicates)	Triclinic $a = 712.3$ pm $b = 784.8$ pm $c = 556.4$ pm $\alpha = 90°\ 5.5'$ $\beta = 101°25'$ $\gamma = 105°\ 44.5'$ $(Z = 4)$ P.G. $\bar{1}$ S.G. P$\bar{1}$	Biaxial (−) $\alpha = 1.712–1.718$ $\beta = 1.721–1.723$ $\gamma = 1.727–1.734$ $\delta = 0.012–0.016$ $2V = 82–83°$ Dispersion weak	5.5 (100) 7.0 (010)	3530–3650	**Habitus:** bladly, columnar, tabular, fibrous. **Color:** blue, white, yellowish, gray, green, or black. **Luster:** vitreous (i.e. glassy), pearly. **Diaphaneity:** translucent to transparent. **Streak:** white. **Fracture:** uneven, brittle. **Luminescence:** pink to red. **Cleavage:** (100) perfect, (010) imperfect. **Twinning:** {100}. **Chemical:** insoluble in strong mineral acids. **Other:** Unfusible, dielectric constant 5.7 to 7.18. **Deposits:** metamorphosed peri-aluminous sedimentary rocks.

11
Minerals, Ores and Gemstones

Table 11.4 (continued)

Mineral name (IMA) [CAS RN] (Synonyms) (Etymology)	Theoretical chemical formula, relative molecular mass, coordinance number, major impurities, mineral class (Strunz)	Crystal system, lattice parameters, Strukturbericht, Pearson symbol, Z, point group, space group, structure type	Optical properties	Mohs hardness (HM)	Density (ρ, kg.m^{-3})	Other mineralogical, physical, and chemical properties
Labradorite (after the Labrador peninsula, Canada)	$(Ca,Na)(Si,Al)_4O_8$ An60 Ab40 $MM = 271.81357$ 3.38wt% Na 8.85wt% Ca 15.88wt% Al 24.80wt% Si 47.09wt% O (Tectosilicates, framework)	Triclinic $a = 815.5$ pm $b = 1284$ pm $c = 1016$ pm $Z = 6$	Biaxial (+) $\alpha = 1.554–1.563$ $\beta = 1.559–1.568$ $\gamma = 1.562–1.573$ $\delta = 0.008–0.010$ $2V = 78–86°$	7	2690	**Habitus:** euhedral crystals, striated. **Color:** white or gray. **Diaphaneity:** translucent to transparent. **Luster:** vitreous (i.e., glassy). **Cleavage:** (001) perfect, (010) good, (110) distinct. **Fracture:** uneven. **Deposits:** magmatic and metamorphic rocks.
Langbeinite	$K_2Mg_2(SO_4)_3$ (Sulfates, chromates, molybdates, and tungstates)	Cubic	Isotropic	n.a.	n.a.	Phosphorescent. Marine evaporite deposits.
Larnite	$Ca_2[SiO_4]$ (Nesosilicates)	Monoclinic $a = 548$ pm $b = 676$ pm $c = 928$ pm $94.55°$ $(Z = 4)$	Biaxial (+) $\alpha = 1.707$ $\beta = 1.715$ $\gamma = 1.730$ $\delta = 0.023$ $2V = 13–14°$ Dispersion Strong	n.a.	3280	—
Laumontite	$CaSi_4Al_2O_{12} \cdot 4H_2O$ Coordinance Ca(6), Si(4), Al(4) (Tectosilicates, zeolite group)	Monoclinic $a = 1475$ pm $b = 1310$ pm $c = 755$ pm $\beta = 111.5°$ $(Z = 4)$ P.G. m S.G. Cm	Biaxial (−) $\alpha = 1.510$ $\beta = 1.520$ $\gamma = 1.520$ $\delta = 0.010$ $2V = 25–45°$	3–4	2300	**Habitus:** prismatic. **Color:** white. **Diaphaneity:** translucent to transparent. **Luster:** vitreous (i.e., glassy). **Cleavage:** (010) perfect, (110) perfect. **Fracture:** uneven. Pyroelectric. **Deposits:** volcanic tuffs.
Lawsonite (after the American mineralogist Prof. A.C. Lawson)	$CaAl_2[Si_2O_7(OH)_2] \cdot (H_2O)$ Coordinance Ca(8), Al(6), Si(4) (Sorosilicates, pair)	Orthorhombic $a = 878.7$ pm $b = 1312.3$ pm $c = 583.6$ pm $(Z = 4)$ P.G. mmm S.G. Ccmm	Biaxial (+) $\alpha = 1.665$ $\beta = 1.674$ $\gamma = 1.685$ $\delta = 0.020$ $2V = 76–87°$ O.A.P. (100) Dispersion	7–8	3050–3120	**Habitus:** prismatic, tabular. **Color:** colorless, pale blue, or grayish blue. **Diaphaneity:** translucent to transparent. **Luster:** vitreous (i.e., glassy), greasy. **Streak:** white. **Cleavage:** (010) perfect, (100) perfect. **Twinning:** {110}. **Fracture:** uneven brittle. **Deposits:** originally described from a crystalline schist associated with serpentine. Also found as a secondary mineral in altered gabbros and diorites.

Name	Formula / composition	Crystal data	Optical	Hardness	Density	Description
(syn., scorzalite: Fe, Ni)	Coordinate Fe(6), Mg(6), Al(6), P(4) (Phosphates, arsenates, and vanadates)	$a = 716$ pm, $b = 726$ pm, $c = 724$ pm, $\beta = 120.67°$, ($Z = 2$), P.G. 2/m, S.G. $P2_1/c$	$\alpha = 1.612$, $\beta = 1.634$, $\gamma = 1.643$, $\delta = 0.031$, $2V = 70°$			translucent. **Luster:** vitreous. **Fracture:** uneven. **Cleavage:** (011) perfect.
Lazurite (syn, lapis lazuli, lasurite, ultramarine) (from the Persian lazward, blue)	$(Na,Ca)_{7-8}(Al,Si)_{12}(O,S)_{24}$ $[(SO_4,Cl,S)_2(OH)_2]$ (Tectosilicates, network)	Cubic, $a = 891$ pm, ($Z = 1$), Sodalite type	Isotropic, $n_D = 1.5$	5.5	2380–2420	**Habitus:** massive, granular. **Color:** dark blue or greenish blue. **Diaphaneity:** translucent. **Luster:** greasy. **Luminescence:** fluorescent. **Cleavage:** (110) imperfect. **Fracture:** conchoidal. **Streak:** light blue.
Lepidocrocite [20344-49-4] (from the Greek, lipis, scale, and krokis, fiber)	FeO(OH), $MM = 88.8517$, 62.85wt% Fe, 36.01wt% O, 1.13wt% H, Coordination Fe(6) (Oxides and hydroxides)	Orthorhombic, $a = 386.8$ pm, $b = 1252.5$ pm, $c = 306.6$ pm, ($Z = 4$), P.G. mmm, S.G. A/2mam, Lepidocrite type	Biaxial (−), $\alpha = 1.94$, $\beta = 2.20$, $\gamma = 2.51$, $\delta = 0.57$, $2V = 83°$, Dispersion weak	5	4090	**Habitus:** scaly, fibrous, bladly, tabular, pulverulent. **Color:** red, yellowish brown, or blackish brown. **Luster:** submetallic. **Diaphaneity:** opaque. **Streak:** dark yellow brown or orange. **Cleavage:** perfect (010), good (001). **Fracture:** uneven. **Deposits:** iron ore deposits.
Lepidolite (from the Greek lepidon, scale, and lithos, stone)	$K(Li,Al)_3(Si,Al)_4O_{10}(F,OH)_2$, $MM = 386.31252$, 10.12wt% K, 3.59wt% Li, 6.98wt% Al, 29.08wt% Si, 0.52wt% H, 49.70wt% O (Phyllosilicates, layered)	Monoclinic, $a = 5.21$, $b = 8.97$, $c = 20.16$, $Z = 4$, Prismatic	Biaxial (−), $\alpha = 1.525$–1.548, $\beta = 1.551$–1.58, $\gamma = 1.554$–1.586, $\delta = 0.029$–0.038, $2V = 25$–$58°$, Dispersion weak	2.5–3	n.a.	**Habitus:** massive, tabular, lamellar. **Color:** pale lilac blue, light red, colorless, or gray. **Cleavage:** (001) perfect. **Diaphaneity:** translucent. **Luster:** vitreous, pearly. **Fracture:** uneven. **Streak:** white. **Deposits:** lithia-bearing pegmatites.
Leucite (syn, amphigene) (from the Greek, leukos, white)	$K[Si_2AlO_6]$, $MM = 218.24724$, 17.91wt% K, 12.36wt% Al, 25.74wt% si, 43.99wt% O, Coordinance K(12), Si(4), Al(4) (Tectosilicates, framework)	Tetragonal (pseudo-cubic), $a = 1343$ pm, $c = 1370$ pm, ($Z = 16$), P.G. 4/m, S.G. $I4_1a$	Isotropic, $n_D = 1.508$–1.511	5.5–6	2470	**Habitus:** crystalline, coarse. **Color:** colorless, white, or gray. **Diaphaneity:** translucent to transparent. **Luster:** vitreous (i.e, glassy). **Streak:** white. **Cleavage:** (110) indistinct. **Twinning:** {100}, {112}. **Fracture:** brittle, conchoidal. **Deposits:** acid volcanic rocks.
Limonite	$FeO \cdot OH \cdot nH_2O$ (Oxides and hydroxides)	Amorphous or cryptocrystalline	Isotropic, $n_D = 2.0$–2.1	4–5.5	2700–4300	
Linnaeite (syn, linneite) (after the Swedish botanist, C Linnaeus)	Co_3S_4, $MM = 305.0636$, 57.95wt% Co, 42.05wt% S (Sulfides and sulfosalts)	Cubic, P.G. 23	Isotropic	4.5–5.5	4800	**Habitus:** crystalline, fine, encrustations, granular. **Color:** white or pinkish white. **Luster:** metallic. **Diaphaneity:** opaque. **Streak:** grayish black. **Cleavage:** (100) imperfect. **Fracture:** uneven.

11

Minerals, Ores and Gemstones

Table 11.4 (continued)

Mineral name (IMA) [CAS RN] (Synonyms) (Etymology)	Theoretical chemical formula, relative molecular mass, coordination number, major impurities, mineral class (Strunz)	Crystal system, lattice parameters, Strukturbericht, Pearson symbol, Z, point group, space group, structure type	Optical properties	Mohs hardness (HM)	Density (ρ, kg.m^{-3})	Other mineralogical, physical, and chemical properties
Livingstonite (after the missionary, D. Livingstone)	$HgSb_4S_8$ $MM = 944.118$ 21.25wt% Hg 51.58wt% Sb 27.17wt% S (Sulfides and sulfosalts)	Monoclinic P.G. 2/m	Biaxial ($-$)	2	4810–4900	**Habitus:** radial, fibrous, columnar. **Color:** steel gray or lead gray. **Luster:** sub-metallic. **Diaphaneity:** opaque to subtranslucent. **Streak:** red. **Cleavage:** (010) perfect, (100) perfect. **Fracture:** uneven.
Magnesiochromite	$MgCr_2O_4$ $MM = 192.29480$ 12.6395wt% Mg 33.2809wt% O 54.0796wt% Cr (Oxides and hydroxides)	Cubic $a = 833.4$ pm Hl$_1$, cF56 ($Z = 8$) S.G. Fd3m Spinel type (Chromite series)	Isotropic $n_D = 2.00$	5.5	4430	
Magnesioferrite	$MgFe_2O_4$ (Oxides and hydroxides)	Cubic $a = 838.3$ pm Hl$_1$, cF56 ($Z = 8$) S.G. Fd3m Spinel type (Magnetite series)	Isotropic $n_D = 2.380$	7.5	4520	Ferromagnetic materials.
Magnesite [546-93-0] (syn, giobertite, bitter spar) (after the Greek city of Magnesia)	$MgCO_3$ $MM = 84.3142$ 28.83wt% Mg 14.25wt% C 56.93wt% O Coordinance Mg(6), C(3) (Nitrates, carbonates, and borates)	Trigonal $a = 463.30$ pm $c = 1501.60$ pm $R\bar{3}^-$ c ($Z = 6$) P.G. $-32/m$ S.G. R-3c Calcite type	Uniaxial ($-$) $\varepsilon = 1.509$–1.563 $\omega = 1.700$–1.782 $\delta = 0.190$-0.218 Dispersion strong	3.5–4.5	2980–3500	**Habitus:** rhombohedral, massive, granular, earthy, fibrous. **Color:** colorless, white, grayish white, yellowish white, or brownish white. **Luster:** vitreous (i.e., glassy). **Diaphaneity:** transparent, translucent, to opaque. **Streak:** white, gray. **Cleavage:** (1011) perfect. **Fracture:** brittle, conchoidal. **Chemical:** decomposes at 990°C giving MgO and readily soluble in diluted acids with evolution of carbon dioxide.
Magnetite [1309-37-1] (syn, lodestone, magnetic iron ore) (from attracted by a magnet)	$Fe_3O_4 = Fe^{II}Fe^{III}_2O_4$ $MM = 231.5386$ 72.36wt% Fe 27.64wt% O Coordinance Fe(4 and 6) (Oxides and hydroxides)	Cubic $a = 839.4$ pm Hl$_1$, cF56 ($Z = 8$) P.G. m3m S.G. Fd3m Spinel type	Isotropic $n_D = 2.42$ $R = 22\%$	5.5–6.5	5201	**Habitus:** octahedral, massive, granular, crystalline. **Color:** black. **Luster:** metallic. **Diaphaneity:** opaque. **Streak:** black. **Cleavage:** none. **Twinning:** [111]. **Fracture:** subconchoidal. **Other:** highly ferromagnetic, electrical resistivity 56 $\mu\Omega$.cm.
Malachite (from the Greek, malache, mallow, in reference to green leaf color)	$Cu_2(CO_3)(OH)_2$ $MM = 221.11588$ 57.48wt% Cu 0.91wt% H 5.43wt% C 36.18wt% O Coordinance Cu(4), C(3)	Monoclinic $a = 950.2$ pm $b = 1197.4$ pm $c = 324.0$ pm $\beta = 98.0°$ ($Z = 4$) P.G. 2/m	Biaxial ($-$) $\alpha = 1.655$ $\beta = 1.875$ $\gamma = 1.909$ $\delta = 0.254$ $2V = 43°$ Dispersion	3.5–4	3700–4050	**Habitus:** botryoidal, stalactitic, massive, fibrous. **Color:** bright green or blackish green. **Luster:** vitreous, silky, or adamantine. **Diaphaneity:** translucent to substranslucent to opaque. **Cleavage:** (-201) perfect (010) fair. **Fracture:** uneven to subconchoidal, brittle. **Streak:** light green. **Chemistry:** readily dissolved by HCl and HNO$_3$ envolving carbon dioxide (i.e., effervescence). **Other:** dielectric constant 6.23 to 4.4. **Deposits:** secondary mineral in the oxidized zones of copper ore deposits, in sandstones. Weathering gives cuprite or azurite.

Mineral	Composition	Crystallography	Optical	Hardness	Density	Description
	$MM = 87.94482$, 62.47wt% Mn, 36.39wt% O, 1.15wt% H, Coordinance Mn(6), (Oxides and hydroxides)	$a = 884$ pm, $b = 523$ pm, $c = 574$ pm, $\beta = 90.0°$, ($Z = 8$), P.G. 2/m, S.G. B2$_1$/d	$\alpha = 2.24$, $\beta = 2.24$, $\gamma = 2.24$, $\delta = 0.29$, $R = 15\%$		4400	...prismatic. Color: black, gray. Luster: submetallic. Diaphaneity: opaque. Streak: red brown. Cleavage: perfect (010), good (110). Twinning: {011}. Fracture: conchoidal.
Manghemite	γ-Fe$_2$O$_3$ (Oxides and hydroxides)	Cubic, $a = 834$ pm, H11, cF56 ($Z = 8$), S.G. Fd3m, Spinel type	Isotropic, $n_D = 2.52$–2.74	5.5–6	5170	**Habitus:** massive, granular, crystalline. **Color:** black. **Luster:** metallic. **Diaphaneity:** opaque. **Streak:** black. **Cleavage:** none. **Fracture:** conchoidal. Ferromagnetic materials.
Marcasite	FeS$_2$, $MM = 119.979$, 46.55wt% Fe, 53.45wt% S, (Sulfides and sulfosalts), Coordinance Fe(6)	Orthorhombic, $a = 444.3$ pm, $b = 542.3$ pm, $c = 338.76$ pm, C18, oP6 ($Z = 2$), S.g. Pnnm, P.G. 222, Marcasite type	Isotropic, $R = 46\%$	6–6.5	4900	**Habitus:** tabular cockscomb aggregate, faces curved. **Color:** white green. **Luster:** metallic. **Diaphaneity:** opaque. **Fracture:** conchoidal. **Cleavage:** (101). **Twinning:** {101}. **Streak:** black. Pyroelectric. **Deposits:** sedimentary, magmatic, metamorphic, and hydrothermal.
Margarite (2M$_1$)	CaAl$_2$(OH)$_2$Si$_2$Al$_2$O$_{10}$, Coordinance Ca(6), Al(6), Al(4), Si(4), (Phyllosilicates, layered)	Monoclinic, $a = 514$ pm, $b = 900$ pm, $c = 981$ pm, 100.8°, ($Z = 4$), P.G. 2/m, S.G. C2/c	Biaxial (–), $\alpha = 1.635$, $\beta = 1.645$, $\gamma = 1.648$, $\delta = 0.013$, 2V = 45°, Dispersion weak	3.5–4.5	3100	**Habitus:** foliated. **Diaphaneity:** transparent to translucent. **Luster:** vitreous. **Cleavage:** (001) perfect. **Fracture:** uneven, flexible. **Streak:** white. Color: gray, yellow.
Marialite	Na$_4$ClSi$_6$Al$_3$O$_{24}$ (Tectosilicates, framework), Coordinance Na(6), Si(4), Al(4)	Tetragonal, $a = 1206.4$ pm, $c = 751.4$ pm, ($Z = 2$), S.G. 4/m, P.G. 14/m, Scapolite type	Uniaxial (–), $\varepsilon = 1.536$, $\omega = 1.540$, $\delta = 0.004$	5–6	2550	**Habitus:** prismatic. **Color:** colorless. **Luster:** vitreous (i.e, glassy). **Diaphaneity:** transparent to translucent. **Fracture:** conchoidal. **Cleavage:** (110). **Streak:** white. Fluorescent under UV-light: yellow, orange.

11

Minerals, Ores and Gemstones

Table 11.4 (continued)

Mineral name (IMA) [CAS RN] (Synonyms) (Etymology)	Theoretical chemical formula, relative molecular mass, coordination number, major impurities, mineral class (Strunz)	Crystal system, lattice parameters, Strukturbericht, Pearson symbol, Z, point group, space group, structure type	Optical properties	Mohs hardness (HM)	Density (ρ, kg.m^{-3})	Other mineralogical, physical, and chemical properties
Massicot	PbO $MM = 223.1994$ 92.83wt% Pb 7.17wt% O (Oxides and hydroxides)	Orthorhombic P.G. 222 Periclase type	Biaxial (+)	n.a.	n.a.	**Habitus:** massive, scaly, earthy. **Color:** yellow or reddish yellow. **Luster:** adamantine. **Deposits:** rare mineral of secondary origin associated with galena. Dimorphous with litharge.
Meionite	$Ca_4CO_3Si_6Al_6O_{24}$ (Tectosilicates, framework) Coordinance Ca(6), Si(4), Al(4)	Tetragonal $a = 1217.4$ pm $c = 765.2$ pm $(Z = 2)$ S.G. 4/m P.G. 14/m Scapolite group	Uniaxial (−) $\varepsilon = 1.558$ $\omega = 1.595$ $\delta = 0.037$	5–6	2760	**Habitus:** prismatic. **Color:** colorless. **Luster:** vitreous (i.e., glassy), **Diaphanity:** transparent to translucent. **Fracture:** conchoidal. **Cleavage:** (110) **Streak:** white. Fluorescent under UV-light: yellow, orange.
Melanterite (syn., green vitrol, pisanite) (from the Greek, melas, black)	$FeSO_4·7H_2O$ $MM = 278.01756$ 20.09wt% Fe 5.08wt% H 11.53wt% S 63.3wt% O (Sulfates, chromates, molybdates, and tungstates)	Monoclinic $a = 1411$ pm $b = 651$ pm $c = 1102$ pm $(Z = 4)$ S.G. P21/c P.G. 2/m	Biaxial (+) $\alpha = 1.47$–1.471 $\beta = 1.477$–1.480 $\gamma = 1.486$ $\delta = 0.015$–0.016 $2V = 86°$ Dispersion none.	2	1890–1900	**Habitus:** efflorescences, encrustations, capillary. **Color:** green, yellow green, brownish black, bluish green, or greenish white. **Luster:** vitreous (glassy). **Diaphanity:** subtransparent to translucent. **Streak:** white. **Cleavage:** (001) perfect, (110) distinct. **Fracture:** conchoidal.
Melilite	$(CaNa)_2(Mg,Fe,Al)[Si_3O_7]$ (Sorosilicates, pair)	Tetragonal $a = 780$ pm $c = 500$ pm $(Z = 2)$ Melilite type	Uniaxial (+/−)	n.a.	n.a.	
Mercury (syn., hydrargyrum, quicksilver) (from the Arabic)	Hg $MM = 200.59$ (Native elements)	Trigonal (Rhombohedral) $a = 300.5$ pm 70.53° A10, hR1 $(Z = 1)$ S.G. R$\bar{3}$m Mercury type	Isotropic	Liquid	13 596	**Deposits:** secondary mineral resulting from oxidation of cinnabar deposits.
Merwinite	$Ca_3Mg[Si_2O_8]$	Monoclinic	Biaxial (+) $\alpha = 1.702$–1.710 $\beta = 1.710$–1.718 $\gamma = 1.718$–1.726 $\delta = 0.008$–0.023	6	3150–3310	

Mineral (name, etymology)	Formula / composition	Crystallography	Optical	Hardness	Density	Properties
(syn, onofrite (Se), guadalcazarite (Zn), saukovite (Cd), from Greek, meta, and cinnabar, similar chemical composition and association with cinnabar)	$MM = 232.656$, 86.22wt% Hg, 13.78wt% S (Sulfides and sulfosalts)	23			7800	**Diaphaneity:** opaque. **Streak:** black. **Fracture:** uneven.
Microcline (syn, amazonite green) (from the Greek, mikron, little and klinetin, to stoop)	$K[Si_3AlO_8]$, $MM = 278.33154$, 14.05wt% K, 9.69wt% Al, 30.27wt% Si, 45.99wt% O, Coordination K(10), Si(4), Al(4), (Tectosilicates, framework)	Triclinic, $a = 857.7$ pm, $b = 1296.7$ pm, $c = 722.3$ pm, $\alpha = 89.7°$, $\beta = 115.97°$, $\gamma = 90.87°$, $(Z = 4)$, P.G. $\bar{1}$, S.G. $P\bar{1}$	Biaxial (−), $\alpha = 1.518$, $\beta = 1.522$, $\gamma = 1.525$, $\delta = 0.007$, $2V = 77–84°$, Dispersion weak	6	2560	**Habitus:** blocky, crystalline, coarse, prismatic. **Color:** white, cream, bright green, or green. **Diaphaneity:** translucent to transparent. **Luster:** vitreous (i.e, glassy). **Cleavage:** (001) perfect, (010) good. **Twinning:** albite (010), pericline [010]. **Fracture:** uneven. **Streak:** white. **Deposits:** granitic pegmatites, hydrothermal and metamorphic rocks.
Millerite (after the British crystallographer William H. Miller)	NiS, $MM = 114.537$, 48.76wt% Fe, 51.24wt% Ni, (Sulfides and sulfosalts), Coordination Ni(5)	Trigonal (Rhombohedral), $a = 961.6$ pm, $c = 315.2$ pm, B13, hR6 $(Z = 3)$, S.G. R3m, P.G. 3m, Millerite type	Uniaxial, $R = 54\%$	3–3.5	5500	**Habitus:** acicular, fibrous. **Color:** brass yellow. **Luster:** metallic. **Diaphaneity:** opaque. **Streak:** greenish gray. **Fracture:** uneven. Electrical resistivity 20 to 40 $\mu\Omega$.cm. Antiferromagnetic.
Minium (syn, red lead oxide) (after the river Minius located in Northwest Spain)	Pb_2PbO_4, $MM = 685.5976$, 90.67wt% Pb, 9.33wt% O, (Oxides and hydroxides)	Tetragonal, P.G. −4, Spinel type	Uniaxial	2.5–3	8200	**Habitus:** scaly, massive, granular, striated. **Color:** light red, brownish red, vivid red, or yellowish red. **Luster:** adamantine. **Diaphaneity:** subtransparent to opaque. **Streak:** yellowish orange. **Cleavage:** (110) perfect, (010) perfect. **Fracture:** earthy. **Deposits:** Oxidized portions of lead ore deposits.
Molybdenite (2H) [1317-33-5] (syn, molybdic ochre) (from the Greek, molybdos, lead)	MoS_2, $MM = 160.072$, 59.94wt% Mo, 40.06wt% S, (Sulfides and sulfosalts), Coordinance Mo(6)	Hexagonal, $a = 316.04$ pm, $c = 1229.50$ pm, C7, hP6 $(Z = 2)$, S.G. P6$_3$/mmc, P.G. 622, Molybdenite type	Uniaxial (−), $\omega = 4.33$, $\varepsilon = 2.03$, $\delta = 2.3$, $R = 18–33\%$	1–1.5	5060	**Habitus:** foliated, massive, disseminated. **Luster:** metallic. **Diaphaneity:** opaque. **Streak:** greenish gray. **Cleavage:** (0001) perfect. **Color:** bluish lead gray or lead gray. **Fracture:** sectile and flexible. **Chemical:** Soluble in conc. acids. Electrical resistivity 0.12 to 7.5 Ω.m.
Monazite (from the Greek, monazeis, to be alone, an allusion to its isolated crystals and their rarity when first found)	(Ce,La,Nd,Th)PO$_4$, Coordination Ce(8), P(4), (Phosphates, arsenates, and vanadates)	Monoclinic, $a = 679.0$ pm, $b = 701$ pm, $c = 646$ pm, $\beta = 104.4°$, $(Z = 4)$, P.G. 2/m, S.G. P2$_1$/n, Crocite type	Biaxial (+), $\alpha = 1.785–1.800$, $\beta = 1.786–1.801$, $\gamma = 1.838–1.851$, $\delta = 0.045–0.075$, $2V = 10–19°$	5–5.5	5150–5300	**Habitus:** crystalline, tabular, prismatic. **Color:** black, gray, brown, red, yellow, green, or orange. **Diaphaneity:** transparent to opaque. **Luster:** adamantine, resinous (Th rich). **Streak:** grayish white. **Cleavage:** (001) distinct, (100) indistinct. **Twinning:** [100] common. **Fracture:** uneven, subconchoidal. **Chemical:** slightly soluble in hot conc. H$_2$SO$_4$. When wetted by H$_2$SO$_4$, colors a Bunsen flame blue-green. Unfusible. **Deposits:** granodiorites, syenites, granitic pegmatites.

11

Minerals, Ores and Gemstones

Table 11.4 (continued)

Mineral name (IMA) [CAS RN] (Synonyms) (Etymology)	Theoretical chemical formula, relative molecular mass, coordinance number, major impurities, mineral class (Strunz)	Crystal system, lattice parameters, Strukturbericht, Pearson symbol, Z, point group, space group, structure type	Optical properties	Mohs hardness (HM)	Density (ρ, kg.m^{-3})	Other mineralogical, physical, and chemical properties
Monticellite	$CaMg[SiO_4]$ Coordinance Ca(6), Mg(6), Si(4) (Nesosilicates)	Orthorhombic $a = 481.5$ pm $b = 1108.4$ pm $c = 637.6$ pm $(Z = 4)$ P.G. mmm S.G. Pbnm (Olivine group)	Biaxial (−) $\alpha = 1.639$–1.653 $\beta = 1.645$–1.664 $\gamma = 1.653$–1.674 $\delta = 0.014$–0.017 $2V = 72.82°$ Dispersion weak	5.5	3080–3570	**Habitus:** crystalline, fine, prismatic. **Color:** colorless or gray. **Diaphaneity:** transparent. **Luster:** vitreous (i.e., glassy). **Streak:** white. **Cleavage:** (010) good, (100) poor. **Twinning:** {031}. **Deposits:** metamorphosed siliceous dolomitic limestones.
Montroydite	HgO $MM = 216.5894$ 92.61wt% Hg 7.39wt% O (Oxides and hydroxides)	Orthorhombic P.G. 222	Biaxial (+) $\alpha = 2.37$ $\beta = 2.50$ $\gamma = 2.65$ $\delta = 0.280$	1.5–2	n.a.	**Habitus:** crystalline, fine. **Color:** reddish orange. **Luster:** adamantine. **Streak:** reddish orange. **Cleavage:** (010) perfect. **Deposits:** oxidized mercury deposits.
Mullite (syn. mullite 3:2)	$Al_4[(SiO_4)]O]_2 \cdot Al_2O_3$ (Nesosilicates)	Orthorhombic $a = 755.7$ pm $b = 768.76$ pm $c = 288.42$ pm $(Z = 3)$	Biaxial (+) $\alpha = 1.649$–1.670 $\beta = 1.642$–1.675 $\gamma = 1.651$–1.690 $2V = 45$–61° O.A.P. (010)	6–7	3150–3260	
Muscovite 2M₁, (syn., isinglass, potash mica, fuchsite; Cr, sericite) (from Latin, vitrum muscoviticum, Muscovy glass, alluding to the Russian province of Muscovy)	$KAl_2[Si_3 Al]O_{10}(OH,F)_2$ $MM = 398.3081$ 9.82wt% K 20.32wt% Al 21.15wt% Si 0.51wt% H 48.20wt% O Coordinance K(6), Al(6), Si(4), and Al(4) (Phyllosilicates, layered)	Monoclinic $a = 520.3$ pm $b = 899.5$ pm $c = 2003.0$ pm $\beta = 94.47°$ $(Z = 4)$ Type 2M₁ mica (Micas group)	Biaxial (−) $\alpha = 1.552$–1.574 $\beta = 1.582$–1.610 $\gamma = 1.587$–1.616 $\delta = 0.34$–0.042 $2V = 30$–47° Dispersion weak	2.5–3 (85 HV)	2770–2880	**Habitus:** massive, lamellar, foliated, micaceous. **Color:** white, gray, silver white, brownish white, or greenish white. **Diaphaneity:** transparent to translucent. **Luster:** vitreous (i.e, glassy). **Streak:** white. **Cleavage:** (001) perfect. **Fracture:** brittle, sectile. **Chemical:** insoluble in strong mineral acids. Unfusible. Dielectric constant 10. **Deposits:** granites and pegmatites.
Natrolite	$Na_2Si_3Al_2O_{10} \cdot 2H_2O$ Coordinance Na(6), Si(4), and Al(4) (Tectosilicates, framework)	Orthorhombic $a = 1830.0$ pm $b = 1863.0$ pm $c = 660.0$ pm $(Z = 8)$ P.G. m2m S.G. Fd2d	Biaxial (+) $\alpha = 1.480$ $\beta = 1.480$ $\gamma = 1.490$ $\delta = 0.012$ $2V = 38$–62°	5–5.5	2230	**Habitus:** acicular. **Color:** colorless, gray. **Diaphaneity:** transparent to translucent. **Luster:** vitreous. **Streak:** white. **Cleavage:** (110) perfect. **Fracture:** uneven.

Name	Chemistry	Crystallography	Optical properties	Mohs hardness	Density (kg/m³)	Description
(syn., nephelite, elaeolite) (from the Greek, nephele, cloud, because it becomes clouded when put in strong acid)	Coordination Na(8), K(9), Si(4), Al(4) (Tectosilicates, framework)		ε = 1.528–1.544, ω = 1.531–1.549, δ = 0.003–0.005	6		...reddish white. Diaphaneity: ... Color: white, gray, brown, brownish gray, or reddish white. Luster: vitreous, greasy. Streak white. Cleavage: {1010} poor. Twinning: {100}, {112}, and {335}. Fracture: subconchoidal. Deposits: silica-poor igneous rocks.
Niccolite (syn., nickeline) (after the old German, nickel, meaning an ore which is not useful)	NiAs MM = 133.61159 43.92wt% Ni 56.08wt% As Traces of Fe, S, Co, Sb, Bi, and Cu (Sulfides and sulfosalts) Coordination Ni(6)	Hexagonal a = 360.9 pm c = 501.9 pm B8, hP4 (Z = 2) S.G. P6$_3$/mmc P.G. 622 Niccolite type	Uniaxial R = 53–56%	5–5.5 (308–642 HV)	7780	Habitus: massive, reniform, columnar. Color: dark tarnish red or pale copper red. Luster: metallic. Diaphaneity: opaque. Streak: brownish black Cleavage: (1010) imperfect, (0001) imperfect. Fracture: uneven, brittle. Chemical: dissolved by the aqua regia. Electrical resistivity 0.1 to 2 mΩ.cm. Deposits: in ore veins with silver, copper, and nickel arsenides and sulfides.
Niter (syn., saltpeter nitre) (after its composition of nitrate and potassium)	KNO$_3$ MM = 101.10324 38.67wt% K 13.85wt% N 47.47wt% O Coordination K(6), N(3) (Nitrates, carbonates, and borates)	Orthorhombic a = 643.1 pm b = 916.4 pm c = 541.4 pm (Z = 4) P.G. mmm S.G. Pmcn Aragonite type	Biaxial (−) α = 1.333, β = 1.505, γ = 1.505, δ = 0.172, 2V = 7°	2	2100	Deposits: efflorescence on cavern walls.
Nitratite (syn., nitratine, soda niter, nitronatrite) (after its composition which contains nitrates)	NaNO$_3$ MM = 84.9947 27.05wt% Na 16.48wt% N 56.47wt% O Coordination Na(6), N(3) (Nitrates, carbonates, and borates)	Trigonal (Rhomboedral) a = 507 pm c = 1682 pm (Z = 6) P.G. 32/m S.G. R3c Calcite type	Uniaxial (−) ε = 1.587, ω = 1.336, δ = 0.251	1.5–2	2240–2290	Habitus: massive. Color: white, reddish brown, gray, or lemon yellow. Luster: vitreous (i.e., glassy). Diaphaneity: transparent to translucent. Cleavage: perfect {0001}. Twinning: {10$\overline{1}$4}. Fracture: uneven, sectile. Deposits: residual water-soluble surface deposits in extremely arid deserts. Nitrates occur in clay-rich caliche deposits replenished by occasional desert thunderstorms which fix N$_2$ from the air.
Norbergite	Mg(OH,F)$_2$·Mg$_2$[SiO$_4$] Coordination Mg(6), Si(4) (Neosilicates)	Orthorhombic a = 470 pm b = 1022 pm c = 872 pm (Z = 4) P.G. mmm S.G. Pbnm (Humite group)	Biaxial (+) α = 1.563–1.567, β = 1.567–1.579, γ = 1.590–1.593, δ = 0.026–0.027, 2V = 44–50°	6–6.5	3200–3320	Habitus: massive, tabular. Color: white, yellow, colorless. Luster: vitreous (i.e., glassy). Diaphaneity: translucent to transparent. Streak: white. Cleavage: {100} poor. Twinning: {001}. Fracture: uneven, brittle. Chemical: attacked by strong mineral acids giving a silica gel. Deposits: dolomites, limestones, skarns.
Northupite	Na$_3$Mg(CO$_3$)$_2$Cl (Carbonates, nitrates and borates)	Cubic	Isotropic n_D = 1.514	3.5–4	2380	Habitus: crystalline-coarse, pyramidal. Color: white, yellow, or gray. Diaphaneity: transparent to translucent. Luster: vitreous (glassy). Deposits: continental evaporite deposits.
Nosean (syn., noselite) (after the German mineralogist, K.W. Nose)	Na$_8$[Al$_6$Si$_6$O$_{24}$](SO$_4$) MM = 1012.38486 18.17wt% Na 15.99wt% Al	Cubic a = 905 pm (Z = 1) Sodalite type	Isotropic n_D = 1.495	5.5	2300–2400	Habitus: massive, granular. Color: white, gray, blue, green, or brown. Luster: vitreous (i.e., glassy), greasy. Luminescence: fluorescent. Streak: bluish white. Cleavage: {110} poor. Twinning: {111}. Fracture: brittle, conchoidal. Deposits: igneous rocks low in silica and rich in alkalis.

Table 11.4 (*continued*)

Mineral name (IMA) [CAS RN] (Synonyms) (Etymology)	Theoretical chemical formula, relative molecular mass, coordination number, major impurities, mineral class (Strunz)	Crystal system, lattice parameters, Strukturbericht, Pearson symbol, Z, point group, space group, structure type	Optical properties	Mohs hardness (HM)	Density (ρ, kg.m^{-3})	Other mineralogical, physical, and chemical properties
Nosean (*continued*)	16.65wt% Si, 0.20wt% H, 3.17wt% S, 45.83wt% O, Traces of Ca, Fe (Tectosilicates, framework)				n.a.	
Oligoclase (syn, sunstone) (from the Greek, oligos, and kasein, little, cleavage)	(Na,Ca)(Si,Al)$_4$O$_8$, An20–Ab80, MM = 265.41986, 6.93wt% Na, 3.02wt% Ca, 12.20wt% Al, 29.63wt% Si, 48.22wt% O (Tectosilicates, framework)	Triclinic, a = 815 pm, b = 1278 pm, c = 850 pm (Z = 5)	Biaxial (+), α = 1.533–1.543, β = 1.537–1.548, γ = 1.542–1.552, δ = 0.009, 2V = 82–86°	7		**Habitus:** euhedral crystals. **Color:** white or gray. **Luster:** vitreous (i.e., glassy). **Luminescence:** fluorescent. **Cleavage:** (001) perfect, (010) good. **Fracture:** uneven. **Streak:** white. **Deposits:** magmatic and pegmatitic rocks.
Olivine (syn, peridot, chrysolite light yellowish green) (after the green color)	(Mg,Fe)$_2$[SiO$_4$] (Nesosilicates)	Orthorhombic	Biaxial (+/−), α = 1.635–1.827, β = 1.651–1.869, γ = 1.670–1.879, δ = 0.035–0.052, 2V = 82–134° Dispersion weak	6.5–7	3220–4390	**Habitus:** massive. **Color:** yellowish green, olive green, greenish black, or reddish brown. **Diaphaneity:** transparent to translucent. **Luster:** vitreous (i.e., glassy). **Cleavage:** (001) good, (010) distinct. **Fracture:** conchoidal, brittle. **Streak:** white. **Deposits:** basic and ultra basic igneous rocks.
Orpiment (from the Latin, auri pigmentum, given by Pliny; an allusion to the vivid golden hue)	As$_2$S$_3$, MM = 246.0412, 60.90wt% As, 39.10wt% S, Traces of Se, Sb, V, and Ge (Sulfides and sulfosalts)	Monoclinic, a = 1149 pm, b = 959 pm, c = 425 pm, D5$_7$, mP20 (Z = 4), S.G. P2$_1$/c, P.G. 2/m, Orpiment type	Biaxial (+), α = 2.40, β = 2.81, γ = 3.02, δ = 0.62, 2V = 76°, R = 29%	1.5–2 (23–52 HV)	3490–3520	**Habitus:** prismatic, massive, fibrous, foliated, flexible crystals. **Color:** lemon yellow, brownish yellow, or orange yellow. **Diaphaneity:** transparent to opaque. **Luster:** resinous. **Streak:** pale yellow. **Cleavage:** (010) perfect. **Fracture:** even, sectile. **Chemical:** dissolved by the aqua regia. **Deposits:** in hydrothermal veins with realgar, stibine and pyrite.
Orthoclase (syn, orthose, adularia) (from the Greek, orthos, right, and kalos, I cleave, an allusion to the mineral's right angle of good cleavage)	K[Si$_3$AlO$_8$], MM = 278.33154, 14.05wt% K, 9.69wt% Al, 30.27wt% Si, 45.99wt% O, Coordinance K(10), Si(4), Al(4) (Tectosilicates, framework)	Monoclinic, a = 862.5 pm, b = 1299.6 pm, c = 719.3 pm, β = 116.01° (Z = 4), P.G. 2/m, S.G. C2/m	Biaxial (−), α = 1.518–1.521, β = 1.523–1.525, γ = 1.526–1.528, δ = 0.005–0.006, 2V = 65–75°, Dispersion strong	6	2560	**Habitus:** prismatic, massive, granular, blocky. **Color:** white, pink, yellow, or red. **Diaphaneity:** transparent to translucent. **Luster:** pearly, vitreous (i.e., glassy). **Streak:** white. **Cleavage:** (001) perfect, (010) good, (110) poor. **Twinning:** Carlsbad [001], Baveno {021}. **Fracture:** uneven. **Deposits:** intrusive and extrusive igneous, and metamorphic rocks.

Name	Formula / Chemistry	Crystallography	Optical	Hardness	S.G.	Description
(…)	Traces of Ca, Fe, Mn, Ni, Cr, Al, and Ti (Inosilicates, single chain)	Orthorhombic a = 908.1 pm b = 1843.1 pm c = 523.8 pm (Z = 16)	Biaxial (+)	5–6	3300–3500	
Palladium (after the discovery of the asteroid, Pallas)	Pd or (Pd,Hg) MM = 106.42 (Native elements)	Cubic a = 389.03 pm Al, $cF4$ (Z = 4) Fm3m Copper type	Isotropic	4.5–5	11,550	Habitus: granular. Color: gray. Diaphaneity: opaque. Luster: metallic.
Pearceite (after the American chemist, R. Pearce)	$Ag_{16}As_2S_{11}$ MM = 2228.4604 77.45wt% Ag 6.72wt% As 15.83wt% S (Sulfides and sulfosalts)	Monoclinic P.G. 2/m	Biaxial	2.5–3	n.a.	Habitus: massive, granular. Color: black. Luster: submetallic. Diaphaneity: opaque. Streak: reddish black. Cleavage: (001) poor. Fracture: uneven.
Pectolite	$Ca_2Na[Si_3O_8](OH)$ Coordinance Ca(6), Na(6), Si(4) (Inosilicates, ribbon)	Triclinic a = 799 pm b = 704 pm c = 702 pm $α$ = 90.05° $β$ = 92.58° $γ$ = 102.47° P.G. $\bar{1}$ S.G. P$\bar{1}$ Pyroxenoid group	Biaxial (−) $α$ = 1.590 $β$ = 1.610 $γ$ = 1.630 $δ$ = 0.04 $2V$ = 35–63°	4.5–5	2900	Habitus: radiating, fibrous. Color: white. Luster: silky. Diaphaneity: transparent to translucent. Streak: white. Fracture: uneven. Cleavage: (100) perfect, (001) distinct. Twinning: {010}.
Pentlandite (after the Irish natural historian, J.B. Pentland)	$(Fe,Ni)_9S_8$ (Sulfides and sulfosalts) Coordinance Ni(6), Fe(6), and Fe (4)	Cubic a = 1009.5 pm D8$_6$, $cF68$ (Z = 4) S.G. Fm–3m P.G. 4-32 Co$_9$S$_8$ type	Isotropic R = 51%	3.5–4	5000	Habitus: massive, granular. Color: light bronze yellow. Diaphaneity: opaque. Luster: metallic. Streak: greenish black. Cleavage: perfect (100), good (111). Fracture: uneven, conchoidal. Electrical resistivity 1 to 11 $\mu\Omega$cm. Deposits: mafic intrusive igneous rocks.
Periclase [1309-48-44] (syn. magnesia) (from the Greek, peri, around, and klao, to cut)	MgO MM = 40.2990 60.31wt% Mg 39.69wt% O Coordinance Mg(6) (Oxides and hydroxides)	Cubic a = 421.17 pm B1, $cF8$ (Z = 4) S.G. Fm3m P.G. 4-32 Rock salt type	Isotropic n_D = 1.736	5.5	3560	Habitus: granular, octahedral crystals. Color: white, gray, or green. Luster: vitreous (i.e., glassy). Diaphaneity: transparent to translucent. Luminescence: fluorescent, long UV-light yellow. Streak: white. Cleavage: (001), (010), (100). Fracture: uneven, brittle, conchoidal. Deposits: contact metamorphism of dolomites and magnesites.
Perovskite or perovskite [12049-50-2] (after the Russian mineralogist, count L.A. Perowski)	$CaTiO_3$ MM = 135.9562 35.22wt% Ti 35.30wt% O 29.48wt% Ca Traces of rare earths, Nb, Ta, Th	Orthorhombic (pseudocubic) a = 536.70 pm b = 764.38 pm c = 544.39 pm E21, $cP5$ (Z = 4)	Biaxial (−) $α$ = 2.34 $β$ = 2.34 $γ$ = 2.34 $δ$ = 0.002 $2V$ = 90°	5.5 (988–1131 HV)	4044	Habitus: faces striated, reniform, pseudocubic, pseudohexagonal. Color: black, reddish brown, pale yellow, yellowish orange. Diaphaneity: transparent to translucent or opaque. Luster: adamantine, submetallic. Streak: light brown. Fracture: subconchoidal. Cleavage: (100). Twinning: {111}. Chemical: attacked by hot H_2SO_4, and HF. Nonfusible: mp 1980° C. Deposits: ultramatic igneous rocks, metamorphosed limestone in contact with mafic igneous rocks.

11

Minerals, Ores and Gemstones

Table 11.4 *(continued)*

Mineral name (IMA) [CAS RN] (Synonyms) (Etymology)	Theoretical chemical formula, relative molecular mass, coordination number, major impurities, mineral class (Stunz)	Crystal system, lattice parameters, Strukturbericht, Pearson symbol, Z, point group, space group, structure type	Optical properties	Mohs hardness (HM)	Density (ρ, kg.m^{-3})	Other mineralogical, physical, and chemical properties
Perowskite or perovskite *(continued)*	Coordinance Ca(12), Ti(6) (Oxides and hydroxides)	P.G. mmm S.G. Pm3m Perowskite type	$R = 16\%$			
Petzite (after the chemist, W. Petz)	Ag_3AuTe_2 MM = 667.90294 32.30wt% Ag 38.21wt% Te 29.49wt% Au (Sulfides and sulfosalts)	Cubic P.G. 23	Isotropic	2.5	8700–9140	**Habitus:** massive, granular. **Color:** iron black or steel gray. **Luster:** metallic. **Diaphaneity:** opaque. **Streak:** grayish black. **Fracture:** brittle, sectile.
Phlogopite (1M) (from the Greek, phlogopos, resembling fire)	$KMg_3(Si_3Al)O_{10}(F,OH)_2$ MM = 417.26002 9.37wt% K 17.47wt% Mg 6.47wt% Al 20.19wt% Si 0.48wt% H 46.01wt% O Coordinance K(12), Mg(6), Si(4) Al(4) (Phyllosilicates, layered)	Monoclinic a = 531 pm b = 923 pm c = 1015 pm β = 95.18° (Z = 2) P.G. 2/m S.G. C2/m Mica type	Biaxial (−) α = 1.53–1.573 β = 1.557–1.617 γ = 1.558–1.618 δ = 0.028–0.045 $2V$ = 0–12° Dispersion weak	2–2.5	2800	**Habitus:** micaceous, scaly, lamellar. **Color:** brown, gray, green, yellow, or reddish brown. **Streak:** white. **Diaphaneity:** transparent to translucent. **Luster:** vitreous, pearly. **Cleavage:** (001) perfect. **Fracture:** uneven. **Twinning:** {310}. **Deposits:** contact and regional metamorphic limestone and dolomites.
Piemontite (from the Piedmont region, Italy)	$Ca_2(Mn,Fe,Al)_3AlO(OH)[SiO_4][Si_2O_7]$ 23.5wt% CaO 24.1wt% Al_2O_3 12.6wt% Fe_2O_3 37.9wt% SiO_2 1.9wt% H_2O (Sorosilicates and nesosilicates)	Monoclinic a = 895.0 pm b = 570.0 pm c = 941.0 pm 115·70 (Z = 2) Epidote group	Biaxial (−) α = 1.732–1.794 β = 1.750–1.807 γ = 1.762–1.829 δ = 0.025–0.088 $2V$ = 64–85° Dispersion strong	6–6.5 (680 HV)	3450–3520	**Habitus:** prismatic according to b, striated, columnar. **Color:** reddish-brown to dark red. **Diaphaneity:** transparent to opaque. **Luster:** vitreous (i.e. glassy). **Streak:** white. **Cleavage:** (001) perfect. **Fracture:** uneven, conchoidal. **Chemical:** insoluble in strong mineral acids, fusible giving a black globule. **Deposits:** regional metamorphic and pegmatite rocks.
Pirssonite	$Na_2Ca(CO_3)_2 \cdot 2H_2O$ (Nitrates, carbonates, and borates)	Orthorhombic	Biaxial (+) α = 1.5 β = 1.5 γ = 1.57 δ = 0.070 $2V$ = 33° Dispersion weak	3	2350	**Color:** colorless or white. **Diaphaneity:** transparent to translucent. **Deposits:** continental evaporite deposits under desert climates.

Name (etymology)	Formula / Class	Hardness	Density	Optical properties	Crystal system	Properties
[7440-06-04] (from Spanish, *platina*, silver)	$MM = 195.08$ Coordination Pt(12) (Native elements)			$n_D = 4.28$ $R = 70\%$	$a = 392.36$ pm Al, cF4 (Z = 4) P.G. m3m S.G. Fm3m Copper type	metallic. **Streak:** grayish white. **Cleavage:** none. **Twinning:** [111]. **Fracture:** hackly, malleable, ductile. **Chemical:** inert in most concentrated mineral acids (e.g., HCl, H_2SO_4, HF, HNO_3), but readily dissolved in aqua regia (i.e., 3 vol. HCl + 1 vol. HNO_3). **Deposits:** mainly in grains and nuggets in alluvial placer deposits.
Plattnerite (after the German metallurgist, K.F. Plattner)	PbO_2 $MM = 239.1988$ 86.62wt% Pb 13.38wt% O (Oxides and hydroxides)	5.5	8500–9630	Uniaxial (–) $\varepsilon = 2.25$ $\omega = 2.35$ $\delta = 0.100$	Tetragonal $a = 392$ pm $c = 430$ pm (Z = 4) P.G. 422	**Habitus:** massive, encrustations. **Color:** black or grayish black. **Luster:** submetallic. **Diaphaneity:** subtranslucent to opaque. **Streak:** chestnut brown. **Fracture:** brittle.
Pollucite (after Pollux, a figure from Greek mythology)	$(Cs,Na)_2Al_2Si_4O_{12}·H_2O$ (Tectosilicates)	6.5	2900	Isotropic $n_D = 1.525$	Cubic	**Habitus:** massive. **Color:** colorless, gray, or white. **Diaphaneity:** transparent. **Luster:** vitreous, dull. **Cleavage:** none. **Streak:** white. **Deposits:** granitic pegmatites.
Polybasite (from the Greek, *poly*, many and *basis*, base, an allusion to the basic character of the compound)	$(Ag,Cu)_{16}Sb_2S_{11}$ (Sulfides and sulfosalts)	2.5–3	4600–5000	Biaxial	Monoclinic P.G. 2/m	**Habitus:** massive, granular, pseudo hexagonal. **Color:** black. **Streak:** reddish black. **Luster:** submetallic. **Diaphaneity:** opaque. **Cleavage:** (001) poor. **Fracture:** uneven.
Polyhalite (from the Greek, *polys*, much and *halos*, salt)	$K_2Ca_2Mg(SO_4)_4·2H_2O$ $MM = 410.81536$ 19.03wt% K 19.51wt% Ca 5.92wt% Mg 0.98wt% H 15.61wt% S 38.95wt% O (Sulfates, chromates, molybdates, and tungstates)	2.5–3.5	2770–2780	Biaxial (–) $\alpha = 1.546-1.548$ $\beta = 1.558-1.562$ $\gamma = 1.567$ $\delta = 0.019-0.02$ $2V = 60-62°$	Triclinic P.G. –1	**Habitus:** massive, lamellar, fibrous. **Color:** white, yellowish white, gray, or flesh pink. **Luster:** vitreous (glassy). **Streak:** white. **Cleavage:** (101) perfect. **Fracture:** conchoidal, brittle. **Deposits:** sedimentary marine evaporite deposits.
Powellite [7789-82-4] (after the American geologist, W. Powell)	$CaMoO_4$ (Sulfates, chromates, molybdates, and tungstates)	3.8	4350	Uniaxial (+) $\varepsilon = 1.971$ $\omega = 1.980$ $\delta = 0.010$	Tetragonal $a = 552.60$ pm $c = 1143.00$ pm (Z = 4) Scheelite type	**Habitus:** euhedral crystals. **Color:** yellow or greenish yellow. **Luster:** adamantine, resinous. **Cleavage:** (111) distinct. **Fracture:** conchoidal, brittle. **Streak:** light yellow.
Proustite (syn. As-ruby silver)	Ag_3AsS_3 (Sulfides and sulfosalts) Coordinance Ag(2), As(3)	2–2.5	5570	Uniaxial (–) $\omega = 2.98$ $\varepsilon = 2.71$ $\delta = 0.17$ $R = 22-26\%$	Trigonal (Rhombohedral) $a = 1081.6$ pm $c = 869.48$ pm (Z = 6) S.G. R3c P.G. 3m	**Habitus:** prismatic, rhombohedral crystals. **Color:** ruby red. **Luster:** adamantine. **Diaphaneity:** translucent to opaque. **Streak:** red. **Fracture:** subconchoidal. **Cleavage:** (101). **Twinning:** {101}, {104}.
Psilomelane (syn. Romanchite) (from Greek, *psilos*, smooth, and *melanos*, black, owing to the common	$Ba(OH)_2Mn^{II}Mn^{IV}_8O_{16}$ $MM = 955.74568$ Coordinance Ba(10), Mn(6) (Oxides and hydroxides)	5–6 (503–813 HV)	3950–4710	Biaxial (n.a.) $R = 22-24\%$	Monoclinic $a = 956$ pm $b = 288$ pm $c = 1385$ pm	**Habitus:** reniform, botryoidal, fibrous, dendritic. **Color:** dark black to dark steel gray. **Luster:** submetallic. **Diaphaneity:** opaque. **Streak:** brown black. **Deposits:** associated with cryptomelane. **Fracture:** uneven to conchoidal, brittle. **Note:** wad is a common name for a low-hardness varieties, while **manganomelane**

11

Minerals, Ores and Gemstones

Table 11.4 (continued)

Mineral name (IMA) [CAS RN] (Synonyms) (Etymology)	Theoretical chemical formula, relative molecular mass, coordinance number, major impurities, mineral class (Strunz)	Crystal system, lattice parameters, Strukturbericht, Pearson symbol, Z, point group, space group, structure type	Optical properties	Mohs hardness (HM)	Density (ρ, kg.m^{-3})	Other mineralogical, physical, and chemical properties
Psilomelane (continued) smooth surface of the concretions)						describes high-density varieties (i.e., above 3000 kg.m^{-3}).
Pyrargyrite [12068-85-8] (syn., dark red silver ore, ruby silver ore) (from the Greek, pyros, and argyros, fire-silver in allusion to color and silver content)	Ag_3SbS_3 $MM = 541.5526$ 59.75wt% Ag 22.48wt% Sb 17.76wt% S Coordinance Ag(2), Sb(2) (Sulfides and sulfosalts)	Trigonal (Rhombohedral) $a = 1105.2$ pm $c = 871.77$ pm $(Z = 6)$ P.G. 3m S.G. R3c	Uniaxial (−) $\varepsilon = 2.881$ $\omega = 3.084$ $\delta = 0.203$ $R = 35–36\%$	2–2.5	5850	**Habitus:** massive, crystalline, prismatic. **Color:** ruby red. **Luster:** adamantine to submetallic. **Diaphaneity:** translucent to opaque. **Streak:** red purple. **Fracture:** subconchoidal, brittle. **Cleavage:** {101}. **Twinning:** {101}, {104}. **Chemical:** readily dissolved by HNO_3 with formation of free S and precipitation of Sb_2O_3. **Deposits:** epithermal veins with other silver-bearing ores.
Pyrite [12068-85-8] (syn., Fool's gold) (from Greek, pyros, fire)	FeS_2 $MM = 119.979$ 46.55wt% Fe 53.45wt% S (Sulfides and sulfosalts) Coordinance Fe(6)	Cubic $a = 541.75$ pm C2, cP12 ($Z = 4$) S.G. Pa3 P.G. 23 Pyrite type	Isotropic $R = 54.5\%$	6–6.5 (1150 HV)	4950–5030 (5011 6)	**Habitus:** faces striated, druse, stalactitic, pyritohedral cubic crystal. **Color:** pale brass yellow. **Luster:** metallic. **Diaphaneity:** opaque. **Fracture:** conchoidal. **Cleavage:** (100) poor, (110) poor. **Twinning:** {110} iron cross. **Streak:** greenish black. Pyroelectric. **Deposits:** sedimentary, magmatic, metamorphic, and hydrothermal.
Pyrochlore (from the Greek, pyros, fire, and chloros, green owing to the green color after pyrolysis)	$(Na,Ca_2U,Th,Ln)(Nb,Ta)_2O_{11}$ $(OH,F)_2 \cdot nH_2O$ Traces of rare earths (Oxides and hydroxides)	Cubic (sometimes metamicte or amorphous) $a = 1037$ pm ($Z = 8$)	Isotropic $n_D = 1.90–2.14$ $R = 14\%$	5–5.5 (514– 764 HV)	3700– 6400	**Habitus:** octahedron, granular, disseminated. **Color:** brown, yellowish brown, yellow, greenish brown, or reddish brown. **Diaphaneity:** translucent to opaque. **Luster:** resinous, greasy or glassy. **Streak:** yellowish brown. **Cleavage:** (111). **Fracture:** uneven to conchoidal. **Chemical:** insoluble in mineral acids. **Other:** radioactive according to U content. Dielectric constant 3.4 to 5.1. **Deposits:** in nepheline-syenite igneous rocks.
Pyrolusite (syn., wad, polianite) (from the Greek, pyros, fire and louein, to wash, because it was used to remove the yellowish color imparted to glass by iron compounds)	MnO_2 $MM = 86.93685$ 63.19wt% Mn 36.81wt% O Traces of rare earths Coordinance Mn(6) (Oxides and hydroxides)	Tetragonal $a = 438.8$ pm $c = 286.5$ pm C4, tP6 ($Z = 2$) P.G. 422 S.G. P42/mnm Rutile group	Uniaxial $R = 30–55\%$	6–6.5 (76– 405 HV)	5234	**Habitus:** reniform, columnar, fibrous, dendritic, or earthy. **Color:** steel gray, iron gray, or bluish gray. **Diaphaneity:** opaque. **Luster:** metallic. **Cleavage:** (110) perfect. **Fracture:** uneven, brittle. **Chemical:** dissolved by HCl, gives a green-blue pearl with molten KOH or NaOH. **Other:** dielectric constant above 81. Electrical resistivity 0.007 to 30 Ω.m. Antiferromagnetic. **Deposits:** in the oxidation zone of manganese ores with rhodonite. In marine sedimentary rocks such as limestone, hydrothermal.
Pyromorphite (syn., green lead ore, campylite, phosphomimetite) (from the Greek, pyros, fire and morfic, form, because, when deep	$Pb_5(PO_4)_3Cl$ $MM = 1356.36678$ 6.85wt% P 76.38wt% Pb 2.61wt% Cl	Hexagonal $a = 997$ pm $c = 733$ pm ($Z = 2$) P.G. 6/m	Uniaxial (−) $\varepsilon = 2.048$ $\omega = 2.058$ $\delta = 0.010$	3.5–4	6850– 7000	**Habitus:** reniform, prismatic, globular. **Color:** green, yellow, brown, grayish white, or yellowish red. **Diaphaneity:** transparent to translucent. **Luster:** adamantine, resinous **Streak:** white. **Cleavage:** (1000) perfect, (1011) imperfect. **Fracture:** subconchoidal, brittle. **Chemical:** dissolved by HNO_3. **Fluorescence:** yellow. **Deposits:** in the weathering zone of lead-zinc ore deposits with andesite, cerussite, hemimorphite,

Coordination Pb(6), P(4) (Phosphates, arsenates, and vanadates)

Mineral	Chemistry	Hardness	S.G.	Optical	Crystallography	Properties
Pyrope (syn., rhodolite) (from the Greek, pyropos, fiery-eyed, an allusion to the red hue)	$Mg_3Al_2(SiO_4)_3$ $MM = 403.12738$ 18.09wt% Mg 13.39wt% Al 20.90wt% Si 47.63wt% O Coordination Mg(8), Al(6), Si(4) (Nesosilicates)	6–7.5	3582	Isotropic $n_D = 1.714$	Cubic $a = 1145.9$ pm $(Z = 8)$ P.G. 432 S.G. Ia3d (Garnet group: Pyralspite series)	**Habitus:** dodecahedral, granular, crystalline, lamellar. **Color:** red or black. **Luster:** vitreous (i.e., glassy). **Streak:** white. **Parting:** (110). **Cleavage:** none. **Fracture:** conchoidal. **Diaphaneity:** transparent to translucent. **Chemical:** soluble in HF. **Deposits:** ultrabasic igneous rocks.
Pyrophyllite (1Tc)	$Al_2Si_4O_{10}(OH)_2$ Coordination Al(6), Si(4) (Phyllosilicates, layered)	1.5–2	2650–2900	Biaxial (–) $\alpha = 1.534–1.556$ $\beta = 1.586–1.589$ $\gamma = 1.596–1.601$ $\delta = 0.045–0.062$ $2V = 52–62°$ Dispersion weak	Triclinic $a = 516$ pm $b = 896$ pm $c = 935$ pm $\alpha = 90.03°$ $\beta = 100.37°$ $\gamma = 89.75°$ $(Z = 2)$ P.G. $\bar{1}$, S.G. P$\bar{1}$	**Habitus:** platy, foliated. **Color:** white, greenish white, or yellowish white. **Diaphaneity:** translucent to opaque. **Luster:** pearly. **Luminescence:** fluorescent. **Streak:** white. **Cleavage:** (001) perfect. **Fracture:** flexible.
Pyrrhotite (syn., magnetic pyrite) (from the Greek, phrrhotes, redness, an allusion to its color)	$Fe_{1-x}S$ (x = 0–0.17) $MM = 85.12065$ 62.33wt% Fe 37.67wt% S (Sulfides and sulfosalts) Coordination Fe(6)	3.5–4.5	4600	Uniaxial $R = 37\%$	Hexagonal $a = 345.2$ pm $c = 576.2$ pm $(Z = 2)$ S.G. P-62c P.G. -62m Defect Niccolite type	**Habitus:** tabular, platy, massive, granular. **Color:** bronze yellow or red. **Diaphaneity:** opaque. **Luster:** metallic. **Cleavage:** (0001) imperfect, (1120) poor. **Fracture:** uneven. **Streak:** gray-black. Electrical resistivity 2 to 160 $\mu\Omega$.cm. **Deposits:** widespread occurrences in igneous and metamorphic rocks. Magnetic.
Quartz (high temperature) [14808-60-7] (from the German, quarz, of uncertain origin)	β-SiO_2 $MM = 60.0843$ 46.74wt% Si 53.26wt% O (Tectosilicates, framework) Coordination Si(4)	7	2530	Uniaxial (+) $\varepsilon = 1.53$ $\omega = 1.54$ $\delta = 0.007$	Trigonal (Hexagonal) $a = 499.9$ pm $c = 545.7$ pm C8, hP9 $(Z = 3)$ S.G. P6$_2$22 (Dextrogyre), and P6$_4$22 (Levogyre) P.G. 622	**Habitus:** stubby bipyramidal. **Color:** colorless. **Luster:** vitreous (i.e., glassy). **Streak:** white. **Twinning:** [102], [302], [201], [112]. **Luminescence:** triboluminescent. **Fracture:** conchoidal. **Chemical:** resistant to strong mineral acids, attacked by HF, transition temperature 867°C.
Quartz (low temperature) [14808-60-7] (syn., rock crystal, smoky quartz: brown to black, amethyst: purple, citrine: yellow) (from the German, quarz, of uncertain origin)	α-SiO_2 $MM = 60.0843$ 46.74wt% Si 53.26wt% O (Tectosilicates, framework) Coordination Si(4)	7	2650	Uniaxial (+) $\varepsilon = 1.543–1.545$ $\omega = 1.552–1.554$ $\delta = 0.009$	Trigonal (Rhombohedral) $a = 491.3$ pm $c = 540.5$ pm $(Z = 3)$ S.G. R3$_2$21 (Dextrogyre), and R3$_2$21 (Levogyre) P.G. 32	**Habitus:** prismatic, massive, crystalline, coarse, druse. **Color:** colorless, yellow, red, or brown. **Diaphaneity:** transparent to translucent. **Luster:** vitreous (i.e., glassy). **Luminescence:** triboluminescent. **Streak:** white. **Cleavage:** (0110) indistinct. **Twinning:** [001]. **Fracture:** conchoidal. **Chemical:** resistant to strong mineral acids, attacked by HF. Transition temperature 573°C. **Deposits:** granitic igneous rocks, metamorphic rocks.

11

Minerals, Ores and Gemstones

Table 11.4 (*continued*)

Mineral name (IMA) [CAS RN] (Synonyms) (Etymology)	Theoretical chemical formula, relative molecular mass, coordination number, major impurities, mineral class (Strunz)	Crystal system, lattice parameters, Strukturbericht, Pearson symbol, Z, point group, space group, structure type	Optical properties	Mohs hardness (HM)	Density (ρ, kg.m^{-3})	Other mineralogical, physical, and chemical properties
Ramsdellite (after the American mineralogist, L.S. Ramsdell who first described the mineral)	MnO_2 $MM = 86.93685$ 63.19wt% Mn 36.81wt% O (Oxide and hydroxides)	Orthorhombic	Biaxial	3	4370	**Habitus:** massive, fibrous, platy. **Color:** steel gray or black. **Diaphaneity:** opaque. **Luster:** metallic. **Fracture:** brittle. **Streak:** brownish black. **Deposits:** manganese deposits with pyrolucite.
Realgar (from the Arabic, rahj al ghar, powder of the mine)	$AsS (= As_4S_4)$ $MM = 106.9876$ 70.03wt% As 29.97wt% S (Sulfides and sulfosalts) Coordination AsS molecules	Monoclinic $a = 929$ pm $b = 1353$ pm $c = 657$ pm $\beta = 106.55°$ B1, mP32 ($Z = 16$) S.G. P21/c P.G. 2/m Realgar type	Biaxial (−) $\alpha = 2.538$ $\beta = 2.684$ $\gamma = 2.704$ $\delta = 0.166$ $2V = 40°$ Dispersion strong	1.5–2 (47– 60 HV)	3560	**Habitus:** prismatic, massive, granular, druse, earthy. **Color:** aurora red, orange yellow, or dark red. **Diaphaneity:** translucent to opaque. **Luster:** resinous, glassy, adamantine. **Streak:** orange red. **Cleavage:** (010), (001), and (100). **Twinning:** {100}. **Fracture:** brittle, sectile. **Chemical:** dissolved by HNO_3, evolves a garlic odor when calcinated and gives sublimated As_2O_3 deposit on cold wall. Electrical resistivity 1 to 150 mΩ.m. **Deposits:** hydrothermal with orpiment. Marcassite and stibnite. Sedimentary rocks.
Rhodocrosite (syn. ponite: rich Fe varieties) (from Greek, Rhodos, pink)	$MnCO_3$ $MM = 114.946949$ 47.79wt% Mn 52.21wt% CO_2 Coordination Mn(6), C(3) (Nitrates, carbonates, and borates)	Trigonal $a = 477.1$ pm $c = 1566.4$ pm ($Z = 6$) P.G. 32/m S.G. R3̄c Calcite type	Uniaxial (−) $\varepsilon = 1.540$–1.617 $\omega = 1.750$–1.850 $\delta = 0.190$–0.230 Dispersion strong	3.5–4	3200–4050	**Habitus:** massive. **Color:** rose-pink, pink, red, brown or brownish yellow, colorless. **Luster:** pearly, vitreous. **Diaphaneity:** transparent to translucent. **Cleavage:** (1011) perfect. **Twinning:** {0112}. **Fracture:** uneven. **Chemical:** dissolved with effervescence in warm dilute acids. On exposure to air develops a brown or black surface alteration layer. **Deposits:** high-temperature metasomatic deposits.
Rhodonite	$(Mn, Ca)[SiO_3]$ Coordination Mn(6), Ca(6), Si(4) (Inosilicates, chain)	Triclinic $a = 768$ pm $b = 1182$ pm $c = 671$ pm $\alpha = 92.35°$ $\beta = 93.95°$ $\gamma = 105.67°$ ($Z = 2$) P.G. 1̄ S.G. P1̄ Pyroxenoid group	Biaxial (+) $\alpha = 1.717$ $\beta = 1.720$ $\gamma = 1.730$ $\delta = 0.013$ $2V = 63$–$76°$ Dispersion weak	4.5–5	3500–3700	**Habitus:** tabular, massive. **Color:** pink, red. **Luster:** vitreous. **Diaphaneity:** transparent to translucent. **Streak:** white. **Fracture:** conchoidal. **Cleavage:** (110) perfect, (001) distinct. **Twinning:** {010}.
Riebeckite (syn. crocidolite: asbestos form) (after the German traveler, E. Riebeck)	$Na_2(Fe,Mg)_3Fe_2Si_8O_{22}(OH)_2$ Inosilicates (double chains)	Monoclinic	Biaxial (−) $\alpha = 1.68$–1.698 $\beta = 1.683$–1.700 $\gamma = 1.685$–1.706 $\delta = 0.005$–0.008	4	3400	**Habitus:** striated, fibrous, massive. **Color:** blue, black, or dark green. **Diaphaneity:** translucent to opaque. **Luster:** vitreous, silky. **Cleavage:** (110) perfect. **Streak:** greenish brown. **Deposits:** magmatic and metamorphic rocks.

Mineral	Formula / Composition	Crystal system / Cell	Optical	Hardness	Density	Properties
Rosenbuschite	$(Ca,Na,Mn)_3(Zr,Ti,Fe)[SiO_4]_2$-$(F,OH)$ (Neosilicates)	Triclinic	Biaxial (+) $\alpha = 1.678–1.680$ $\beta = 1.687–1.688$ $\gamma = 1.705–1.708$ $\delta = 0.027–0.028$ $2V = 68–78°$	5–6	3310–3380	
Rutile [13463-67-7] (from the Latin, rutilus, meaning reddish)	TiO_2 $MM = 79.8788$ 59.94wt% Ti 40.06wt% O Coordinance Ti(6) (Oxides and hydroxides)	Tetragonal $a = 459.37$ pm $c = 296.18$ pm C4, tP6 (Z = 2) P.G. 422 S.G. P4/mm Rutile type	Uniaxial (+) $\varepsilon = 2.605–2.613$ $\omega = 2.899–2.901$ $\delta = 0.286–0.296$ Dispersion strong	6–6.5	4230–4250	**Habitus:** acicular, prismatic, massive. **Color:** reddish brown, yellowish brown, black or bluish violet, inclusion in quartz. **Diaphaneity:** transparent, translucent, opaque. **Luster:** adamantine. **Streak:** grayish black. **Fracture:** uneven. **Cleavage:** (111). **Twinning:** {101}, {301}. **Chemical:** insoluble in water, slightly soluble in HCl, HNO₃, sol. HF and in hot H_2SO_4, or $KHSO_4$. Attacked by molten Na_2CO_3. Electrical resistivity 29 to 910 Ωm.
Sanidine (syn. anorthoclase) (from the Greek, sanis, little plate, and idos, to see)	$(K,Na)(Si,Al)_4O_8$ Coordinance K(10), Si(4), Al(4) (Tectosilicates, framework)	Monoclinic $a = 856.2$ pm $b = 1303.6$ pm $c = 719.3$ pm $\beta = 116.58°$ (Z = 4) P.G. 2/m S.G. C2/m	Biaxial (−) $\alpha = 1.518–1.527$ $\beta = 1.522–1.532$ $\gamma = 1.525–1.534$ $\delta = 0.006–0.007$ $2V = 80–85°$	6	2560	**Habitus:** blocky, prismatic, massive, granular. **Color:** colorless, white, gray, yellowish white, or reddish white. **Diaphaneity:** transparent to translucent. **Luster:** vitreous, pearly. **Cleavage:** (001) perfect, (010) good. **Twinning:** Carlsbad {001}. **Fracture:** uneven. **Streak:** white. **Deposits:** acid volcanic igneous rocks.
Sapphirine	$(Mg,Fe)_2Al_2O_6[SiO_4]$ (Neosilicates)	Monoclinic $a = 996$ pm $b = 2860$ pm $c = 985$ pm $110.5°$ (Z = 8)	Biaxial (−) $\alpha = 1.701–1.725$ $\beta = 1.703–1.728$ $\gamma = 1.705–1.732$ $\delta = 0.005–0.007$ $2V = 50–114°$ O.A.P. (010)	7.5	3400–3580	
Scheelite [7790-75-2] (after the Swedish chemist, K.W. Scheele)	$CaWO_4$ $MM = 287.9256$ 13.92wt% Ca 63.85wt% W 22.23wt% O Coordinance Ca(4), W(4) (Sulfates, chromates, molybdates, and tungstates)	Tetragonal $a = 524.2$ pm $c = 1137.20$ pm (Z = 4) P.G. 4/m S.G. I4₁/a Scheelite type	Uniaxial (+) $\varepsilon = 1.918–1.920$ $\omega = 1.934–1.937$ $\delta = 0.016–0.017$	4.5–5	6060–6110	**Habitus:** massive, granular, disseminated, tabular, columnar, bipyramidal or pseudo octahedrons. **Color:** colorless, white, pale yellow, greenish, brownish yellow, or reddish yellow. **Luster:** vitreous (i.e., glassy), greasy or subadamantine. **Diaphaneity:** transparent to translucent. **Luminescence:** fluorescent under short UV light, bright bluish white, sometimes pale yellow (traces of Mo). **Streak:** white. **Cleavage:** (101) distinct, (112) poor. **Twinning:** {110}. **Fracture:** uneven, brittle. **Chemical:** soluble in HCl or HNO₃ giving a yellow solid residue of WO₃, soluble in NH₄OH. The soln. in HCl gives a deep blue color when a crystal of pure Sn, or Zn is added. **Other:** dielectric constant 3.5 to 5.75. Melting point 1620°C. **Deposits:** high-temperature quartz veins, in metamorphism contact halo of intrusive granitic igneous rocks, in detritic sedimentary rocks near granites.
Schorl	$NaFe_3Al_6(OH)_4B_3O_9[Si_6O_{18}]$ Coordinance Na(6), Fe(6), Al(6), B(3), Si(4) (Cyclosilicates, ring)	Trigonal $a = 1646$ pm $c = 715$ pm (Z = 3) P.G. 3m S.G. R3m Tourmaline group	Uniaxial (−) $\varepsilon = 1.668$ $\omega = 1.639$ $\delta = 0.029$	7–7.5	3270	**Habitus:** prismatic. **Color:** white. **Luster:** resinous. **Diaphaneity:** transparent to translucent. **Streak:** white. **Cleavage:** (101) poor, (110) poor. **Twinning:** {101}. **Fracture:** subconchoidal.

11
Minerals, Ores and Gemstones

Table 11.4 (continued)

Mineral name (IMA) [CAS RN] (Synonyms) (Etymology)	Theoretical chemical formula, relative molecular mass, coordination number, major impurities, mineral class (Strunz)	Crystal system, lattice parameters, Strukturbericht, Pearson symbol, Z, point group, space group, structure type	Optical properties	Mohs hardness (HM)	Density (ρ, kg.m^{-3})	Other mineralogical, physical, and chemical properties
Scorodite	$Fe(AsO_4)\cdot 4H_2O$ $MM = 266.82556$ 20.92wt% Fe 28.08wt% As 47.97wt% O 3.03wt% H Coordination Fe(6), As(4) (Phosphates, arsenates, and vanadates)	Orthorhombic $a = 1043$ pm $b = 896$ pm $c = 1015$ pm ($Z = 8$) P.G. mmm S.G. P2$_1$/cab Scorodite type	Biaxial (+) $\alpha = 1.784$ $\beta = 1.796$ $\gamma = 1.814$ $\delta = 0.030$ $2V = 54°$	3–4	3200	**Habitus:** prismatic, dipyramidal. **Color:** green, brown. **Streak:** white. **Diaphaneity:** transparent to translucent. **Luster:** resinous. **Fracture:** uneven. **Cleavage:** (120) good, (100) perfect, (010) poor.
Senarmontite (after the French mineralogist, H.H. de Senarmot)	Sb_2O_3 $MM = 275.4988$ 88.39wt% Sb 11.61wt% O (Oxides and hydroxides)	Cubic	Isotropic $n_D = 2.087$	2	5200–5300	**Habitus:** euhedral crystals, massive-granular, encrustations. **Color:** white, colorless, or gray. **Luster:** adamantine. **Diaphaneity:** transparent to translucent. **Streak:** white. **Cleavage:** (111) imperfect. **Fracture:** uneven. **Deposits:** oxidation of stibnite and other antimony minerals.
Shortite	$Na_2Ca_2(CO_3)_3$ (Nitrates, carbonates, and borates)	Orthorhombic	Biaxial	n.a.	n.a.	Fluorescent
Siderite (from Greek, sideros, iron)	$FeCO_3$ $MM = 115.8539$ 62.1wt% FeO 37.9wt% CO$_2$ Coordination Fe(6), C(3) (Nitrates, carbonates, and borates)	Trigonal (Rhombohedral) $a = 468.87$ pm $c = 1537.3$ pm ($Z = 6$) P.G. 32/m S.G. R$\bar{3}$c Calcite type	Uniaxial (−) $\varepsilon = 1.575$–1.637 $\omega = 1.782$–1.875 $\delta = 0.207$–0.242 Dispersion strong	4.5–5	3500–3960	**Habitus:** rhombohedral, crystalline, coarse, stalactitic, massive. **Color:** yellow brown. **Luster:** vitreous (i.e., glassy). **Diaphaneity:** transparent, translucent, to opaque. **Streak:** white. **Cleavage:** (1014) perfect. **Twinning:** {0118}. **Fracture:** brittle, subconchoidal. **Chemical:** readily dissolved in diluted acids with evolution of carbon dioxide. **Deposits:** sedimentary rocks.
Sillimanite (syn., viridine: green fibrolite; acicular) (after the American chemist and mineralogist, B. Silliman)	$Al_2O[SiO_4] = Al_2SiO_5$ $MM = 162.04558$ 33.30wt% Al 17.33wt% Si 49.37wt% O Coordination Al(6), Si(4), Al(4) Traces of Fe, Mn (Nesosubsilicates)	Orthorhombic $a = 748.43$ pm $b = 767.30$ pm $c = 577.11$ pm ($Z = 4$) P.G. mmm S.G. Pbnm	Biaxial (+) $\alpha = 1.653$–1.661 $\beta = 1.658$–1.662 $\gamma = 1.673$–1.684 $\delta = 0.020$–0.023 $2V = 21$–$30°$ O.A.P. (010) Dispersion strong	6.5–7.5	3240	**Habitus:** fibrous, prismatic, acicular. **Color:** colorless, white, yellowish or green. **Fracture:** splintery, brittle. **Diaphaneity:** transparent to translucent. **Luster:** vitreous (i.e., glassy). **Cleavage:** (010) perfect. **Fluorescence:** bluish white. **Chemical:** insoluble in strong mineral acids, decomposed by molten Na$_2$CO$_3$. When heated in an aqueous solution of Co(NO$_3$)$_2$ gives a blue color (Thénard blue). **Other:** dielectric constant 9.29. **Deposits:** metamorphosed peri-aluminous sedimentary rocks. Gneiss and shales.
Silver [7440-22-4] (syn., argentum)	Ag $MM = 107.8682$ Coordination Ag(12) (Native elements)	Cubic $a = 408.56$ pm Al, cF4 ($Z = 4$) P.G. m3m S.G. Fm3m Copper type	Isotropic $n_D = 0.181$ $R = 94\%$	2.5–3	10,506	**Habitus:** octahedral, dendritic. **Color:** silver white. **Diaphaneity:** opaque. **Luster:** metallic. **Streak:** gray. **Cleavage:** none. **Twinning:** {111}. **Fracture:** hackly, malleable, ductile. **Chemical:** readily dissolved in nitric acid, HNO$_3$.

Mineral	Chemical composition	Crystallography	Optical	Hardness	Density	Properties
Sinhalite	$MgAlBO_4$ MM = 126.095138 19.28wt% Mg 21.40wt% Al 8.57wt% B 50.75wt% O Coordinance Mg(6), Al(6), B(4) (Nitrates, carbonates, and borates)	Orthorhombic a = 432.8 pm b = 987.8 pm c = 567.5 pm (Z = 4) P.G. mmm Olivine type	Biaxial (+) α = 1.670 β = 1.700 γ = 1.710 δ = 0.04 2V = 55°	6.5–7	3420	**Habitus:** prismatic. **Color:** white, yellow. **Diaphaneity:** transparent to translucent. **Luster:** vitreous. **Streak:** green-blue. **Cleavage:** good (010). **Fracture:** conchoidal.
Skutterudite	$(Co,Ni)As_3$ (Sulfides and sulfosalts) Coordinance Co(6), Ni(6)	Cubic a = 820 pm D02, cI32 (Z = 8) S.G. Im-3 P.G. m-3 CoAs3 type	Isotropic R = 54%	5.5–6	6100–6800	**Habitus:** skeletal, cubic crystals. **Color:** tin white. **Luster:** metallic. **Diaphaneity:** opaque. **Streak:** black. **Fracture:** uneven. **Cleavage:** (100), (111). **Twinning:** [112]. Electrical resistivity 5 to 400 $\mu\Omega$.cm.
Smithsonite [3486-35-9] (syn. galmei, calamine, zinc spar) (after the English mineralogist, J. Smithson)	$ZnCO_3$ MM = 125.3992 52.15wt%Zn 9.58wt% C 38.28wt% O Coordinance Zn(6), C(3) (Nitrates, carbonates, and borates)	Trigonal (Rhombohedral) a = 465.28 pm c = 1502.8 pm (Z = 6) P.G. -32/m S.G. R-3c Calcite type	Uniaxial (−) ε = 1.625 ω = 1.848 δ = 0.225	4.5	4450	**Habitus:** massive, botryoidal, reniform, earthy. **Color:** grayish white, dark gray, green, blue, or yellow. **Diaphaneity:** transparent to translucent. **Luster:** vitreous (i.e., glassy), pearly. **Streak:** white. **Cleavage:** (10$\bar{1}$1) perfect. **Fracture:** subconchoidal, brittle. **Chemical:** readily attacked by strong mineral acids with evolution of carbon dioxide.
Sodalite (from its chemical composition)	$Na_8[Al_6Si_6O_{24}]Cl_2$ MM = 933.75868 19.70wt% Na 17.34wt% Al 18.05wt% Si 3.80wt%, Cl 41.12wt% O Coordinance Na(7), Si(4), Al(4) Traces of K and Ca (Tectosilicates, framework)	Cubic a = 891 pm (Z = 1) P.G. 43m S.G. P43m (Sodalite type)	Isotropic n_D = 1.483–1.487	5.5–6	2270–2330	**Habitus:** massive, granular, disseminated. **Color:** azure blue, white, yellow, pale pink, colorless, gray, or green. **Diaphaneity:** transparent to translucent. **Luster:** vitreous (i.e., glassy), greasy. **Streak:** white. **Cleavage:** (110) good. **Twinning:** [111]. **Fracture:** conchoidal. **Deposits:** volcanic tuffs and volcano-clastic sediments.
Spessartine (syn., spessartite) (after the locality, Spessart, Germany)	$Mn_3Al_2(SiO_4)_3$ MM = 495.02653 33.29wt% Mn 10.90wt% Al 17.02wt% Si 38.78wt% O Coordinance Mn(8), Al(6), Si(4) (Nesosilicates)	Cubic a = 1.162 pm (Z = 8) P.G. 432 S.G. 1a3d Garnet group (Pyralspite series)	Isotropic n_D = 1.805	6.5–7.5	4180	**Habitus:** massive, crystalline, lamellar. **Color:** red or brownish red. **Diaphaneity:** transparent to translucent. **Luster:** vitreous, resinous. **Streak:** white. **Parting:** (110). **Fracture:** Subconchoidal. **Deposits:** magmatic, metamorphic, and pegmatic rocks.
Sphalerite [1314-98-3] (syn., zinc blende, mock, lead ore, black Jack, false galena) (from the Greek, sphaleros, misleading)	ZnS MM = 97.456 67.10wt% Zn 32.90wt% S Traces Fe (Sulfides and sulfosalts) Coordinance Zn(4)	Cubic a = 540.93 pm B3, cF8 (Z = 4) S.G. F-43m P.G. -43m Blende type	Isotropic n_D = 2.369 R = 17.5–19%	3.5–4 (198 (HV))	4089	**Habitus:** tetrahedral crystals, granular, colloform. **Color:** brown, yellow, orange, red, green, or black. **Luster:** resinous, metallic, greasy. **Diaphaneity:** transparent to opaque. **Luminescence:** fluorescent and triboluminescent. **Streak:** brownish white. **Cleavage:** (110). **Fracture:** uneven, conchoidal. **Twinning:** [111]. **Chemicals:** attacked by strong mineral acids, HCl, or HNO_3 with evolution of H_2S and yellow precipitate of sulfur. Infusible. Electrical resistivity 2.7 mΩ.m. **Deposits:** veins in igneous, sedimentary, and metamorphic rocks.

11
Minerals, Ores and Gemstones

Table 11.4 (continued)

Mineral name (IMA) [CAS RN] (Synonyms) (Etymology)	Theoretical chemical formula, relative molecular mass, coordination number, major impurities, mineral class (Strunz)	Crystal system, lattice parameters, Strukturbericht, Pearson symbol, Z, point group, space group, structure type	Optical properties	Mohs hardness (HM)	Density (ρ, kg.m^{-3})	Other mineralogical, physical, and chemical properties
Sphene (syn, titanite) (from the Greek, sphen, coin)	CaTi[SO$_4$](O,OH,F)] 28.6wt% CaO 40.8wt% TiO$_2$ 30.6wt% SiO$_2$ Coordination Ca(7), Ti(6), Si(4) (Nesosilicates)	Monoclinic a = 656 pm b = 872 pm c = 744 pm 119.72° (Z = 4) P.G. 2/m S.G. C2/c	Biaxial (+) α = 1.843–1.950 β = 1.870–2.034 γ = 1.943–2.110 δ = 0.100–0.192 2V = 17–40° Dispersion strong O.A.P. (010)	5	3450–3550	**Habitus:** wedge-shaped crystals, tabular, prismatic. **Color:** yellow, brown, white, greenish, gray. **Luster:** adamantine, resinous, greasy. **Diaphaneity:** transparent to opaque. **Cleavage:** (110) perfect. (100) poor. **Fracture:** uneven, conchoidal. **Twinning:** {100}, {221} lamellar. **Chemical:** insoluble in HCl, but decomposed by hot concentrated H$_2$SO$_4$. Fusible. **Deposits:** igneous rocks, granitic, pegmatites.
Spinel (syn, ruby spinal, balas ruby, red rubicelle) from Latin, spina, thorn, in allusion to sharply pointed crystals	MgAl$_2$O$_4$ MM = 142.26568 17.08wt% Mg 37.93wt% Al 44.98wt% O Coordination Mg(4), Al(6) (Oxides and hydroxides)	Cubic a = 808.0 pm Hl1, cF56 (Z = 8) P.G. m3m S.G. Fd3m Spinel type	Isotropic n_D = 1.719	7.5–8	3583	**Habitus:** euhedral crystals, massive, granular. **Color:** colorless, red, blue, green, or brown. **Luster:** vitreous (i.e., glassy). **Diaphaneity:** transparent, translucent, opaque. **Streak:** grayish white. **Cleavage:** (111) poor. **Twinning:** {111}. **Fracture:** conchoidal or uneven.
Spodumene (syn, pink: kunzite)	LiAl[Si$_2$O$_6$] Coordination Li(6), Al(6), and Si(4) (Inosilicates, chains)	Monoclinic a = 952 pm b = 832 pm c = 525 pm 110.46° (Z = 4) P.G. 2/m S.G. C2/m	Biaxial (+) α = 1.650 β = 1.660 γ = 1.670 δ = 0.020 2V = 60–80° Dispersion weak	6.5–7	3150	**Habitus:** euhedral prismatic crystals. **Color:** colorless, pink. **Luster:** vitreous (i.e., glassy). **Diaphaneity:** transparent, translucent. **Streak:** white. **Cleavage:** 110 perfect. **Twinning:** {100}. **Fracture:** uneven. **Chemical:** insoluble in strong mineral acids. Nevertheless when heated above 1100°C it transforms to beta spodumene which is readily attacked by hot concentrated H$_2$SO$_4$. **Deposits:** granitic pegmatites.
Spurrite	2Ca$_2$[SiO$_4$]·CaCO$_3$	Monoclinic	Biaxial (−) α = 1.637–1.641 β = 1.672–1.676 γ = 1.676–1.681 δ = 0.039–0.040 2V = 35–41°	5	3010	
Stannite (syn, tin pyrites, bell metal ore) (from the Latin, stannum, tin)	Cu$_2$FeSnS$_4$ MM = 429.913 12.99wt% Fe 29.56wt% Cu 27.61wt% Sn 29.83wt% S (Sulfides and sulfosalts)	Tetragonal a = 546 pm c = 1072 pm H26, tI16 (Z = 2) S.G. I42m	Uniaxial	3.5–4	4400	**Habitus:** massive, euhedral crystals. **Color:** steel gray or olive green. **Diaphaneity:** opaque. **Luster:** metallic. **Cleavage:** (110) poor. **Fracture:** uneven. **Streak:** black. Electrical resistivity 1.2 to 570 mΩ.m.

Mineral (name / etymology)	Chemical composition	Crystallography	Optical properties	S.G.	Hardness	Description
Staurolite (syn, staurolite) (from the Greek, stauros, cross, and lithos, stone, in allusion to the common cross-shaped twins of the crystals)	$(Fe, Mg, Zn)_2(Al, Fe)_9[O_6|Si_4O_{22}](OH)_2$; 27–29wt% SiO_2; 53–54wt% Al_2O_3; 1–3wt% Fe_2O_3; 11–12wt% FeO; 2–3wt% MgO; Coordinance Al(6), Si(4), Fe(4) (Nesosubsilicates)	(pseudo-orthorhombic); $a = 790$ pm; $b = 1665$ pm; $c = 563$ pm; $90.0°$; (Z = 2); P.G. 2/m; S.G. C2/m	Biaxial (+); $\alpha = 1.739$–1.747; $\beta = 1.745$–1.753; $\gamma = 1.752$–1.761; $\delta = 0.012$–0.014; $2V = 82$–$90°$; O.A.P. (100); Dispersion weak	3740–3830	7–7.5	Habitus: tabular, prismatic, massive. Color: reddish brown, brownish black, or yellowish brown. Diaphaneity: translucent to opaque. Luster: vitreous (i.e., glassy), dull. Streak: gray. Twinnings: common in cross. Cleavage: (001) distinct. Twinning: (031), and (231). Fracture: subconchoidal. Chemical: attacked by hot conc. H_2SO_4. Unfusible. Other: dielectric constant 6.80. Deposits: metamorphosed aluminous sedimentary rocks.
Stephanite (syn, brittle silver ore) (after the Austrian engineer, A. Stephan)	Ag_5SbS_4; MM = 789.355; 68.33wt% Ag; 15.42wt% Sb; 16.25wt% S (Sulfides and sulfosalts)	Orthorhombic; P.G. mm2	Biaxial	6250	2–2.5	Habitus: pseudo hexagonal, tabular, massive. Color: iron black. Diaphaneity: opaque. Luster: metallic. Streak: black. Fracture: subconchoidal. Cleavage: (010) imperfect, (021) poor.
Stibnite (syn, antimonite, antimony glance, gray antimony, stibium) (from the Greek, stimmi or stibi, antimony, hence to the Latin, stibium)	Sb_2S_3; $M = 339.698$; 71.68wt% Sb; 28.32wt% S (Sulfides and sulfosalts); Coordinance Sb(7)	Orthorhombic; $a = 1122.9$ pm; $b = 1131.0$ pm; $c = 383.89$ pm; $D5_{11}$, oP20 (Z = 4); S.G. Pccn; P.G. 222; Stibnite type	Biaxial (−); $\alpha = 3.194$; $\beta = 4.046$; $\gamma = 4.303$; $\delta = 1.110$; $2V = 26°$; $R = 25$–38%	4630	2	Habitus: prismatic, faces striated, granular. Color: lead gray, bluish lead gray, steel gray, or black. Diaphaneity: opaque. Luster: Metallic. Streak: blackish gray. Cleavage: (010) perfect. Fractus: subconchoidal.
Stilbite	$(Ca, Na)Si_7Al_2O_{18}\cdot 7H_2O$; Coordinance Ca(6), Na(6), Si(4) and Al(4). (Tectosilicates, framework)	Monoclinic; $a = 1364$ pm; $b = 1824$ pm; $c = 1127$ pm; $129.16°$; (Z = 4); P.G. 2/m; S.G. C2/m	Biaxial (?); $\alpha = 1.490$; $\beta = 1.500$; $\gamma = 1.500$; $\delta = 0.010$; $2V = 30$–$50°$	2150	3.5–4	Habitus: prismatic, striated, curved crystals. Color: gray. Diaphaneity: translucent to transparent. Luster: pearly. Streak: gray. Cleavage: (010) distinct, (001) poor, (101) poor. Fracture: subconchoidal.
Stishovite	SiO_2; MM = 60.0843; 46.74wt% Si; 53.26wt% O (Tectosilicates, framework); Coordinance Si (4)	Tetragonal; $a = 417.9$ pm; $c = 266.49$ pm (Z = 2); C4, tP6 (Z = 2); P.G. 422; S.G. P4/mnm; Rutile type	Uniaxial (+); $\varepsilon = 1.826$; $\omega = 1.799$; $\delta = 0.027$	4300	6	Habitus: prismatic. Color: colorless. Luster: vitreous (i.e., glassy). Streak: white. Twinning: [011]. Fracture: conchoidal. Chemical: resistant to strong mineral acids, attacked by HF and molten alkali hydroxides.
Stromeyerite (after the German chemist, F. Stromeyer)	AgCuS; MM = 203.4802; 32.13wt% Cu; 53.01wt% Ag; 15.76wt% S	Orthorhombic; P.G. 222	Biaxial	6000–6300	2.5–3	Habitus: granular, massive, pseudo hexagonal. Color: steel gray. Luster: metallic. Diaphaneity: opaque. Streak: steel gray. Fracture: conchoidal. Deposits: copper-silver veins where silver replaces copper in bornite.

11

Minerals, Ores and Gemstones

Table 11.4 (continued)

Mineral name (IMA) [CAS RN] (Synonyms) (Etymology)	Theoretical chemical formula, relative molecular mass, coordinance number, major impurities, mineral class (Strunz)	Crystal system, lattice parameters, Strukturbericht, Pearson symbol, Z, point group, space group, structure type	Optical properties	Mohs hardness (HM)	Density (ρ, kg.m^{-3})	Other mineralogical, physical, and chemical properties
Strontianite [1633-05-2] (after Strontian, a Scottish town)	$SrCO_3$ $MM = 147.6292$ 59.35wt% Sr 8.14wt% C 32.51wt% O (Sulfates, chromates, molybdates, and tungstates)	Orthorhombic $a = 602.9$ pm $b = 841.4$ pm $c = 510.7$ pm $(Z = 4)$ P.G. mmm S.G. Pmcn Aragonite type	Biaxial (−) $\alpha = 1.516$–1.520 $\beta = 1.664$–1.667 $\gamma = 1.666$–1.669 $\delta = 0.149$–0.150 $2V = 7$–10° Dispersion weak	3.5	3720	**Habitus:** pseudohexagonal, columnar, massive, granular, acicular, spadelike. **Color:** white, yellowish gray, greenish gray, or bluish white. **Luster:** vitreous (i.e., glassy). **Diaphaneity:** transparent to translucent. **Cleavage:** (110) good. **Fracture:** uneven, conchoidal, brittle. **Streak:** white. **Chemical:** decomposes at 1494°C giving off SrO and CO_2. Soluble in strong mineral acids with evolution of CO_2.
Sulfur or sulfur (From Sanskrit, sulvere and Latin, sulfurium)	S_8 $MM = 256.528$ Coordinance S(2) (Native elements)	Orthorhombic $a = 1046.46$ pm $b = 1286.60$ pm $c = 2448.60$ pm A16, $oF128$ ($Z = 128$ S or 16 S$_8$) P.G. 222 S.G. Fddd Sulfur type	Biaxial (+) $\alpha = 1.958$ $\beta = 2.038$ $\gamma = 2.245$ $\delta = 0.290$ $2V = 68.58°$ Dispersion weak	1.5–2.5	2068	**Habitus:** massive, reniform, stalactitic. **Color:** yellow, yellowish brown, or gray. **Diaphaneity:** transparent to translucent. **Luster:** resinous. **Streak:** white. **Cleavage:** (101), (110). **Fracture:** sectile. **Chemical:** highly soluble in carbon disulfide CS_2. **Deposits:** volcanic exhalations and bacterial reduction of sulfates in sediments.
Sylvanite (after Transylvania)	$(Au,Ag)_2Te_4$ (Sulfides and sulfosalts)	Monoclinic $a = 896$ pm $b = 449$ pm $c = 1462$ pm $(Z = 4)$ P.G. 2/m	Biaxial	1.5–2	7900–8300	**Habitus:** prismatic, skeletal, platy. **Color:** yellowish silver white or white. **Luster:** metallic. **Diaphaneity:** opaque. **Streak:** steel gray. **Cleavage:** (010) perfect. **Fracture:** uneven.
Sylvite (syn. sylvinite) [7447-40-7] (after the Dutch chemist, Sylvia de la Boe (1614–1672))	KCl $MM = 74.551$ 52.45wt% K. 47.55wt% Cl (Halides) Coordinance K(6)	Cubic $a = 629.31$ pm B1, cF8 ($Z = 4$) S.G. Fm3m P.G. 4-32 Rock salt type	Isotropic $n_D = 1.490$	2.0	1988	**Habitus:** massive, cubic euhedral crystals, fibrous. **Color:** white, yellowish white, reddish, white, bluish white, or brownish white. **Luster:** vitreous, greasy. **Diaphaneity:** transparent to translucent. **Streak:** white. **Cleavage:** (100), (010), (001). **Fracture:** uneven, brittle, sectile. **Chemical:** soluble in water, the solution colors the flame of a bunsen violet. Bitter taste. Fusible (778°C).
Talc (2M1) (syn., steatite: massive, soapstone, kerolite) (from the Arabic)	$Mg_3Si_4O_{10}(OH)_2$ $MM = 379.26568$ 19.23wt% Mg 29.62wt% Si 0.53wt% H 50.62wt% O Coordinance Mg(6), Si(4) (Phyllosilicates, layered)	Monoclinic $a = 528.7$ pm $b = 915.8$ pm $c = 1895$ pm $\beta = 99.50°$ $(Z = 4)$ P.G. m S.G. Cc Type 2M1, mica	Biaxial (−) $\alpha = 1.539$–1.550 $\beta = 1.589$–1.594 $\gamma = 1.589$–1.600 $\delta = 0.037$–0.050 $2V = 6$–30°	1	2580–2830	**Habitus:** foliated, scaly, massive. **Color:** pale green, white, gray white, yellowish white, or brownish white. **Diaphaneity:** translucent. **Luster:** vitreous (i.e, glassy), pearly. **Streak:** white. **Cleavage:** (001) perfect. **Fracture:** uneven. **Deposits:** hydrothermal alteration of non-aluminous magnesian silicates.

Name (etymology)	Formula / composition	Density	Hardness	Optical properties	Crystal system	Properties
Tapiolite (named after the god Tapio of (Finnish mythology))	$(Fe,Mn)(Ta,Nb)_2O_6$ $MM = 513.7392$ 70.44wt% Ta 10.87wt% Fe 18.69wt% O (Oxides, and hydroxides)	8170	6–6.5	Biaxial	Tetragonal $a = 475.4$ $c = 922.8$ ($Z = 2$)	Habitus: granular, massive. Color: black. Diaphaneity: opaque. Luster: metallic. Streak: brown. Cleavage: (110) imperfect. Fractures: uneven. Deposits: pegmatites and alluvial deposits.
Tenorite (syn., melaconite, melanochalcite) (after the Italian botanist, M. Tenore)	CuO $MM = 79.5454$ 79.89wt% Cu, 20.11wt% O	6500	3.5–4	Biaxial	Triclinic P.G. -1	Habitus: scaly, earthy, massive. Color: black. Luster: earthy (dull). Diaphaneity: opaque. Streak: black. Cleavage: none. Fracture: conchoidal. Deposits: secondary copper mineral.
Tephroite (from the Greek tephros, ash gray after its color)	Mn_2SiO_4 (Nesosilicates)	4110–4390	6.5	Biaxial (+) $\alpha = 1.759$ $\beta = 1.797$ $\gamma = 1.86$ $\delta = 0.101$ $2V = 78°$	Orthorhombic $a = 476$ pm $b = 102$ pm $c = 598$ pm ($Z = 4$) P.G. mmm Olivine group	Habitus: massive-granular, granular, prismatic. Color: gray, olive gray, flesh pink, or reddish brown. Luster: vitreous-greasy. Diaphaneity: transparent to translucent. Streak: gray. Cleavage: 010 indistinct. Fracture: brittle-conchoidal. Deposits: contact metamorphism of manganese-bearing rocks.
Terlinguaite (after Terlingua, Texas, USA)	Hg_2ClO (Halides)	8700	2–3	Biaxial (−) $\alpha = 2.35$ $\beta = 2.64$ $\gamma = 2.66$ $\delta = 0.310$ $2V = 26°$ Dispersion strong	Monoclinic	Habitus: triated, prismatic. Color: yellow or green. Luster: adamantine. Diaphaneity: transparent to translucent. Cleavage: (101) perfect. Deposits: oxidized portions of mercury deposits.
Tetradymite (from the Greek, tetradymos, four-fold)	Bi_2Te_2S $MM = 705.2268$ 59.27wt% Bi 36.19wt% Te 4.55wt% S (Sulfides and sulfosalts)	7550	1.5–2	Uniaxial	Trigonal	Habitus: lamellar, granular, pseudo hexagonal. Color: steel gray or yellow gray. Diaphaneity: opaque. Luster: metallic. Cleavage: (0001) perfect. Fracture: uneven. Streak: steel gray.
Tetrahedrite (derived from its crystal form)	$(Cu,Fe)_{12}Sb_4S_{13}$ (Sulfides and sulfosalts)	4600–5200	3.5–4	Isotropic $R = 24\%$	Cubic P.G. 43m $a = 1034$ pm ($Z = 2$) S.G. = 143m	Habitus: massive-granular, massive. Color: steel gray or black. Luster: metallic. Diaphaneity: opaque. Streak: black. Fracture: uneven.
Tetrahedrite (from its tetrahedral crystal form)	$(Cu,Fe)_2Sb_4S_{13}$ (Sulfides and sulfosalts) Coordinance Cu(3), Cu(4)	4900–5100	3.5–4	Isotropic $n_D = 2.720$ $R = 24\%$	Cubic $a = 1027$ pm ($Z = 2$) S.G. 143 m P.G. 43 m Tetrahedrite type	Habitus: tetrahedral crystals, granular, massive. Color: steel gray or black. Diaphaneity: opaque. Luster: metallic. Streak: brown, black. Cleavage: None. Twinning: {111}. Fracture: uneven, subconchoidal. Electrical resistivity 0.3 to 30,000 Ω.m.
Thorianite (from the presence of thorium)	ThO_2 (Oxides and hydroxides)	10,000	6	Isotropic	Cubic	Habitus: pseudo hexagonal, granular. Color: black or brown. Diaphaneity: opaque. Luster: metallic. Cleavage: (100), (010), (001) poor. Fracture: conchoidal, brittle. Streak: black. Radioactive. Deposits: pegmatites and alluvial deposits.

11
Minerals, Ores and Gemstones

Table 11.4 (*continued*)

Mineral name (IMA) [CAS RN] (Synonyms) (Etymology)	Theoretical chemical formula, relative molecular mass, coordination number, major impurities, mineral class (Strunz)	Crystal system, lattice parameters, Strukturbericht, Pearson symbol, Z, point group, space group, structure type	Optical properties	Mohs hardness (HM)	Density (ρ, kg.m^{-3})	Other mineralogical, physical, and chemical properties
Thorite (syn. orangite) (from the Scandinavian God, Thor)	$(Th,U)SiO_4$ $MM = 324.1212$ 71.59wt% Th 8.67wt% Si 9.74wt% O (Nesosilicates)	Tetragonal $a = 711.7$ pm $c = 629.5$ pm $(Z = 4)$	Uniaxial (−) $\varepsilon = 1.78–1.82$ $\omega = 1.79–1.84$ $\delta = 0.010–0.020$	5	5350	**Habitus:** prismatic, granular, massive. **Color:** reddish brown, black, or orange. **Diaphaneity:** transparent to translucent. **Luster:** resinous. **Cleavage:** (110) poor. **Fracture:** conchoidal. **Streak:** light brown. **Deposits:** augite-syenite rocks.
Tiemannite (after C.W.F. Tiemann)	$HgSe$ $MM = 279.55$ 71.75wt% Hg 28.25wt% Se (Sulfides and sulfosalts)	Cubic P.G. Diploidal	Isotropic	2.5	8190–8470	**Habitus:** euhedral crystals, granular. **Color:** dark lead gray. **Luster:** metallic. Diaphaneity: opaque. **Streak:** grayish black. **Cleavage:** none. **Fracture:** brittle.
Titanite (sphene) (after titanium)	$CaTiO[SiO_4]$ (Nesosilicates) Coordinance Ca(7), Ti(6), Si(4)	Monoclinic $a = 656$ pm $b = 872$ pm $c = 744$ pm $\beta = 119.72°$ $(Z = 4)$ P.G. 2/m S.G. C2/c	Biaxial (+) $\alpha = 1.84–1.95$ $\beta = 1.87–2.034$ $\gamma = 1.943–2.11$ $\delta = 0.103–0.160$ $2V = 20–56°$	5–5.5	3480	**Habitus:** massive, lamellar, euhedral crystals. **Color:** reddish brown, gray, yellow, green, or red. **Diaphaneity:** transparent to opaque. **Luster:** adamantine, resinous. **Cleavage:** (110) distinct, (100), (112) imperfect. **Fracture:** subconchoidal. **Streak:** reddish white. **Deposits:** magmatic, metamorphic and hydrothermal rocks.
Tobernite	$Cu(UO_2)_2(VO_4)_2 \cdot 8H_2O$ Coordinance Cu(6), U(2), P(4) (Phosphates, arsenates, and vanadates)	Tetragonal $a = 70$ pm $c = 206.7$ pm P.G. 4/mmm S.G. I4/mmm $(Z = 4)$	Uniaxial (−)	2–2.5	3250	
Topaz (syn. pycnite; yellowish white) (after the locality, Topasos Island, in the Red Sea)	$Al_2[SiO_4](F,OH)_2$ Coordinance Al(6), Si(4) (Nesosubsilicates)	Orthorhombic $a = 464.9$ pm $b = 879.2$ pm $c = 839.4$ pm $(Z = 4)$ P.G. mmm S.G. Pbnm	Biaxial (+) $\alpha = 1.606–1.629$ $\beta = 1.609–1.631$ $\gamma = 1.616–1.638$ $\delta = 0.009–0.011$ $2V = 48–68°$ O.A.P. (010)	8	3490–3570	**Habitus:** crystalline, prismatic, massive, granular. **Color:** colorless, pale blue, yellow, yellowish brown, or red. **Luster:** vitreous (i.e., glassy). **Diaphaneity:** transparent. **Streak:** white. **Cleavage:** (001) perfect. **Fracture:** uneven grading. **Luminescence:** fluorescent under short UV-light: golden yellow, and long UV-light: cream. **Chemical:** slightly attacked by hot conc. H_2SO_4. **Other:** dielectric constant 6.09 to 7.4. **Deposits:** pegmatites and high-temperature quartz veins. Cavities in granites and rhyolites.
Tremolite (syn. asbestos) (after Tremola Valley, South side	$Ca_2(Mg,Fe)_5Si_8O_{22}(OH)_2$ $MM = 417.3278$ 9.60wt% Ca	Monoclinic $a = 984.00$ pm $b = 180.52$ pm	Biaxial (−) $\alpha = 1.599–1.620$ $\beta = 1.612–1.630$	5–6	3000–3100	**Habitus:** acicular, columnar, massive, fibrous, granular. **Color:** colorless, white, gray, or greenish gray. **Luster:** vitreous (i.e., glassy), pearly. **Diaphaneity:** transparent to translucent. **Luminescence:** fluorescent under short UV-light: yellow,

Name	Composition	Crystallography	Optical	Hardness	Density	Description
Alps)	46.01wt% O Coordinance Ca(8), Mg(6), Si(4) (Inosilicates, double chain)	(Z = 2) P.G. 2/m S.G. C2/m Tremolite type	2V = 65–86° Dispersion weak			...acids, quite unfusible giving a greenish globule. **Deposits:** contact metamorphism of Ca-rich rocks.
Trevorite	$NiFe_2O_3$ (Oxides and hydroxides)	Cubic $a = 843$ pm HI$_1$, cF56 ($Z = 8$) C10, Fd3m Spinel type	Isotropic $n_D = 2.30$	5.5	5260	
Tridymite [15468-32-3]	SiO_2 $MM = 60.0843$ 46.74wt% Si 53.26wt% O (Tectosilicates, framework) Coordinance Si (4)	Hexagonal $a = 504.63$ pm $c = 825.63$ pm ($Z = 4$) C10, hP12 ($Z = 4$) P.G. 6 mmm P6$_3$/mmc	Uniaxial (+) $\varepsilon = 1.475$ $\omega = 1.479$	6–7	2280	**Habitus:** wedge shaped. **Color:** colorless. **Streak:** white. **Diaphaneity:** transparent to translucent. **Luster:** vitreous (i.e., glassy). **Fracture:** conchoidal. **Twinning:** {110}. Transition temperature 1470°C.
Triphylite	$Li(Fe,Mn)PO_4$ Coordinance Li(8), Fe(8), and P(4) (Phosphates, arsenates, and vanadates)	Orthorhombic $a = 601$ pm $b = 468$ pm $c = 1036$ pm ($Z = 4$) P.G. mmm S.G. P2$_1$/mcn	Biaxial (−) $\alpha = 1.680$ $\beta = 1.680$ $\gamma = 1.690$ $\delta = 0.01$ 2V = 0–56°	5–5.5	3500–5500	**Habitus:** coarse massive. **Color:** blue green. **Streak:** white gray. **Diaphaneity:** transparent to translucent. **Luster:** vitreous, resinous. **Fracture:** subconchoidal. **Cleavage:** (001) perfect, (010) good.
Trona (Arabic origin, meaning natron)	$Na_3(CO_3)(HCO_3) \cdot 2H_2O$ $MM = 226.0262$ 30.51wt% Na 2.23wt% H 10.63wt% C 56.63wt% O (Nitrates, carbonates, and borates)	Monoclinic P.G. 2/m	Biaxial (−) $\alpha = 1.412$ $\beta = 1.492$ $\gamma = 1.540$ $\delta = 0.128$ 2V = 72° Dispersion strong	2.5	2110–2170	**Habitus:** fibrous, columnar, massive. **Color:** yellowish white, gray, or white. **Luster:** vitreous (glassy). **Diaphaneity:** translucent. **Streak:** white. **Cleavage:** (100) perfect, (111) indistinct, (001) indistinct. **Fracture:** subconchoidal.
Turquoise (syn. callaite) (after Turkey from where it was brought to Europe)	$CuAl_6(PO_4)_4(OH)_8 \cdot 4(H_2O)$ $MM = 813.44052$ 19.90wt% Al 7.81wt% Cu 15.23wt% P 1.98wt% H 55.07wt% O Coordinance Cu(6), Al(6), P(4) (Phosphates, arsenates, and vanadates)	Triclinic $a = 748$ pm $b = 995$ pm $c = 768$ pm $\alpha = 111.65°$ $\beta = 115.38°$ $\gamma = 69.43°$ ($Z = 1$) P.G. $\bar{1}$ S.G. P$\bar{1}$	Biaxial (+) $\alpha = 1.610$ $\beta = 1.615$ $\gamma = 1.650$ $\delta = 0.040$ 2V = 40–44° Dispersion strong	5–6	2700	**Habitus:** concretionary, reniform, massive, encrustations. **Color:** light blue, apple green, or greenish blue. **Diaphaneity:** translucent to opaque. **Luster:** resinous, waxy. **Streak:** pale bluish white. **Cleavage:** (001) perfect, (010) good. **Fracture:** subconchoidal, brittle.
Ulexite (syn. natroborocalcite) (after the German chemist, G.L. Ulex)	$NaCaB_5O_6(OH)_6 \cdot 5(H_2O)$ Coordinance Na(6), Ca(9), B(3, and 4) (Nitrates, carbonates, and borates)	Triclinic $a = 873$ pm $b = 1275$ pm $c = 670$ pm	Biaxial (+) $\alpha = 1.491$ $\beta = 1.505$ $\gamma = 1.520$	2.5	1950–1960	**Habitus:** acicular, fibrous, capillary. **Color:** colorless or white. **Luster:** silky. **Diaphaneity:** transparent to translucent. **Streak:** white. **Cleavage:** (010) perfect, (110) perfect. **Fracture:** uneven brittle.

11
Minerals, Ores and Gemstones

Table 11.4 (*continued*)

Mineral name (IMA) [CAS RN] (Synonyms) (Etymology)	Theoretical chemical formula, relative molecular mass, coordinance number, major impurities, mineral class (Strunz)	Crystal system, lattice parameters, Strukturbericht, Pearson symbol, Z, point group, space group, structure type	Optical properties	Mohs hardness (HM)	Density (ρ, kg.m^{-3})	Other mineralogical, physical, and chemical properties
Ulexite (*continued*)		$\alpha = 90.27°$ $\beta = 109.13°$ $\gamma = 105.12°$ ($Z = 2$) P.G. $\bar{1}$ S.G. $P\bar{1}$	$\delta = 0.029$ $2V = 73–78°$			
Ullmanite (after the German chemist and mineralogist, J.Ch. Ullmann)	NiSbS $MM = 212.506$ 27.62wt% Ni 57.29wt% Sb 15.09wt% S (Sulfides and sulfosalts)	Cubic $a = 588$ pm FO$_1$, cP12 ($Z = 4$) S.G. P2$_1$3 P.G. 23 Ullmannite type	Isotropic $R = 42\%$	5–5.5	6700	**Habitus:** tetrahedral crystals, granular, massive. **Color:** steel gray or silvery white. **Luster:** metallic. **Diaphaneity:** opaque. **Cleavage:** [100], [010], [001]. **Fracture:** brittle uneven. **Streak:** grayish black.
Ulvöspinel	Fe$_2$TiO$_4$ (Oxides and hydroxides)	Cubic $a = 853$ pm H1$_1$, cF56 ($Z = 8$) S.G. Fd3m Spinel type	Isotropic $n_D \approx 2.16$	5.5	4780	
Uraninite	UO$_2$ $MM = 270.0277$ Coordinance U(2) (Oxides and hydroxides)	Cubic $a = 546.82$ pm C1, cF12 ($Z = 4$) P.G. 432 S.G. Fm3m Fluorite type	Isotropic $R = 14\%$	5–6	10,970	**Habitus:** cubic, massive. **Color:** black. **Diaphaneity:** opaque. **Luster:** metallic. **Streak:** brown, black. **Cleavage:** (100). **Other:** electrical resistivity 1.5 to 200 Ω.m. Radioactive.
Uranophane (syn. uranotile) (from Greek, *uran* and *phanos*, to appear)	Ca(UO$_2$)$_2$SiO$_3$(OH)$_2$·5H$_2$O $MM = 586.36418$ 6.84wt% Ca 40.59wt% U 9.58wt% Si 2.06wt% H 40.93wt% O (Inosilicate, chain)	Monoclinic	Biaxial (−) $\alpha = 1.643$ $\beta = 1.666$ $\gamma = 1.669$ $\delta = 0.026$ $2V = 38°$ Dispersion strong	2.5	3900	**Habitus:** radial, earthy, massive, fibrous. **Color:** yellow. **Diaphaneity:** translucent. **Luster:** vitreous (i.e., glassy). **Cleavage:** (100) perfect. **Fracture:** uneven. **Streak:** yellowish white. **Deposits:** alteration product of gummite.

Name	Composition	Crystallography	Optical	Hardness	Density	Properties
Uvarovite (after Count S.S. Uvarov, Russian statesman and ardent amateur mineral collector)	$Ca_3Cr_2(SiO_4)_3$, MM = 500.4755, 24.0 wt% Ca, 20.78wt% Cr, 16.84wt% Si, 38.36wt% O, Coordinance Ca(8), Cr(6), Si(4) (Neosilicates)	Cubic, a = 1200 pm (Z = 8), P.G. 432, S.G. Ia3d, Garnet group (Ugrandite series)	Isotropic, n_D = 1.860	6.5–7.5	3900	**Habitus**: dodecahedral crystals. **Color**: green. **Diaphaneity**: transparent to translucent. **Luster**: vitreous (i.e., glassy). **Streak**: white. **Fracture**: subconchoidal. **Deposits**: metamorphosed chromite deposits.
Valentinite (syn. antimony bloom) (after the German alchemist, B. Valentinus)	Sb_2O_3, MM = 291.4982, 83.53wt% Sb, 16.47wt% O (Oxides and hydroxides)	Orthorhombic, D5₈, oP20 (Z = 8), S.G. Pbmn, P.G. 222	Biaxial (–), α = 2.18, β = 2.35, γ = 2.35, δ = 0.170	2.5–3	5600–5800	**Habitus**: euhedral crystals, divergent, striated. **Color**: white gray, or yellowish gray. **Luster**: adamantine. **Streak**: white. **Cleavage**: (110) perfect, (010) distinct. **Fracture**: uneven. **Deposits**: occurs as an oxidation product of various antimony minerals.
Vanadinite	$Pb_5(VO_4)Cl$, MM = 1186.3921, 87.33wt% Pb, 4.29wt% V, 5.39wt% O, 2.99wt% Cl, Coordinance Pb(6), V(4) (Phosphates, arsenates, and vanadates)	Hexagonal, a = 1033 pm, c = 735 pm (Z = 2), P.G. 6/m, S.G. P6₃/m, Apatite type	Uniaxial (–), ε = 2.350, ω = 2.416, δ = 0.066	3	6900	**Habitus**: prismatic, hollow prisms, encrustation. **Color**: red orange. **Streak**: white yellow. **Diaphaneity**: translucent. **Luster**: resinous. **Fracture**: subconchoidal.
Variscite (syn. strengite: Fe)	$Al(PO_4)\cdot 2H_2O$, MM = 157.983, 17.08wt% Al, 19.61wt% P, 60.76wt% O, 2.55wt% H, Coordinance Al(6), P(4) (Phosphates, arsenates, and vanadates)	Orthorhombic, a = 987 pm, b = 957 pm, c = 852 pm (Z = 8), P.G. mmm, S.G. P2₁/cab, Scorodite type	Biaxial (–), α = 1.55–1.56, β = 1.57–1.58, γ = 1.58–1.59, δ = 0.03, 2V = 48–54°	4	2500	**Habitus**: prismatic. **Color**: green, yellow. **Streak**: white. **Diaphaneity**: transparent to translucent. **Luster**: resinous. **Fracture**: subconchoidal, splintery. **Cleavage**: (010) good.
Vesuvianite (syn. idocrase, cyprine) (from Greek, eidos, appearance, krasis, mixture, owing to the resemblance of its crystals to other species, from Vesuvius volcano, Italy)	$Ca_{10}(Mg,Fe)_2Al_4[Si_2O_7]_2[SiO_4]_5(OH)_4F_4$, Coordinance Ca(8), Mg(6), Fe(6), Al(6), Si(4) (Neosilicates and sorosilicates)	Tetragonal, a = 1560 pm, c = 1180 pm (Z = 4), P.G. 4/mmm, S.G. Pnmc	Uniaxial (–), ε = 1.700–1.746, ω = 1.703–1.752, δ = 0.01–0.008, Dispersion strong	6–7	3330–3430	**Habitus**: prismatic. **Color**: reddish brown to green, emerald green (Cr-rich), reddish brown (Ti-rich), pale yellow (Cu-rich). **Luster**: vitreous (i.e., glassy), greasy. **Diaphaneity**: transparent to opaque. **Cleavage**: (110), (100), (001). **Fracture**: uneven. **Chemical**: attacked by strong mineral acids after calcination. **Deposits**: contact metamorphism.
Villiaumite [7681-49-4] (after the French traveller, Villium)	NaF, MM = 41.98817, 54.75wt% Na, 45.25wt% F (Halides)	Cubic, a = 463.42 pm, B1, cF8 (Z = 4), S.G. Fm3m, Rock salt type	Isotropic, n_D = 1.327	2.5	2785	**Habitus**: massive, granular. **Color**: crimson red, dark cherry red, or colorless. **Luster**: vitreous (i.e., glassy). **Diaphaneity**: transparent. **Luminescence**: fluorescent. **Streak**: pinkish white. **Cleavage**: (100), (010), (001). **Fracture**: brittle. **Deposits**: nepheline-syenite rocks. Soluble in water. Fusible (mp 996°C).

11
Minerals, Ores and Gemstones

Table 11.4 (continued)

Mineral name (IMA) [CAS RN] (Synonyms) (Etymology)	Theoretical chemical formula, relative molecular mass, coordination number, major impurities, mineral class (Strunz)	Crystal system, lattice parameters, Strukturbericht, Pearson symbol, Z, point group, space group, structure type	Optical properties	Mohs hardness (HM)	Density (ρ, kg.m^{-3})	Other mineralogical, physical, and chemical properties
Violarite	FeNi$_2$S$_4$ MM = 301.475 18.52wt% Fe 38.94wt% Ni 42.24wt% S (Sulfides and sulfosalts) Coordination Ni(6), Fe(4)	Cubic a = 946.4 pm H1$_1$, cF56 (Z = 8) S.G. Fd-3 P.G. 4-32 Spinel type	Isotropic R = 45%	4.5–5.5	4500–4800	**Habitus:** octahedral crystals. **Color:** violet gray. **Diaphaneity:** opaque. **Luster:** metallic. **Streak:** gray. **Cleavage:** perfect (111). **Twinning:** {111}. **Fracture:** uneven, conchoidal.
Vivianite	Fe$_3$(PO$_4$)$_2$·8H$_2$O MM = 596.57180 28.08wt% Fe 15.58wt% P 53.64wt% O 2.70wt% H Coordination Fe(6), P(4) (Phosphates, arsenates, and vanadates)	Monoclinic a = 1008 pm b = 1343 pm c = 470 pm β = 104.5° (Z = 2) P.G. 2/m S.G. C2/m	Biaxial (+) α = 1.579 β = 1.603 γ = 1.633 δ = 0.054 2V = 83°	2	2580	**Habitus:** reniform, lamellar, fibrous. **Color:** colorless green. **Streak:** white **Diaphaneity:** transparent to translucent. **Luster:** pearly, vitreous. **Fracture:** sectile. **Cleavage:** (010) perfect, (100) perfect.
Wavellite	Al$_3$(PO$_4$)$_2$(OH)$_3$·5H$_2$O MM = 506.957507 15.97wt% Al 18.33wt% P 63.12wt% O 2.58wt% H Coordination Al(6), P(4) (Phosphates, arsenates, and vanadates)	Orthorhombic a = 962 pm b = 1736 pm c = 699 pm (Z = 4) P.G. mmm S.G. P2$_1$/cmn	Biaxial (+) α = 1.525 β = 1.534 γ = 1.552 δ = 0.027 2V = 72°	3–4	2360	**Habitus:** globular, radiating, spherolitic. **Color:** white, yellow, green. **Streak:** white. **Diaphaneity:** translucent. **Luster:** vitreous. **Fracture:** subconchoidal. **Cleavage:** (101) perfect, (010) perfect.
Willemite (syn. Troostite) (after Willem I, King of the Netherlands)	Zn$_2$SiO$_4$ MM = 222.8631 58.68wt% Zn 12.60wt% Si 28.72wt% O (Nesosilicates)	Trigonal (Rhombohedral) a = 1394 pm c = 931 pm (Z = 18) P.G. $\bar{3}$	Uniaxial (+) ω = 1.691–1.72 ε = 1.719–1.73 δ = 0.010–0.028	5.5	3900–4200	**Habitus:** prismatic, massive-granular, massive. **Color:** white, yellow, green, reddish brown, or black. **Luster:** vitreous-resinous. **Diaphaneity:** transparent to translucent to opaque. **Streak:** white. **Luminescence:** fluorescent, green under short UV wavelength. **Cleavage:** (0001) poor, (1120) poor. **Deposits:** main ore mineral of franklin, a metamorphosed zinc orebody. **Fracture:** uneven.
Witherite [513-77-9] (after the English natural historian, W. Withering)	BaCO$_3$ MM = 197.336 77.54wt% Ba 4.49wt% C 17.96wt% O Coordination Ba(6), C(3) (Nitrates, carbonates, and borates)	Orthorhombic a = 643.0 pm b = 890.4 pm c = 531.4 pm (Z = 4) S.G. Pmcn P.G. mmm	Biaxial (-) α = 1.529 β = 1.676 γ = 1.677 δ = 0.148 2V = 16° Dispersion weak	3.5	4290–4300	**Habitus:** pseudo hexagonal, columnar, globular, reniform, fibrous. **Color:** colorless, milky white, grayish white, pale yellowish white, or pale brownish white. **Luster:** vitreous (i.e., glassy). **Diaphaneity:** transparent to translucent. **Streak:** white. **Fracture:** subconchoidal. **Cleavage:** (010), (110) distinct. **Twinning:** [110]. **Chemical:** decomposes at 1555°C giving off CO$_2$. Soluble in diluted HCl with evolution of CO$_2$.

Name	Chemical composition	Crystallography	Optical properties	Hardness	Density	Properties
Wolframite [13870-24-1] (from the German, Wolfram, name for tungsten)	$Fe_{0.5}Mn_{0.5}WO_4$, MM = 303.2291, 9.06wt% Mn, 9.21wt% Fe, 60.63wt% W, 21.10wt% O. Traces of V, Nb, Ta, Sc, Ti, Mo, Al, and In. (Sulfates, chromates, molybdates, and tungstates)	Monoclinic, a = 478.2 pm, b = 573.1 pm, c = 498.2 pm, 90.57° (Z = 2), Wolframite type (Ferberite-Hubnerite)	Biaxial (+), α = 2.20-2.26, β = 2.22-2.32, γ = 2.30-2.42, δ = 0.10-0.16, 2V = 73-90°, R = 16%, Dispersion none	4-4.5 (258-657 HV)	/510	**Habitus:** prismatic, lamellar, tabular, massive, granular. **Color:** reddish and brownish black to iron black. **Luster:** resinous to submetallic. **Diaphanity:** transparent to opaque. **Streak:** reddish brown. **Cleavage:** {010} perfect. **Fracture:** uneven, brittle. **Chemical:** attacked by conc. and hot HCl and H_2SO_4. **Other:** dielectric constant 12.51. Magnetic. **Deposits:** high-temperature quartz veins, in greisen, pegmatites and skarns.
Wollastonite	$Ca[SiO_3]$, Coordinance Ca(6), Si(4) (Inosilicates, chain)	Triclinic, a = 794 pm, b = 732 pm, c = 707 pm, α = 90.03°, β = 95.37°, γ = 103.43° (Z = 4), P.G. $\bar{1}$, S.G. $P\bar{1}$, Pyroxenoid group	Biaxial (−), α = 1.620, β = 1.632, γ = 1.634, δ = 0.014, 2V = 39°, Dispersion weak	4.5-5	3100	**Habitus:** prismatic, needlelike. **Color:** white. **Luster:** silky. **Diaphanity:** transparent to translucent. **Streak:** white. **Cleavage:** {100} perfect, (001) distinct. **Twinning:** {010}. **Fracture:** uneven.
Wulfenite [10190-55-3] (syn., yellow lead ore) (after the Austrian mineralogist, F.X. Wulfen)	$PbMoO_4$, MM = 367.1376, 26.13wt% Mo, 56.44wt% Pb, 17.43wt% O. Coordinance Pb(6), Mo(4) (Sulfates, chromates, molybdates, and tungstates)	Tetragonal, a = 543.5 pm, c = 1211.0 pm (Z = 4), P.G. 4/m, S.G. $I4_1/a$, Scheelite type	Uniaxial (−), ε = 2.283, ω = 2.404, δ = 0.121	2.9	6750-7000	**Habitus:** tabular, massive, granular, bipyramidal pseudooctahedron. **Luster:** resinous, greasy, or adamantine. **Diaphanity:** subtransparent to subtranslucent. **Color:** orange-yellow, waxy yellow, yellowish gray, olive green, or brown. **Streak:** yellowish white. **Cleavage:** {101} imperfect. **Twinning:** {001}. **Fracture:** subconchoidal, brittle.
Wurtzite (syn., radial blende, HT ZnS) (after the French chemist, Ch.A. Wurtze)	ZnS, MM = 97.456, 67.10wt% Zn, 32.90wt% S. Traces Fe (Sulfides and sulfosalts) Coordinance Zn(4)	Hexagonal, a = 382.30 pm, c = 625.65 pm, B4, hP4 (Z = 2), S.G. $P6_3mc$, P.G. 6 mm, Wurtzite type	Uniaxial (+), ε = 2.356, ω = 2.378, δ = 0.022, R = 19%	3.5-4	4030	**Habitus:** pyramidal, radial, tabular, colloform. **Color:** orange red, light brown or dark brown. **Luster:** resinous. **Streak:** brown-yellow. **Cleavage:** (1010), (0001). **Fracture:** uneven. **Chemical:** attacked by strong mineral acids, such as HCl, or HNO_3, with evolution of H_2S and yellow precipitate of sulfur. Infusible.
Xenotime (from the Greek, xenos, foreign, and time, honor, owing to the rarity and the small size of its crystals)	YPO_4, MM = 183.87721, 48.35wt% P, 16.84wt% P, 34.80wt% O. Traces of U, Th, Zr, Si, and rare earths. Coordinance Y(6), P(4) (Phosphates, arsenates, and vanadates)	Tetragonal, a = 688.5 pm, c = 598.2 pm, $I4_1/amd$ (Z = 4), P.G. 4/mmm, S.G. $I4_1/amd$, Zircon type	Uniaxial (+), ε = 1.720-1.721, ω = 1.816-1.827, δ = 0.095-0.107	4.5-5	4750	**Habitus:** radial, prismatic, massive, and granular. **Color:** yellowish brown, greenish brown, gray, reddish brown, or brown. **Luster:** vitreous (i.e., glassy), greasy or resinous. **Diaphanity:** translucent to opaque. **Streak:** pale brown. **Cleavage:** (100) perfect. **Fracture:** uneven, splintery. **Chemical:** insoluble in strong mineral acids.

11

Minerals, Ores and Gemstones

Table 11.4 *(continued)*

Mineral name (IMA) [CAS RN] (Synonyms) (Etymology)	Theoretical chemical formula, relative molecular mass, coordinance number, major impurities, mineral class (Strunz)	Crystal system, lattice parameters, Strukturbericht, Pearson symbol, Z, point group, space group, structure type	Optical properties	Mohs hardness (HM)	Density (ρ, kg.m^{-3})	Other mineralogical, physical, and chemical properties
Zincite [1314-13-2] (syn, red zinc oxide) (from the German, zink)	(Zn,Mn)O MM = 81.3894 80.34wt% Zn 19.66wt% O (Oxides and hydroxides) Coordinance Zn(4)	Trigonal (Hexagonal) a = 660.4 pm c = 597.9 pm B4, hP4 (Z = 2) S.G. P63mc P.G. 6mm Wurtzite type	Uniaxial (+) ε = 2.013 ω = 2.029 δ = 0.016	4–4.5	5560	**Habitus:** massive, fibrous, granular, disseminated. **Color:** deep red or yellowish orange. **Diaphaneity:** translucent to translucent. **Luster:** submetallic, subadamantine. **Streak:** yellowish orange. **Cleavage:** (0001). **Parting:** [110]. **Fracture:** subconchoidal. **Deposits:** metamorphosed weathered ore deposit.
Zircon [10101-52-7] (syn, hyacinthe: orange, red, malacon: white) (from Arabic, zar, gold, and gum, color)	Zr[SiO$_4$] MM = 183.3071 49.77wt% Zr 15.32wt% Si 34.91wt% O Traces of Hf, U, Th Coordinance Zr(4), Si(4) (Nesosilicates)	Tetragonal a = 660.4 pm c = 597.9 pm (Z = 4) P.G. 4/mmm S.G. I4$_1$/amd Zircon type	Uniaxal (+) ε = 1.923–1.960 ω = 1.968–2.015 δ = 0.042–0.065 Dispersion very strong	7.5	4600–4700 (3900–4200 metamicte)	**Habitus:** prismatic, tabular, crystalline. **Color:** brown, reddish brown, colorless, gray, green. **Diaphaneity:** transparent, translucent, opaque. **Luster:** adamantine. **Streak:** white. **Fracture:** uneven. **Cleavage:** (110). **Twinning:** {111}. **Chemical:** dissociates at 1540°C in SiO$_2$ and ZrO$_2$. Insoluble in HCl and HNO$_3$, slightly soluble in conc. H$_2$SO$_4$, readily dissolved by HF. Radioactive owing to the isomorphic substitution of Zr(IV) cations by U(IV) and Th(IV).
Zoisite (syn, tanzanite: blue, thulite: pink or red, anyolite: green) (after the Austrian natural scientist, S.Von Zoïs)	Ca$_2$Al$_3$O(OH)[SiO$_4$][Si$_2$O$_7$] MM = 427.37572 18.76wt% Ca 12.63wt% Al 19.71wt% Si 0.24wt% H 48.67wt% O Coordinance Ca(6, 9), Fe(6), Si(4) (Sorosilicates and nesosilicates)	Orthorhombic a = 1615.0 pm b = 558.1 pm c = 1006.0 pm (Z = 4) P.G. 2/m S.G. P2$_1$/m Epidote group	Biaxial (–) α = 1.685–1.705 β = 1.688–1.710 γ = 1.697–1.725 δ = 0.004–0.008 2V = 0–60° Dispersion strong	6–6.5 (680 HV)	3150–3270	**Habitus:** prismatic, striated, columnar. **Color:** gray, apple green, brown, blue, or rose red. **Diaphaneity:** transparent to translucent. **Luster:** vitreous (i.e, glassy), pearly. **Streak:** white. **Cleavage:** (100), perfect, (001) imperfect. **Twinning:** {100}. **Fracture:** uneven. **Chemical:** insoluble in strong mineral acids, fusible giving a white globule. **Deposits:** regional metamorphic and pegmatite rocks.

Astrakhanite = bloedite
Aventurine = quartz + mica
Badenite = bismuth + safflorite + modderite
Baikalite = diopside
Barkevikite = Fe-hornblende
Barsanovite = eudialyte
Baryte = barite
Bastonite = biotite
Bauxite = hydroxides and oxides of Al and Fe
Bellite = mimetite
Belorussite = byelorussite
Bentonite = montmorillonite
Bertonite = bournonite
Binnite = tennantite
Bisbeeite = plancheite-chrysocolla
Blackjack = sphalerite
Blanchardite = brochantite
Bleiglanz = galena
Blende = sphalerite
Blockite = penroseite
Bloodstone = chalcedony
Borickite = delvauxite
Boronatrocalcite = ulexite
Brandisite = clintonite
Braunbleierz = pyromorphite
Bravoite = Ni-pyrite
Breislakite = vonsenite (ludwigite, fibrous ilvaite)
Breunnerite = Fe-magnesite
Brocenite = Ce-fergusonite
Broggerite = Th-uraninite
Bromlite = alstonite
Bromyrite = bromargyrite
Bronzite = Fe-enstatite
Buratite = aurichalcite
Byssolite = actinolite-tremolite
Cabrerite = Mg-annabergite
Cairngorm = smoky quartz
Calamine = hemimorphite
Calciumlarsenite = esperite
Californite = vesuvianite
Campylite = P-mimetite
Canbyite = hisingerite
Carbonado = black diamond
Carbonytrine = Y-tengerite
Carborundum = synthetic moissanite
Carnelian = carneol (cornaline, agate, quartz)
Carneol (carnelian, cornaline) = red chalcedony (quartz)
Carpathite = karpatite (coronene)
Carphosiderite = hydrogeno-jarosite
Caryocerite = Th-melanocerite
Cathophorite = brabantite
Catoptrite = katoptrite
Cenosite = kainosite
Centrallasite = gyrolite

(*From page 409*)

Cerargyrite = chlorargyrite
Chalcedony = quartz
Chalcolite = torbernite
Chalcosine = chalcocite
Chalcotrichite = cuprite
Challantite = ferricopiapite
Chalmersite = cubanite
Chalybite = siderite
Chathamite = Fe-skutterudite
Chavesite = monetite
Chengbolite = moncheite
Chessylite = azurite
Chiastolite = andalusite
Chile saltpeter nitratine (soda niter, nitronatrite)
Chileloeweite = humberstonite
Chloanthite = nickelskutterudite
Chlorastrolite = pumpellyite
Chlormanasseite = altered koenenite
Chloromelanite = jadeite
Chloropal = nontronite
Chlorotile = agardite
Christensenite = tridymite
Christianite = anorthite or phillipsite
Chromrutile = redledgeite
Chrysolite = olivine
Chrysoprase = green chalcedony (quartz)
Cinnamon stone = hessonite (grossular)
Cirrolite = attacolite (bearthite, lazulite, cyanite)
Citrine = yellow quartz
Cleavelandite = albite
Cliftonite = graphite
Clinobarrandite = Al-phosphosiderite
Clinoeulite = Mg-clinoferrosilite
Clinostrengite = phosphosiderite
Cobaltoadamite = adamite
Cobaltocalcite = sphaerocobaltite
Coccinite = moschelite
Coccolite = diopside
Cocinerite = chalcocite + silver
Collophanite = carbonated apatite
Colophonite = andradite
Columbite = ferrocolumbite (magnocol, maganocol)
Comptonite = thomsonite
Connarite = garnierite
Copperas = melanterite
Corindon = corundum
Cornaline = agate (quartz)
Corundophilite = clinochlore
Corynite = Sb-gersdorffite
Coutinite = Nd-lanthanite
Crestmoreite = tobermorite + wilkeite
Crocidolite = riebeckite
Csiklovaite = tetradymite + galenobismutite + bismuthinite
Cyanite = kyanite

Cyanose = chalcanthite
Cymatolite = muscovite + albite
Cymophane = chrysoberyl
Cyrtolite = zircon
D'ansite = dansite
Dahllite = carbonated hydroxylapatite
Dakeite = schroeckingerite
Damourite = muscovite
Danaite = Co-arsenopyrite
Daphnite = Mg-chamosite
Dashkesanite = K-hastingsite
Davisonite = apatite + crandallite
Dehrnite = carbonate-fluorapatite
Delatorreite = todorokite
Delessite = Mg-chamosite
Delorenzite = tanteuxenite
Deltaite = crandallite + hydroxy-apatite
Deltamooreite = torreyite
Demantoide = andradite
Dennisonite = davisonite (apatite + crandallite)
Desmine = stilbite
Destinezite = diadochite
Dewalquite = ardennite
Deweylite = clinochrysotile-lizardite
Diabantite = Fe-clinochlore
Diallage = diopside
Dialogite = rhodochrosite
Diatomite = tripolite (opaline silica)
Dipyre = scapolite
Disthene = kyanite
Djalmaite = uranmicrolite
Donatite = chromite + magnetite
Dornbergite (doernbergite) = bottinoite
Doverite = synchysite
Droogmansite = kasolite
Dubuissonite = montmorillonite
Duporthite = talc + chlorite
Durdenite = emmonsite
Dysanalyte = Nb-perovskite
Dysodile = resin
Eardleyite = takovite
Eastonite phlogopite + serpentine
Ekbergite = wernerite
Elaterite = mineral rubber
Electrum = Au–Ag alloy
Eleonorite = beraunite
Ellestadite = fluorellestadite-hydroxyellestadite
Ellsworthite = uranpyrochlore
Embolite = Br-chlorargyrite or Cl-bromargyrite
Emerald = beryl
Endellite (hydrohalloysite) = halloysite
Endlichite = As-vanadinite
Enigmatite = aenigmatite
Epidesmine = stilbite

Epiianthinite = schoepite
Eschynite = aeschynite
Eucolite = eudialyte
Fahlore = tetrahedrite-tennantite
Fassaite = diopside or augite
Femolite = Fe-molybdenite
Fengluanite = isomertieite
Ferchevkinite = Fe-chevkinite
Fernandinite = bariandite + roscoelite + gypsum
Ferridravite = povondraite
Ferrifayalite = laihunite
Ferrithorite = thorite + Fe hydroxide
Ferutite = davidite
Fibrolite = sillimanite
Flint (silex)= quartz
Fluorichterite = richterite
Fluorspar = fluorite
Forbesite = annabergite + arsenolite
Forcherite = opal
Fowlerite = Zn-rhodonite
Francolite = carbonated fluoroapatite
Freirinite = lavendulan
Fuchsite = Cr-muscovite
Fuggerite = melilite
Genevite = vesuvianite (idocrase)
Genthite = garnierite
Geyserite = opal
Gilpinite = johannite
Ginzburgite = roggianite
Giobertite = magnesite
Girasol = opaline quartz
Glagerite = halloysite
Glaserite = apthitalite
Glauber's salt = mirabilite
Glaukolite (glavcolite) = scapolite
Glockerite = lepidocrocite
Goongarrite = heyrovskyite
Gorgyite = goergeyite
Goshenite = beryl
Grammatite = tremolite
Grandite = grossular-andradite
Griffithite = Fe-saponite
Grunlingite (Gruenlingite) = joseite + bismuthinite
Gummite = secondary uranium oxides
Hackmanite = sodalite
Hatchettite = hydrocarbons mixture
Hatchettolite = uranpyrochlore
Heliodor = beryl
Heliotrope = chalcedony or plasma = quartz
Hemafibrite = synadelphite
Hematoide = quartz + goethite-hematite
Hessonite = grossular
Heubachite = Ni-heterogenite
Hexagonite = Mn-tremolite

Hexastannite = stannoidite
Hiddenite = spodumene
Hjelmite = tapiolite + pyrochlore
Hokutolite = Pb-barite
Hortonolite = Mg-Mn-fayalite
Hoshiite = Ni-magnesite
Huehnerkobelite = alluaudite or ferroalluaudite
Humboldtilite = melilite
Hyacinth = orange zircon
Hyacinthe de Compostelle = amethyste (quartz)
Hyalite = opal
Hyaloallophane = allophane + hyalite
Hyalosiderite = fayalite
Hydrargillite = gibbsite
Hydrated halloysite = endellite
Hydrogrossular = hibschite-katoite
Hydrohalloysite = endellite
Hydromica = brammallite, hydrobiotite, illite
Hydrophilite = antarcticite or sinjarite
Hydrotroilite = colloidal hydrous ferrous sulfide
Idocrase = vesuvianite
Iglesiasite = cerussite + smithsonite
Indicolite = indigolite = elbaite
Iodobromite = I-bromargyrite
Iodyrite = Iodargyrite
Iolite = cordierite
Iozite = wuestite
Iridosmine = iridium–osmium alloy
Iron-cordierite = sekaninaite
Iserine (nigrine) = illmenite + rutile
Isoplatinocopper = hongshiite
Isostannite = kesterite-ferrokesterite
Jade = jadeite (nephrite)
Jasper = quartz
Jefferisite = vermiculite
Jeffersonite = Mn Zn acmite or augite
Jelletite = andradite
Jenkinsite = Fe-antigorite
Johnstrupite = mosandrite
Josephinite = awaruite, kamacite, taenite, tetrataenite
Kallilite = ullmannite
Kamarezite = brochantite
Kammererite (kaemmererite) = Cr-clinochlore
Karafveite = monazite
Karpinskyite = leifite + clay
Kasoite = celsian
Katayamalite = baratovite
Keeleyite = zinkenite
Keilhauite = yttrian titanite
Kellerite = Cu-pentahydrite
Kennedyite = armalcolite or pseudobrookite
Kerchenite = metavivianite
Kerolite = talc
Kertschenite = oxidation product of vivianite

Khlopinite = Ta-samarskite
Klaprothite = lazulite
Kleberite = pseudorutile
Klipsteinite = altered rhodonite
Knebelite = Mn-fayalite
Knipovichite = Cr-alumohydrocalcite
Knopite = perovskite
Kolskite = lizardite + sepiolite
Koppite = pyrochlore
Kotschubeite = Cr-clinochlore
Kramerite = probertite
Kularite = Ce-monazite
Kunzite = spodumene
Kurskite = carbonated fluorapatite
Lapis lazuli = lazurite
Lapparentite = khademite (rostite)
Laubmannite = dufrenite + kidwellite + beraunite
Lavrovite = diopside
Lazarevicite = As-sulvanite
Lehiite = apatite + crandallite
Leonhardtite = starkeyite
Lepidomelane = Fe-biotite
Lesserite = inderite
Lettsomite = cyanotrichite
Leuchtenbergite = clinochlore
Leucochalcite = olivenite
Leucoxene = alterartion of illmenite
Leverrierite = kaolinite + illite
Lewistonite = carbonated fluorapatite
Lievrite = illvaite
Lingaitukuang = brabantite
Liujinyinite = uytenbogaardtite
Lotrite = pumpellyite
Lovchorrite = mosandrite
Lunijianlaite = cookeite + pyrophyllite
Lunnite = pseudomalachite
Lusakite = Co-staurolite
Lusungite = goyazite
Lydian stone (basanite, touchstone) = quartz
Macconnellite = mcconnellite
Mackintoshite = thorogummite
Magnophorite = Ti-K-richterite
Maitlandite = thorogummite
Malacon = zircon
Mangan neptunite = manganneptunite
Manganocalcite = calcite
Manganomelane = manganese oxides
Manganophyllite = Mn-biotite
Marignacite = Ce-pyrochlore
Mariposite = Cr-phengite
Marmatite = Fe-sphalerite
Martite = hematite pseudomorph on magnetite
Maskelynite = glass of plagioclase composition
Mauzeliite = Pb-romeite

Medmontite = chrysocolla + mica
Melaconite = tenorite
Melanite = Ti-andradite
Melanochalcite = tenorite (chrysocolla, malachite)
Melinose = wulfenite
Melnikovite = greigite
Menilite = opal
Meroxene = biotite
Merrilite = whitlockite
Mesitite = Fe-magnesite
Mesotype = natrolite (mesolite, scolecite)
Metahalloysite = halloysite
Metastrengite = phosphosiderite
Metauranopilite = meta-uranopilite
Minette = iron hydroxides and oxides
Miomirite = Pb-davidite
Mispickel = arsenopyrite
Mizzonite = marialite-meionite
Moganite = fine crystalline quartz
Mohsite = Pb-crichtonite
Monheimite = smithsonite
Monsmedite = voltaite
Montdorite = biotite
Montesite = Pb-herzenbergite
Morencite = nontronite
Morganite = beryl
Morion = quartz
Mossite = tantalite-tapiolite
Muchuanite = altered molybdenite
Mushketovite = magnetite pseudomorph on hematite
Nanekevite = Ba-orthojoaquinite
Nasturan = pitchblende
Nemalite = Fe-brucite
Nenadkevite = uraninite + boltwoodite
Neotype = barytocalcite
Nephrite = actinolite
Nevyanskite = iridosmine
Niccolite = nickeline
Nickeliron = kamacite (taenite, tetrataenite)
Nigrine (Iserine) = illmenite + rutile
Nimesite = brindleyite
Niobite = ferrocolumbite
Niobozirconolite = Nb-zirkelite
Nitroglauberite = darapskite + nitratine
Nitrokalit = niter (saltpeter, nitre, saltpetre)
Nitronatrite = nitratine (soda niter)
Nocerite = fluoborite
Nuevite = samarskite
Nuttalite = wernerite
O'danielite = odanielite
Obruchevite = Y-pyrochlore
Yellow ochre = limonite
Octahedrite = anatase
Oellacherite = Ba-muscovite

Oligiste = hematite
Oligonite = Mn-siderite
Olivine = peridot
Onofrite = Se-metacinnabar
Onyx = layered chalcedony (quartz)
Orthite = allanite
Orthose = orthoclase
Osmiridium = iridium–osmium alloy
Outremer = ultramarine (lazurite)
Ozocerite = hydrocarbon mixture
Pageite = vonsenite
Paigeite = hulsite
Panabase = tetrahedrite
Pandaite = Ba-pyrochlore
Pandermite = priceite
Paranthine = wernerite
Partridgeite = bixbyite
Paternoite = kaliborite
Paulite = hypersthene
Pennine (penninite) = clinochlore
Pericline = albite
Peridot = forsterite
Peristerite = albite
Perthite = intergrowth orthoclase + plagioclase
Phacolite = chabazite
Pharaonite = microsommite
Phengite = muscovite
Piazolite = hydrogrossular
Picotite = Cr-spinel
Picrochromite = magnesiochromite
Pinite = altered cordierite
Pisanite = Cu-melanterite
Pisekite = monazite
Pistacite = epidote
Pitchblende = uraninite
Plasma = green quartz
Platiniridium = iridium–platinium alloy
Pleonaste = Fe-spinel
Plessite = kamacite-taenite
Plumbago = graphite
Plumosite = boulangerite
Polianite = pyrolusite
Polyadelphite = andradite
Porcelainite = mullite
Prase = green quartz
Priorite = aeschynite
Pseudowavellite = crandallite
Psilomelane = romanechite
Ptilolite = mordenite
Pycnite = topaz
Pyralspite = garnet subgroup
Quercyite = carbonate apatite
Rashleighite = Fe-turquoise
Resinite = opal

Rhodusite = Mg-riebeckite
Rijkeboerite = Ba-microlite
Ripidolite = Fe-clinochlore
Risorite = fergusonite
Rock salt = halite
Roepperite = fayalite or tephroite
Rostite = khademite
Rozhkovite = Pd-auricupride
Rubellane = altered biotite
Rubellite = elbaite
Ruby = corundum
Ruby silver = proustite, pyrargyrite
Ruthenosmiridium = iridium-osmium-ruthenium alloy
Sagenite = rutile
Sal ammoniac = salmiac
Salite = sahlite = diopside
Salmiac = salammoniac (sal ammoniac)
Saltpeter = niter (nitre, saltpetre, nitrokalit)
Samiresite = Pb-uranpyrochlore
Sapphire = corundum
Sard = brown chalcedony
Saukovite = Cd-metacinnabar
Saussurite = zoisite (scapolite)
Schefferite = Mn-aegirine
Scheibeite = phoenicochroite
Schizolite = Mn-pectolite
Schoenite = picromerite
Schuchardtite = clinochlore (vermiculite)
Schwatzite (Schwazite) = Hg-tetrahedrite
Schweizerite = serpentine
Selenite = gypsum
Sericite = muscovite
Seybertite = clintonite
Sheridanite = clinchlore
Sideretine = pitticite
Siderochrome = chromite
Siserskite (syserskite) = iridosmine
Smaltite = skutterudite
Smoky quartz = brown quartz
Sobotkite = Al-saponite
Soda felspar = albite
Soda niter = nitratine (nitronatrite)
Sophiite = sofiite
Spartalite = zincite
Specularite = hematite
Sphene = titanite
Staffelite = carbonated flurapatite
Staringite = cassiterite + tapiolite
Stassfurtite = boracite
Steatite = talc
Steinsalz = Halite
Stibine = stibnite (antimonite)
Strahlstein = actinolite •
Succinite = amber

Sukulaite = Sn-microlite
Sylvinite = halite + sylvite
Syssertskite = osmium
Taaffeite = musgravite
Tagilite = pseudomalachite
Tanzanite = zoisite
Tarasovite = mica-smectite M
Tarnowitzite (tarnowskite) = Pb-aragonite
Tavistockite = apatite
Tellurbismuth = Te-bismuthite
Ternovskite = Mg-riebeckite
Teruelite = dolomite
Thulite = zoisite
Thuringite = Fe-chamosite
Tibiscumite = allophane
Tiger eye = quartz + crocidolite
Tincal = borax
Toddite = columbite + samarskite
Topazolite = andradite
Treanorite = allanite
Triphane = spodumene
Troostite = Mn-willemite
Trudellite = chloraluminite + natroalunite
Tsavolite = tsavorite = grossular
Turgite = turjite (hematite)
Turnerite = monazite
Ufertite = davidite
Ugrandite = garnet subgroup
Ultramarine = lazurite
Ulvite = ulvospinel
Uralite = amphibole
Uranite = autunite
Uranotile = uranophane
Uranotile beta = uranophane beta
Vegasite = Pb-jarosite
Verdelite = tourmaline
Vernadskite = antlerite
Vibertite = bassanite
Viridine = Mn-andalusite
Voltzite = wurtzite
Vorobievite = beryl
Vredenburgite = jacobsite + hausmannite
Vulpinite = anhydrite
Wad = manganese oxides
Walchowite = resinoid
Warrenite = owyheeite (jamesonite)
Westgrenite = Bi-microlite
Wiikite = Y-pyrochlore + euxenite
Wilkeite = apatite (fluorellestadite)
Williamsite = antigorite
Withmamite = piemontite
Wolframite = huebnerite-ferberite
Wood Tin = cassiterite
Yttroorthite = Y-allanite

Zeiringite = aragonite + aurichalcite
Zigueline = chalcopyrite + cuprite + limonite + cinnabar
Zinconine = hydrozincite
Zinkblende = sphalerite

References

[1] Mohs F (1824) Grundriss der mineralogie.
[2] Staples LF (1964) Friedrich Mohs and the scale of hardness. J Geol Educat: 98–101.
[3] Ridgeway RR, Ballard AH, Bailey BL (1933) Trans Electrochem Soc 63: 267.
[4] Povarennikh AS (1962) A fifteen division Mohs scale of hardness. Zap Ukr Otd Vses Mineralog Obshchestva Akad Nauk Ukr SSSR 1: 67–74.
[5] Povarennikh AS (1964) Necessary revisions to be made in the Mohs scale of hardness. Dopovidi Akad Nauk Ukr SSSR 6: 804–806.

Further Reading

Anthony JW, Bideaux RA, Bladh KW et al. (1990) Handbook of mineralogy, Vol. 1: Elements, sulfides, sulfosalts. Mineral Data Publishing.
Anthony JW, Bideaux RA, Bladh KW et al. (1995) Handbook of mineralogy, Vol. II: Silica, silicates. Mineral Data Publishing.
Aubert G, Guillemin C, Pierrot R (1978) Précis de minéralogie. Masson, Paris.
Babushkin VI, Matveyev, GM, Mchedlov-Petrossyan OP (1985) Thermodynamics of silicates. Springer-Verlag, New York.
Bardet M (1975) Le diamant, 2 vols. Éditions du BRGM, Orléans.
Batemann AM (1950) Economic mineral deposits. New York.
Bates RL (1960) Geology of the industrial rocks and minerals. Harper and Brothers Publishers, New York.
Bayliss P, Erd DC, Mrose ME et al. (1986) Mineral powder diffraction file, data book. International Centre for Diffraction Data.
Berry LG (ed.) (1983) Mineralogy: Concepts, descriptions, determinations, 2nd edn. Freeman and Co., San Francisco.
Blackburn WH, Dennen WH (1997) Encyclopedia of mineral names: special publication of the Canadian Mineralogist. Mineralogical Association of Canada, Ottawa, ON, Canada.
Bordet P (1968) Précis d'optique cristalline appliqué á l'identification des minéraux. Masson & Cie, Paris.
Caillére S, Hénin S, Rautureau M (1982) Minéralogie des argiles, Tome I: Structure et propriétés physico-chimiques, 2nd edn. Masson, Paris.
Caillére S, Hénin S, Rautureau M (1982) Minéralogie des argiles, Tome 2: Classification et nomenclature, 2nd edn. Masson, Paris.
Carmichael RS (1989) Practical handbook of physical properties of rocks and minerals. CRC Press, Boca Raton, FL.
Cavenago Bignami S (1964) Gemmologia. Edizioni Enrico Hoepli, Milano.
Clark AM (1993) Hey's mineral index: mineral species, varieties and synonyms, 3rd edn. Chapman & Hall, New York.
Criddle AJ, Stanley CJ (1993) The quantitative data file for ore minerals. Chapman & Hall, New York.
Dana JD (1944) Dana's system of mineralogy, 7th edn. John Wiley and Sons, New York.
Dana E, Salisbury F, William E (1949) A texbook of mineralogy, 4th edn. John Wiley and Sons, New York.
Deer WA, Howie RA, Zussman J (1962) Rock-forming minerals (5 vols); vol.1: Ortho- and ring-silicates; vol. 2: Chain silicates; vol.3: Sheet silicates, vol. 4: Framework silicates; vol. 5: Non-silicates. Longman, London.
Deer WA, Howie RA, Zussman J (1966) An introduction to the rock-forming minerals. Longman, London.
Deer WA, Howie RA, Zussman J (1992) An introduction to the rock-forming minerals, 2nd edn. Longman Scientific and Technical, Harlow, Essex.
Denayer ME (1955) Tableaux de pétrographie. Editions Lamarre-Poinat, Paris.
Dixon (1979) Atlas of economic mineral deposits. Chapman & Hall, New York.
Donnay JDH, Ondik HM (1973) Crystal data determinative tables, 3rd edn. Vol.2, Inorganic compounds. Joint Committee on Powder Diffraction Standards (JCPDS), Swarthmore, PA.
Embrey PG, Fuller JP (1983) A manual of new mineral names 1892–1978. British Museum. London.
Feklichev VG (1992) Diagnostic constants of minerals. CRC Press, Boca Raton, FL.
Fischesser R (1955) Données des principales espéces minérale. Éditions Sennac, Paris.
Fleischer M, Mandarino J (1995) Glossary of mineral species 1995. The Mineralogical Record Inc.
Gaines RV, Skinner HC, Foord EE et al. Dana's new mineralogy, 8th edn. John Wiley and Sons, New York.
Grill E (1963) Minerali industriali e minerali delle rocce. Edizioni Enrico Hoepli, Milano.
Hey MH (1963) First appendix to the second edition of an index of mineral species and varieties arranged chemically, 2nd edn. The British Museum, London.

Hey MH (1974) A second appendix to the second edition of an index of mineral species and varieties arranged chemically, 2nd edn. Trustees of the British Museum, London.

Hey MH (1975) An index of mineral species arranged chemically, 2nd edn. British Museum, London.

Jones AP, Williams CT, Wall F (1996) Rare earth minerals: chemistry, origin and ore deposits. Chapman & Hall, New York.

Kipfer A (1974) Mineralindex. Ott Verlag.

Klein C, Hurlbut CS (1985) Manual of mineralogy, 20th edn. John Wiley and Sons, New York.

Lapadu-Hargues P (1954) Précis de minéralogie. Masson & Cie, Paris.

Lafitte P (1957) Introduction á l' étude des roches métamorphiques et des gîtes métallifères. Masson & Cie, Paris.

Liebau F (1985) Structural chemistry of silicates. Springer-Verlag, Berlin.

Lindgren W (1940) Mineral deposits. New York.

Mange M, Maurer H (1992) Heavy minerals in color. Chapman & Hall, New York.

Manning DAC (1995) Introduction to industrial minerals. Chapman & Hall, New York.

Milovsky AV, Kononov OV (1985) Mineralogy. Mir Editions, Moscow.

Nickel EH, Nichols MC (1991) Mineral reference manual. Van Nostrand Reinhold, New York.

Parfenoff A, Pomerol C, Tourenq J (1970) Les minéraux en grain: méthodes d'étude et détermination. Masson and Cie, Paris.

Phillips WR, Griffen DT (1981) Optical mineralogy, the nonopaque minerals. Freeman, New York.

Picot P, Johan Z (1982) Atlas of ore minerals. BRGM, Orléans.

Raguin E (1961) Géologie des gîtes minéraux, 2nd edn. Masson & Cie, Paris.

Ramdohr P (1960) Die erzmineralien und ihre verwaschsungen. Akademie Verlag, Berlin.

Ramdohr P (1969) The ore minerals and their intergrowths, 3rd edn. Pergamon Press, New York.

Ramdohr P, Strunz H (1960) Handbuch der mineralogie. Akademie Verlag, Berlin.

Ramdohr P, Strunz H (1978) Klockmanns lehrbuch der mineralogie, 16th edn. Enke Verlag, Berlin.

Read PG (1988) Dictionary of gemmology 2nd edn. Butterworth, Oxford.

Roberts WL, Rapp GR, Weber J (1974) Encyclopedia of minerals. Van Nostrand Reinhold, New York.

Roberts L, Rapp GR, Cambell TJ (1990) Encyclopedia of minerals, 2nd edn. Van Nostrand Reinhold, New York.

Roubault M (1958) Géologie de l'uranium. Masson & Cie, Paris.

Roubault M (1960) Les minerais uranifères français (3 vols). Presses Universitaires de France (PUF), Paris.

Roubault M, Fabries, J, Touret J et al. (1963) Détermination des minéraux des roches au microscope polarisant. Editions Lamarre-Poinat, Paris.

Routhier P (1963) Les gisements métallifères (2 vols). Masson & Cie, Paris.

Strunz H (1966) Mineralogische tabellen. Akademische Verlags-Gesellschaft, Geest und Portig, Liepzig.

Strunz H (1982) Mineralogische tabellen, 8th edn. Geest und Portig.

Winchell AN (1959) Elements of optical mineralogy, 4th edn. John Wiley, New York.

Zoltai T, Stout JA (1984) Mineralogy, concepts and principles. Macmillan Publishing Company, New York.

Zoltai T, Stout J (1985) Mineralogy, concepts and principles. Burges Publishing Company, Minneapolis, MN.

12 Rocks and Meteorites

12.1 Introduction

Rocks represent the overall geological materials constituting the Earth's crust (lithosphere), which are commonly made from an aggregate of crystals of one or more minerals and/or glass. From a wide geological point of view, rocks can be either solid (e.g., granite, limestone, rock salt, and ice), or fluid (e.g. sand and volcanic ashes), or liquid (bitumen and oil), or gaseous (natural gas and hydrothermal fluids). The important discipline of Earth sciences which studies the rock formation processes, chemical composition and physical properties, is named **petrology** (from Latin, petrus, stone) sometimes called **lithology** in the old textbooks (from Greek, lithos, stone), while **petrography** sensu-stricto only classifies rocks and can be understood as simple taxonomy. There are several reasons to study, identify, and measure properties of the different types of rocks present in the Earth's crust. First of all, rock materials contain valuable mineral ore deposits and can also contain fossil fuels (e.g., oil, coal, and natural gas). Hence understanding of the different types of rocks is necessary in order to locate and recover these valuable economical resources. Secondly, from a civil engineering point of view, the knowledge of the physical properties of different rocks used in construction is important to select the most appropriate building materials. Actually, some rock types are more susceptible to slope failure (landslides) or structural failure (disintegration under pressure) than others. Consequently, it is necessary to know the characteristics of the underlying rocks when doing any major civil engineering construction. Thirdly, from an agricultural point of view, rocks are the basic geological material from which all soils are formed. Hence, the rock chemical composition strongly influences the nature of the soil and the types of vegetation which the soil can support. Finally, from an environmental point of view, rock type also influences the flow of water, a major necessity of life, above and below the ground surface.

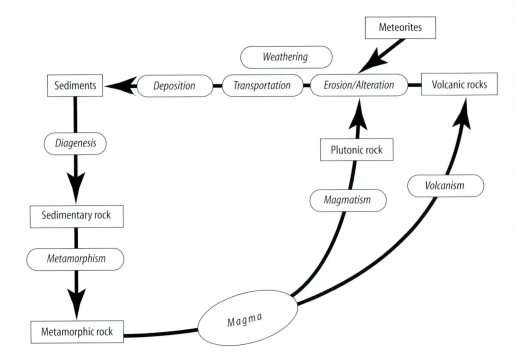

12.2 Types of Rocks

Rocks can be classified into one of four categories on the basis of their formation process: (1) **igneous** or **magmatic rocks** are formed by the cooling and solidification of a molten silicate bath (magma), (2) **sedimentary rocks** are produced by the weathering/erosion process (physical erosion and chemical alteration) of pre-existing rocks. After that, raw materials undergo three successive processes: (i) the transportation of degradation material by several media (water, wind or ice), (ii) the deposition as sediment, (iii) **diagenesis**[1] after which the sediment is cemented or not into a sedimentary rock. (3) **Metamorphic rocks** are rocks whose original form has been modified chemically and physically as a result of nature, high pressure, and hot fluids or a combination. Metamorphic rocks may form from the sedimentary, or previous metamorphic rocks. Finally, (4) **meteorites** are extraterrestrial materials coming from the solar system which are continually falling on Earth. The processes which produce the four general rock types, and the relationships between them are summarized in the rock cycle depicted in the figure.

12.3 Igneous Rocks

Igneous rocks (sometimes called **magmatic** or **endogeneous** rocks) are rocks resulting from the solidification on cooling of a molten silicate material called **magma** and occur in a wide variety of forms of different shapes and sizes. The magma is characterized by: (i) a chemical composition which is essentially a silicate melt generated by melting deep within the Earth's crust, (ii) a high melting temperature usually ranging between 500°C to 1500°C, and (iii) a mobility, i.e., ability to flow. Magma forms at depths of about 15 to 25 km, where temperatures are in the range 500–1500°C and lithostatic pressure around 1 GPa (10 kbar). The types of

---

[1] Diagenesis or lithification is the set of physical (e.g., pressure, temperature), chemical (e.g., dissolution, precipitation), or biological (e.g., fermentation) parameters which transform the unconsolidated sediment into a final rock.

igneous rocks that form from this magma depend generally on three factors: (1) original chemical composition of the magma, (2) the temperature at which the cooling begins, and finally (3) the cooling rate. According to the cooling depth, igneous rocks can be classified into two major subdivisions. (i) **Plutonic** or **intrusive rocks** are igneous rocks formed from a slow cooling rate of a magma as it rises through the Earth's crust forming very large crystalline bodies called **batholiths**. Hence, this means that the crystal size is medium or coarse, and the rock exhibits a so-called typical phaneritic texture (e.g., granite, syenite, or gabbro). Intrusive rocks occur in a variety of deposit forms. Vertical sheets of igneous rock are called **dykes**. Horizontal sheets, parallel or near parallel to layering are known as **sills**. Fatter pods of crystalline rock are called **laccoliths**. (ii) **Volcanic** or **extrusive rocks** are igneous rocks obtained from a rapid cooling rate of a magma formed if it reaches the surface of the Earth. Extrusive rocks occur as lava flows and pyroclastic ashes or debris that is ejected into the air during eruptions and are entirely related to volcanoes. These kinds of rocks often occur in characteristic volcanic cones. Submarine lava flows form characteristic pods called pillows. Therefore the rock texture exhibits a partially crystallized or totally amorphous texture called aphanitic and hyaline respectively (basalt, obsidian glass, and pumice).

Magma is a molten silicate medium which originates inside the Earth's crust probably due to the partial melting of the deep lithospheric material. Petrologists have identified two main classes of magmas from which igneous rocks are generally derived. **Hypersiliceous or felsic magmas** are silicate melts having a high silica content (above 65wt% SiO_2) and a low melting temperature range, usually 500–900°C. Due to these two chief characteristics these silica-rich magmas are highly viscous and hence move slowly and solidify slowly before reaching the Earth's surface. This type of magma leads principally to the formation of plutonic rocks (e.g., granite). By contrast, **hyposiliceous or mafic magmas** are silicate melts having a low silica content (45–52wt% SiO_2) and a high melting temperature range, usually 1100–1500°C. Due to these two chief characteristics these magmas are highly fluid and hence move quickly upward to the Earth's surface (as volcanoes) and they lead to the formation of lavas and pyroclastic products. As a general rule, owing to the initial chemical composition of the deep lithosphere, mother magmas are initially hyposiliceous melts; however, due to the fractional crystallization of ferromagnesian minerals (e.g., olivine, pyroxenes) occurring during cooling modifies their composition and becoming hypersiliceous. Igneous rocks that are rich in calcium, iron, and magnesium and relatively silica poor form from mafic magma. These rocks are generally dark in color and contain alkaline elements. Some common mafic igneous rocks include basalt, gabbro, and andesite. Rocks that contain relatively high quantities of sodium, aluminum and potassium and contain more than 65% silica originate from felsic magmas. The rocks created from felsic magma include granite and rhyolite. These rocks are light in color and are acidic in nature. Moreover, a silica-poor magma which solidifies at or near the Earth's surface would give the aphanitic rock basalt and would be composed of a very finely crystalline aggregate of the minerals peridot, pyroxenes, and plagioclase. If the same silica-poor magma were instead cooled more slowly at some depth within the crust, the overall chemical composition of the resulting rock would be the same as that of the basalt, but the rock's coarsely crystalline texture would instead classify it as a gabbro. Basalts and gabbros are considered extrusive and intrusive equivalents because while rhyolite has an extrusive texture and granite has an intrusive texture, they both have the same silica-poor chemical composition.

12

Rocks and Meteorites

12.3.1 Classification of Igneous Rocks

The petrographic classification of igneous rocks is essentially based on the following three main characteristics: their actual mineralogical composition, their texture and their chemical composition. The texture of an igneous rock is principally a function of the cooling rate of the mother magma, while mineralogical composition is both a function of chemistry and cooling history of a magma. Several quantitative parameters or indices are commonly used in order to help in the classification of igneous rocks.

12.3.1.1 Crystal Morphology and Dimensions

Table 12.1 Crystal dimensions

Order of magnitude of crystal size	Obsolete designation	Modern designation
Decimeter (dm)	Megablastes	Megacrystals
Centimeter (cm)	Porphyroblastes	Porphyrocrystals
Millimeter (mm)	Phaneroblastes	Phanerocrystals
Inframillimeter	Phenoblastes	Phenocrystals
Submillimeter	Spherolites, microlites	Spherocrystals, microcrystals
Micrometer (μm)	Crystalites	Cryptocrystals
Nonvisible	Mesostase	Glass, hyaline, amorphous

Table 12.2 Crystal development

Development	Obsolete designation		Modern designation	Example
Regular geometrical shape	Automorphous	Idiomorphous	Euhedral	
	Subautomorphous	Hypidiomorphous	Subhedral	
No regular geometrical shape	Xenomorphous	Allotriomorphous	Anhedral	

Table 12.3 Crystal proportion

Crystal proportion	Designation (Latin root)	Designation (Greek root)
Similar crystal dimensions	Equigranular	Isogranular
Different crystal dimension	Inequigranular	Heterogranular

Table 12.4 Crystal external shapes

External shape	Designation	Example
Grain	Massive	Quartz
Plate	Tabular	Pyroxenes
Flake	Lamellar	Micas
Prism	Columnar	Beryl, sillimanite
Fibrous, needlelike	Acicular	Rutile, asbestos

Table 12.5 Mineral composition

Category	Example
Essential minerals	Quartz, feldspars (i.e., orthoclase, Na and Ca plagioclases), olivine, pyroxenes, and amphiboles
Accessory minerals	Micas (i.e., biotite and muscovite)
Less common minerals	Rutile, titanite, apatite, beryl, tourmaline, zircon

12.3.1.2 Mineralogy

Igneous rocks are essentially composed of the six major rock-forming silicate minerals (i.e., constitute about 95% of all igneous rocks), which include peridots, pyroxenes, amphiboles, feldspars, micas, and quartz. Other minerals may also be present, but they usually make up only a small fraction of the rock. These less abundant and less common minerals are often called accessory minerals (e.g., micas). The **modal composition** corresponds to the actual mineralogical composition measured statistically under the polarizing microscope with a thin section of the rock, while the **normal composition** CIPW[1] is the ideal mineralogical composition calculated from the quantitative chemical analysis of the rock. The former allows the mineralogy of partially or hyaline crystallized rocks to be estimated, i.e., if the mother magma was able to solidify slowly.

N.L. Bowen devised the following crystallization series to explain the origin of all igneous rock types from a single parent magma. He observed that minerals crystallized at different temperatures, and reasoned that all igneous rocks could form from the parent basaltic magma through reaction and removal of certain crystals during cooling. Hence he summarized the temperature sequence of mineral crystallization in what is called Bowen's reaction series. Initial mafic molten silicate liquids tend to crystallize minerals at relatively high temperatures to form mafic or ferromagnesian minerals rich in iron, magnesium, and calcium. These minerals (olivine, pyroxene, and calcium-rich plagioclase feldspar) tend to react with the silica-enriched melt to form the next mineral in the sequence and then crystallize in a specific sequence according to their melting point. Intermediate and silicic liquids crystallize at lower temperatures to form minerals which are richer in silica, sodium, and potassium. These pale-colored minerals (quartz, sodium-rich plagioclase feldspar, potassium feldspar) also crystallize in a specific sequence.

Olivine
Mg pyroxene   Calcic plagioclase
Mg–Ca pyroxene   Calcic-alkaline plagioclase
Amphibole   Alkaline-calcic plagioclase
Biotite   Alkaline plagioclase
Potassium feldspar
Muscovite
Quartz

12

Rocks
and
Meteorites

Table 12.6 Crystallization sequence (Rosenbuch)

Early minerals	Apatite, zircon, sphene
	Pyroxenes, amphiboles, biotite
	Alkaline plagioclases
	Neutral plagioclases
	Acidic plagioclases
	Albite, orthose
Late minerals	Quartz

Table 12.7 Minerals according to density (E. Lacroix)

Density (d, kg.m^{-3})	Class
2770	Coupholithes
>2770	Barylites

12.3.1.3 Coloration

The coloration of igneous rocks is a dimensionless quantity introduced by the French geologist Élie de Beaumont. It corresponds to the surface fraction of white minerals in a thin section of the rock which is accurately determined using statistical counting methods. From a practical point of view, the coloration is determined under the microscope observing a thin section of the igneous rock under nonpolarized light. However, although this technique gives good results with igneous rocks having coarse grain size, igneous rocks having a fine crystal size, with individual crystals too small to identify need a less exact method. In this particular case, coloration is obtained from the calculated mineralogical composition of the rock (modal analysis) using the two following formulae: for saturated rocks, coloration = 100 – %(quartz + feldspars), while for subsaturated rocks coloration = 100 – %(feldspars + feldspathoids). This method generally associates higher silica content with lighter rock color. Lighter-colored rocks are felsic or silicic which means that they contain abundant feldspars and quartz. Darker-colored, mafic rocks are richer in mafic minerals such as olivine, pyroxene, and amphibole. These mafic minerals are also called ferro-magnesian minerals because they contain relatively large amounts of iron and magnesium. Mafic igneous rocks like basalt and gabbro also contain significant amounts of plagioclase feldspars. Ultramafic rocks are composed entirely or almost entirely of mafic minerals. Rocks with color and composition between mafic and felsic such as andesite are considered intermediate. Glassy rocks can be an exception to this generalization about color.

Table 12.8 Coloration (E. Beaumont)

Designation	Fraction of white minerals (%)	Examples
Hololeucocrates	95–100	Leucogranite
Leucocrates	65–95	Granite
Mesocrates	35–65	Diorite
Melanocrates	5–35	Gabbro
Holomelanocrates	0–5	Pyroxenite

12.3.2 Texture

Rock texture is the overall appearance of a rock based of the size and arrangement of its interlocking crystals. Crystal size is the most important aspect of igneous texture. Amongst the several texture varieties, three main classes can be identified: phaneritic, aphanitic, and glassy textures.

Phaneritic igneous rocks are formed from a slow cooling rate of a mother magma (i.e., far from the Earth's surface), possess a coarse grain size, with visible grains (1–20 mm), and exhibit a so-called typical phaneritic texture (e.g., granite, syenite, or gabbro). In a few cases, intrusive igneous rocks can have a distinctly mixed crystal size with a so-called **porphyroid texture**. In this case scattered, prominent, extremely coarse crystals often called **mega-crystals** are surrounded by a groundmass of medium or coarse crystals. Porphyritic textures indicate that the magma stopped at some depth where the larger crystals formed, before migrating to the surface where it erupted. The magma may also have been supersaturated with the coarse crystalline mineral phase and these phaneritic rocks exhibit a **pegmatitic texture** and are called pegmatites. The texture possesses large crystals greater than 2–5 cm and usually above 12 cm in diameter. Sometimes, exceptionally large crystals may be several meters in size. These usually form in the latest stage of cooling of water-rich magmas, and represent the accumulated volatiles from the magma.

Aphanitic and porphyritic igneous rocks – aphanitic igneous rocks are also crystalline. They form by rapid cooling of lava at or near the Earth's surface. As a result all or most of the crystals are so small that individual crystals cannot be distinguished. These tiny crystals, called **microlites** have a fine crystal size. Most aphanitic igneous rocks are also called extrusive or volcanic rocks because they are formed at or near the surface of the Earth and are often associated with volcanoes. Sometimes aphanitic or glassy rocks contain scattered coarse or medium crystals called **phenocrystals** which are surrounded by a groundmass composed of fine crystals and/or glass. A rock with such a mixed crystal size has a **porphyritic texture**. This texture is sometimes formed by a two stage cooling history. Initially, a magma rising through the crust begins to cool slowly at depth forming the phenocrysts. Then the magma erupts onto the surface where the remaining liquid cools rapidly forming the matrix. In other cases, the phenocrystals represent crystals which had a much faster growth rate during cooling than did the fine crystals (e.g., porphyritic rhyolite).

Glassy and hyaline igneous rocks are the result of a rapidly solidified magma, and because it is a molten liquid cooled with a high cooling rate the random microscopic organization of a magma is fixed and avoids the crystallization process. Therefore, by contrast with other igneous rocks, which are crystalline, these rapidily quenched materials exhibit an amorphous or vitreous (glassy) aspect. Nevertheless, due to thermodynamic instability of glasses, some devitrification processes occur and particular textures are often present such as cracks (spherolitic) and bubbles (perlitic).

Table 12.9 Chief textures of igneous rocks

Main texture type	Cooling rate	Definition	Varieties and facies
Phaneritic	Slow	All the minerals have a medium or coarse crystal size	Malgachitic, pegmatitic, pegmatoidic, rapa-kiwic, porphyroidal, granular, isogranular, saccharoidal, homogeneous, Dent de Cheval, orbicular, aplitic, cataclastic, graphic, amygdalar
Micro-phaneritic	Slow/ moderate	All the minerals have a small crystal size	Porphyritic, aphanitic, phaneritic, graphic, ophitic, poecilitic, intersertale, intergranular, lamproporphyric, doleritic
Aphanitic or microlitic	Rapid	The crystals are so small (microlites) that individual crystals cannot be distinguished without magnification and are surrounded by a glassy matrix. Aphanitic textures form primarily when cooling rates are fast, such as in lava flows	Porphyritic, trachytic
Hyaline, glassy or vitreous	High	The glassy or vitreous texture with no crystals usually indicates that the magma cooled extremely quickly and/or that it was so viscous that ions could not migrate to form seed crystals. Most glasses are related to pyroclastic igneous rocks	Vitreous, amorphous, spherolitic, perlitic, fluidale, brechiform, vacuolaire, vesicular

12
Rocks and Meteorites

Table 12.10 Crystallinity

Glass fraction (%)	Category	Example
0–5	Holocrystalline	Granite
5–55	Hypocrystalline	Basalt
55–95	Hypophyaline	Pumice
95–100	Holohyaline	Obsidian

12.3.2.1 Chemistry

The chemical composition of an igneous rock is also an important parameter in its classification. The chemical composition of a rock may be expressed by the types of minerals present and their relative abundances or in the rock color. Rocks may also be analyzed chemically using quantitative chemical analysis techniques to determine the relative proportions of chemical elements present. These chemical abundances can be used directly to classify igneous rocks. The chemical composition of the mother magma and, to a lesser extent, that of the country rock largely controls the types of minerals which may be formed.

Table 12.11 Abundance of chemical elements

Category	Examples
Major chemical elements	O, Si, Al, Fe, Mg, K, Na, Ca
Minor chemical elements	H, F, Cl
Dispersed chemical elements	Ti, P
Rare chemical elements	U, Th, Zr, Hf

Table 12.12 Acidity (silica content)

Silica content (wt% SiO_2)	Category	Examples
66–100	Acid igneous rocks	Granite
52–66	Neutral igneous rocks	Syenite
45–52	Mafic igneous rocks	Gabbro
0–45	Ultramafic igneous rocks	Peridotites

Table 12.13 Saturation

Category	Example
Sursaturated	Granite
Saturated	Syenite
Subsaturated	Peridotites

Table 12.14 Alkalinity

Category	Criteria
Alkaline igneous rocks	$n(Na_2O) + n(K_2O) > n(Al_2O_3)$ $n(Na_2O) + n(K_2O) > 1/6(SiO_2)$
Non-alkaline igneous rocks	$n(Na_2O) + n(K_2O) < n(Al_2O_3)$ $n(Na_2O) + n(K_2O) < 1/6(SiO_2)$

Table 12.15 Feldspar Index

Category	F-Index (%)
Alkaline igneous rocks	80–100
Subalkaline igneous rocks	60–80
Monzonitic igneous rocks	40–60
Subplagioclasic igneous rocks	20–40
Holoplagioclasic igneous rocks	0-20

12.3.3 General Classification of Igneous Rocks

The actual mineralogical composition of an igneous rock can be based on the microscopic surface area fraction estimation of mineral species present in a thin section of a rock sample. The easiest classification scheme for the igneous rocks uses first the coloration of the rock described previously (i.e., fraction of dark minerals), then the most abundant rock-forming minerals in the rock, and finally the textures as the basis for naming. There are eight common names to remember, four intrusive rocks and four extrusive rocks. From the most acid to the most alkaline, the intrusive rocks are: granite, diorite, syenite, and gabbro. The corresponding extrusive rock names with similar composition and mineralogy are: rhyolite, trachyte, andesite, and basalt. The remaining names are applied to ultrabasic rocks with a very low silica content: peridotite, pyroxenite, and anorthosite. A very simple classification of igneous rocks is reported in Table 12.18. However, for a more quantitative identification of igneous rocks apart from ultramafic rocks, it is recommended to refer to the most comprehensive and accurate classification of igneous rocks developed in 1976 by Streckeisen.[2] This classification is based on a modal analysis. The rock mineralogical analysis is located in a double triangle "QAPF", where the vertices represent quartz (Q), the alkali feldspars (A), the feldspathoids (F), and the potassic feldspars (P) respectively.

12.3.4 Vesicular and Pyroclastic Igneous Rocks

As volcanic rocks cool, but before they completely solidify, gas originally dissolved in the silicate melt separates from the liquid to evolve bubbles called vesicles which result in holes in the completely solidified porous and low-density pyroclastic rock (vesicularity). It is this gas which helps to produce explosive volcanic eruptions. Explosive volcanic eruptions provide a group of extrusive igneous rocks called **pyroclastic** rocks which are formed by the ejection of molten rock into the atmosphere. Pumice is an example of a light-colored, very vesicular, and glassy pyroclastic igneous rock. Scoria is similar to pumice, but it is colored, often red to black, and more dense. Scoria is sometimes sold as lava rock for landscaping and for gas grills.

Table 12.16 Average chemical composition of igneous common rocks (wt%)

Igneous rock	SiO_2	TiO_2	Al_2O_3	Fe_2O_3	FeO	MnO	MgO	CaO	Na_2O	K_2O	H_2O^{+a}	H_2O^-	P_2O_5	CO_2
Andesite	57.94	0.87	17.02	3.27	4.04	0.14	3.33	6.79	3.48	1.62	0.83	0.34	0.21	0.05
Anorthosite	50.28	0.64	25.86	0.96	2.07	0.05	2.12	12.48	3.15	0.65	1.17	0.14	0.09	0.14
Basalt	49.20	1.84	15.74	3.79	7.13	0.20	6.73	9.47	2.91	1.10	0.95	0.43	0.35	0.11
Basanite	44.30	2.51	14.70	3.94	7.50	0.16	8.54	10.19	3.55	1.96	1.20	0.42	0.74	0.18
Dacite	65.01	0.58	15.91	2.43	2.30	0.09	1.78	4.32	3.79	2.17	0.91	0.28	0.15	0.06
Diorite	57.48	0.95	16.67	2.50	4.92	0.12	3.71	6.58	3.54	1.76	1.15	0.21	0.29	0.10
Dolerite	50.18	1.14	15.26	2.86	8.05	0.19	6.78	9.41	2.56	1.04	1.46	0.43	0.27	0.18
Dunite	38.29	0.09	1.82	3.59	9.38	0.71	37.94	1.01	0.20	0.08	4.59	0.25	0.20	0.43
Earth's crust[b]	60.18	1.06	15.61	3.14	3.88	n.a.	3.56	5.17	3.91	3.19	n.a.	n.a.	0.30	n.a.
Gabbro	50.14	1.12	15.48	3.01	7.62	0.12	7.59	9.58	2.39	0.93	0.75	0.11	0.24	0.07
Granite	71.30	0.31	14.32	1.21	1.64	0.05	0.71	1.84	3.68	4.07	0.64	0.13	0.12	0.05
Granodiorite	66.09	0.54	15.73	1.38	2.73	0.08	1.74	3.83	3.75	2.73	0.85	0.19	0.18	0.08
Hawaiite	47.48	3.23	15.74	4.94	7.36	0.19	5.58	7.91	3.97	1.53	0.79	0.55	0.74	0.04
Latite	61.25	0.81	16.0	3.28	2.07	0.09	2.22	4.34	3.71	3.87	1.09	0.57	0.33	0.19
Monzonite	62.60	0.78	15.65	1.92	3.08	0.10	2.02	4.17	3.73	4.06	0.90	0.19	0.25	0.08
Mugearite	50.52	2.09	16.71	4.88	5.86	0.26	3.20	6.14	4.73	2.46	1.27	0.87	0.75	0.15
Nephelenite	40.60	2.66	14.33	5.48	6.17	0.26	6.39	11.89	4.79	3.46	1.65	0.54	1.07	0.60

Nepheline syenite	54.99	0.60	20.96	2.25	2.05	0.15	0.77	2.31	8.23	5.58	1.30	0.17	0.13	0.20
Norite	50.44	1.00	16.28	2.21	7.39	0.14	8.73	9.41	2.26	0.70	0.84	0.13	0.15	0.18
Obsidian	73.84	–	13.00	1.82	0.79	–	0.49	1.52	3.82	3.92	0.53	–	–	–
Peridotite	42.26	0.63	4.23	3.61	6.58	0.41	31.24	5.05	0.49	0.34	3.91	0.31	0.10	0.30
Phonolite	56.19	0.62	19.04	2.79	2.03	0.17	1.07	2.72	7.79	5.24	1.57	0.37	0.18	0.08
Pumice	70.38	–	15.82	1.42	1.50	–	0.48	1.56	3.70	4.10	–	3.62	–	–
Pyroxenite	46.27	1.47	7.16	4.27	7.18	0.16	16.04	14.08	0.92	0.64	0.99	0.14	0.38	0.13
Rhyodacite	65.55	0.60	15.04	2.13	2.03	0.09	2.09	3.62	3.67	3.00	1.09	0.42	0.25	0.21
Rhyolite	71.30	0.28	13.27	1.48	1.11	0.06	0.39	1.14	3.55	4.30	1.10	0.31	0.07	0.05
Syenite	58.58	0.84	16.64	3.04	3.13	0.13	1.87	3.53	5.24	4.95	0.99	0.23	0.29	0.28
Tephrite	47.80	1.76	17.00	4.12	5.22	0.15	4.70	9.18	3.69	4.49	1.03	0.22	0.63	0.02
Tonalite	61.52	0.73	16.48	1.83	3.82	0.08	2.80	5.42	3.63	2.07	1.04	0.20	0.25	0.14
Trachyandesite	58.15	1.08	16.70	3.26	3.21	0.16	2.57	4.96	4.35	3.21	1.25	0.58	0.41	0.08
Trachybasalt	49.21	2.40	16.63	3.69	6.18	0.16	5.10	7.90	3.96	2.55	0.98	0.49	0.59	0.10
Trachyte	61.21	0.70	16.96	2.99	2.29	0.15	0.93	2.34	5.47	4.98	1.15	0.47	0.21	0.09

[a] Losses on ignition at 120°C.
[b] Clarke FW, Washington HS (1924) The composition of the earth's crust. US Geol Survey, Profess Paper 127.

12
Rocks and Meteorites

Table 12.17 Deposit depth location

Depth location in the Earth's crust	Rock family	Texture
Surface (e.g., lava flows, volcanoes)	Volcanic igneous rocks (vulcanites)	Hyaline, microlitic
Mid-depth (e.g., veins)	Hypovolcanic igneous rocks Periplutonic igneous rocks	Aphanitic, doleritic, porphyritic
Deep deposits (e.g., intrusives)	Plutonic igneous rocks (plutonites)	Phaneritic, phorphyroid, pegmatitic

12.4 Sedimentary Rocks

Sedimentary rocks are reaction products resulting from the interaction of the atmosphere and hydrosphere on the Earth's crust. The formation of sedimentary rocks always follows several steps grouped under four chief successive processes: (i) **weathering/erosion** (involving physical wear/erosion and chemical alteration) is the process of breakdown of pre-existing rocks (igneous, metamorphic or sedimentary) at the Earth's surface by the action of wind, water, and ice, (ii) **transportation**, during which the mobile raw materials (e.g., colloids, salts, minerals, and rock particles) are carried by several possible natural agents (water, wind or ice) to the site of deposition, (iii) **sedimentation** or **deposition**, the process of particles falling from suspension or being precipitated from ions in solutions to form layers of sediments, (iv) **diagenesis** or **lithification**, the process after which the sediments are buried, and are compacted and cemented or not into a sedimentary rock, due to the increasing pressure and temperature encountered during subsidence. But none of these processes works in isolation. The original constituents of the lithosphere, i.e., minerals of igneous rocks, are to a large extent thermodynamically unstable with respect to atmosphere and hydrosphere. They have been formed at both high temperature and pressure, and cannot be expected to remain chemically stable under the very different conditions encountered at the Earth's surface. Therefore, except for quartz which is highly resistant, all the minerals tend to alter. The altered rock rapidly disintegrates under the mechanical effect of erosion and its constituents are transported and redeposited as sediments. For instance, lithification of shale occurs as a result of the compaction and cementation of wet mud. Randomly oriented clay particles in the deposited sediment are reoriented as water is expelled during compaction. The compaction results from the increased load of newly deposited sediment. As the water content is reduced the pore-waters become more concentrated and cement is deposited from solution. Splitting surfaces form normal to the loading direction and parallel to the orientation of the platy clay minerals. The process of volume reduction and water expulsion is called consolidation. If the fluid cannot be expelled the sediments remain unconsolidated. The rate of consolidation is controlled by the permeability of the sediment. Both porosity (volume of voids) and permeability (rate of fluid transfer) are drastically reduced by compaction and cementation. Amongst the sedimentary rocks, nine sedimentary rocks are important volumetrically, they are: (1) petroleum and coal, (2) ironstone, (3) bauxite, (4) rock salt, (5) phosphate rock, (6) sandstone, (7) limestone, (8) dolomite, and (9) mudstone (shale). These very common rocks contain a very limited set of five or six rock forming minerals from the following group: quartz, calcite, dolomite, kaolinite, illite, goethite, and boehmite.

12.4.1 Sediments

Sediments are commonly subdivided into three types. (i) Clastic or detrital sediments comprise particles of various sizes carried in suspension by wind, water, or ice. Sand is an example of a clastic sediment. (ii) Chemical or precipitated sediments are carried in aqueous solution as ions. Calcite is an example of a chemical precipitate. (iii) Organic or biogenic sediments are

Table 12.18 Simple classification of igneous rocks

Coloration	Feldspar group	Specific gravity	Saturated (acid) (SiO₂ > 60wt%)		Unsaturated (alkaline) (45wt% < SiO₂ < 60wt%)		Substated ultrabasic (mafic) (45wt% > SiO₂)		
			Both quartz and feldspars	Feldspars	Feldspars and feldspathoids	Feldspathoids	Peridots	Pyroxenes	Amphiboles
Hololeucocrate (i.e., white)	Orthoclases (orthose, microcline)	2.65 to 2.80	**Granite** (71.3)	**Syenite** (58.58)	**Nepheline syenite** (54.99)	**Ijolite** (n.a.)			
Leucocrate			Rhyolite (72.82)	Trachyte (61.21)	Phonolite (56.19)	Nephelenite (n.a.)			
	Na-Plagioclases (albite oligoclase, andesine) An0–50	2.80 to 2.90	**Tonalite** (61.52)	**Diorite** (57.48)	**Essexite** (n.a.)	**Missourite** (n.a.)			
Mesocrate			Dacite (65.01)	Andesite (57.94)	Tephrite (47.80)	Nephelenite (40.60)			
	Ca-Plagioclases labrador, bytownite, anorthite) An50–100	2.80 to 3.25	**Quartz gabbro** (n.a.)	**Gabbro** (50.14)	**Theralite** (n.a.)	**Melteigite** (n.a.)			
Melanocrate			Tholeite (n.a.)	Basalt (49.20)	Basanite (44.30)	n.a.	**Peridotite** (42.26) Kimberlite	**Pyroxenite** (46.27)	**Amphibolite**
Holomelanocrate (i.e., dark)	Non-present or rare	3.5 to 4.3							

Bold type indicates phaneritic rocks.
SiO₂ wt% in parentheses.

12
Rocks
and
Meteorites

Table 12.19 Common pyroclastic rocks

Pyroclastic	Description
Scoria	Pyroclastic volcanic rock made of fragments of rock and ash cemented together by a glassy matrix. It may also resemble a sedimentary conglomerate or breccia, except that rock fragments are all fine-grained ingneous or vesicular and dark in color; brown, black, or dark red; similar to vesicular basalt but fully riddled with holes to form a spongy mass.
Pumice	Pumice is sponge-like pyroclastic volcanic rock, with a very low density (below that of water), and light in color (white to gray), it may be glassy or dull, and fully riddled with holes. It is used as an abrasive (e.g., pumice stone, lava soap).
Tuff	Tuff is a pyroclastic igneous rock with a fragmental texture and is often friable (loosely held together). Tuff may be composed of volcanic ash (tiny shards of volcanic glass), pumice fragments (see above), pieces of obsidian (solid volcanic glass without significant vesicles), and relatively dense rock fragments.
Obsidian	Dark amorphous igneous rocks with a rhyolitic chemical composition, conchoidal fracture and brittle, sometimes devitrification figures.

Table 12.20 Classification of pyroclastic rocks[a]

Fragment or clast size (mm)	Pyroclast or fragment type	Consolidated rock name
>265	Block, bomb	Pyroclastic breccia
64–264	Bomb	Agglomerate
2–64	Lapilli	Lapilli tuff
< 2	Coarse ash grain	Coarse ash tuff
1/16	Fine ash (dust)	Fine ash tuff

[a]From Le Maitre RW, Bateman P, Dudek A (1989) A classification of igneous rocks and glossary of terms. Blackwell Scientific Publications, Oxford.

Table 12.21 Classes of sedimentary rocks

Group	Examples
Residual sedimentary rocks	Laterites, bauxite, and residual clays
Detritic or terrigeneous rocks	Sand, sandstones, breach conglomerate, pelites
Carbonated sedimentary rocks	Limestones, dolomites, and marl
Evaporitic sedimentary rocks	Rock salt, gypsum, and anhydrite
Phosphatic sedimentary rocks	Phosphates, phosphorites
Carbonaceous and organic sedimentary rocks	Peat, coals (e.g., lignites, anthracites), petroleum, bitumen, natural gas
Ferrous sedimentary rocks	Ironstone, taconite
Siliceous	Diatomite, radiolarites, spongites, flint, cherts

precipitated or accumulated by biological agents. Many micro-organisms promote the precipitation of calcite to form biogenically precipitated calcareous muds. Sediments are classified on the basis of the origin, size, and mineralogical composition of the particles. They are produced by the action of weathering and erosion that break down pre-existing rocks by physical and chemical processes. The sediment is then transported by wind, water, or ice to the site of deposition. The character of the sediment is determined by the extent of weathering and the type and distance of transportation. Some sediments weather in situ with little or no transport (e.g., laterites) giving residual sedimentary rocks. Others may be transported over large distances from mountain top to ocean. The transport agents wind, water, and ice generate distinctive sediments that can be identified by the extent of particle abrasion and the degree of sorting. Particle size is an important factor in determining many important physical rock properties including strength, porosity, permeability, density, and many others. It also determines the name of the sedimentary rock type for clastic rocks.

12.4.2 Residual Sedimentary Rocks

This category of sedimentary rocks comprises rocks formed by the degradation materials located near or at the site of weathering of the pre-existing rocks. It consists mainly of the degradation products depleted by the leaching action of water. Hence soluble cations (e.g., Na, K, Ca, Mg) are removed, while insoluble cations such as iron and aluminum associated with clays and silica remain in the materials. Amongst them (e.g., residual clays, terra rossa), the most important economically is bauxite.

12.4.3 Detritic or Clastic Sedimentary Rocks

Detritic or clastic sedimentary rocks represent 80–90% of all the sedimentary rocks. They can be separated into two main classes. (i) The **clastic** or **terrigenous** rocks (i.e., detritic sensu stricto) are formed from the erosion of pre-existing rocks. They contain 50% clastic elements and are classified according to the grain size of the particles (clasts) that are cemented together to form the consolidated rock. (ii) The **pyroclastic** sedimentary rocks are formed from the volcanic clasts (see Section 12.3). The strength of cementation is often an important characteristic in engineering terms. Well-cemented quartz sandstones can be very strong mechanically, whereas friable uncemented sandstones are relatively weak rocks. Siltstones, mudstones, and shales are usually weak rocks because of the dominance of platey clay minerals that provide little frictional resistance. Moreover, conglomerates and sandstones exhibit a relatively high volumes of void fraction (porosity) and are economically important as aquifers for water supplies and reservoir rocks for gas and petroleum.

12.4.4 Chemical Sedimentary Rocks

Chemical sedimentary rocks are classified according to the predominant minerals precipitated to form the rock and their texture. They are deposited by both chemical precipitation of a salt from an aqueous saturated solution rather than release from suspension, and the biological action of micro-organisms. They can be arranged in five classes according to their chemistry: carbonated (e.g., limestone), siliceous (e.g., chert), evaporitic (e.g., gypsum), ferrous (e.g., ironstone), and phosphated (e.g., phosphorite). In Table 12.23 are listed some of the more common chemical sedimentary rocks. Calcite, silica, collophane, and iron oxides are the main cements that bind sedimentary rocks. Iron oxides (e.g., goethite, limonite) in very small amounts can be responsible for the red, orange, and green coloration in sedimentary rocks. Fine-grained rocks such as shales and mudstones usually appear dark gray to black owing to the

Table 12.22 Detritic sedimentary rocks

Class	Particle diameter (d, mm)	Unconsolidated material or particle type	Consolidated (i.e., cemented) rock type	
Rudites	> 256	Boulders	Very coarse conglomerate and breccia	**Conglomerate** has rounded rock fragments, while **breccia** has angular rock fragments.
	64–256	Cobble	Coarse conglomerate and breccia	
	2–64	Pebbles	Conglomerate and breccia	
Arenites	0.62–2	Coarse sands	Coarse sandstone	**Sandstone**: quartz predominant, visible grains, often thickly bedded, depositional structures such as cross-bedding common. **Arkose**: sandstone with more than 25wt% feldspar grains.
	02–0.6	Medium sands	Medium sandstone	
	0.06–0.2	Fine sands	Fine sandstone	
Lutites	0.004–0.06	Silts	Siltstone	Quartz predominant, grains barely visible, gritty feel.
	< 0.004	Clays	Mudstone and shale	**Mudstone**: thick beds > 1 cm blocky, fine, mud, no particles discernable, may show polygonal dried mud-cracks, composed predominantly of clay minerals and very fine quartz. **Shale**: laminated mudstone, fissile, splits into thin sheets.

presence of sulfide minerals (e.g., pyrite, chalcopyrite). Black shales may contain significant amounts of organic matter and carbonaceous material. There are a large number of these rocks that form deposits of economic significance (e.g., ironstone, phosphorite, dolomite, rock salt, and potash).

12.4.5 Biogenic Sedimentary Rocks

Organic or biogenic sedimentary rocks are the result of biological processes. They may be clastic accumulations of animal skeleton debris (e.g., limestone), biologically catalyzed precipitates (e.g., ironstones), obtained from the diagenesis of organic matter coming from marine-originating carbon-rich sediments (e.g., oil), fluvial vegetal matter (e.g., coals), or alteration products of siliceous organisms (e.g., chert).

Coals originated from the diagenesis in anaerobic conditions (i.e., in the absence of air) of the remains of trees, ferns, mosses, vines, and other forms of plants, accumulated under water, which flourished in huge swamps, marshlands, and bogs in the paleozoic era during prolonged periods of humid rainforest climate and abundant rainfall. The precursor of coal was peat, which was formed in the early stage of diagenesis by bacterial and chemical action on the plant debris. Subsequent action of heat and lithostatic pressure during diagenesis metamorphosed the peat into various types of coals.

Petroleum and natural gas originated from degradation under bacterial and chemical action of microorganisms debris in anaerobic conditions in seawater or brackish waters. Petroleum accumulates over geological times in complex underground geological formations called reservoirs made of porous sedimentary rocks (e.g., sandstones) surrounded by overlying and underlying strata of impervious rocks (e.g., clays, rock salt). Petroleum is a brownish green to

Table 12.23 Chemical sedimentary rocks

Class	Rock name	Texture	Mineral composition	Description
Carbonated	Limestone	Clastic	Calcite	Calcite fragments and calcite cement. White, gray, or bluish in color. Fizzes strongly with dilute HCl.
	Oolitic limestone	Clastic		Rounded calcite ooliths bounded by a calcite cement. Can be slightly dolomitized.
	Dolomitic lime-stone	Clastic	Calcite and dolomite	Calcite fragments and calcite cement with significant alteration to the magnesium-bearing carbonate dolomite. Reacts with dilute HCl.
	Dolomite or dolostone	Clastic	Dolomite	Carbonate almost completely transformed to dolomite. Often yellowish or pinkish in color. Reacts weakly with dilute HCl.
Siliceous	Travertin or siliceous tuff	Amorphous	Silica	Rocks made from 50wt% silica precipitated from saturated aqueous solutions or metasomatic reaction. Diatomite is made from accumulation of siliceous skeletons of micro-organisms.
	Calcedoine, jasper	Glassy		
	Diatomite, spongolite, radiolarite	Clastic		
Evaporitic	Rocksalt	Crystalline	Halite	Halite, interlocking cubic crystals.
	Potash	Crystalline	Halite, sylvite, and carnallite	Halite with sylvite, interlocking cubic crystals, sometime contains orange-to-red carnallite crystals.
	Rock gypsum	Crystalline	Gypsum	Gypsum, commonly interlocking prismatic or fibrous crystals. Usually white or light gray.
Phosphated	Phosphate rock	Clastic	Collophanite	Pisoliths and organic debris cemented by collophaite-rich material.
	Phosphorite		Calcite and collophanite	
Ferrous	Ironstone, faconite	Crystalline	Siderite, goethite, limonite	

black liquid with an extremely complex chemical composition. Actually, it is a mixture of hydrocarbons (mainly alkanes) as well as compounds containing nitrogen, oxygen, and sulfur. Most petroleum contains traces of nickel and vanadium.

12.4.6 Chemical Composition

The chemical composition of sedimentary rocks is shown in Table 12.25.

12.5 Metamorphic Rocks

Metamorphism is the sum of deep subsurface processes, working below the weathering zone, that result in the partial or complete recrystallization of a pre-existing rock (called protolith), with the production of new structures, new minerals, deformation and rotation of mineral grains, recrystallization of initial minerals as larger grains, and production of strong brittle

Table 12.24 Carbonaceous sedimentary rocks

Type	Carbon content (wt% C)	Bulk density (kg.m^{-3})	Gross caloric value (MJ.kg^1)	Description
Peat	55–70	870	20.9	Light brown, high porosity.
Crude oil	83–87	810–985	38.5 MJ.dm^{-3}	Petroleum is a brownish green to black liquid.
Natural gas	–	–	38.4 MJ.m^{-3}	Mainly methane with other alkanes and a minute amount of hydrogen sulfide.
Lignite	70–75	640–860	23.9–25.5	Dark brown with ligneous debris.
Coal	85–92	670–910	25.5–30.1	Dark, matt.
Bituminous coal	80–95	670–910	24.4–32.6	Friable grayish brown solid.
Anthracite	92–95	800–930	30.2–32.5	High carbon coal that approaches graphite in structure and composition. It is hard, compact, dark and shiny, with generally a conchoidal fracture. It is difficult to ignite and burns with a smokeless blue flame.
Graphite	95–100	1800–2200	> 32.5	Pure carbon in lamellar crystals.

rocks or anisotropic rocks weak in shear. Metamorphism is induced in solid rocks as a result of pronounced changes in three factors: (i) temperature, T, (ii) pressure, P, and (iii) chemistry of surrounding and reactive fluids. But in any case metamorphism does not imply melting and only consists of a subsolidus recrystallization (from a metallurgical point of view). The lower limit of metamorphic temperatures is stated as 150°C in order to make a clear distinction from the diagenetic process. The upper limit is the melting temperature when a magma forms (anatexy process).

The heat may originate from several sources, (i) chiefly the increase of temperature with depth (geothermal gradient). Actually, the geothermal gradient affects sediments during subsidence and is an important factor in regional metamorphism. The order of magnitude of several geothermal gradients are listed in Table 12.27 (page 498). (ii) The contiguous magmatic intrusions (e.g., plutons), veins, and lava flows during volcanic activity are responsible for contact metamorphism. (iii) Additional sources of heat are exoenergetic transformations, friction losses during tectonic activity (e.g., mylonites), impact of meteorites (e.g., impactites), lightning in desert areas (e.g., fulgurites), and more recently human activity (e.g., atomic explosions). Pressure may be resolved in two ways: (i) lithostatic (hydrostatic) or uniform pressure, which leads to change in volume of the overlying pile of rock, and (ii) oriented pressure or mechanical stress, which leads to change of shape or distortion. Finally, the action of chemically reactive hot fluids is a most important factor in metamorphism since they do not alter the initial chemical composition, i.e., they promote reaction by dissolution; on the other hand when mass balance is strongly modified, it is called metasomatism. The type of metamorphic rock is determined by the parent rock (protolith) and the P–T conditions

12.5.1 Metamorphic Rock Classification

First of all, metamorphic rocks can be divided into two groups on the basis of the pressure and temperature conditions of their formation. **Regional metamorphic rocks** (e.g., schists, mica schists, gneiss) are generated mainly by pressure and stresses in the roots of mountain belts, while **contact metamorphic rocks** (e.g., quartzite, marble, skarns) are generated mainly by

Table 12.25 Average chemical composition of sedimentary rocks (wt%)[a]

Igneous rock	SiO_2	TiO_2	Al_2O_3	Fe_2O_3	FeO	MgO	CaO	Na_2O	K_2O	H_2O	P_2O_5	CO_2	SO_3	BaO	C
Sediment	57.95	0.57	13.39	3.47	2.08	2.65	5.89	1.13	2.86	3.23	0.13	5.38	0.54	trace	0.66
Sandstone	78.33	0.25	4.77	1.07	0.30	1.16	5.50	0.45	1.31	1.63	0.08	5.03	0.07	0.05	trace
Limestone	5.19	0.06	0.81	0.54	–	7.89	42.57	0.05	0.33	0.77	0.04	41.54	0.05	trace	trace
Shale	58.10	0.65	15.40	4.02	2.45	2.44	3.11	1.30	3.24	5.00	0.17	2.63	0.64	0.05	0.80

[a]Data from Pettijohn FJ (1949) Sedimentary rocks. Harper and Brothers, New York, p. 82.

12
Rocks
and
Meteorites

Table 12.26 Designation according to protolith

Protolith type	Designation prefix	Example
Sedimentary	Para-	Para-schist
Igneous	Ortho-	Ortho-basalt
Metamorphic	Meta-	Meta-gneiss

Table 12.27 Geothermal gradients

Country rocks	Geothermal gradient		
	Practical range	$(°C.km^{-1})$	$(°C.mile^{-1})$
Granitic	1°C per 60–100m	10 to 17	16 to 27
Sedimentary basins	1°C per 33 m	30	50
Coal deposits	1°C per 20–25 m	40 to 50	51 to 64
Volcanic area	1°C per 10–15 m	67 to 100	107 to 161

temperature increases at the boundaries of igneous intrusions, sometimes called thermal metamorphic rocks. A second common subdivision of metamorphic rocks is based on their texture which is determined by the parent rock and the temperature and pressure conditions. In general, metamorphism can lead to two classes of texture: (i) the non-foliated rocks which exhibit a recrystallized texture but no preferred mineral orientation (e.g., marble, quartzite), and (ii) the foliated rocks (e.g., mica schist, gneiss) having a strong mineral orientation and/or mineral banding or layering. There are a limited number of common metamorphic rock types which are listed in Table 12.28.

12.5.2 Metamorphic Grade

As the degree of metamorphism increases, new minerals become stable and crystallize, while unstable minerals disappear. The minerals present in metamorphic rocks are thus precise

Table 12.28 Most common types of metamorphic rocks

Metamorphic rocks	Protolith (parent rock)	Texture	Description
Marble	Limestones	Non-foliated	Interlocking, often coarse, calcite crystals, little or no porosity
Quartzite	Sandstones	Non-foliated	Interlocking almost fused quartz grains, little or no porosity
Hornfels	Shales	Foliated	Shales baked by igneous contact form with very hard fine-grained rocks
Gneiss	Coarse-grained rocks	Foliated	Dark and light bands or layers of aligned minerals
Schists	Fine-grained rocks	Foliated	Mica minerals, often crinkled or wavy
Skarns	Calcareous rocks	Non-foliated	Contact metamorphism and alteration by hot fluids
Slates	Shales, clays, and muds	Foliated	Prominent splitting surfaces

Table 12.29 Common metamorphic facies as a function of temperature and pressure

Pressure range/ temperature range	Low temperature (0 to 400°C)	Medium temperature (400 to 600°C)	High temperature (600 to 1000°C)	Ultrahigh temperature (1000 to 1200°C)
Low pressure (0 to 500 MPa)	Zeolites facies (diagenesis)	Hornblende facies	Pyroxene facies	Sanidinite facies
Medium pressure (500 to 1000 MPa)	Prehnite and pumpellyite facies	Green schist facies	Amphibolite facies	Granulite facies
High pressure (above 1 GPa)	Glaucophane and lawsonite facies		Eclogitic facies	

indicators of the pressure and temperature conditions at the time of the last recrystallization. Metamorphic grade is a scale of metamorphic intensity which uses indicator minerals as geothermometers and geobarometers. For instance, a particular sequence such as slate–phyllite–schist indicates metamorphic rocks of increasing grade. The corresponding indicator minerals are chlorite, biotite, and garnet respectively. The transition from chlorite grade to biotite grade is the first appearance of chlorite.

12.5.3 Metamorphic Facies

Metamorphic facies is a more sophisticated extension of the grade concept to include pressure (geobarometry) as well as temperature (geothermometry) to the interpretation of metamorphic rocks. The indicator minerals become groups of minerals or mineral assemblages that characterize a particular region of pressure and temperature.

12.6 Meteorites

12.6.1 Definitions

Meteorites are extraterrestrial materials coming from the solar system which are continually falling on Earth. Meteorite sizes commonly range from the finest dust particles (several μm) up to those that are several kilometers in diameter if we consider asteroids, which appear to be similar to meteorites in many aspects. Meteorites consist essentially of a Ni–Fe alloy, or of ferromagnesian silicates (e.g., olivine, pyroxenes), feldspars, plagioclases, and sulfides (e.g., troilite), or a mixture of them. Many systems of classification have been devised in the past but owing to their aspect and mineralogical and chemical composition, they can be grouped and classified as follows in three main groups: entirely stony meteorites (**litholites** or **aerolites**), entirely metallic meteorites (**siderites**), and a combination of the two previous types called **lithosiderites** (**siderolites**). This modern classification is depicted in Table 12.30. Glassy meteorites known as tektites are also found and will be presented at the end of this section.

12.6.2 Modern Classification of Meteorites

See **Table 12.30** (pages 500–504).

Table 12.30 Modern classification of meteorites

Percentage of all falls	Category	Description and distinguishing features	Chondrule character and/or letter designation
		1. Aerolites or **stony meteorites** possess a chemical composition reflecting solar abundances of nonvolatile elements and are subdivided as achondrites and chondrites.	
85		**1.1. Chondrites** are stony meteorites which contain minerals, mainly olivine, pyroxenes, and also feldspars; in addition small nickel–iron grains with 0.2 to 3 mm particle size, called chondrules are always present. Further divisions are based on chemical trends and identify the following chondrite subgroups: (i) carbonaceous, (ii) ordinary, (iii) enstatite, and (iv) other chondrites. Still further divisions based on petrologic type gives us types 1–7; type 3 chondrites have remained unaltered, with lower types experiencing progressive aqueous alteration and higher types experiencing progressive thermal or shock alteration. Type 7 chondrites are transitional to an achondrite. Chondrules are found in all petrologic types except types 1 and 7 where alteration has left them indistinct from the matrix.	
		1.1.1. Carbonaceous chondrites are primitive and rare undifferentiated meteorites composed of silicate chondrules set in a fine-grained silicate matrix which contain carbon compounds including long-chain hydrocarbons and amino acids similar to those used in protein synthesis in living organisms and basic building blocks of life. Within the matrix can be found calcium–aluminum silicate inclusions that represent the earliest material that condensed from the hot nebula, while certain isotopes originated in interstellar grains that predate the formation of the solar system. Carbonaceous chondrites formed in an oxygen-rich environment with most metal combined into silicates, sulfides, water or other oxides. They formed on the smaller asteroids that retain the oldest record of the solar nebula, containing solar abundances of non volatile elements. Carbonaceous chondrites have been divided according to the ratio Ca/Si into individual chemical subgroups including the CI, CM, CR, CO, CK, CV, and CH groups, along with the three-member Coolidge-grouplet and a few rare ungrouped members. The discovery of new and unique CCs helps us to continually revise the record of processes occuring in the early solar system. Carbonaceous chondrites have a large abundance of opaque mineral-rich porphyritic chondrules. Moreover, they have a CI-normalized mean-refractory–lithophile abundance ratio of 1.00–1.35 and a fine-grained matrix/chondrule modal abundance ratio of 0.5–7.0. Finally, carbonaceous chondrites have an abundance of isotopically heterogeneous refractory inclusions $\sim$0.5–5.0vol%.	
		The **CM subgroup** contains meteorites which are friable and have a low water content.	Sparse CM2
		The **CI subgroup** contains meteorites which are friable and have a high water content.	Absent CI
		The **CV subgroup** contains up to 20wt% H_2O locked into hydrated minerals, iron-rich olivine, with calcium and aluminum inclusions. The CV3 subgroup has recently been further divided into three subgroups: (1) reduced (CV3R), (2) oxidized-bali (CV3OxB), and (3) oxidized-allende (CV3OxA).	Sparse CV2 Abundant CV3 Distinct CV4 Less Distinct CV5

The CO subgroup contains meteorites which have minute amounts of chondrules.

1.1.2. Ordinary chondrites are composed of varying ratios of olivine and pyroxenes with spherical chondrules that represent unmelted condensates of the presolar nebula. This group is subdivided according to ratio of iron to silicone (Fe/Si) into: H (olivine-bronzite), L (olivine-hypersthene), and LL (amphoterite). The H subgroup, having the lowest oxygen content, formed nearest the sun, with the L's and LL's at increasing heliocentric distances. The petrologic types of the ordinary chondrites range from 3 to 7 since no aqueous alteration took place on their parent bodies. Ordinary chondrites have few opaque mineral-rich porphyritic chondrules. Ordinary chondrites have a CI-normalized mean-refractory-lithophile abundance ratio of 0.77–0.82. Ordinary chondrites have a fine-grained matrix-chondrule modal abundance ratio of 0.3. Ordinary chondrites have a negligible amount of isotopically heterogeneous refractory inclusions.

Abundant CO3
Distinct CO4

The **H subgroup** chondrites with H for high iron content (12 to 27wt% Fe metal) are also called bronzite chondrites (Mg/Si = 0.97).

Abundant H3
Distinct H4
Less Distinct H5
Indistinct H6
Melted H7

The **L subgroup** chondrites with L for low iron (5 to 10% Fe metal) are also called hypersthene chondrites (Mg/Si = 0.92).

Abundant L3
Distinct L4
Less Distinct L5
Indistinct L6
Melted L7

The **LL subgroup** chondrites with LL for both a low iron content of 20wt% along with a low metal content of only 2wt% are also called amphoterite. The chief minerals are bronzite, olivine, and to a lesser extent, oligoclase (Mg/Si = 0.92).

1.1.3. Enstatite chondrites are highly reduced with all of the iron visible as metal or troilite (FeS). The silicate consists mainly of the iron-free pyroxene, enstatite. As with the ordinary chondrites, a subdivision is made according to the ratio of iron to silicon (Fe/Si) into two subgroups. The EL (Mg/Si = 0.73) and EH (Mg/Si = 0.88) chondrites are from separate parent bodies. From studies of rare-gas fractionation patterns, some researchers believe they may have formed inside the orbit of Venus, while the identification of E-type asteroids in the inner asteroid belt suggests this was their location of origin. Enstatite chondrites have a low to moderate abundance of opaque mineral-rich porphyritic chondrules. Enstatite chondrites have a CI-normalized mean-refractory-lithophile abundance ratio of roughly 0.6. Enstatite chondrites have a negligible amount of isotopically heterogeneous refractory inclusions.

Distinct E4
Less Distinct E5
Indistinct E6
Melted E7

79

1

12
Rocks
and
Meteorites

Table 12.30 (continued)

Percentage of all falls	Category	Description and distinguishing features	Chondrule character and/or letter designation
		1.1.4 Other chondrites	
		B chondrites – this is a newly designated grouplet of four meteorites which are members of the CR clan. The bencubbinites consist of Bencubbin, Weatherford, HaH 237, and GRO95551. The group has a metal–silicate chondritic composition with highly reduced silicates and over 50% Fe–Ni. Cryptocrystalline chondrules are present, as are CAIs in HaH 237. Oxygen isotopes suggest a very close relationship between the bencubbinite grouplet and the CR and CH chondrites.	
		R chondrites – this group of 12 meteorites, formerly known for the Carlisle Lakes specimen, is now known for the only fall of the group, Rumuruti. The group is highly oxidized, olivine-rich, and metal-poor. R chondrites have a high degree of Fe oxidation. They differ greatly in oxidation state, oxygen isotope composition, and mineralogy from ordinary, carbonaceous, or enstatite chondrites, or silicate inclusions in IAB and IIE siderites. The parent body was originally highly unequilibrated but was subsequently thermally metamorphosed and impact-melted to a moderate degree. Most members are highly brecciated and contain implanted solar wind gases. R chondrites have few opaque mineral-rich porphyritic chondrules. R chondrites have a CI-normalized mean-refractory–lithophile abundance ratio of 0.85. R chondrites have a fine-grained matrix–chondrule modal abundance ratio of 1.6. R chondrites have essentially no amount of isotopically heterogeneous refractory inclusions.	
		K chondrites – the type specimen of this chondrite grouplet, Kakangari, along with two other members, have unique petrologic, bulk chemical, and O isotopic characteristics that distinguish them from other chondrite groups. The grouplet also does not fit into the existing systematics of the E, O, R, or C chondrites as their characteristics relate to heliocentric distance of formation. K chondrites therefore represent a unique, primitive, parent asteroid.	
		F chondrites – forsterite chondrites are intermediate in composition, mineralogy, and oxidation state between the H-group ordinary and enstatite chondrites. They represent a highly unequilibrated distinct chondritic suite that underwent nebula condensation/accretion before colliding with the aubrite parent body. The highly-shocked chondritic fragments were incorporated into the aubrite meteorites Cumberland Falls and ALH78113 forming a breccia.	
8		**1.2 Achondrites** – all members of this classification originated on chondritic bodies that underwent igneous melting and recrystallization. Their parent bodies were large enough to melt and segregate the denser metals from the lighter silicates, generally forming a metallic core, a magnesium-rich mantle, and a calcium-rich crust. Of the various achondrites, three are believed to have originated on the asteroid 4 Vesta. These represent the brecciated surface materials (howardites), the extrusive basalts (eucrites), and the plutonic cumulates (diogenites). In addition, 13 meteorites comprising three groups originated on Mars (eight shergottites, four nakhlites (including one orthopyroxenite), and one chassignite), and 13 meteorites found are of lunar origin. The winonaites formed in the same nebula locality as that in which the iron group IAB silicate inclusions	

...angrites, brachinites, acapulcoites, lodranites, ureilites, and the aubrites. Achondrites represent about 8% of all meteorite falls. Achondrites contain no chondrules, but are not chemically homogeneous; the major minerals are pyroxene, olivine, and feldspars.

Aubrites		
Angrite		
Ureilites		
Subgroup HED		
	Howardites, bucrite–diogenite mix	AHOW
	Eucrites, anorthite–pigeonite	AEUC
	Diogenites, hypersthene	ADIO
Subgroup SNC (= Shergotty–Nakha–Chassigny)		
	Shergottites, basaltic	AEUC
	Nakhlites, diopside–olivine	ACANOM
	Chassignite, olivine	ACANOM

5.7

2. Siderites or iron meteorites – these meteorites are made of a crystalline iron–nickel alloy. Scientists believe that they resemble the outer core of the Earth. This is a varied group of meteorites composed mainly of Fe–Ni metal with small amounts of other minerals. Most were formed in the cores of differentiated asteroids, although some probably formed in small melt pods distributed around smaller parent bodies. They are separated into distinct chemical groups based on their trace element contents, each representing an origin on a unique asteroid. The determining factors are groupings of meteorites with similar ratios of trace elements to nickel. Generally, the higher the Roman numeral of the classification, the lower the concentration of trace elements (chemical classification). Chemical classification is important because it suggests that certain iron meteorites share a common origin or were formed under similar conditions. Siderites can also be classified by their internal structure (the Widmanstätten bandwidth), influenced by bulk nickel content, into hexahedrites, octahedrites, and ataxites (structural classification).

Hexahedrites > 50 mm		H
Octahedrites		
	Coarsest 3.3–50 mm Ogg	
	Coarse 1.3–3.3 mm Og	
	Medium 0.5–1.3 mm Om	
	Fine 0.2–0.5 mm Of	
	Finest < 0.2 mm Off	
	Plessitic < 0.2 mm (kamacite spindles) Opl	
	Ataxites (no structure)	D

12
Rocks and Meteorites

Table 12.30 (continued)

Percentage of all falls	Category	Description and distinguishing features	Chondrule character and/or letter designation
1.5–2.8	**3. Lithosiderites or stony iron meteorites** – these meteorites are mixtures of iron–nickel alloys and minerals such as schreibersite (Fe,Ni,Co)$_3$P, troilite (FeS), cohenite (Fe$_3$C), and graphite (C). Scientists believe that they are like the material that would be found where the Earth's core meets the mantle. Polishing of a cut surface with etching produces different structures: Neumann lines, i.e., parallel lines, that cross each other under various angles mostly formed under mechanical pressure; and Wittmanstätten patterns. There is not yet a consensus for the origin of the siderites, and different theories currently exist to explain their formation. The standard theory calls for a large differentiated asteroid that underwent igneous activity to produce a basaltic crust. A large impactor mixed molten metal with the cooler silicates and was rapidly cooled. This was followed by burial in a deep regolith where slow cooling proceeded until excavation and delivery to Earth. Another theory has the basaltic crust of a molten parent body founder and sink through the mantle to the metallic core where mixing occurred. Subsequent collisions exposed this stony-iron layer and delivered fragments to Earth. There is also a theory that calls for the collisional disruption and gravitational reassembly of an asteroid to explain the mixing observed.	
		Pallasites – these meteorites are mixtures of olivine and Fe-Ni metal that formed deep in the core–mantle boundary of a small, differentiated asteroid. As the overlying cumulate olivine cooled and contracted, the still slightly molten metal was injected into the crystalline olivine forming a continuous matrix. Later collisions exposed this layer and delivered samples to Earth. There are three compositional clusters representing separate parent bodies.	PAL
		Mesosiderites – the mesosiderites are complex assemblages of Fe-Ni metal with orthopyroxene, plagioclase, olivine, and eucritic clasts. They range from little recrystallized to melted (subgroups I–IV), which cooled slowly at great depth. Silicate material was mixed with the viscous Fe-Ni metal, cooling to form silicated irons. Most belong to the three non-magmatic groups IAB, IIICD, and IIE. Group IIE silicated irons are related to the H chondrites, while the unique silicated iron Steinbach is related to group IVA irons.	MES
		Lodranites iron, pyroxene, olivine	LOD
		Siderophyre iron, orthopyroxene	IVA-ANOM

Weathering Grade (A)

- W1 – minor oxide rims around metal and troilite minor oxide veins.
- W2 – moderate oxidation of metal, about 20–60% being affected.
- W3 – heavy oxidation of metal and troilite, 60–95% being replaced.
- W4 – complete (>95%) oxidation of metal and troilite, but no alteration of silicates.
- W5 – beginning alteration of mafic silicates, mainly along cracks.
- W6 – massive replacement of silicates by clay minerals and oxides.

Weathering Grade (B)

- A – minor rustiness; rust haloes on metal particles and rust stains along fractures are minor.
- B – moderate rustiness; large rust haloes occur on metal particles and rust stains on internal fractures are extensive.
- C – severe rustiness; metal particles have been mostly stained by rust throughout.
- E – evaporite minerals visible to the naked eye.

Fracturing Scale

- A – minor cracks; few or no cracks are conspicuous to the naked eye and no cracks penetrate the entire specimen.
- B – moderate cracks; several cracks extend across exterior surfaces and the specimen can be readily broken along the cracks.
- C – severe cracks; specimen readily crumbles along cracks that are both extensive and abundant.

Shock Stage

- S1 – unshocked, peak shock pressure < 5 GPa.
- S2 – shocked, peak shock pressure 5–10 GPa.
- S3 – shocked, peak shock pressure 10–20 GPa.
- S4 – shocked, peak shock pressure 20–35 GPa.
- S5 – shocked, peak shock pressure 35–55 GPa.
- S6 – very strongly shocked, peak shock pressure 55–75 GPa.
- Note that whole rock impact melting occurs at 75–90 GPa.

12.6.3 Tektites, Impactites, and Fulgurites

Tektites, or glassy meteorites, consist of a silica-rich glass (70wt% SiO_2) similar to some hyaline igneous rocks such as obsidian. They exhibit an unusual chemical composition which

Table 12.31 Names of tektites according to geographical locations

Continent	Region	Name
Africa	Ivory Coast	Ivoirites
	Libya	Tectites
North America	Texas	Bediasites
	Georgia	Georgite
Latin America	Mexico, Peru, Columbia	Americanites
Australasia	Philippines	Rhizalites
	Thailand	Australasites
Europe	Czech and Slovak Republics	Moldavites
Russia	Kazakstan	Irgizites

comprises a high content of SiO_2, Al_2O_3, K_2O, and CaO, and a low content of MgO, FeO, Fe_2O_3, and Na_2O. They are found generally as small rounded masses in areas that preclude a volcanic origin. Nevertheless, by contrast with meteorites, tektites have not been seen to fall and their origin is still not well known. On the other hand, impactites and fulgurites are certainly terrestrial materials produced by a sudden high-energy impact or explosion (e.g., atomic bomb), or lightning in desert areas.

12.7 Physical Properties of Common Rocks

Engineers not only need to have a basic understanding of the physical, chemical, and mineralogical characteristics of rocks and soils but also of their response to applied loads, the flow of fluids, and other environmental stresses. Therefore, knowledge of the mechanical behavior of rocks and soils and their fluid flow characteristics as porous media are the important data for geotechnical and civil engineers. The materials in geological engineering are rocks and soils. At the surface, projects include: building as foundations, highways and railroads, dams and reservoirs, slopes and landslides. Below the surface, projects in which the mechanical properties of rocks are important include the construction of: mine shafts, levels, raises and adits, tunnels, storage caverns, and disposal chambers. On the other hand, fluids exert a very strong influence on the mechanical behavior of rocks and soils. The engineering properties of rocks are influenced by a large number of geological factors. Mineralogy and particle-contacts control strength on a small scale; tectonic deformation, igneous activity, and metamorphism all result in substantial changes in the mechanical behavior of rocks through recrystallization and fracturing. The increase in sediment load during diagenesis combined with cementation and filling pores results in increased strength, and decreased porosity, and hence permeability. In general rocks become stronger and less porous and permeable as they get older. Recent sediments are normally weaker than ancient rocks with similar lithology and mineralogy. Rocks and soils with a level of compaction corresponding to their present burial depth are said to be normally consolidated. Where erosion has occurred, rocks may be compacted much more than expected for their current depth of burial. These rocks and soils are said to be overconsolidated. Rocks that have not compacted to the expected extent for their depth of burial, perhaps because fluids could not escape, are said to be underconsolidated. Underconsolidated rocks are often associated with high fluid pressures (overpressure). An overpressure is a pressure in excess of the pressure predicted from the normal hydrostatic gradient.

References

[1] Cross W, Iddings JP, Pirsson, IV et al. (1912) A quantitative chemico-mineralogical classification and nomenclature of igneous rocks. J Geology 10: 555–690.
[2] Streckeisen AL (1976) Classification of the common igneous rocks by means of their chemical composition: a provisional attempt. Neues Jahrbuch fur Mineralogie, Monatshefte H1, H1–15.

Further Reading

Carzzi A (1953) Pétrographite des roches sédimentaires. F Rouge, Lausanne.
Johanssen A (1932, 1937, 1938, 1939) A descriptive petrography of igneous rocks. University of Chicago Press, Chicago.
Jung J (1959) Classification modale des roches éruptives: utilisant les données fournies par le compteur de points. Masson, Paris.
Jung J (1977) Précis de pétrographie, 3rd edn. Masson, Paris.

Table 12.32 Some physical properties of selected rocks

Name	Density (ρ, kg.m^{-3})	Thermal conductivity (k, W.m^{-1}.K^{-1})	Specific heat capacity (c_p, kJ.kg^{-1}.K^{-1})	Electrical volume resistivity (ρ, Ω.cm)
Sedimentary rocks				
Anthracite	1400–1700	0.26	1260	10^2–10^9
Asphalt	1100–1500	0.06	920	n.a.
Bauxite	2550	n.a.	n.a.	n.a.
Chalk	1900–2800	0.2–0.92	n.a.	n.a.
Clay, soft shale	2200–2700	1673	837	10^2–10^3
Coal	1400–1800	0.26	1089–1548	n.a.
Conglomerate	2000–2700	n.a.	n.a.	n.a.
Craie	1200–2400	n.a.	n.a.	n.a.
Diatomite	200–300	0.05	n.a.	n.a.
Dolomite	2760–2840	n.a.	n.a.	n.a.
Flint	2500–2800	n.a.	n.a.	n.a.
Gravel (dry)	1400–1700	n.a.	n.a.	n.a.
Gypsum	2200–2320	0.753–1.297	1088	n.a.
Lignite	1100–1400	n.a.	n.a.	n.a.
Limestone (hard)	2200–2760	1.67–2.15	907–921	10^6
Limestone (soft)	1200–2200	0.92–1.67	907–921	10^3–10^5
Marl	1800–2600	n.a.	n.a.	n.a.
Mud (river, moist)	1440	n.a.	n.a.	n.a.
Oil (crude)	870	n.a.	n.a.	10^8–10^{11}
Peat	840	n.a.	n.a.	n.a.
Phosphate rock	3200	n.a.	n.a.	n.a.
Rock salt	2100–2200	n.a.	n.a.	10^5–10^6
Sand (dry)	1600–1700	0.27–0.33	753–799	10^3–10^5
Sandstone (hard)	2140–2900	1.30	745	10^4–10^5
Sandstone (soft)	1600–2100	2.90	963	n.a.
Shale	2600–2900	n.a.	n.a.	n.a.
Soil	1120–1700	0.52	1840	n.a.
Metamorphic rocks				
Asbestos	2000–2800	0.07	n.a.	n.a.
Gneiss	2700–2900	n.a.	n.a.	10^4–10^{10}
Marble	2600–2850	2.80	879	n.a.
Mica schist	2400–3200	n.a.	n.a.	n.a.
Quartzite	2300–2700	5.38	711–1105	n.a.
Slate	1500–3200	1.4–2.5	n.a.	5×10^4 to 7×10^4
Igneous rocks				
Amphibolite	2900–3000	n.a.	n.a.	n.a.
Andesite	2000–2900	n.a.	n.a.	n.a.
Basalt	2800–3200	0.92–2.0	n.a.	n.a.
Basanite	2600–3200	n.a.	n.a.	n.a.
Diabase	2800–3100	n.a.	n.a.	n.a.

12

Rocks
and
Meteorites

Table 12.32 (continued)

Name	Density (ρ, kg.m^{-3})	Thermal conductivity (k, W.m.$^{-1}$.K^{-1})	Specific heat capacity (c_p, kJ.kg^{-1}.K^{-1})	Electrical volume resistivity (ρ, Ω.cm)
Diorite	2700–3000	n.a.	n.a.	n.a.
Dolerite	2700–3000	n.a.	n.a.	n.a.
Gabbro	2800–3100	n.a.	n.a.	n.a.
Granite	2640–2760	2.51–3.97	837–1088	10^4–10^9
Granodiorite	2680–3000	n.a.	n.a.	n.a.
Grenatite	3500–4300	n.a.	n.a.	n.a.
Microdiorite	2700–3000	n.a.	n.a.	n.a.
Microgabbro	2700–3100	n.a.	n.a.	n.a.
Microgranite	2500–2700	n.a.	n.a.	n.a.
Microsyenite	2700–2900	n.a.	n.a.	n.a.
Nepheline syenite	2600–2700	n.a.	n.a.	n.a.
Obsidian and tachylite	2300–2850	n.a.	n.a.	n.a.
Peridotite	3100–3450	n.a.	n.a.	n.a.
Pumice	390–1100	n.a.	n.a.	n.a.
Rhyolithe	2300–2500	n.a.	n.a.	n.a.
Syenite	2600–2900	n.a.	n.a.	n.a.
Trachyte	2200–2700	n.a.	n.a.	n.a.
Volcanic slag	1100–2000	n.a.	n.a.	n.a.
Ice	917	1.8–2.2	2093	3×10^8

Le Bas MJ, Streckeisen AL (1991) The IUGS systematic of igneous rocks. J Geol Soc 148: 825–833.

Lliboutry L (1964, 1965) Traité de glaciologie: tomes 1 et 2. Masson & Cie, Paris.

Michel-Levy A (1889) Structures et classification des roches eruptives. Baudry Editeur, Paris.

Milner HB (1940) Sedimentary petrology. Th. Murby, London.

Moorhouse WW (1959) The study of rocks in thin sections. Harper, New York.

Pettijohn, FJ (1949) Sedimentary rocks. Harper and Brothers, New York.

Raguin E (1970) Pétrographie des roches plutoniques dans leur cadre géologique. Masson, Paris.

Raguin E (1976) Géologie du granite, 3rd edn.. Masson, Paris.

Shelley D (1993) Igneous and metamorphic rocks under the microscope. Chapman & Hall, New York.

Turner FJ, Verhoogen J (1967) Igneous and metamorphic rocks. Springer-Verlag, Berlin.

Twenhofel WH (1950) Principles of sedimentation. McGraw-Hill, New York.

Williams H, Turner FJ, Gilbert CM (1954) Petrography. Freeman, San Francisco.

13 Timbers and Woods

13.1 General Description

Timber can be considered as a typical natural composite material with a highly anisotropic structure. Indeed, this structure has two chief directions, radial and longitudinal, corresponding to its botanical organization. Furthermore, superimposed to these two degrees of variability are local effects such as growing conditions. For classifications, the terms hardwood and softwood have no relation to the actual mechanical hardness of the wood. It is only a broad botanical distinction. **Hardwoods** are generally broad-leaved deciduous trees which carry their seeds in seedcases (i.e., Angiosperms), such as ash, balsa, beech, greenheart, oak, obeche, and maple, while **softwoods** are generally coniferous trees (i.e., Gymnosperms) such as douglas fir, yellow pine, larch, spruce, hemlock, red cedar, and yew. From a structural-botanical point of view, wood contains many cells. These cells have different functions depending on their location in the tree. Inner cells, located in the **heartwood**, are mostly dead and provide mechanical support for the tree and in which the reverse materials, e.g., starch, have been removed or converted into resinous substances, Heartwood is generally darker than sapwood, although the two are not always clearly differentiated. Cells located in the **sapwood** store nutrients and act a conduits for water. Only the **cambium**, one-cell-thick layer, located just beneath the bark, contains new growing cells allowing the tree to grow, and subdivides the new wood from bark cells. This creates the rings each year. Hence, wood is considered from a strict mechanical point of view, as a complex fiber-reinforced composite composed of long, unidirectionally aligned tubular cellulosic polymer cells in a polymer matrix made of **lignine**. **Cellulose** is a naturally occurring carbohydrate and a thermoplastic polymer; it is arranged in long chains to form a framework. A bundle of these long chains is enclosed by both hemicellulose, a short polymer, and lignine, an organic cement that bonds these bundles, or microfibrils, together. Many of these unidirectionally aligned microfibrils compose the inner cell structure. Wrapped around the core is the cell wall consisting of more microfibrils, except that they are randomly oriented.

13.2 Properties of Woods

Physical properties of woods strongly depend on the moisture content (water mass fraction). Owing to this strong dependence, it is advisable when using timber in some particular applications (e.g., marine, chemical process industry, foodstuffs) to take figures at maximum moisture content when data are available. Hence, specification of timber is not a simple matter of identification of the species of tree from which the material has been cut. As a general rule, accurate determination of a timber species from small samples of wood is quite impossible by macroscopic examination alone and always requires an accurate microscopic identification by a botanical expert, in particular with tropical timber species.

Moisture content is a dimensionless quantity which consists of the mass fraction of water contained in the wood, expressed as a percentage of the mass of the oven-dry wood (wt%). Wood is commonly grouped in five classes according to the moisture content: (i) **Air-dried** – wood having an average moisture content of 25wt% or lower, with no material over 30wt% (ii) **Green** – freshly sawn wood or wood that essentially has received no formal drying. (iii) **Kiln-dried** – dried in a kiln or by some other refined method to an average moisture content specified or understood to be suitable for a certain use. Kiln-dried lumber can be specified to be free of drying stresses. (iv) **Partly air-dried** – wood with an average moisture content between 25 and 45wt%, with no material over 50wt%. (v) **Shipping dry** – lumber partially dried to prevent stain or mold in brief periods of transit, preferably with the outer 3 mm dried.

Density and specific gravity – these two physical quantities may be related to important wood attributes such as mechanical strength, shrinkage, paper-forming properties, and cutting forces required in machining. In assessing the use potential of a species, density is a physical quantity which often receives initial attention. Basic specific gravity (noted as oven-dry values in common tables) is the dimensionless ratio of wood density to the density of water at its maximum density ($3.98°C$) and is calculated from the oven-dry mass and volume in the green condition. This may range from less than 0.34 for balsam poplar (*Populus balsamifera*), to about 0.88 for live oak (*Quercus virginiana*). Density is accurately calculated from mass and volume ratios, when the woods are green or when air dry, usually at a moisture content of 12wt% water. Densities may range from 320 to 1100 $kg.m^{-3}$ balsam to ebony. Two factors control the density of wood: (i) the species of the wood, and (ii) the moisture content. Hardwoods such as oak, elm, and maple have higher densities than softwoods such as pine, spruce, and cedar. Although a live tree contains large amounts of water, when the tree is cut the moisture content depends on the surrounding relative humidity. The higher the humidity, the more water will remain contained in the dead wood. Water content and the type of tree control the density which, in turn, controls the mechanical properties (Table 13.1, pages 512-516).

Mechanical properties – the common mechanical properties reported for woods are, according to the standard ASTM D 143: the hardness in N, the work to maximum load in $kJ.m^{-3}$, Young's modulus (modulus of elasticity) in GPa, the bending strength (modulus of rupture) in MPa, the compression strength (maximum crushing strength) both parallel and perpendicular to the grain (stress at proportional limit) in MPa, and finally the shear stress parallel to the grain in MPa.

Drying and shrinkage – the response of individual woods to air-drying and kiln-drying is noted as well as the absence or presence of degradation due to checking, warp, or collapse. percentage of shrinkage values (i.e., volumetric, radial, tangential) from the green to oven-dry condition (0wt% moisture content) or green to various air-dry conditions (6, 12 or 20wt% moisture content) are the most common properties reported.

Durability – resistance of the wood to attack by decay fungi, insects, and marine borers is another important characteristic to consider when selecting a wood for marine applications and outdoor service. As a general rule wood kept constantly dry does not decay. Further, if it is kept continuously submerged in water even for long periods of time, it is not decayed significantly by the common decay fungi regardless of the wood species or the presence of sapwood. Bacteria and certain soft-rot fungi can attack submerged wood but the resulting deterioration is very slow. A large proportion of wood in use is kept so dry at all times that it lasts indefinitely. Moisture and temperature, which vary greatly with local conditions, are the principal factors affecting the degradation rate.

The strength of a wood depends on its density; higher density, higher strength. When wood is drying, almost no change in strength is observed until the amount of water drops below 30wt%. A piece of wood can carry different loads in different directions (anisotropic behavior). For instance, a specimen of wood can carry a much greater load in the longitudal direction (with the grain) than it can in the radial or tangential directions (against or across the grain). The modulus of elasticity is also highly anisotropic. The modulus of elasticity perpendicular to the grain is 1/20th of the modulus parallel to the grain. Mechanical properties also depend on imperfections of the wood. Knots in the wood can decrease the tensile strength. The use of plywood is a good way to reduce the anisotropic behavior of wood. Thin layers of wood, called plies are stacked atop each other aligning the grains perpendicular with each new layer. A thermosetting phenolic resin is spread between the layers of plies and set under pressure.

13.3 Applications

Despite its continuing use in building and marine engineering since ancient times, timber has declined considerably as an engineering material since the beginning of the twentieth century. Nevertheless, in the chemical process industries (CPI), oak, pine, redwood, and cypress are mainly used for corrosion applications such as cooling towers and storage tanks. On the other hand, filter-press frames, structural members of buildings, and barrels are also sometimes made of wood. There is no better wood, and the selection depends on the exposure. Containers must be kept wet or the staves will shrink, warp, and leak. A number of manufacturers offer wood impregnated to resist acids or alkalis or the effect of high temperatures. Apart from its use as a construction material, since the beginning of humanity wood has been a valuable fuel owing to its desirable properties: (i) readily available, (ii) renewable, (iii) relatively inexpensive, (iv) easy to store and handle, and (v) relatively non-polluting (see Table 13.2).

Table 13.2 Heating value of common woods

Heating value	High	Medium	Low
Type	Beech, hornbeam, yew, oak, ash, birch, haw-thorn, hazel, plane, apple	Elm, cherry, sycamore, cedar, douglas, fir, wal-nut, larch	Poplar, spruce, alder, pine, willow

13.4 Wood Performance in Various Corrosives

While wood is fairly inert chemically, it is readily dehydrated by concentrated solutions, and hence shrinks badly. It is also slowly hydrolyzed by acids and alkalis, especially when hot. Strong acids and dilute alkalis attack wood. In tank construction, if sufficient shrinkage takes place to allow crystals to form between the staves, it becomes very difficult to make the tank tight again. A strong drawback of timber construction material is that wood is also subject to biological attack (e.g., bacteria, marine borers). Therefore, in order to reduce both chemical and biological attack, timber structures need to be treated (e.g., impregnated or coated) with special mixtures such as waxes, plastic resins, or others depending on applications (e.g., marine, foodstuffs). For example, in marine applications, in order to be immune to borers, wood is treated by a high-pressure impregnation with a creosote or copper–chromium–arsenic water-borne mixture. The use of biocides should be also considered. A number of manufacturers offer wood impregnated to resist acids or alkalis or the effect of high temperatures.

Table 13.1 Mechanical properties of selected woods (12wt% moisture)

Usual name	Botanical Latin name (genus and species)	Category[a]	Radial shrinkage (%) (tangential)[b]	Specific gravity[c]	Bulk density (ρ, kg.m^{-3})	Young's modulus (||) (E, GPa)	Hardness (N)	Work to maximum load (WML, KJ.m^{-3})	Module of rupture or static bending strength (MOR, MPa)	Compressive or crushing strength (||) (MPa)	Compressive or crushing strength ($\perp$) (MPa)	Shear strength (||) (MPa)
Afromosia	*Pericopsis elata*	H	6.4 (10.7)	0.57	689	12.45	6926	n.a.	133.76	71.36	n.a.	n.a.
Ash (blue)	*Fraxinus quadrangulata*	H	3.9 (6.5)	0.58	651	9.65	9024	99.29	95.15	48.13	9.79	14.00
Ash (white)	*Fraxinus americana*	H	4.9 (7.9)	0.64	603–673	11.99	5871	114.46	106.26	51.09	8.00	13.17
Aspen (quaking)	*Populus tremula*	H	(9.2)	0.380	417	8.136	1557	52.40	57.92	29.30	2.55	5.86
Balsa	*Ochroma*	H	n.a.	n.a.	120	n.a.	n.a.	n.a.	n.a.	n.a.	n.a.	n.a.
Balsam poplar	*Populus balsamifera*	H	3.9 (9.2)	0.34	0.68	7.59	1334	34.48	46.89	27.72	2.07	6.41
Basswood (American)	*Tilia americana*	H	6.6 (9.3)	0.37	400–417	10.07	1824	49.64	59.99	32.61	2.55	6.83
Beech (American)	*Fagus grandifolia*	H	5.1 (11.0)	0.64	560–720	11.86	5782	104.12	102.74	50.33	6.96	13.86
Berlinia, ebiara	*Berlinia* spp.	H	4.4 (8.9)	0.58	705	8.75–10.82	6038	n.a.	91.01–118.58	53.02–55.16	n.a.	n.a.
Birch (silver)	*Betula populifolia*	H	n.a.	n.a.	n.a.	11.86	n.a.	n.a.	n.a.	n.a.	n.a.	n.a.
Birch (yellow)	*Betula alleghaniensis*	H	7.2 (9.2)	0.66	660–689	13.91	8347	104.1	114.45	56.33	6.95	8.68
Black wattle	*Acacia molissima*	H	n.a.	0.64	721	14.34	7770	105.3	120.66	60.67	7.54	n.a.
Bluegum (southern)	*Eucalyptus globulus*	H	8.0 (12.2)	0.80	817–977	16.34–20.34	11,455	n.a.	114.45–146.16	68.53–82.73	n.a.	n.a.
Boxelder	*Acer negundo*	H	3.9 (7.4)	0.457	513	5.998	3203	64.5	35.992	34.13	5.38	9.38

Common name	Species											
Bucida (oxhorn)	Bucida buceras	H	4.4 (7.9)	0.93	1.105	13.79	10,390	n.a.	106.17	n.a.	n.a.	n.a.
Cedar (western red)	Cedrela toona	S	2.4 (5.0)	0.34	368	7.7–8.0	n.a.	n.a.	51.71	31.44	3.17	5.92
Cedar (white)	Cedrela orientalis	S	n.a.	n.a.	300	4.8	n.a.	n.a.	n.a.	n.a.	n.a.	n.a.
Cherry (black)	Prunus serotina	H	3.7 (7.1)	0.53	560	10.3	7548	n.a.	84.8	49.02	4.8	6.55
Chesnut (American)	Castanea dentata	H	2.7 (5.4)	0.43	481	8.48	2401	44.82	59.30	36.68	4.28	7.45
Cottonwood (black)	Populus trichocarpa	H	3.6 (8.6)	0.35	384	8.76	1557	46.20	58.60	31.03	2.07	7.17
Cottonwood (eastern)	Populus deltoides	H	3.9 (9.2)	0.43	449	9.45	1912	51.02	58.61	33.85	2.62	6.41
Cypress	Taxodium distichum	S	3.8 (6.2)	0.48	482–512	9.9	2264	n.a.	73.08	43.85	5.37	6.89
Danta, kotibe	Nesogordonia papaverifera	H	5.4 (8.2)	0.65	800	10.90–11.65	9502	n.a.	128.24–136.52	65.16–69.29	n.a.	n.a.
Douglas fir (coast)	Pseudostuga taxifolia	S	4.8 (7.6)	0.51	512–545	13.4	3152	n.a.	85.49	49.91	5.52	7.99
Ebony	Diospyros kurzii	H	n.a.	n.a.	978	n.a.	n.a.	n.a.	n.a.	n.a.	n.a.	n.a.
Elm (American)	Ulmus americana	H	4.2 (9.5)	0.55	551–658	9.24	3692	89.66	81.36	38.06	4.76	10.41
Elm (rock)	Ulmus thomasii	H	4.8 (8.1)	0.63	660–795	10.62	9101	132.34	102.5	48.61	8.48	13.24
Eucalyptus	Eucalyptus diversicolor	H	n.a.	n.a.	829	n.a.	n.a.	n.a.	n.a.	n.a.	n.a.	n.a.
Eucalyptus	Eucalyptus hemilampra	H	n.a.	n.a.	1058	n.a.	n.a.	n.a.	n.a.	n.a.	n.a.	n.a.
Hairi	Alexa imperatricis	H	4.0 (8.5)	0.505	513	10.89	n.a.	n.a.	73.02	38.75	n.a.	n.a.
Hemlock (eastern)	Tsuga canadensis	S	3.0 (6.8)	0.43	430–449	8.3	2220	n.a.	61.36	37.30	4.48	7.31

Table 13.1 (continued)

Usual name	Botanical Latin name (genus and species)	Category[a]	Radial shrinkage (%) (tangential)[b]	Specific gravity[c]	Bulk density (ρ, kg.m^{-3})	Young's modulus (∥) (E, GPa)	Hardness (N)	Work to maximum load (WML, KJ.m^{-3})	Module of rupture or static bending strength (MOR, MPa)	Compressive or crushing strength (∥) (MPa)	Compressive or crushing strength (⊥) (MPa)	Shear strength (∥) (MPa)
Hemlock (western)	Tsuga heterophylla	S	4.3 (7.9)	0.44	440–465	11.3	2398	n.a.	77.91	49.02	3.79	8.61
Hickory (pignut)	Hicoria alba	H	7.0 (10.5)	0.77	770–836	15.58	n.a.	209.61	138.59	63.65	13.65	14.82
Iroko	Chlorophora excelsa	H	7.0 (10.5)	n.a.	800	n.a.	n.a.	n.a.	n.a.	n.a.	n.a.	n.a.
Ironwood	Rhamnidium ferrum	H	n.a.	n.a.	1077	n.a.	n.a.	n.a.	n.a.	n.a.	n.a.	n.a.
Juniper (alligator)	Juniperus deppeana steud.	H	2.7 (3.6)	0.51	577	4.48	5160	44.82	46.49	28.41	11.72	n.a.
Lapacho, bethabara, ipe	Tabebuia spp.	H	6.6 (8.0)	0.85–0.97	1057–1201	20.75–23.10	n.a.	n.a.	173.75–193.05	89.70–96.52	n.a.	n.a.
Larch (western)	Larix occidentalis	S	4.5	0.59	590–609	12.9	3685	n.a.	90.32	52.68	6.76	9.38
Magnolia	Magnolia virginiana	H	4.7–9.1 (8.3)	0.45	465	11.5	n.a.	n.a.	70.4	39.8	3.9	11.8
Mahogany	Swietenia spp.	H	3.5 (4.8)	0.51	460–545	10.3	3552	n.a.	79.01	46.88	7.58	8.48
Maple (black)	Acer nigrum	H	4.8 (9.3)	0.57	641	11.17	5249	86.19	91.70	46.06	7.03	12.55
Maple (red)	Acer rubrum	H	4.0 (8.2)	0.54	609	11.31	4226	86.19	92.39	45.09	6.90	12.76
Maple (silver creek)	Acer saccharinum	H	3.0 (7.2)	0.47	529	7.86	3114	57.23	61.37	35.99	5.102	10.21
Maple (sugar)	Acer saccharum	H	4.9 (9.5)	0.63	676–704	12.62	6450	113.77	108.94	53.99	10.14	16.07

Maple (bigleaf)	Acer macrophyllum	H	3.7 (7.1)	0.48	545	9.99	3781	53.78	73.78	41.03	5.17	11.93
Oak (chestnut)	Quercus prinus	H	4.0 (8.2)	0.66	737	10.96	5026	75.85	91.70	47.09	5.79	10.27
Oak (northern) (red)	Quercus rubra	H	4.0 (8.2)	0.66	705	12.55	5737	100.67	99.98	48.54	7.79	12.27
Oak (post)	Quercus stellata	H	4.0 (8.2)	0.67	753	10.41	6049	91.01	91.01	45.51	9.86	12.69
Oak (red) scarlet	Quercus coccinea	H	4.0 (8.2)	0.67	753	13.17	6227	103.43	119.97	57.44	7.72	13.03
Oak (white)	Quercus alba	H	5.3 (9.0)	0.68	769	12.27	8800	102.05	104.80	51.30	7.38	13.79
Oboto	Mammea africana	H	6.5 (10.0)	0.53–0.70	657–865	14.34–14.62	n.a.	n.a.	138.58–160.65	68.26–77.22	n.a.	n.a.
Padauk (African)	Pterocarpus soyauxii	H	n.a. (n.a.)	n.a.	n.a.	n.a.	n.a.	n.a.	n.a.	n.a.	n.a.	n.a.
Pine (Eastern) white	Pinusi strobus	S	2.6 (6.1)	0.37	370–432	8.5–10.06	1687	n.a.	66.88	34.75	3.24	7.17
Pine patula	Pinus patula	S	4.1 (7.9)	n.a.	481–609	8.34–12.82	n.a.	n.a.	82.74–97.91	37.92–50.33	n.a.	n.a.
Pine (ponderosa)	Pinus ponderosa	S	3.9 (6.3)	0.40	420–449	8.89	2046	48.95	64.81	36.68	4.00	7.79
Pine (red)	Dracydium cupressium	S	3.8 (7.2)	0.47	470–497	11.2	2486	n.a.	75.84	41.85	4.14	8.34
Pine (shortleaf)	Pinus echinata	S	4.4 (7.7)	0.54	540–577	12.13	3064	n.a.	90.32	50.12	5.65	9.58
Pine (western white)	Pinus monticola	S	2.6 (5.3)	0.42	420–433	10.1	1865	n.a.	41.85	34.75	3.24	7.17
Poplar (yellow)	Populus balsamifera	H	4.2 (7.6)	0.43	331–465	10.9	2398	n.a.	69.84	38.20	8.21	3.72
Red cedar (Australian)	Toona spp.	S	3.8 (6.3)	0.52	512	8.96	4604	n.a.	73.08	23.99	n.a.	n.a.

Table 13.1 (continued)

Usual name	Botanical Latin name (genus and species)	Category[a]	Radial shrinkage (%) (tangential)[b]	Specific gravity[c]	Bulk density (ρ, kg.m^{-3})	Young's modulus (II) (E, GPa)	Hardness (N)	Work to maximum load (WML, KJ.m^{-3})	Module of rupture or static bending strength (MOR, MPa)	Compressive or crushing strength (II) (MPa)	Compressive or crushing strength ($\perp$) (MPa)	Shear strength (II) (MPa)
Redwood	*Sequoia sempervirens*	S	2.6 (4.4)	0.35	384–449	8.14–9.2	2135	n.a.	69	28.95–42.4	2.90–4.83	5.51–120.75
Rosewood (Indian)	*Dalbergia latifolia*	H	2.7 (5.8)	0.70	849	11.45–12.27	14,080	n.a.	116.66–120.75	63.57–65.16	n.a.	n.a.
Silver birch	*Betula papyrifera*	H	n.a.	n.a.	n.a.	n.a.	n.a.	n.a.	n.a.	n.a.	n.a.	n.a.
Silver fir (Pacific)	*Abies amabilis*	S	4.4 (9.2)	0.43	433	12.13	1913	64.12	75.84	44.19	3.10	8.41
Spruce (sitka)	*Picea rubra*	S	4.3 (7.5)	0.42	413–448	10.8	2264	n.a.	7032	38.68	3.99	7.92
Spruce (white)	*Picea alba*	S	4.7 (8.2)	0.45	449	9.2	2131	n.a.	67.57	37.71	3.17	7.44
Sycamore	*Platanus occidentalis*	S	n.a.	n.a.	539	n.a.	n.a.	n.a.	n.a.	n.a.	n.a.	n.a.
Teak	*Tectona grandis*	H	n.a.	n.a.	582	n.a.	n.a.	n.a.	n.a.	n.a.	n.a.	n.a.
Thuja	*Thuja occidentalis*	H	n.a.	n.a.	315	n.a.	n.a.	n.a.	n.a.	n.a.	n.a.	n.a.
Tupelo (black)	*Nyssa sylvatica*	H	4.4 (7.7)	0.55	550–561	8.3	5585	n.a.	66.19	38.06	6.41	9.24
Walnut (black)	*Juglans nigra*	H	5.2 (7.1)	0.56	560–609	11.6	4484	n.a.	100.66	52.26	6.96	9.44
Willow (black)	*Salix nigra*	H	3.3 (8.7)	0.37	408–417	5.5	n.a.	78.5	32.95	14.3	1.3	8.8

Note: average mechanical properties are given for a moisture content of 12wt%.

[a]H = hardwoods, S = softwoods.
[b]From green to oven-dry condition based on initial condition when green.
[c]Oven-dry to green volume ratio.

14 Building and Construction Materials

14.1 Portland Cement

14.1.1 Introduction and History of Cement

In England, in the 1700s, it was noticed that certain particular types of limestone containing clay minerals and silica could be calcined and that the product, after grinding and mixing with clear water, would set to a hard cement. This new type of cement was stronger than the cements in previous use at that time, such as the pozzolan cement. Another important advantage was also noticed by the first users, namely that it sets under water and hence could be used for piers, lighthouse foundations, and canal locks. In reference to this type of cement, the mother limestones were designated as **hydraulic limes** or **water limes**. Further investigations demonstrated that a mixture of pure limestone with clay and silica sand also produced a **natural cement** having these valuable properties. Later, in 1824, an Englishman, Joseph Aspdin observed that by calcinating at high temperature a limestone, called Portland stone, extensively used at that time as dimension stone in Great Britain, it was possible to obtain a cement of superior quality, especially strength, in comparison with natural cement. It was the beginning of the well-known **Portland cement**. Since the 19th century Portland cement has been indispensable for civil engineering applications. In these applications Portland cement is the main ingredient in a castable or moldable mixture of cement with water and aggregates.

14.1.2 Raw materials for Portland Cement

The basic raw materials used for the manufacture of Portland cement depend upon availability at the quarry near the cement plant location, and are commonly limestones, shales, marl, chalk, clays, and sand. However, the particular intimate mixture must have an overall composition with 80wt% of low-magnesium calcium carbonate, $CaCO_3$ (such as limestone, marl, or chalk), and about 20wt% of clay (in form of clays, shale, or slag). This chemical composition expressed as oxide in

517

percentage by weight is roughly 75wt% CaO and 25wt% SiO_2. However, another important requirement regarding limestones for the manufacture of Portland cement, is that they should contain no more than 3wt% MgO (i.e., 5wt% $MgCO_3$). Therefore, this obviously excludes dolomites and dolomitic limestones and imposes a narrow selection for carbonate sedimentary rocks.

14.1.3 Processing of Portland Cement

In making Portland cement, the raw materials such as limestone or shale are supplied by raw silos and intimately mixed with sand creating a raw meal, and are transported to the raw ball mill by a belt conveyor in order to be crushed, proportioned under close chemical control and ground to a fine powder by either a wet or a dry process. Actually, there are four grinding processes which are commonly used to manufacture Portland cement. These range from the dry process, through the semi-dry, and semi-wet, to the wet process. The selection of the appropriate process is determined by the composition of raw material available at the plant location, especially its moisture content. For instance, wet or semi-wet processes are used for chalk or clay owing to their higher levels of moisture, while a dry process is used for dry materials (low moisture content) such as limestones. The dry process is the more modern. After grinding, the powdered material is then preheated at 260°C and precalcined at 900°C in order to initiate the chemical reactions. The material is then transported to a rotary kiln. Actually, powder is fed into the upper end of a slightly inclined long rotary kiln, rotating at 3.5 rpm, which is a cylindrical steel reactor vessel, 16 ft (4.8 m) outside diameter and 290 ft (87 m) long lined with refractory bricks. In a typical cement plant the kiln is rated at 4650 tonnes per day. The charge moves gradually down the kiln under gravity, toward the lower end, where a high heat is produced by combustion of coal, oil, or gas (sometimes scrap tires). During calcination, the maximum temperature can reach 1450°C, and in some regions the charge is partially melted, and it emerges as a vitreous (i.e., glassy) material, **clinker**, mainly composed of calcium silicates and aluminates in a nodular product. After rapid cooling, the clinker is mixed with 2 to 4wt% gypsum, $CaSO_4 \cdot 2H_2O$, in order to regulate the setting time, and the mixture is ground in a finish ball mill to a fine powder. The resulting powder is known as **Portland cement**. The cement is then stored in large silos prior to be despatch either in (i) bulk quantities by road or by rail or (ii) in sealed bags packed onto pallets. The average chemical composition of Portland cement is given in Table 14.1. A typical cement plant produces about 1,500,000 tonnes of cement type I and II per year and 1 tonne of clinker is used to make approximately 1.1 tonnes of Portland cement.

Table 14.1 Chemical composition of Portland cement

Formula	Average mass fraction (x, wt%)
SiO_2	21.8 to 21.9
Al_2O_3	4.9 to 6.9
Fe_2O_3	2.4 to 2.9
CaO	63.0 to 65.0
MgO	1.1 to 2.5 (max. 3.0)
SO_3	1.7 to 2.6
Na_2O	0.2
K_2O	0.4
H_2O	1.4 to 1.5

Table 14.2 Common letter designation of cement oxide components

Oxide (common name)	Formula	Symbol
Silicon dioxide (silica)	SiO_2	S
Aluminum oxide (alumina)	Al_2O_3	A
Iron (III) oxide (limonite)	Fe_2O_3	F
Iron (II) oxide	FeO	f
Calcium oxide (lime)	CaO	C
Magnesium oxide (magnesia)	MgO	M
Sulfur trioxide	SO_3	$\bar{S}$
Sodium oxide (soda)	Na_2O	N
Potassium oxide (potash)	K_2O	K
Loss on ignition (water)	H_2O	W, LOI

14.1.4 Portland Cement Chemistry

In cement chemistry, it is common to represent the chemical formula of compounds involved in the calcination and hydratation reactions by a capital letter abbreviation of oxides. These standard symbols are shown in Table 14.2. For instance, the dicalcium silcate or belite can be written C_2S. Therefore, it is possible to represent cement composition in a ternary phase diagram C–S–A. But most commercial cement compositions are restricted to the subsystem C–C_5A_3–C_2S.

During Portland cement processing, several chemical transformation stages can be clearly identified. During **calcination** (high-temperature firing in air), the calcium carbonate from the limestone and sometimes marl give off carbon dioxide producing free calcium oxide, or quicklime, CaO:

$$CaCO_3 \rightarrow CaO + CO_2 \uparrow .$$

At the same time, clay materials and sand release silica, SiO_2, alumina, Al_2O_3, iron sesquioxide, Fe_2O_3, and lose their constitutive water. On melting, these oxides according to the following chemical reaction produce four definite stoichiometric synthetic phases:

$$3CaO + SiO_2 \rightarrow Ca_3(SiO_5) \equiv C_3S$$
$$2CaO + SiO_2 \rightarrow Ca_2(SiO_4) \equiv C_2S$$
$$3CaO + Al_2O_3 \rightarrow Ca_3Al_2O_6 \equiv C_3A$$
$$4CaO + Al_2O_3 + Fe_2O_3 \rightarrow Ca_4Al_2Fe_2O_{10} \equiv C_4AF.$$

Therefore, almost all Portland cements contain the same five main mineral compounds or phases: (i) **alite** with the chemical formula $Ca_3(SiO_5) \equiv C_3S$ gives to the cement most of its early strength, (ii) **belite** with the chemical formula $Ca_2(SiO_4) \equiv C_2S$ hydrates more slowly than alite and provides to the concrete its late strength, (iii) tricalcium aluminate, $Ca_3Al_2O_6 \equiv C_3A$ is responsible for the workability of the mortar and acts as fluxing agent assisting the melting during calcination, (iv) **gypsum**, $CaSO_4 \cdot 2H_2O \equiv Ca\bar{S}H_2$, prevents too-rapid setting (flash setting), and finally (v), the ferrite phase referred to as **tetracalcium aluminoferrite** $Ca_4Al_2Fe_2O_{10} \equiv C_4AF$ has no significant hydraulic properties and owing to its iron oxide content provides the gray color to cement. Nevertheless, along with the previous main components, several other compounds can be found in the cement in minute amounts. The impurities are for instance: magnesium and alkali metal salts, ashes and sulfur chemicals from the fuel, incomplete reaction products, and weathering products of precursors during storage.

The curing and hardening process of cement consists of complex hydratation reactions during which new phases appear. These reactions begin as the spaces between cement particles

14

Building &
Construction
Materials

are filled with water and are of sol-gel reaction type. Actually, when the **hydration** process takes place, a thin gel layer develops onto each cement particle. The gel consists mainly of hydrated calcium aluminates and precipitated calcium hydroxide, $Ca(OH)_2$ (portlanditec) in the lime saturated water between the grains. The rate of the reaction is controlled by gypsum which acts as a retardant. At this stage, the setting is not sufficient to insure sufficient strength. However, after 5 h, owing to the hydration of calcium silicates hardening produces a little strength and the process continues until the network of microfibrils grow and interconnect. Amongst them only the two minerals C_3S and C_2S react with gypsum and added water during the hydration process giving the required strength properties of Portland cement. As a general rule, cement properties can be obtained after a curing of 28 days.

14.1.5 Portland Cement Nomenclature

ASTM has defined five types of Portland cement in its standard C150–84. Amongst them, type I is made in the greatest quantity.

14.2 Aggregates

Aggregates are various irregularly shaped materials with two or more size distributions such as coarse and fine which are mined from quarries. They give to the concrete its necessary volume and strength. As a rule of the thumb, 1 m^3 of concrete contains 2 tonnes of both gravel and sand.

14.2.1 Coarse Aggregates

The materials commonly used as coarse aggregate are for instance crushed stones such as limestone, basalt, diabase, granite, gravel, slag, or other hard inert material with similar

Table 14.3 ASTM Portland cement types

ASTM Type	Description	Compressive strength after 28 days (MPa)	Applications
Type I	Normal or ordinary Portland cement (NPC)	42	General uses, and hence used where no special properties are required.
Type II	Modified Portland cement (MPC)	47	Low heat generation during the hydration process. More resistant to sulfate attack than previous type. Used in structures with large cross-sections and for drainage pipes where sulfate levels are low.
Type III	Rapid-hardening Portland cement (RHPC)	52	Used when high strengths are required after short periods of curing.
Type IV	Low-heat Portland cement (LHPC)	34	Less heat generation during hydration than type II. Used for mass concrete construction where large heat generation could create issues. The tricalcium aluminate content must be maintained below 7wt%.
Type V	Sulfate-resisting Portland cement (SRPC)	41	High sulfate resistance, and hence special cement used when severe sulfate attack is possible.

characteristics. In some particular applications requiring high density, such as counter-weights, dry docks or nuclear radiation shields, the following materials can be used as heavyweight coarse aggregates: crushed cast iron scrap, heavy minerals and ores such as hematite, ilmenite, and barite. By contrast lightweight aggregates essentially use pumice, lava, slag, shales, cinders from coal, and coke. However, in all cases, aggregate grains must be clean, durable, and free from organic matter and alkali. Grain size must range between 5 mm (below sieve no. 4) and the coarsest size allowed for the structure, but 28 mm seems to be a maximum.

14.2.2 Fine Aggregate

The material commonly used as fine aggregates is silica sand. It should be clean, hard, and, like coarse aggregate free from organic matter and alkali metal compounds. Sometimes ground stones, slag or other hard materials can replace silica sand partially or totally. Grain sizes must be not less than 95wt% passing through sieve no. 4, not less than 10wt% passing through sieve no. 50 and finally no more than 5wt% passing through sieve no. 100.

14.3 Mortars and Concrete

14.3.1 Definitions

Concrete and mortar are composite materials which chiefly consist of two main components: (i) a matrix made of a hardened cement in which, (ii) irregularly shaped aggregates, with two or more size distributions (e.g., coarse and fine) are dispersed. As general rule, a **mortar** is a mixture containing fine aggregates, i.e., with a maximum size of 2 mm (e.g., sand), and having a cement/fine aggregate/water mass ratio of 1:3:0.5, while **concrete** is a mixture made with fine and coarse aggregates i.e., exhibiting a minimum size of 5 mm (e.g., gravel, crushed stones), and having a cement/fine/coarse aggregate/water mass ratio of 1:2:3:0.5.

When additional structural material such as steel reinforcing bars are added, the concrete is defined as **steel reinforced concrete**, while when prestressed cables are inserted, concrete is defined as **prestressed concrete**. As a general rule Portland cement should always be used for reinforced concrete, for mass concrete and concretes servicing under water. Moreover, many concretes prepared with normal Portland cement show very little gain in compressive strength after 28 days.

The concrete mixture can be proportioned in numerous ways: (i) arbitrary selection based on experience and common practice such as one part of cement, two parts of fine aggregate and four parts of coarse aggregates, (ii) proportioning on the basis of the water/cement ratio, (iii) combining materials on the basis of either the voids in the aggregates or mechanical-analysis curves in order to obtain the concrete with a maximum density for a given cement content.

14.3.2 Degradation Processes

Concrete, like other materials commonly submitted to weathering such as natural rocks or manmade construction materials shows a degradation as a function of service life; the degradation strongly depends both on the cement type and environmental conditions. As a general rule, the major degradation processes are: sulfate attack, freezing and thawing, corrosion of reinforcing bars, thermal stresses, acid attack, and phase change.

14.4 Ceramics for Construction

The ceramics used in standard construction projects are of three types: (i) **fired ceramics** (e.g. brick usually called common brick or facing brick) and (ii) **cast** or (iii) **formed hydraulic cement structures** (e.g. regular poured or placed concrete, or precured cement block or cinder block, precast concrete shapes, occasionally glass block or similar ceramic). However, none of these materials is intended specifically for service under severe chemical exposure, and all are designed for institutional, residence, or similar construction, and other buildings not normally subject to chemical spills. But today, with the acid rain phenomenon and corrosives included in off-gases from various processes, and from the incineration of industrial wastes, they may be exposed to conditions well beyond those of past years.

14.5 Building Stones

14.5.1 Limestones and Dolomites

Limestone is a general name for a wide variety of calcareous sedimentary rocks made mainly of calcium carbonate, while dolomites refers to calcareous rocks with a magnesium carbonate content above 45wt%. Limestones are mildly dense (i.e., 2150–2500 kg.m^{-3}), hard rocks with a low compressive strength (28 to 50 MPa) compared to basalt and granite, while dolomites have a density ranging between 2800 and 2900 kg.m^{-3} and exhibit similar mechanical properties to limestones. They are extensively used as building stones, as fluxes in steelmaking, and for the manufacture of lime (CaO), magnesia, and dolime (MgO + CaO) in the chemical industry.

14.5.2 Sandstones

Sandstones are consolidated, siliceous detritic sedimentary rocks. Usually, the main component is quartz grains cemented by amorphous silica, with occasionally feldspars, mica, and clays. Sandstone density ranges from 2240 to 2650 kg.m^{-3} depending on the porosity, and a higher compresive strength of 70–90 MPa which is superior to that of limestones and dolomites. Sandstones are mainly used as building stones, and silica-rich varieties (more than 99wt% SiO_2) such as quartzite are used as source of silica in glass making, and in metallurgy for ferroalloy preparation.

14.5.3 Basalt

Basalt is dark-brown phaneritic volcanic igneous rock (see Chapter 12 for a precise petrological definition). Its main components are microcrystals of alkali feldspars, pyroxenes, and olivine embedded in a volcanic glass matrix. The quantitative chemical analysis usually falls into the following ranges: 45–48wt% SiO_2, 14–16wt% Al_2O_3, 12–14wt% (Fe_3O_4, Fe_2O_3, FeO), 10–12wt% CaO, 8wt% MgO, 6wt% (K_2O, Na_2O), and 2wt% TiO_2, with traces of Mn and S. Basalt is a dense (2880–3210 kg.m^{-3}), hard (Mohs hardness 5.5–6) rock, not subject to absorption. Moreover, it exhibits a high compressive strength of 150 MPa, and possesses a modulus of elasticity between 10–12 GPa. Its coefficient of linear thermal expansion is 0.6–0.8 μm.m^{-1}.K^{-1}, and its thermal operating limit is 500°C. The material is stated to have excellent resistance to acids except hydrofluoric to which it has only limited resistance, and to a wide range of alkalis and salts. It also has excellent abrasion resistance and finds its greatest use in the lining of hoppers and chutes where both abrasion and chemical resistance are required. Bricks are made from basalt by melting it at 1250°C and casting it in molds. It is made in Europe in the form of bricks, cylinders for lining pipe, and special sectional shapes for lining all kinds of equipment.

Table 14.4 Physical properties of natural building stones

Building stone	Density (ρ, kg.m^{-3})	Compressive strength (MPa)	Thermal conductivity (k, W.m^{-1}.K^{-1})	Specific heat capacity (c_p, J.kg^{-1}.K^{-1})
Sandstones	2240–2650	70–90	2.51–2.90	745–960
Limestones	2150–2500	28–50	0.92-2.15	810–920
Dolomites	2800–2900	30–55	2.92	920
Granite	2630–2800	160–240	2.51–2.79	775–840
Basalt	2880–3210	150	2.60	950

14.5.4 Granite

Granite and to a lesser extent granodiorite are coarse-grained plutonic igneous rocks (see Chapter 12 for a precise petrological definition) primarily composed of silicate minerals such as quartz and feldspars with often small amounts of accessory minerals such as micas. Therefore silica and alumina are the major components of granite. Granite is a mildly dense (2630–2800 kg.m^{-3}), very hard and compact rock (e.g., crushing strength about 160–240 MPa) with a low absorption. It exhibits a very low thermal conductivity, and a coefficient of thermal expansion close to that of acid brick. It has been used in construction from prehistoric times, and where ancient structures have best survived the effects of time and weathering, they have often been made of granite. It is an important building stone because of its durability and corrosion resistance; a few decades ago a major steel mill experimented with it as the sole construction material for a continuous pickler, and owing to its fine polish it is used for mill rods in the pulp and paper industry.

Table 14.5 Physical properties of building materials

Building materials	Density (ρ, kg.m^{-3})	Specific heat capacity (c_p, J.kg^{-1}.K^{-1})	Thermal conductivity (k, W.m^{-1}.K^{-1})
Brick alumina, fused (96wt% alumina) (22% porosity)	2900	753.10	3.0962
Brick, chrome (32wt% Cr_2O_3)	3200	627.60	1.1715
Brick, chrome magnesite	3000	753.10	2.0920
Brick, diatomaceous earth	440	795.00	0.0877
Brick, diatomaceous earth (high burn)	590	795.00	0.2259
Brick, diatomaceous earth (molded)	610	795.00	0.2427
Brick, diatomaceous earth (fused at 1100°C)	600	795.00	0.2218
Brick, diatomaceous earth (850°C)	440	795.00	0.0921
Brick, Egyptian fire (64–71wt% silica)	950	732.20	0.3138
Brick, forsterite (58wt% MgO, 38wt% SiO_2) (20% porosity)	2760	795.00	1.0042
Brick, fused alumina (96wt% alumina) (22% porosity)	2900	753.10	3.0962
Brick, hard fired silica (94–95wt% silica)	1800	753.10	1.6736
Brick, high alumina (53wt% alumina) (20% porosity)	2330	753	1.3807
Brick, high alumina (83wt% alumina) (28% porosity)	2570	753	1.5062
Brick, high alumina (87wt% alumina) (22% porosity)	2850	753	2.9288
Brick, kaolin insulating (heavy)	430	774	0.2510
Brick, kaolin insulating (light)	300	774	0.0837
Brick, magnesite (86wt% MgO) (17.8% porosity)	2920	837	3.6819
Brick, magnesite (87wt% MgO)	2530	837	3.8493
Brick, magnesite (89wt% MgO)	2670	837	3.4727
Brick, magnesite (90wt% MgO) (14.5% porosity)	3080	837	4.9371
Brick, magnesite (93wt% MgO) (22.6% porosity)	2760	837	4.8116
Brick, masonry, medium	2000	837	0.7113
Brick, fireclay	2000	753	1.0042
Brick, fireclay, missouri	645	960	1.0042
Brick, normal fireclay (22% porosity)	1980	732	1.2970
Brick, siliceous (25% porosity)	1930	753	0.9372
Brick, siliceous fireclay (23% porosity)	2000	753	1.0878
Brick, sillimanite (22% porosity)	2310	711	1.4644
Brick, stabilized dolomite (22% porosity)	2700	837	1.6736
Brick, vermiculite	485	837	0.1674
Calcium oxide (packed powder)	1700	753	0.3180
Calcium oxide (pressed)	3030	753	13.8070
Carbon brick, fired	1470	707	3.5982
Chalk	1540	921	0.8368
Concrete, 1–4 dry	2300	657	0.7531
Concrete, cinder	1600	657	0.3347
Concrete, lightweight	950	657	0.2092

Table 14.5 (continued)

Building materials	Density (ρ, kg.m^{-3})	Specific heat capacity (c_p, J.kg^{-1}.K^{-1})	Thermal conductivity (k, W.m^{-1}.K^{-1})
Concrete, stone (1–2–4 mix)	2300	880	1.460
Diabasic glass (artificial)	2400	753	1.1715
Dolomite	2700	921	2.9288
Granite	2650	837	2.5104
Gypsum	2320	1088	1.2970
Limestone (dense, dry)	2500	810	1.6736
Limestone	2650	921	2.1500
Marble	2680	879	2.5104
Plaster, building (molded, dry)	1250	1088	0.4310
Quartz sand (dry)	1600	753	0.3347
Quartz sand (wet) (4–23wt% water)	1700	753	1.6736
Sand, northway (4–10wt% water)	1700	837	0.8368
Sand, quartz (wet (4–23wt% water)	1700	753	1.6376
Sandstone	2300	963	2.5104
Sandstone (heavy)	2600	963	4.1840
Sandstone (light)	2200	745	1.8410
Soil (average)	1300	1046	0.8368
Soil, clay (wet)	1500	2929	1.5062
Soil, fine quartz flour (dry)	880	745	0.1674
Soil, fine quartz flour (21wt% water)	1820	1464	2.2175
Soil, loam (dry)	1200	837	0.2511
Soil, loam (4–27wt% water)	1600	1046	0.4184
Soil, sandy (8wt% water)	1750	1004	0.5858
Soil, sandy dry	1650	795	0.2636
Vermiculite insulating powder	270	837	0.1213
Vermiculite, expanded (light)	220	753	0.0711
Vermiculite, expanded (heavy)	300	753	0.0690

14
Building &
Construction
Materials

15

Appendices

15.1 Periodic Chart of the Elements

Table 15.1 Mendeleev's periodic chart of the elements

IA(1)	IIA(2)	IIIB(3)	IVB(4)	VB(5)	VIB(6)	VIIB(7)	VIIIB(8, 9, 10)			IB(11)	IIB(12)	IIIA(13)	IVA(14)	VA(15)	VIA(16)	VIIA(17)	VIII(18)
Alkali metals	Alkaline-earth metals													Pictinides	Chalcogenes	Halogens	Noble gases
^{1}H																	^{2}He
^{3}Li	^{4}Be				Transition metals							^{5}B	^{6}C	^{7}N	^{8}O	^{9}F	^{10}Ne
^{11}Na	^{12}Mg	R.E.	Reactive and refractory metals				Ferrous metals and PGMs			Precious and noble metals		^{13}Al	^{14}Si	^{15}P	^{16}S	^{17}Cl	^{18}Ar
^{19}K	^{20}Ca	^{21}Sc	^{22}Ti	^{23}V	^{24}Cr	^{25}Mn	^{26}Fe	^{27}Co	^{28}Ni	^{29}Cu	^{30}Zn	^{31}Ga	^{32}Ge	^{33}As	^{34}Se	^{35}Br	^{36}Kr
^{37}Rb	^{38}Sr	^{39}Y	^{40}Zr	^{41}Nb	^{42}Mo	^{43}Tc	^{44}Ru	^{45}Rh	^{46}Pd	^{47}Ag	^{48}Cd	^{49}In	^{50}Sn	^{51}Sb	^{52}Te	^{53}I	^{54}Xe
^{55}Cs	^{56}Ba	^{57}La	^{72}Hf	^{73}Ta	^{74}W	^{75}Re	^{76}Os	^{77}Ir	^{78}Pt	^{79}Au	^{80}Hg	^{81}Tl	^{82}Pb	^{83}Bi	^{84}Po	^{85}As	^{86}Rn
^{87}Fr	^{88}Ra	^{89}Ac	^{104}Rf	^{105}Db	^{106}Sg	^{107}Bh	^{108}Hs	^{109}Mt				Fusible metals					

Transfermium elements

Lanthanides and rare earths	^{57}La	^{58}Ce	^{59}Pr	^{60}Nd	^{61}Pm	^{62}Sm	^{63}Eu	^{64}Gd	^{65}Tb	^{66}Dy	^{67}Ho	^{68}Er	^{69}Tm	^{70}Yb	^{71}Lu
Actinides (uranides, curides)	^{89}Ac	^{90}Th	^{91}Pa	^{92}U	^{93}Np	^{94}Pu	^{95}Am	^{96}Cm	^{97}Bk	^{98}Cf	^{99}Es	^{100}Fm	^{101}Md	^{102}No	^{103}Lr

Alkali metals: Li, Na, K, Rb, Cs, Fr
Alkaline-earth metals: Be, Mg, Ca, Sr, Ba, Ra
Actinides (uranides: Ac, Th, Pa, U, Np, Pu and curides: Am, Cm, Bk, Cf, Es, Fm, transfermium elements)
Rare earth and lanthanides: Sc, Y, La (ceric and yttric)
Reactive and refractory metals: Ti, Zr, Hf, Nb, Ta, Mo, W, Re
Platinum group metals: Ru, Rh, Pd, Os, Ir, Pt
Precious and noble metals: Ag, Au
Ferrous metals: Fe, Co, Ni
Heavy metals: Zn, Cd, Hg, Sn, Pb
Fusible metals: Cd, Hg, Ga, In, Tl, Sn, Pb, Sb, Bi
Pictinides: N, P, As, Sb, Bi
Chalcogenes: O, S, Se, Te, Po
Halogens: F, Cl, Br, I, As
Rare and noble gases: He, Ne, Ar, Kr, Xe, Rn

15.2 Selected Physical Properties of the Elements

These are shown in **Table 15.2a** (pages 530–537).

Table 15.2b Physical properties of water at room temperature (i.e. 293.15 K)

Property (symbol/SI unit)	Value
Density (ρ, kg.m^{-3})	997.045
Melting point (mp, °C)	0
Boiling point (bp, °C)	100
Dynamic viscosity (μ, mPa.S)	1.002
Surface tension (γ, mN.m^{-1})	72.88
Thermal conductivity (k, W.m^{-1}.K^{-1})	0.5983
Coefficient of volumic thermal expansion (β, 10^{-9} K^{-1})	2.07
Critical temperature (T_c, K)	647.13
Critical pressure (P_c, MPa)	22.048
Critical density (ρ_c, kg.m^{-3})	325
Critical molar volume (v, dm^{-3}.mol)	0.056
Specific heat capacity, isobaric (c_P, J.kg^{-1}.K^{-1})	4177.4
Latent specific enthalpy of fusion (h_f, kJ.kg^{-1})	333.9
Latent specific enthalpy of vaporization (h_v, kJ.kg^{-1})	2258.3
Vapor pressure (Π, Pa)	3167.21
Sound longitudinal velocity (v_l, m.s^{-1})	1509
Electrical resistivity (ρ, MΩ.cm)	18.2
Dielectric permittivity (ε_r)	80.2
Refractive index (n_D)	1.33300

15.3 Geochemical Classification of the Elements

Table 15.3 Geochemical classification of the elements[a]

Classes	Description	Examples
Lithophiles	Affinity to silicate materials	O, Si, Al, Mg, Ca, Na, K, Ti, Zr, Hf, Nb, Ta, W, Sn, U
Siderophiles	Affinity to iron	Fe, Co, Ni, PGMs
Chalcophiles	Affinity to sulfur forming sulfides, sulfosalts, and chalcogenides	Cu, Fe, Co, Ni, Hg, Cd, Os, Ir, Pt, Ru, Rh, Pd, Zn, Re, As, Sb, Se, Te
Hydrophiles	Affinity to water, and aqueous solutions (i.e., brines, geothermal fluids)	H, O, Na, K, Li, Cl, F, Mg
Atmophiles	Gaseous elements	H, O, N, He, Ar, rare gases
Biophiles	Animals and plants	C, H, O, N, P

[a]Goldschmidt (1937) J Chem Soc 55.

15.4 Historical Names of the Elements

Table 15.4 Obsolete and historical names of the elements

Obsolete name (symbol)	IUPAC name	Obsolete name (symbol)	IUPAC name
Actinon (An)	Radon	Erythronium	Vanadium
Alabamine	Astatine	Ferrum	Iron
Aluminium	Aluminum	Glucinium (Gl)	Beryllium
Argentum	Silver	Hydrargyrum	Mercury
Arsenicum	Arsenic	Illinium (Il)	Promethium
Aurum	Gold	Kalium	Potassium
Azote (Az)	Nitrogen	Masurium (Ma)	Technetium
Caesium	Cesium	Mischmetal	Cerium impure
Cassiopeium	Lutetium	Natrium	Sodium
Celtium (Ct)	Hafnium	Plumbum	Lead
Columbium (Cb)	Niobium	Stannum	Tin
Cuprum	Copper	Stibium	Antimony
Didynium (Dm)	Neodymium, praseodymium	Sulfur	Sulphur
Ekaaluminium	Gallium	Thoron (Tn)	Radon
Ekacaesium	Francium	Virginium (Vi)	Francium
Emanation (Em)	Radon	Wolfram	Tungsten

Table 15.5 Mononuclidic elements (isotopes)

^{9}Be, ^{19}F, ^{23}Na, ^{27}Al, ^{31}P, ^{45}Sc, ^{55}Mn, ^{59}Co, ^{75}As, ^{89}Y, ^{103}Rh, ^{127}I, ^{133}Cs, ^{141}Pr, ^{159}Tb, ^{165}Ho, ^{169}Tm, ^{197}Au, ^{209}Bi

15.5 Cost of the Pure Elements

See **Table 15.6** (pages 538–539).

15.6 Crystallography and Crystallochemistry

15.6.1 Direct Space Lattice Parameters

A crystal is a periodic array of ordered entities (e.g., ions, atoms, molecules) in three dimensions. The repeating unit is imagined to be a unit cell whose volume and shape are designated by the three vectors representing the length and direction of the cell edges as a three-unit vector of translation. Therefore, a space lattice is defined by either the three-unit lattice vectors a, b, and c or the set of the six lattice parameters: a, b, c, α, β, and γ, where the last three quantities represent the plane angles between the cell edges. The International Union of Crystallography (IUCr) has now standardized this notation as follows:

$$\alpha \equiv \text{mes } (b, c) \text{ and plane A} \equiv (b, c)$$
$$\beta \equiv \text{mes } (c, a) \text{ and plane B} \equiv (b, c)$$
$$\gamma \equiv \text{mes } (a, b) \text{ and plane C} \equiv (b, c)$$

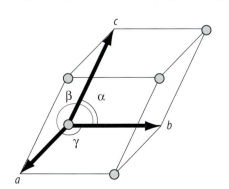

Table 15.2a Properties of the elements

Element name (IUPAC)	Chemical abstract registry number [CARN]	Symbol (IUPAC)	Atomic number (Z)	Atomic relative mass (1995) ($^{12}C = 12,000$)	Electronic configuration (ground state)	Spectral term	Electronegativity (Pauling)	Crystal space lattice	Space group (Hermann–Mauguin)	Pearson symbol	Strukturbericht and structure type	Lattice parameters (pm)	Transition temperature (α to β) (°C)
Actinium	[7440-34-8]	Ac	89	[227]	[Rn]$6d^24s^1$	$^2D_{3/2}$	1.10	fcc	Fm3m	$cF4$	A1 (Cu)	$a = 531.11$	n.a.
Aluminum	[7429-90-5]	Al	13	26.981538(2)	[Ne]$3s^23p^1$	$^2P_{1/2}$	1.61	fcc	Fm3m	$cF4$	A1 (Cu)	$a = 404.96$	None
Americium	[7440-35-9]	Am	95	[243]	[Rn]$5f^76d^07s^2$	$^8S_{7/2}$	1.30	hcp	P6$_3$/mmc	$hP2$	A3 (Mg)	$a = 346.80$ $c = 1124.00$	1074
Antimony	[7440-36-0]	Sb	51	121.760(1)	[Kr]$4d^{10}5s^25p^3$	$^4S_{3/2}$	2.05	Rhomb.	R-3m	$hR2$	A7 (α-As)	$a = 336.90$ $b = 533.00$	None
Argon (g)	[7440-37-1]	Ar	18	39.948(1)	[Ne]$3s^23p^6$	$^1S_0$	n.a.	fcc	Fm3m	$cF4$	A1 (Cu)	$a = 531.09$	−189.2
Arsenic (α-)	[7440-38-2]	As	33	74.92160(2)	[Ar]$3d^{10}4s^24p^3$	$^4S_{3/2}$	2.18	Rhomb.	R-3m	$hR2$	A7 (α-As)	$a = 413.18$ $\alpha = 54°10'$	228
Astatine (α-)	[7440-68-8]	At	85	[210]	[Xe]$4f^{14}5d^{10}6s^26p^5$	$^3P_{3/2}$	2.20	n.a.	n.a.	n.a.	n.a.	n.a.	n.a.
Barium	[7440-39-3]	Ba	56	137.327(7)	[Xe]$6s^2$	$^1S_0$	0.89	bcc	Im3m	$cI2$	A2 (W)	$a = 501.30$	370
Berkelium	[7440-40-6]	Bk	97	[247]	[Rn]$5f^86d^07s^2$	$^6H_{15/2}$	1.30	n.a.	n.a.	n.a.	n.a.	n.a.	n.a.
Beryllium (α-)	[7440-41-7]	Be	4	9.012182(3)	[He]$2s^2$	$^1S_0$	1.57	hcp	P6$_3$/mmc	$hP2$	A3 (Mg)	$a = 228.59$ $c = 358.32$	1254
Bismuth	[7440-69-9]	Bi	83	208.98038(2)	[Xe]$4f^{14}5d^{10}6s^26p^3$	$^4S_{3/2}$	2.02	Rhomb.	R-3m	$hR2$	A7 (α-As)	$a = 474.60$ $\alpha = 57.23°$	None
Bohrium	n.a.	Bh	107	[262]	[Rn]$5f^{14}6d^57s^2$	n.a.	n.a.	n.a.	n.a.	n.a.	n.a.	n.a.	n.a.
Boron (β-)	[7440-42-8]	B	5	10.811(7)	[He]$2s^22p^1$	$^2P_{1/2}$	2.04	Rhomb.	R-3m	hR 105	β-B	$a = 1014.5$ $\alpha = 65°12'$	None
Bromine (liq., Br$_2$)	[7726-95-6]	Br	35	79.904(1)	[Ar]$3d^{10}4s^24p^5$	$^3P_{3/2}$	2.96	Orthorh.	Cmca	$oC8$	A14 (I$_2$)	$a = 668.00$ $b = 449.00$ $c = 874.00$	−7.25
Cadmium	[7440-43-9]	Cd	48	112.411(8)	[Kr]$4d^{10}5s^2$	$^1S_0$	1.69	hcp	P6$_3$/mmc	$hP2$	A3 (Mg)	$a = 297.94$ $b = 561.86$	None
Calcium (α-)	[7440-70-2]	Ca	20	40.078(4)	[Ar]$4s^2$	$^1S_0$	1.00	fcc	Fm3m	$cF4$	A1 (Cu)	$a = 558.84$	443
Californium	[7440-71-3]	Cf	98	[251]	[Rn]$5f^{10}6d^07s^2$	$^5I_8$	1.30	n.a.	n.a.	n.a.	n.a.	n.a.	n.a.
Carbon (diam.)	[7782-40-3]	C	6	12.0107(8)	[He]$2s^22p^2$	$^3P_0$	2.55	cubic	F-3md	$cF8$	A4 (diam.)	$a = 356.69$	n.a.
Carbon (graph.)	[7440-44-0]	C	6	12.0107(8)	[He]$2s^22p^2$	$^3P_0$	2.55	hex.	P6$_3$/mmc	$hP4$	A9 (graph.)	$a = 246.16$ $c = 670.90$	n.a.
Cerium (β-)	[7440-45-1]	Ce	58	140.116(1)	[Xe]$5d^16s^24f^1$	$^3H_4$	1.12	dhcp	P6$_3$/mmc	$hP4$	A3' (α-La)	$a = 368.10$ $c = 1185.7$	61
Cesium	[7440-46-2]	Cs	55	132.90545(2)	[Xe]$6s^1$	$^2S_{1/2}$	0.79	bcc	Im3m	$cI2$	A2 (W)	$a = 614.00$	n.a.
Chlorine (g, Cl$_2$)	[7782-50-5]	Cl	17	35.4527(9)	[Ne]$3s^23p^5$	$^3P_{3/2}$	3.16	Orthorh.	Cmca	$oC8$	A14 (I$_2$)	$a = 624.00$ $b = 448.00$ $c = 826.00$	−100.97
Chromium	[7440-47-3]	Cr	24	51.9961(6)	[Ar]$3d^54s^1$	$^7S_3$	1.66	bcc	Im3m	$cI2$	A2 (W)	$a = 288.46$	n.a.
Cobalt (ϵ-)	[7440-48-4]	Co	27	58.933200(9)	[Ar]$3d^74s^2$	$^4F_{9/2}$	1.88	hcp	P6$_3$/mmc	$hP2$	A3 (Mg)	$a = 250.70$ $c = 406.90$	440, 1120
Copper	[7440-50-8]	Cu	29	63.546(3)	[Ar]$3d^{10}4s^1$	$^2S_{1/2}$	1.90	fcc	Fm3m	$cF4$	A1 (Cu)	$a = 361.51$	None
Curium	[7440-51-9]	Cm	96	[247]	[Rn]$5f^76d^17s^2$	$^9D_2$	1.30	n.a.	n.a.	n.a.	n.a.	n.a.	n.a.
Dubnium	n.a.	Db	105	[262]	[Rn]$5f^{14}6d^37s^2$	$^3F_{3/2}$	n.a.	n.a.	n.a.	n.a.	n.a.	n.a.	n.a.
Dysprosium (α-)	[7429-91-6]	Dy	66	162.50(3)	[Xe]$5d^06s^24f^{10}$	$^5I_8$	1.22	hcp	P6$_3$/mmc	$hP2$	A3 (Mg)	$a = 359.15$ $c = 565.01$	1381
Einsteinium	[7429-92-7]	Es	99	[252]	[Rn]$5f^{11}6d^07s^2$	$^5I_{15/2}$	1.30	n.a.	n.a.	n.a.	n.a.	n.a.	n.a.
Erbium	[7440-52-0]	Er	68	167.26(3)	[Xe]$5d^06s^24f^{14}$	$^3H_6$	1.24	hcp	P6$_3$/mmc	$hP2$	A3 (Mg)	$a = 355.92$ $c = 558.50$	1367

Density (d, kg·m⁻³) (298.15 K)	Young's or elastic modulus (E, GPa)	Coulomb or shear modulus (G, GPa)	Bulk or compression modulus (K, GPa)	Poisson's ratio (ν)	Electrical resistivity (ρ, μΩ·cm) (293.15 K)	Temperature coefficient electrical resistivity (TCR, 10^{-3} K⁻¹)	Melting point (mp, °C)	Boiling point (bp, °C)	Thermal conductivity (k, W·m⁻¹·K⁻¹) (300 K)	Specific heat capacity (c_p, J·kg⁻¹·K⁻¹) (300 K)	Coefficient linear thermal expansion (α, 10^{-6} K⁻¹) (0–100°C)	Latent heat of fusion (l_f, kJ·mol⁻¹)	Thermal neutron capture cross-section (σ_{th}, 10^{-28} m²)	Thermal neutron mass absorption coefficient ((μ/r), cm²·g⁻¹)	Mass magnetic susceptibility (cm, 10^{-9} kg⁻¹·m³)
10,060	n.a.	n.a.	n.a.	n.a.	n.a.	n.a	1046.9	3196.9	12	119.8	14.9	14.2	810	0.79000	n.a.
2698.9	70.26	26.2	75.2	0.345	2.6548	4.50	660.4	2466.9	237	903	23.5	10.47	0.233	0.00300	7.80
13,670	n.a.	n.a.	n.a.	n.a.	68	n.a	993.9	2606.9	est. 10	n.a.	n.a.	14.4	74	n.a.	51.50
6696	54.7	20.7	n.a.	0.330	41.7	5.10	630.7	1634.9	24.3	205	8.5	19.89	5.4	0.01600	-10.90
1.784	n.a.	n.a.	n.a.	n.a.	n.a.	n.a.	-189.4	-185.9	0.0177	520	n.a.	1.21	0.65	0.00600	-6.00
5778	n.a.	n.a.	n.a.	n.a.	26	n.a	816.9	615.9	50	329	4.7	27.8	4.3	0.02000	-3.90
n.a.	n.a.	n.a.	n.a.	n.a.	n.a.	n.a.	n.a.	n.a.	1.7	n.a.	n.a.	23.8	n.a.	n.a.	n.a.
3594	12.8	4.86	n.a.	0.280	33.2	n.a	728.9	1636.9	18.4	285	18.1	7.66	1.3	0.00270	11.30
14,790	n.a.	n.a.	n.a.	n.a.	n.a.	n.a	n.a.	n.a.	est. 10	n.a.	n.a.	n.a.	710	n.a.	n.a.
1847.70	318	156	11	0.075	3.56	9.00	1277.9	2969.9	194–210	1886	11.5	12.22	0.0092	0.00030	-12.60
9747	34	12.8	n.a.	0.330	106.8	4.60	271.4	1559.9	7.87	123	13.4	10.89	0.034	0.00060	-17
n.a.	n.a.	n.a.	n.a.	n.a.	n.a.	n.a.	n.a.	n.a.	n.a.	n.a.	n.a.	n.a.	n.a.	n.a.	n.a.
2340	440	n.a.	n.a.	n.a.	6500	n.a	2299.9	3657.9	27.6	1107	5.0	22.6	755	24.00000	-8.70
3122.60	n.a.	n.a.	n.a.	n.a.	7.8 x 10^{18}	n.a	-7.3	58.8	0.122	947	n.a.	10.55	6.8	0.02000	-4.90
8650	62.6	24	51	0.300	6.83	4.30	321.0	764.9	96.8	231	29.8	6.41	2450	14.00000	-2.30
1550	19.6	7.9	17.2	0.310	3.43	4.57	838.9	1483.9	200	632	22.3	8.36	0.43	0.00370	13.80
n.a.	n.a.	n.a.	n.a.	n.a.	n.a.	n.a.	n.a.	n.a.	n.a.	n.a.	n.a.	n.a.	2900	n.a.	n.a.
3513	980	n.a.	n.a.	n.a.	10^{11}	n.a.	3820	5100	990–2320	509	1.2	105	0.0035	0.00015	-6.20
2260	n.a.	n.a.	n.a.	n.a.	1375	n.a.	5.701, 1960d	n.a.	n.a.	709	n.a.	105	0.0035	0.00015	-6.20
8240	33.6	13.5	21.5	0.248	82.8	8.70	798.9	3425.9	11.4	192	8.5	5.23	0.6	0.00210	220.00
1873	1.7	0.65	n.a.	0.295	20	4.80	28.4	678.5	35.9	242	97.0	2.09	29	0.07700	-2.80
3.214	n.a.	n.a.	n.a.	n.a.	10^{9}	n.a.	-101.0	-34.0	0.0089	956	n.a.	6.41	35.5	0.33000	-7.20
7190	279	115.3	160.2	0.210	12.7	2.14	1856.9	2671.9	93.7	459.8	6.2	20.9	3.1	0.02100	44.50
8900	211	82	181.5	0.320	6.24	6.60	1454.9	2869.9	99.2	421	13.4	15.5	37.2	0.21000	Ferro-magnetic
8960	129.8	48.3	137.8	0.343	1.673	4.30	1083.5	2566.9	401	494	16.5	13.02	3.78	0.02100	-1.08
13,300	n.a.	n.a.	n.a.	n.a.	n.a.	n.a.	n.a.	n.a.	est. 10	n.a.	n.a.	n.a.	60	n.a.	n.a.
n.a.	n.a.	n.a.	n.a.	n.a.	n.a.	n.a.	n.a.	n.a.	n.a.	n.a.	n.a.	n.a.	n.a.	n.a.	n.a.
8551	61.4	24.7	40.5	0.247	92.6	1.19	1411.9	2561.9	10.7	170.5	9.9	17.2	920–1100	2.00000	5450.00
n.a.	n.a.	n.a.	n.a.	n.a.	n.a.	n.a	n.a	n.a.	n.a.	n.a.	n.a.	n.a	160	n.a.	n.a.
9066	69.9	28.3	44.4	0.237	87	2.01	1528.9	2862.9	14.5	168	12.2	17.2	160–170	0.36000	3770.00

Table 15.2a *(continued)*

Element name (IUPAC)	Chemical abstract registry number [CARN]	Symbol (IUPAC)	Atomic number (Z)	Atomic relative mass (1995) ($^{12}C = 12.000$)	Electronic configuration (ground state)	Spectral term	Electronegativity (Pauling)	Crystal space lattice	Space group (Hermann–Mauguin)	Pearson symbol	Strukturbericht and structure type	Lattice parameters (pm)	Transition temperature (α to β) (°C)
Europium	[7440-53-1]	Eu	63	151.964(1)	$[Xe]5d^06s^24f^7$	$^8S_{7/2}$	n.a.	b.c.c.	Im3m	cI2	A2 (W)	$a = 458.27$	None
Fermium	[7440-72-4]	Fm	100	[257]	$[Rn]5f^{12}6d^07s^2$	$^3H_6$	1.30	n.a.	n.a.	n.a.	n.a.	n.a.	n.a.
Fluorine (g, F_2)	[7782-41-4]	F	9	18.9984032 (5)	$[He]2s^22p^5$	$^3P_{3/2}$	3.98	Monocl.	C2/c	mC8	α-F_2	$a = 550.00$ $b = 338$ $c = 728$ $\beta = 102.7°$	–227. 60
Francium	[7440-73-5]	Fr	87	[223]	$[Rn]7s^1$	$^2S_{1/2}$	0.70	n.a.	n.a.	n.a.	n.a.	n.a.	n.a.
Gadolinium (α-)	[7440-54-2]	Gd	64	157.25(3)	$[Xe]5d^16s^24f^7$	$^9D_2$	1.20	hcp	P6₃/mmc	hP2	A3 (Mg)	$a = 363.36$ $c = 578.10$	1235
Gallium	[7440-55-3]	Ga	31	69.723(1)	$[Ar]3d^{10}4s^24p^1$	$^2P_{1/2}$	1.81	Orthorh.	Cmca	oC8	A11 (α-Ga)	$a = 451.86$ $b = 765.70$ $c = 452.58$	None
Germanium	[7440-56-4]	Ge	32	72.61(2)	$[Ar]3d^{10}4s^24p^2$	$^3P_0$	2.01	cubic	F-3md	cF8	A4 (Diamond)	$a = 565.74$	None
Gold	[7440-57-5]	Au	79	196.96655(2)	$[Xe]5d^{10}6s^14f^{14}$	$^2S_{1/2}$	2.54	f.c.c.	Fm3m	cF4	A1 (Cu)	$a = 407.82$	None
Hafnium	[7440-58-6]	Hf	72	178.49(2)	$[Xe]5d^26s^24f^{14}$	$^3F_2$	1.30	hcp	P6₃/mmc	hP2	A3 (Mg)	$a = 319.46$ $c = 505.10$	1743
Hassium	n.a.	Hs	108	[265]	$[Rn]5f^{14}6d^67s^2$	n.a.	n.a.	n.a.	n.a.	n.a.	n.a.	n.a.	n.a.
Helium (g)	[7440-59-7]	He	2	4.002602(2)	$1s^2$	$^1S_0$	None	hcp	P6₃/mmc	hP2	A3 (Mg)	$a = 347.00$	–269. 20
Holmium	[7440-60-0]	Ho	67	164.93032(2)	$[Xe]5d^06s^24f^{11}$	$^4I_{15/2}$	1.23	hcp	P6₃/mmc	hP2	A3 (Mg)	$a = 357.78$ $c = 561.78$	
Hydrogen (g, H_2)	[1333-74-0]	H	1	1.00794(7)	$1s^1$	$^2S_{1/2}$	2.20	f.c.c.	Fm3m	cF4	A1 (Cu)	$a = 533.80$	–271. 90
Indium	[7440-74-6]	In	49	114.818(3)	$[Kr]4d^{10}5s^25p^1$	$^2P_{1/2}$	1.78	Tetragonal	I4/mmm	tI2	A6 (In)	$a = 325.30$ $c = 494.70$	None
Iodine (s, I_2)	[7553-56-2]	I	53	126.90447(3)	$[Kr]4d^{10}5s^25p^5$	$^3P_{3/2}$	2.66	Orthorh.	Cmca	oC8	A14 (I_2)	$a = 726.97$ $b = 479.03$ $c = 979.42$	None
Iridium	[7439-88-5]	Ir	77	192.217(3)	$[Xe]5d^76s^24f^{14}$	$^4F_{9/2}$	2.20	f.c.c.	Fm3m	cF4	A1 (Cu)	$a = 383.92$	None
Iron	[7439-89-6]	Fe	26	55.845(2)	$[Ar]3d^64s^2$	$^5D_4$	1.83	b.c.c.	Im3m	cI2	A2 (W)	$a = 286.65$	914, 1391
Krypton	[7439-90-9]	Kr	36	83.80(1)	$[Ar]3d^{10}4s^24p^6$	$^1S_0$	n.a.	f.c.c.	Fm3m	cF4	A1 (Cu)	$a = 581.00$	–193
Lanthanum (α-)	[7439-91-0]	La	57	138.9055(2)	$[Xe]5d^16s^24f^0$	$^2D_{3/2}$	1.10	dhcp	P6₃/mmc	hP4	$\alpha3'$ (α-La)	$a = 377.40$ $c = 1217.10$	868
Lawrencium	[22537-19-5]	Lr	103	[262]	$[Rn]5f^{14}6d^17s^2$	$^2D_{5/2}$	n.a.	n.a.	n.a.	n.a.	n.a.	n.a.	n.a.
Lead	[7439-92-1]	Pb	82	207.2(1)	$[Xe]4f^{14}5d^{10}6s^26p^2$	$^3P_0$	1.80	f.c.c.	Fm3m	cF4	A1 (Cu)	$a = 495.02$	None
Lithium (β-)	[7439-93-3]	Li	3	6.941(2)	$[He]2s^1$	$^2S_{1/2}$	0.98	b.c.c.	Im3m	cI2	A2 (W)	$a = 350.93$	–195
Lutetium	[7439-94-3]	Lu	71	174.967(1)	$[Xe]5d^16s^24f^{14}$	$^2D_{3/2}$	1.27	hcp	P6₃/mmc	hP2	A3 (Mg)	$a = 350.52$	None
Magnesium	[7439-95-4]	Mg	12	24.3050(6)	$[Ne]3s^23p^0$	$^1S_0$	1.31	hcp	P6₃/mmc	hP2	A3 (Mg)	$a = 320.87$ $b = 520.90$	Noned
Manganese	[7439-96-5]	Mn	25	54.938049(9)	$[Ar]3d^54s^2$	$^6S_{5/2}$	1.55	Cubic	I-43m	cI58	α-Mn	$a = 891.39$	710, 1090, 1136
Meitnerium	n.a.	Mt	109	[266]	$[Rn]5f^{14}6d^77s^2$	n.a.	n.a.	n.a.	n.a.	n.a.	n.a.	n.a.	n.a.
Mendelevium	[7440-11-1]	Md	101	[258]	$[Rn]5f^{13}6d^07s^2$	$^2F_{7/2}$	1.3	n.a.	n.a.	n.a.	n.a.	n.a.	n.a.
Mercury (liq.)	[7439-97-6]	Hg	80	200.59(2)	$[Xe]4f^{14}5d^{10}6s^26p^0$	$^1S_0$	1.90	Rhomb.	R-3m	hR1	$\alpha10$ (α-Hg)	$a = 300.50$ $\alpha = 70.53°$	–38. 836

Density (d, kg.m⁻³) (298.15 K)	Young's or elastic modulus (E, GPa)	Coulomb or shear modulus (G, GPa)	Bulk or compression modulus (K, GPa)	Poisson's ratio (ν)	Electrical resistivity (ρ, $\mu\Omega$.cm) (293.15 K)	Temperature coefficient electrical resistivity (TCR, 10^{-3} K⁻¹)	Melting point (mp, °C)	Boiling point (bp, °C)	Thermal conductivity (k, W.m⁻¹.K⁻¹) (300 K)	Specific heat capacity (c_p, J.kg⁻¹.K⁻¹) (300 K)	Coefficient linear thermal expansion (α, 10^{-6} K⁻¹) (0–100°C)	Latent heat of fusion (L_f, kJ.mol⁻¹)	Thermal neutron capture cross-section (σ_{th}, 10^{-28} m²)	Thermal neutron mass absorption coefficient ((μ_t/ρ), cm².g⁻¹)	Mass magnetic susceptibility (cm, 10^{-9} kg⁻¹.m³)
5243	18.2	7.9	8.3	0.152	90	n.a.	821.9	1596.9	13.9	182.3	35.0	10.5	4300–4600	6.00000	276.00
n.a.	n.a.	n.a.	n.a.	n.a.	n.a.	n.a.	n.a.	n.a.	n.a.	n.a.	n.a.	n.a.	5800	n.a.	n.a.
1.696	n.a.	n.a.	n.a.	n.a.	n.a.	n.a.	−219.6	−188.1	0.0279	1648	n.a.	5.10	0.0096	0.00020	n.a.
n.a.	n.a.	n.a.	n.a.	n.a.	n.a.	n.a.	26.9	676.9	est. 15	n.a.	n.a.	n.a.	n.a.	n.a.	n.a.
7901	54.8	21.8	37.9	0.259	134	1.76	1312.9	3265.9	10.6	235.9	9.4	15.5	49000	73.00000	Ferro-magnetic
5907	9.81	6.67	n.a.	0.470	27.795	n.a.	29.8	2402.9	40.6	371	18.3	5.594	2.9	0.01500	−3.00
5323	79.9	29.6	n.a.	0.320	46, 000	n.a.	937.5	2829.9	59.9	322	5.8	36.8	2.2	0.01100	−1.50
19,320	78.5	26	171	0.420	2.35	4.00	1064.4	2806.9	317	129	14.1	12.78	98.7	0.17000	−1.80
13,310	141	56	109	0.260	35.1	4.40	2229.9	5196.9	23	147	6.0	24.07	104	0.20000	5.30
n.a.	n.a.	n.a.	n.a.	n.a.	n.a.	n.a.	n.a.	n.a.	n.a.	n.a.	n.a.	n.a.	n.a.	n.a.	n.a.
0.1785	n.a.	n.a.	n.a.	n.a.	n.a.	n.a.	−272.2	−273.2	0.152	5193	n.a.	0.021	0.007	0.00010	−5.90
8795	64.8	26.3	40.2	0.231	81.4	1.71	1473.9	2694.9	16.2	164.9	11.2	17.2	65	0.15000	5490.00
0.08988	n.a.	n.a.	n.a.	n.a.	n.a.	n.a.	−259.1	−273.2	0.1815	14310	n.a.	0.117	0.332	0.11000	−24.80
7310	10.6	3.68	n.a.	0.450	8.37	5.20	156.2	2079.9	81.6	233	24.8	3.27	194	0.60000	−1.40
4930	n.a.	n.a.	n.a.	n.a.	1.3 x 10^{15}	n.a.	113.6	184.4	0.449	429	n.a.	15.78	6.2	0.01800	−4.50
22, 420	528	209	371	0.260	4.71	4.50	2409.9	4129.9	147	130	6.8	26	425	0.80000	2.30
7874	208.2	81.6	169.8	0.291	9.71	6.50	1534.9	2749.9	80.2	447	11.8	15.2	2.56	0.01500	Ferro-magnetic
3.7493	n.a.	n.a.	n.a.	n.a.	n.a.	None	−156.6	−152.3	102.4	248	n.a.	1.64	25	0.13000	−4.40
6145	36.6	14.3	27.9	0.280	57	2.18	920.9	3456.9	13.5	195.1	4.9	8.37	8.98	0.02300	11.00
n.a.	n.a.	n.a.	n.a.	n.a.	n.a.	n.a.	n.a.	n.a.	n.a.	n.a.	n.a.	n.a.	n.a.	n.a.	n.a.
11,350	16.1	5.59	45.8	0.440	20.648	4.20	327.5	1739.9	35.3	129	29.0	4.81	0.171	0.00030	−1.50
534	4.91	4.24	n.a.	0.360	8.55	4.35	180.5	1346.9	84.7	3572	56.0	4.60	0.045	n.a.	6.30
9840	68.6	27.2	47.6	0.261	58.2	n.a.	1662.9	3394.9	16.4	154	125.0	19.2			
1738	44.7	17.3	35.6	0.291	4.45	4.25	648.9	1089.9	156	1024	23.7	8.79	0.063	0.00100	6.90
7440	191	79.5	n.a.	0.240	144	n.a.	1243.9	1961.9	7.82	480	21.7	14.7	13.3	0.08300	121.00
n.a.	n.a.	n.a.	n.a.	n.a.	n.a.	n.a.	n.a.	n.a.	n.a.	n.a.	n.a.	n.a.	n.a.	n.a.	n.a.
n.a.	n.a.	n.a.	n.a.	n.a.	n.a.	n.a.	n.a.	n.a.	n.a.	n.a.	n.a.	n.a.	n.a.	n.a.	n.a.
13,546	None	None	None	None	94.1	1.00	−38.9	356.6	8.34	138	62.0	2.324	374	0.63000	−2.10

Table 15.2a (continued)

Element name (IUPAC)	Chemical abstract registry number [CARN]	Symbol (IUPAC)	Atomic number (Z)	Atomic relative mass (1995) ($^{12}C = 12,000$)	Electronic configuration (ground state)	Spectral term	Electronegativity (Pauling)	Crystal space lattice	Space group (Hermann–Mauguin)	Pearson symbol	Strukturbericht and structure type	Lattice parameters (pm)	Transition temperature (α to β) (°C)
Molybdenum	[7439-98-7]	Mo	42	95.94(1)	$[Kr]4d^55s^1$	$^7S_3$	2.16	b.c.c.	Im3m	cI2	A2 (W)	$a = 314.70$	None
Neodymium (α-)	[7440-00-8]	Nd	60	144.24(3)	$[Xe]5d^06s^24f^4$	$^5I_4$	1.14	dhcp	P6₃/mmc	hP4	α3' (α-La)	$a = 365.82$ $c = 1179.66$	863
Neon (g)	[7440-01-9]	Ne	10	20.1797(6)	$[He]2s^22p^6$	$^1S_0$	1.14	Cubic	Fm3m	cF4	A1 (Cu)	$a = 446.20$	-248.59
Neptunium	[7439-99-8]	Np	93	237.0482	$[Rn]5f^46d^17s^2$	$^6L_{11/2}$	1.36	Orthorh.	Pnma	oP8	(α-Np)	$a = 472.30$ $b = 488.70$ $c = 666.30$	280
Nickel	[7440-02-0]	Ni	28	58.6934(2)	$[Ar]3d^84s^2$	$^3F_4$	1.91	f.c.c.	Fm3m	cF4	A1 (Cu)	$a = 352.38$	358
Niobium	[7440-03-1]	Nb	41	92.90638(2)	$[Kr]4d^45s^1$	$^6D_{1/2}$	1.60	b.c.c.	Im3m	cI2	A2 (W)	$a = 330.04$	None
Nitrogen (g, N₂)	[7727-37-9]	N	7	14.00674(7)	$[He]2s^22p^3$	$^4S_{3/2}$	3.04	Cubic	Pa3	cP8	α-N	$a = 566.1$	-237.54
Nobelium	[10028-14-5]	No	102	[259]	$[Rn]5f^{14}6d^07s^2$	$^1S_0$	1.3	n.a.	n.a.	n.a.	n.a.	n.a.	n.a.
Osmium	[7440-04-2]	Os	76	190.23(3)	$[Xe]5d^66s^24f^{14}$	$^5D_4$	2.20	hcp	P6₃/mmc	hP2	A3 (Mg)	$a = 273.41$ $c = 431.98$	None
Oxygen (g, O₂)	[7782-44-7]	O	8	15.9994(3)	$[He]2s^22p^4$	$^3P_2$	3.44	Monocl.	C2m	mC4	α-O	$a = 540.3$ $b = 342.9$ $c = 508.6$ $\beta = 132.53°$	-249.38
Palladium	[7440-05-3]	Pd	46	106.42(1)	$[Kr]4d^{10}5s^0$	$^1S_0$	2.20	f.c.c.	Fm3m	cF4	A1 (Cu)	$a = 389.03$	None
Phosphorus (P₄)	[7723-14-0]	P	15	30.973761(2)	$[Ne]3s^23p^3$	$^4S_{3/2}$	2.19	Cubic	I43m	n.a.	P(white)	$a = 718$	None
Platinum	[7440-06-4]	Pt	78	195.078(2)	$[Xe]5d^96s^14f^{14}$	$^3D_3$	2.28	f.c.c.	Fm3m	cF4	A1 (Cu)	$a = 392.36$	None
Plutonium	[7440-07-5]	Pu	94	[244]	$[Rn]5f^66d^07s^2$	$^7F_0$	1.28	Monocl.	P2₁/m	mP16	(α-Pu)	$a = 618.30$ $b = 482.20$ $c = 1096.30$ $\beta = 101.79°$	122
Polonium	[7440-08-6]	Po	84	[209]	$[Xe]4f^{14}5d^{10}6s^26p^4$	$^3P_2$	2.00	Cubic	Pm3m	cP1	A$_h$ (α-Po)	$a = 336.60$	54
Potassium	[7440-09-7]	K	19	39.0983(1)	$[Ar]4s^1$	$^2S_{1/2}$	0.82	b.c.c.	Im3m	cI2	A2 (W)	$a = 533.4$	None
Praseodymium (α-)	[7440-10-0]	Pr	59	140.90765(2)	$[Xe]5d^06s^24f^3$	$^4I_{9/2}$	1.13	dhcp	P6₃/mmc	hP4	A3' (α-La)	$a = 367.21$ $c = 1183.26$	795
Promethium (α-)	[7440-12-2]	Pm	61	[145]	$[Xe]5d^06s^24f^5$	$^6H_{5/2}$	n.a.	dhcp	P6₃/mmc	hP4	A3' (α-La)	$a = 365.00$ $c = 1165.00$	890
Protoactinium	[7440-13-3]	Pa	91	231.03588(2)	$[Rn]5f^26d^17s^2$	$^4K_{11/2}$	1.50	Tetragonal	I4/mmm	tI2	A$_a$ (α-Pa)	$a = 392.21$ $c = 323.8$	1170
Radium	[7440-14-4]	Ra	88	[226]	$[Rn]7s^2$	$^1S_0$	0.89	b.c.c.	Im3m	cI2	A2 (W)	$a = 514.80$	n.a.
Radon	[10043-92-2]	Rn	86	[222]	$[Xe]4f^{14}5d^{10}6s^26p^6$	$^1S_0$	n.a.	fcc	n.a.	n.a.	n.a.	n.a.	n.a.
Rhenium	[7440-15-5]	Re	75	186.207(1)	$[Xe]5d^56s^24f^{14}$	$^6S_{5/2}$	1.90	hcp	P6₃/mmc	hP2	A3 (Mg)	$a = 276.09$ $c = 445.8$	None
Rhodium	[7440-16-6]	Rh	45	102.90550(2)	$[Kr]4d^85s^1$	$^4F_{9/2}$	2.28	f.c.c.	Fm3m	cF4	A1 (Cu)	$a = 380.32$	None
Rubidium	[7440-17-7]	Rb	37	85.4678(3)	$[Kr]5s^1$	$^2S_{1/2}$	0.82	b.c.c.	Im3m	cI2	A2 (W)	$a = 570.50$	None
Ruthenium	[7440-18-8]	Ru	44	101.07(2)	$[Kr]4d^75s^1$	$^5F_5$	2.20	hcp	P6₃/mmc	hP2	A3 (Mg)	$a = 270.58$ $c = 428.16$	None
Rutherfordium	n.a.	Rf	104	[261]	$[Rn]5f^{14}6d^27s^2$	$^3F_2$	n.a.	n.a.	n.a.	n.a.	n.a.	n.a.	n.a.
Samarium (α-)	[7440-19-9]	Sm	62	150.36(3)	$[Xe]5d^06s^24f^6$	$^7F_0$	1.17	Rhomb.	R-3m	hR3	C19 (α-Sm)	$a = 899.6$ $\alpha = 23° \ 13'$	922

Density (d, kg.m^{-3}) (298.15 K)	Young's or elastic modulus (E, GPa)	Coulomb or shear modulus (G, GPa)	Bulk or compression modulus (K, GPa)	Poisson's ratio (ν)	Electrical resistivity (ρ, $\mu\Omega$.cm) (293.15 K)	Temperature coefficient electrical resistivity (TCR, 10^{-3} K^{-1})	Melting point (mp, °C)	Boiling point (bp, °C)	Thermal conductivity (k, W.m^{-1}.K^{-1}) (300 K)	Specific heat capacity (c_p, J.kg^{-1}.K^{-1}) (300 K)	Coefficient linear thermal expansion (α, 10^{-6}K^{-1}) (0–100°C)	Latent heat of fusion (L_f, kJ.mol^{-1})	Thermal neutron capture cross-section (σ_{th}, 10^{-28} m^2)	Thermal neutron mass absorption coefficient ((μ/ρ), cm^2.g^{-1})	Mass magnetic susceptibility (cm, 10^{-9} kg^{-1}.m^3)
10 220	324.8	125.6	261.2	0.293	5.2	4.35	2616.9	4611.9	138	251	5.1	35.6	2.6	0.00900	11.70
7007	41.4	16.3	31.8	0.281	64	1.64	1020.9	3067.9	16.5	190.3	6.7	7.14	49	0.11000	480.00
0.89994	n.a.	n.a.	n.a.	n.a.	n.a.	None	−248.7	−246.1	0.0493	1030	n.a.	0.324	0.04	0.00600	−4.10
20 250	n.a.	n.a.	n.a.	n.a.	122	n.a.	639.9	3901.9	6.3	n.a.	n.a.	9.46	180	n.a.	
8902	199.5	76	177.3	0.312	6.844	6.92	1452.9	2731.9	90.7	471	13.3	17.16	37.2	0.02600	Ferro-magnetic
8570	104.9	37.5	170.3	0.397	12.5	2.60	2467.9	4741.9	53.7	265	7.2	29.3	1.15	0.00400	28.10
1.2506	None	None	None	None	None	None	−209.9	−195.8	0.02958	1041	n.a.	0.72	1.91	0.04800	−10.00
n.a.	n.a.	n.a.	n.a.	n.a.	n.a.	n.a.	n.a.	n.a.	n.a.	n.a.	n.a.	n.a.	n.a.	n.a.	n.a.
22,590	559	223	373	0.250	8.12	4.10	3053.9	5026.9	87.6	130	4.6	29.3	15	0.02300	0.60
1.429	None	None	None	None	None	None	−218.4	−183.0	0.02674	920	n.a.	0.445	0.00028	0.00001	1335.00
12 020	121	43.6	187	0.390	10.8	4.20	1551.9	3139.9	71.8	244	11.0	16.7	6.9	0.02300	65.70
1820	n.a.	n.a.	n.a.	n.a.	10^{17}	None	44.4	279.9	0.235	770	n.a.	2.64	0.18	0.00200	−11.30
21, 450	170	60.9	276	0.390	10.6	3.92	1771.9	3826.9	71.6	134.4	9.0	19.7	10	0.02000	12.20
19, 840	87.5	34.5	225	0.180	146	n.a.	640.9	3231.9	6.74	142	55.0	2.9	1.7	n.a.	31.40
9320	n.a.	n.a.	n.a.	n.a.	40	n.a.	253.9	961.9	20	n.a.	n.a.	10	0.5	n.a.	n.a.
862	3.175	1.3	n.a.	0.350	7.2	5.70	63.7	773.9	102.4	757	83.0	2.39	2.1	0.01800	6.70
6773	37.3	14.8	28.8	0.281	68	1.71	930.9	3511.9	12.5	193	6.8	11.3	11.4	0.02900	423.00
7220	46.0	18	33	0.280	est. 50	n.a.	1167.9	2726.9	est. 17.9	est. 185	n.a.	12.6	8000	n.a.	n.a.
15, 370	n.a.	n.a.	n.a.	n.a.	17.7	n.a.	1839.9	4000.0	est. 47	n.a.	n.a.	16.7	500	n.a.	32.50
c. 5000	None	None	None	None	100	n.a.	699.9	1139.9	est. 18.6	n.a.	n.a.	7.15	20	n.a.	n.a.
9.73	n.a.	n.a.	n.a.	n.a.	n.a.	None	−71.2	−61.8	est. 0.00364	n.a.	n.a.	2.7	0.7	n.a.	n.a.
21, 020	466	181	334	0.260	19.3	4.50	3179.9	5626.9	47.9	136	6.6	33.5	90	0.16000	4.60
12, 410	379	147	276	0.260	4.51	4.40	1965.9	3726.9	150	243	8.5	22.6	145	0.63000	13.20
1532	2.35	0.91	0.3	0.300	12.5	4.80	39.1	687.9	58.2	363	9.0	2.198	0.38	0.00300	2.60
12, 370	432	173	286	0.250	7.6	4.10	2309.9	3899.9	117	238	9.6	23.7	2.6	0.00900	5.40
n.a.	n.a.	n.a.	n.a.	n.a.	n.a.	n.a.	n.a.	n.a.	n.a.	n.a.	n.a.	n.a.	n.a.	n.a.	n.a.
7536	49.7	19.5	37.8	0.274	88	1.48	1076.9	1790.9	13.3	181	n.a.	8.92	5900	47.00000	111.00

Table 15.2a (continued)

Element name (IUPAC)	Chemical abstract registry number [CARN]	Symbol (IUPAC)	Atomic number (Z)	Atomic relative mass (1995) ($^{12}C = 12,000$)	Electronic configuration (ground state)	Spectral term	Electronegativity (Pauling)	Crystal space lattice	Space group (Hermann–Mauguin)	Pearson symbol	Strukturbericht and structure type	Lattice parameters (pm)	Transition temperature (α to β) (°C)
Scandium (α-)	[7440-20-2]	Sc	21	44.955910(8)	[Ar]$4s^2 3d^1$	$^2D_{3/2}$	1.36	hcp	P6$_3$/mmc	hP2	A3 (Mg)	$a = 330.90$ $c = 527.30$	950
Seaborgium	n.a.	Sg	106	[263]	[Rn]$5f^{14} 6d^4 7s^2$	n.a.	n.a.	n.a.	n.a.	n.a.	n.a.	n.a.	n.a.
Selenium (γ-)	[7782-49-2]	Se	34	78.96(3)	[Ar]$3d^{10} 4s^2 4p^4$	$^3P_2$	2.55	Hex.	P3$_1$21	hP3	A8 (γ-Se)	$a = 436.59$ $c = 495.37$	None
Silicon	[7440-21-3]	Si	14	28.0855(3)	[Ne]$3s^2 3p^2$	$^3P_0$	1.90	Cubic	F-3md	cF8	A4 (Diamond)	$a = 543.06$	None
Silver	[7440-22-4]	Ag	47	107.8682(2)	[Kr]$4d^{10} 5s^1$	$^2S_{1/2}$	1.93	f.c.c.	Fm3m	cF4	A1 (Cu)	$a = 408.57$	None
Sodium	[7440-23-5]	Na	11	22.989770(2)	[Ne]$3s^1 3p^0$	$^2S_{1/2}$	0.93	b.c.c.	Im3m	cI2	A2 (W)	$a = 429.06$	None
Strontium	[7440-24-6]	Sr	38	87.62(1)	[Kr]$5s^2$	$^1S_0$	0.95	f.c.c.	Fm3m	cF4	A1 (Cu)	$a = 608.40$	235, 540
Sulphur (α-)	[7704-34-9]	S	16	32.066(6)	[Ne]$3s^2 3p^4$	$^3P_2$	2.58	Orthorh.	Fddd	oF128	α16 (α-S)	$a = 104.64$ $b = 1286.60$ $c = 2448.60$	93.55
Tantalum	[7440-25-7]	Ta	73	180.9479(1)	[Xe]$5d^3 6s^2 4f^{14}$	$^4F_{3/2}$	1.50	b.c.c.	Im3m	cI2	A2 (W)	$a = 330.30$	None
Technetium	[7440-26-8]	Tc	43	9898	[Kr]$4d^5 5s^2$	$^6S_{5/2}$	1.90	hcp	P6$_3$/mmc	hP2	A3 (Mg)	$a = 273.80$ $c = 439.30$	n.a.
Tellurium	[13494-80-9]	Te	52	127.60(3)	[Kr]$4d^{10} 5s^2 5p^4$	$^3P_2$	2.10	Hex.	P3$_1$21	hP3	A8 (γ-Se)	$a = 445.66$ $c = 592.64$	None
Terbium (α-)	[7440-27-9]	Tb	65	158.92534(2)	[Xe]$5d^0 6s^2 4f^9$	$^6H_{15/2}$	1.20	hcp	P6$_3$/mmc	hP2	A3 (Mg)	$a = 360.55$ $c = 569.66$	1289
Thallium	[7440-28-0]	Tl	81	204.3833(2)	[Xe]$4f^{14} 5d^{10} 6s^2 6p^1$	$^2P_{1/2}$	1.62	hcp	P6$_3$/mmc	hP2	A3 (Mg)	$a = 345.66$ $c = 552.48$	230
Thorium	[7440-29-1]	Th	90	232.0381(1)	[Rn]$6d^2 7s^2$	$^3F_2$	1.30	f.c.c.	Fm3m	cF4	A1 (Cu)	$a = 508.42$	1360
Thulium	[7440-30-4]	Tm	69	168.93421(2)	[Xe]$5d^0 6s^2 4f^{13}$	$^2F_{7/2}$	1.25	hcp	P6$_3$/mmc	hP2	A3 (Mg)	$a = 353.75$ $c = 555.40$	None
Tin (β-)	[7440-31-5]	Sn	50	118.710(7)	[Kr]$4d^{10} 5s^2 5p^2$	$^3P_0$	1.96	Tetragonal	I4$_1$/amd	tI4	A5 (β-Sn)	$a = 648.92$	13
Titanium (α-)	[7440-32-6]	Ti	22	47.867(1)	[Ar]$3d^2 4s^2$	$^3F_2$	1.54	hcp	P6$_3$/mmc	hP2	A3 (Mg)	$a = 295.11$ $c = 468.43$	882
Tungsten	[7440-33-7]	W	74	183.84(1)	[Xe]$5d^4 6s^2 4f^{14}$	$^5D_0$	2.36	b.c.c.	Im3m	cI2	A2 (W)	$a = 316.52$	None
Ununnilium	n.a.	Uun	110	[269]	[Rn]$5f^{14} 6d^8 7s^2$	n.a.	n.a.	n.a.	n.a.	n.a.	n.a.	n.a.	n.a.
Unununium	n.a.	Uuu	111	[272]	[Rn]$5f^{14} 6d^{10} 7s^1$	n.a.	n.a.	n.a.	n.a.	n.a.	n.a.	n.a.	n.a.
Uranium	[7440-61-1]	U	92	238.0289(1)	[Rn]$5f^3 6d^1 7s^2$	$^5L_6$	1.38	Orthorh.	Cmcm	oC4	A20 (α-U)	$a = 285.37$ $b = 586.95$ $c = 495.48$	662, 770
Vanadium	[7040-62-2]	V	23	50.9415(1)	[Ar]$3d^3 4s^2$	$^4F_{3/2}$	1.63	b.c.c.	Im3m	cI2	A2 (W)	$a = 302.78$	None
Xenon	[7040-63-3]	Xe	54	131.29(2)	[Kr]$4d^{10} 5s^2 5p^6$	$^1S_0$	n.a.	f.c.c.	Fm3m	cF4	A1 (Cu)	$a = 635.00$	−185
Ytterbium	[7040-64-4]	Yb	70	173.04(3)	[Xe]$5d^0 6s^2 4f^{14}$	$^1S_0$	1.11	hcp	P6$_3$/mmc	hP2	A3 (Mg)	$a = 387.99$ $c = 638.59$	−3
Yttrium	[7040-65-5]	Y	39	88.90585(2)	[Kr]$4d^1 5s^2$	$^2D_{3/2}$	1.22	hcp	P6$_3$/mmc	hP2	A3 (Mg)	$a = 364.82$ $c = 573.18$	1485
Zinc	[7040-66-6]	Zn	30	65.39(2)	[Ar]$3d^{10} 4s^2$	$^1S_0$	1.65	hcp	P6$_3$/mmc	hP2	A3 (Mg)	$a = 266.48$ $c = 494.69$	None
Zirconium	[7040-67-7]	Zr	40	91.224(2)	[Kr]$4d^2 5s^2$	$^3F_2$	1.33	hcp	P6$_3$/mmc	hP2	A3 (Mg)	$a = 323.21$ $c = 514.77$	862

Density (d, kg.m⁻³) (298.15 K)	Young's or elastic modulus (E, GPa)	Coulomb or shear modulus (G, GPa)	Bulk or compression modulus (K, GPa)	Poisson's ratio (ν)	Electrical resistivity (ρ, $\mu\Omega$.cm) (293.15 K)	Temperature coefficient electrical resistivity (TCR, 10^{-3} K⁻¹)	Melting point (mp, °C)	Boiling point (bp, °C)	Thermal conductivity (k, W.m⁻¹.K⁻¹) (300 K)	Specific heat capacity (c_p, J.kg⁻¹.K⁻¹) (300 K)	Coefficient linear thermal expansion (α, 10^{-6} K⁻¹) (0–100°C)	Latent heat of fusion (l_f, kJ.mol⁻¹)	Thermal neutron capture cross-section (σ_{th}, 10^{-28} m²)	Thermal neutron mass absorption coefficient ((μ/ρ)$_{th}$, cm².g⁻¹)	Mass magnetic susceptibility (cm, 10^{-9} kg⁻¹.m³)
2989	74.4	29.1	56.6	0.279	56.2	n.a.	1540.9	2830.9	15.8	567.4	10.2	15.9	27.2	0.25000	88.00
n.a.	n.a.	n.a.	n.a.	n.a.	n.a.	n.a.	n.a.	n.a.	n.a.	n.a.	n.a.	n.a.	n.a.	n.a.	n.a.
4790	58	n.a.	n.a.	n.a.	106	n.a.	216.9	685.0	2.04	321	37.0	6.28	11.7	0.05600	–4.00
2329	113	39.7	n.a.	0.420	100,000	n.a.	1409.9	2354.9	148	712	7.6	50.66	171	0.00200	–1.60
10,500	82.7	30.3	103.6	0.367	1.587	4.10	962.0	2211.9	429	235	19.1	11.09	63.6	0.20000	–2.30
971.2	6.8	2.53	n.a.	0.340	4.2	5.50	97.8	883.0	141	1228	68.9	2.64	0.53	0.00700	6.40
2540	15.7	6.03	12	0.280	13.2	n.a.	768.9	1383.9	35.3	737	100.0	8.4	1.2	0.00500	–2.50
2070	n.a.	n.a.	n.a.	n.a.	2×10^{23}	None	112.9	444.7	0.269	706	74.3	1.235	0.52	0.00550	–6.20
16,654	185.7	69.2	196.3	0.342	12.45	3.50	2995.9	5424.9	57.5	140	6.5	24.7	20.5	0.04100	10.70
11 500	n.a.	n.a.	n.a.	n.a.	22.6	n.a.	2171.9	4876.9	0.206	708	n.a.	23.81	22	n.a.	34.20
6240	47.1	16.7	n.a.	0.180	43,600	n.a.	449.6	989.9	2.35	202	27.0	17.6	5.4	0.01300	–3.90
8229	55.7	22.1	38.7	0.261	114	n.a.	1355.9	3122.9	11.1	172	7.0	16.3	23	0.09000	13 600.00
1,850	7.9	2.7	28.5	0.450	18	5.20	303.5	1456.9	46.1	130	30.0	4.3	3.4	0.00600	–3.00
11,720	78.3	30.8	54	0.260	13	4.00	1749.9	4786.9	54	118	11.2	19.0	7.4	0.01000	5.30
9321	74	30.5	44.5	0.213	67.6	1.95	1544.9	1946.9	16.8	160	11.6	18.4	105	0.25000	1990.00
7298.40	49.9	18.4	58.2	0.357	11	4.60	232.0	2269.9	66.6	227	23.5	7.08	0.63	0.00200	–3.10
4540	120.2	45.6	108.4	0.361	42	3.80	1659.9	3286.9	21.9	537.8	8.4	17.5	6.1	0.04400	40.10
19,300	411	160.6	311	0.280	5.65	4.80	3406.9	5656.9	174	132	4.5	35.2	18.4	0.03600	3.90
n.a.	n.a.	n.a.	n.a.	n.a.	n.a.	n.a.	n.a.	n.a.	n.a.	n.a.	n.a.	n.a.	n.a.	n.a.	n.a.
n.a.	n.a.	n.a.	n.a.	n.a.	n.a.	n.a.	n.a.	n.a.	n.a.	n.a.	n.a.	n.a.	n.a.	n.a.	n.a.
8,950	175.8	73.1	97.9	0.200	30.8	3.40	1132.4	3744.9	27.6	116	5–23	15.5	7.57	0.00500	21.50
6160	127.6	46.7	158	0.365	24.8	3.90	1886.9	3376.9	30.7	498	8.3	16.74	5.06	0.03300	62.80
6.897	None	None	None	None	n.a.	n.a.	–111.9	–107.1	0.00569	1583	n.a.	3.10	25	0.08300	–4.30
965	23.9	9.9	30.5	0.207	29	1.30	823.9	1192.9	34.9	145	25.0	9.20	35	0.07600	5.90
469	63.5	25.6	41.2	0.243	57	2.71	1521.9	3337.9	17.2	298	10.8	11.43	1.28	0.00600	66.60
133	104.5	41.9	69.4	0.249	5.916	4.20	419.6	906.9	116	389	25.0	7.28	1.1	0.00550	–2.21
506	98	35	89.8	0.380	42.1	4.40	1851.9	4376.9	22.7	278	5.9	19.3	0.184	0.00660	16.60

15
Appendices

Table 15.6 Cost of pure elements and some alloys (1998)

Material	Purity (wt%)	Cost ($US tr oz^{-1})	Cost ($US lb^{-1})	Cost ($US kg^{-1})
Aluminum	99.50	0.045	0.67	1.48
Antimony	99.99	0.045	0.65	1.43
Barium	99.70	12.44	181.44	400.00
Beryllium	99.50	22.43	327.04	721.00
Bismuth	99.99	0.24	3.50	7.72
Boron	99.00	155.52	2267.96	5000.00
Calcium	99.90	0.28	4.02	8.86
Cadmium	99.90	0.02	0.27	0.61
Cesium	99.98	630.87	9200	20,283
Cerium	99.90	3.89	56.70	125.00
Chromium	99.90	0.51	7.37	16.26
Cobalt	99.50	1.30	19.00	41.89
Copper	99.9990	0.09	1.27	2.80
Dysprosium	99.9	9.33	136.08	300.00
Erbium	99.9	20.22	294.84	650.00
Europium	99.00	233.28	3401.94	7500.00
Ferrochromium	65.00	0.049	0.71	1.565
Ferromaganese	76–78	0.02	0.34	0.75
Ferromolybdenum	65–70	0.17	2.43	5.35
Ferroniobium	65–70	0.44	6.35	14.00
Ferrosilicon	75.00	0.03	0.41	0.90
Ferrovanadium	80.00	0.55	8.03	17.70
Ferrotungsten	70–75	0.26	3.80	8.37
Gadolinium	99.00	15.09	219.99	485
Gallium	100.00	13.22	192.77	424.99
Germanium	99.99	38.88	566.99	1250
Gold 10 Kt	41.67	162.08	2363.72	5211.10
Gold 14 Kt	58.33	226.92	3309.20	7295.54
Gold 18 Kt	75.00	291.75	4254.69	9379.98
Gold 20 Kt	83.33	324.17	4727.43	10,422.20
Gold 24 Kt	99.995	389.00	5672.92	12,506.64
Hafnium	97.00	50.20	732.09	1614
Holmium	99.00	311.03	4535.92	10 000
Indium	99.99	8.55	124.69	274.89
Iridium	99.999	415.01	6052.28	13,343
Iron	99.99	0.03	0.45	1.00
Lanthanum	99.00	155.52	2267.96	5000
Lead	99.90	0.016	0.24	0.53
Lithium	99.80	2.97	43.27	95.40
Lutecium	99.00	2332.76	34,019.43	75,000
Magnesium	99.80	0.13	1.85	4.08
Mercury	100.00	0.158	2.30	5.08

Table 15.6 (continued)

Material	Purity (wt%)	Cost ($US tr oz^{-1})	Cost ($US lb^{-1})	Cost ($US kg^{-1})
Molybdenum	99.95	4.06	59.26	130.65
Niobium-zirconium	99.00	7.95	116.00	255.74
Neodymium	99.00	31.10	453.59	1000.00
Nickel	99.00	0.157	2.29	5.05
Niobium	99.90	6.88	100.40	221.34
Osmium	99.999	399.99	5833.20	12,860.00
Palladium	99.999	359.99	5249.88	11,574
Platinum	99.999	398.18	5806.76	12,801.72
Potassium	99.90	2.80	40.82	90.00
Praseodymium	99.00	70.00	1020.83	2250.55
Rhenium	99.90	27.99	408.23	900.00
Rhodium	99.999	890.00	12,979.17	28,614.17
Rubidium	99.80	622.07	9071.85	20,000.00
Ruthenium	99 999	40.99	597.83	1318
Samarium	99.99	155.52	2267.96	5000.00
Selenium	99.00	0.27	4.00	8.82
Silicon	99.90	0.06	0.88	1.94
Silver	99.99	5.19	75.74	167
Sodium	99.90	2.05	29.94	66.00
Strontium	99.95	311.03	4535.92	10,000.00
Tantalum	99.90	14.33	209.00	460.77
Tellurium	99.50	1.51	22.00	48.50
Terbium	99.00	933.10	13 607.77	30,000.00
Terbium	99.90	1555.17	22 679.62	50,000.00
Thallium	99.00	2.74	39.92	88.00
Thorium	99.90	150.00	2187.57	4822.76
Tin	99.90	0.17	2.48	5.47
Titanium (high purity)	99.99	3.51	51.26	113.00
Titanium alloy Ti–0.25Pd	–	7.63	111.29	245.35
Titanium alloy Ti–6A1–4V	–	1.53	22.33	49.24
Titanium C.P. Grade 2	99.80	1.44	21.00	46.30
Tungsten	99.90	19.20	280.00	617.29
Vanadium	99.00	50.00	729.17	1607.54
Ytterbium	99.90	622.07	9071.85	20,000.00
Yttrium	99.90	6.86	100.00	220.46
Zinc	100.00	0.033	0.48	1.06
Zircadyne® 702	99.00	3.43	50.00	110.23
Zirconium	99.80	5.91	86.18	190

There are seven possible space lattices which entirely describe crystalline materials, either inorganic or organic. These are called the seven crystal systems (cubic, tetragonal, hexagonal, trigonal, orthorhombic, monoclinic, and triclinic).

15.6.2 Symmetry Elements

Table 15.7 Symmetry element notations

Symmetry element	Notation International Hermann–Mauguin	Old Schonflies	Symmetry operation
Center	$\bar{1}$	C_i	Center of inversion
Reflection plane (mirror)	m	C_s	Single reflection plane of symmetry
n-fold rotation axis	n	C_n	n-fold rotation axis with $n = 2, 3, 4,$ and 6, the angle of rotation, A, expressed in radians is given by the following relation: $A(\text{rad}) = 2\pi/n$
Inversion axis	n	C_{ni}	Vertical n-fold rotation axis followed by an inversion by a symmetry center lying on the axis ($\bar{2} = m, \bar{3}, \bar{4} = 4, \bar{6}$)
Glide plane	$a, b, c,$ n, d	–	Reflection in a plane followed by a translation according to a vector parallel to the plane: Translation in the a direction: a Translation in the b direction: b Translation in the c direction: c Translation in the $1/2(a + b)$ or face diagonal direction: n Translation in the $1/2(a + b + c)$ or volume diagonal direction: d
Screw axis	n_m	–	Vertical n-fold axis, followed by a translation parallel to the axis
Rotary-reflection axis	$\tilde{n}$	S_n	Point group with an n-fold axis of rotary reflection

Table 15.8 Five Platonician regular polyhedrons

Regular polyhedron	Face	Volume	Surface area	No. of faces	No. of edges	No. of vertices
Tetrahedron	Equilateral triangle	$a^3\sqrt{2}/12$	$a^2\sqrt{3}$	4	6	4
Octahedron	Equilateral triangle	$a^3\sqrt{2}/3$	$2a^2\sqrt{3}$	8	12	6
Hexahedron (cube)	Square	a^3	$6a^2$	6	12	8
Pentagonal dodecahedron	Regular pentagon	n.a.	n.a.	12	30	20
Icosahedron	Equilateral triangle	n.a.	$5a^2\sqrt{3}$	20	30	12

15.6.3 The Seven Crystal Systems

See **Table 15.9** (page 541).

Table 15.9 The seven crystal systems

Crystal system	Synonyms, old names	Symbol	Geometrical description	Symmetry Hermann–Mauguin (Schoenflies–Fedorov)	Lattice parameters (IUCr) (edge length, interaxial angles)
Cubic	Isometric	C (c)	Cube	$m3m$ (O_h)	$a = b = c$ $\alpha = \beta = \gamma = \pi/2$ rad
Hexagonal		H (h)	Upright prism with a regular hexagonal base	$6/mmm$ (D_{6h})	$a = b \neq c$ $\alpha = \beta = \pi/2$ rad and $\gamma = 2\pi/3$ rad
Tetragonal	Quadratic	T (t)	Upright prism with a square base	$4/mmm$ (D_{4h})	$a = b \neq c$ $\alpha = \beta = \gamma = \pi/2$ rad
Rhombohedral	Trigonal	R (h)	Prism with faces of identical lozenges	$3m$ (D_{3d})	$a = b = c$ $\alpha = \beta = \gamma \neq \pi/2$ rad
Orthorhombic	Ortho-gonal	O (o)	Upright prism with a rectangular basis	mmm (D_{2h})	$a \neq b \neq c$ $\alpha = \beta = \gamma = \pi/2$ rad
Monoclinic	Clino-rhombic	M (m)	Inclined prism with a rectangular base	$2/m$ (C_{2h})	$a \neq b \neq c$ $\alpha = \gamma = \pi/2$ rad and $\beta > 2\pi/3$ rad
Triclinic	Anorthic	T (a)	Uneven prism	1 (C)	$a \neq b \neq c$ $\alpha \neq \beta \neq \gamma \neq \pi/2$ rad

15.6.4 Conversion of Rhombohedral to Hexagonal Lattice

The rhombohedral unit cell is defined by three equal length unit translations, a, and the plane angle between them, α. The rhombohedral lattice parameters can be converted to hexagonal by using the two following equations:

$$a_{\mathrm{H}} = 2a_{\mathrm{R}}\sin(\alpha/2)$$
$$c_{\mathrm{H}} = 3[a_{\mathrm{R}}^2 - a_{\mathrm{H}}^2/3]^{1/2}$$

15.6.5 The 14 Bravais Space Lattices

See **Table 15.10** (page 542).

15.6.6 Close-Packed Arrangements Characteristics

See **Table 15.11** (page 542).

15.6.7 The 32 Classes of Symmetry

There are 10 elements of symmetry in crystals. These 10 symmetry operators can be combined in 32 ways to produce the 32 point groups (Table 15.13, page 544).

Table 15.10 The 14 Bravais space lattices

Crystal system	Bravais space lattice	ASTM notation	Hermann–Mauguin symbol[a]	Pearson notation
Cubic	Primitive cell	C	P	cP1
	Body centered	B	I	cI2
	Face centered	F	F	cF4
Hexagonal	Primitive cell	H	P	hP2
Tetragonal	Primitive cell	T	P	tP1
	Body centered	U	I	tI2
Rhombohedral	Primitive cell	R	R	hR1
Orthorhombic	Primitive cell	O	P	oP1
	Base centered	Q	A, B, C	oA2
	Body centered	P	I	oI2
	Face centered	S	F	oF4
Monoclinic	Primitive cell	M	P	mP1
	Base centered	N	A, B, C	mP2
Triclinic	Primitive cell	A	P	aP1

[a] **P** primitive, **I** body centered (from German, Innercentrum), **F** face-centered (from German, Flaschencentriert), A, B, C faces orthogonal to lattice vectors *a*, *b* and *c* respectively.

Table 15.11 Characteristics of close-packed arrangements

Parameters	Simple cubic	Body-centered cubic	Face-centered cubic	Hexagonal close packed
Notation	cs, *P*	bcc, *I*	fcc, *F*	hcp
Unit cell volume	a^3	a^3	a^3	$a^2 c \sqrt{3}/2$
Number of entities per unit cell	1	2	4	2
Primitive cell volume	a^3	$a^3/2$	$a^3/4$	$a^2 c \sqrt{3}/2$
Number of first neighboring entities (coordinance number)	6	8	12	12
Number of second neighboring entities	12	6	6	12
Smallest distance between first neighbors	a	$a\sqrt{3}/2 \cong 0.866a$	$a\sqrt{2}/2 \cong 0.707a$	a
Smallest distance between second neighbors	$a\sqrt{2} = 1.414a$	a	a	$a\sqrt{3}$
Packing fraction	$\pi/6 \cong 0.524$	$\pi\, a\sqrt{3}/8 \cong 0.680$	$\pi a \sqrt{2}/6 \cong 0.740$	$\pi\, a\sqrt{2}/6 \cong 0.740$

Table 15.12 Schoenflies–Fedorov point group notation

Notation	Description
C_n	Point group with a single n-fold rotation axis
C_{nh}	Point group with a single vertical n-fold rotation axis, together with a horizontal mirror plane
C_{nv}	Point group with a single vertical n-fold rotation axis, together with n vertical planes
D_n	Point group with a single vertical n-fold rotation axis, together with two-fold rotation axis perpendicular to it
V	Alternative symbol to D_2
O	Holohedral cubic point group
T	Tetartohedral cubic point groups
S_n	Point group with an n-fold axis of rotary reflection
i	Center of inversion
s	Single plane of symmetry
d	Diagonal reflection plane, bisecting the angle between two horizontal axes

15.6.8 Strukturbericht Designations

See **Tables 15.14–15.19** (pages 546–551).

15.6.9 The 230 Space Groups

See **Tables 15.20–15.26** (pages 552–557).

15.6.10 Crystallographic Calculations

15.6.10.1 Theoretical Crystal Density

The theoretical density, ρ, expressed in kg.m^{-3}, of a crystal having a number Z of entities with atomic (or molecular) molar mass M, expressed in kg.mol^{-1}, placed in a space lattice structure having a unit cell of volume V, expressed in m^3 is given by the following equation, where N_A is Avogadro's number (6.0221367×10^{23} mol^{-1}):

$$\rho = \frac{ZM}{N_A V}.$$

15.6.10.2 Lattice Point and Vector Position

A lattice point, $\{M\}$, which describes the position of a microscopic entity (e.g., electrons, ions, atoms, molecules or clusters), is located in the crystal space lattice by giving the number of unit translations, along each of the three distinct translation directions, by which it is removed from the point $\{Q\}$ as fixed origin. Therefore, each lattice point is entirely and simply described by a set of three coordinates (u, v, w) or by the single position vector V:

$$V = \mathbf{OM} = u \cdot \mathbf{a} + v \cdot \mathbf{b} + w \cdot \mathbf{c}.$$

Note: sometimes the lattice point coordinates are denoted by the designation: $\cdot$ uvw $\cdot$ (e.g. $\cdot$ 320 $\cdot$).

Table 15.13 The 32 classes of symmetry

Crystal system	Hermann–Mauguin	Schoenflies-Fedorov	Crystal morphology (names of classes according to Von Groth)	Typical mineral	Class no.
Cubic	$m3m$	O_h	Cubic hexaoctahedral (= holohedral)	Galena, PbS	32
	$\bar{4}3m$	T_d	Cubic hexatetrahedral (= tetrahedral)	Sphalerite, ZnS	31
	$m3$	T_h	Cubic dyakis-dodecahadral (=diploidal, or pyritohedral)	Pyrite, FeS_2	30
	432	O	Cubic pentagonal icositetrahedral (= gyroidal, or plagihedral)	Cuprite, Cu_2O	29
	23	T	Cubic tetrahedral-pentagonal dodecahedral (= tetartohedral)	Ullmanite, NiSSb	28
Hexagonal	$6/mmm$	D_{6h}	Dihexagonal-dipyramidal (= holohedral)	Beryl, $Be_3A1_2Si_6O_{18}$	27
	$6mm$	C_{6v}	Dihexagonal-pyramidal (= hemimorphic)	Greenockite	26
	$6/m$	C_{6h}	Hexagonal-dipyramidal (= pyramidal)	Apatite, $Ca_5(PO_4)_3(F,OH,CI)$	25
	622	D_6	Hexagonal trapezohedral (= trapezohedral)	Kalsilite	24
	6	C_6	Hexagonal pyramidal (= tetartohedral)	Nepheline	23
	$\bar{6}m2$	D_{3h}	Ditrigonal-dipyramidal (= trigonal holohedral)	Benitoite	22
	$\bar{6}$	C_{3h}	Trigonal-dipyramidal	Not found in minerals	19
Trigonal (rhombohedral)	$\bar{3}m$	C_{3d}	Hexagonal scalenohedral (= ditrigonal pyramidal, holohedry)	Calcite, $CaCO_3$	21
	$3m$	C_{3v}	Ditrigonal-pyramidal (= **hemimorphic** hemihedry)	Tourmaline	20
	32	D_3	Trigonal-trapezohedral	α-Quartz, SiO_2	18
	$\bar{3}$	$S_5 = C_{3I}$	Trigonal-rhombohedral	Dolomite	17
	3	C_3	Trigonal-pyramidal (= tetartohedry)	Not found in minerals	16
Tetragonal	$4/mmm$	D_{4h}	Ditetragonal-dipyramidal (= holohedry)	Zircon, $ZrSiO_4$	15
	$4mm$	C_{4v}	Ditetragonal-pyramidal (= hemimorphic hemihedry)	Diaboleite	14
	$4/m$	C_{4h}	Tetragonal-dipyramidal (= paramorphic hemihedry)	Scheelite, $CaWO_4$	13
	422	D_4	Tetragonal-trapezohedral (= enantiomorphic hemihedry)	Phosgenite	12
	$\bar{4}2m$	$V_4 = D_{2d}$	Tetragonal scalenohedral (= sphenoidal, hemihedry of 2nd sort)	Chalcopyrite, $CuFeS_2$	11
	4	C_4	Tetragonal-pyramidal (= tetartohedry)	Wulfenite	10
	$\bar{4}$	S_4	Tetragonal-disphenoidal (= ogdohedry)	Cahnite	9
Orthorhombic	mmm	$V_h = D_{2h}$	Orthorhombic-dipyramidal (= holohedral)	Baryte, $BaSO_4$	8

Table 15.13 (continued)

Crystal system	Hermann–Mauguin	Schoenflies–Fedorov	Crystal morphology (names of classes according to Von Groth)	Typical mineral	Class no.
	mm2	C_{2v}	Orthorhombic-pyramidal (= hemimorphic hemihedry)	Topaz	7
	222	$V = D_2$	Orthorhombic-disphenoidal (= enantiomorphic hemihedry)	Sulfur, S_8	6
Monoclinic	2/m	C_{2h}	Rhomboidal prismatic (= holohedry)	Gypsum, $CaSO_4$	5
	m	$C_{h1} = C_s$	Monoclinic domatic (= clinohedral, hemihedry)	Clinohediite	4
	2	C_2	Monoclinic sphenoidal (= hemimorphic hemihedry)	Tartaric acid	3
Triclinic	-1	C_i	Triclinic pinacoidal (= holohedry)	$CuSO_4$, axinite	2
	1	C_1	Triclinic asymmetric (= Pedial, hemihedry)	Not found in minerals	1

15.6.10.3 Scalar Product

The scalar product between two vectors is a scalar quantity represented as $V_1 \cdot V_2$ and is defined by the following equation:

$$V_1 \cdot V_2 = |V_1| \times |V_2| \cos(V_1, V_2) = |V_1| \times |Vc_2| \cos \theta$$

where θ is the plane angle, expressed in radians, measured counterclockwise between the two vectors. Introducing the set of the six vector coordinates, is possible to express the scalar product analytically as:

$$V_1 \cdot V_2 = [u_1 u_2 a^2 + v_1 v_2 b^2 + w_1 w_2 c^2 + (u_1 v_2 + v_1 u_2)ab \cos \gamma +$$
$$(u_1 w_2 + v_1 u_2)ac \cos \beta + (w_1 v_2 + v_1 w_2)bc \cos \alpha]$$

Finally, the scalar product can be also written as a matrix product:

$$V_1 \cdot V_2 = (u_1 v_1 w_1) \cdot \begin{bmatrix} a \cdot a & a \cdot b & a \cdot c \\ b \cdot a & b \cdot b & b \cdot c \\ c \cdot a & c \cdot b & c \cdot c \end{bmatrix} \cdot \begin{bmatrix} u_2 \\ v_2 \\ w_2 \end{bmatrix}$$

15.6.10.4 Vector or Cross Product

The vector product between two vectors is a vector quantity represented as $V_1 \times V_2$ or $V_1 \wedge V_2$ and is defined by the following equation:

$$V_1 \times V_2 = |V_1| \times |V_2| \sin(V_1, V_2) = |V_1| \times |V_2| \sin \theta$$

where θ is the plane angle, expressed in radians, measured counterclockwise between the two vectors. Introducing the set of the six vector coordinates, is possible to express the vector product analytically as:

$$V_1 \times V_2 = [(v_1 w_2 - w_1 v_2)b \times c + (u_2 w_1 - u_1 w_2)c \times a + (u_1 v_2 - u_2 v_1)a \times b].$$

Table 15.14 Strukturbericht designations for pure elements (A type)

Desig-nation	Typical example (Mineral name)	Crystal system	Hermann –Mauguin	Pearson
A_a	α-Protoactinium	Tetragonal	$14/mmm$	$tI2$
A_b	β-Uranium	Tetragonal	$P4nm$	$tP30$
A_c	α-Neptunium	Orthorhombic	$Pmcn$	$oP8$
A_d	β-Neptunium	Tetragonal	$P42_1$	$tP4$
A_e	β-TiCu$_3$	Orthorhombic	$Cmcm$	$oC4$
A_f	HgSn$_{10}$	Hexagonal	$P6/mmm$	hPI
A_g	γ-Boron	Tetragonal	$P\bar{4}n2$	$tP50$
A_h	α-Polonium	Cubic	$Pm3m$	cPI
A_i	β-Polonium	Rhombohedral	$R\bar{3}m$	tRI
A_k	α-Selenium	Monoclinic	$P2_1/n$	$mP32$
A_l	β-Selenium	Monoclinic	$P2_1/a$	$mP32$
A1	Copper	Cubic fcc	$Fm3m$	$cF4$
A2	Tungsten	Cubic bcc	$Im3m$	$cI2$
A3	Magnesium	Hexagonal hcp	$P6_3/mmc$	$hP2$
A4	Diamond	Cubic	$Fd3m$	$cF8$
A5	β-Tin, white	Tetragonal	$14/amd$	$tI4$
A6	Indium	Tetragonal	$F4/mmm$	$tF4$
A7	α-Arsenic	Rhombohedral	$R\bar{3}m$	$hR2$
A8	γ-Selenium	Trigonal	$P3_221$	$hP3$
A9	Graphite	Hexagonal	$P6_3/mmc$	$hP4$
A10	α-Mercury	Rhombohedral	$R\bar{3}m$	$hR1$
A11	α-Gallium	Orthorhombic	$Cmca$	$oC8$
A12	α-Manganese	Cubic	$1\bar{4}3m$	$cI58$
A13	β-Manganese	Cubic	$P4_13$	$cP20$
A14	Iodine (I$_2$)	Orthorhombic	$Pm3n$	$cP8$
A15	β-tungsten, or Cr$_3$Si	Cubic	$Pm3n$	$cP8$
A16	α-Sulfur (S$_4$)	Orthorhombic	$Fddd$	$oF128$
A17	Phosphorus (Black)	Orthorhombic	Cmca	$oC8$
A19	Polonium	Monoclinic	n.a.	n.a.
A20	α-uranium	Orthorhombic	$Cmcm$	$oC4$

Table 15.15 Strukturbericht designations for binary compounds (AX type)

Desig-nation	Typical example (Mineral name)	Crystal system	Hermann–Mauguin	Pearson
B_a	CoU	Cubic	12_13	$cI16$
B_b	ζ-AgZn	Hexagonal	$P\bar{3}$	$hP9$
B_c	CaSi	Orthorhombic	$Cmmc$	$oC8$
B_d	η-NiSi	Orthorhombic	$Pbnm$	$oP8$
B_e	CdSb	Orthorhombic	$Pbca$	$oP16$
B_f	CrB	Orthorhombic	$Cmcm$	$oC8$
B_g	MoB	Tetragonal	$14/amd$	$tI16$
B_h	WC	Hexagonal	$P6/mmm$	$hP2$
B_i	γ'-MoC	Hexagonal	$P6_3/mmc$	$hP8$
B_k	BN	Hexagonal	$P6_3/mmc$	$hP4$
B_l	AsS (realgar)	Monoclinic	$P2_1n$	$mP32$
B_m	TiB	Orthorhombic	$Pnma$	$oP8$
B1	Halite, rocksalt, NaCl	Cubic	$Fm3m$	$cF8$
B2	CsCl	Cubic	$Pm3m$	$cP2$
B3	ZnS (sphalerite)	Cubic	$F\bar{4}3m$	$cF8$
B4	ZnS (Wurzite)	Hexagonal	$P6_3mc$	$hP4$
B8$_1$	α-NiAs	Hexagonal	$P6_3/mmc$	$hP4$
B8$_2$	β-Ni$_2$In	Hexagonal	$P6_3/mmc$	$hP4$
B9	HgS (cinnabar)	Hexagonal	$P3_121$	$hP6$
B10	LiOH	Tetragonal	$P4/nmm$	$tP4$
B11	PbO (massicot)	Tetragonal	$P4/nmm$	$tP4$
B13	NiS (Millerite)	Hexagonal	$R3m$	$hR6$
B16	GeS	Orthorhombic	$Pnma$	$oP8$
B17	PtS (Cooperite)	Tetragonal	$P4_2/mmc$	$tP4$
B18	CuS (Covellite)	Hexagonal	$P6_3/mmc$	$hP12$
B19	AuCd	Orthorhombic	$Pmcm$	$oP4$
B20	FeSi	Cubic	$P2_13$	$cP8$
B26	CuO	Monoclinic	n.a.	n.a.
B27	FeB	Orthorhombic	$Pbnm$	$oP8$
B29	SnS	Orthorhombic	$Pmcn$	$oP8$
B31	MnP	Orthorhombic	$Pcnm$	$oP8$
B32	NaTl	Cubic	$Fd3m$	$cF16$
B34	PdS	Tetragonal	$P4_2/m$	$tP16$
B35	CoSn	Hexagonal	$P6/mmm$	$hP6$
B37	TlSe	Tetragonal	$14/mcm$	$tI16$

Table 15.16 Strukturbericht designations for ternary compounds (A_2X or AX_2 type)

Designation	Typical example (Mineral name)	Crystal system	Hermann–Mauguin	Pearson
C_a	Mg_2Ni	Hexagonal	$P6_222$	$hP18$
C_b	Mg_2Cu	Orthorhombic	$Fddd$	$oF48$
C_c	$ThSi_2$	Tetragonal	$14/amd$	$tI12$
C_e	$CoGe_2$	Orthorhombic	Aba	$oA24$
C_g	ThC_2	Monoclinic	$C2/c$	$mC12$
C_h	Cu_2Te	Hexagonal	$P6/mmm$	$hP6$
C_k	$LiZn_2$	Hexagonal	$P6_3/mmc$	$hP3$
C_1	CaF_2 (fluorite)	Cubic	$Fm3m$	$cF12$
$C1_b$	$MgAgAs$	Cubic	$F\bar{4}3m$	$cF12$
$C2$	FeS_2 (pyrite)	Cubic	$Pa3$	$cP12$
$C3$	Cu_2O (cuprite)	Cubic	$Pn3m$	$cP6$
$C4$	TiO_2 (rutile)	Tetragonal	$P4_2/mmm$	$tP6$
$C5$	TiO_2 (anatase)	Tetragonal	n.a.	n.a.
$C6$	CdI_2	Hexagonal	$P\bar{3}m1$	$hP3$
$C7$	MoS_2 (molybdenite)	Hexagonal	$P6_3/mmc$	$hP6$
$C8$	SiO_2 (quartz)	Hexagonal	n.a.	n.a.
$C9$	SiO_2 (β-cristoballite)	Cubic	n.a.	n.a.
$C10$	SiO_2 (β-tridymite)	Hexagonal	n.a.	n.a.
$C11_a$	CaC_2	Tetragonal	$14/mmm$	$tI6$
$C11_b$	$MoSi_2$	Tetragonal	$14/mmm$	$tI6$
$C12$	$CaSi_2$	Rhombohedral	$R\bar{3}m$	$hR6$
$C14$	$MgZn_2$	Hexagonal	$P6_3/mmc$	$hP12$
$C15$	$MgCu_2$	Cubic	$Fd3m$	$cF24$
$C15_b$	$AuBe_5$	Cubic	$F\bar{4}3m$	$cF24$
$C16$	Al_2Cu	Tetragonal	$14/mcm$	$tI12$
$C18$	FeS_2 (marcasite)	Orthorhombic	$Pnnm$	$oP6$
$C19$	α-Sm	Hexagonal	$R\bar{3}m$	$hR3$
$C21$	TiO_2 (brookite)	Orthorhombic	n.a.	n.a.
$C22$	Fe_2P	Hexagonal	$P2\bar{6}m$	$hP9$
$C23$	$PbCl_2$	Orthorhombic	$Pnma$	$oP12$
$C28$	$HgCl_2$	Orthorhombic	n.a.	n.a.
$C29$	SrH_2	Orthorhombic	n.a.	n.a.
$C32$	AlB_2	Hexagonal	$P6/mmm$	$hP3$
$C33$	Bi_2Te_3S	Hexagonal	$R\bar{3}m$	$hR5$
$C34$	$AuTe_2$ (calaverite)	Monoclinic	$C2/m$	$mC6$
$C35$	$CaCl_2$	Orthorhombic	n.a.	n.a.
$C36$	$MgNi_2$	Hexagonal	$P6_3/mmc$	$hP24$
$C37$	Co_2Si	Orthorhombic	$Pbnm$	$oP12$
$C38$	Cu_2Sb	Tetragonal	$P4/nmm$	$tP6$
$C40$	$CrSi_2$	Hexagonal	$P6_222$	$hP9$
$C42$	SiS_2	Orthorhombic	$Icma$	$oI12$
$C43$	ZrO_2 (baddeleyite)	Monoclinic	$P2_1C$	$mP12$

Table 15.16 (continued)

Designation	Typical example (Mineral name)	Crystal system	Hermann–Mauguin	Pearson
C44	GeS_2	Orthorhombic	$Fdd2$	$oF72$
C46	$AuTe_2$ (krennerite)	Orthorhombic	$Pma2$	$oP24$
C49	$ZrSi_2$	Orthorhombic	$Cmcm$	$oC12$
C54	TiS_2	Orthorhombic	$Fddd$	$oF24$

Table 15.17 Strukturbericht designations for quaternary compounds (A_3X or AX_3 type)

Designation	Typical example (Mineral name)	Crystal system	Hermann–Mauguin	Pearson
$D0_a$	β-$TiCu_3$	Orthorhombic	$Pmmn$	$oP8$
$D0_b$	γ-Ag_3Ga	Hexagonal	$P\bar{3}$	$hP9$
$D0_c$	U_3Si	Tetragonal	$14/mcm$	$tI16$
$D0_d$	Mn_3As	Orthorhombic	$Pmmn$	$oP16$
$D0_2$	$CoAs_3$	Cubic	$Im3$	$cI32$
$D0_3$	BiF_3 or $BiLi_3$	Cubic	$Fm3m$	$cF16$
$D0_9$	ReO_3 or Cu_3N	Cubic	$Pm3m$	$cP4$
$D0_{11}$	Fe_3C	Orthorhombic	$Pnma$	$oP16$
$D0_{18}$	Na_3As	Hexagonal	$P6_3/mmc$	$hP8$
$D0_{19}$	Mg_3Cd	Hexagonal	$P6_3/mmc$	$hP8$
$D0_{20}$	$NiAl_3$	Orthorhombic	$Pnma$	$oP16$
$D0_{21}$	Cu_3P	Hexagonal	$P\bar{3}cl$	$hP24$
$D0_{22}$	$TiAl_3$	Tetragonal	$14/mmm$	$tI8$
$D0_{23}$	$ZrAl_3$	Tetragonal	$14/mmm$	$tI16$
$D0_{24}$	$TiNi_3$	Hexagonal	$P6_3/mmc$	$hP16$

Table 15.18 Strukturbericht designations for compounds (A_4X or AX_4 type)

Designation	Typical example (Mineral name)	Crystal system	Hermann–Mauguin	Pearson
$D1_3$	$BaAl_4$	Tetragonal	$14/mmm$	$tI10$
$D1_a$	$MoNi_4$	Tetragonal	$14/m$	$tI10$
$D1_b$	UAl_4	Orthorhombic	$Imma$	$oI20$
$D1_c$	$PtSn_4$	Orthorhombic	$Aba2$	$oC20$
$D1_d$	$PtPb_4$	Tetragonal	$P4/nbm$	$tP10$
$D1_e$	UB_4	Tetragonal	$P4/mbm$	$tP20$
$D1_f$	Mn_4B	Orthorhombic	$Fddd$	$oF40$
$D1_g$	B_4C	Rhombohedral	$R\bar{3}m$	$tR15$

15
Appendices

Table 15.19 Strukturbericht designations for other compounds

Desig-nation	Typical example (Mineral name)	Crystal system	Hermann–Mauguin	Pearson
$D2_a$	$TiBe_{12}$	Hexagonal	$P6/mmm$	$hP13$
$D2_b$	$ThMn_{12}$	Tetragonal	$I4/mcm$	$tI26$
$D2_c$	U_6Mn	Tetragonal	$I4/mcm$	$tI28$
$D2_d$	$CaCu_5$	Hexagonal	$C6/mmm$	$hC6$
$D2_e$	$BaHg_{11}$	Cubic	$Pm3m$	$cP36$
$D2_f$	UB_{12}	Cubic	$Fm3m$	$cF52$
$D2_g$	Fe_8N	Tetragonal	$I4/mmm$	$tI18$
$D2_h$	Al_6Mn	Orthorhombic	$Cmcm$	$oC28$
$D2_1$	CaB_6	Cubic	$Pm3m$	$cP7$
$D2_3$	$NaZn_{13}$	Cubic	$Fm3c$	$cFI12$
$D5_a$	U_3Si_2	Tetragonal	$P4/mbm$	$tP10$
$D5_b$	Pt_2Sn_3	Hexagonal	$P6/mmc$	$hR10$
$D5_c$	Pu_2C_3	Cubic	$I\bar{4}3d$	$cI40$
$D5_1$	$\alpha\text{-}Al_2O_3$	Rhombohedral	$R\bar{3}c$	$hR10$
$D5_2$	La_2O_3	Hexagonal	$P\bar{3}m1$	$hP5$
$D5_3$	Mn_2O_3	Cubic	$Ia3$	$cI80$
$D5_8$	Sb_2S_3	Orthorhombic	$Pbnm$	$oP20$
$D5_9$	Zn_3P_2	Tetragonal	$P4/nmc$	$tP40$
$D5_{10}$	Cr_3C_2	Orthorhombic	$Pbnm$	$oP20$
$D5_{13}$	Ni_2Al_3	Hexagonal	$C\bar{3}m1$	$hC5$
$D7_a$	Ni_3Sn_4	Monoclinic	$C2/m$	$mC14$
$D7_b$	Ta_3B_4	Orthorhombic	$Immm$	$oI14$
$D7_1$	Al_4C_3	Rhombohedral	$R\bar{3}m$	$hR7$
$D7_2$	Co_3S_4	Cubic	$Fd3m$	$cF56$
$D7_3$	Th_3P_4	Cubic	$I\bar{4}3d$	$cI26$
$D8_a$	Th_6Mn_{23}	Cubic	$Fm3m$	$cFI16$
$D8_b$	V_3Ni_2	Tetragonal	$P4/mnm$	$tP30$
$D8_c$	$Mg_2Cu_6Al_5$	Cubic	$Pm3m$	$cP39$
$D8_d$	Co_2Al_9	Monoclinic	$P2_1/a$	$mP22$
$D8_e$	$Mg_{32}(Al, Zn)_{49}$	Cubic	$Im3m$	$cI162$
$D8_f$	Ir_3Sn_7	Cubic	$Im3m$	$c140$
$D8_g$	Mg_5Ga_3	Orthorhombic	$Ibam$	$oI28$
$D8_h$	W_2B_5	Hexagonal	$P6_3/mmc$	$hP14$
$D8_i$	Mo_2B_5	Rhombohedral	$R\bar{3}m$	$hR7$
$D8_k$	Th_7S_{12}	Hexagonal	$P6_3/m$	$hP19$
$D8_l$	Bi_3Cr_5	Tetragonal	$I4/mcm$	$tI32$
$D8_m$	Si_3W_5	Tetragonal	$I4/mcm$	$tI32$
$D8_1$	Fe_3Zn_{10}	Cubic	$Im3m$	$cI52$
$D8_2$	Cu_5Zn_8	Cubic	$I\bar{4}3m$	$cI52$
$D8_3$	Cu_9Al_4	Cubic	$P\bar{4}3m$	$cP52$
$D8_4$	$Cr_{23}C_6$	Cubic	$Fm3m$	$cFI16$
$D8_5$	Fe_7W_6	Rhombohedral	$R\bar{3}m$	$hR13$

Table 15.19 (continued)

Desig-nation	Typical example (Mineral name)	Crystal system	Hermann–Mauguin	Pearson
$D8_6$	$Cu_{15}Si_4$	Cubic	$I\bar{4}3d$	$cI76$
$D8_8$	Mn_5Si_3	Hexagonal	$P6_3/mcm$	$hP16$
$D8_9$	Co_9S_8	Cubic	$Fm3m$	$cF68$
$D8_{10}$	Cr_5Al_8	Rhombohedral	$R\bar{3}m$	$hR26$
$D8_{11}$	Co_2Al_5	Hexagonal	$P6_3/mcm$	$hP28$
$D10_1$	Cr_7C_3	Hexagonal	$P31c$	$hP80$
$D10_2$	Fe_3Th_7	Hexagonal	$P6_3/mcm$	$hP20$
$E0_1$	PbC1F	Tetragonal	$P4/nmm$	$tP6$
$E0_7$	FeAsS	Monoclinic	$B2_1/d$	$mB24$
$E1_a$	$MgCuAl_2$	Orthorhombic	$Cmcm$	$oC16$
$E1_b$	$AuAgTe_4$ (sylvanite)	Monoclinic	$P2/c$	$mP12$
$E1_1$	$CuFeS_2$ (chalcopyrite)	Tetragonal	$I\bar{4}2d$	$tI16$
$E2_1$	$CaTiO_3$ (perowskite)	Cubic	$Pm3m$	$cP5$
$E2_4$	Sn_2S_3	Orthorhombic	$Pnma$	$oP20$
$E3$	Al_2CdS_4	Tetragonal	$I\bar{4}$	$tI14$
$E9_a$	Al_7FeCu_2	Tetragonal	$P4/mnc$	$tP40$
$E9_b$	$FeMg_3Al_8Si_6$	Hexagonal	$P\bar{6}2m$	$hP18$
$E9_c$	Mn_3Al_9Si	Hexagonal	$P6_3/mmc$	$hP26$
$E9_d$	$A1Li_3N_2$	Cubic	$Ia3$	$cI96$
$E9_e$	$CuFe_2S_3$ (cubanite)	Orthorhombic	$Pnma$	$oP24$
$E9_3$	Fe_3W_3C	Cubic	$Fd3m$	$cF112$
$F0_1$	NiSSb (ullmanite)	Cubic	$P2_13$	$cP12$
$F5_a$	$KFeS_2$	Monoclinic	$C2/c$	$mC16$
$F5_1$	$CrNaS_2$	Rhombohedral	$R\bar{3}m$	$hR4$
$F5_6$	CuS_2Sb	Orthorhombic	$Pnma$	$oP16$
$H1_1$	Al_2MgO_4 (spinel)	Cubic	$Fd3m$	$cF56$
$H2_4$	Cu_3S_4V	Cubic	$P\bar{4}3m$	$cP8$
$H2_5$	$AsCu_3S_4$	Orthorhombic	$Pmn2_1$	$oP16$
$H2_6$	$FeCu_2SnS_4$ (stannite)	Tetragonal	$I\bar{4}2m$	$tI16$
$L1_a$	Pt_3Cu	Cubic	$Fm3c$	$cF32$
$L1_0$	CuAu	Tetragonal	$C4/mmm$	$tC4$
$L1_2$	Cu_3Au	Cubic	$Pm3m$	$cP4$
$L2_a$	δ-TiCu	Tetragonal	$P4/mmm$	$tP2$
$L2_1$	$AlCu_2Mn$	Cubic	$Fm3m$	$cF16$
$L2_2$	Sb_2Tl_7	Cubic	$Im3m$	$cI54$
L'_1	Fe_4N	Cubic	$Pm3m$	$cP5$
L'_2	Martensite	Tetragonal	$I4/mmm$	$tI3$
$L'2_b$	ThH_2	Tetragonal	$I4/mmm$	$tI6$
$L'3$	Fe_2N	Hexagonal	$P6_3/mmc$	$hP3$
$L'6_0$	$CuTi_3$	Tetragonal	$P4/mmm$	$tP4$
$L'6$	No name	Tetragonal	$F4/mmm$	$tF4$

15

Appen-dices

Table 15.20 Triclinic space groups

Ordered number	Space group (Hermann–Mauguin)
001	$P1$
002	$P\bar{1}$

Table 15.21 Monoclinic space groups

Ordered number	Space group (Hermann–Mauguin)
003	$P2$
004	$P2_1$
005	$C2$
006	Pm
007	Pc
008	Cm
009	Cc
010	$P2/m$
011	$P2_1/m$
012	$C2/m$
013	$P2/c$
014	$P2_1/c$
015	$C2/c$

Table 15.22 Orthorhombic space groups

Ordered number	Space group (Hermann–Mauguin)
016	$P222$
017	$P222_1$
018	$P2_12_12$
019	$P2_12_12_1$
020	$C222_1$
021	$C222$
022	$F222$
023	$I222$
024	$I2_12_12_1$
025	$Pmm2$
026	$Pmc2_1$
027	$Pcc2$
028	$Pma2$

Table 15.22 Orthorhombic space groups

Ordered number	Space group (Hermann–Mauguin)	Ordered number	Space group (Hermann–Mauguin)
029	$Pca2_1$	052	$Pnna$
030	$Pnc2$	053	$Pmna$
031	$Pmn2_1$	054	$Pcca$
032	$Pba2$	055	$Pbam$
033	$Pna2_1$	056	$Pccn$
034	$Pnn2$	057	$Pbcm$
035	$Cmm2$	058	$Pnnm$
036	$Cmc2_1$	059	$Pmmn$
037	$Ccc2$	060	$Pbcn$
038	$Amm2$	061	$Pbca$
039	$Abm2$	062	$Pnma$
040	$Ama2$	063	$Cmcm$
041	$Aba2$	064	$Cmca$
042	$Fmm2$	065	$Cmmm$
043	$Fdd2$	066	$Cccm$
044	$Imm2$	067	$Cmma$
045	$Iba2$	068	$Ccca$
046	$Ima2$	069	$Fmmm$
047	$Pmmm$	070	$Fddd$
048	$Pnnn$	071	$Immm$
049	$Pccm$	072	$Ibam$
050	$Pban$	073	$Ibca$
051	$Pmma$	074	$Imma$

15

Appen-
dices

Table 15.23 Tetragonal space groups

Ordered number	Space group (Hermann–Mauguin)
075	$P4$
076	$P4_1$
077	$P4_2$
078	$P4_3$
079	$I4$
080	$I4_1$
081	$P\bar{4}$
082	$I\bar{4}$
083	$P4/m$
084	$P4_2/m$
085	$P4/n$
086	$P4_2/n$
087	$I4/m$
088	$I4_1/a$
089	$P422$
090	$P42_12$
091	$P4_122$
092	$P4_12_12$
093	$P4_222$
094	$P4_22_12$
095	$P4_322$
096	$P4_32_12$
097	$I422$
098	$I4_122$
099	$P4mm$
100	$P4bm$
101	$P4_2cm$
102	$P4_2nm$
103	$P4cc$
104	$P4nc$
105	$P4_2mc$
106	$P4_2bc$
107	$I4mm$
108	$I4cm$
109	$I4_1md$
110	$I4_1cd$
111	$P\bar{4}2m$
112	$P\bar{4}2c$
113	$P\bar{4}2_1m$
114	$P\bar{4}2_1c$
115	$P\bar{4}m2$

Table 15.23 (continued)

Ordered number	Space group (Hermann–Mauguin)
116	$P\bar{4}c2$
117	$P\bar{4}b2$
118	$P\bar{4}n2$
119	$I\bar{4}m2$
120	$I\bar{4}c2$
121	$I\bar{4}2m$
122	$I\bar{4}2d$
123	$P4/mmm$
124	$P4/mcc$
125	$P4/nbm$
126	$P4/nnc$
127	$P4/mbm$
128	$P4/mnc$
129	$P4/nmm$
130	$P4/ncc$
131	$P4_2/mmc$
132	$P4_2/mcm$
133	$P4_2/nbc$
134	$P4_2/nnm$
135	$P4_2/mbc$
136	$P4_2/mnm$
137	$P4_2/nmc$
138	$P4_2/ncm$
139	$I4/mmm$
140	$I4/mcm$
141	$I4_1/amd$
142	$I4_1/acd$

Table 15.24 Trigonal space groups

Ordered number	Space group (Hermann–Mauguin)
143	$P3$
144	$P3_1$
145	$P3_2$
146	$R3$
147	$P\bar{3}$
148	$R\bar{3}$
149	$P312$
150	$P32_1$
151	$P3_112$
152	$P3_121$
153	$P3_21_2$
154	$P322_1$
155	$R32$
156	$P3m1$
157	$P31m$
158	$P3c1$
159	$P31c$
160	$R3m$
161	$R3c$
162	$P\bar{3}1m$
163	$P\bar{3}1c$
164	$P\bar{3}m1$
165	$P\bar{3}c1$
166	$R\bar{3}m$
167	$R\bar{3}c$

Table 15.25 Hexagonal space groups

Ordered number	Space group (Hermann–Mauguin)
168	$P6$
169	$P6_1$
170	$P6_5$
171	$P6_2$
172	$P6_4$
173	$P6_3$
174	$P\bar{6}$
175	$P6/m$
176	$P6_3/m$
177	$P622$
178	$P6_122$
179	$P6_522$
180	$P6_222$
181	$P6_422$
182	$P6_322$
183	$P6mm$
184	$P6cc$
185	$P6_3cm$
186	$P6_3mc$
187	$P\bar{6}m2$
188	$P\bar{6}c2$
189	$P\bar{6}2m$
190	$P\bar{6}2c$
191	$P6/mmm$
192	$P6/mcc$
193	$P6_3/mcm$
194	$P6_3/mmc$

Table 15.26 Cubic space groups

Ordered number	Space group (Hermann–Mauguin)
195	$P23$
196	$F23$
197	$I23$
198	$P2_13$
199	$I2_13$
200	$Pm\bar{3}$
201	$Pn\bar{3}$
202	$Fm\bar{3}$
203	$Fd\bar{3}$
204	$Im\bar{3}$
205	$Pa\bar{3}$
206	$Ia\bar{3}$
207	$P432$
208	$P4_232$
209	$F432$
210	$F4_132$
211	$I432$
212	$P4_332$
213	$P4_132$
214	$I4_132$
215	$P\bar{4}3m$
216	$F\bar{4}3m$
217	$I\bar{4}3m$
218	$P\bar{4}3n$
219	$F\bar{4}3c$
220	$I\bar{4}3d$
221	$Pm\bar{3}m$
222	$Pn\bar{3}n$
223	$Pm\bar{3}n$
224	$Pn\bar{3}m$
225	$Fm\bar{3}m$
226	$Fm\bar{3}c$
227	$Fd\bar{3}m$
228	$Fd\bar{3}c$
229	$Im\bar{3}m$
230	$Ia\bar{3}d$

15

Appendices

Finally, the vector product can be also written as a matrix product:

$$V_1 \times V_2 = \begin{bmatrix} a & b & c \\ u_1 & v_1 & w_1 \\ u_2 & v_2 & w_2 \end{bmatrix}.$$

15.6.10.5 Mixed Product and Cell Multiplicity

The mixed product between three vectors is a scalar quantity represented as (V_1, V_2, V_3) and is defined to be equal to:

$$V_1 \cdot (V_2 \times V_3) = (V_1 \times V_2) \cdot V_3.$$

The vector product can be also written as a matrix product:

$$(V_1, V_2, V_3) = \underbrace{\begin{bmatrix} u_1 & v_1 & w_1 \\ u_2 & v_2 & w_2 \\ u_2 & v_3 & w_3 \end{bmatrix}}_{m \text{ multiplicity of the cell}} (a, b, c).$$

The **multiplicity of the cell**, m, is a dimensionless physical quantity equal to the number of entities (e.g., electrons, ions, atoms, molecules) contained in the crystal lattice structure.

Table 15.27 Cell multiplicity

Class	Multiplicity	Name
Single unit cell	$m = 1$	Primitive cell
Multiple cell	$m = 2$	Double cell
	$m = 3$	Triple cell
	$m = 4$	Quadruple cell

Important note: the rigorous deduction of entities (e.g., ions, atoms, molecules) contained inside the unit cell only depends to their particular locations in the crystal space lattice. Actually, (i) entities located on the corners are counted as eighth (1/8), because they are shared by eight other neighboring cells, (ii) entities located on the edges of the lattice are counted as fourth (1/4) because they are shared by four neighboring cells, (iii) while entities located at the faces of the cell are counted as half (1/2) because they are shared by two adjacent cells. (iv) Finally, entities located inside the cell space lattice are counted as unity (1). Therefore the multiplicity, m, of the cell can be easily calculated reporting the number, N, of entities in each particular locations (i.e., corners, edges, faces, interior).

$$m = N_{inside} + \frac{N_{faces}}{2} + \frac{N_{edges}}{4} + \frac{N_{corners}}{8}.$$

15.6.10.6 Unit Cell Volume

The unit cell volume is given by the following general equation which is calculated from the mixed product of the three lattice vectors:

$$V_{unit\ cell} = (a, b, c) = abc\sqrt{1 - \cos^2\alpha - \cos^2\beta - \cos^2\gamma + 2\cos\alpha\cos\beta\cos\gamma}.$$

Table 15.28 Space lattice volume

System	Volume
Cubic	$V_C = a^3$
Tetragonal	$V_Q = a^2 c$
Hexagonal	$V_H = a^2 c \sqrt{3}/2 = 0.866 a^2 c$
Rhombohedral	$V_R = a^3 (1 - 3\cos^2\alpha + 2\cos^3\alpha)^{1/2}$
Orthorhombic	$V_O = abc$
Monoclinic	$V_M = abc\sin\beta$
Triclinic	$V_T = abc (1 - \cos^2\alpha - \cos^2\beta - \cos^2\gamma + 2\cos\alpha\cos\beta\cos\gamma)^{1/2}$

15.6.10.7 Plane Angle Between Two Lattice Planes

One is also occasionally interested in computing the angle between planes. The angle, φ, between the plane with Miller's indices (h_1, k_1, l_1) and the plane with Miller's indices (h_2, k_2, l_2). The basic equation allowing calculation of this angle is (see coefficients s_{ii} in Table 15.30 on page 561):

$$\cos\varphi = \frac{d_{h_1 k_1 l_1} \cdot d_{h_2 k_2 l_2}}{v^2} [s_{11}h_1 h_2 + s_{22}k_1 k_2 + s_{33}l_1 l_2 + s_{23}(k_1 l_2 + k_2 l_1)$$
$$+ s_{13}(l_1 h_2 + l_2 h_1) + s_{12}(h_1 k_2 + h_2 k_1)].$$

15.6.11 Interplanar Spacing

See **Tables 15.30–15.31** (page 561).

15.6.12 Reciprocal Lattice Unit Cell

See **Table 15.32** (page 561).

15.7 Properties of Liquid Metals

See **Tables 15.33-15.34** (pages 562–64).

15.8 Properties of Molten Salts

See **Table 15.35** (page 565).

15.9 Electrochemical Galvanic Series

See **Table 15.36** (page 566).

Table 15.29 Plane angle between lattice planes

System	Plane angle
Cubic	$\cos\varphi = \dfrac{h_1 h_2 + k_1 k_2 + l_1 l_2}{\sqrt{(h_1^2 + k_1^2 + l_1^2)(h_2^2 + k_2^2 + l_2^2)}}$
Tetragonal	$\cos\varphi = \dfrac{\dfrac{h_1 h_2 + k_1 k_2}{a^2} + \dfrac{l_1 l_2}{c^2}}{\sqrt{\left(\dfrac{h_1^2 + k_1^2}{a^2} + \dfrac{l_1^2}{c^2}\right)\left(\dfrac{h_2^2 + k_2^2}{a^2} + \dfrac{l_2^2}{c^2}\right)}}$
Hexagonal	$\cos\varphi = \dfrac{h_1 h_2 + k_1 k_2 + \dfrac{h_1 k_2 + h_2 k_1}{2} + \dfrac{3a^2 l_1 l_2}{4c^2}}{\sqrt{\left(h_1^2 + k_1^2 + h_1 k_1 + \dfrac{3a^2 l_1^2}{4c^2}\right)\left(h_2^2 + k_2^2 + h_2 k_2 + \dfrac{3a^2 l_2^2}{4c^2}\right)}}$
Rhombohedral	$\cos\varphi = \dfrac{(h_1 h_2 + k_1 k_2 + l_1 l_2)\sin^2\alpha + 2(h_1 k_1 + k_1 l_1 + h_1 l_1)[\cos^2\alpha - \cos\alpha]}{\sqrt{((h_1^2 + k_1^2 + l_1^2)\sin^2\alpha + (k_1 l_2 + k_2 l_1 + l_1 h_2 + l_2 h_1 + h_1 k_2 + k_1 h_2)[\cos^2\alpha - \cos\alpha])((h_2^2 + k_2^2 + l_2^2)\sin^2\alpha + 2(h_2 k_2 + k_2 l_2 + h_2 l_2)[\cos^2\alpha - \cos\alpha])}}$
Orthorhombic	$\cos\varphi = \dfrac{\dfrac{h_1 h_2}{a^2} + \dfrac{k_1 k_2}{b^2} + \dfrac{l_1 l_2}{c^2}}{\sqrt{\left(\dfrac{h_1^2}{a^2} + \dfrac{k_1^2}{b^2} + \dfrac{l_1^2}{c^2}\right)\left(\dfrac{h_2^2}{a^2} + \dfrac{k_2^2}{b^2} + \dfrac{l_2^2}{c^2}\right)}}$
Monoclinic	$\cos\varphi = \dfrac{\dfrac{h_1 h_2}{a^2} + \dfrac{k_1 k_2 \sin^2\beta}{b^2} + \dfrac{l_1 l_2}{c^2} - \dfrac{2h_1 l_1 \cos\beta}{ac} - \dfrac{(h_2 l_1 + h_1 l_2)\cos\beta}{ac}}{\sqrt{\left(\dfrac{h_1^2}{a^2} + \dfrac{k_1^2 \sin^2\beta}{b^2} + \dfrac{l_1^2}{c^2} - \dfrac{2h_1 l_1 \cos\beta}{ac}\right)\left(\dfrac{h_2^2}{a^2} + \dfrac{k_2^2 \sin^2\beta}{b^2} + \dfrac{l_2^2}{c^2} - \dfrac{2h_2 l_2 \cos\beta}{ac}\right)}}$
Triclinic	See general formula

Table 15.30 General formula for the interplanar spacing

$$\frac{1}{d_{hkl}} = \sqrt{\frac{s_{11} \cdot h^2 + s_{22} \cdot k^2 + s_{33} \cdot l^2 + 2s_{12} \cdot hk + 2s_{23} \cdot kl + 2s_{13} \cdot hl}{V}}$$

with $V = abc(1 - \cos^2\alpha - \cos^2\beta - \cos^2\gamma + 2\cos\alpha\cos\beta\cos\gamma)^{1/2}$

$s_{11} = b^2c^2 \sin^2\alpha$	$s_{12} = abc^2(\cos\alpha\cos\beta - \cos\gamma)^2$
$s_{22} = a^2c^2 \sin^2\alpha$	$s_{23} = a^2bc(\cos\beta\cos\gamma - \cos\alpha)$
$s_{33} = a^2c^2 \sin^2\alpha$	$s_{31} = ab^2c(\cos\gamma\cos\alpha - \cos\beta)$

Table 15.31 Interplanar spacing according to crystal lattice

System	Interplanar spacing
Cubic	$\dfrac{1}{d_{hkl}} = \sqrt{\dfrac{h^2 + k^2 + l^2}{a^2}}$
Tetragonal	$\dfrac{1}{d_{hkl}} = \sqrt{\dfrac{h^2 + k^2}{a^2} + \dfrac{l^2}{c^2}}$
Hexagonal	$\dfrac{1}{d_{hkl}} = \sqrt{\dfrac{2(h^2 + hk + k^2)}{3a^2} + \dfrac{2l^2}{3c^2}}$
Rhombohedral	$\dfrac{1}{d_{hkl}} = \sqrt{\dfrac{(h^2 + k^2 + l^2)\sin^2\alpha + 2(hk + kl + hl)(\cos^2\alpha - \cos\alpha)}{a^2(1 - 3\cos^2\alpha + 2\cos^3\alpha)}}$
Orthorhombic	$\dfrac{1}{d_{hkl}} = \sqrt{\dfrac{h^2}{a^2} + \dfrac{k^2}{b^2} + \dfrac{l^2}{c^2}}$
Monoclinic	$\dfrac{1}{d_{hkl}} = \sqrt{\dfrac{h^2}{a^2 \sin^2\beta} + \dfrac{k^2}{b^2} + \dfrac{l^2}{c^2 \sin^2\beta} - \dfrac{2hl\cos\beta}{ac\sin^2\beta}}$
Triclinic	See general formula

Table 15.32 Definition of the reciprocal unit lattice

The three reciprocal vectors are a^*, b^*, and c^* defined by the nine relations below

$a.a^* = 1$	$b.a^* = 0$	$c.a^* = 0$
$a.b^* = 0$	$b.b^* = 1$	$c.b^* = 0$
$a.c^* = 0$	$b.c^* = 0$	$c.c^* = 1$

Note: A condensed notation used by crystallographers is as follows: $a_i.b_j = \delta_{ij}$, where δ_{ij} is the Kronecker operator (i.e., for $i = j$, $\delta_{ij} = 1$ and for $i \neq j$, $\delta_{ij} = 0$). On the other hand, a slightly different notation is used in solid state physics: $a_i.b_j = 2\pi\delta_{ij}$.

15

Appendices

Table 15.33 Physical properties of liquid metals at the melting point

Metal	Melting point (mp, °C)	Volume expansion on melting (%)	Density (ρ, kg.m^{-3})	Dynamic viscosity (η, mPa.s)	Surface tension (γ, mJ.m^{-2})	Thermal conductivity (k, W.m^{-1}.K^{-1})	Specific heat capacity (c_p/J.kg^{-1}.K^{-1})	Electrical resistivity (ρ, $\mu\Omega$.cm)
Aluminum	660.0	6.5	2385	1.30	914	94.03	1080	24.25
Antimony	630.5	0.8	6483	1.22	367	21.8	258	114
Arsenic	817	10	5220	n.a.	n.a.	n.a.	n.a.	210.0
Barium	727	n.a.	3321	n.a.	224	n.a.	228	133
Beryllium	1283	n.a.	1690	n.a.	1390	n.a.	3480	45
Bismuth	271	−3.35	10,068	1.80	378	17.1	146	129
Boron	2077	n.a.	2080	n.a.	1070	n.a.	2910	210
Cadmium	321	4.0	8020	2.28	570	42	264	33.7
Calcium	865	n.a.	1365	1.22	361	n.a.	775	25
Cerium	804	n.a.	6685	2.88	740	n.a.	250	126.8
Cesium	28.6	2.6	1854	0.68	69	19.7	280	37
Chromium	1875	n.a.	6280	n.a.	1700	n.a.	780	31.6
Cobalt	1493	3.5	7760	4.18	1873	n.a.	590	102
Copper	1083	4.2	8000	4.0	1285	165.6	495	20
Gadolinium	1312	n.a.	7140	n.a.	810	n.a.	213	27.8
Gallium	29.8	−3.2	6090	2.04	718	25.5	398	26
Germanium	934	−5.1	5600	0.73	621	n.a.	404	67.2
Gold	1063	5.1	17,360	5.0	1140	104.44	149	31.25
Hafnium	1943	n.a.	11,100	7.90	1630	n.a.	n.a.	218
Indium	156.6	2.0	7023	1.89	556	42	259	32.3
Iridium	2443	n.a.	20,000	n.a.	2250	n.a.	n.a.	n.a.
Iron	1536	3.5	7015	5.5	1872	n.a.	795	138.6
Lanthanum	930	n.a.	5955	2.45	720	21	57.5	138
Lead	327	3.5	10,678	2.65	468	15.4	152	94.85
Lithium	180.5	1.65	525	0.57	395	46.4	4370	24
Magnesium	651	4.12	1590	1.25	559	78	1360	27.4
Manganese	1241	1.7	5730	n.a.	1090	89.7	838	40
Mercury	−38.87	3.7	13,691	2.10	498	6.78	142	90.5
Molybdenum	2607	n.a.	9340	n.a.	2250	n.a.	570	60.6
Neodymium	1024	n.a.	6688	n.a.	689	n.a.	232	126
Nickel	1454	4.5	7905	4.90	1778	n.a.	620	85
Niobium	2468	n.a.	7830	n.a.	1900	n.a.	n.a.	105
Osmium	2727	n.a.	20,200	n.a.	2500	n.a.	n.a.	n.a.
Palladium	1552	n.a.	10,490	n.a.	1500	n.a.	n.a.	n.a.
Platinum	1769	n.a.	19,000	n.a.	1800	n.a.	178	73
Plutonium	640	−2.5	16,640	6.0	550	n.a.	n.a.	133
Potassium	63.5	2.55	827	0.51	111	53	820	136.5
Praeseodymium	935	n.a.	6611	2.80	n.a.	n.a.	238	138
Rhenium	3158	n.a.	18,800	n.a.	2700	n.a.	n.a.	145

Table 15.33 (continued)

Metal	Melting point (mp, °C)	Volume expansion on melting (%)	Density (ρ, kg.m^{-3})	Dynamic viscosity (η, mPa.s)	Surface tension (γ, mJ.m^{-2})	Thermal conductivity (k, W.m^{-1}.K^{-1})	Specific heat capacity (c_p/J.kg^{-1}.K^{-1})	Electrical resistivity (ρ, $\mu\Omega$.cm)
Rhodium	1966	n.a.	10,800	n.a.	2000	n.a.	n.a.	n.a.
Rubidium	38.9	2.5	1437	0.67	83	33.4	398	22.83
Ruthenium	2427	n.a.	10,900	n.a.	2250	n.a.	n.a.	8.4
Selenium	217	15.8	3989	24.8	106	0.3	445	10^6
Silicon	1410	−10	2510	0.94	865	n.a.	1040	75
Silver	960.7	3.8	9346	3.88	903	174.8	283	17.25
Sodium	96.5	2.5	927	0.68	195	89.7	1386	9.64
Strontium	770	n.a.	2480	n.a.	303	n.a.	354	58
Tantalum	2977	n.a.	15,000	n.a.	2150	n.a.	n.a.	118
Tellurium	451	4.9	5710	2.14	180	2.5	295	550
Thorium	1691	n.a.	10,500	n.a.	978	n.a.	n.a.	172
Tin	232	2.3	7000	1.85	544	30	250	47.2
Titanium	1685	n.a.	4110	5.2	1650	n.a.	700	172
Tungsten	3377	n.a.	17,600	n.a.	2500	n.a.	n.a.	127
Uranium	1133	n.a.	17,900	6.5	1550	n.a.	161	63.6
Vanadium	1912	n.a.	5700	n.a.	1950	n.a.	780	67.8
Zinc	419	4.7	6575	3.85	782	49.5	481	37.4
Zirconium	1850	n.a.	5800	8.0	1480	n.a.	367	153

15.10 Hardness Scales

For common hardness scales used in mineralogy (e.g., Mohs, Rosival, Ridgeway), see Table 11.3 (page 404) and Table 15.37 (pages 567–568).

15.11 UNS Standard Alphabetical Designation

The **Unified Numbering System** (UNS) (Table 15.38, page 569) is the accepted alloy designation system in North America and worldwide for commercially available metals and alloys.[1] The UNS is managed jointly by the **American Society for Testing and Materials** (ASTM) and the **Society of Automotive Engineers** (SAE). The standard code designation consists of five digits following the prefix letter identifying the alloy family. Generally, UNS designations are simply expansions of the former designations (i.e., AISI, AA, CDA, etc.).

Table 15.34 Maximum operating temperature (°C) of metallic container materials for corrosive liquid metals under inert atmosphere (A = attacked)

Molten metal or alloy	AISI 316L	Ti	Zr	Hf	Nb	Ta	Mo	W	Ag	Au	Pt	Rh	Ir
Ag	n.a.	n.a.	n.a.	n.a.	n.a.	1200	n.a.	n.a.	n.a.	n.a.	n.a.	n.a.	n.a.
Al	A	750	n.a.	n.a.	n.a.	A	n.a.	n.a.	A	A	n.a.	n.a.	A
Bi	A	n.a.	n.a.	n.a.	n.a.	560	n.a.	n.a.	A	A	A	A	470
Ca	n.a.	n.a.	n.a.	n.a.	n.a.	1200	n.a.	n.a.	n.a.	n.a.	n.a.	n.a.	A
Cd	A	450	n.a.	n.a.	n.a.	n.a.	n.a.	n.a.	n.a.	n.a.	n.a.	n.a.	A
Ga	A	400	n.a.	n.a.	n.a.	400	400	n.a.	A	A	n.a.	n.a.	230
Hg	n.a.	150	n.a.	n.a.	n.a.	600	600	n.a.	A	A	A	550	550
In	A	n.a.	n.a.	n.a.	n.a.	n.a.	n.a.	n.a.	A	n.a.	A	n.a.	360
K	n.a.	n.a.	600	600	n.a.	900	n.a.	n.a.	A	A	n.a.	260	260
Li	540	750	1000	1000	1000	1000	1200	1200	A	A	n.a.	n.a.	380
Mg	n.a.	850	A	A	1000	1150	n.a.	n.a.	A	A	n.a.	n.a.	A
Na	n.a.	600	600	600	n.a.	900	n.a.	n.a.	A	A	n.a.	290	290
NaK	n.a.	n.a.	600	600	n.a.	900	n.a.	n.a.	A	A	n.a.	n.a.	n.a.
Pb	n.a.	600	n.a.	n.a.	n.a.	1000	850	n.a.	A	A	A	A	n.a.
Sb	A	n.a.	n.a.	n.a.	n.a.	n.a.	n.a.	n.a.	n.a.	n.a.	n.a.	n.a.	n.a.
Sn	A	600	n.a.	n.a.	n.a.	n.a.	n.a.	n.a.	A	n.a.	A	n.a.	n.a.
Th–Mg	n.a.	n.a.	n.a.	n.a.	850	1000	n.a.	n.a.	n.a.	n.a.	n.a.	n.a.	n.a.
U	n.a.	n.a.	n.a.	n.a.	1400	1450	n.a.	n.a.	n.a.	n.a.	n.a.	n.a.	n.a.
Zn	A	750	A	A	450	500	n.a.	n.a.	A	A	n.a.	n.a.	A

15.12 Fuel Energy Content

See **Table 15.39** (page 570).

15.13 Natural Radioactivity

During the formation of the Earth, 4.65 billion years ago, along with stable nuclides, several radionuclides were formed. Those that were radioactive with a half-life too short with respect to the Earth's formation obviously disappeared. On the other hand, those with a half-life the same order of magnitude or greater to that of the formation of Earth are mainly responsible for the natural radioactivity of the Earth's crust materials (i.e., ice, sea and ocean water, minerals, ores, rocks, and soils). Today, over 60 radionuclides occur in nature, and they can be grouped in three main categories (Tables 15.40-15.42, pages 571-572): (i) **primordial radionuclides**, i.e., radionuclides present since the formation of the Earth, (ii) **cosmogenic radionuclides** or **cosmonuclides**, i.e., radionuclides formed by nuclear interaction between primary and secondary cosmic radiations (cosmic rays) and the upper atmosphere nuclides, and (iii) **artificial radionuclides**, i.e., radionuclides enhanced or produced due to human activities (e.g., atmospheric testing of nuclear weapons, nuclear power reactors, and industries involved in the

Table 15.35 Physical properties of selected pure molten salts at the melting point

Molten salt chemical formula	Melting point (mp, °C)	Boiling point (bp, °C)	Molar latent heat of fusion (L_m, kJ.mol^{-1})	Density (ρ, kg.m^{-3})	Dynamic viscosity (η, mPa.s)	Surface tension (γ, mN.m^{-1})	Electrical conductivity (ρ, S.cm^{-1})
$BaCl_2$	962	1560	16.7	3829	4.506	170	n.a.
BaF_2	1368	2260	28.5	4163	n.a.	n.a.	n.a.
$BeCl_2$	415	482	8.66	1443	n.a.	n.a.	0.00087
BeF_2	552	1169	4.77	n.a.	n.a.	n.a.	n.a.
$CaCl_2$	775	1936	28.5	2085	3.010	151	2.342
CaF_2	1418	2533	29.7	2498	n.a.	n.a.	4.100
$CsCl$	645	1297	15.90	2737	n.a.	88 (700)	1.270
CsF	703	1231	21.8	3649	n.a.	106	n.a.
$CsOH$	272	990	4.56	n.a.	n.a.	n.a.	n.a.
K_2CO_3	898	dec.	27.6	1897	n.a.	n.a.	2.178
K_2SO_4	1069	1670	34.39	1889	n.a.	143	1.940
KCl	771	1437	26.6	1527	1.094	98	2.350
KF	858	1502	28.3	1910	n.a.	138	4.602
$KHSO_4$	197	n.a.	n.a.	2079	n.a.	n.a.	0.141
KNO_3	337	400	10.1	1870	2.705	105	0.820
KOH	406	1327	8.00	1714	2.300	n.a.	n.a.
Li_2CO_3	737	d.1300	4.1	1825	n.a.	n.a.	n.a.
Li_2SO_4	859	n.a.	7.3	2004	n.a.	225	n.a.
$LiCl$	613	1383	19.9	1502	n.a.	137	6.1714
LiF	848	1673	26.8	1810	n.a.	252	8.780
$LiNO_3$	253	n.a.	24.9	1785	5.500	n.a.	1.336
$LiOH$	471	1626	20.88	n.a.	n.a.	n.a.	n.a.
$MgCl_2$	714	1412	43.1	1680	n.a.	n.a.	1.183
MgF_2	1263	2270	58.2	2404	n.a.	n.a.	n.a.
Na_2CO_3	858	n.a.	29.64	1972	n.a.	n.a.	3.222
Na_2SO_4	884	2227	23.6	2069	n.a.	193	2.370
Na_3AlF_6	1273	n.a.	107.28	n.a.	7.000	130	2.997
$NaCl$	801	1465	28.1	1556	1.463	116	3.870
$NaCN$	563	n.a.	8.79	n.a.	n.a.	n.a.	1.270
NaF	996	1704	32.7	1948	n.a.	185	4.970
$NaHSO_4$	315	n.a.	n.a.	n.a.	n.a.	n.a.	0.168
$NaNO_3$	307	d.500	n.a.	1900	3.156	116	1.360
$NaOH$	323	1388	6.60	1783	4.000	n.a.	2.827
$SrCl_2$	874	1250	15.9	2727	n.a.	168	2.285
SrF_2	1477	2460	28.5	3432	n.a.	n.a.	n.a.

Sources: (i) Janz GJ (1967) Molten salts handbook. Academic Press, New York.
(ii) Lovering DG, Gale RJ (1983, 1984, 1990) Molten salts techniques, Vols. 1–4. Plenum Press, New York. (iii) Barin I, Knacke O (1973) Thermodynamical properties of inorganic substances Springer-Verlag, Berlin.

Table 15.36 Electrochemical galvanic series

Metal or alloy (decreasing nobility)
Iridium
Platinum
Gold
Graphite
Titanium (passive)
Silver
Zirconium
Stainless steels AISI 316, 317 (passive)
Stainless steels AISI 304 (passive)
Nickel (passive)
Monels
Bronzes
Copper
Brasses
Nickel (active)
Naval brass
Tin
Lead
Stainless steels AISI 316, 317 (active)
Stainless steels AISI 304 (active)
Cast irons
Steel or iron
Aluminum 2024
Cadmium
Aluminum
Zinc and zinc alloys
Magnesium alloys
Magnesium

Table 15.37 Conversion between several hardness scales (approx.)

Vickers hardness (HV)	Brinell hardness (3000 kg mass, WC 10 mm ball) (HB)	Rockwell hardness			Scleroscope hardness number
		A Scale (60 kg mass, diamond cone indenter) (HRA)	B Scale (100 kg mass, diamond cone indenter) (HRB)	C Scale (150 kg mass, diamond cone indenter) (HRC)	
940	–	85.6	–	68.0	97
920	–	85.3	–	67.5	96
900	–	85.0	–	67.0	95
880	767	84.7	–	66.4	93
860	757	84.4	–	65.9	92
840	745	84.1	–	65.3	91
820	733	83.8	–	64.7	90
800	722	83.4	–	64.0	88
780	712	–	–	–	87
780	710	83.0	–	63.3	85
772	698	82.6	–	62.5	83
746	684	82.2	–	61.8	81
730	682	82.2	–	61.7	80
720	670	81.8	–	61.0	79
700	656	81.3	–	60.1	78
697	653	81.2	–	60.0	77
674	647	81.1	–	59.7	76
653	638	80.8	–	59.2	75
648	630	80.6	–	58.8	–
644	627	80.5	–	58.7	–
633	601	79.8	–	57.3	–
612	578	79.1	–	56.0	–
595	555	78.4	–	54.7	–
565	534	77.8	–	53.5	–
544	514	76.9	–	52.1	–
528	495	76.3	–	51.0	–
513	477	75.6	–	49.6	–
484	461	74.9	–	48.5	–
471	444	74.2	–	47.1	–
458	429	73.4	–	45.7	–
446	415	72.8	–	44.5	–
423	401	72.0	–	43.1	–
412	388	71.4	–	41.8	–
395	375	70.6	–	40.4	–
382	363	70.0	–	39.1	–
372	352	69.3	–	37.9	–
363	341	68.7	–	36.6	–

Table 15.37 (continued)

Vickers hardness (HV)	Brinell hardness (3000 kg mass, WC 10 mm ball) (HB)	Rockwell hardness			Scleroscope hardness number
		A Scale (60 kg mass, diamond cone indenter) (HRA)	B Scale (100 kg mass, diamond cone indenter) (HRB)	C Scale (150 kg mass, diamond cone indenter) (HRC)	
350	331	68.1	–	35.5	–
336	321	67.5	–	34.3	–
327	311	66.9	–	33.1	–
318	302	66.3	–	32.1	–
310	293	65.7	–	30.9	–
302	285	65.3	–	29.9	–
294	277	64.6	–	28.8	–
286	269	64.1	–	27.6	–
279	262	63.6	–	26.6	–
272	255	63.0	–	25.4	–
260	248	62.5	–	24.2	–
254	241	61.8	100.0	22.8	–
248	235	61.4	99.0	21.7	–
243	229	60.8	98.2	20.5	–
235	223	–	97.3	20.0	–
230	217	–	96.4	18.0	–
220	212	–	95.5	17.0	–
214	207	–	94.6	16.0	–
210	201	–	93.8	15.0	–
205	197	–	92.8	–	–
200	192	–	91.9	–	–
196	187	–	90.7	–	–
194	183	–	90.0	–	–
190	179	–	89.0	–	–
185	174	–	87.8	–	–
180	170	–	86.8	–	–
176	167	–	86.0	–	–
170	163	–	85.0	–	–
163	156	–	82.9	–	–
159	149	–	80.8	–	–
150	143	–	78.7	–	–
145	137	–	76.4	–	–
139	131	–	74.0	–	–
130	126	–	72.0	–	–
126	121	–	69.8	–	–
120	116	–	67.6	–	–
110	111	–	65.7	–	–

Table 15.38 UNS metals and alloys alphabetical designation

UNS designation	Description
AXXXXX	Aluminum and aluminum alloys
CXXXXX	Copper and copper alloys
DXXXXX	Specified mechanical properties steels
EXXXXX	Rare-earth and rare-earth-like metals and alloys
FXXXXX	Cast irons and cast steels
GXXXXX	AISI and SAE carbon and alloy steels
HXXXXX	AISI and SAE H-steels
JXXXXX	Cast steels
KXXXXX	Miscellaneous steels and ferrous alloys
LXXXXX	Low-melting metals and alloys
MXXXXX	Miscellaneous nonferrous metals and alloys
NXXXXX	Nickel and nickel alloys
PXXXXX	Precious metals and alloys
RXXXXX	Reactive and refractory metals and alloys
SXXXXX	Heat and corrosion resistant stainless steels
TXXXXX	Tool steels, wrought, and cast
WXXXXX	Welding filler metals
ZXXXXX	Zinc and zinc alloys

nuclear fuel cycle). About 68% of the total amount of radioactivity on Earth is of natural origin (i.e., both primordial and cosmogenic radionuclides), while 32% is due to the former human activities. As a general rule, a material is arbitrarily defined as **radioactive** when it exhibits a specific radioactivity greater than 74 kBq.kg^{-1} (i.e., 2 nCi.g^{-1}). The **primordial radionuclides** are usually divided into two groups: (i) those that occur individually (i.e., non-series) and decay directly to a stable nuclide such as: ^{50}V, ^{87}Rb, ^{113}Cd, ^{115}In, ^{123}Te, ^{138}La, ^{142}Ce, ^{144}Nd, ^{147}Sm, ^{152}Gd, ^{174}Hf, ^{176}Lu, ^{187}Re, ^{190}Pt, ^{192}Pt, ^{209}Bi, and (ii) those that occur in decaying chains (i.e., **nuclear series**) and decay to a stable isotope of lead through a sequence of radionuclides of wide-ranging half-lives. They are typically long lived, with half-lives often on the order of hundreds of millions of years. There are three naturally occurring decay series, headed by the radionuclides ^{238}U, ^{235}U, and ^{232}Th, and one artificial series headed by ^{237}Np. These series are commonly called the uranium series, the actinium series, the thorium series, and the neptunium series respectively.

15.14 Scientific and Technical Societies

See **Table 15.43** (pages 573–577).

15

Appen-
dices

Reference

[1] Society of Automotive Engineers (SAE) (1998). Metals and alloys in the unified numbering system, 7th edn. ASTM/SAE.

Table 15.39 Common fuel energy content

Fuel	Heat of combustion	Unit
Solid fuels		
Anthracite coal	34.2	$MJ.kg^{-1}$
Bituminous coal	23.2	$MJ.kg^{-1}$
Charcoal	33.7	$MJ.kg^{-1}$
Coke	30.2–33.14	$MJ.kg^{-1}$
Lignite coal	21.3–23.2	$MJ.kg^{-1}$
Nylon	30.7	$MJ.kg^{-1}$
Paper	17.6	$MJ.kg^{-1}$
Peat	16.7–20.9	$MJ.kg^{-1}$
Polyethylene scrap	44.6	$MJ.kg^{-1}$
Polypropylene scrap	49.0	$MJ.kg^{-1}$
PVC	26.6	$MJ.kg^{-1}$
Wood	20.0	$MJ.kg^{-1}$
Liquid fuels		
Coal tar	37.7	$MJ.kg^{-1}$
Crude oil	38.5–43.7	$MJ.kg^{-1}$
Gasoline JP–4	45.6	$MJ.kg^{-1}$
Natural gas liquefied	25.2	$MJ.dm^{-3}$
Gaseous fuels (at 273.15 K and 101,325 Pa)		
Acetylene	55	$MJ.m^{-3}$
Butane	117.23	$MJ.m^{-3}$
Butene	115.13	$MJ.m^{-3}$
Carbon monoxide	12.04	$MJ.m^{-3}$
Coke oven gas	22.0	$MJ.m^{-3}$
Ethane	65.52	$MJ.m^{-3}$
Ethylene	59.08	$MJ.m^{-3}$
Hydrogen	12.12	$MJ.m^{-3}$
Hydrogen sulfide	23.86	$MJ.m^{-3}$
Methane	37.68	$MJ.m^{-3}$
Natural gas	38.4	$MJ.m^{-3}$
Produced gas	5.6	$MJ.m^{-3}$
Propane	93.9	$MJ.m^{-3}$
Propene	87.11	$MJ.m^{-3}$
Synthetic coal gas	10.8	$MJ.m^{-3}$
Water gas	11.5	$MJ.m^{-3}$

Table 15.40 The four nuclear series

Mass number parity (A)	Series name	Header radionuclide ($T_{1/2}$; E_α, MeV)	End stable nuclide	Gaseous radioelement (emanation, old symbol)
$4n$	Thorium 232	^{232}Th (14.05 Gy; 4.08)	^{208}Pb	^{220}Rn (thoron, Tn)
$4n + 1$	Neptunium 237 (artificial)	^{237}Np (2.14 My; 4.96)	^{209}Bi	None
$4n + 2$	Uranium 238 – radium	^{238}U (4.468 Gy; 4.19)	^{206}Pb	^{222}Rn (radon, Rn)
$4n + 3$	Uranium 235 – actinium	^{235}U (703.7 My; 4.6793)	^{207}Pb	^{223}Rn (actinon, An)

Important note: The three stable nuclide isotopes of lead are the ultimate end decaying nuclides of three natural series, and hence are called **radiogenic** lead nuclides in contrast with the naturally occurring isotope ^{204}Pb.

Table 15.41 Non-series primordial radionuclides

Chemical element	Relative atomic mass (^{12}C = 12.000)	Radionuclides (s)	Stable nuclides	Relative isotopic abundance (a, at%)	Half-life ($T_{1/2}$, y)	Decay type, maximum energy	Specific radioactivity (A_m, kBq.kg^{-1})
Potassium	39.0983	^{40}K	^{40}Ca, ^{40}Ar	0.0117	1.277×10^9	β^- (1.32MeV) 89.28%, (EC, β+) 10.72%	30.996
Vanadium	50.9415	^{50}V	^{50}Cr, ^{50}Ti	0.2500	1.4×10^{17}	β^-, EC	0.0000046
Rubidium	85.4678	^{87}Rb	^{87}Sr	27.8350	4.88×10^{10}	β^-(0.273)	882.743
Cadmium	112.4110	^{113}Cd	^{113}In	12.2200	9.3×10^{15}	β^-(0.59)	0.00155
Indium	114.8180	^{115}In	^{115}Sn	95.7100	4.4×10^{14}	β^-(0.496)	0.2505879
Tellurium	127.6000	^{123}Te	^{123}Sb	0.9080	1.3×10^{13}	EC (0.051)	0.0724031
		^{130}Te	^{130}Xe	33.7990	2.5×10^{21}	2β	0.000000014
Lanthanum	138.9055	^{138}La	^{138}Ce, ^{138}Ba	0.0902	1.06×10^{11}	β^-(1.04)34%, CE(1.75) 66%	0.8103017
Neodymium	144.2400	^{144}Nd	^{140}Ce	23.8000	2.29×10^{15}	α(1.83)	0.00953
Samarium	150.3600	^{147}Sm	^{143}Nd	15.0200	1.06×10^{11}	α(2.15)	124.651
		^{148}Sm	^{144}Nd	11.3000	7.00×10^{15}	α(1.96)	0.0014201
		^{149}Sm	^{145}Nd	13.8000	1.00×10^{16}	α	0.0012140
Gadolinium	157.250	^{152}Gd	^{148}Sm	0.2000	1.1×10^{14}	α(2.24)	0.00153
Lutetium	174.9670	^{176}Lu	^{176}Hf, ^{176}Yb	2.5900	3.80×10^{10}	β^-(1.02)97%, CE 3%, γ	51.526
Hafnium	178.4900	^{174}Hf	^{170}Yb	0.1620	2.00×10^{15}	α(2.55)	0.0000600
Rhenium	186.2070	^{187}Re	^{187}Os	62.6000	4.56×10^{10}	β^-(0.0025)	975.17
Osmium	190.2300	^{186}Os	^{182}W	1.5800	2.00×10^{15}	α(2.75)	0.0005493
Platinum	195.0780	^{190}Pt	^{186}Os	0.0100	6.5×10^{11}	α(3.18)	0.0104314
Thorium	232.0381	^{232}Th	^{208}Pb	100.000	1.40×10^{10}	α(4.081)	4071.723
Uranium	238.0289	^{234}U	^{208}Pb	0.0055	2.45×10^5	α(4.856)	12,474.77
		^{235}U	^{207}Pb	0.7200	7.04×10^8	α(4.6793)	568.324
		^{238}U	^{206}Pb	99.2745	4.46×10^9	α(4.190)	12,369.116

15

Appendices

Table 15.42 Major cosmogenic radionuclides

Cosmonuclide	Half-life ($T_{1/2}$)	Decay type and maximum energy (E_m, MeV)
$^3T(^3H)$	12.43 years	β^- (0.018)
^7Be	53.29 days	CE (0.477)
^{10}Be	1.5×10^6 years	β^- (0.555)
^{14}C	5730 years	β^- (0.15648)
^{22}Na	2.605 years	β^+ (2.842) 90%, EC 10%, γ
^{26}Al	7.4×10^5 years	β^+ (4.003) 82%, EC
^{32}Si	172 years	β^- (0.227)
^{32}P	14.262 days	β^- (1.710)
^{33}P	25.56 days	β^- (0.249)
^{35}S	87.44 days	β^- (0.1674)
^{36}Cl	3.01×10^5 years	β^- (0.709)
^{39}Cl	55.6 min	β^- (1.50)
^{39}Ar	269 years	β^- (0.57)
^{81}Kr	2.29×10^5 years	EC (0.287)

Further Reading

Crystallography and Crystallochemistry

Borchard Ott W (1995) Crystallography, 2nd edn. Springer-Verlag, Heidelberg.

Collective (1910) Struktürbericht, Zeitschrift für Kristallographie (7 vols). Akademische Verlagsgesellschaft, Leipzig.

Collective (1983–1992) International tables for crystallography (vols A, B, and C). Riedel, Dordrecht, NL.

Fedorov ES (1892) Zusammenstellung der Kristallographischen Resultate. Zs. Krist 20.

Giacovazzo C, Monaco HL, Viterbo D et al. (1992) Fundamentals of crystallography. OUP/IUCr, London.

Groth P (1921) Elemente der Physikalischen und Chemischen Krystallographie. R. Oldenbourg, München/Berlin.

Janecke E (1940) Kurzgefasstes Handbuch aller Legierung. Verlag Von Robert Kiepert.

McGillavry CH, Riecki GD (1952–1962) International tables of X-ray crystallography (Vols I, II, and III). Kynock Birmingham.

Pearson WB (1958 and 1967) Handbook of lattice spacings and structures of metals and alloys, vols 1 and 2. Pergamon Press, New York.

Schoenflies A (1891) Kristallsysteme und Kristallstructur. Leipzig.

Van Meershe M, Feneau-Dupont J (1984) Introduction à la cristallographie et à lachemie structurale. Éditions Peeters, Louvain-la-Neuve.

Table 15.43 US and international professional societies

Acronym	Professional society	Address
AA	Aluminum Association	900 19th Street, N.W., Washington, DC 20006, USA Telephone: 1 (202) 862 5100 Internet: http://www.aluminum.org
ACA	American Crystallographics Association	P.O. Box 96, Ellicott Station, Buffalo NY 14205–0096, USA Telephone: 1 (716) 856 9600 Fax: 1 (716) 852 4846 Internet: http://www.hwi.buffalo.educ/ACA
ACI	American Concrete Institute	P.O. Box 19150, Detroit, MI 48219, USA Telephone: 1 (313) 930 9277 Fax: 1 (313) 930 9088 E-mail: service@cssinfo.com Internet: http://www.cssinfo/info/aci.html
ACS	American Chemical Society	1155 16th Street, N.W., Washington, DC 20036, USA Telephone: 1 (202) 872 4600 Internet: http://www.acs.org
ADA	American Dental Association	211E, Chicago Avenue, Chicago, IL 60611, USA Telephone: 1 (312) 440 2500 Fax: 1 (312) 440 2800 Internet: http://www.ada.org
AESF	American Electroplaters and Surface Finishers Society	12644 Research Parkway, Orlando FL 32826-3298, USA Telephone: 1(407) 281 6441 Fax: 1 (407) 281 6446 Internet: http://www.aesf.org
AGA	American Gas Association	400 N. Capitol Street, Washington, DC 20001, USA Telephone: 1 (202) 824 7000 Fax: 1 (202) 824 7115 Internet: http://www.aga.org
AGU	American Geophysical Union	2000 Florida Avenue N.W., Washington, DC 20009-1277, USA Telephone: 1 (202) 462 6900 Fax: 1 (202) 328 0566 E-mail: service@kosmos.agu,org Internet: http://www.agu.org
AIAA	American Institute of Aeronautics and Astronautics	Suite 500, 1801 Alexander Bell Drive, Reston, VA 20191-4344, USA Telephone: 1 (703) 264 75 00 Fax: 1 (703) 264 75 51 Internet: http://www.aiaa.org
AIChE	American Institute of Chemical Engineers	3 Park Avenue, New York, New York, 10016-5901, USA Telephone: 1 (212) 591 7338 Internet: http://www.aiche.org
AIE	American Institute of Engineers	1018 Appian Way, El Sobrante, CA 94803-3142, USA Telephone: 1 (510) 223 8911 Fax: 1(888) 868 9243 E-mail: aie@members-aie.org Internet: http://www.members-aie.org
AIME	American Institute of Mining, Metallurgical, and Petroleum Engineers	3 Park Avenue, New York, NY 10016, USA Telephone: 1 (212) 419 7676 Fax: 1 (212) 419 7671 Internet: http://www.idis.com/aime

Table 15.43 (continued)

Acronym	Professional society	Address
AIP	American Institute of Physics	One Physics Ellipse, College Park, MD 20740-3843, USA Telephone: 1 (301) 209 3100 Fax: 1 (301) 209 0843 Internet: http://www.aip.org
AISI	American Iron and Steel Institute	1101 17th Street NW, Washington D.C., 20036, USA Telephone: 1 (412) 281 6323 Internet: http://www.steel.org
ANS	American Nuclear Society	555 North Kennington Avenue, La Grange Park, IL 60526,USA Telephone: 1 (708) 352 6611 Fax: 1 (708) 579 0499 E-mail: nucleus@ans.org Internet: http://www.ans.org
API	American Petroleum Institute	1220L Street NW, Washington D.C., 20005, USA Telephone: 1 (202) 682 8000 Internet: http://www.api.org
APS	American Physical Society	One Physics Ellipse, College Park, MD 20740-3844, USA Telephone: 1 (301) 209 3200 Fax: 1 (301) 209 0865 E-mail: opa@aps.org Internet: http://aps.org
ASCE	American Society of Civil Engineers	1015 15th Street, N.W. Suite 600, Washington, DC 20005, USA Telephone: 1 (202) 789 2200 Fax: 1 (202) 289 6797 Internet: http://www.asce.org
ASM	American Society for Metals	9639 Kinsman Road, Materials Park, OH 44073–0002, USA Telephone: 1 (440) 338 5151 Fax: 1 (440) 338 4634 Internet: http://www.asm-intl.org
ASME	American Society of Mechanical Engineers	3 Park Avenue, New York, NY 10016–5990 USA Telephone: 1 (212) 705 7722 Internet: /http://www.asme.org
ASNE	American Society for Naval Engineers	1452 Duke Street Alexandria, Virginia 22314–3458, USA Telephone: 1 (703) 836 6727 Fax: 1 (703) 836 7491 Internet: http://www.jhuapl.edu/ASNE
ASTM	American Society for Testing and Materials	100 Bau Harbor Drive W., Conshohocken, PA 19428–2959 USA Telephone: 1 (202) 862 5100 Internet: http://www.astm.org
AVS	American Vacuum Society	120 Wall Street – 32 floor, New York, NY 10005, USA Telephone: 1 (212) 248 0200 Fax: 1 (212) 248 0245 Internet: http://www.vacuum.org
AWS	American Welding Society	550 NW LeJeune Road, P.O. Box 351040, Miami, Florida, FL-33126, USA Telephone: 1 (305) 443 9353 Fax: 1 (305) 443 7559 Internet: http://www.amweld.org

Table 15.43 *(continued)*

Acronym	Professional society	Address
CDA	Copper Development Association	260 Madison Avenue, New York, NY 10016, USA Telephone: 1 (212) 251 7200 Fax: 1 (212) 251 7234 E-mail: The-Copper-Page@cda.copper.org Internet: http://www.cda.org
ECS	Electrochemical Society	10 South Main Street, Pennington NJ 08534-2896, USA Telephone: 1 (609) 737 1902 Fax: 1 (609) 737 2743 E-mail: ecs@electrochem.org Internet: http://www.electrochem.org
EPRI	Electric Power Research Institute	3412 Hillview Avenue, Palo Alto, CA 94304-1395, USA Telephone: 1 (650) 855 2000 Internet: http://www.epri.com
GI	Gold Institute	1112 16th Street, N.W., Suite 240, Washington D.C. 20036, USA Telephone: 1 (202) 835 0185 Fax: 1 (202) 835 0155 E-mail: info@goldinstitute.org
GSA	Geological Society of America	3300 Penrose Place, Boulder, CO 80301, USA Telephone: 1 (303) 447 2020 Fax: 1 (303) 447 1133 E-mail: web@geosociety.org Internet: http://www.geosociety.org
IEEE	Institute of Electrical and Electronics Engineers	445 Hoes Lane, PO Box 1331, Piscataway, NJ 08855-0459, USA Telephone: 1 (732) 981 0060 Fax: 1 (732) 981 0225 Internet: http://www.ieee.org
ILZRO	International Lead-Zinc Research Organization Inc.	2525 Meridian Parkway, Post Office Box 12036, Research Triangle Park, NC 27709-2036, USA Telephone: 1 (919) 361 4647 Fax: 1 (919) 361 1957 E-mail: rputnam@ilzro.org Internet: http://www.ilzro.org
IMA	International Magnesium Association	1303 Vincent Place, Suite One, McLean VA 22101 USA Telephone: 1 (703) 442 888 Fax: 1 (703) 821 1824 E-mail: ima@bellatlantic.net Internet: http://www.intlmag.org
IMOA	International Molybdenum Association	Unit 7 Hackford Walk, 119–123 Hackford Road, London SW9 0QT, UK Telephone: 44 0171 582 2777 Fax: 44 0171 582 0556 E-mail: ITIA_IMOA@compuserve.com Internet: http://www.imoa.org.uk
IPMI	International Precious Metals Institute	4400 Bayou Blvd., Suite 18 Pensacola, FL32503-1908, USA Telephone: 1 (850) 476 1156 Fax: 1 (850) 476 1548 E-mail: ipmi@pond.com Internet: http://www.ipmi.org

15

Appendices

Table 15.43 *(continued)*

Acronym	Professional society	Address
ITA (TDA)	International Titanium Association (former Titanium Development Association)	1871 Folsom Street, Suite 200 Boulder, CO 80302-5714, USA Telephone: 1 (303) 443 7515 Fax: 1 (303) 443 4406 E-mail: afitz@titanium.net Internet: http://www.titanium.org
ITIA	International Tungsten Industry Association	Unit 7 Hackford Walk, 119–123 Hackford Road, London SW9 0QT, UK Telephone: 44 0171 582 2777 Fax: 44 0171 582 0556 E-mail: ITIA_IMOA@compuserve.com Internet: http://www.itia.org.uk
ITRI	International Tin Research Institute	Kingston Lane, Uxbridge, Middx UB8 3PJ, UK Telephone: 44 (0) 1895 272406 Fax: 44 (0) 1895 251841 http://www. itri.co.uk/index.htm
IZA	International Zinc Association	Fax: 32 2 776 0089 Internet: http://www.iza.com
LDA	Lead Development Association International	42 Weymount Street, London WIN 3LQ, UK Telephone: 44 0171 499 8422 Fax: 44 0171 493 1555 Internet: http://www.ldaint.org
NACE	National Association of Corrosion Engineers	1440 South Creek Drive, Houston, TX 77084-4906, USA Telephone: 1 (281) 228 6200 Fax: 1 (281) 579 6694 Internet: http://www.nace.org
NiDI	Nickel Development Institute	214 King Street W., Suite 510, Toronto, Canada M5H356 Telephone: 1 (416) 591 7999 Fax: 1 (416) 591 7987 Internet: http://www.nidi.org
NIST (NBS)	National Institute for Science and Technology (former NBS)	Building 225, Room B162 Washington DC., 20234, USA Telephone: 1 (301) 975 6478 Internet: http://www.nist.org
OSA	Optical Society of America	2010 Massachusetts Avenue, NW, Washington, DC 20036 USA Telephone: 1 (202) 223 8130 Fax: 1 (2020) 223 1096 Internet: http://www.osa.org
SAE	Society for Automotive Engineers	400, Commonwealth Drive, Warrendale, PA., 15096-0001, USA Telephone: 1 (724) 776 4841 Fax: 1 (724) 776 5760 Internet: http://www.sae.org
SI	Salt Institute (SI)	700 N. Fairfax Street, Suite 600 Fairfax Plaza, Alexandria, VA 22314-2040, USA Telephone: 1 (703) 549 4648 Fax: 1 (703) 548 2194 Internet: http://www.saltinstitute.org
SME	Society of Manufacturing Engineers	One SME Drive, P.O. Box 930, Dearborn, MI 48121–0930, USA Telephone: 1 (313) 271 1500 Internet: http://www.sme.org

Table 15.43 (continued)

Acronym	Professional society	Address
SNAME	Society of Naval Architects and Marine Engineers	601 Pavonia Avenue, Jersey City, NJ 07306 USA Telephone: 1 (201) 798 4800 Fax: 1 (201) 798 4975 Internet: http://www.sname.org
SPE	Society of Petroleum Engineers	P.O. Box 833836, Richardson, TX 75083-3836, USA Telephone: 1 (972) 952 9393 Fax: 1 (972) 952 9435 Internet: http://www.spe.org
SVC	Society of Vacuum Coaters	71 Pinon Hill Place N.E., Albuquerque, NM 87122-1407, USA Telephone: 1 (505) 856 7188 Fax: 1 (505) 856 6716 E-mail: svcinfo@svc.org Internet: http://www.svc.org
TIC	Tantalum Niobium International Study Center	Washington Street, 40, Brussels B-1050, Belgium Telephone: (02) 649 51 58 Fax (02) 649 64 47 E-mail: tantniob@agoranet.be Internet: http://www.tanb.org
TMS	The Mineral, Metals, and Materials Society	184 Thorn Hill Road, Warrendale, PA 15086 USA Telephone: 1 (724) 776 9000 Fax: 1 (724) 776 3770 E-mail: robinson@tms.org Internet: http://www.tms.org
UI	Uranium Institute	12th Floor, Bowater House West, 114 Knightsbridge, London SWIX 7LJ, UK Telephone: 44 0171 225 0303 Fax: 44 0171 225 0308 E-mail: ui@uilondon.org. Internet: http://www.uilondon.org
USGS	US Geological Survey	807 National Center, Reston, VA 20192, USA Internet: http://www.usgs.gov
ZI	Zinc Institute	292 Madison Avenue, New York, NY 10017, USA Telephone: 1 (212) 578 4750 Fax: 1 (212) 578 4750

15

Appen-
dices

16 Bibliography

Only the general bibliographical references not previously listed in individual chapters are included here. For instance, references about semiconductors and superconductors are listed at the end of Chapters 4 and 5 respectively, while general references about metallurgy or comprehensive series in material sciences are listed in this section.

16.1 General Desk References

16.1.1 Mathematics, Statistics, and Units

Abramowitz M, Stegun IA (1972) Handbook of mathematical functions, with formulas, graphs, and mathematics tables. Dover Publications, New York.

Beyer WH (1991) CRC standard mathematical tables and formulae, 29th edn. CRC Press, Boca Raton, FL.

Cardarelli F (1999) Scientific unit conversion: practical guide to metrication, 2nd edn. Springer-Verlag, London.

Sachs L (1984) Applied statistics: a handbook of techniques, 2nd edn. Springer-Verlag, New York.

Tallarida RJ (1992) Pocket book of integrals and mathematical formulas, 2nd edn. CRC Press, Boca Raton, FL.

16.1.2 Physics and Chemistry

ACS (1992) Reagent chemicals, 8th edn. American Chemical Society, New York.

Adamson AW (1990) Physical chemistry of surfaces, 5th edn. John Wiley, New York.

AIP (1989) A physicist's desk reference, physics vade mecum, 2nd edn. American Institute of Physics (AIP) Press, New York.

Ash M, Ash I (1992) Handbook of plastics compounds, elastomers, and resins: an international guide by category, trade names, composition and suppliers. VCH Publishers, New York.

Condon EU, Odishaw H (1958) Handbook of physics. McGraw-Hill Company, New York.

Craver CD (1977) Desk book of infrared spectra, 2nd edn. Coblentz Society.

Dean J (ed.) (1999) Lange's handbook of chemistry, 15th edn. McGraw-Hill, New York.

Driscoll WG (1978) Handbook of optics. Optical Society of America, McGraw-Hill, New York.

Emsley J (1991) The elements, 2nd edn. Clarendon Press, Oxford.

Gray, DE (ed.) (1972) American Institute of Physics AIP-handbook, 3rd edn. McGraw-Hill, New York.
Greenwood NN, Earnshaw A (1984) Chemistry of the elements. Pergamon Press, New York.
Grigoriev IS, Meilikhov EZ (eds.) (1996) Handbook of physical quantities. CRC Press, Boca Raton, FL.
JANAF (1986) Thermochemical tables, 3rd ed, 2 vols. American Chemical Society, Washington, DC.
Kaye GWC, Laby TH (1995) Tables of physical and chemical constants, 16th end. Longman, New York.
Lide D (ed.) (1998) CRC handbook of chemistry and physics, 78th edn. CRC Press, Boca Raton FL.
Lide DR (1995) Handbook of organic solvents. CRC Press, Boca Raton, FL.
Lide DR, Milne GWA (1994) Handbook of data on common organic compounds. CRC Press, Boca Raton, FL.
Lide DR, Milne GWA (1995) Names, synonyms, and structure of organic compounds. CRC Press, Boca Raton, FL.
Merck (1989) The Merck index, 11th edn. Merck and Co, Rahway.
Mills I, Cvitas T, Homann K, Kallay N, Kuchitsu K. (1993) Quantities, units and symbols in physical chemistry, 2nd edn. IUPAC/Blackwell Scientific Publications, Oxford.
Milne GWA (1995) CRC handbook of pesticides. CRC Press, Boca Raton, FL.
Nuclear Data Group (1973) Nuclear level schemes A = 45 through A = 257. Academic Press, New York.
Perry DL, Phillips SL (1995) Handbook of inorganic compounds. CRC Press, Boca Raton, FL.
Pourbaix M (1963) Atlas d'équilibres électrochimiques á 25°C. Gauthiers Villars & Cie, Paris.
Schmidt E, Grigull U (1982) Properties of water and stream in SI units: 0–800°C and 0–1000 bar. Springer-Verlag, Heidelberg.
Schweitzer P (1991) Corrosion resistance tables: metals, nonmetals, coatings, mortars, plastics, elastomers and linings, and fabrics, 3rd edn. Vols. A, B and C. Marcel Dekker, New York.
Smithsonian physical tables 9th rev. edn. (1954) Smithsonian Institution Press, Washington.
Wagman DD (ed.) (1982) The NBS tables of chemical thermodynamic properties. ACS and AIP, New York.

16.1.3 Engineering

Amiss J (1992) Machinery's handbook guide, 24th edn. Industrial Press.
Ashby MF (1992) Materials selection in mechanical design. Pergamon Press, Oxford.
Avallone E, Baumeister T (1997) Mark's standard handbook for mechanical engineers, 10th edn. McGraw-Hill, New York.
Beitz W, Kuttner K-H (1995) Dubbel's handbook of mechanical engineering. Springer-Verlag, London.
Braun AM, Maurette MT, Oliveros E (1991) Photochemical technology. Wiley, New York.
Eshbach OW, Souders M (1974) Handbook of engineering fundamentals, 3rd edn. John Wiley and Sons, New York.
Fink D, Beaty H (1994) Standard handbook for electrical engineers, 13th edn., McGraw-Hill, New York.
Ganic E, Hicks T (1990) The McGraw-Hill handbook of essential engineering information and data. McGraw-Hill, New York.
Hine F (1985) Electrode processes and electrochemical engineering. Plenum Press, New York.
Incropera, FP, Witt DP (1985) Fundamentals of heat and mass transfer, 2nd edn. John Wiley, New York.
Karassik I, Krutzsh W, Frazer W, Messina J (eds) (1986) Pump handbook, 2nd edn. McGraw-Hill, New York.
Linden D (1984) Handbook of batteries and fuel cells. McGraw-Hill, New York.
Lowenheim FA (1974) Modern electroplating, 3rd edn. Wiley, New York.
Orfeuil M, Robin A (1987) Electric process heating: technologies, equipment and applications. Batelle Press, Columbus, OH.
Perry RH, Green DW (eds) (1998) Perry's chemical engineer's handbook, 7th edn. McGraw-Hill, New York.
Pletcher D, Walsh FC (1990) Industrial electrochemistry, 2nd edn. Chapman & Hall, London.
Wendt, S (1998) Electrochemical engineering. Springer-Verlag, Heidelberg.

16.1.4 Metallurgy

Barett CS (1952) Structure of metals, 2nd edn. McGraw-Hill, New York.
Bousfield B (1992) Surface preparation and microscopy of materials. John Wiley & Sons, Chichester.
Brandes EA, Brook GB (1992) Smithell's metal reference handbook, 7th edn. Butterworth, London.
Bringas JE (1995) The metals black book, vol. 1 Ferrous metals, 2nd edn. CASTI Publishing, Edmonton, Canada.
Bringas JE (1995) The metals red book, vol. 2 Nonferrous metals. CASTI Publishing, Edmonton, Canada.
Cahn RW, Haasen P (1995) Physical metallurgy, 4th edn. North-Holland, New York.
Dieter GE (1984) Mechanical metallurgy, 3rd edn. McGraw-Hill, New York.
Habashi F (1997) Handbook of extractive metallurgy (vols I, II, III, and IV). VCH Weinheim.
Hume-Rothery W, Raynor GV (1954) The structure of metals and alloys, 3rd edn. The Institute of Metals, London.

Miner DF, Seastone JB (eds) (1955) Handbook of engineering materials. John Wiley & Sons, New York.
Pearson WB (ed.) (1959) Handbook of lattice spacing and structures of metals, vol. I. Pergamon Press, New York.
Pearson WB (ed.) (1972) Crystal chemistry and physics of metals and alloys. Wiley, New York.
Shackelford JF, Alexander W (ed) (1991) The CRC materials science and engineering handbook. CRC Press, Boca Raton, FL.
Smith WF (1981) Structure and properties of engineering alloy. McGraw-Hill, New York.
Vander Voort GF (1984) Metallography: principles and practice. McGraw-Hill, New York.
Woldman NE, Gibbons RC (eds) (1979) Woldman's engineering alloys, 6th edn, ASM, Metals Park, OH.

16.1.5 Phase Diagrams

American Society for Metals (1990) Binary alloy phase diagrams, 2nd edn. (3 vols) ASM, Ohio Park, OH.
American Society for Metals (1995) Handbook for ternary alloy phase diagrams (10 vols). ASM, Ohio Park, OH.
Champion P, Guillet L, Poupeau P (1981) Diagrammes de phases des matériaux cristallins. Masson, Paris.
Hansen M, Anderko K (1958) Constitution of binary alloys, 2nd edn. McGraw-Hill, New York.
Haughton LJ, Prince A (1956) The constitutional diagrams of alloys: a bibliography, 2nd edn. Monograph and Reports Series No. 2, Institute of Metals.
Hulgren R, Desai PD, Hawkins DT, Gleiser M, Kelley KK (1973) 4.1 Selected values of the thermodynamics properties of the elements. University of California, Berkeley/ASM.
Hulgren R, Desai PD, Hawkins DT, Gleiser M, Kelley KK (1973) 4.2 Selected values of the thermodynamics properties of binary alloys. University of California, Berkeley/ASM.
Janecke E (1940) Kurzgefasstes Handbuch aller Legierung. Verlag Von Robert Kiepert.
Levin EM, McMurdie, Hall FP (1956) Phase diagrams for ceramists, Part I. American Ceramic Society.
Levin EM, McMurdie, Hall FP (1956) Phase diagrams for ceramists, Part II. American Ceramic Society.
Moffat WG (1978) The handbook of binary phase diagrams. General Electric.
Prince A (1978) Multicomponent alloy constitution bibliography 1955–1973. The Metals Society, London.
Tamas F, Pal I (1970) Phase equilibria spatial diagrams. Iliffe Books.

16.2 Dictionaries and Encyclopediae

Ash M, Ash I (ed.) (1994) Gardner's chemical tradenames dictionary, 10th edn. VCH Publishers, New York.
Ash M, Ash I (1991) Concise encyclopedia of industrial chemical additives. Edward Arnold, London.
Ash M, Ash I (1989) The condensed encyclopedia of surfactants. Edward Arnold, London.
Ballentyne DWG, Lovett DE (1980) A dictionary of named effects and laws. In: Chemistry, physics and mathematics, 4th edn. Chapman and Hall, London.
Becher P (1990) Dictionary of colloid and surface science. Marcel Dekker, New York.
Besancon RM (ed.) Encyclopedia of physics, 3rd edn. Van Nostrand Reinhold, New York.
Considine DM (ed.) (1984) Van Nostrand Reinhold encyclopedia of chemistry, 4th edn. Van Nostrand Reinhold, New York.
De Sola R (1991) Abbreviation dictionary, 8th edn. CRC Press, Boca Raton, FL.
Dictionary of inorganic compounds, 5 vols. Chapman & Hall, London (1992).
Dictionary of organic compounds, 6th edn. Chapman & Hall, London (1996).
Flugge S (ed.) (1955–1988) Handbuch der Physik, 55 vols, 2nd edn. Springer-Verlag, Berlin.
Grayson M, Eckroth D (1985) Kirk–Othmer concise encyclopdia of chemical technology. Wiley, New York.
Howard P, Neal M (1992) Dictionary of chemical names and synonyms. Lewis, New York.
Howard PH, Neal MW (1992) Dictionary of chemical names and synonyms. CRC Press, Boca Raton, FL.
Kajdas C, Harvey SSK, Wilusz E (1991) Encyclopedia of tribology. Elsevier, Amsterdam.
Kroschwitz JI, Howe-Grant M (1991) Encyclopedia of chemical technology, 4th edn. Wiley, New York.
Lapedas DN (1978) McGraw-Hill dictionary of physics and mathematics. McGraw-Hill, New York.
Lerner R, Trigg GL (1991) Encyclopedia of physics 2nd edn. VCH Publishers, New York.
Lerner RG, Trigg GL (ed.) (1983) Concise encyclopedia of solid state physics. Addison Wesley, Reading, MA.
Lewis RJ, Sax NI (1992) Hawley's condensed chemical dictionary, 12th edn. Van Nostrand Reinhold.
McGraw-Hill encyclopedia of science and technology, 7th edn., 20 vols. McGraw-Hill New York (1992).
Parker S (1994) Dictionary of scientific and technical terms, 5th edn. McGraw-Hill, New York.
Snell FD, Hilton L (1966–1974) Encyclopedia of industrial chemical analysis, 20 vols. Interscience, New York.

16.3 Comprehensive Series

Table 16.1 Comprehensive series in material sciences

Advances in Materials Science and Engineering (11 volumes), Pergamon Press, New York, (1989–1994). [1] Brook R (ed.) Concise encyclopedia of advanced ceramic materials (1991), [2] Kelly A (ed.) Concise encyclopedia of composite materials (1994), [3] Evetts J (ed.) Concise encyclopedia of magnetic and superconducting materials (1992), [4] Cahn R, Lifshin E (ed.) Concise encyclopedia of materials characterization (1992), [5] Bever MB (ed.) Concise encyclopedia of materials economics, policy, and management (1992), [6] Williams D (ed.) Concise encyclopedia of medical and dental materials (1994), [7] Carr DD (ed.) Concise encyclopedia of mineral resources (1989), [8] Corish PJ (ed.) Concise encyclopedia of polymer processing and applications (1991), [9] Mahajan S, Kimerling LC (ed.) Concise encyclopedia of semiconducting materials and related technologies (1992), [10] Schniewind AP (ed.) Concise encyclopedia of wood and wood-based materials (1989), [11] Moavenzadeh F (ed.) Concise encyclopedia of building and construction materials (1990).

ASM Handbook of Metal Series 9th ed (20 volumes), American Society of Metals (ASM), Ohio Park, OH (1984–1996). [1] Properties and selection: irons, steels and high-performance alloys (1990), [2] Properties and selection: nonferrous alloys and special-purpose materials (1991), [3] Alloys phase diagrams (1992), [4] Heat treating (1991), [5] Surface engineering (1994), [6] Welding, brazing and soldering (1993), [7] Powder metallurgy (1984), [8] Mechanical testing (1985), [9] Metallography and microstructures (1985), [10] Materials characterization (1986), [11] Failure analysis and prevention (1986), [12] Fractography (1987), [13] Corrosion (1987), [14] Forming and forging (1988), [15] Casting (1988), [16] Machining (1984), [17] Nondestructive evaluation and quality control (1989), [18] Friction, lubrication, and wear technology (1992), [19] Fatigue and fracture (1996), [20] Materials selection (1994).

Chemistry and Physics of Carbon: A Series of Advances (25 volumes), Walker PL Jr (ed.), Marcel Dekker, New York (1966–1996).

Corrosion Technology Series (7 volumes), Schweitzer PA, Marcel Dekker, New York (1989–1995).

DECHEMA Corrosion Handbook, Corrosive Agents and Their Interaction with Materialss (12 volumes), Behrens D (ed.), VCH Weinheim (1987–1993). [1] Acetates, aluminium chloride, chlorine and chlorinated water, fluorides, potassium hydroxides, steam, sulfonic acids (1987), [2] Aliphatic aldehydes, ammonia and ammonium hydroxide, sodium hydroxide, soil (1988), [3] Acid halides, amine salts, bromides, bromines, carbonic acid, lithium hydroxide (1988), [4] Alkanecarboxylic acids, formic acid, hot oxidizing gases, polysols (1989), [5] Aliphaic amines, alkaline earth chlorides, alkaline earth hydroxides, fluorine, hydrogen fluoride and hydrofluoric acid, hydrochloric acid and hydrogen chloride (1989), [6] Acetic acid, alkanes, benzene and benzene homologues, hydrogen chloride (1990), [7] Aliphatic ketones, ammonium salts, atmosphere, potassium chloride (1990), [8] Sulfuric acid (1991), [9] Methanol, sulfur dioxide (1991), [10] Carboxylic acids, esters, drinking water, nitric acid) (1992), [11] Chlorine dioxide, seawater (1992), [12] Chlorinated hydrocarbon, phosphoric acid (1993).

International Critical Tables of Numerical Data, Physics, Chemistry, and Technology (7 volumes), The National Reserach Council, McGraw-Hill, New York (1926–1933).

Landolt and Bornsteins's Numerical Data and Functional Relationships in Science and Technology, Springer-Verlag, Berlin (1961–1999). **Volume I**: Atomic and molecular physics. Part 1: Atoms and ions (1950), Part 2: Molecules I (Nuclear structure) (1951), Part 3: Molecules II (Electron shells) (1951), Part 4: Crystals (1955), Part 5: Atomic nuclei and elementary particles (1952). **Volume II**: Properties of matter in its aggregated states. Part 1: Mechanical-thermal properties of states (1971), Part 2: Equilibria except fusion equilibria, Part 2a: Equilibria vapor-condensate and osmotic phenomena (1960), Part 2b: Solution equilibria I (1962), Part 2c: Solution equilibria II (1964), Part 3: Fusion equilibria and interfacial phenomena (1956), Part 4: Caloric quantities of state (1961), Part 5a: Transport phenomena I (Viscosity and diffusion) (1969), Part 5b: Transport phenomena II (Kinetics, homogenous gas equilibria), (1968), Part 6: Electrical properties (1959), Part 7: Electrical properties II (Electrochemical systems) (1960), Part 8: Optical constants (1962), Part 9: Magnetic properties I (1962), Part 10: Magnet properties II (1967). **Volume III**: Astronomy and geophysics (1952). **Volume IV**: Technology. Part 1: Material values and mechanical behavior of non-metals (1955), Part 2: Material values and behavior of metallic industrial materials, Part 2a: Principles; testing methods; iron industrial materials (1964), Part 2b: Sinter materials; heavy metals (except special industrial materials) (1964), Part 2c: Light metals; special industrial materials; semiconductors; corrosion (1965), Part 3: Electrical engineering; light technology, x-ray technology (1957), Part 4: Heat technology, Part 4a: Methods of measurement, thermodynamic properties of homogenous materials (1967), Part 4b: Thermodynamic properties of mixtures, combustion; heat transer (1972), Part 4c: Absorption equilibria of gases in liquids, Part 4c1: Absorption of liquids of low vapor pressure (1976), Part 4c2: Absorption in liquids of high vapor pressure (1980). The **New Series** is divided into seven main subject groups: Group I: Nuclear and particle physics, Group II: Atomic and molecular physics, Group III: Crystal and solid state physics, group IV: Macroscopic and technical properties of matter, Group V: Geophysics and space research, Group VI: Astronomy, astrophysics, and space research, Group VII: Biophysics.

Table 16.1 (continued)

Materials Engineering Series A (9 volumes), Marcel Dekker, New York, 1992–1996. [1] Richerdson, DW. Modern ceramic engineering, properties, processing, and use in design, 2nd. ed (1992), [2] Murray GT. Introduction to Engineering materials: behavior, properties and selection (1993), [3] Liebermann HH. Rapidly solidified alloys; processes-structures-properties-applications (1993), [4] Belitskus DL. Fiber and whisker reinforced ceramics for structural applications (1993), [5] Speyer, RF. Thermal analysis of materials (1994), [6] Jahanmir S. Friction and wear of ceramics (1994), [7] Ochiai S. Mechanical properties of metallic composites (1994), [8] Lee BI, Pope BJA. Chemical processing of ceramics (1994), [9] Cheremisinoff NP, Cheremesinoff PN. Handbook of advanced materials testing (1995).

Materials Science and Technology, a Comprehensive Treatment (19 volumes) Cahn RW, Hassen, P, Kramer BJ (eds), VCH Weinheim (1991–1996). [1] Structure of solids (1992), [2A] Characterization of materials (1992), [2B] Characterization of materials (1994), [3A] Electronic and magnetic properties of metals and ceramics (1992), [3B] Electronic and magnetic properties of metals and ceramics (1994), [4] Electronic structure and properties of semi-conductors (1991), [5] Phase transformation in materials (1991), [6] Plastic deformation and fracture of materials (1992), [7] Constitution and properties of steels, (1992), [8] Structure and properties of nonferrous alloys (1994), [9] Glasses and amorphous materials (1991), [10A] Nuclear materials (1994), [10B] Nuclear materials (1994), [11] Structure and properties of ceramics (1994), [12] Structure and properties of polymers (1993), [13] Structure and properties of composite (1993), [14] Medical and dental materials, [15] Processing of metals and alloys (1991), [16] Processing of semi-conductors (1995), [17] Processing of ceramics (1995), [18] Processing of polymers (1996), [19A] Corrosion of materials, [19B] Corrosion of materials.

Ternary Alloys, A Comprehensive Compendium of Evaluated Constitutional Data and Phase Diagrams (13 volumes), Petzow G, Effenberg G (eds), VCH, Weinheim (1988–1995). [1] Ag–Al–Au to Ag–Cu–P, [2] Ag–Cu–Pb to Ag–Zn–Zr, [3] Al–Ar–O to Al–Ca–Zn, [4] Al–Cd–Ce to Al–Cu–Ru, [5] Al–Cu–S to Al–Gd–Sn, [6] Al–Gd–Tb to Al–Mg–Sc, [7] Al–Mg–Se to A1–Ni–Ta, [8] Al–Ni–Tb to Al–Zn–Zr, [9, 10] All As systems, [11] As–Ir–K to As–Yb–Zn, [12] Au systems, [13] Cu systems.

Thermal Expansion Data Articles, Taylor D. [I] Binary oxides with the sodium chloride and wurtzite structures, MO, Br Ceram Trans J 83 (1984) 5–9, [VIII] Complex oxides, ABO_3, the perowskites, Br Ceram Trans J 84 (1985) 181–188, [X] Complex oxides, ABO_4, Br Ceram Trans J 85 (1986) 146–155, [XI] Complex oxides, A_2BO_5, and the garnets, Br Ceram Trans J 86 (1987) 1–6, [XII] Complex oxides, $A_2 BO_6$, A_2BO_7, $A_2B_2O_7$, plus complex aluminates, silicates and analoguous compounds, Br Ceram Trans J 87 (1988) 39–45, [XIII] Complex oxides with chain, ring, and layer structures and the apatites, Br Ceram Trans J 87 (1988) 87–95, [XIV] Complex oxides with the sodalite and nasicon framework structure, Br Ceram Trans J 90 (1991) 64–69, [XV] Complex oxides with the leucite structure and frameworks based on the six-membered ring of tetraedra, Br Ceram Tran J 90 (1991) 197–204.

Thermophysical Properties of Matter (14 volumes), Touloukian YS, Kirby RK, Taylor RE, Lee TYR (eds), New York, IFI/Plenum (1970–1977). [1–3] Thermal conductivity, [4–6] Specific heat capacity, [10] Thermal radiative properties, and thermal diffusivity, [11] Absolute and dynamic viscosity, [12, 13], Coefficients of therma expansion [14] index.

Traité des Matériaux (20 volumes), Gerl M, Ilschner B, Mercier, J-P Mocellin A (eds), Presses Polytechniques Romandes, Lausanne, Switzerland (1990–1996). [1] Introduction à la science des matériaux, [2] Caractérisation chimique, physique et microstructurale des matériaux, [3] Caractérisation des matériaux par rayons X, électrons et neutrons, [4] Qualité et fiabilitté des matériaux: essais normalisés, [5] Thermodynamique chimique des é tats de la matiére, [6] Transformations de phases, [7] Phénomènes de transport: application à l'élaboration et au traitement des matériaux, [8] Propriétés physique des matériaux, [9] Déformation et résistance des matériaux, [10] Modélisation numérique en science et technologie des matériaux, [11] Métaux et alliages: propriétés, technologie et applications, [12] Corrosion et chimie des surfaces des matériaux, [13] Chimie des polyméres: synthèse, réactions et dégradation, [14] Composites à matrice organique, [15] Les céramiques: principes et méthodes d'élaboration, [16] Physique et technologies des semi-conducteurs, [17] Matériaux de l'an 2000: bilan et perspective.

Index

Note: Page numbers in *italic* type refer to tables.

ac magnetic permeability, 283
acanthite, 193, *410*
acceptors, 248, 250
acrylic, 377
acrylonitrile butadiene styrene (ABS),
    373–4, *382, 383*
actinolite, *410*
aegirine, *410*
aggregates, 520–1
    coarse, 520
    fine, 521
aging of ferroelectric materials, 305–7
akermanite, *410*
albite, *410*
alite, 519
alkali metals, 75–84
    properties, *76–7*
    see also specific alkali metals
alkaline-earth metals, 94
    properties of, *95–6*
    see also specific alkaline-earth
        metals
allanite, *411*
allotropism, 2
almandine, *411*
alpha-iron, 3
alps, *461*
alumina, 47, *339*
aluminum
    alloying, 48
    description, 45–6
    electrowinning/scrap recycling, 47
    further reading, 69
    general properties, 45–6
    history, 46
    industrial preparation, 47–8
    industrial uses and applications, 51
    metal, 47
    minerals, 46–7
    natural occurrence, 46–7
    ores, 46–7

processing, 47–8
refining, 48
secondary production, 48
aluminum alloys, 48–50
    cast, 49, *56–7*
    further reading, 69
    industrial uses and applications, 51
    standard designations, 49–50,
        49–50
    temper designation, *50*
    wrought, 49, *52–4*
aluminum carbide, *346, 347*
aluminum diboride, *342, 343*
aluminum dodecaboride, *342, 343*
aluminum mononitride, *352, 353*
aluminum sesquioxide, *358, 359*
alunite, *411*
Alvite, 154
amalgam, 193
amber, *411*
amblygonite, 79, *411*
American Society for Testing and
        Materials (ASTM), 563
aminoplastics, 377
analcime, *411, 412*
anatase, 121, *412*
andalusite, *412*
andesine, *412*
andradite, *412*
anglesite, *413*
anhydrite, 108, *413*
anisotropic materials, 404–5
ankerite, *413*
annabergite, *413*
anodes
    carbon, 326
    dimensionally stable, *328*
    DSA type, 327–8
    DSA-Cl$_2$, 330
    DSA-O$_2$, 330
    for cathodic protection, *332*

for oxygen evolution in acidic
        media, 328–31
    impressed current, *333*
    lead and lead alloy, 325–6
    lead dioxide, 329
    manganese dioxide, 329
    materials, 324–31
    materials for oxygen evolution in
        acid media, *324*, 328–31
    materials used in batteries and fuel
        cells, *321*
    noble metals, 325
    platinized, 326–7
    precious metals, 325
    tantalum, 326–7
    titanium, 326–7
anodic protection, cathodes for,
        331–2, *332*
anorthite, *413*
antiferromagnetic compounds, Néel
        temperature, *282*
antiferromagnetic elements, Néel
        temperature, *281*
antiferromagnetic materials, 280–1
antigorite, *414*
antimony, *414*
antlerite, *414*
apatite, 108, *414*
aphanitic igneous rocks, 485
aphthitalite, *414*
aquamarine, 97
aragonite, 108, *415*
argentite, 193
arrest points, 6
arsenic, *415*
arsenopyrite, *415*
atacamite, *415*
atomic orbitals, linear combination
        of, 245
atomic polarizability, 294
augelite, *415*

augite, 415, 416
austenite gamma, 6
austenitic stainless steels, 16, 18
    physical properties, 20–1
autnite, 416
azurite, 416

baddeleyite, 146, 154, 416
band theory of bonding, 245
barite, 112, 416
barium
    description, 111
    general properties, 111
    history, 111–12
    industrial uses and applications,
        112
    minerals, 112
    natural occurrences, 112
    ores, 112
    preparation, 112
    processing, 112
    properties of, 95–6
barium titanate, lattice structure,
        305
basalt, 522
bastnaesite, 219, 220, 235, 416
    hydrochloric acid digestion
        process, 222
batholiths, 481
batteries, electrode material, 319–20
bauxite, 46, 47
Bayer process, 47
belite, 519
bertrandite, 97, 417
beryl, 97, 417
    mining and mineral dressing, 98
beryllium
    description, 97
    further reading, 241
    general properties, 97
    history, 97
    industrial preparation, 98
    industrial uses and applications,
        98–9
    metal, 98
    minerals, 97
    natural occurrences, 97
    ores, 97
    processing, 98
    properties of, 95–6
    world metal producers, 99
beryllium boride, 342, 343
beryllium carbide, 346, 347
beryllium diboride, 342, 343
beryllium hemiboride, 342, 343
beryllium hexaboride, 342, 343
beryllium hydroxyde, 98
beryllium monoboride, 342, 343
beryllium monoxide, 360, 361
beryllium nitride, 352, 353
beta-iron, 3
B-H magnetization curve, 281–3
binary compounds, Strukturbericht
        designations, 547
bindheimite, 417
binding energy, 315

biogenic sedimentary rocks, 494–5
biotite, 417
birefringence, 405
bischofite, 417
bismuth, 417
black iron, 10
blast furnace, 4
blende, 59
Bloch walls, 303
blodite, 417
boehmite, 47, 418
Boltzmann distribution, 249
Bolzano process, 102
bonding, band theory of, 245
boracite, 418
borax, 86, 418
bornite, 418
boron, 342, 343
boron carbide, 348, 349
boron mononitride, 352, 353
boulangerite, 418
bournonite, 419
bradleyite, 419
bravais space lattices, 542
brines, lithium carbonate from, 81
brochantite, 419
bromoargyrite, 193, 419
brookite, 121, 419
brucite, 101, 419
building and construction materials,
        517–25
    physical properties of, 524–5
    see also specific materials
building stones, 522–3, 523
butadiene acrylonitrile rubber, 385
butyl rubber, 379–80, 382, 383
bytownite, 419, 420

calamine, 59
calaverite, 420
calcination, 519
calcite, 420
calcium
    description, 103–8
    further reading, 242
    general properties, 103–8
    history, 108
    industrial preparation, 109
    industrial uses and applications,
        109–10
    minerals, 108
    natural occurrences, 108
    ores, 108
    processing, 109
    properties of, 95–6
calcium-based chemicals, 109–10
calcium monoxide, 360, 361
calomel, 420
capacitance, 292
    temperature coefficient, 292–3
capacitors
    electrostatic energy stored in, 293
    of different geometries, 293
    parallel electrode, 293
carbon anodes, 326
carbon–iron phases, 6

carbon product properties, 341
carbon steels, 6, 9–11
    designation, 12
    physical properties, 13–14
carbonaceous sedimentary rocks, 496
carnallite, 89, 91, 101, 420
carnotite, 155, 420
cassiterite, 421
cast irons, 6–7
    alloyed, 11
    classification, 8–9
    high-silicon, 9
cast stainless steels, 22
cast steels, 15
Castner cells, 86
cathodes
    for anodic protection, 331–2, 332
    materials for hydrogen evolution
        in acid media, 323
    materials used in batteries and fuel
        cells, 321
cathodic protection, anodes for, 332
cathodoluminescence, 405
CdTe, 260
celestine, 110
celestite, 110, 421
cell multiplicity, 558, 558
cellulose, 509
cellulose acetate, 382, 383
cellulose acetate-butyrate, 382, 383
cellulose acetate-butyrate-
        proprionate, 382, 383
cellulosics, 376
celsian, 421
cement, history, 517
ceramics
    advanced, 339–40, 340
    definitions, 337
    for construction, 522
    further reading, 365
    physical properties, 342–65
    traditional, 337–8, 338
cerium, 217
cerium dioxide, 360, 361
cerussite, 421
cervantite, 421
cesium
    description, 92–3
    general properties, 92–3
    history, 93
    industrial preparation, 93
    industrial uses and applications,
        93–4
    minerals, 93
    natural occurrences, 93
    ores, 93
    processing, 93
    properties, 76–7
    world metal producers, 94
chabasite, 421, 422
chalcanthite, 422
chalcocite, 422
chalcopyrite, 422
charge carriers, 249
charge transfer transitions (CTTs),
        400

chemical reactivity of minerals, 407
chemistry, bibliography, 579–80
Chilian saltpeter, 86
chloanthite, *422*
chlorinated polyvinylchloride
    (CPVC), 373, *382, 383*
chlorite, *422*
chloritoid, *423*
chloroargyrite, 193, *423*
chlorofluorinated polyethylene, *382,
    383*
chloroprene rubber (CPR), 380
chlorosulfonated polyethylene
    (CSM), 380
chondrodite, *423*
chromite, *423*
chromium boride, *342, 343*
chromium carbide, *348, 349*
chromium diboride, *342, 343*
chromium disilicide, *356, 357*
chromium herminitride, *352, 353*
chromium monoboride, *342, 343*
chromium mononitride, *352, 353*
chromium oxide, *360, 361*
chromium silicide, *356, 357*
chrysoberyl, 97, *423*
chrysocolla, *423*
cinnabar, *424*
clastic sedimentary rocks, 493
Clausius–Mosotti equation, 294
clays, 337
clinohumite, *424*
clinozolsite, *424*
close-packed arrangements, *542*
C–Mn steels, 9
cobalt alloys, properties, *42*
cobaltite, *424*
coercitive force, 281
coercive electric field strength, 305
coercive magnetic field, 281
coercivity, 281
coesite, *424*
colemanite, *425*
coloradoite, *425*
columbite, *425*
compounds, Strukturbericht
    designations, *549, 550–1*
concrete
  definition, 521
  degradation processes, 521
conduction band, 245
conductors, definition, 246
construction materials *see* building
    and construction materials
contact metamorphic rocks, 496
copper, *425*
  description, 51
  electrorefining by-product, 200
  general properties, 51
  hydrometallurgical process, 54–5
  industrial preparation, 54–5
  minerals, 54
  natural occurrence, 54
  ores, 54
  processing, 54–5
  pyrometallurgical process, 55

copper alloys, 55–8
  cast, 56, 58, *64*
  categories, *58*
  UNS designation, 55–6
  wrought, 56–8, *60–3*
Copper Development Association,
    Inc. (CDA), 55–6
cordierite, *425*
corona mechanism, 302
corrosion protection, electrodes,
    331–2
corrosion resistance
  metallic container materials, 564
  niobium, 156–7
  platinum group metals, 212,
    214–16
  tantalum, 165
  titanium alloys, 128, *140–3*
  woods, 511
  zirconium, *151*
  zirconium alloys, *151*
corundum, *425*
cosmogenic radionuclides, 572
cost of elements and alloys, 538–9
cotunnite, *426*
Coulomb's law, 291
covellite, *426*
cristobalite, *426*
critical points, 3
crocoite, *426*
cross product, 545–58
cryolite, *426*
cryolithe, 86
crystal
  definition, 395
  development, 482
  dimensions, 482
  external shapes, 482
  field theory (CFT), 400
  form, 399–400
  habit, 399
  lattice, interplanar spacing
    according to, *561*
  morphology, 482
  properties, 405
  proportion, 482
  structure, 405
  symmetry, 541, *544–5*
  systems, *541*
crystallinity, 486
crystallization process, Czochralski
    (CZ), 260
crystallization sequence, 483
crystallochemistry, 529–59
  further reading, 572
crystallography, 529–59
  further reading, 572
cubanite, *426*
cubic space groups, *557*
cuprite, *427*
Czochralski crystallization process
    (CZ), 260

datolite, *427*
delta-iron, 3
demagnetization curve, 281

demantoid, *413*
detritic sedimentary rocks, *493, 494*
diagenesis, 480
diamagnetic materials, 278
diamond, *348, 349, 427*
diaspore, *427*
dichroism, 405
dielectrics, 291–311
  absorption, 296
  behavior, 300–2
  breakdown mechanisms, 302–3
  breakdown voltage, 296
  class I, 308
  class II, 308
  classification, 307–8
  further reading, 311
  heating, 297–8
  linear, 308
  losses, 296–7
    frequency dependence, 301–2
  permittivity, 291
  physical quantities, 291
  polarization, 295
  properties of, 308, *309–11*
diffusion, 251
diopside, *427*
dioptase, *427*
dipole orientation, 300–1
dispersion, 405
dissipation factor, 297
dolomite, 101, 108, *428*, 522
donors, 248, 250
dopants, 247
Downs cells, 86, *86*
dravite, *428*
DSA type anodes, 327–8
DSA-Cl$_2$ anodes, 330
DSA-O$_2$ anodes, 330
ductile (nodular) cast irons, 8
duplex stainless steels, 16, 21
  physical properties, *22*
dykes, 481
dysprosium, 217
dysprosium oxide, *360, 361*

Ebonex, 329–30
eddy current losses, 283
eglestonite, *428*
elastomers, 371, 378–81
  further reading, 393
  properties of, *382–91*
elbaite, *428*
electric current density, 315
electric dipole moment, 294–5, 298
electric field, 304
  strength, 294
electric flux density, 294
electrical classification of solids,
    246–7
electrical materials, 313–35
electrical resistivity, temperature
    coefficient, 298–9
electrical susceptibility, 295–6
electrochemical equivalence, 319
electrochemical galvanic series, *566*

electrode materials, 319–32
  for batteries and fuel cells, 319–20
  for corrosion protection and
      control, 331–2
  for electrolytic cells, 320–4
  requirements, 322
electrolyte, 321
electrolytic cells, electrode materials
      for, 320–4
electrolyzer, 321
electromechanical coupling, 303
electromigration, 250–1
electron-emitting materials, 314–15
electron work function, 315
electronic dielectric breakdown, 302
electronic energy, 246
electronic polarization, 300
electrostatic energy stored in a
      capacitor, 293
electrum, 193, 428
elements
  geochemical classification, 528
  historical names, 528–9
  mononuclidic, 529
  periodic chart, 527
  properties of, 530–7
  Strukturbericht designations,
      546–7
  symmetry, 540
embolite, 428
enargite, 428
energy band gap, 245, 246
engineering, bibliography, 580
enstatite, 429
epichloridrin rubber, 382, 383
epidote, 429
epoxy resins, 378, 382, 383
epsomite, 429
erbium, 217
erythrite, 429
ethenic polymers, 371–5
ethylene chlorotrifluoroethylene, 382,
      383
ethylene propylene diene rubber, 382,
      383
ethylene propylene rubber, 380, 382,
      383
ethylene tetrafluoroethylene, 382, 383
eudialyte, 146, 429
europium, 217
europium oxide, 360, 361
euxenite, 158

falcondoite, 429
Faraday's constant, 319
faujasite, 429
fayalite, 430
Feldspar index, 487
feldspars, 108
ferberite, 430
fergusonite, 430
Fermi level, 245
ferrimagnetic materials, 281
ferrite alpha, 6
ferrite delta, 6

ferritic stainless steels, 16
  physical properties, 19
ferroelectric domains, 303
ferroelectric hysteresis loop, 304
ferroelectric materials, 303–5, 308
  aging of, 305–7
  properties of, 306–7
ferromagnetic compounds,
      properties of, 280
ferromagnetic elements, properties
      of, 279
ferromagnetic ferrites and garnets,
      properties of, 280
ferromagnetic materials, 279, 281–9
ferrous metals and alloys, 1–43
  see also specific ferrous metals and
      alloys
float zone (FZ) method, 261
fluorelastomers, 381
fluorescence, 405
fluorinated ethylene propylene, 375,
      382, 383
fluorinated polyolefins
      (fluorocarbons), 374
fluorite, 108, 430
fluorspar, 108
forsterite, 430
francium, 94
franklinite, 59, 431
fuel cells, electrode material, 319–20
fuel energy content, 570
fulgurites, 505–6
furane plastics, 378

GaAs, 249, 260
gadolinium, 217
gadolinium oxide, 360, 361
gahnite, 431
galaxite, 431
galena, 431
gamma-iron, 3
gangue
  common minerals, 396
  definition, 396
GaP, 260
GaSb, 260
gaylussite, 431
gehlenite, 431
gemstones, definition, 396
geochemical classification of
      elements, 528
germanium
  applications and industrial uses,
      259–60
  description, 258–9
  general properties, 258–9
  history, 259
  natural occurrence, 259
  preparation, 259
gersdorffite, 431, 432
gibbsite, 47, 432
Ginzburg–Landau theory, 266
glasses
  definitions, 340–64
  further reading, 365–8
  physical properties, 364, 366–9

glassy igneous rocks, 485
glauberite, 432
glauconite, 432
glaucophane, 432
goethite, 432
gold, 196–203, 433
  caratage, 196–7
  carbon in pulp process, 199
  cementation method, 199
  copper electrorefining by-product,
      200
  cyaniding process, 199
  description, 196–7
  general properties, 196–7
  gravity separation method, 199
  history, 197
  industrial preparation, 198–200
  industrial uses and applications,
      200
  metal and alloy world producers,
      203
  mineral dressing and mining, 198
  minerals, 197–8
  natural occurrences, 197–8
  ores, 197–8
  placer mining, 198, 199
  processing, 198–200
  refining, 199–200
  vein or lode mining, 198
gold alloys, 196–203
  caratage, 196–7
  industrial uses and applications,
      200
  physical properties, 201–2
goslarite, 433
granite, 523
graphite, 260, 339, 348, 349, 433
gray cast iron, 7
greenalite, 433
Grimm–Sommerfeld rule, 249
grossular, 433
gummite, 433
gypsum, 108, 433, 519
habitus, 399
hafnium
  acid cleaning, 115
  description, 153
  general properties, 153
  history, 153–4
  industrial preparation, 154
  industrial uses and applications,
      154
  machining characteristics, 116
  minerals, 154
  natural occurrences, 154
  ores, 154
  processing, 154
  properties of, 117–19
  pyrophoric properties, 116
  world producers, 154

hafnium alloys, 153–4
hafnium diboride, 342, 343
hafnium dioxide, 360, 361
hafnium disilicide, 356, 357
hafnium monocarbide, 348, 349

hafnium mononitride, *354, 355*
Hafnon, 154
halite, 86, *434*
Hall effect, 251–2
Hall-Heroult process, 47–8
hanksite, *434*
hard magnetic materials, 283, 284
  properties of, *286–8*
hardness of minerals, 402–3, *404*
hardness scales
  conversion, *567–8*
  Mohs, 402–3
hardwoods, 509
haussmannite, *434*
hauyne, *434*
heartwood, 509
heat-resistant alloys, 15
heat-resistant stainless steels, 22
heavy spar, 112
hedenbergite, *434*
hematite, 2, *435*
hemimorphite, 59
hercynite, *435*
hessite, *435*
heulandite, *435*
hexagonal lattice, conversion of
        rhombohedral lattice to, 541
hexagonal space groups, *556*
high-carbon steels, 10
high-density polyethylene (HDPE),
        372
high-strength low-alloy (HSLA)
        steels, 24
historical names of elements, *529*
holmium, 217
horn silver, 193
hot dip galvanizing, 65
huebnerite, *435*
humite, *435*
Hunter process, 123–4
hyacinth, 146
hyaline igneous rocks, 485
hydrogen evolution in acid media,
        cathode materials for, *323*
hysteresis loop, 281–3

igneous rocks
  pyroclastic, 487
  texture, *485*
illite, 337, *436*
ilmenite, 121, 122, 147, *436*
impactites, 505–6
InAs, 249, 260
indicatrix, 404
initial magnetic permeability, 283
InP, 260
InSb, 260
insulation resistance, 298
insulators, 291–311
  definition, 246
  further reading, 311
  physical properties, 298–300
  properties of, 308, *309–11*
internal discharge, 303
International Annealed Copper
        Standard (IACS), 51

International Union of
        Crystallography (IUCr), 529
interplanar spacing
  according to crystal lattice, *561*
  general formula, *561*
intrinsic mechanism, 303
iodargyrite, *436*
iodoargyrite, 193
ionic polarization, 300
iridium, 210–11, *436*
  physical and chemical properties,
        206–8
iron
  critical points, 2–3
  crystallographic phase
        transformation, 3
  crystallography, 2–3
  general properties and description,
        1–6
  history, 2
  minerals, 2
  mining and mineral dressing, 3–4
  natural occurrence, 2
  ores, 2–4
  pure iron grades, 5
iron-based superalloys, 27
iron–carbon diagram, 6
iron–carbon phases, 6
ironmaking, 4–5
isotopes, *529*
isotropic materials, 404

jacinth, 146
jacobsite, *436*
jadeite, *436*
jargon, 146

kainite, 101, *437*
kaolinite, 337, *437*
kernite, *437*
kiesserite, 101
krennerite, *437*
Kroll process, 122–3
kyanite, *437*

labradorite, *438*
laccoliths, 481
langbeinite, 101, *438*
lanthanides, 217
  discovery milestones, *219*
lanthanum, 217
lanthanum dicarbide, *348, 349*
lanthanum dioxide, *360, 361*
lanthanum hexaboride, *342, 343*
larnite, *438*
laterite ores, nickel oxide from, 33
lattice planes, plane angle between,
        559, 560
laumontite, *438*
lawsonite, *438*
lazulite, *439*
lazurite, *439*
lead
  anodes, 325–6
  description, 68–9
  further reading, 74

general properties, 68–9
physical properties, *70–1*
lead alloys
  anodes, 325–6
  further reading, 74
  physical properties, *70–1*
lead dioxide anodes, 329
leakage electric current, 299–300
lepidocrocite, *439*
lepidolite, 79, 91, *439*
leucite, 91, *439*
leucoxene, 122, 147
lignine, 509
limestones, 522
limonite, 2, *439*
linear combination of atomic
        orbitals, 245
linnaeite, *439*
liquid metals
  maximum operating temperature
        of metallic container
        materials for, *564*
  physical properties at melting
        point, *562–3*
lithium
  abundances in different geological
        materials, *80*
  applications and industrial uses, *83*
  description, 75–9
  further reading, 240–1
  general properties, 75–9
  history, 79
  industrial preparations, 80–2
  industrial uses and applications,
        82–4
  metal, 81–2
  minerals, 79
  natural occurrence, 79
  ores, 79
  processing, 80–2
  properties of, *76–7*
  world metal producers, 84, *84*
lithium carbonate
  from brines, 81
  from silicate ores, 80–1
lithology, 479
livingstonite, *440*
loss tangent, 297
low-alloy steels, 11–15
  designation, *12*
  physical properties, *13–14*
low-carbon steels, 10
low-density polyethylene (LDPE),
        371–2
luminescence of minerals, 405–6
lutetium, 217

machining steels, 25
macroscopic electric dipole moment,
        295
macroscopic magnetic moment, 276
magma, 480, 481
  hypersiliceous or felsic, 481
  hyposiliceous or mafic, 481
magnesiochromite, *440*
magnesioferrite, *440*

magnesite, 101, *440*
magnesium
  aluminothermic reduction, 102
  description, 99–100
  electrolytic reduction, 101–2
  further reading, 241–2
  general properties, 99–100
  history, 100
  industrial preparation, 101–2
  industrial uses and applications,
     103, *107–8*
  minerals, 100–1
  natural occurrences, 100–1
  nonelectrolytic processes, 101
  ores, 100–1
  processing, 101–2
  properties of, *95–6*
  refining, 102
  silicothermic reduction, 102
  thermochemical reduction, 102
  world producers, 103
magnesium alloys
  further reading, 241–2
  physical properties, *104–6*
  standard ASTM designation, *103*
magnesium monoxide, *360*, *361*
Magnetherm process, 102
magnetic dipole moment, 275–6
magnetic energy density stored, 277
magnetic energy losses, 281
magnetic field, 277
magnetic field strength, 275
magnetic flux density, 276–7
magnetic induction, 276–7
magnetic materials, 275–90
  classification, 278–81
  further reading, 289–90
  selection, 289
magnetic moment, 275
magnetic permeability, 276–7, 283
magnetic physical quantities, 275–7
  and conversion factors, *277*
magnetic shielding, 289
magnetism, minerals, 405
magnetite, 2, *440*
magnetization, 276
magnetization curve, 281–3
malachite, *440*, *441*
malleable cast irons, 7–8
manganese dioxide anodes, 329
manganite, *441*
manghemite, *441*
marcasite, 2, *441*
margarite, *441*
marialite, *441*
martensitic stainless steels, 16
  physical properties, *17–18*
massicot, *442*
material sciences, comprehensive
    series, *582–3*
mathematics, bibliography, 579
maximum kinetic energy, 315
maximum magnetic permeability,
    283
medium-carbon steels, 10
meionite, *442*

Meissner–Ochsenfeld effect, 270
melamine formaldehyde, *382*, *383*
melanterite, *442*
melilite, *442*
mercury, *442*
Merryl–Crowe process, 199
merwinite, *442*
metacinnabar, *443*
metallic character, 246
metallic container materials,
    corrosion resistance, 564
metalloids, 247
metallurgy, bibliography, 580–1
metamorphism, 495
meteorites, 480
  classification, 499–505, *500–4*
  definitions, 499
microcline, 89, *443*
microscopic electric dipole moment,
    294–5
microscopic magnetic dipole
    moment, 275–6
millerite, *443*
mineraloids, definition, 395
minerals
  chemical composition, 399
  chemical formulae, 399
  chemical reactivity, 407
  classifications, 398–9
  cleavage, 401
  color variations, 400
  crystallographic properties, 399
  definition, 395
  density, 402
  diaphaneity, 400–1
  fracture, 401
  further reading, 477–8
  hardness, 402–3, *404*
  industrial, *397–8*
    definition, 396
  interaction of electromagnetic
    radiation, 400
  luminescence, 405–6
  luster, 401
  magnetism, 405
  optical properties, 403–5
  parting, 401
  properties, 396–8, 408–9, *410–66*
  specific gravity, 402
  streak, 402
  Strunz classification, 407–8
  synonyms, 409–77
  tenacity, 402
  transmission of light, 400–1
  see also specific minerals
minium, *443*
mixed product, 558
Mohs scale of hardness, 402–3
molecular polarizability, 294
molten salts, physical properties at
    melting point, *565*
molybdenite, 177, *443*
molybdenum, 176–85
  acid cleaning, *115*
  cleaning, 182
  descaling, *182*

  description, 176–7
  electrical discharge machining
    (EDM), 182
  etching, 182, *182*
  general properties, 176–7
  history, 177
  industrial preparation, 177
  industrial uses and applications,
    182–5, *183–5*
  joining, 179–80
  machining, *116*, 180–2
  metalworking, 179
  minerals, 177
  natural occurrences, 177
  ores, 177
  pickling, 182, *182*
  processing, 177
  properties, *117–19*
  pyrophoric properties, *116*
  welding, 179–80
  world metal producers, *185*
molybdenum alloys, 176–85
  descaling, *182*
  etching, *182*
  industrial uses and applications,
    182–5, *183–5*
  pickling, *182*
  properties, 177–9, *178*
molybdenum boride, *344*, *345*
molybdenum diboride, *344*, *345*
molybdenum disilicide, *356*, *357*
molybdenum hemiboride, *344*, *345*
molybdenum hemicarbide, *348*, *349*
molybdenum herminitride, *354*, *355*
molybdenum monoboride, *344*, *345*
molybdenum monocarbide, *348*, *349*
molybdenum mononitride, *354*, *355*
monazite, 147, 219, 220, 235, *443*
  alkali digestion, 221
  sulfuric acid digestion process, 221
monoclinic space groups, *552*
mononuclidic elements, *529*
monticellite, *444*
montmorillonite, 337
montroydite, *444*
mortars, definition, 521
mullite, *339*, *444*
muscovite, *444*

natrolite, 86, *444*
natural radioactivity, 564–9
natural rubber, 378–9, *382*, *383*
neodymium, 217
Neoprene, 380
nepheline, 46, *445*
niccolite, *445*
nickel
  description, 32
  general properties, 32
  history, 32
  industrial preparation, 33–4
  metal, 33
  minerals, 32–3
  natural occurrence, 32–3
  ores, 32–3
  physical properties, *36*

nickel (*continued*)
  processing, 33–4
  refining processes, 33
nickel alloys, 34
  main classes, *35–6*
  Ni-Ti shape memory, 34–6
    properties, *42*
  physical properties, *36, 37–41*
nickel oxide
  from laterite ores, 33
  from sulfide ores, 33
niobite, 157
niobium
  acid cleaning, *115*
  anodic electroetching, 163
  carbothermic reduction, 159
  cleaning, 162
  corrosion resistance, 156–7
  degreasing, 162
  description, 156–7
  further reading, 243
  general properties, 156–7
  history, 157
  industrial preparation, 158–9
  industrial uses and applications,
    164
  joining, 162
  machining, *116*, 159–62
  metal, 159
  metallothermic reduction, 159
  metalworking, 159
  minerals, 157–8
  natural occurrences, 157–8
  ores, 157–8
  pickling, 163
  processing, 158–9
  properties of, *117–19, 160–1*
  pyrophoric properties, *116*
  welding, 162
  world metal producers, *163*
niobium alloys
  further reading, 243
  properties of, *160–1*
niobium diboride, *344, 345*
niobium hemicarbide, *348, 349*
niobium monoboride, *344, 345*
niobium monocarbide, *350, 351*
niobium mononitride, *354, 355*
niobium pentoxide, *360, 361*
niobium silicide, *356, 357*
niobium-tantalum concentrates, 158
niter, 90, *445*
nitratite, *445*
nitrile rubber (NR), 379
noble metals, 192–203
  anodes, 325
  see also specific noble metals
nonferrous metals and alloys
  common, 45–74
  less common, 75–244
  see also specific nonferrous metals
    and alloys
nonmetallics
  color variations, 400
  industrial, *398*
    definition, 396

norbergite, *445*
northupite, *445*
nosean, *445, 446*
nuclear fuel cycle, 233
nuclear series, *571*
nylon, 375

Ohm's law, 266
oligoclase, *446*
olivine, *446*
ores
  definition, 395
  see also specific metals/ores
orpiment, *446*
orthoclase, 46, 89, *446*
orthoferrosilite, *447*
orthorhombic space groups, *552, 553*
osmium, 210
  physical and chemical properties,
    206–8
oxygen evolution in acid media,
    anodes for, *324*, 328–31

palladium, 209–10, *447*
  physical and chemical properties,
    206–8
parallel electrode capacitor,
    capacitance, 293
paramagnetic materials, 278–9
PbTe, 260
P-E diagram, 304
pearceite, *447*
pectolite, *447*
pelletizing, 4
pentlandite, *447*
perfluorinated alkoxy, 375, *385*
periclase, *447*
periodic chart of elements, *527*
permittivity of a vacuum, 291
perowskite (or perovskite), 122, *447,
    448*
petalite, 79
petrography, 479
petzite, *448*
phaneritic igneous rocks, 484
phase diagram, bibliography, 581
phase stabilizers, titanium, *127*
phenakite, 97
phenocrystals, 485
phenol formaldehyde, *385*
phenolics, 377
phlogopite, *448*
phosphorescence, 405
photocathode materials, 315–18, *319*
photoelectric effect, 315, 318
photoelectric quantum yield, 315
photoemission, 318
physics, bibliography, 579–80
Pidgeon's process, 102
piemontite, *448*
piezoelectric materials, 303
pirssonite, *448*
pitchblende, 112–13
plain carbon steels, 9
plane angle between lattice planes,
    559, 560

platinized anodes, 326–7
platinum, 211–12, *449*
  physical and chemical properties,
    206–8
platinum alloys, properties of, *212–13*
platinum group metal alloys,
    properties of, *212–13*
platinum group metals, 203–17
  corrosion resistance, 212, *214–16*
  further reading, 243
  general overview, 203–4
  industrial applications and uses,
    217
  minerals, 204
  natural occurrences, 204
  ores, 204
  physical and chemical properties,
    206–8
  world producers, *218*
  see also specific metals
platinum labware, cleaning, 212
platonician regular polyhedrons, *540*
plattnerite, *449*
pleochroism, 405
point groups, *543*
polarization, 300–1
  frequency dependence, 301
pollucite, 91, 93, *449*
polyacetals, 375–6
polyacrylic butadiene rubber, *385*
polyamide nylon 4,6, *385*
polyamide nylon 6, *385*
polyamide nylon 6,6, *385*
polyamide nylon 6,10, *385*
polyamide nylon 6,12, *385*
polyamide nylon 11, *385*
polyamide nylon 12, *385*
polyamide-imide, *385*
polyamides (PA), 375
polyaramide, *385*
polyarylate resins, *385*
polybasite, *449*
polybenzene-imidazole, *385*
polybutadiene rubber, 379, *385*
polybutadiene terephtalate, *385*
polybutylene, 372, *385*
polycarbonates, 376, *385*
polychloroprene, 380, *385*
polyether ether ketone, *386, 387*
polyether imide, *386, 387*
polyether sulfone, *386, 387*
polyethylene, 371–2, *386, 387*
polyethylene naphtalate, *386, 387*
polyethylene oxide, *386, 387*
polyethylene terephtalate, *386, 387*
polyhalite, *449*
polyhydroxybutyrate (biopolymer),
    386, 387
polyimides (PI), 376
polyisoprene, *386, 387*
  trans-1,4-polyisoprene rubber
    (PIR), 379
polymers, 371–93
  ASTM standards, *392–3*
  classes, 371
  further reading, 393

physical properties, 381–93
physical quantities, *392–3*
see also specific polymers
polymethyl methacrylate (PMMA),
    374, *386, 387*
polymethyl pentene, *386, 387*
polymethyl penthene, 373
polymide, *386, 387*
polymorphism, 3
polyolefins, 371–5
polyoxymethylene, *386, 387*
polyphenylene oxide, 377, *386, 387*
polyphenylene sulfide, 377, *388, 389*
polypropylene, 372, *388, 389*
polysiloxanes, 381
polystyrene, 373, *388, 389*
polysulfide rubber, 380, *388, 389*
polysulfone, 376, *388, 389*
polytetrafluoroethylene, 374, *388, 389*
polytrifluorochloroethylene, 374–5,
    388, 389
polyurethane, 377, *388, 389*
polyvinyl acetate, 373, *388, 389*
polyvinyl alcohol, *388, 389*
polyvinyl chloride, 372–3, *388, 389*
polyvinyl fluoride, *388, 389*
polyvinylidene chloride (PVDC), 373
polyvinylidene fluoride, 375, *388, 389*
porcelain brick, 338
porphyritic igneous rocks, 485
Portland cement, 517
    chemical composition, *518*
    chemistry, 519–20, *519*
    nomenclature, 520, *520*
    processing, 518
    raw materials, 517–18
    types, *520*
potassium
    description, 88–9
    further reading, 241
    history, 89
    industrial preparation, 90
    industrial uses and applications,
        90, *90*
    minerals, 89
    natural occurrence, 89
    ores, 89
    processing, 90
    properties of, *76–7*, 88–9
potassium heptafluorotantalate, 167
powellite, 177, *449*
praseodymium, 217
precious metals, 192–203
    anodes, 325
    see also specific metals
precipitation-hardening (PH)
    stainless steels, 16, 21
    physical properties, *23*
primordial radionuclides, *571*
promethium, 217
propylene-vinylidene hexafluoride,
    390
proustite, *449*
psilomelane, *449, 450*
pulling crystal growth technique, 260
pyrargyrite, *450*

pyrite, *450*
pyrochlore, 166, *450*
pyroclastic rocks, 487, *492*
pyrolusite, *450*
pyromorphite, *450*
pyrope, *451*
pyrophyllite, *451*
pyrrhotite, *451*

quartz, *451*
quaternary compounds,
        Strukturbericht
        designations, *549*
radioactivity, natural, 564–9
radionuclides, 564–9
    cosmogenic, *572*
    primordial, *571*

radium
    description, 112
    general properties, 112
    history, 112
    industrial preparation, 113
    industrial uses and applications,
        113
    natural occurrences, 113
ramsdellite, *452*
rare-earth metals, 217–23
    description, 217–19
    further reading, 244
    general properties, 217–19
    history, 219–20
    hydrometallurgical concentration,
        221
    industrial applications and uses,
        224–5
    industrial preparation, 220–3
    industrial uses and applications,
        223
    liquid-liquid extraction process, 223
    metallothermic reduction, 222
    minerals, 220
    mining and mineral dressing, 221
    natural occurrence, 220
    ore concentration, 221
    ores, 220
    processing, 220–3
    refining, 222
    world suppliers, *225*
    see also specific rare-earth metals
rare-earth oxide prices, *225*
reactive and refractory metals *see*
        refractory and reactive
        metals (RMs)
realgar, *452*
reciprocal unit lattice, *561*
refractories
    definitions, 337
    further reading, 365
    physical properties, *342–65*
    properties, *341*
refractory and reactive metals (RMs)
    cleaning, 114
    common properties, 114
    corrosion resistance, *120*
    descaling, 114, *115*

etching, 114, *115*
general overview, 113–15
machining characteristics, *116*
pickling, 114
properties of, *117–19*
pyrophoricity, 114–15, *116*
    see also specific refractory and
        reactive metals
regional metamorphic rocks, 496
relative dielectric permittivity, 292
remanence, 281
remanent magnetic induction, 281
remanent polarization, 304
retentivity, 281
rhenium, 188–92
    description, 188–91
    general properties, 188–91
    history, 191
    industrial preparation, 192
    minerals, 192
    natural occurrences, 192
    ores, 192
    processing, 192
    uses and applications, 192
rhenium alloys, 188–92
    uses and applications, 192
rhodium, 206–9
    physical and chemical properties,
        206–8
rhodizite, 91, 93
rhodocrosite, *452*
rhodonite, *452*
rhombohedral lattice, conversion to
        hexagonal lattice, 541
Richardson–Dushman equation, 315
riebeckite, *452*
rock salt, 86
rocks, 479–508
    endogeneous, 480
        further reading, 508
    igneous, 480–7
        chemical composition, 486,
            486–9
        classification, 481–4, *491*
        coloration, 484
        general classification, 487
        mineralogy, 483
        texture, 484–6
    industrial, *397–8*
    magmatic, 480
    metamorphic, 480, 495–9
        classification, 496–9
        metamorphic facies, 499, *499*
        metamorphic grade, 498–9
        types, *498*
    physical properties, 506, *506–8*
    plutonic or intrusive, 481
    pyroclastic, *492*
    sedimentary, 480, 490–5
        chemical, 493–4
        chemical composition, *497*
        classes, *492*
        residual, 493
    types, 480
    volcanic or extrusive, 481
    see also specific types

roscoelite, 155
rosenbuschite, *453*
rubbers, 378–81
  see also specific types
rubidium
  description, 91
  general properties, 91
  history, 91
  industrial preparation, 91–2
  industrial uses and applications, 92
  minerals, 91
  natural occurrences, 91
  ores, 91
  processing, 91–2
  properties of, *76–7*
  world chief producers, 92
ruthenium, 205
  physical and chemical properties,
    206–8
ruthenium tetraoxide, 330
rutile, 121, 122, 133, 147, *453*

salpeter, 90
samarium, 217
samarium oxide, *362, 363*
sandstones, 522
sanidine, *453*
sapphirine, *453*
saturation magnetic induction, 281
saturation polarization point, 304
scalar product, 545
scandium, 217
scheelite, *453*
Schoenflies–Fedorov point group
    notation, *543*
schorl, *453*
scientific and technical societies,
    573–7
scorodite, *454*
secondary emission, 318, *320*
secondary emission coefficient, 318
sediments, 490–3
Seebeck coefficient, 313
Seebeck effect, 313–14
Seebeck electromotive force, 313
semiconductors, 245–64
  applications, *252*
  classes, 247–9
  compound, 249
  definition, 246
  extrinsic-doped, 247–8
  intrinsic or elemental, 247
  manufacturing, 260
  miscellaneous materials, 260
  monocrystal growth, 260–1
  n-type, 248, 263
  p-type, 248, 263
  p-n junction, 263–4
  physical properties, 254–7
  physical quantities, 249–50
  stoichiometric compounds, 249
  transport properties, 250–2
  wafer processing, 260–3
  wafer production, 261–3
  see also germanium; silicon
semimetals, 247

senarmontite, *454*
separator, 322
shape memory Ni-Ti alloys, 34–6
  properties, *42*
shortite, *454*
sialons, *339*
siderite, *454*
silica, *339*
silica brick, 338
silicate ores, lithium carbonate from,
    80–1
silicium dioxide, *362, 363*
silicon
  applications and industrial uses, 258
  cost, 258
  description, 252–3
  general properties, 252–3
  history, 253
  natural occurrence, 253
  preparation, 253–8
silicon carbide, *339*
silicon hexaboride, *344, 345*
silicon monocarbide, *350, 351*
silicon nitride, *354, 355*
silicon tetraboride, *344, 345*
silicone rubber, 381, *390*
sillimanite, *454*
silver, 192–6, *454*
  description, 192–3
  further reading, 243
  general properties, 192–3
  history, 193
  industrial preparation, 194
  industrial uses and applications,
    194–6
  minerals, 193–4
  natural occurrences, 193–4
  ores, 193–4
  processing, 194
silver alloys, 192–6
  further reading, 243
  industrial uses and applications,
    194–6
  physical properties, *195*
sinhalite, *455*
skutterudite, *455*
smithsonite, 59, *455*
Society of Automotive Engineers
    (SAE), 563
soda ash, 86
sodalite, *455*
sodium
  description, 84–5
  further reading, 241
  general properties, 84–5
  history, 86
  industrial applications and uses, *88*
  industrial preparation, 86–7
  industrial uses and applications, 87
  minerals, 86
  molten salt electrowinning, *86*
  natural occurrence, 86
  ores, 86
  processing, 86–7
  properties, *76–7*
  world metal producers, 87, *87*

soft magnetic materials, 283–4
  properties of, *285*
softwoods, 509
solids, electrical classification, 246–7
space charge polarization, 301
space groups
  cubic, *557*
  hexagonal, *556*
  monoclinic, *552*
  orthorhombic, *552, 553*
  tetragonal, *554*
  triclinic, *552*
  trigonal, *555*
space lattices, 540
  parameters, 529–59
  volume, *559*
spessartine, *455*
sphalerite, 59, *455*
sphene, 122, *456*
spinel, 329, *456*
spodumene, 79, *456*
spontaneous polarization, 303
spurrite, *456*
stabilizing elements, *6*
stainless steels
  general description, 15–16
  melting process, 22–3
  processing, 22–3
  see also specific types
stannite, *456*
statistics, bibliography, 579
staurolite, *457*
stephanite, *457*
stibnite, *457*
stilbite, *457*
stishovite, *457*
stromeyerite, *457*
strontianite, 110, *458*
strontium
  description, 110
  general properties, 110
  history, 110
  industrial uses and applications,
    111
  minerals, 110
  natural occurrences, 110
  ores, 110
  preparation, 111
  processing, 111
  properties of, *95–6*
Strukturbericht designations
  binary compounds, *547*
  compounds, *549, 550–1*
  elements, *546–7*
  quaternary compounds, *549*
  ternary compounds, *548–9*
styrene-butadiene rubber, 379, *390*
styrene-butadiene styrene rubber,
    390
sulfide ores, nickel oxide from, 33
sulfur, *458*
superalloys, iron-based, 27
superconductors, 265–73
  basic theory, 269–70
  BCS theory, 269–70
  further reading, 273

general description, 265–6
high critical temperature, 268–9
high-magnetic-field applications, 272–3
history, 271–2
industrial uses and applications, 272–3
low-magnetic-field applications, 273
Meissner–Ochsenfeld effect, 270
organic, 269
type I, 266, *267*
type II, 266–7, *268*
supercooled liquid, 340
surface electrical resistivity, 299
sylvanite, *458*
sylvinite, 89
sylvite, *458*
symmetry classes, 541, *544–5*
symmetry elements, *540*
synthetic isoprene rubber, *390*

talc, *458*
tantalum, 163–76
acid cleaning, *115*
cladding, 171–4
cleaning, 171
coating, 171–4
corrosion resistance, 165
degreasing, 171
description, 163–6
further reading, 243
general properties, 163–6
history, 166
hydriding-dehydriding process, 167
industrial preparation, 166–7
industrial uses and applications, 174, *175–6*
joining, 170
machining, *116*, 170
metal, 167
metalworking, 168
minerals, 166
natural occurrences, 166
ores, 166
powder, 167
processing, 166–7
properties of, *117–19*
pyrophoric properties, *116*
welding, 170
world metal producers, 176
tantalum alloys, 163–76
further reading, 243
physical properties, *169*
tantalum anodes, 326–7
tantalum diboride, *344, 345*
tantalum disilicide, *356, 357*
tantalum hemicarbide, *350, 351*
tantalum herminitride, *354, 355*
tantalum monoboride, *344*
tantalum monocarbide, *350, 351*
tantalum mononitride, *354, 355*
tantalum pentoxide, *362, 363*
tantalum silicide, *356, 357*
tapiolite, 157, *459*

tektites, 505–6, *505*
tenorite, *459*
tephroite, *459*
terbium, 217
terlinguaite, *459*
ternary compounds, Strukturbericht designations, *548–9*
tetracalcium aluminoferrite, 519
tetradymite, *459*
tetragonal space groups, *554*
tetrahedrite, *459*
thermal discharge, 302
thermal instability, 302
thermal mechanism, 302
thermionic emission, 315
thermionic properties of materials, *318*
thermocouple materials, 313–14
physical properties, *316–17*
properties of, 314
thermocouples, 314
types and common uses, *314*
thermoelectronic emission, 315
thermoluminescence, 405
thermoplastics, 371–7
properties of, *382–91*
thermosets, 371, 377–8
properties of, *382–91*
Thiokol, 380
thorianite, *459*
thorite, *460*
thorium
caustic soda digestion process, 236
concentration, 235–6
conversion
of thorium nitrate to, 237
to thoria, 237
to thorium tetrachloride, 237
to thorium tetrafluoride, 237
description, 234
further reading, 244
general properties, 234
history, 234
hydrometallurgical concentration processes, 236
industrial preparation, 235
industrial uses and applications, *238*
metal, 237
minerals, 235
mining and mineral dressing, 235
natural occurrence, 235
ores, 235
processing, 235
properties of, *226–7*
refining, 236
sulfuric acid digestion process, 236
thorium dicarbide, *350, 351*
thorium dioxide, *362, 363*
thorium disilicide, *358, 359*
thorium hexaboride, *344, 345*
thorium monocarbide, *350, 351*
thorium mononitride, *354, 355*
thorium nitride, *354, 355*
thorium tetraboride, *346, 347*
thullium, 217

tiemannite, *460*
timber, 509–16
general description, 509
see also woods
tin, further reading, 74
tin alloys
further reading, 74
physical properties, *72–3*
titanite, 122, *460*
titanium
acid cleaning, *115*
annealing, 129
bending, 129
blasting, 133
casting, 129
chemical composition, *124*
cleaning, 133
commercially pure, 124–5
corrosion resistance, 119–21
degreasing, 133
descaling, 133
description, 115–21
designations, *124*
etching, 133
further reading, 242
general properties, 115–21
grinding, 133
history, 121
industrial preparation, 122–4
industrial uses and applications, 133, *144*
joining, 129–33
machining, *116*, 129
metalworking, 128–9
minerals, 121–2
mining and mineral dressing, 122
natural occurrences, 121–2
ores, 121–2
phase stabilizers, *127*
physical properties of, *126*
pickling, 133
processing, 122–4
properties of, *117–19*
punching, 129
pyrophoric properties, *116*
refining, 124
shearing, 129
superplastic forming, 129
trade names of, *125*
world producers, *145*
titanium alloys
alpha-beta-titanium alloys, 127
alpha-titanium alloys, 127
annealing, 129
ASTM designation, 128
bending, 129
beta-titanium alloys, 127
casting, 125, 129
chemical composition, *130–2*
corrosion resistance, 119–21, 128, *140–3*
description, *134–5*
electrical properties, *138–9*
electron-beam melting, 125
further reading, 242
general description, 125

titanium alloys (*continued*)
    industrial uses and applications,
        133, *133*
    joining, 129–33
    machining, 129
    manufacturing, 125
    mechanical properties, *136–7*
    metallurgical classification, 127
    metalworking, 128–9
    near-alpha-titanium alloys, 127
    plasma melting, 125
    properties, 125
    punching, 129
    shearing, 129
    superplastic forming, 129
    thermal properties, *138–9*
    uses, *139*
    vacuum arc remelting, 125
titanium anodes, 326–7
titanium diboride, *344*
titanium dioxide, *362, 363*
titanium disilicide, *356, 357*
titanium monobride, *345*
titanium monocarbide, *350, 351*
titanium mononitride, *354, 355*
titanium tetrachloride, 122
titanium trisilicide, *358, 359*
titanomagnetite, 122
tobernite, *460*
tool steels, 25
    AISI designation, *26–7*
    physical properties, *28–31*
topaz, *460*
transition temperatures, 3
tremolite, *460*
trevorite, *461*
triboluminescence, 405
trichroism, 405
triclinic space groups, *552*
tridymite, *461*
trigonal space groups, *555*
triphyllite, 79, *461*
trona, *461*
tungsten, 186–92
    acid cleaning, *115*
    description, 186
    further reading, 243
    general properties, 186
    history, 186–7
    industrial preparation, 187–8
    industrial uses and applications,
        188
    machining characteristics, *116*
    minerals, 187
    natural occurrences, 187
    ores, 187
    processing, 187–8
    properties of, *117–19*
    pyrophoric properties, *116*
    world metal producers, *190–1*
tungsten alloys, 186–92
    further reading, 243
    industrial uses and applications,
        188
    physical properties, *189*
tungsten dinitride, *354, 355*

tungsten disilicide, *358, 359*
tungsten hemiboride, *346, 347*
tungsten hemicarbide, *350, 351*
tungsten heminitride, *354, 355*
tungsten monoboride, *346, 347*
tungsten monocarbide, *350, 351*
tungsten mononitride, *354, 355*
tungsten silicide, *358, 359*
turquoise, *461*

ulexite, *461, 462*
ullmanite, *462*
ultrahigh-molecular-weight
        polyethylene (UHMWPE),
        372
ultrahigh-strength steels, 24
    mechanical properties, *25*
ulvospinel, *462*
Unified Numbering System (UNS)
        standard alphabetical
        designation, 563, *569*
unit cell volume, 558
United Numbering System for Metals
        and Alloys (UNS), 55
units, bibliography, 579
unplasticized polyvinyl chloride, *390*
unsaturated polyester, *390*
uranides, 223–38
    see also thorium; uranium
uraninite, *462*
uranium
    concentration by leaching, 230–1
    crushing and grinding, 230
    description, 223–8
    enrichment, 232
    fluorination of UF_4 to UF_6, 232
    further reading, 244
    general properties, 223–8
    history, 228
    hydrofluorination of UO_2 to UF_4,
        232
    industrial preparation, 230–3
    industrial uses and applications, 233
    metal, 233
    mineral dressing and mining,
        229–30
    minerals, 228–9
    natural occurrence, 228–9
    ores, 228–9
    preparation of UO_2, 232–3
    processing, 230–3
    properties of, *226–7*
    recovery from leach liquors, 231
    reduction of UO_3 to UO_2, 232
    refining, 231–2
uranium carbide, *350*
uranium diboride, *346, 347*
uranium dicarbide, *352, 353*
uranium dioxide, *362, 363*
uranium disilicide, *358, 359*
uranium dodecaboride, *346, 347*
uranium monocarbide, *352, 353*
uranium mononitride, *356, 357*
uranium nitride, *356, 357*
uranium silicide, *358, 359*
uranium tetraboride, *346, 347*

uranophane, *462*
urea formaldehyde, *390*
uvarovite, *463*

valence band, 245
valentinite, *463*
vanadinite, 155, *463*
vanadium
    aluminothermic reduction, 156
    carbothermic reduction, 156
    description, 154–5
    general properties, 154–5
    history, 155
    industrial preparation, 155–6
    industrial uses and applications,
        156
    minerals, 155
    natural occurrences, 155
    ores, 155
    processing, 155–6
vanadium alloys, 154
vanadium diboride, *346, 347*
vanadium disilicide, *358, 359*
vanadium hemicarbide, *352, 353*
vanadium monocarbide, *352, 353*
vanadium mononitride, *356, 357*
vanadium silicide, *358, 359*
variscite, *463*
vector product, 545–58
vein deposits, definition, 396
vesicular igneous rocks, 487
vesuvianite, *463*
villiaumite, *463*
violarite, *464*
Viton fluoroelastomers, 381
vivianite, *464*
volume electrical resistivity, 298

wavellite, *464*
Weiss domains, 303
white cast iron, 7
willemite, *464*
witherite, 112, *464*
wolfram *see* tungsten
wolframite, *465*
wollastonite, *465*
woods, 509–16
    applications, 511
    corrosion resistance, 511
    density and specific gravity, 510
    drying and shrinkage, 510
    durability, 510–11
    heating value, *511*
    mechanical properties, 510,
        512–16
    moisture content, 510
    properties of, 510–11
    see also timber
work hardening, 15
wulfenite, 177, *465*
wurzite, *465*

xenotime, 147, 220, *465*

ytterbium, 217
yttria, 219

yttrium, 217
yttrium oxide, *362*, *363*

zeolites, 108
zinc
  description, 59
  general properties, 59
  history, 59, 68
  hydrometallurgical process, 59–65
  industrial applications and uses, 65
  industrial preparation, 59–65
  industrial uses and applications, 69
  metal, 59
  minerals, 59, 69
  natural occurrence, 59, 69
  ores, 59, 69
  physical properties, *66–7*
  processing, 59–65
  pyrometallurgical process, 65
zinc alloys, physical properties, *66–7*
zincite, *466*
zinnwaldite, 91
zircon, 146, 147, 154, *340*, *466*
zirconia, *340*

zirconium, *364*
  acid cleaning, *115*
  cleaning, 152
  corrosion resistance, *151*
  descaling, 152
  description, 135–46
  etching, 152
  further reading, 242
  general properties, 135–46
  history, 146
  industrial preparation, 147–8
  industrial uses and applications,
      152, *152*
  machining, 151–2
  machining characteristics, *116*
  minerals, 146–7
  mining and mineral dressing,
      147
  natural occurrences, 146–7
  ores, 146–7
  physical properties, *149–50*
  pickling, 152
  processing, 147–8
  properties of, *117–19*
  pyrophoric properties, *116*

  welding, 151–2
  world metal producers, *153*
zirconium alloys
  chemical composition, *148*
  cleaning, 152
  corrosion resistance, *151*
  descaling, 152
  etching, 152
  further reading, 242
  industrial uses and applications,
      152, *152*
  machining, 151–2
  nuclear grades, 148, *148*
  physical properties, *149–50*
  pickling, 152
zirconium diboride, *346*, *347*
zirconium dioxide, *362*, *363*, *365*
zirconium dioxide PSZ, *364*, *365*
zirconium dioxide TTZ, *364*
zirconium dioxide TZP, *364*
zirconium disilicide, *358*, *359*
zirconium dodecaboride, *346*, *347*
zirconium monocarbide, *352*, *353*
zirconium mononitride, *356*, *357*
zoisite, *466*